Diseases and Pests of Fibre Crops

Diseases and Pests of Fibre Crops
Identification, Treatment and Management

Subrata Biswas

Principal Scientist
ICAR-Central Research Institute for Jute and Allied Fibres
Nilganj, Berrackpore, Kolkata - 700 120
West Bengal

CRC Press is an imprint of the
Taylor & Francis Group, an **informa** business

NEW INDIA PUBLISHING AGENCY
New Delhi – 110 034

First published 2021
by CRC Press
2 Park Square, Milton Park, Abingdon, Oxon, OX14 4RN

and by CRC Press
6000 Broken Sound Parkway NW, Suite 300, Boca Raton, FL 33487-2742

© 2021, New India Publishing Agency

CRC Press is an imprint of Informa UK Limited

The right of Subrata Biswas to be identified as author of this work has been asserted by him in accordance with sections 77 and 78 of the Copyright, Designs and Patents Act 1988.

Reasonable efforts have been made to publish reliable data and information, but the author and publisher cannot assume responsibility for the validity of all materials or the consequences of their use. The authors and publishers have attempted to trace the copyright holders of all material reproduced in this publication and apologize to copyright holders if permission to publish in this form has not been obtained. If any copyright material has not been acknowledged please write and let us know so we may rectify in any future reprint.

All rights reserved. No part of this book may be reprinted or reproduced or utilised in any form or by any electronic, mechanical, or other means, now known or hereafter invented, including photocopying and recording, or in any information storage or retrieval system, without permission in writing from the publishers.

For permission to photocopy or use material electronically from this work, access www.copyright.com or contact the Copyright Clearance Center, Inc. (CCC), 222 Rosewood Drive, Danvers, MA 01923, 978-750-8400. For works that are not available on CCC please contact mpkbookspermissions@tandf.co.uk

Trademark notice: Product or corporate names may be trademarks or registered trademarks, and are used only for identification and explanation without intent to infringe.

Print edition not for sale in South Asia (India, Sri Lanka, Nepal, Bangladesh, Pakistan or Bhutan).

British Library Cataloguing-in-Publication Data
A catalogue record for this book is available from the British Library

Library of Congress Cataloging-in-Publication Data
A catalog record has been requested

ISBN: 978-1-032-02459-2 (hbk)

Preface

Plant fibres are being used by man for several thousand years. The fibre crops, like, cotton (*Gossypium* spp.), jute (*Corchorus capsularis* and *C. olitorius*), mesta/kenaf (*Hibiscus cannabinus*), roselle (*H. sabdariffa*), sunnhemp (*Crotalaria juncea*), sisal (*Agave sisalana*), ramie (*Boehmeria nivea*), flax (*Linum usitatissimum*) and hemp (*Canabis sativa*), share the great economic importance in different countries. These crops, based on the type and origin of fibres, may be classified into three major groups, such as, (1) the plants producing fibres from stem bark or more precisely, from phloem tissues (e.g. jute, mesta, roselle, sunnhemp, ramie, flax and hemp); (2) the plants producing fibres from the whole vascular bundles of leaf (e.g. Sisal) and (3) the plants that produce fibres from a single elongated cell of fruit or seed (e.g. cotton). The fibres, which producing from phloem tissues, are known as bast fibres. The bast fibres are composed of lignocelluosic substances and mostly derived by retting only, except, for ramie which requires degumming after retting by chemical treatment.

As regards the fibre production scenario in the world , the major fibre crop producing countries are Bangladesh, Bhutan, China, India, Myanmar, Nepal, Pakistan and Thailand for Jute; Argentina, China, Cuba, Egypt, Guatamala, Hewti, India, Indonesia, Iran, Italy, Malaysia, Mozambique, New Guina, Peru, Russia, South Africa, Spain, Thailand, U.S.A and Zimbabwe for Kenaf; Australia, Brazil, Caribbean countries, China, Egypt, Hawaii, India, Jamaica, Mali, Mexico, Philippines, Senegal, South Africa, Sudan, Tanzania, Thailand, U.S.A. for Roselle; Bangladesh, Brazil, Chilie, China, Hungary, India, Poland, Romania, Russia and Turkey for Sunnhemp; Brazil, China, Philippines, South Korea and Taiwan for Ramie; Argentina, Belgium, Canada, France, Germany, Holland, Italy, Russia, Switzerland, U.K. and Ukraine for Flax; Brazil, Chile, China, France, Holland and South Korea for Hemp; Angola, Brazil, China, Kenya, Madagascar, Morocco, Mozambique, South Africa, Tanzania and Venezuela for Sisal and African tropics, Australia, China, Egypt, India and United States for Cotton. Cotton is referred to as "King of Fibres" and "White Gold". It is the most important fibre crops in the world and plays a pivotal role in socio-economic development of India which is the second largest producer of cotton, producing 352 lakh bales as in 2015-16. Jute, also called as "Golden Fibre" for its natural lustre, is the second most important fibre crop. It is an important bast fibre produced mostly in Indo-Bangladesh region. India is the largest producer of jute, producing 114 lakh bales as in 2013-14. Kenaf fibre is produced mainly in India and China, followed by Bangladesh. The world kenaf production in 2008 was estimated at 272.000 tonnes. In the Indian subcontinent, especially in the Ganges Delta region, roselle is cultivated for plant fibres. However, this crop is grown for various other purposes in different countries. China and Thailand are the largest producers of different products of roselle and control much of the world supply. The sunnhemp cultivation is now dropped down with the present annual production of its fibre is around 18 thousand ton in India. The global production of flax fibre is 809258 tonnes as in 2016. France is the largest producer of flax fibre, produced 587,047 tonnes fibres in 2016, followed by Belgium with 87,162 tonnes. The world's best flax fibre comes from Belgium and adjoining countries. In India, the manufacturer of linen fabrics import the flax fibres from European countries and does not utilize the flax produced in India. Sisal is the major source of vegetal hard fibres. Processed sisal fibres are used to make twine, cord, carpets, bags and hats. Global production of sisal fibre is amounted to 281 thousand tonnes as in 2013 of which Brazil produced 150,584 tonnes, Tanzania approximately 34,875 tons, Kenya 28,000 tonnes, Madagascar 18,950 tonnes and China 16,500 tonnes. Venezuela contributes 4,826 tonnes with a smaller amount coming from Morocco, South Africa, Mozambique, and Angola. As regards the ramie production, FAO estimated world production of ramie green plant as 280 000 tonnes in 2005, of which almost all was grown in China. The high fibre quality, such as, smooth, long and excellent tensile strength, is the reason why ramie is widely cultivated in China. In China, ramie is the second most important fibre crop, with its growth acreage and quantity of fibre production second only to cotton. Hemp is grown specifically for the industrial uses of its derived products. It is one of the fastest growing plants. Hemp fibre due to its qualities such as excellent strength, toughness and capability to hold oil is widely used for making canvases. The largest producer of hemp fibre is France, producing 48,264 tonnes fibres annually and that closely followed by China with 44,000 tonnes of annual production.

Diseases and insect pests have a serious impact on the economic output of the fibre crops. The fibre plant pathogens and crop-feeding insects are integral part of agroecosystems in these crops, where they have coevolved with these crops over a period of thousands years. A cascade of mutual and complex interactions exists between the cultivated fibre crops and their pests and diseases. As such, the fibre crops are affected by various diseases and insect pests and these cause considerable yield losses. In addition to yield losses caused by diseases, these new elements of complexity also include post-harvest quality losses. Crop losses due to pests and pathogens are direct, as well as indirect; they have a number of facets, some with short-, and others with long-term consequences.

An estimate was made on potential and actual losses despite the current crop protection practices with cotton and other crops for the period 2001-03 on a regional basis (19 regions) as well as for the global total. Insect pests that feed on plant structures directly produced yield, such as growing tips and fruiting structures, are generally the greatest problem in a cotton crop. The total global potential loss due to pests was 26-29% for cotton, while, overall losses due to animal pests and pathogens were 18 and 16%, respectively. Incidence of different diseases and insect pests is the major limiting factors in productivity improvement of fibre crops, except sisal, which is to some extent resistant to those. Crop losses due to harmful organisms can be substantial and may be prevented, or reduced, by crop protection measures. The fibre crop protection against plant diseases and insect pests, in particular, has an obvious role to play in meeting the growing demand for fibre quality and quantity. The crop protection is now primarily focused on protecting the crops from yield losses due to diseases and insect pests. Still, the problem remains as challenging today as in the 20th century.

A comprehensive knowledge on the diseases and pests of fibre crops is essential to adopt any management method in combating their attacks. In this regard, vast research findings have already been achieved, but these are in very scattered way, either as short research articles or technical bulletin or somewhat in aggregated manner in a few chapters of a book on fibre crops or along with other crops. In certain cases, crop wise diseases and pests are described in books and in website documents, but these are mostly dealt with a few aspects of descriptions. Practically, there is no book at present which comprehensively and exclusively dealt with the diseases and insect pests of fibre crops. This book, entitled, 'Diseases and Pests of Fiber Crops', presents the major and alarming diseases and pests of different fibre crops, ignoring a few mere occurrences of pests. It covers major disease and pest damages with the methods to combat them in fibre crop cultivations. The diseases and pests are properly described giving emphasis on both morphological and molecular characteristics of pathogens and biology of different insect pests. The latest and most up-to-date knowledge on these aspects which acquired from diverse, complex, contemporary scientific discoveries in the field of fibre crop diseases and insect pests are compiled and presented in this book. This is a reference book; however the subject matters are described in textbook format which attempt to provide undergraduate, postgraduate and research personnel a means to acquire deeper knowledge on these subjects. Plant pathologists, entomologists and agricultural research scientists, and in academia, may find much of great use in this book. This book is written in 8 (eight) major chapters, each representing a type fibre crop, except for chapter 2 (two) which deals with both Mesta (kenaf) and Roselle for their similarities in disease and insect pest attacks. Each of the eight chapters is again subdivided into 2 or 3 (only for Chapter 2) subchapters to deal with different types of diseases and pests separately. In writing this book, the author has used various research findings of many scientists from different decades and computed in his own way of presentation. Practically, without their contributions, this book would never be possible to write. The author is very much thankful to them whose works are used in this book. The author is thankful to Dr. P.G. Karmakar, Ex Director of ICAR-Central Research Institute for Jute and Allied Fibres, Barrackpore for extending him research facilities. In addition, the author wishes to give thanks to the present Director and Scientists colleague of this institute for their heartfelt supports in all of his activities, which encouraged the author to write this book. The author is also thankful to his wife Mrs. Lila Biswas, for her support and encouragement.

Subrata Biswas

Contents

Preface v

1. **Jute (*Corchorus olitorius, C. capsularis*) -Malvales: Malvaceae** 1
   A. Diseases 1
   1. Bacterial leaf blight 1
   2. Anthracnose 3
   3. Black band disease 6
   4. Leaf spot 8
   5. Powdery Mildew 10
   6. Soft Rot 14
   7. Sooty Mould/ Sooty mould of pods 18
   8. Stem Gall 21
   9. Stem rot/Stem & root rot/ Charcoal rot 23
   10. Yellow mosaic disease 30
   11. Root knot disease 32
   B. Pests 33
   1. Grey weevil /Ash weevil 33
   2. Hairy caterpillar/ bihar hairy caterpillar / jute hairy caterpillar 35
   3. Mealy Bugs 38
   4. Mites 46
   5. Jute semilooper/Angled gem 53
   6. Jute stem-weevil 57

2. **Mesta/Kenaf & Roselle (*Hibiscus cannabinus; H. sabdariffa*) Malvales: Malvaceae** 61
   A. Mesta diseases 61
   1. Anthracnose 61
   2. Charcoal rot 63
   4. Eye rot 66
   5. Leaf spots 69
   6. Root rots/wilts 73
   7. Tip rot 77

    8. White stem rot/cottony soft rot/ white mould 78
    9. Viral diseases 81
    10. Root knot nematode disease 85
  B. Roselle diseases 88
    1. Bacterial Wilt 88
    2. Black leg 89
    3. Foot and stem rot 92
    4. Foliar blight /Leaf blight 97
    5. Leaf fleck 98
    6. Leaf spots 100
    7. Boeremia (Phoma) leaf blight 105
    8. Powdery mildews 107
    9. Seedling blights 111
    10. Stem rot/Southern blight 115
    11. Vascular wilt 116
    12. Watery leaf Spot 119
    13. Virus diseases 121
  C. Mesta (kenaf & roselle) pests 126
    1. Flea Beetle 126
    2. Mesta Hairy caterpillar 128
    3. Mealybugs 130
    4. Spiral borer 133
    5. Mesta stem weevil 135

3. Sunnhemp (*Crotalaria juncea*) -Fabales: Fabaceae 139
  A. Diseases 139
    1. Anthracnose 139
    2. Leaf spot 141
    3. Powdery mildews 143
    4. Rust 146
    5. Root and stem rot 147
    6. Twig blight 149
    7. Wilts 151
    8. Seedling diseases (blight, damping off, rot & wilt) 157
    9. Virus diseases 159
  B. Pests 165
    1. Crotalaria pod borer 165
    2. Eriophid mite 167
    3. Sunnhemp flea beetle 169

4. Hairy caterpillar	171
5. Stem grinder	173
6. Sunnhemp mirid	175
7. Top shoot borer	176

4. Ramie (Boehmeria nivea)- Rosales: Urticaceae ... **181**

A. Diseases ... 181
   1. Alternaria leaf spot ... 181
   2. Angular leaf spot ... 183
   3. Anthracnose ... 186
   4. Brown root rot ... 192
   5. *Curvularia* leaf blight ... 194
   6. White cane rot ... 197
   7. White fungus disease ... 200
   8. Ramie mosaic ... 203
   9. Nematode diseases ... 205

B. Pests ... 219
   1. Indian red admiral caterpillar ... 219
   2. Six-spotted zigzag ladybird ... 222
   3. Leaf beetle ... 224
   4. Leaf roller ... 225
   5. Mealy bug/ Pink hibiscus mealybug ... 230
   6. Ramie moth ... 231
   7. Termite ... 234
   8. White grubs ... 236

5. Hemp (*Cannabis sativa*)- Rosales Cannabaceae ... **239**

A. Diseases ... 239
   1. Anthracnose ... 239
   2. Bacterial blight /leaf spot of hemp ... 244
   3. *Xanthomonas* leaf spot ... 246
   4. Downy mildew ... 248
   5. *Fusarium* wilt ... 251
   6. Grey mould ... 253
   7. Powdery mildew ... 258

B. Pests ... 263
   1. Aphids ... 263
   2. European Corn Borer ... 273
   3. Hemp flea beetle ... 276

    4. Hemp borer 278
    5. Mites 282

6. Flax (*Linum usitatissimum*)- Malpighiales: Linaceae 287
  A. Diseases 287
    1. Anthracnose 287
    2. Basal stem blight 292
    3. Pasmo 294
    4. Powdery mildews 299
    5. Rust 304
    6. *Sclerotiana* stem rot 309
    7. Seedling blight 314
    8. Stem break 323
    9. Wilt 327
    10. Other minor diseases 335
    11. Virus/ virus like diseases 338
  B. Pests 345
    1. Aphid 345
    2. Army cutworm 348
    3. Linseed Blossom Midge 351
    4. Flax Bollworm 354
    5. Cutworms 356
    6. Flea Beetles 363
    7. Foliar caterpillars (climbing cutworms) 367
    8. Thrips 369

7. Sisal (*Agave sisalana*)Asparagales: Asparagaceae 377
  A. Diseases 377
    1. Anthracnose 377
    2. Bole rot 380
    3. Korogwe leaf spot 383
    4. Zebra disease 385
  B. Pests 390
    1. Agave weevil 390

8. Cotton (*Gossypium hirsutum, G. barbadense, G. arboretum, G. herbaceum*)- Malvales: Malvaceae 395
  A. Diseases 395
    1. Bacterial Disease: Angular leaf spot 395
    2. Alternaria leaf spot 399

3. Boll rot 403
4. *Corynespora* leaf blight 407
5. Grey mildew 409
6. *Mycosphaerella* (Cercospora) leaf spot 413
7. *Paramyrothecium* (Myrothecium) leaf spot 415
8. Root rot 417
9. Rust disease 422
10. *Stemphylium* leaf spot 429
11. Tobacco Streak virus disease 431
12. Vascular (*Fusarium*) Wilt 433
13. *Verticillium* wilt 438
B. Pests 442
1. Cotton Aphid/Melon aphid 442
2. Bollworms 446
3. Dusky cotton bug 455
4. Weevils 458
5. Cotton leaf roller 463
6. Cotton mealybug 466
7. Midges 471
8. Mirid bugs 477
9. Pink bollworm 487
10. Red cotton bugs 494
11. Spider Mites 500
12. Spotted bollworm 506
13. Stink bugs 515
14. Thrips 524
15. White fly 529

References **535**

Index **671**

1
Jute (*Corchorus olitorius, C. capsularis*) -Malvales: Malvaceae

A. Diseases

1. Bacterial leaf blight

Occurrence & severity status

The disease Bacterial leaf blight, also known as 'Bacterial leaf spot', occurs on white jute (*Corchorus capsularis*) and tossa jute (*C. olitorius*) in seed crops of jute at the ICAR-CRIJAF research farm, Barrackpore, West Bengal, India (Biswas *et al.*, 2013, 2014), in Formosa (Elliott, 1951) and Sudan (Sabet, 1957). The incidence of the disease varies from 2-5% (Borker and Yamlembam, 2017).

Symptoms

The disease initially appears on leaves as small brown circular spots of 2-5mm diameter (in tossa jute) or angular brown spots (in white jute), In most cases, the spots are surrounded by a yellow halo and the angular spots coalesce at later stage show larger blighted areas on lamina (Borker and Yamlembam, 2017). The lesions on the stems are elongated; sometimes the spots girdle the stem. The disease on the capsules produces small sunken brown spots. Defoliation and death of the plants may result in severe infections. In the later stages of disease, brown sunken spots were found on the green capsules (Sabet, 1957; Biswas *et al.*, 2013, 2014).

Causal organism: *Xanthomonas campestris* (Pammel 1895) Dowson, 1939

[**N.B.** The bacterium is generally classified in two different types based on host plants they attacked and that is considered as pathovar. For example, the pathogen of white jute (*C. campestris*) is classified as *Xanthomonas campestris* pv. *capsularii*, while, the pathogen of tossa jute (*C. olitorius*) is named as *Xanthomonas campestris* pv. *olitorii* (Biswas *et al.*, 2013, 2014). However, the validity of pathovar is questionable in some cases (Bila *et al.*, 2013), particularly the latter one is yet to be recognized]

Synonyms

= *Bacillus campestris* Pammel, 1895

= *Pseudomonas campestris* (Pammel) Smith, 1897

= *Bacterium campestris* (Pammel) Smith, 1897

= *Phytomonas campestris* (Pammel) Bergey *et al.*, 1923

Morphological/Morphophysiolgical characteristics

Bacterial colonies on nutrient agar medium appear as pale yellow cream in color (Biswas *et al.*, 2013). These are convex, mucoid, slimy, glistering and individual colony round in shape. The bacteria is an aerobic, Gram-negative and rod shaped, measuring 0.4-1.0µ x 0.7-2.0µ in size, motile by a single polar flagellum. These are positive in several biochemical tests, like, KOH string assay, H_2S production and catalase activity, while, they are negative to oxidase test and nitrate reduction with oxidative metabolism of glucose (Naqvi *et al.*, 2013). The bacteria produce an extrcellular polysaccharide known as Xanthan gum.

Molecular characteristics

PCR, sequencing and subsequent BLASTn analysis are used as a molecular technique for identifying the pathogen (Biswas *et al.*, 2014). PCR with *Xanthomonas campestris* specific primers NZ8F3/NZ85R3 generates an amplicon of 530 bp from the bacteria attacked tossa jute (Biswas *et al.*, 2013). Biswas *et al.* (2013) has observed that the amplicons from the bacterial strain JB-CO-13 shows 100% identity with *X. Campestris* pv. *olitorii* (EU285213). Nucleotide span and ORF finder (NCBI) analysis indicates the 530-bp PCR amplicon coded part of a gyrase B gene has 100% identity with a translated gene product (Protein ID: ABX84334) (Biswas *et al.*, 2013).

Predisposing facters

The disease favours 25-30°C.

Disease cycle

The primary source of the disease inoculum is infected plant parts or seeds. It can survive in plant debris in soil up to 2 years, but not more than 6 weeks in free soil. The bacteria get entry through the hydathodes, wounds caused by insects or occasionally through stomata. The bacteria manifests in the diseased plant parts and dispersed by rain water splashes, mechanical equipments under wet and humid conditions.

Control measures

The antagonist bacteria, like, *Bacilus subtilis* (strain BB as found by Wulff *et al.*, 2002) or *Pseudomonas fluoroscens* (strain ID-3 as found by Naqvi *et al.*, 2013) may be applied in case of jute disease management, since these are found effective in brassica and sesame, respectively, against the same bacterial species.

Spraying Copper oxychloride (@ 1525g a.i./ha) or cupric hydroxide (@540-810g a.i./ha) solution (keeping in mind that higher concentration cupric hydroxide as phytotoxic) may be effective to manage the disease chemically in jute, since, the chemicals were found effective against the same pathogen species in pepper in Israel (Azaizeh and Bashan, 1984)

2. Anthracnose

Occurrence & severity status

The jute anthracnose [(Fig. 1A-B)] was first observed in 1938 at Kumamoto and Shizuoka Prefecture in Japan (Ikata and Yoshida, 1940). Now, it is found to occur in white jute (*Corchorus capsularis* L.) in most of the major jute growing countries like, India (Ghosh, 1956), Bangladesh (Islam and Ahmed, 1964), China (Niu *et al.*, 2016a), etc. In China, it is considered as a most serious disease, particularly, in Fujian, Henan, Guangxi and Zhejiang provinces, where, yield was reduced by 20% on average, and in certain cases it reached up to 50%. (Niu *et al.*, 2016) . In India, the disease was first observed prior to 1952 at Chinsurah in an epidemic form in an exotic variety, 'Jap Rey', of capsularis. The disease appears in late July and remains as most widespread in late August. In case of *Corchorus olitorius*, the disease is particularly virulent in Assam (especially prevalent where fertiliser dosage is over 60 Kg N/ha) affecting over 60 % of the crop, but mortality due to disease is low. It has been suggested that the Indian strain of the pathogen is different a form Malayalam strain (Ghosh, 1956). The Indian varieties of *C. capsularis* and almost all *C. olitorius* varieties are reported to be immune to anthracnose. Further, it is believed that the disease has found its way from Far Eastern countries. It has been shown that non-branching types of capsularis of Japanese, Chinese and Formosan origin are more susceptible than the Indian branching types. Whether the fungus is an indigenous strain or a new introduction is not definitely known.

Fig. 1 (A-B). Anthracnose of jute caused by *Colletotrichum corchori*

Symptoms

Lesions appear on stems, leaves and pods as brown to black, not sunken but with definite in outline. Initially, it appears as small brownish-black spots (black in olitorius), covering the stem, which coalesce to form cankerous tissues. Later, the disease causes chlorotic regions with black brown sunken necrotic pits on the surfaces of stems. In late stages, the diseased plants undergo defoliation, dieback and blight, successively. Stems break at infected points and die, when such tissues yield the weight of the crown or to the strong wind. In certain cases, the pods of *C. olitorius* shrivel due to fungal attack and also, the disease affects fibre yield, when the lesions coalesce to form a big canker. Sometimes, girdling of the stem and shredding the fibre make anthracnose a major threat to jute fiber production and quality (Ikata and Yoshida,1940; Niu *et al.*, 2016a).

Causal organism

***Colletotrichum corchori* Ikata& I. Tanaka1940** (accepted name, NARO Gene bank)

[Annals of Phytopathological Society of Japan, 10: 148 (1940)]

Synonyms

= *Colletotrichum corchorum-capsularis* sp. nov. Niu *et al.*2016

[*Scientific Reports* **6**, Article number: 25179 (2016); doi:10.1038/srep25179]

= *Colletotrichum corchorum* IKATA *et*TANAKA *nov.* sp. (Ikata and Tanaka)

[Japanese Journal of Phytopathology Vol. 10 (1940-1941) No. 2-3 P 141-149 (no Latin diagnosis)]

[N.B. Prior to the polyphasic identification of *Colletotrichum* species, *C. gloeosporioides* and *C. corchorum* were generally recognized as the most important jute pathogens worldwide (Ikata and Yoshida, 1940; Purkayastha and Sen-Gupta, 1975). The pathogen, *C. gloeosporioides,* has been found to cause anthracnose disease in Tossa jute (*C. olitorious*) in West Bengal (India). Purkayastha and Sen Gupta (1973) studied on conidial germination and appressoria formation]

Morphological/Morphophysiolgical characteristics

The fungus on PDA medium develops colonies @ 6.5–10.5 mm/day in diameter at 28 °C. After 7 days, it becomes greyish white to dark gray mycelium and dense, concentric, circular conidia masses. Acervuli black, superficial, scattered; stroma patelliform, 100-350μ in diameter by 25-50μ high; setae several to abundant, originating from margin of stroma, yellowish brown to black and becoming lighter toward the apex, 2 to 5 septate, 36-117μ long by 3.6-5.0μ wide. Conidiophores simple, hyaline, arising from stroma, 15-35μ long by 3-4μ wide; conidia abundant, non-septate, hyaline, curved, falcate, fusiform, bluntly tapered to each end, (12) 14.5 - 19.5 (25) x (3.6) 3.5 -4(6.0)μm with 16-22 × 4μ (differently 18.3–26.3 × 2.7–4.3 μm with mean size 22.6 × 3.62μ) as the most common size. Appressoria are, brown, ovoid to ellipsoidal, edge entire, (6.8) 6.5 -11(12.5) x (6.0) 5.5 - 8.5 (9.8) μm, sometimes becoming Complex (Ikata andYoshida , 1940; Wijesekara, 2005; Niu *et al.*, 2016b).

Molecular characteristics

The molecular analyses was performed on all of the *Colletotrichum* strains isolated from jute, which included 2 strains of *C. gloeosporioides* complex and 5 strains with curved conidia (Niu *et al.*, 2016b). A phylogram was constructed to identify the strains in the *C. gloeosporioides* species complex. The 2 strains were confidently identified as *C. fructicola* and *C. Siamense.* with 100% bootstrap support, based on the combined datasets of partial CAL, GAPDH, GS and TUB2 sequence analysis. The other 5 strains with curved conidia did not cluster with any currently known species based on these 4 molecular markers. Therefore, a further 6 gene regions (ACT, CAL, GAPDH, GS, ITS and TUB2) of these five strains were sequenced and phylogenetic relationships were predicted using parsimony and Bayesian methods. However, these 5 strains did not cluster well with any other *Colletotrichum* species in the 6 gene phylogenetic tree. The morphological and culture characteristics were closest to the species *C. corchorum*. So these 5 strains were described by Niu *et al.* as *C. corchorum-capsularis* sp. Nov (Niu *et al.*, 2016b).

Predisposing factors

The optimum temperature for growth of the fungus is 30°C. The disease is seed-borne, the fungus mycelium exists in the seed and spores are adhering on the external part of seeds.

It spreads from the stem during hot, humid days (temperature above 33°C, humidity 84% and above, high dose of N) of late July and is most widespread in late August.

Disease cycle

Disease is disseminated by seed, soil and air as conidia or over wintering mycelia. The infection occurs on host where it forms conidia in sporodochia. These conidia repeatedly infect the host tissue.

Control measures

For the management of this disease following measures are suggested:

- The growing of disease resistant varieties like D 154 and JRC 212
- Crop rotation with rice, wheat, etc. in case of this disease affected areas
- Soil amelioration with lime (where pH is lower than 6.2) @ 22 Kg/ha
- Seed treament with Vitavex-200/Provax-200 @ 4g/1 kg seed
- Copper oxychloride (50% Cu) at 0.75% conc., should be sprayed at the initial stage of infection and followed by 2-3 sprays at intervals of 7 days.

3. Black band disease

Occurrence & severity status

The disease was first described as 'black band' by Shaw in 1921. The disease is considered to be of minor importance and occurs only in few places (Rangaswami and Mahadevan, 1998). It generally occurs sporadically in *Terai* or red soil areas where soil fertility and moisture are low. Both white and tossa jutes are prone to be infected by this disease. It is prevalent in mid-July, when the plants are matured (Shaw, 1921). In Japan at Okinawa, it is known by the name, "kurogare-byo^'" of Jew's marrow, and occurs on tossa jute (nalta jute in Japanese) in March (Sato *et al.*, 2002, 2008). It occurs in all major seed growing countries, however, its occurrence in epidemic form reported once from Rangpur and Kishoreganj (earlier East Pakistan, at present Bangladesh).

After 1948, "black band" as a major disease has not been reported (Wadud and Ahmed 1962; Punithalingam 1980; Sutton 1980a; Mukerjee and Mishra 1988).

Symptoms

The lesion appears on stems just above the ground level, at first, as a small black or blackish brown patch, which gradually enlarges and eventually, encircles the stem making a black band around. The surface of the affected portion of stems becomes warty, due to development of pycnidia all over the stem. Black coloured sooty spore masses develop over the warts of the affected stem parts during high humid condition and that adhere to the fingers on touching. Later, wilting of leaves, rapid defoliation, desiccation and death of plants occur (Shaw 1921; Mukerjee and Mishra 1988; Sato *et al.*, 2008).

Causal organism: *Lasiodiplodia theobromae* **(Pat.) Griffon & Maubl. 1909**

(Ref. Sato *et al.*, 2008)

Synonyms

= *Botryodiplodia theobromae* Pat. 1892

= *Diplodia corchori* Syd. & P. Syd. 1916

Morphological/Morphophysiolgical characteristics

The fungal mycelium is hyaline, profusely branched, sparsely septate, 2.0-4.8 µm in diameter. At later stages, the mycelium turns greenish to brown and ultimately black with close septation. The pycnidia on the host are black, ostiolate, stromatic, erumpent and confluent measuring 87.7-300.8 µm in diameter (mostly 130–200 x 120–160 µm in size). Pycnidia are produced in stromata as single or often compounded in the group of 2-6 and aggregated in botryose clusters (Patel, 1971). The pycnidia are elongated flask shaped, usually papillate with prominent ostiole. Conidia are initially hyaline, unicellular, ellipsoid to oblong, thick walled with granular contents. On maturity the conidia become 2-celled, dark brown, dolliform to shortly cylindrical, with longitudinal striations, usually 20-30 X 10-15 (16) µm in size (Ainsworth *et al.*, 1973; Punithalingam, 1980; Cedeno *et al.*, 1995; Sato *et al.*, 2008).

The fungus forms fruiting structure, stromata, in pure culture. These stromata produced on malt agar and on oatmeal agar are similar in both macroscopic and microscopic structure to those formed on naturally infected plant. Both the one celled, hyaline, 'immature' conidia and the one-septate, brown coloured, 'mature' conidia germinate readily in sterile distilled water. At suitable temperatures, light in the blue region of the visible spectrum with wavelengths 400–520 mµ are effective in the photo-induction of stromata (Alasoadura, 1970). When the fungus is cultured on PDA in the dark, it grows optimum at 30°C. On PDA at

25°C under black light after 16 days, the colonies consisted of olive to grayish aerial mycelia and numerous black pycnidia with black reverse view. The morphology of the pathogenic fungus on the host plant is almost the same as that observed on PDA, except for pycnidia, which are immersed in stromata beneath the host epidermis and smaller in size (Sato *et al.*, 2008).

Molecular characteristics

Polymerase chain reaction amplification of total genomic DNA of different isolates of *L. theobromae* generates numerous polymorphisms among the isolates, which indicate high genetic variability (Sangeetha *et al.*, 2012).

Predisposing factors

Low fertility and soil moisture enhance the disease occurrence and severity.

The optimum temperature for occurrence of the disease is 30°C.

Disease cycle

The conidia may survive in seeds and in soil. These are disseminated by wind and water from decaying portions to the new host tissue. They may also be disseminated by insects. Conidia on seeds remain viable for 4 months, while, as mycelium for 1 yr. (Punithalingam, 1976). On host tissue, they develop infection and produce conidia for secondary infection.

Control measures

- Early sown crops can escape the attack of the disease
- Two times irrigations, coupled with maintenance of proper soil fertility and spraying of copper oxychloride @ 0.5% or mancozeb @ 0.2% concentration are effective to control the disease. Otherwise, early harvest is recommended to check the spread of this disease.
- Alternaively, Spraying of Mancozeb @ 2g/ litre water for 2-3 times or spraying of Mancozeb along with roughing and Malathion spray is also effective to control black band (Ashraf-Uz-Zaman, 2004).
- Spraying of garlic bulb extract may be effective. Since, it inhibits the spore germination and mycelial growth of this causal fungus (Ahmed and Sultana, 1984).

4. Leaf spot

Occurrence & severity status

The disease occurs both on tossa (*Corchorus olitorius*) and white (*Corchorus capsularis*) jutes. In addition, it occurs in several other *Corchorus* spp, like,

C. pilolobus Link, *C. tridens* L., etc, of Malvaceae (earlier Tiliaeace) family. It is mainly distributed in Bangladesh (Miah, 1974), Brazil, Cuba, India, China, Nepal, Pakistan, Philippines, South Africa and Taiwan. (Chupp, 1954; Hsieh and Goh, 1990). In India, the occurrence of *C. corchori* in jute was first reported by Chaudhury in the year 1948.

Symptoms

The disease on leaf develops angular or irregular spots which are vein-limited, 0.5-2 mm wide (differently, 3-14mm in diameter according to Chupp, 1954), often confluent to form larger spots, dark brown, paler on lower surface. Fruiting is amphigenous, but more abundant on lower surface of leaf. Stromata, when present, are small, dark brown (Chupp, 1954; Hsieh and Goh, 1990). On capsules, most predominantly affecting one week old capsule (Khan and Fakir, 1993), the disease appears as brown amphigenous spots that in certain cases named as 'Fruit mould' (Mukherjee and Bosu, 1973) or 'Sooty Mould' (Ghosh, 1983). The disease also affects the seeds in capsule, developing light brown flimsy seeds that either do not germinate or if germinate, they do not survive (Miah, 1974).

Causal organism: *Cercospora corchori* Sawada, 1019

[Report of the Department of Agriculture Government Research: 19 Institute of Formsa: 37]

Morphological/Morphophysiolgical characteristics

Stroma is dark brown, consisting conidiophores on its surface. Conidiophores arise singly or (2) 4-7 (10) in a fascicle, rarely as many as 20 spreading stalks, pale to medium brown or olivaceous brown, paler at the apex, sparingly septate (0-3 septate), not branched, rarely geniculate, almost straight, subtruncate at tip, with thickened conidial scars, (24)25-60 (150) x 4-5(6) µm in size. Conidia are hyaline to subhyaline, acicular to obclavate, straight to curved, indistinctly multiseptate, and acute at the apex, truncate or subtruncate at the base with a thickened hilum, (25) 32-90 (240) x (2) 3-5 µm. (Chupp, 1954; Mukherjee and Bosu, 1973; Ghosh, 1983; Hsieh and Goh, 1990).

Predisposing factors

Hot and humid conditions favour the disease

Disease cycle

The pathogen is seed borne. It can disperse through seeds as mycelia or by air as conidia from the site of infection. Seed borne pathogen may survive in jute plant residues in soil. During favourable condition it may produce spores that infect new hosts.

Control measures

Diseases caused by *Cercospora* species can be controlled by spraying Carbendazim (Bavistin 50%) @ 1g a.i./l.

5. Powdery Mildew

Occurrence & severity status

The disease is of minor importance in fibre yielding crop. However, it seriously affects the seed crop, where it attacks in late August up to the stage of harvesting of seeds in all areas of capsularis and olitorius jute seed production fields. Its occurrence on Jute in India was reported in 1943 (VaradaRajan., 1943).

Symptoms

The disease appears at the pod stage. A white to ash coloured powdery mass of spores with mycelia cover the leaves, branches and pods. Leaves later turns brown, wither and drop off. Infected fruits do not develop in normal way and are discoloured. Most of the seeds shrivel and lose germinability (Ghosh, 1983).

Causal organism: Podosphaera fuliginea (Schltdl.) U. Braun & S. Takam., (2000)

[Preferably, known as *Podosphaera xanthii* (Castagne) U. Braun & Shishkoff (2000) { Schlechtendalia, 4: 31}]

Synonyms

= *Oidium erysiphoides* Fr., (1832) (Anamorph)

= *Sphaerotheca fuliginea* (Schltdl.) Pollacci, (1913)

= *Sphaerotheca fuliginea f. fuliginea* (Schltdl.) Pollacci, (1911)

= *Sphaerotheca fuliginea var. fuliginea* (Schltdl.) Pollacci, (1911)

Morphological/Morphophysiolgical characteristics

Anamorph: Mycelium is superficial, uninucleate, conidiophore erect, 4-7 septate, 80-146 µm long, foot cell cylinderical followed by 1-4 barrel-shaped cell. Conidia are in chain (4-7), hyaline, ellipsoidal, more or less cylindrical in shape, contained irregular shaped fibrosin bodies, size 21-37 x 15-23 µm (average 34.5 x 20.0 µm), L/B ratio 1.28-2.27 (range). Conidia are germinated from its apical and basal part or from the side wall, germ tube simple or forked (Hirata, 1942, 1955; Pawar *et al.*, 2009).

Teleomorph: Chasmothecia (Cleistothecia, perithecia) are formed on both abaxial and adaxial surfaces of leaves. They are yellow to brown when young and turned black at maturity, globose. Its diameter is 90-113 μm (av. 99.3 μm), differently 105-155 μm (Pawar et al., 2009), Chasmothecial wall cell is large and variable in shape, 35.8 x 22.5 μm, and it has appendages variable in number. The appendage is myceloid, hyaline but dark brown at the base, its length is variable, usually as long as the diameter of the chasmothecium and has 1 to 3 septa. Ascus is single, ovate in shape, hyaline, double walled, 52-76 x 56-118 μm (average 71.3x 55.3 μm) in size, contains 8 ascospores when matured. Ascospore is ellipsoidal, single celled and hyaline, its size is 16-18 x 18–24 μm, differently, 14.8 x 18.5 μm (Hirata, 1942, 1955;Braun and Takamatsu, 2000; Pawar et al., 2009; Endo et al., 2012).

Molecular characteristics

Strains of *Podosphaera fuliginea* (=*Podosphaera xanthii*, *Sphaerotheca fuliginea*) which is mostly known as one of the causal agents of powdery mildew of cucurbits show sexual compatibility heterothallism in all isolates. Frequency of the two mating types differs significantly in the population. DNA polymorphism is determined both by restriction fragment length polymorphism (RFLP) of the ribosomal internal transcribed spacers (ITS) and 5.8S DNA amplified by the polymerase chain reaction and by random amplified polymorphic DNA (RAPD). For all 11 restriction enzymes, the strains present an identical pattern of ITS RFLP. RAPD analysis, using 22 primers which provided reproducible patterns, reveals a relatively low degree of polymorphism. Furthermore, cluster analysis based on RAPD data (152 markers) does not separate groups within the species *P. fuliginea*. There is no association between virulence, mating type, geographical and host origin and RAPD patterns. The lack of association between phenotypic and molecular markers and the close fit to linkage equilibrium for the characters examined suggest that recombination may play a role in populations of *S. fuliginea* (Bardin et al., 1997).

In a study of Lee (2012b) the molecular phylogenetic analysis of a Korean strain of *Podosphaera xanthii* was determined by rDNA sequence analysis by amplifying the internal transcribed spacer (ITS) region including the 5.8S rDNA and 28S rDNA with the ITS1F (5'-CTTGGTCATTTAGAGGAAGT-3') and LR5F (5'-GCTATCCTGAGGGAAAC-3') primer set (Lee et al., 2011; Lee, 2012a). He found that each of the rDNA ITS and 28S sequences of the strains which included NCBI Gene Bank accessions showed 100% identity values with the related strains of *P. Xanthii*. Generally, for phylogenetic analysis of the Erysiphaceae, mainly rDNA nucleotide sequence data are used. rDNA sequence data are especially useful for identifying powdery mildew fungi in the absence of teleomorphs (Okomoto et al., 2002; Lee, 2012a). The powdery mildew fungi are split into 5 major lineages based on nuclear ribosomal DNA sequences (Mori et al., 2000; Braun et al., 2006). Recently, *Podosphaera* has been divided into 2 clades: clade 1, consisting of the section *Podosphaera* on Prunus (*P. tridactylas*) and subsection Magnicellulatae; and clade 2, composed of the other member of section *Podosphaera* and subsection *Sphaerotheca* based on ITS and 28S rDNA sequence analyses (Takamatsu *et al.*, 2010). Although these data are sometimes insufficient to confirm identifications, some fragmentary molecular data including ITS rDNA sequence are becoming increasingly available and can be useful in complementing other data. P. xanthii is widely distributed and has been designated variously as *Sphaerotheca fuliginea*, *Sphaerotheca fusca*, and *Podosphaera fusca*. Although, separation of P. xanthii from the P. fusca group based on morphological properties of teleomorph was proposed earlier, it is now considered that the natural variation of morphological characters within a species make it difficult to assess the phylogenetic status at the species level. Currently, many mycologists have considered *P. xanthii* (*fusca*) as a synonym of *P. fusca* (*xanthii*) although this separation has remained controversial. Lee (2012b) reported that the molecular classification system is a useful tool for construction of the phylogenetic tree with regard to *Podosphera* species, showing that the two species belong to a distinct Xanthii/Fusca Group. Several recent workers used molecular basis with different primer sets to confirm the infection of *P. xanthii*. For example, CosmeBojorques, *et al.* (2011) used S1/S2 primer sets to confirm the *P. xanthii* infection of greenhouse cucurbits in California. Sujatha *et al.* (2015) used powdery mildew-specific ITS universal primer pair PN23/PN34 and a group of primer sets S1/S2, G1/G2 and L1/L2 which are specific to the ITS regions of *Podosphaera xanthii*, *Golovinomyces cichoracearum*, and *Leveillula taurica*, respectively, to confirm the identity *P. Xanthii* infection in sesame.

Predisposing factors

The disease coincides with the drop in night temperature in October. Foggy weather is favourable for its growth. The temperatures 16° to 32°C and high humid conditions favour the powdery mildew development.

Disease cycle

The disease is disseminated by seed, soil and air from the living plant tissues of various crops and weed hosts which are year-round availabe. *Podosphaera fuliginea* uses haustoria to gain access to the leaf epidermal cells. The fungus is usually spread through mycelium from infected plant or through chasmothecia (ascocarps). The special structure, chasmothecia, allow overwinter survival of the species that causes the disease in different other crops and weeds. Spores, which are the primary means of dispersal, are carried by wind to new hosts. Although humidity is required, the spores of this powdery mildew species can germinate and infect in the absence of free water. In fact, germination is inhibited by water on plant surfaces for extended periods. The conidia of the fungus which produced on the infected region are spread through the air and thus can cause repeated infection. The mycelium can also overwinter in the buds of infected plants.

Control measures

This pathogen is able to develop resistance to systemic fungicides. Most of the systemic fungicides are at risk for resistance development because they have single-site mode of action. Thus modification of one gene in the pathogen may be enough to enable the pathogen to resist the action of the fungicide. The protectant fungicides are not at risk for resistance development because they have multi-site mode of action. So, in most cases protectant fungicides are used to manage the disease chemically. The disease can be checked by lime-sulphur, dusting (mixing @ 3:1 ratio) especially in the early hours of the morning (when dew effectively arrests particles) and by ensuring clean cultivation. Lime sulphur dusting or spraying at early stage of infection is very effective measure to control the spread of disease. Irrigation is believed to be helpful in building resistance (Ghosh, 1983). The disease can also be controlled by spraying powdered sulphur @ 40kg/hactre but this is not economical for a crop like jute (Rangaswami and Mahadevan, 2006). In some cases, spraying of sulphur @2-2.5kg/h is recommended for controlling the disease. The best sulphur product to use for powdery mildew control is wettable sulphur that is specially formulated with surfactants similar to those in dishwashing detergent. Spraying of wettable sulphur (Thiovit 80 WG) @ 32g/10 litres water is also suggested to apply for controlling the disease. The fungicide program should consist of an alternation among systemic fungicides with different modes of action (eg a Q_O inhibitor fungicides (Q_O I Strobilurin fungicides), namely, azoxystrobin (Amistar), fenamidone(Reason), fluoxastrobin (Disarm), kresoxim methyl (Sovran), pyraclostrobin (Cabrio) and trifloxystrobin (Compass; Flint), are alternated with Chlorothalonil (aka) non-systemic fungicide (Bravo, Echo, Daconil) or DMI (DeMethylation Inhibitors) fungicides (viz., Propiconazloe (Tilt), myclobutanil (Rally), tebuconazole (Folicur), triflumizole (Procure), mycobutanil (Nova)] or alternatively, mixing these with protectant fungicides (McGrath, 2004). However, the best measure is by growing of resistant cultivars and much effort is required to develop that.

A study on the powdery mildew of squash caused by *Sphaerotheca fuliginea* in Egypt reveals that before or after infection- spraying of potassium salt (KCl, KH_2PO_4 or K_2CO_3) solution @ of 750 µg/ml decreased infection, as compared with the fungicide Prochloraz (Master 25%), suggesting biochemical changes are associated with induced systemic resistance (Muhanna *et al.*, 2011).

6. Soft Rot

Occurrence & severity status

The disease is widely distributed in tropical and subtropical countries, as it requires high temperature and a rainy season to thrive. It occurs in most of the Jute growing countries. In India, it is most prevalent in Assam and North Bengal (Ghosh and George, 1955). It attacks both species of jute, *C. capsularis* and *C. olitorius*. It appears in mid-july and gains in severity by harvesting time (Ghosh and George, 1955). Apart from jute, the disease affects many plants belonging to nearly 100 families.

Symptoms

The disease first appears at the basal region of the stem, just above the ground level as wet, soft brown lesions with profuse white fluffy mycelial growth. The lesions progress towards roots and cause decay. Later, the white fluffy mycelia turn compact and contain mustard-shaped white (initially) or dark brown (on maturation) sclerotia. The peeling-off bark exposes rusty brown and damaged fibre layer. The plant breaks off at the points of infection.

Causal organism: *Athelia rolfsii* (Curzi) C.C. Tu & Kimbr., 1978 [Botanical Gazette Crawfordsvile, 139: 460]

[N.B. The fungus is mostly described as *Sclerotium rolfsii* (considered as a synonym). The recent accepted name of the fungus is *Athelia rolfsii* (Curzi) C.C. Tu & Kimbr. In 1932, Mario Curzi discovered that the teleomorph (spore-bearing state) was a corticioid fungus and accordingly placed the species in the form genus *Corticium*. With a move to a more natural classification of fungi, *Corticium rolfsii* was transferred to *Athelia* in 1978. *Sclerotium rolfsii* is the anamorphic stage of the pathogen and the teliomorph stage i.e. the sexual stage is rarely observed. The teleomorph *Athelia rolfsii* is a basiomycete classified in the order: Atheliales.]

Synonyms

= *Corticium rolfsii* Curzi

= *Pellicularia rolfsii* (Curzi) E. West

= *Botryobasidium rolfsii* (Curzi) Venkatar.

= *Sclerotium rolfsii* Sacc. (anamorph)

Morphological/Morphophysiolgical characteristics

The teliomorph of *S. rolfsii* (*Athelia rolfsii*) is rarely observed on hosts or in culture. The fungus produces effused basidiocarps that are smooth and white.

Microscopically, they consist of ribbon-like hyphae with clamp connections. Basidia are club-shaped, 4 to 6 μm x 7 to 14 μm in size, bearing four basidiospores. Basidiospores are ellipsoid, hyaline, smooth, measuring 4-7 μm x 3-5 μm (differently 1.0 to 5 μm x 5 to 12 μm) (Aycock, 1966; Roberts et al., 2014). Small, brownish sclerotia (hyphal propagules) are also formed, arising from the hyphae .

The anamorph (*Sclerotium rolfsii*) is a necrotrophic, soilborne fungus, causing disease in susceptible host plants. It produces abundant white mycelium on infected plants and in culture. Advancing mycelium and colonies often grow in a distinctive fan-shaped pattern and the coarse hyphal strands may appear as ropy structure. Cells are hyaline with thin cell walls and sparse cross walls. Main branch hyphae may have clamp connections on each side of the septum (Aycock, 1966). The fungus grows well on a variety of media and can utilize a wide range of carbon and nitrogen compounds. On agar media, linear growth is rapid at 30° C.; in liquid media best growth occurs between 20° and 30° C (Abeygunawardena and Wood, 1957). In agar plate culture, sclerotia are not formed until the mycelium covers the plate. Sclerotia begin as small tufts of white mycelium that form spherical sclerotia 0.5 to 1.5 mm in diameter. They become tan to dark brown on maturation due to presence of melanin-like pigment (Chet *et al.*, 1967). Young sclerotia often exude droplets of clear to pale yellowish fluids. Mature sclerotia are hard, slightly pitted, and have a distinct rind. Although most sclerotia are spherical, some are slightly flattened or coalesce with others to form an irregular sclerotium. *S. rolfsii* does not form asexual fruiting structures or spores. (Aycock, 1966; Roberts *et al.*, 2014)

Molecular characteristics

In order to identify the *Athelia rolfsii* (= *Sclerotium rolfsii*), Mahadevakumar *et al.* (2016) isolated genomic DNA of *Sclerotium rolfsii* causing Fruit rot of *Cucurbita maxima* and amplified internal transcribed region of ribosomal DNA using universal primers ITS1- 5 - CGGATCTCTTGGTTCTGGCA-3 and ITS4 - 5 -GACGCTCGAACAGGCATGCC-3 . In their studies the rDNA sequence results showed 99-100 % similarity with reliable sequence of *S. rolfsii* retrieved from GenBank (AB075298.1 and JF966208.1), which confirmed the identity of the pathogen. Also, the phylogenetic tree constructed based on Neighbour-Joining [NJ] method using MEGA6 software showed that the representative sequence (GenBank accession number KJ002764.1) from the their study shared a common clade with the ITS sequences of GenBank (KP412466.1 and JF966208.1). In this way the identity of the pathogen was further confirmed (Mahadevakumar *et al.*, 2016). Similarly, the identity of *S. rolfsii* associated with leaf blight disease of *Psychotria nervosa* (wild coffee) was confirmed by Mahadevakumar and Janardhana (2016).

However, there is considerable genetic variability among the isolates collecting from different regions and hosts. The genetic variability among the virulent isolates of *S. rolfsii* collecting from Groundnut was studied by Prasad *et al.* (2010). They used molecular techniques like RAPD, ITS-PCR and RFLP. The RAPD banding pattern reflected the genetic diversity among the isolates by formation of two clusters. A total of 221 reproducible and scorable polymorphic bands ranging approximately as low as 100 bp to as high as 2500 bp were generated with five RAPD primers. The amplified ITS region of rDNA with the specific ITS1 and ITS4 universal primers produced approximately 650 to 700 bp polymorphic bands in all the isolates and that confirmed the isolates were *S. rolfsii*. The ITS-RFLP studies indicated that there was no polymorphism in restriction banding pattern among the isolates with the restriction endonucleases used (Prasad *et al*, 2010). Similar genetic variability amongst the different isolates of *S. rolfsii* in RAPD-PCR analysis was observed by several earlier researchers (Almeida *et al.*, 2001; Kokub *et al.*, 2007; Punja and Sun, 2001). Prasad *et al.* (2010) observed the polymorphic and distinguishable banding pattern for the DNA fragments amplified with five random primers signifying genetic diversity among all eight isolates of *S. rolfsii* as studied. Parvin *et al.* (2016) used DNA fingerprinting by RAPD analysis to select virulent strains of *S. rolfsii* casing diseases in egg plant and tomato. They identified most virulent strain is S8 which collected from tomato of Thakurgaon.

The UPGMA based dendrogram generated for the *S. rolfsii* isolates reveals that they were divided into two main clusters, A and B which were further subdivided into subclusters (Rasu *et al.*, 2013). Perez *et al.* (1998) observed that the isolates collected from different fields clustered into different groups in the dendrogram based on pair-wise dissimilarities study. Four distinct groups of *S. rolfsii* isolates based on DNA polymorphisms for 21 random primers were observed by Saude *et al.* (2004).

Predisposing factors

The disease is favoured by hot, humid weather and jute plant debris. The pathogen can grow in a wide range of temperature (8°C to 40°C). However, the mycelial growth is most rapid at 30° C (Abeygunawardena and Wood, 1957). The sclerotia are produced as the temperature rises from 27°C to 35°C. The fallen leaves of jute while come in contact with infested soil encourage the multiplication of the pathogenic fungus. The disease requires high temperature and a rainy season to thrive. However, temperature and moisture may interact in influencing survival of *S. rolfsii* under field conditions (Beute and Rodriguez-Kabana 1981). In one such study, high moisture plus high temperature adversely affected survival of sclerotia of *S. rolfsii* (Beute and Rodriguez-Kabana 1981).

Disease cycle

Primarily, the disease is spread by sclerotia. The sclerotia spread to uninfected areas by wind, water, animals, and soil. Mycelium is also carried to new places by infected seeds. Sclerotia serve as overwintering structures (Xu, 2008). They germinate in the presence of oxygen; hence, they survive better in top layer of soil. Sclerotia germinate either by hyphal or eruptive germination (Punja and Grogan 1981). Eruptive germination means that aggregates of white mycelium burst out of the sclerotial rind. An external food source is not required for this type of germination. Sclerotia can germinate eruptively only after being induced by dry conditions or volatile compounds (Punja and Grogan 1981). Sclerotia can germinate hyphally .In this case, growth of individual hypha from sclerotia occurs in response to availability of exogenous nutrients (Punja and Damiani 1996). The hyphae can directly penetrate the susceptible plant tissues. During infection, *S. rolfsii* secretes oxalic acid and tissue degrading enzymes such as cellulase (Aycock 1966; Punja and Damiani 1996). Oxalic acid is corrosive to tissues of plants (Ghaffar 1976). Oxalic acid can combine with calcium in plant tissues, removing it from association with the pectic compounds in plant cell walls, lowering cell wall pH, and thereby favouring activity of the cell wall-degrading enzymes endopolygalacturonase and cellulase (Deacon 2006). Massive mycelial growth on plant tissues produces large quantities of oxalic acid that facilitate penetration of hyphae into tissue (Aycock 1966). Oxalic acid and tissue-degrading enzymes work together to break down cell walls, resulting in tissue maceration. The pathogens then absorb nutrients from the macerated tissue (Aycock 1966). Maceration interrupts transport of water and nutrients in plant tissues, thereby causing wilting, yellowing, and necrosis (Bateman and Beer 1965). The slerotia formed on the infected tissue infects new hosts.

Control measures

Sclerotium rolfsii is difficult to manage as it has broad host range and the sclerotia can survive in soil for many years. However, there are certain measures, which can reduce the disease incidence and losses caused by the pathogen. These are as follows:

Cultural

Deep ploughing or burning of jute stubble kills the fungus by exposing the lower surface. Deep ploughing reduces the germination rate of sclerotia. Soil depth of 8 cm or more prevents the germination of sclerotia . However deep ploughing can lead to more even and uniform spread of sclerotia in the whole field thus increasing the disease incidence.

Biological

Trichoderma harzianum and *Gliocladium virens* can be used as effective biological control agents against *S. rolfsii* .Further, aplication *Pseudomonas fluorescens* (10^8cfu ml^{-1}) may be able to protect *S. rolfsii* infection as observed in case of greenhouse peanut (Ganesan and Gnanamanickam, 1987).

Chemical

Spraying copper oxychloride (50% Cu) on the basal region and ground checks the spread of the disease. Alternatively, spraying of mancozeb @ 2g/litre water for 2-3 times at the base of the plant may apply to manage the disease.

However, for effective management, integration of cultural practices with chemical and biological control is needed for *S. rolfsii*.

7. Sooty Mould/ Sooty Mould of Pods

Occurrence & severity status

The presence of sooty mould fungi usually indicates that a plant has become affected by a sap-sucking pest. Sooty moulds do not attack the plant directly, but their growth is unsightly and can reduce plant vigour by preventing photosynthesis. However, in jute 'Sooty moulds' are caused by the fungi that are pathogenic to jute (Fakir, 1977; Khan and Fakir, 1993). Probably, amphigenous spots are described as 'Sooty Mould'. So, the naming of such as a disease is, so far, questionable. In jute, as reported, "Sooty Mould" occurs in seed crops, more particularly in the north-east and mid-eastern regions. It is a very common disease in a seed crop. Both the two species (*C. olitorius* and *C. capsularies*) of jute are attacked by the disease. The diseased pods show drastic reduction of viable seeds (Bhattacharya, 2013). The disease was earlier described as 'Fruit Mould' by Mukherjee and Bosu in 1973. The sooty moulds caused by *Cercospora corchori* Syd. (*Cercospora corchori* Sawada, 1919?) and *Cercospora cassicola* (Berk & Curt.) Wei was reported earlier to cause considerable damage to the capsules and fruit of both species of jute, rendering the bulk of seed useless (Mukherjee and Ghosh, 1960; Mukherjee and Bosu, 1973; Ghosh, 1983).

[**N.B.** It is to be noted that both the species are not general sooty mould fungi. They are pathogenic to jute and various other crops, causing leaf spots. The fungus *Corynespora cassiicola* causes severe leaf spot disease on more than 70 host plant species. Generally, sooty mould is caused by the fungi belonging to the genera *Alternaria, Aureobasidium, Antennariella, Capnodium, Cladosporium, Fumago, Limacinula, Meliola* and *Scorias*]

Symptoms

The infected pods show brown or sooty mycelial growth on them (Fig. 2). Such pods are shrivelled and poor seed setting. In severe attack pods may be of empty seeds (Bhattacharya, 2013).

Fig. 2. Sooty mould of jute, showing amphigenous spots on pods

Causal organism: *Corynespora cassiicola* (Berk.& M.A. Curtis) C.T. Wei, (1950)

[Notes on Corynespora. Mycol. Papers 34: 1-10]

[**N.B.** *Cercospora corchori* Syd. was named earlier as a sooty mould fungus of jute (Ghosh, 1983). It is probably, *Cercospora corchori* Sawada, 1919. This is a seed borne fungal pathogen in jute and is the most predominant in one week old growing jute capsules (Khan and Fakir, 1993). The fungus causes leaf spot (?) disease in Jute (*Corchorus olitorius* L). Cercospora *corchori* has hyaline, acicular conidia, which differed from other species of *Cercospora* on this host (Sawada, 1919). Conidia are 32-117µm x 3-5µm with 2-7 septa; Conidiophore 24-79µm x 4-6µm with 0-3 septa (Mukherjee and Bosu, 1973; Ghosh, 1983). In this book this fungus is described under Leaf Spot disease of Jute]

Synonyms

= *Cercospora melonis* Cooke, Gard. Chron., (1896)

= *Cercospora vignicola* E. Kawam., (1931)

= *Corynespora melonis* (Cooke) Sacc., (1913)

= *Corynespora vignicola* (E. Kawam.) Goto, (1950)

= *Helminthosporium cassiicola* Berk. & M.A. Curtis [as 'cassiaecola'], (1868)

= *Helminthosporium papaya* Syd., (1923)

= *Helminthosporium vignae* Olive{?}, in Olive, Bain & Lefebvre, (1945)

= *Helminthosporium vignicola* (E. Kawam.) Olive, Mycologia 41: 355 (1949)

Morphological/Morphophysiolgical characteristics

The fungus grows and sporulates well in the potao dextrose agar (PDA) medium at temperature of 25°C (Tsay and Kuo, 1991; Ahmed et al., 2013). The fungal colony on PDA medium appears as effuse, gray to light olivaceous green at immature stage and that turns brown to dark blackish brown at maturity, often hairy or velvety. Conidophore is 178.8-185.7μm x 6.88μm in size. Conidia are formed solitary or in chains of 2 - 6, very variable in shape, obclavate to cylindrical, straight or curved, pale olivaceous brown or brown, smooth, with 4 - 20 pseudoseptae, mostly 7-16 (i.e., 8-17 celled as described by Ghosh, 1983) and measured 41.2 - 219.7μ long and 9.2 - 21.5μ wide (Ahmed et al., 2013), differently, the conidial size is 103.2-237.4μm x 12.0-18.9 μm (Mukherjee and Ghosh, 1960; Ghosh, 1983).

Molecular characteristics

The genetic variability in different isolates of *C. cassiicola* collected from diverse hosts and locations in Sri Lanka and Australia was assessed by analyzing total fungal DNA, and using restriction fragment length polymorphism (RFLP) analysis of the internal transcribed spacer (ITS) region of ribosomal DNA and random amplified polymorphic DNA-polymerase chain reaction (RAPD-PCR). Amplified ITS fragments from all *C. cassiicola* isolates exhibited an identical size, and restriction analysis with seven different restriction endonucleases revealed identity in all of the detected DNA fragments. This finding of high genetic relationship was further supported by the cloning and DNA sequencing of the ITS2 region from one Sri Lankan and one Australian isolate. However, RAPD-PCR profiles generated by 15 oligonucleotide decamer primers revealed significant polymorphism between groups of organisms. Genetic relationships among the isolates were determined by cluster analysis of the RAPD-PCR data and seven different RAPD groups were identified. Isolates showed strong correlations between the assigned RAPD group and the location and host plant genotype from which the isolate was collected. Correlations were also observed between the RAPD group, growth of the isolate and pathogenicity on different plant hosts (Silva et al., 1998).

Predisposing factors

The disease is promoted by heavy dew or rain and accelerated by cloudy conditions (Bhattacharya, 2013). The optimum temperature for spore germination and mycelial growth ranged between 25 and 30°C. The fungus requires a nutrient source on which to grow, and this is most commonly the honeydew excreted by a number of sap-sucking pests (e.g. aphids, mealybugs, scale insects, whiteflies). Droplets of honeydew are shed by these pests and fall onto surfaces below where they are feeding. Occasionally, sooty mould growth develops on sugary, sticky exudates produced by the leaves of the plant itself.

Control measures

It is sensitive to copper, a spray of copper oxychloride (50% Cu) at 0.75% conc., is recommended on seed multiplication farms. Further, it is evident that spraying of endosulfan against semilooper gives capsules completely free from sooty mould (Ghosh, 1983). Prophylactic spray of Bordeaux mixture is also **suggested to manage sooty mould (Desai** *et al.*, 1997). However, it should be noted that chemical control of sooty mould itself is not needed. Synthetic insecticides such as the organophosphates acephate (orthene), malathion, or diazinon can be used in severe cases for the control of sap-sucking pest responsible for the honeydew on which the mould is growing.

8. Stem Gall

Occurrence & severity status

The disease was first reported from Banaras (Vanaras) by Lingappa in the year 1955. Later it was reported from Maharashtra (Patil, 1962), West Bengal , Tripura and other places (Das and SenGupta,1964; Prakash and Ghosh , 1964). It is found on olitorius (tossa) jute in the Terai and low-lying areas which facilitated by alkaline soil, water logging at early stages for at least 3 days and in the crops of late sowing (Lingappa, 1955). The incidence of the disease is sporadic and the disease has been reported as absent in the main jute growing tracts of India and Bangladesh (Prakash and Ghosh, 1964). The disease is reported as localized in Uttar Pradesh of India and is of minor importance (Ghosh, 1983).

Symptoms

The disease appears as galls that develop on young stems. The galls are initially green but subsequently turn pink to brown and eventually become rusty brown. The severely affected plants wither and die. The galls are measured as 0.5-2.0 mm long, protruding 1-2mm from surface of stem. In some cases these are coalesced and formed larger blisters with lacerated fibre layer (Ghosh, 1983).

The transverse section through gall of an infected stem shows disintegration in phloem and cortical tissues and a large number of bright golden yellow resting spores. Galls are formed due to hypertrophy of host cells and there is no evidence of abnormal cell multiplication (Ghosh, 1983).

Causal organism: *Physoderma corchori* **Lingappa, 1955** [Mycologia 47(1): 109 (1955)]

Morphological/Morphophysiolgical characteristics

The fungal species is a chytrid, consisting of both monocentric eucarpic thallus (pl.-thalli) and an endobiotic polycentric thallus. Monocentric thallus is epibiotic, primary zoosporangium generally anchored with endobiotic rhizoids that confined to a single host cell. Eucarpic thallus is ephemeral. Primary zoosporangium is probably operculate and produces primary zoospores. The zoosporangia are internally proliferous, can produce a second round of zoospores after releasing the first one. Secondary zoospores are ovoid, 3.5-6 µm with blunt ends, a flagellum of 18 µm long and an eccentric refractive globule near the anterior end. At later stage of the season, the zoospores develop into endobiotic polycentric thalii (sing. thallus). The polycentric endobiotic thallus is a rhizomycellium, profusely branched, tenous portions 1.5-2 µm diameter, intercalary enlargements are one to two celled, spherical or spindle shaped. Resting spore is formed from the endobiotic thallus. Resting spore is 1.5 µm thick walled, dark brown coloured, subspherical or globose, 19-30 µm in diameter. (Lingappa, 1955; Sparrow, 1960; Dayal, 1997).

Predisposing factors

The disease occurs in marshy land it is mostly confined to the submerged portion of hosts. Thus, it requires free water at least an inch of standing water to initiate the infection of a host plant. Once the plant is infected, then high humidity, dew or rain is sufficient to keep the infection going through the season. Resting spores germinate at pH 7.2 and not at below of that, thus alkaline soil is suitable for the disease (Prakash and Ghosh, 1964).

Disease cycle

The fungus produces resting spores which are formed from the intercalary cells. These resting spores are the primary source of infection. They germinate to produce zoospores during jute growing season and that infect the host under favourable conditions. Once the plant is infected, secondary infection may take place by both the primary zoospores and secondary zoospores formed at gall. The liberated zoospores infect new host cell repeatedly and in this fashion several generations take place. At later stage, the zoospores begin to develop endobiotic polycentric thallus. This endobiotic thallus often cases extensive damage by infecting many host cells with the highly branched fine rhizoids. Resting spores which produced acts for overwinter survival.

Control measures

To check the spread of this disease, farmers should restrict for olitorius jute to high land, deep plough the field and ensure timely sowing. Capsularis (white) jute is tolerant to the disease and may grow the disease prone areas.

9. Stem rot/Stem & root rot/ Charcoal rot

Occurrence & severity status

The disease [(Fig. 3(A-D) was initially described by Shaw (1912) as seedling disease of jute in Pusa (India). He named the pathogen as *Rhizoctonia solani* Kuhn. Later, Sawada (1916) reported stem-rot of jute from Formosa. He found that the fungus had a pycnidial phase, and named it *Macrophoma corchori*. Ashby (1927) commented on that in pure culture minute black sclerotia were developed which Sawada (1916) might thought as immature pycnidia (Ashby, 1927). Shaw in 1924 changed the name of his previously reported fungus, *R. Solani*, to *Macrophoma corchori* Shaw that constantly associated with black sclerotia in the cortex of jute. Ashby (1927) showed the sclerotial stage of the fungus causing root-rot of jute to be *Rhizoctonia batatocola* (Taub) Butler. However, after comparing all the available cultures and carrying out inoculation tests, he found that the pycnidial stage and named the fungus as *Macrophomina phaseoli* (Maubl.) Ashby. Presently, the causal fungus is again renamed as *Macrophomina phaseolina*. The fungus is a global devastating necrotrophic pathogen, causing charcoal rot disease. It infects more than 500 plant hosts (Wyllie, 1998; Islam *et al.*, 2012) including major fibre crops, like, jute & cotton (De *et al.*, 1992; Aly *et al.*, 2007). Although, it has a wide host range, *Macrophomina* is a monotypic genus. This pathogen can cause severe crop losses. In India and Bangladesh, stem rot/charcoal rot disease causes substantial loss of fibre yield of jute (De *et al.*, 1992). In Bangladesh, the disease severity is very high causing about 10% yield loss. The stem rot disease occurs widely in almost all jute growing countries. In India its severity is more in Assam and West Bengal, particularly in North Bengal. In south Bengal, Hoogly, Burdwan, North-24 parganas and Southern parts of Nadia districts are more affected areas of the disease.

Symptoms

The disease occurs sporadically in field and attacks jute plants both in the seedling and adult stages. In case of seedling attack, the infected one shows reddish brown discoloration at the collar region extending up the stem, that may turn brown to black and eventually, the plant dies with the symptoms of soft and

Fig. 3 (A-D). Stem rot disease of jute caused by *Macrophomina phaseolina*. **A.** Initial symptom of the disease in young plants, showing epinasty and wilting; **B.** Attack on root of the wilting plants; **C-D**. Typical stem rot/charcoal rot symptom on the stem of mature jute plants

rotted. The disease is primarily restricted to the intercellular spaces of the cortex of the primary roots. Consequently, adjacent cells collapse and heavily infected plantlets may die. Dark lesions appear on the epicotyls and hypocotyls followed by seedling death due to obstruction of xylem vessels. In plants, the pathogen causes red to brown lesions on roots and stems with production of dark mycelia and black microsclerotia (Abawi and Pastor-Corrales, 1990). In case of young plant, during hot summer season, the diseased plant shows wilt symptom without any apparent discolouration at stem, although on uprooting the plants, dark spots due to presence of numerous dark microsclerotia of infection region is visible at root or/and collar region. At latter stage, the infected plant turns brown and dried. Usually typical charcoal rot symptom develops at later in the season. At that time, the stem appears as brown to black in colour producing typical rot symptoms with a few brown to dark wilted leaves at top or bare head. Sloughing of cortical and tap root tissues occur simultaneously with characteristic black speckled appearance due to the presence of microsclerotia in vascular tissues and pith. Necrotic spots may also appear on the tender leaves at the top of plants and/or from the edges of leaves, due topresence of a translocatable toxin (s) [(–)-botryodiplodin (Siddiqui *et al.*, 1979; Ramezani *et al.*, 2007) and, differently, that with phaseolinone (Dhar *et al.*, 1982; Bhattacharya *et al.*, 1994). The toxins are sometimes responsible for rot symptoms without showing other signs.The infected seeds are brown without any lusture and light in weight. Microsclerotia are developed within the capsule and on the seeds (Chan and Sackston 1973; Dhingra and Sinclair 1974; Mehak, 2017). Pycnidia are rarely produced on the host whereas mycelia and sclerotia are abundantly formed (Knox-Davies, 1966).

In this contrast, it is also reported that as rotting of the leaves and hypocotyl of young seedlings advances, abundant pycnidial formation takes place. Further, the diseased and wilted plants with bare dead stalks in the field contain abundant pycnidia all over. Pycnidia are usually embedded in the epidermis of the stem being visible as small black dots. Numerous pycnidia are also reported to cover the entire lamina and the petiole with pycnidial ostiole on the upper surface of the leaf. Pycnidia occur on the outer surface of the capsule and sporadically on the seeds (Mehak, 2017).

[N.B. The abundance in pycnidial formation under natural condition needs further attention for jute. Since, so far, pycnidial formation under natural condition is scanty (Shaw in 1924) although in culture medium supplemented with jute plant parts, the formation of pycnidia is profuse, which usually embedded in the epidermis of the stem being visible as small black dots with erumpent ostioles].

Causal organism: *Macrophomina phaseolina* **(Tassi) Goid. 1947** [Annali della Sperimentazione Agaria, 1(3): 457]

Synonyms

= *Botryodiplodia phaseoli* (Maubl.) Thirum, 1983

= *Dothiorella phaseoli* (Maubl.) Petr. & Syd., 1927

= *Macrophomina philippinensis* Petr., 1923

= *Macrophoma phaseolina* Tassi, 1901

= *Macrophomina phaseoli* var. *indica* Moniz & V.P. Bhide, 1963

= *Macrophoma phaseoli* Maubl. 1905

= *Macrophoma corchori* Sawada, 1916

= *Macrophoma phaseoli* (Maubl.) Ashby, 1927

= *Rhizoctonia bataticola* (Taub.) Butler, 1925

[**N.B.** A teleomorph was described as *Orbilia obscura* by Ghosh, Muckerji & Basak 1964 (Holliday, 1980) but this has not been confirmed (Mihail 1992).]

Morphological/Morphophysiolgical characteristics

Macrophomina phaseolina is an anamorphic fungus (Crous *et al.*, 2006). The fungal hypha is colourless, 3-8µ broad, septate, branched. It frequently produces dark minute sclerotia (microsclerotia) inside the host tissue. Microsclerotia are fromed from the specialized highly compressed 50-200 hyphal cells called monoloids which joined together by melanin pigment (Beas-Fernández *et al.*, 2006; Aboshosha *et al.*, 2007; PartridgeMehak, 2017). The fungus also forms sub-epidermal, oval, black pycnidia with erumpent ostiole. Pycnidia are generally larger than microsclerotia and are present in linear rows on the outside of stems, whereas charcoal rot microsclerotia form throughout (inside) the taproot and lower stem. Ostiole is central, circular and surrounded by dark brown thick-walled cells. Pycnidia are 100-200µ in diameter with multicellular wall. Conidiophores are numerous, hyaline, simple, straight to slightly curved with tapering apex, 10-14 x 2.5-3.5µm in size. Conidia are borne singly at the tip of conidiophores, uni-cellular, hyaline, smooth-walled, ovoid to oblong, straight or slightly curved, 14-33 x 6-12µm (PartridgeMehak, 2017).

The fungus on PDA is characteristically grey to black in colour. Colony is with even margin, and abundant, fluffy aerial mycelium. The mycelia are buff, turning vinaceous buff to pale olivaceous grey, with dense, black sclerotial masses. Colony grows optimum at 30 36°C, and keeps itself still growing at 40°C (Sarr *et al.*, 2014). Pycnidia are rarely produced in the culture while it can be induced in culture by near UV irradiation for 12 h and darkness for 12 h with alternating cycle at 20-24°C. The pycnidia produced in the culture are typically dark brown to black, sub-globose to lageniform with the diameter of around 300μm and composed of several layers of cells. The inner layer is hyaline while the outermost layer of cells is dark brown to black in colour (Punithalingam, 1982; Kaur et al., 2012). Production of pycnidia is not common on potato dextrose agar (PDA) medium, but can be produced on media containing plant parts or oilseed extracts. In vitro production of large numbers of pycnidia has been accomplished with the using of sterilized plant parts (Goth and Ostazeshi, 1965; Dhingra and Sinclair, 1978; Ma et al., 2010). Different isolates of M. phaseolina may show morphological variability in terms of growth, colour, radial growth, sclerotial size, pycnidium production (Dhingra and Sinclair, 1973a, 1978; Pearson et al., 1986; Adam, 1986; Atiq et al., 2001; Riaz et al., 2007; Iqbal, 2010)

Molecular characteristics

In recent years, molecular characterization is considered as an important tool for accurate identification of the fungus. This has been aided with the use of nucleic acid–based molecular techniques, such as the use of species-specific oligonucleotide primers for polymerase chain reaction (PCR) and digoxigenin (DIG) labeled DNA probes, both are prepared based on sequences of the internal transcribed spacers (ITS) (Babu et al., 2007). Internal transcribed spacers (ITS) and 18S rRNA are the most conserved genes used to identify the fungus (Babu et al., 2010b).

The housekeeping gene 18S rRNA sequence is an ideal marker to identify the fungi at genus level but has some limitations to discriminate the intra species of fungi. The sequence of hyper-variable regions like V2, V4, V7, and V9 play an important role in identification of fungi. The region V7 has been utilized to discriminate various strains by phylogenetic analysis (Wu et al., 2015). Khan et al. (2017) reported that they have identified M. phaseolina strains by 18S rRNA and analyzed for genetic diversity by using random amplified polymorphic DNA (RAPD) markers. However, species-specific primers and probes are mostly used for the identification and detection of M. Phaseolina at molecular level by exploiting rDNA gene cluster as a target. The primers, MpKF1 (52 - CCG CCA GAG GAC TATCAA AC-32) and MpKR1 (52 -CGT CCG AAG CGA GGTGTA TT-32) designed from the conserved sequences of the ITS region are highly specific and yielded a specific 350 bp products. Since, this 350 bp amplicon is absent from other soil-borne pathogens, so, it can be used for the species-specific identification for M. Phaseolina (Babu et al., 2007; Kaur et al., 2012). Jana et al. (2003) used random amplified polymorphic DNA (RAPD) markers, the primer OPA-13 (5'-CAGCACCCAC-3') in producing fingerprint profiles that clearly distinguish between the different isolates of *M. phaseolina*. UPGMA analysis was also done to classify different isolates into five major groups (Jana *et al.*, 2003).

Predisposing factors

The pathogen is seed-borne (Varadarajan and Patel, 1943). So the infected seed is one of the most important factors for the dissemination of the disease in different areas. The infected seeds do not germinate or produce seedlings that die soon after emergence; however, they help the establishment of the pathogen in new areas. The pathogen is also soil borne. The pathogenic fungus as a form of sclerotia (microsclerotia) may survive in association with plant residues or in soil for prolonged periods (Cook *et al.*, 1973; Ilyas and Sinclair, 1974; Seikh and Ghaffar, 1979). At high temperature (30°C or above) and at water stress condition the pathogen grows well and damage the plants severely. The disease is most pronounced when plants are stressed by adverse environmental factors such as drought and heat stress during the reproductive growth stages (Smith and Wyllie, 1999). Thus, high temperature and low moisture favour the disease development (Aegerter *et al.*, 2000). The germination of microsclerotia, also, occurs frequently in the temperature range of 28–35°C (Mihail, 1989).

Disease cycle

The disease is monocyclic and the fungus survives as microsclerotia in the soil and on infected plant debris. The microsclerotia produced in the roots and stem tissues of its hosts serve as the primary source of inocula and can persist in the soil for three years (Dhingra and Sinclair, 1977), but, according to Short *et al.* (1980) these can survive to around 15 years. Microsclerotia germinate on the root surface, germ tubes form appresoria that again form infection pegs to penetrate the host epidermal cell walls by mechanical pressure and enzymatic digestion or through natural openings (Bowers and Russin, 1999). The hyphae grow first intercellularly in the cortex and then intracellularly through the xylem colonizing the vascular tissue. Once in the vascular tissue *M. phaseolina* spreads through the taproot and lower stem of the plant producing microsclerotia that plug the vessels (Wyllie, 1988). The mechanical plugging of the xylem vessels by microsclerotia and toxin production leads to disease development.

Management/Control measures

Host resistance

Most of the jute varieties cultivated in India are more or less susceptible to the disease. Meena *et al* (2015) from ICAR-Central Research Institute for Jute and Allied Fibres (Barrackpore) reported that the tossa jute Acc No. OIN-431 is tolerant and none of the capsularis germplasms shows enough resistance to stem rot. However, a wild jute species (*C. trilocularis*) is known to be resistant to *Macrophomina phaseolina* (Biswas *et al.*, 1968).

Cultural control

Colonization of roots by *M. phaseolina* can be lower by irrigation. However, root colonization of the pathogen still occurs in irrigated system and when the soil moisture becomes predominantly low (e.g. under drought conditions), colonization by *M. phaseolina* can result in the production of microsclerotia, which will increase the level of inoculum for subsequent host crops (Smith *et al.*, 2014). Flooding the field for several weeks reduces the inoculum present in soil. The mycelium of *M. Phaseolina* cannot survive more than seven days in wet soil, while, microscerotia required to die about seven to eight weeks in wet soil (Abawi and Pastor Corrales, 1990). Hence, it is required to manage the fields for avoiding drought stress. Rotation with crops that have relatively low susceptibility to the pathogen or non-host and that required standing water for growth, such as, paddy (rice), for 1 to 2 years in the fields with a history of charcoal rot, may be very effective in reducing the disease in jute. An examination of the cropping system also reveals that *M. phasolina* disease in jute is significantly low when Jute-paddy-potato and jute-paddy-groundnut patterns are used. Reduced seeding rates may also be helpful to manage drought stress and charcoal rot (Smith *et al.*, 2014).

Biological control

Soil application of *T. viride* thrice (on 7, 15 & 30 DAS) was found best in controlling the disease in jute and giving minimum per cent disease incidence as compared to control (Srivastava *et al.*, 2010).

Chemical control

Application of chemical measures in controlling the disease under field condition of the crop, like, jute is less possible. However, seed treatment with benomyl (Benlate) and carboxin (Vitavax) may be effective to control the seed borne pathogens, hence, in reducing the disease (Abawi and Pastor Corrales, 1990). An integrated management system with the use of 50% N: P: K of recommended fertilizers + seed treatments with *Azotobacter* and phosphorus solubilising bacteria (PSB) @ 5g/Kg+ *Trichoderma viride* (seed treatment @ 5g/Kg of seed and soil application @ 2Kg/ha at 21DAS) + *Psuedomonas fluorescens* spray @ 0.2% at 45DAS has been found effective to minimize the disease of jute under field conditions (Meena *et al.*, 2014)

10. Yellow mosaic disease

Occurrence & severity status

Jute yellow mosaic disease [(Fig. 4 (A-B)] was first reported from Bengal (India) by Finlow in 1917, but the etiology was not known for many years, until Ha *et al.* (2006, 2008). They identified two begomoviruses, namely, *Corchorus yellow vein virus* (CoYVV) and *Corchorus golden mosaic virus* (CoGMV), from the diseased plants in Vietnam. Subsequently, the yellow mosaic disease was identified as a cause of CoGMV from India (Ghosh *et al.*, 2008, 2012). This identification of virus has provided further evidence that CoGMV is widely distributed in Asia. The disease causes drastic reduction in the quality and yield of jute (Ghosh *et al.*, 2011). It has been considered to be one of the most important limiting factors of jute cultivation. The incidence of the disease has been found to be around 50% on some of the leading *C. capsularis* cultivars in India. In Bangladesh, which is one of the major jute growing countries (Kundu, 1951), yellow mosaic disease is a major limiting factor for white jute (*C. capsularis* L.) cultivation. In a study, the occurrence of Yellow mosaic disease of jute (*C. capsularis*) was observed on several plants in different jute growing regions in April 2013 (Hasan *et al.*, 2014).

Fig. 4 (A-B). Yellow mosaic disease in jute. **A.** Disease showing yellow mosaic with crinkling and malformation of leaves. **B.** Disease showing yellow mosaic on young plant.

Symptoms

Symptoms on jute generally include crinkled, leathery and malformed leaves with a yellow mosaic of varying intensity. The yellow flakes appear on the lamina of young leaves of CoGMV infected plants and that gradually increase in size to form green and chlorotic intermingled patches (Hasan *et al.*, 2015). The infected plant becomes dwarf due to remarkable alteration of biochemical components (Ghosh *et al.*, 2011) and produces less fibre. The virus (CoGMV) infection also reduces the levels of chlorophyll, total protein, catalase, peroxidase and esterase (Ghosh *et al.*, 2011).

Causal organism: Virus, *Corchorus golden mosaic virus* (CoGMV) [Geminiviridae : Begomovirus]

Virus

The virus is bipartite possesses DNA A, which is a typical New World bipartite begomovirus genome organization that lacks AV2 open reading frame (AV2ORF) which is present in Old World begomoviruses and DNA B genomic components (Ghosh, 2008; Hassan *et al.*, 2015). It is found that the initial symptom as small spots appears on leaf lamina at 15 days of post inoculation, which intensifies further as typical yellow mosaic symptoms at 20 days of post inoculation in jute plants only when both the genomic components (DNA A and DNA B) are co-inoculated. While, the DNA A genomic component alone fails to develop disease in jute (Hasan and Sano, 2014; Hasan *et al.*, 2015).

The complete nucleotide sequences of DNA A of three different isolates are found 2676 nt and 2687 nt in length, while that of DNA B is ranged from 2658nt to 2668 nt (Hasan *et al*, 2015). The DNA A nucleotide sequences of Bangladesh isolates share identity either with Vietnam isolate or with Indian isolate (Hasan *et al.*, 2015).

Transmission

Like other begomoviruses, CoGMV can be transmitted by whitefly and the efficiency of its transmission ranges from 20 to 60% for infestation with 3 and 10 viruliferous whiteflies (Ghosh *et al.*, 2008). The mechanical transmission of the yellow mosaic disease has not been demonstrated, presumably due to the presence of a large amount of mucilage and phenolics in jute plant. Begomoviruses can be transmitted by the whitefly (*Bemisia tabaci*) and sometimes by mechanical inoculation (King *et al.*, 2012).

Management/Control measures

A field study on the management of jute yellow mosaic virus disease through cultural practices indicates that top dressing of nitrogenous fertiliser is very effective in terms of increasing yield (3.05 t/ha) and that is better than both rouging with field sanitation and Malathion 57 EC spray. The top dressing of nitrogenous fertiliser also shows the highest benefit-cost ratio (4.84) (Mahmud *et al.*, 2014).

The disease is suggested to manage by the prophylactic spray of Folidol (Ebos @ 1 ml/5 l of water) or Endrin (20 EC; @ 7.5 ml/5 l of water) for killing the vector. Recently, an economical prophylactic spray of Rogor (0.5%), Metasystox (0.0 2%) or Dimecron (0.05%) is being recommended. However, it should be remembered that the white fly (*B. tabaci*) has enormous ability to develop resistance against systemic fungicides. So, the fungicides of different mode of actions should be applied simultaneously or alternatively. It is also found (by the author from an unpublished work) that the use of *Clitoria ternatea* (Butterfly pea) grown in earthen pots and placing in jute fields for white fly trapping and monitoring the disease is very suitable, since, the white fly has more affinity to this trap plant.

11. Root Knot Disease

Occurrence & severity status

Several Nematode species are found to cause damage on both *Corchorus capsularis* and *C. olitorius* jute (Chaturvedi and Khera, 1979), however, amongst them, *Meloidogyne* spp. causing root knot disease are most important. The root knot disease is one of the important diseases in both premier jute growing countries like, India (Singh and Ghosh, 1981) and Bangladesh (Talukder, 1974). In Bangladesh, the prevalency of Root-knot is very high, damaging plants by different *Meloidogyne* spp. along with 14 other plant parasitic nematodes (Timm and Ameen, 1960). In India, also, 47 nematode species were recorded from jute rhizosphere (Chaturvedi and Khera, 1979). The root knot disease is caused by different species of *Meloidogyne* at different places, depending upon the occurrence of the species in that agroclimatic conditions. However, The two species, *M. javanica* and *M. incognita* are mostly found in jute of both the countries (Timm and Ameen, 1960; Chaturvedi and Khera, 1979).

Symptoms

The root knot nematodes affect jute at various growth stages (Talukder, 1974; Ahmed, 1977). The symptoms appear as stunted plants, sometimes wrinkled, with chlorotic or yellowish and variegated leaves and on uprooting large number different sized galls (knots) are found on short crumpled roots and at later stage the plants wilt (Chaturvedi and Khera, 1979; Ghosh, 1983). At initial stage of infection, the root cell walls are thickened; the cell cytoplasm is vacuolated and contains many nuclei. Hyperplasia occurs around giant cells, and heavily infested roots consist of 2 cork layers (Chaturvedi and Khera, 1979).

Causal organisms: ***Meloidogyne javanica*** **and** ***Meloidogyne incognita***

(**N.B.** Further details are in the following the chapters of Mesta/Kenaf and Ramie diseases)

Control measures

In order to prevent the perpetuation and increase of nematodes in jute fields, ploughing after each harvest with provision of drainage is suggested to expose nematode infected roots to the sun. Further, at least one non-host crop, like, rice is to be incorporated in jute based cropping system (Ghosh, 1983). Amendment of soil with old jute seed powder, potash and sulphur is found effective to check rootknot infection in jute (Begum *et al.*, 1991). Moreover, growing sunhemp as a trap crop, in lines in the jute field is found effective in reducing knot (gall) formation in jute plants (Haque *et al.*, 2008).

The nematicides, Cadusafos (Rugby) and Sodium tetrathiocarbonate (Enzone), at 8ppm (a.i.) concentration are effective to control root knot disease caused by Meloidogyne javanica (Soltani *et al.*, 2013).

B. Pests

1. Grey weevil /Ash weevil

Occurrence & severity

This weevil is a polyphagous pest attacking several cultivated crops in different states of India viz., Assam, Bengal, Himachal Pradesh, Orissa, Jammu and Kashmir, Karnataka, Punjab, Tamil Nadu and Uttar Pradesh, as well as in the neighbouring countries, like, Bangladesh, Myanmar and Srilanka (Marshall, 1916; Ramamurthy and Ghai, 1988; Azam, 2007). Earlier it was considered as a minor pest, but during the recent years it has emerged as one of the major pests of jute (Selveraj *et al.*, 2016). The Grey Weevil attacks tossa jute (*C. olitorius*), while white jute (*C. capsularis*) is considered as immune to the pest attack, since it does not damage white jute at any stage of plant growth (Mallick *et al.*, 1980). It generally attacks jute crops when the plant age becomes to 35-40 days during the month of April to May. The atmospheric relative humidity and rainfall have positive correlation with the incidence of the pest (Rahaman and Khan, 2012a). Rainfall favours proliferation and incidence of the weevil. The temperature ranging 35-39° C with relative humidity 85-94% after one or two pre-monsoon showers is the most congenial condition for the grey weevil attacks (Selvaraj *et al.*, 2016).

Scientific name & order: *Myllocerus discolour* **Boheman, 1834 (Curculionidae: Coleoptera)**

Types of damage

Adult weevils notch on the leaves and leaf margins for feeding. They prefer apical unopened leaves to feed. On feeding they make irregular holes on leaf which later causes defoliation in tossa jute at early stages of plant growth. The grubs are voracious eater of soft adventitious roots and rootlets, causing stunted growth of plant (Butani, 1979).

Biology

Egg: The female lays about 300 eggs singly or in groups in soil. Eggs are Small, ovoid and cream coloured.

Grub: Grubs are small and apodous (legess). These are C shaped, creamy white body and with brown heads. Its development completes underground.

Pupa: The weevil pupates in earthen cocoons prepared by the mature grubs in soil. Pupation takes about one week time period.

Adult: Adult weevil is gray with its back as ferruginous brown, with patches of fawn-coloured scaling and mottled black. The head is black, with rectangular snout, bearing 12 segmented, geniculate, laterally arranged antenna. The first segment of antenna is the longest. Eyes are black, prominent, laterally present at the base of rostrum. Elytra are prominent covering the abdomen completely and that covered by green scales completely. Legs are similar in size, covered by brownish hairs and scales, bearing a pair of claws at the tip. Mesothorax is rectangular and hairy.

Life cycle

There is generally one generation in the course of a year. The female lays about 300 eggs singly or in groups in soil. Hatching takes place after three to seven days of incubation. The grubs feed on roots and become full grown in about 10 days. The pupation takes place in an earthen cell at a depth of 5 to 7 cm. Pupal period usually varies from five to seven days. Sometimes pupae remain in the soil in diapause until the next year. The newly hatched weevil is grey.

Management/Control measures

Host resistance

The tossa jute cv. JRO 7835 is to some extent tolerant to *M. discolour* attack, while, JRO 524, JRO 632 and JRO 878 are susceptible (Rahaman and Khan, 2012b).

Cultural control

After harvest the land should be ploughed well so that the pupae are exposed, providing perching of bamboo top for sitting the birds. Light traps may be used to destroy the adult weevil.

Biocontrol

Application of neem cake @ 500 kg/ha at the time of last ploughing is suggested.

Chemical Control

In endemic areas, it is needed to apply carbofuran 3 G @ 15 kg/ha on 15 days after planting. Spraying of carbaryl 50 WP @ 3g + wettable sulphur 2g/litre may also give better control. Control may be also achieved by the use of any chlorinated hydrocarbons.

2. Hairy caterpillar/ bihar hairy caterpillar / jute hairy caterpillar

Occurrence & severity

Bihar hairy caterpillar is a sporadic pest, widely distributed in Afghanistan, Bhutan, Bangladesh, China, India, Myanmar, Nepal and Pakistan (CPC 2004). In India, it is a serious pest in West Bengal, Bihar, Madhya Pradesh, Uttar Pradesh, Punjab, Manipur, and other states. The insect is highly polyphagous and attacks a variety of bast fibre crops, including jute, and several other economically important crop plants. Its incidence exhibits positive relationship with the atmospheric humidity (Rahaman and Khan, 2012c). The hairy catterpilar causes severe economic damage in jute (Gupta and Bhattacharya, 2008) and that considered as one of the major pests of bast fibre crops, like jute (*Corchorus* spp.) and Mesta (*Hibiscus* spp) of which jute is more preferred host.. The insect pest is also a major constraint in Bangladesh where it is commonly known as bicha or shoapoka. In West Bengal, it may cause a total foliage loss up to 20–30% in jute. In a recent study, it is found that the hairy caterpillar causes yield loss up to 30% in jute (Bandyopadhyay *et al.* 2014). Selvaraj *et al.* (2016) also reported that the infestation of bihar hairy caterpillar in jute under West Bengal agroclimatic conditions is more regular and so severe that the situation may reach the status of outbreaks. Of the two types of cultivated jutes, tossa jute (*C. olitorius*) is found more susceptible to *Spilosoma obliqua* than the white jute (*C. capsularis*) (Pandit, 1985).

Types of damage

Only caterpillars are harmful to the jute crop. They eat almost all the leaves including growing points. At initial stage, the young gregarious larvae feed on chlorophyll on the under surface of the same leaf giving a membranous appearance. Then they attack another leaf. They scrape the green part of the leaves excluding midrib and cross veins so that the leaves appeared as net or web of brownish-yellow in colour. At later stages, the grown up larvae eat the leaves from the margin. The fed up leaves defoliate as skeleton and in severe cases, only stems are left behind, thus, in turn, causing the loss of yield.

Scientific name & order: *Spilosoma obliqua* **Walker, 1855; (Lepidoptera: Arctiidae)**

Synonyms

= *Spilarctia obliqua* (Walker)

= *Diacrisia obliqua*

Subspecies

- *Spilarctia obliqua obliqua* (Walker, 1855) (south-eastern Afghanistan, northern Pakistan, India, Bhutan, Bangladesh, Burma)
- *Spilarctia obliqua montana* (Guérin-Méneville, 1843) (southern India)

Biology

Egg: Female lays eggs in cluster (consisting each with about 75-174 eggs on different jute species; Gotyal *et al.*, 2015) on lower surface of leaves. One female lays about 500-550 eggs in 6 days. The eggs are spherical in shape but slightly flattened at one side. Freshly laid eggs are creamy white in colour and that turned pale yellow with age. The size of eggs is about 0.25mm in diameter. When the eggs are about to hatch, they turned blackish due to development of larval head inside (Ward and Kalleshwaraswamy, 2017).

larva: Larva shows colour variation from greenish to brownish. The mature larval body is with black head and broad transverse dark and pale bands and tufts of yellow long hairs that are dark at both ends. During the larval developmental period, the caterpillar moults five times and has six larval instars. The first instar larva is sluggish in nature, translucent light yellow body colour and with dark head. In second instar the larval body is also same in colour of that of first instar, but, it grows faster. The body turns wider than head and with prominent setae and tuercules. The third instar larva is morphologically similar to that of second instar. The larva becomes light yellow in colour with black patches on the anterior and posterior regions of the body. The spiracles are seen as black in colour. The larva of fourth instar is yellow with longitudinal stripes. Head and prothoracic shield are dark brown. Black patches are present on the anterior and posterior regions. Fifth instar larva is almost similar to that of fourth instar, except in size. The head capsule width is 0.4-0.54μ. The sixth instar larvae is the fully grown caterpillar with head capsule width about 0.45-0.70μ and similar in appearance as described mature larva earlier (Ward and Kalleshwaraswamy, 2017).

Pupa: The mature larva pupates in a thin silken cocoon by interwoven shed hairs of the larvae under dry debris, foliage and soil close to the plants. The pupa is elongated oval in shape. The eyes and antennal case are prominent (Ward and Kalleshwaraswamy, 2017).

Adult: Medium sized brown moth with a red (crimson) abdomen. Adult moth is reddish brown with black spots. Both the wings are pinkish and possess black spots. The head thorax and abdomen are distinct. The antennae and legs are light brown. Two long segmented filiform antennae are located dorsally on the head and close to compound eyes. Both and female moth are almost similar in appearance, except that the size of the male is smaller and with sharply tapered abdomen (Ward and Kalleshwaraswamy, 2017). The wingspan is about 23 mm.

Life cycle: Adults survive for a minimum period of 6 days of which, pre oviposition period takes 3 days and oviposition period remains for 3 days (Nath and Singh, 1996). However, under favourable condition the longevity of adults is more and so on for oviposition period that may increase up to 6 days. Egg incubation period varies from 6.5-10.5 days, the shortest period is in October-November and Longest one is in December-January. The larval period varies from 24 days (june-July) to 44 days (December-January). The total period of life cycle (egg to adult) takes only 39.6 days in June-July and that is prolonged up to 70 days during December to January (Debraj and Singh, 2010). The moth passes through several generations per year. The pupae may overwinter in cocoons in sheltered places or in the soil (Kabir and Khan, 1969).

Management/Control measures

Cultural and manual control

The insect pest can be minimized by deep summer ploughing that will facilitate in exposing soil borne pupae to sun for killing. It is also suggested to avoid pre monsoon sowing, to use optimum seed rate and to provide adequate plant spacing in order to escape the insect attack. Hand picking of egg masses and early instar larvae and killing them by burning or kerosinized water is considered to be the easiest method of control.

Light Trap: It is suggested to install one light trap (200W mercury vapour lamp) per hectare to catch the adults of some nocturnal pests such as hairy caterpillar (positively phototropic).

Biological control

Spraying insect repellent, like, Azadirachtin (0.1%), may be effective to manage the pest in field, since in an *in vitro* test it was found as very effective by exhibiting 69.78% feeding inhibition of 3rd instar larvae of jute hairy caterpillar (Chowdhury *et al.*, 2012). Bandyopadhyay *et al.* (2014) suggested to apply the combination of microbe [like, *Bacillus thuringiensis* var. *kurstaki* (Delfin) as used by the authors] and Azadiractin spray in a given integrated pest management system with reference to hairy caterpillar of jute.

Chemical control

Application insecticide solution of chlorpyriphos 20 EC @ 1.5 lit/ha or trizophos 40 EC @ 0.8 Lit/ha or quinalphos 25 EC @ 1.5 lit/ha or dusting Chlorpyriphos 1.5% DP or quinalphos 1.5% @ 25kg/ha when the population is likely to reach 10/m row length (ETL) is suggested to manage the pest chemically (TNAU Agritech Portal, ND1).

[**N.B.** Use of host resistance which is the most important method to manage insect pests, the work in this line in jute is still at infancy. A certain level of resistance against the hairy caterpillar has been found in two wild jute varieties (*Corchorus tridens* and *C. aestuans*), by showing adverse larval development, growth, survival, pupation and adult emergence of *S. obliqua* (Gotyal *et al.*, 2013).]

3. Mealy Bugs

Several mealy bug species are found to affect jute plants in different jute growing areas. However, except a few reports their severity under farmers' field condition is yet to determine. In most cases they are considered as minor pest of jute and occur sporadically in certain areas, although, recently a new jute pest, cotton mealy bug (*Phenacoccus solenopsis* Tinsley) is found to attack jute severely in the research farm of Central Jute and Allied Fibre Research Institute, Barrackpore (West Bengal, india). The mealy bugs, *Ferrisia virgata*, *Nipaecoccus viridis* (Newstead) (= *Pseudococcus filamentosus* var. *corymbatus*) and *Phenacoccus solenopsis* Tinsley are found to affect jute in diierent places (Das, 1948; Dutt and Ganguli, 1956; Satpathy *et al.*, 2009, Mani and Satpathy, 2016). Another species of mealy bugs, *Maconellicoccus hirsutus (= Phenacoccus hirsutus)*, although reported earlier as a minor pest of jute, their existence on jute plants is probably doubtful.

i) Striped mealybug: *Ferrisia virgata* Cockerell, 1893 (Pseudococcidae: Hemiptera)

Occurrence and severity

Originally the mealy bug (*Ferrisia* spp.) is a New World genus native to the Americas (Williams and Watson, 1988). However, this stripe mealy bug (*Ferrisia virgata*) is spread to the Old World and it can now be found in tropical and temperate climates across the globe (Kosztarab, 1996; Williams, 2004; Williams and Watson, 1988). In the tropical areas, this mealy bug is widely distributed and highly polyphagous. Apart from jute, it also attacks cotton amongst fibre crops. The striped mealy bug was first recorded on tossa jute (*C. olitorius*) in Dacca (Now in Bangladesh) in 1944 (Das, 1947, 1948; Kundu *et al.*, 1959).

Type of Damage

The mealy bug colonies often occur at the growing points, around the stem nodes, on the undersides of leaves and on the fruit. Heavy infestations are conspicuous because of the white waxy secretions. Damage is caused by the nymphs that remain congregated in large numbers and suck up the sap of the leaf petiole, stem and pod. Repeated attacks by gregarious nymphs cause a crust to develop on stem, which hinders separation of fibre bundles during retting and thus forms "barky fibre". In case of seed crops the damage is more by decreasing pod setting. The affected pods become deformed and remain abortive. The pest appears at late in the season before the initiation of the flower buds.

Biology

Egg: Eggs are pale yellow, cylindrical with round ends, about 60-80 eggs are laid by a female @ 3-5 eggs per day and that deposited singly.

Nymph: There are three nymphal instars in stripe mealy bug. First and second instar nymphs are light yellow with six antennal segments (Highland, 1956). Third instar nymphs have seven antennal segments. Male begins to differentiate from females in the third instar when the body colour turns dark and body shape resembles to adult form. Wing buds are developed in third instar males. Faint dorsal stripes and small caudal tassels gradually become apparent in females with each moult. Total nymphal duration in females is about 43 and 93 days at 29 and 17°C., respectively, while that in male's elapses by 25 days.

Adult: Adult females are oval, greyish yellow, (2.0) 4-4.5 mm long (Kaydan and Gullan, 2012), with two longitudinal, submedian, interrupted dark stripes on the dorsum showing through the waxy secretion. The dorsum also bears numerous, straight, crystalline (glassy) threads of wax that are extended laterally from the body. Two posterior waxy tails or tassels with a length half that of mealy bug body are also present. The mealy bug does not possess an ovisac (Ferris, 1950). Longevity of the adult female was 36-53 days, and that of the male 1-3 days (Ghose and Paul, 1972).

Life Cycle: *Ferrisia virgata* reproduces sexually, with each female mating only once. Parthenogenetic reproduction is also of common occurrence in indoor rearing. Life cycle requires about 6-7 weeks in warmer season and that may extend to 13-14 weeks in low temperature (Cerson, 2016). Eggs are laid in groups beneath the body on a pad of cottony wax filaments (Kaydan and Gullan, 2012) over a period of 20-29 days (Schmutterer, 1969). Each female is ovoviviparous. Female lay about 60–80 eggs (Awadallah *et al.*, 1979) that hatch within 30 minutes of being laid (Ghose and Paul, 1972) or differently, after 3-4 hours (Schmutterer, 1969) or within few hours (Awadallah *et al.*, 1979). Female and male nymphs moult 3 and 4 times (4rth moult is from puparia), and the development period varies from 26-47 and 31-57 days, respectively (Ghose and Paul, 1972). At the end of third instar the female moults into the sexually mature adult form. Males begin constructing a cocoon at the onset of second instar and continue feeding and maturing while inside (Highland, 1956). The third instar in males is a prepupa stage after which a puparia is formed. The male reaches sexual maturity upon the moult from puparia to the winged adult stage. There are three to five generations per year (Ammar *et al.*, 1979; Awadallah *et al.*, 1979; Highland, 1956).

Control measures

Several chemical insecticides are found effective to control the mealy bug. Satisfactory control may be obtained by spraying diazinon, malathion and dimethoate at their effective doses (Schmutterer, 1969). Prothiofos, either alone or with mineral oil may give better control of *F. virgata* (Villiers and Stander, 1978). Permethrin is also another insecticide highly effective to control this mealy bug (Price, 1979).

As regards the biocontrol measure, several hymenopterous parasites and predators are found to be associated with *F. virgata*. However, their numbers in nature are considerably low compared to the relatively high population of the pest (Awadallah *et al.*, 1979)

ii) Spherical mealybug: *Nipaecoccus viridis* Newstead, 1894 (Hemiptera: Pseudococcidae)

Synonym

= *Pseudococcus filamentosus* var. *corymbatus* Green, 1922

Occurrence and severity

This mealy bug attacks over 100 plant species in more than 30 families. In jute, it was first observed in the year 1954 in 24-Parganas, West Bengal. Further, it is reported that this mealy bug causes serious damage both in white and tossa jutes (Dutt and Ganguli, 1956).

Type of Damage

The infestation by this pest arrests the vertical growth of the stem and after making a loop that recovers and reverts to its normal direction of growth. The attacked region becomes deep green in colour along with swelling and shortening of internodes. Leaves arising from the attacked regions become deformed and the attacked region appears as bushy. The first and second instar nymphs mostly induce this kind of malformation, while the later instars or the adults are not associated with such damage. Apical meristem of the growing tip is mostly preferred by them for feeding. Hypertrophy of cells in a radial direction of different tissues, like, the cortex, xylem and pith occurs. An increase in the number of cell layers is also noticed (Dutt and Ganguli, 1956, Kundu *et al.*, 1959).

Biology

Egg: Eggs are dark purple and laid by the female in a yellowish to white ovisac formed by wax threads. Large females may lay more than 1000 eggs.

Nymph: The female nymph and two early male nymphal instars resemble the female, but are smaller and purple in colour. Later they become darker and covered with wax.

Pupa: The male pupa is light-brown to purple and develops in a loose cottony-white cocoon.

Adult: Adult females are round or broadly oval, somewhat flattened dorsoventrally, about 4 mm long and 3mm in width, with body colour being black, purple to blue green and with thick white or pale yellow wax. The body segmentation is visible prior to oviposition and covered by wax. The young ones covered in mealy white wax with short projecting filaments arranged around the margin. In high densities waxy secretions, may appear as a continuous layer of wax which will obscure individual mealybugs. Wax may turn yellow in older infestations. Ovipositing females become covered in abundant white waxy threads that form a smooth globular ovisac. The wax threads are very elastic and if the ovisac is grasped and pulled with the fingers, it can be drawn out for 150 mm or more. The body contents are purple and this can be observed when individuals are squashed.

Adult males have well-developed legs, antennae and genitalia, one pair of simple wings and no mouthparts. They are very short-lived (Cilliers and Bedford, 1978; Annecke and Moran, 1982; Hattingh *et al.*, 1998).

Life Cycle

Female passes through 3 nymphal stages, often moving a little after each moult to change their feeding sites. They live about 50 days and die soon after depositing all eggs. The male develops through two larval instars, prepupal and pupal stages. It stops feeding in the former stage and seeks a pupation site, away from the colony, often in aggregates. Female develops little faster than that of the male (19-20 days). Males live up to three days and die after copulation. The mealybug reproduces sexually. Females begin to secrete an ovisac about a day after mating, and 3-7 days later begin to lay eggs, for a period of 6-15 days. Egg survival is adversely affected by high relative humidity (above 60) and temperatures below 20°C. The emerging crawlers move away from the hatching site and aggregate in cracks on stems, twigs, leaves or young fruits. They may crawl long distances and can easily be dispersed by winds. At all stages they feed on host phloem and parenchyma (Ghosh and Ghose, 1989; Ghose and Ghosh, 1990).

Control measures

A predator, *Scymnus* (*Pullus*) *pallidicollis* has been found to be very efficient in checking the pest. .

Nicotine sulphate 1.5 per cent and Folidol 0.005 per cent (a.i.) are found effective for control (Kundu *et al.*, 1959)

iii) Cotton mealybug: *Phenacoccus solenopsis* Tinsley (Hemiptera: Pseudococcidae)

Occurrence and severity

Recently, for the first time in May 2009 cotton mealy bug, *Phenacoccus solenopsis* Tinsley [(Fig.5 (A-B)] has been recorded on tosa jute (*C.olitorius* cv. JRO-524) in research farm of ICAR-Central Jute and Allied Fibre Research Institute, Barrackpore. Later, it was found that it attacked both tossa and white jute (*C. capsularis*) in someparts of Southern West Bengal, India.

Fig. 5 (A-B). Cotton mealybug, *Phenacoccus solenopsis*, in jute. **A.** Cotton mealy bug on jute. **B.** Symbiotic association of ants with cotton mealy bug in jute.

At first, it was observed on olitorius jute (cv. JRO-524) in considerable limit (60-80% infestation with an average intensity of 2-3 in 0-4 scale) in the year 2009 at CRIJAF, Barrackpore. Later, another outbreak of mealybug on jute was observed in 2011. The infestation of mealy bug on tossa jute particularly in the early crop growth stage of the plant (40-65 days old) during intermittent stretches of dry period has been found to occur in many parts of South Bengal. In a few areas the infestation was too high to control the pest (Satpathy *et al.*, 2009, 2013; Gotyal *et al.*, 2014).

Type of Damage

The damage is mostly caused by attacking the plant during early (40-65 days of sowing) vegetative phase to feed the cell sap by the nymphs of mealy bug. This exhibits symptoms of distorted and bushy shoots, crinkled and/ or twisted bunchy leaves, and stunted plants that dry completely in severe cases. Apical meristem is the most susceptible part of the plant. The vertical growth of plant is arrested with shortened internodes. The plant gives bushy appearance. Repeated attacks on the stem cause the development of crust due to which fibre bundles resist separation at the time of retting, resulting in the formation of 'barky fibre". Late season infestation during reproductive stage of the crop results in reduced plant vigour and early crop senescence. The infestation of mealy bug on tossa jute becomes severe particularly during the early crop growth period in presence of intermittent stretches of dry period (Gotyal *et al.*, 2014).

Biology

Egg: Eggs are whitish yellow, semi-transparent, oval to oblong in shape and loosely remain inside the white thread like cottony mass. Length and breadth of eggs are about 0.35 mm and 0.20 mm, respectively.

Nymph: The female nymph moults three times, whereas male moults four times to attain maturity. Newly emerged crawlers are light yellow in colour with light red eyes, three pairs of legs and a pair of seven segmented filiform antennae. The colour of nymphs appeared to grayish yellow within two to three days after hatching. First instar nymphs are oblong in shape and yellow in colour devoid of mealy scale cover on body. Duration of first instar nymph is about 5 to 9 days. The second instar nymph is to some extent larger than the first instar nymph. It is whitish yellow in colour and oblong in shape. It secretes a waxy material on its body thus a white mealy covering appears and the body margin shows rudimentary waxy filaments. Duration of second instar nymph is about 4 to 10 days. The third instar nymph is bigger in size than the earlier instars. Pair of antennae and smoky brownish legs are clearly visible whereas the eyes remain covered with mealy secretion and not clearly visible. The secretion of white waxy material covers whole body except the either side of dorsal region of metathorasic and few abdominal segments along the mid dorsal line of the body. Duration of third instar nymphs is about 4 to 6 days. Male nymph passes through the fourth instar and the duration of that varies from 5 to 7 days (Nikam *et al.*, 2010; Vennila *et al.*, 2010). Thus, the developmental period of male is longer in nymphal stages as compared to that of females. The total nymphal period of male and female is 21-29 days and 18-25 days, respectively.

Adult

Female adult is wingless, oblong in shape and light to dark yellow in colour ventrally. Dorsally the whole body is well segmented, covered with waxy deposition except at the posterior abdominal region where blackish stripes on either side of mid dorsal line are visible. The deposition of waxy material along the margin of the body is so intense that waxy filament clearly visible on the outer margin. It possesses a pair of brownish, short, filiform eight segmented antennae and three pairs of brownish red legs. The newly emerged female adult is about 1.92 to 2.80 mm in length and 0.70 to 1.00 mm in breadth, while, the length and breadth of gravid female are 4.1 to 4.7 mm and 2.8 to 3.0 mm, respectively.

Male adult is smaller than females, slender, delicate, smoky white in colour, possessing single pair of delicate, long and slender antennae, three pairs of brownish legs, two pairs of terminal filaments and one pair of well-developed mesothoracic wings. The abdominal region is pale yellow. The second pair of wings is modified as hamulohalters (Power *et al.*, 2017)

Life Cycle

The mealy bug reproduces sexually. The gravid female of *P. solenopsis* lay about 150-600 number of whitish yellow eggs within a white cottony ovisac which initially remain underneath the body of the mother and later, the female move leaving the ovisac on the leaf surface. Incubation period of eggs varies from 30 to 56 minutes. The emerged nymphs (crawlers) crawl out of thread like pouches to leaf surface within a day. The crawler is very active, fast moving and remains on the lower surface of leaf near the midrib or petiole. It passes through different instars. At the third instar, the nymph becomes sluggish and appears encased in loose waxy white covering called cocoon from which only the male adult emerges after 5-7 day and passes through the fourth instar. The total nymphal period of male and female is 21-29 days and 18-25 days, respectively. The adult female passes through pre-oviposition, oviposition and postoviposition periods which are about 4-9 days, 8-18 days and 3-10, respectively, depending upon the seasonal variations. The longevity of female adult is about 30-50 days, while, adult male survives only 3-5 days. (Akintola and Ande, 2008; Aheer *et al.*, 2009b; Arve, 2009; Dhawan and Saini, 2009; Nikam *et al.*, 2010; Vennila *et al.*, 2010; Power *et al.*, 2017)

Management/Control measures

Cultural control

Monitoring of the pest and natural enemies is an effective measure to determine need, time and type of management practices to adopt for mealy bug control. The level of parasitization and abundance of natural enemies are to be ascertained on the basis of regular monitoring prior to the insecticide application. Field sanitation is one of the effective cultural practices to keep the mealy bug under controlled condition. The infested plants or plant parts should be removed and destroyed from the farm to check the further incidence of mealy bugs. Balanced use of fertilizers also keeps the pests under check. (Gotyal *et al.*, 2014). If the infestation is going to cause economic damage the chemical control measure may be adopted.

Chemical Control

Dusting of methyl parathion 2 per cent, spraying of profenophos 50 EC or chlorpyriphos 25 EC or quinalphos 25 EC help to reduces the pest population (Joshi *et al.*, 2010). Methomyl, imidacloprid, thiamethoxam and chlorpyrifos also show very much effective against *P. solenopsis* giving 92.3 to 80.4% reduction of the insect population (Zahi *et al.*, 2016)

Biocontrol

Although, there are certain controversies about the efficacy of bioagents in mealy bug control under field condition, here a few important studies are mentioned below:

The naturally occurring parasite, *Aenasius bambawalei* Hayat, is a potential bioagent of *P. solenopsis*. The parasitoid completes its life cycle on the mealybug leaving the mummified body along with the exit hole behind. The parasitization efficiency of the parasitoid from field collected mealybugs was found 57.2 per cent (Kumar *et al.*, 2009).

Apart from this there are several other predators of this mealy bug. Amongst them, *Brumus suturalis* Fabricus, *Chelomenes sexmaculatia* Fabricus, *Cryptolaemus montrouzieri* (Muslant) and *Chrysoperla carnea* (Stephens) are most common predators (Tanwar *et al.*, 2007; Radadia *et al.*, 2008). Sattar *et al.* (2007) found that the *C. carnea* mostly preferred the first instar nymphs of the mealy bug and consumed up to 1604 mealy bugs per day. However, Moore (1988) stated that despite the frequent use of predators, only the coccinellid, *C. montrouzieri* can be considered successful. However, the survivability of *C. montrouzieri* during adverse atmospheric condition and against natural enemies is questionable. It cannot survive in winter and several earlier attempts resulted negligible achievements. Since, it was observed that the lady bird beetle, *Cryptolaemus montrouzieri* was an effective control used at the vineyards in southern India in the 1970s. However failure was occurred as *Maconellicoccus hirsutus* could not survive in winter and attacked by several parasites and predators. It is attacked worldwide by 21 parasites and 41 predators (Sagarra and Peterkin, 1999).

As regards the efficacy of pathogenic fungi in mealy bug control, it is found that *Verticillium lecanii* (Zimmermann), *Metarrhizium anisoplae* Metschnikow and *Beauveria bassiana* (Vuillemin) are most effective fungal pathogens of the pest (Tanwar *et al.*, 2007 and Radadia *et al.*, 2008).

Considering the earlier research findings, Joshi *et al.* (2010) suggested that *Aenasius bambawalei* Hayat, *Anagyruska mali* Mani, *Cryptolaemus montrouzieri* (Mulsant), *Chrysoperla carnea* (Stephens), *Verticillium lecanii* (Zimmermann) and *Beauveria bassiana* (Vuillemin) are the effective biological control agents in managing the infestation of the pest.

4. Mites

i) Red mite/ Red spider mite

Occurrence & severity

Red mite, also known as red spider mite, is an important pest of jute, particularly of white jute (*Corchorus capsularis*) in India (Misra, 1913 ; Das, 1948). The mite is polyphagous, hence its distribution is widespread. In India, it is widely distributed and considered as a serious pest of tea in north-eastern India (Harrison, 1937; Awasthy and Venkatakrishna, 1977). It attacks jute, *C. capsularis*, in India (Das, 1959) and that, so far, happens in neighbouring countries also. The incidence of this mite is of sporadic in nature but occasionally it causes serious damage to the crop. It is found that, the intermittent shower followed by dry spell with high humidity favours multiplication of the mite faster (Tripathi and Ram, 1971). Further, prolonged dry weather during the early part of the flushing season normally increases the red spider incidence. It prefers bright sun and unshaded areas and in that condition the mites attack plants are more severely (Das, 1959). The red spider mites live under a cover of web that they spin as a protection against inclement weather. The pest occurs in severe form from March to June but with the monsoon rains it practically disappears. A second, light, attack may, however, develop in September or October.

Types of damage

Red spider mite normally attacks the upper surface of the mature leaves in which the sap is not flowing freely. In a severe infestation, particularly under conditions of dry weather, the lower surface and the young leaves are almost equally attacked. Both adults and nymphs residing under the web suck the sap from upper surface. Due to continuous feeding by both adult and nymphs, the affected leaves become leathery, turn yellowish brown that at later become bronzy and drop off prematurely.

Scientific name & order: *Oligonychus coffeae* **Nietner, 1861 (Acari: Tetranychidae)**

Biology

Egg: The egg is ovoid or spherical, smooth, with a slight depression on the exposed top side and flattened on the lower surface (Das 1959; Rao 1974). A short hair-like process arises from the upper pole and bends in a form of hook. The eggs are blood red to chrome red glossy and shining, but change to light orange before hatching. Eggs are laid singly on the upper surface of the leaf along the mid-rib and veins. They are laid one at a time ranging from 4 to 6 per day (Das, 1959). Hatching period is 2-7 days. When the larva emerges, the egg shells breaks into two halves with bottom half remaining attached to the leaf surface. The remnant shell is white transparent (Roy *et al.*, 2014).

Larva: The larva is very small in size, almost round and it has six (three pairs) legs. A freshly hatched larva is yellowish orange but the colour subsequently changes to pale orange. The idiosoma later shows a greenish appearance with consumption of food. The larva after passing through the inactive nymphochrysalis (quiescent) stage developed into the first nymphal stage called protonymph. Larval period is 1-3 days.

Nymphs

Protonymph: It is oval shaped body and characterised by the presence of four pairs of legs The anterior legs are pale crimpson, whereas the posterior pairs are deep reddish brown. Protonymph enters the second quiescent stage called deutochrysalis and then develops into deutonymph. Protonymphal period is 2-3 days.

Deutonymph: The deutonymph is like protonymph but larger in size. At this stage, both the sexes can be differentiated. The female is larger and broader than the male. Its abdomen is rounded at the posterior end, where as the abdomen of male is pointed at the posterior end. Deutonymph then underwent the quiescence period. Adult mite emerged from the third quiescent stage called teleiochrysalis. Deutonymphal period is 1-2 days.

Adult: The minute adult mites are just visible to the naked eye. Their body color differs from orange-red to dark red, with black patches. The adults are sexually distinguishable. Males are smaller in size with a narrow body and tapering end whereas females are larger and broad in shape. In females, the legs and front portion of the body (propodosoma) are bright crimson, whereas the abdomen is dark purplish brown. The male is the same body color as the female except that the tip of its abdomen is crimson. The legs, particularly the ûrst pair, in males are longer than those of the female. The tip of the aedeagus is bent to the ventral surface at right angles (Das 1959).

Both males and females are sexually mature on emergence. Males emerge earlier and wander about in search of female deutonymphs (Das 1959). Pairing takes place immediately on emergence of the female. In most cases several males are found to surround a quiescent female deutonymph which starts admitting males (Das 1959).

Males are short-lived (4-5 days) whereas females may live for about 3 weeks during summer or for a couple of months during winter (Rao, 1974). The female dies soon after the last egg is laid. In general, the mites survive for 20-30 days (Roy *et al.*, 2014).

Life cycle

The life cycle is composed of egg, larva, protonymph, deutonymph and adult with quiescent stages in between two successive active stages (Roy *et al.*, 2014). The duration of the life-cycle varies with the season depending on the temperature and humidity. In May and June the life-cycle is completed in 9·4–12 days outdoors, while in the cold weather it may take as much as 28 days. The maximum length of life of a female has been found to be 29 days indoors. The males usually die within four or five days. Parthenogenetic reproduction may take place under induced conditions, the progeny being all males. On average, there are 15–16 generations each year but these overlap, no winter diapause is reported and the red mite can be found in all stages of development at almost any time throughout the year. Hu and Wang (1965) reported that under laboratory conditions, the mites passed through 22 generations in a year.

Management/ Control measure

Cultural

Field sanitation is very much required to manage the mite. The weed plant (*Physalis minima*) is an alternate host, so, weeding is required.

Use of botanicals

Neem oil (Azadirachtin) in various commercial formulations is recommended to manage the red spider mite (Muraleedharan, 2006). Several other plant extracts, such as, water extracts of a common weed, *Clerodendrum viscosum*, and *Melia azadirachta* are found promising in controlling red spider mite populations under field conditions (Roy *et al*, 2010b, 2011b). Acaricidal activity is also noticed with the different solvent extracts of *Polygonum hydropiper* (Samarah *et al.*, 2006; Roy *et al.*, 2011a).

Use of Natural enemies

There are several naturally occurring insect predators, such as coccinellid and staphylinid larvae, lacewing larvae, and mite predators (Roy *et al.*, 2014). These predatory insects attack eggs and other stages of the red spider, often keeping it considerably in check (Das, 1959). The naturally occurring mite predators are most important species of the families Phytoseiidae and Stigmaeidae (Roy *et al.*, 2014). However, the population of natural enemies is comparatively low in most cases.

Use of Fish extract

Fish extract (1:100) exhibited the highest mortality of nymphs of red spider mite. Mortality of nymphs of red spider mite was found to be highest with Fish extract at 1:100 dilution (83.33%). The fish extraction was prepared by mixing 80 kilograms of different parts of fish along with 50 litres cow urine, 15kg cow dung and 100 litres of water. The mixture was kept for 7 days in a plastic drum installed underground. *Polygonum hydropiper* extract at 15:200 caused significant mortality on nymphs and adult of red spider mite whereas Azadirachtin 5% was at par with fish extract and *P. hydropiper* extract (Bhuyan *et al.*, 2017).

Chemical control

Dicofol (0.04%) is effective to manage the mite on jute.

ii) Yellow mite

Occurrence & severity

The yellow mite/jute mite/broad mite is an important pest of diverse crops in tropical and subtropical regions. The mite was first described by Banks (1904) as *Tarsonemus latus* in Washington, D.C., USA (Denmark 1980). This species has a large host range and is distributed worldwide and is known by a number of common names. In India and Sri Lanka it is called the yellow tea mite, while those in Bangladesh call it the yellow jute mite. In some European countries it is called the broad spider. In parts of South America it is called the tropical mite or the broad rust mite. In respect of jute, this mite is a very destructive pest and causes damage to both fibre and seed crops (Das and Roychoudhori, 1979). In recent years, the yellow mite in jute is more regular and so severe that the situation may reach the status of outbreaks (Selvaraj *et al.*, 2016). This pest attacks both the cultivated species of jute, *Corchorus capsularis* and *C. olitorius*. Initial infestation of jute by the mite occurs in mid-May and the population reaches a peak at the end of June and again during the last week of July. Dry periods are most suitable for rapid multiplication of the mite, while damp weather and heavy rainfall are unfavourable. However, a recent study indicates that Yellow mite incidence shows positive association with atmospheric relative humidity (Rahaman and Khan, 2012a).

Types of damage

Both adults and nymphs are injurious. They mostly feed on the apical leaves on dorsal side and thus cause curling and crumpling of lamina backwards along the midrib. The affected younger leaves become discoloured. The natural green colour of the leaves turns into brown with change of shape due to curling (Das and Singh, 1985). Loss of nutrition in young plant is appeared due to sucking of saps. The height of plant becomes stunted and that causes significant yield loss (Nair, 1986, Pradhan and Saha, 1997). The malformation of terminal leaves is associated with the toxin of mite's saliva. The toxic saliva causes twisted, hardened and distorted growth in the terminal part of the plant (Baker 1997). The yellow mites are usually seen on the newest leaves. Owing to the tiny size of the pest, it is hardly seen through naked eyes but the damage is obvious. Early crops sown during March and April are liable to be more infested by mite than the late sown crops. It is important, that the yellow mite does not vector any known plant virus diseases (Waterhouse and Norris, 1987; Higa and Namba, 1970).

Scientific name & order: *Polyphagotarsonemus latus* Banks, 1904 (Acari: Tarsonemidae)

Synonym

= *Tarsonemus latus* Banks, 1904

Biology

Egg: The egg is colourless, translucent and oval in shape and slightly flattened (Lavoipierre, 1940). The exposed translucent surface is covered with five or six rows of 29 to 37 white bumps called tubercles. The egg is about 0.07-0.08 mm long (Denmark 1980, Peña and Campbell 2005, Baker 1997). It is usually laid singly on the underside of new growing leaves (Hill, 1983b). Egg usually hatches in 2 to 3 days.

Larva: Larva is very small, pear-shaped (Hill, 1983b) and has three pairs of legs. Just after hatching the larva is slow moving and with translucent body that appearing white due to presence of minute ridges (Pena and Campbell, 2005).

As it grows up its size becomes 0.1-0.2mm long. The female becomes yellowish green or dark green in color and male is yellowish brown (Waterhouse and Norris, 1987). Then the quiescent stage appears as an immobile, engorged larva (Baker, 1997). The larval period lasts for 2-3 days.

Nymph: After 2-3 days, the larva becomes a quiescent nymph that is clear and pointed at both ends. The nymphal stage lasts about a day. Nymphs are usually found in depressions on the leaves, although female nymphs are often carried about by males (Peña and Campbell 2005). In this quiescent period, there is no feeding. Sexes are similar in appearance, except for the fourth pair of legs. On males the fourth pairs of legs are enlarged, while, that on female, the legs are reduced and whip-like (Lavoipierre, 1940). The nymphal (quiescent) stage lasts 2 to 3 days After the quiescent stage, the mites become active, at which time , mating occurs.

Adult: The adult mites are very small and difficult to see without a hand lens. Adult yellow mites are elliptical in shape, but slightly wider at the front than the rear (Brown and Jones, 1983). Females are about 1.5 mm long and males are slightly shorter and more broad (Lavoipierre, 1940). Both male and female mites are light, translucent yellowish green. A pale white stripe runs longitudinally down the back of the female. They have 4 pairs of whitish legs, but the fourth pair of the female adult is greatly reduced. Females live for about 10 days and each female lays about 22-50 eggs at an average of 2 to 5 eggs per day (Hill, 1983b; Brown and Jones, 1983). Without fertilization, females produce eggs that hatch into only male progeny (Waterhouse and Norris, 1987).

Life cycle

The yellow/broad mite has a short and rapid life cycle living between 5 and 13 days. This mite has four stages in its life cycle: egg, larva, nymph and adult. Adult female lays 22 to 50 eggs (averaging two to five per day) on the undersides of leaves in about 8 to 13 days period and then dies. Adult male may live for five to nine days. The unmated female lays eggs that hatch into males, while, mated female usually lays four female eggs for every male egg. The egg hatches within two or three days and the larva emerges from the egg to start feeding. Larva is slow moving and does not disperse far. After two or three days, the larva develops into a quiescent larval (nymph) stage. Female nymph becomes attractive to the male that picks up the female and carries to the new foliage. Male and female both are very active, but the male apparently accounts for much of the dispersal of a yellow mite population in their frenzy to carry the female nymph to new leaves. When females emerge from the quiescent stage, males immediately mate with them (Annonymous, 2016; Baker, 1997; Peña and Campbell 2005).

Management/Control measures

Cultural control

There is no specific recommendation to control jute mites by means of cultural practices. However, field sanitation, timely sowing, use of adequate spacing between plants, use of balanced NPK fertilization and regular crop monitoring are the essential tools for managing any pest attack in jute.

Biological Control

Throughout the world, specific natural enemies are not known and there have been no attempts at biological control. However, many locally occurring general mite predators give satisfactory control in many areas (Wilkerson 2005, Peña and Campbell 2005, Fan and Petitt 1994, Peña et al. 1996). Broad mites are satisfactorily controlled by introduced general predators (Waterhouse and Norris, 1987).

Use of Botanicals

The botanical oils, like, Neem oil and Mahogany oil, are found effective to minimize the yellow mite infestation in jute by 60.55 and 55.89%, respectively (Rahaman et al., 2016). In addition, application of neem oil is reported to increase jute plant height by 24.64% and base diameter by 27.87%, giving higher yield (Yasmin et al., 2013).

Chemical measure

The acaricide, Abamectin (Ambush 1.8 EC), has been found highly effective showing reduction of 80.25% yellow mite infested plant (Rahaman et al., 2016). Piao et al. (1999) have found maximum percent mortality of *P. latus* on jute after application of abamectin (Ambush 1.8 EC) @ 0.5 ml/l. The hatching percentage of eggs is also reduced by the application of Ambush 1.8 EC @ 4 ml/l (Watson et al., 1985).

Foliar spray of Endosulfan provides superior control of yellow mite. Dicofol is effective, but may require a second application a week later in severe attacks (Hill, 1983). Sulphur is effective but requires 2 -3 weeks to achieve control (Swaine, 1971; Brown and Jones, 1983). Elsewhere, the following insecticides are reported to control this pest. Chlorpyrifos, methamidophos, monocrotophos, phosphamidon, acephate or carbosulfan (Dhandapani and Jayaraj, 1982). Dimethoate or dinocap are also recommended for yellow mite control in certain country (Waterhouse and Norris, 1987).

5. Jute semilooper/Angled gem

Occurrence & severity

The jute semilooper or angeled gem has been regarded as the most important pest of jute for many years (Lefroy, 1906, 1907; Woodhouse, 1913; Chowdhury, 1933; Patel and Ghosh, 1940). The insect occurs in the entire jute growing region all over the world (India, Bangladesh, Myanmar, Sri Lanka and in parts of Africa). In West Bengal (India), jute semilooper (*Anomis sabulifera* Guen) is found as a major pest of jute, causing economic damage (Rahman and Khan, 2012a). During the years 1950s there were regular outbreaks of the insect in West Bengal, reducing fibre yield by 14-40% (Kundu *et al.*, 1959). In Assam, Orissa and Uttar Pradesh also this pest is very destructive. In Bangladesh, the pest is found to cause 2-3 maunds (80-120kg) fibre yield loss per acre (Kabir, 1975). It is estimated that infestations of the insect up to 15% do not cause any appreciable loss of fibre yield, while, 20-30% infestation causes moderate yield loss and 40-60% infestation is liable to decrease yield by 24.58-25.56% (Chatterjee *et al.*,1988). The jute plants subjected to a single attack at the age of 40 days suffer less damage whereas those infested repeatedly during 60 to 90 days of age suffer the maximum damage (Tripathi and Ram,1972). The tossa jute (*C. olitorious*) is more susceptible than the white jute (*C. capsularis*). The incidence of semilooper is negatively correlated with the maximum temperature but had positive significant association with minimum temperature, morning relative humidity (RH) and afternoon RH (Rahman and Khan, 2012a). Pre monsoon rains followed by drought condition are congenial for the outbreak of semilooper and may lead up to 50% loss of crop (Dutt, 1958a). The semilooper attack is severe on half-grown plants which are one metre high. They camouflage but are easily noticed when they crawl by producing a loop in the middle. The second generation is the most damaging and sometimes up to 90 per cent of the leaves may be eaten up. Generally, the top 7-9 leaves are damaged and plant growth is adversely affected, resulting in a considerable reduction in the yield of fibre (Dutt, 1958a; Sing and Das, 1979).

Types of damage

Fruits, growing seedlings, leaves and seeds are mostly affected by the caterpillars and adults as well. They externally feed on the plant parts leading to dieback, chlorosis and reduction of harvest. The larvae, just after hatching, feed on the epidermal tissue and mesophyll of the lower surface of young leaf, keeping the upper epidermal membrane intact (Das, 1948). The first sign of its attack is seen at the top buds of the plant, since, the apical buds are the most susceptible to damage (Dutt, 1958a). Later, the infestation extends downwards in an almost regular sequence up to the tenth or eleventh leaf (Kabir, 1975) and never beyond the fifteenth one from top. The intensity of attack tends to vary inversely with the age of the foliage, about 81% of attack being limited to the seventh fully opened leaf below the top. Lefroy (1907) reported that holes are made by *A. sabulifera* feeding on the body of the leaves and their margins, giving the leaves at the top of each shoot a characteristic feeding appearance. [(Fig. 6 (A-B)] In cases of severe attack the growing points are eaten and destroyed, the stems are totally defoliated, and internodes are shortened and weakened at the place of attack. As a result of excessive feeding the whole plant becomes dwarf. The larvae are very active in eating jute leaves voraciously, particularly in the crown, where the tender leaves are completely destroyed. In case large leaf, the softer portion of the lamina is devoured, leaving only the skeleton of the harder main and secondary veins. Branching caused by damage to the apical bud tangles the fibres and makes their extraction difficult (Ghosh, 1983). The pest feeds on the developing seeds inside the pods. A single large hole is made in the spherical capsules of *C. capsularis*, while, two holes are prepared in the long cylindrical pods of *C. olitorius* (Singh and Das, 1979).

Fig. 6 (A-B). Jute semilooper (*Anomis sabulifera*) attack in jute

Scientific name & order: *Anomis sabulifera* **Guenée, 1852 (Noctuidae: Lepidoptera)**

Synonyms

= *Gonitis sabulifera* Guenee (1852)

= *Gonitis propingua* Butler (1884)

= *Gonitis marginata* Holland (1894)

Biology

Egg: Egg is small, transparent and has the appearance of a water droplet. The eggs are laid singly and fecundity is 28-247 eggs/female (Senapati and Ghose, 1991). The egg stage remains for about 2 days.

larva: The larva of semilooper is Light green to dark greenish in colour, creeping in nature, easily hides over the green leaves of the jute plant. Its head is slightly yellowish in colour. The elongated larva moves with loop like formation, hence it is called semilooper. The loop is formed because its five pairs of sucker feet are not well- developed, and when it moves it humps its back into an arch like structure. There are small warts over its dorsal side and the lateral sides possess yellow stripes. Each segment bears short hair on small white ringed black papillae. The full grown stage attains a length of 4 cm. The larval stage is consisted of 5 instars and lasts for 16 days.

Pupa: The pupa is inactive, soft bodied and brownish soil like colour. It can be seen naked in the soil as no pupal chamber is formed. The larva shrinks to about 3 cm in length at the end of pupal stage. The pupal stage continues 6 days (Gaikwad and Pawar, 1983)

Adult: The adult has a typical moth like structure. The adult moth is of dull earth brown color, stout build and with brownish wings. The wings are yellowish in colour, with black spots. Wingspan is about 32-38mm. The females are larger than males. Antennae of male are ciliated. Antemedial line of fore wings is bent outwards between vein 1 and inner margin. The postmedial line is incurved beyond the cell. It has diffused black on the antemedial line of fore wings and between postmedial and sub-marginal lines. A small orbicular spot is usually present and tow specks are conjoined into a reniform spot (Hampson, 1892). These moths are not visible during day time but come out after the sun sets.

Life cycle

The insect pest passes winter in soil in the pupal stage and the adult moths appear in May-June, when the crop begins to grow in the field. They lay eggs singly on the underside of young leaves. There may be several eggs in one leaf. A female may lay more than 150 eggs which look like water droplets. Egg period lasts for 2-3 days. The tiny green caterpillars, on emerging with three pairs of tiny sucker feet, feed on the apical leaves and buds. Subsequently fourth pair of legs develops, the fifth pair of legs does not well developed and the larva attains full length in about 17 days after five moults. The pupation takes place on the plant or in the soil. In summer, the pupae emerge in about a week, but those, which diapause, spend the entire winter in that stage. The life-cycle is completed in about one month and several generations are completed in a year.

Management/Control measures

Use of host resistance

Although, it is known that tossa jute (*C. olitorius*) varieties are more susceptible than white jute (*C. capsularis*) varieties (Kabir, 1975), till date literature regarding the host resistance in jute against semilooper is scanty (CRIJAF, 1987-1989).

Cultural Control

Early sowing before second week of April with less fertilization may be effective to escape the pest attack. Since, certain earlier experimental data of ICAR-CRIJAF, West Bengal, indicated that sowing before the second week in April reduced the incidence of *A. sabulifera* and resulted in better yields (Dutt, 1952a; Tripathi and Ram, 1971; CRIJAF, 1990-1992), while higher levels of fertilizer increased infestation by *A. sabulifera* (CRIJAF, 1986). Further, as a preventive measure ploughing the infested fields after harvest is suggested to kill the pupae that remained in soil. Dislodging the caterpillars into kerosenized water by drawing a rope across the young crop may also be effective to minimize early infestation. Setting up light trap 1-2 nos/acre and killing the trapped moths is also suggested to manage the pest.

Biological Control

Encouraging predatory birds to feed on the larvae of *A. sabulifera* by fixing perches for them in the field was the earliest biological control recommendation (ICJC, 1946). This method is still valid but is no longer practised (Ghosh, 1983). Spraying of *Bacillus thuringiensis* spores has been found effective in increasing the mortality of jute semilooper both under laboratory (Das and Singh, 1976) and field (Chatterjee, 1965) conditions. In a field trial with *B. thuringiensis*, the mortality of *A. sabulifera* on jute plants was found up to 96% (CRIJAF, 1977). Spores of the fungus *Beauveria bassiana* may also be used as a bioagent to control *A. sabulifera*, since; this parasitic fungus was effective under laboratory condition (Pandit and Som, 1988). Nuclear polyhedrosis virus (NPV) was also found effective to cause mortality as high as 80% of jute semilooper (Ishaque and Kabir, 1967) and it's apply as a spray has been suggested as a very promising control method.

Chemical Control

Several insecticides are found effective in different investigations. These insecticides may be used alternately in avoiding the development of resistance in the insect. Spraying of Endosulfan 35EC (0.075%) in 500 l water per ha thrice at 15 days interval from mid June or at first appearance of the pest was the earlier suggested very effective chemical measure. However, this chemical is now banned in various countries, including India. So the other insecticides which are found effective may be used alternately to manage the pest. These are: (1)Spraying with Spinosad 45 % SC @ 1 ml/5 lit, (2) Chlorpyriphos 20 % EC @ 2.5 ml/lit, (3) Lufenuron 5.8 % EC @ 1 ml/lit, (4) Quinalphos 25 % EC @ 2 ml/lit, (5) Carbaryl 50 % WDP @ 2.5 gm/lit, (6) Triazophos 40 % EC @ 1 ml/lit, (7) Acephate 75 % WP @ 0.75 gm/lit and (8) Profenophos 50 % EC @ 1 ml/lit of water.

6. Jute stem-weevil

Occurrence & Severity

The pest occurs in almost all jute growing areas. In Bangladesh and West Bengal (India), the stem weevil is one of the major harmful pests of jute causing economic damage. In West Bengal, nearly 5% jute plants are found as infested by the pest. The insect also occurs in serious form on jute in Bihar and U.P. There are numerous overlapping generations of weevil in the field. The weevil exists throughout the entire crop season. The crop raised either during the month of April or in the first week of May is liable to stem weevil attack. Moreover, the crop grown with high nitrogen fertilizer and sown early suffers more. The weevil preferably fed on *Corchorus olitorius* than on *C. capsularis* (Tripathi and Bhattacharya, 1968). The rainfall has negative impact on the incidence of jute stem weevil (Das and Singh 1977; Rahman and Khan, 2010, 2012a). Although, cloudy damp weather associated with low daytime temperatures of both soil and air are congenial for incidence and multiplication of the jute stem weevil (Dutt, 1958b; Chatterjee *et al.*, 1978). The average yield loss of crop due to apion was estimated at 18% in India (Tripathi and Bhattacharya, 1968; Singh, 1981)

Types of damage

The nodal region below leaf base is the most ideal place for the insect damage, whereas the nodal region opposite to leaf-base is less ideal place for the pest. The internodes of the stem and petioles are sometimes attacked by the pest but no incidence is found in the leaf lamina. Adult weevil feeds on tender leaves making pinholes and attacks top shoot for egg laying. The top shoots dry up; as a result side branches develop and reduce the fibre length. The main damage to the quality of fibre is caused by weevil made oviposition holes. A female may make a number of holes before laying eggs and damages numerous stems in her life time. Grub damages the fibre bundles and as a result of injury, mucilaginous and gummy substances produced in the stem and produced knots on the fibre. This knot does not dissolve during retting and persists through the process of washing and drying. The fibre with knots is called knotty fibre.

Further, during the time of retting of jute, some of the fibres are torn out at the point of the knots, resulting decreased production (Dasgupta and Ghosh, 2008). This knot on the fibre is considered as a major defect and degrades the fibre (Das, 1948; Tripathi, 1967). The destruction of tissues by the grub also causes wilting of the crown leaves just above the seat where the insect is concealed. The withered crown leaves turn black and drop off (Ghosh, 1983).

Scientific name, family & order: *Apion corchori* **Marshall (Apionidae: Coleoptera)**

Biology

Egg: Egg is oval, very small measuring about 0.43 x0.33 mm in size, and glistening white. The egg hatches out into grub in about 3 days (Ghosh, 1983). Eggs are laid singly in the made up hole on jute stem.

Grabs: The grub is creamy white with light brown head, measuring 1-2mm in size, body is wrinkled and remains within the stem

Pupa: It measures about 2.07 mm in length and 1.08 mm in breadth.

Adult: Tiny in size about 2-3mm long, brown or dull black coloured weevil with very conspicuous snout, body is covered with minute white setae. Adults may live for about 120-200 days. A single female lays about 500-600 eggs during its life span (Ghosh, 1983).

Life cycle

Female weevil lays about 500-600 eggs within its oviposition period of around 120-124 days. The female lays its egg singly in a hole that punctured by itself at the node in the stem just below the young crown leaves of young plants (Ghosh, 1983). The egg hatches out into grub in about 3 days. The grabs remain in jute stem and starts feeding on tissues inside bark. The grabs move downwards making tunnel within stem and after 10-12 days of feeding in jute stem the grabs pupate in a rough chamber inside the stem. The pupal period lasts for 4-6 days. Adult weevil emerges through the exit hole of tunnel made earlier by the grab/mother adult. The normal life-cycle completes within 15-24 days. A number of overlapping generations are completed during the jute season. The adult passes the winter in bushes, shrubs and hedges, and start laying eggs on the new crop next year (Prasad, 2007; Arora *et al.*, 2012).

Management/Control measures

Use of host resistance

Although, a few varieties of both white and tossa jutes show certain degree tolerance, however, their use in any management programme against stem weevil under field conditions is negligible. The *C. capsularis* cv. NDC-2005-7, JRC 321 and *C. olitorius* cv. JRO-2360 are to some extent resistant to the pest attacks (Das and Singh, 1977; Das and Singh, 1986; Yadav and Yadav, 2010; Yadav and Singh, 2012). However, these are not popular amongst most of the jute farmers of West Bengal (India) and Bangladesh.

Cultural methods

Balanced fertilization with N.P.K., timely removal of weeds and timely thinning are recommended in order to minimize the weevil attack by cultural methods (Tripathi and Bhattacharya, 1969). The pest is suppressed by collecting and destroying the stubble after harvest, and by removing Bon okra (*Triumfetta rhomboidea*) and other shurbs and bushes from the field.

Chemical control

Effective control of the pest is very crucial as because the insect passes there most of the life inside the stem and thus escapes from the direct contact of the applied pesticides (Das and Singh, 1986). However, infested plants should be sprayed with any of the recommended insecticides. Carbaryl (50WP) @ 2.5kg in 625 L of water per ha; Endosulfan 0.075% a.i. (Das and Singh, 1977) and Endrin (0.3%) at an interval of 15 days (Tripathi and Bhattacharya, 1970) are very effective chemical pesticides to minimize the pest attacks.

2
Mesta/Kenaf & Roselle (*Hibiscus cannabinus*; *H. sabdariffa*) Malvales: Malvaceae

A. Mesta diseases

1. Anthracnose

Occurrence & severity status

Mesta (*Hibiscus cannabinus*) is very much prone to be attacked by anthracnose disease (Singh, 2014). It is very destructive disease in India (Ghosh, 1983). The disease also occurs in South Korea with overall disease incidence of 20%. (Kwon *et al.*, 2015).

Symptoms

Symptoms on kenaf plants include dark brown spots on the leaves, petioles and stems. New leaves and shoots are most susceptible to infection (Kwon *et al.*, 2015). Initially the terminal bud is attacked, where stipules and young leaves are affected and develop necrotic spots and withers (Singh, 2014). Mature lesions on fully expended leaves are dark brown and often irregular in shape, and not vein delimited. Later, the infected leaves fall off; Flowers and seed capsules are also affected by the disease (Singh, 2010). The stems are affected in patches. On stem the lesions are elongated and black, which later form cavities and in severe cases, stem-dieback or wilt occurs (Singh, 2010; Kwon *et al.*, 2015).

Causal organism: *Colletotrichum hibisci* **Pollacci , 1888 (1896?)**
[AttiIst. bot. R. Univ. Pavia, 2 Sér. 5: 16]

Synonym

(?) = *Colletotrichum hibisci-cannabini* Sawada [in Ghosh, 1983)

Morphological/Morphophysiolgical characteristics

The causal fungus (named as *Colletotrichum* sp. by the authors) produced pinkish-gray colonies on PDA medium, which later became gray to dark gray. Subsequently, pink conidial masses formed under artificial light. The conidia were single-celled, fusiform, and measured 8 to 17 × 3 to 4 μm. The appressoria on water agar were pale to dark brown, clavate, and measured 8 to 11 × 4 to 6 μm. (Kwon *et al.*, 2015).

Molecular characteristics

Phylogenetic analysis reveals that the *Colletotrichum* sp. infecting kenaf is grouped separately from *C. acutatum*, based on molecular data (Kwon *et al.*, 2015).

Predisposing facters

The optimum temperature for growth of this pathogen is 25-28°C, and pH 5.8-6.5. It is usually inactive in dry season but during favourable conditions it causes anthracnose disease (Sharma and Kulshrestha, 2015). High temperature and high humidity favour the growth of the pathogen. Germenation of spores requires high relative humidity.

Disease cycle

The fungal colonises may remain on dead twigs and injured plant tissues and forms an abundance of acervuli and conidia. Conidia can spread by rain splash or overhead irrigation. The spores infect the host tissue by penetrating the host cuticle. After penetration, infection follows the hemibiotrophic mode of infection where, biotrophic and necrotrophic phases are sequentially occurred (Sharma and Kulsherestha, 2015). Later, necrotrophic secondary hyphae develop and spread to kill the host cell (Munch *et al.*, 2008). This tissue is subsequently colonized, acervuli are formed, thus completing the pathogen's life cycle. Dead wood and plant debris are primary sources of inocula. Fruits with quiescent infections remain asymptomatic before harvesting.

Control measures

A proper spray of copper oxychoride (50% copper fungicide at the rate of 3kg per/ha) checks this disease to a great extent (Singh, 2010). However, Seed treatment with thiram @ 1.25 mg/kg of seed along with prophylactic spray of Copper oxycloride at 0.075% (a.i.) may give better result as suggested (Singh, 2014).

Azoxystrobin is one of the strobilurin class fungicide has been found very effective to inhibit mycelial growth completely of *Colletotrichum* spp. It also suppresses the development of both panicle and leaf anthracnose and is very effective against mango anthracnose (Sundravadana *et al.*, 2006). This fungicide may be used against kenaf anthracnose.

2. Charcoal rot

Occurrence & severity status

The disease is also known as Root rot or Collar rot. The Kenaf (*Hibiscus cannabinus* L.) plant is found to be affected by charcoal rot disease in South Carolina where it is a potential crop (Blake *et al.*, 1994). The disease is also found to be increased in its intensity in greenhouse-grown kenaf (*H. cannabinus* L.) seedlings in Taiwan (Tu and Cheng, 1971).

Symptoms

The fungus infects roots and lower part of stem. The plants at seedling stage show wilting symptoms due to rotting at root and collar regions. However, at later stage in the season, the charcoal rot symptom is developed, by showing dark coloured stem and bare headed and wilted plant (Fig. 7).

Fig. 7. Charcoal rot disease in Mesta

Causal organism: *Macrophomina phaseolina* **(Tassi) Goidanich**

Predisposing factors

The disease is predisposed to high temperature and moisture stress. The hot and dry weather favours the disease. Infected seeds are the source of the disease in field. However the pathogen is mainly soil borne. Being a soil-inhabiting pathogen, many environmental and soil factors are responsible for the development of disease

[**N.B.** The causal fungus along with its control has been described in details in the preceding chapter with jute stem rot disease]

3. Collar rot

Occurrence & severity status

The disease occurs worldwide in the tropical, subtropical, and warm temperate regions

Symptoms

The disease initially appears as water soaked lesion on the crown (collar region) and lower stem tissue near the soil line. Subsequently, it forms a dense white cottony mycelial growth on the lesion. In advancement of disease, the mycelia on the stem form large number of sclerotia and the middle of the mycelial mat become compressed and pale yellowish. The sclerotia are spherical, initially white and cottony but, later hard and tan to dark brown coloured, smooth textured and 0.5-1.5 mm diameter in size. Plant foliage become pale and wilted. The stem area near the soil level becomes soft and pulpy and, thus, the stem breaks down at the point of infection at late stage.

Causal organism: *Athelia rolfsii* **(Curzi) C.C. Tu & Kimbr., 1978**

[**N.B.** The synonym *Sclerotium rolfsii* (anamorph) is mostly used to describe the disease, although, *Athelia rolfsii* (teliomorph) is the accepted name of the fungus (Krik, 2018)]

Synonyms

= *Botryobasidium rolfsii* (Curzi) Venkatar. 1950

= *Corticium rolfsii* (Sacc.) Curzi, 1932

= *Pellicularia rolfsii* (Curzi) E. West, 1947

= *Sclerotium rolfsii* Sacc, 1911

Morphological/Morphophysiolgical characteristics

Athelia rolfsii (=Sclerotium rolfsii) is a necrotrophic, soilborne fungal plant pathogen. The pathogen at its anamorph stage produces abundant white mycelium on the infected plants and in culture. It produces two types of hyphae (Aycock, 1966). In one type, the mycelium is coarse, straight, with large cells (2-9 µm x 150-250 µm) and have two clamp connections at each septation, but may exhibit branching in place of one of the clamps. In second type, branching is common in the slender hyphae (1.5-2.5 µm in diameter) that tend to grow irregularly and lack clamp connections. Slender hyphae are often observed to penetrate the substrate (Aycock, 1966). Advancing mycelium and colonies grow in a distinctive fan-shaped pattern and the coarse hyphal strands appear somewhat ropy. Its cells are hyaline with thin walls and sparse cross walls. Main branch hyphae may have clamp connections on each side of the septum.

In agar plate culture, sclerotia are formed when the mycelium covers the plate. Sclerotia begin as small tufts of white mycelium that become spherical sclerotia with 0.5 to 1.5 mm in diameter. Sclerotia become dark as they mature, changing tan to dark brown in colour. Young sclerotia often exude droplets of clear to pale yellowish fluids. Mature sclerotia are hard, slightly pitted, and have a distinct rind. Although most of the sclerotia are spherical, some are slightly flattened or coalesce with others to form an irregular sclerotium. The fungus does not form asexual fruiting structures or spores; (Ridge and Shew, 2014).

The teliomorph is rarely observed on hosts or in culture. The fungus produces basidia on an exposed hymenium, and basidium produces four haploid basidiospores. The appressed hymenium develops in small, thin, irregular patches. The basidia are clavate, 4 to 6 µm x 7 to 14 µm in size. Basidiospores are hyline, 1.0 to 5 µm x 5 to 12 µm in size (Ridge and Shew, 2014)

Molecular characteristics

In order to confirm the identity of *Athelia rolfsii*, Mahadevakumar et al. (2015) aplified genomic DNA and internal transcribed region of ribosomal DNA with the use of ITS1-ITS4 universal primers The amplified PCR product (550bp) were sequenced in both direction with the use of nBLAST. They reported that the rDNA query sequence of their material showed 100 % similarity to the representative sequence of the GenBank (KP412466.1 and JF966208.1). However, different isolates of the fungus show different genomic structures. For example, in a study of Sharma (2007), six isolates showed a length variation in the ITS amplified region of 5.8s rRNA gene.

Predisposing factors

Atmospheric high temperatures (27 to 35°C), humid conditions, and acidic soil are favourable for the fungus and disease development. Germination of sclerotia occurs at pH 3-7 and it is inhibited at pH above 7(Ridge and Shew, 2014).

Disease cycle

Sclerotia serve as the overwintering structures and primary inoculum for disease. Sclerotia may exist free in the soil near the surface or in association with plant debris (BackmanandBrenneman, 1984). The fungus may also survive as mycelium in crop debris. Sclerotia as well as the mycelium of plant debris are disseminated through the movement of infested soil and contaminated farm tools and machinery. They infect new plants at the place where they are moved (Ridge and Shew, 2014). On infection, they cause disease and produce mycelia and sclerotia on the host surface which again act as inocula.

Management/Control measures

Cultural control

Certain fertilization regimes, such as high calcium levels and ammonium type fertilizers, may suppress disease under low disease pressure. So, the non-acidifying fertilizers such as calcium nitrate or calcium ammonium nitrate (CAN) can be used to prevent acidifying soil and creating conditions conducive to disease development. Close plant spacing and over-irrigation should be avoided, since, those promote disease development and injury during cultivation (Ridge and Shew, 2014). Sclerotia buried deep in the soil may survive for a year or less. Thus, deep plowing serves as a cultural control tactic by burying sclerotia deep in the soil.

Chemical control

Spraying of hexaconazole or difenconazole at 0.0125% at the base of plants is an effective chemical control measure for the disease (Sharma, 2007).

4. Eye Rot

Occurrence & severity status

The disease occurs mostly in tropical areas of Asia, particularly, in the Indian sub-continent. Its occurrence in mesta was first reported from West Bengal (India) by Mukherjee and Basak in the year 1969. Nowadays, the disease is considered as of economic importance, damaging both quality and quantity of mesta fibre (Mukherjee and Basu, 1974). It is in increasing trend, by spreading into new areas. Apart from mesta, the disease also affects both kind of cultivated jutes (*C. olitorius* and *C. capsularis*) and Congo jute (*Urena lobata* L.). Its occurrence in other fibre crops, like, roselle, suunhemp and ramie, has also been reported (Mukherjee and Basu, 1974)

Symptoms

The disease affects both stem and leaves and developed necrotic brown small spots initially. Later, the infected plant shows eye shaped dark brown patches on the stem and leaves. In some cases, particularly during high humid conditions, concentric rings are formed on the brown necrotic patches which expand rapidly (Pelayo-Sánchez, 2017). Finally, the infected stem brakes at the place of infection (Singh, 2010) and the leaves fall off.

Causal organism: *Paramyrothecium roridum* (Tode) L. Lombard & Crous, 2016

[Persoonia, 36: 211; Krik, 2018]

Synonyms

= *Dacrydium roridum* (Tode) Link, (1809)

= *Myrothecium advena* Sacc. (1908)

= *Myrothecium advena* var *terricola* H.Q. Pan & T.Y. Zhang (2014)

= *Myrothecium roridum* Tode (1790)

= *Myrothecium roridum* var. *apiculatum* Haware & Pagvi (1971)

= *Myrothecium roridum* var. *eichhorniae* Ponnappa (1970)

= *Myrothecium roridum* var. *roridum* Tode (1790)

= *Myrothecium roridum* var. *violae* Lobik (1928)

Morphological/Morphophysiolgical characteristics

Sporodochia on leaf or stem are sessile or slightly stipitate, black, polymorphic in surface view, convex, surrounded by white tufts of mycelia (Han *et al.*, 2014). Spore mass are wet when young, drying hard and shiny black. Marginal hyphae are contorted, hyaline, usually tapering towards blunt apex, branched, septate. No external mycelium is found on plant material. Conidiophores arise directly from the mycelium in culture or from epidermal cells of the host. In case of latter one, the basal cells of the conidiophores are short and closely compacted forming a stromatic layer with the host epidermal cells, hyaline, cylindrical, branched below and then into two or three branches bearing the phialides. Phialides are closely compacted together to form subhymenial layer, septate. Phialides are in whorls of 2-5 at apex of conidiophore branches, mostly cylindrical, rarely slightly clavate, hyaline, some dilute olivaceous, occasionally the apex darkened, closely compacted into parallel rows forming a dense hymenial layer. Conidia are rod shaped, ends rounded, one end rarely slightly truncate, smooth walled, hyaline to olivaceous, black in mass, guttulate, 4.5-11 (av. 7.2) x 1.5-2.5 (av. 1.8) μm (Fitton and Holliday, 1970).

Cultures on PDA medium are pinkish white (Han *et al.*, 2014), floccose, wrinkled and often raised in the centre. Reverse view is pinkish buff to light pinkish cinnamon. Sporulation is spreadings throughout the colony often in concentric zones, consisting of small groups of conidiophores forming rudimentary sporodochia. Sporodochial surface is white at first then olivaceous black, shiny and wet. Hyphae smooth walled, hyaline, rarely branched, septate (Fitton and Holliday, 1970).

Molecular characteristics

In order to identify *Paramyrothecium roridum* (= *M. roridum*) on molecular basis, the fungal genomic DNA was amplified in PCR with the use of universal primers ITS1 and ITS4 . The sequence of a 473 bp fragment of the ITS rDNA PCR product was determined as DUCC4002 ITS rDNA sequence which showed 100% sequence similarity with that of *M. roridum* of Genbank accession (AJ302001). The full-length ITS nucleotide sequence of DUCC4002 had been used for Phylogenetic analysis. A phylogentic tree was constructed using the neighbor-joining method with 1,000 bootstrap replications. In the phylogenetic tree, the ITS rDNA sequence of DUCC4002 was grouped with the known *M. roridum* sequences retrieved from GenBank (Kwon *et al.*, 2014). In another study using a Wizard Genomic Purification Kit, it was found that the sequence of the ITS region exhibited 100% identity over 561 bp with another *M. roridum* sequence of GenBank (JF343832) (Pelayo-Sanchez *et al.*, 2017). Hong et al (2013) also observed that their query rDNA sequence derived from the amplified product of rDNA ITS using primer pairs, ITS1 and ITS4, showed 100% similarity with the sequence that retrieved from *M. roridum* isolates (No. BB71015 & BBA 67679 having Genbank Acc. numbers, AJ302001 and AJ301995, respectively).

Predisposing factors

The heavily infested field soil as a result of the continuous cropping of susceptible host acts as an important predisposing factor for *P. roridum* infection. Warm and wet weather conditions favour disease outbreak (Bruton and Fish, 2012). The temperature range for the growth of the *P. roridum* is from 15° C to 35° C with an optimum at about 30° C. The severity of *P. roridum* is associated with the build up of high humidity in the field (Tomar *et al.*, 2009). Older plants are more susceptible than young ones.

Disease cycle

Spores are easily mixed in water and they are spread by irrigation or rain splash, workers activity and wind. It infects the new host where it forms spores for further infection.

Control measures

Earlier, Copper fungicide was used to suggest in controlling the disease. Later, in a study it was found that Benlate (0.01%) and Dithane C-90 (0.05%) were most effective, while, Copper oxychloride was not so promising (Mukherjee and Basu, 1974). Recently, due to advancement in fungicide research several new fungicides are showing effective against the disease. Such as, Chipco 26019 (iprodione) at 2 lb per 100 gal of water, Compass O (trifloxystrobin) at 1 oz per 100 gal of water, Daconil 2787 (chlorothalonil) at 1.5 lb per 100 gal of water, Heritage (azoxystrobin) at 4 oz per 100 gal of water, Medallion (fludioxonil) at 2 oz per 100 gal of water, and Systhane (myclobutanil) at 2 oz per 100 gal of water are found significantly more effective in the control of *Paramyrothecium* incited disease than copper at 1.5 lb per 100 gal of water. (McMillan, 2010).

5. Leaf spots

i) Leaf spot (*Cercospora malayensis*)

Occurrence & severity status

The leaf spot disease of kenaf (*Hibiscus cannabinus*) is reported to occur in Cambodia, China, South Africa, Tanzania, Zambia, Zimbabwe (Farr and Rossman, 2017) and Korea (Park *et al.*, 2017). The possible presence of the disease in other countries, including India, cannot be over ruled. Since, the presence of the causal fungus in India and other countries was reported earlier (Chupp and Sherf, 1960). It is very common disease during humid season in kenaf and other *Hibiscus* spp, including *H. Sabdariffa* (Farr and Rossman, 2017). In Iksanand Namwon areas of Korea, the disease was very high, showing 50% disease incidence during the month of September in the years 2013 and 2014 (Park *et al.*, 2017).

Symptoms

The disease appears as spots on leaf surface [(Fig. 8 (A-B)]. The spots are circular to irregular, typically vein-limited, and may reach up to 10 mm in diameter. The spots are initially brown to reddish brown; later, turning pale brown, irregular shaped, necrotic spots, with purplish margins and greyish patches of amphigenous growth of the fungus. Ultimately the necrotic areas become shrivelled and crack (little, 1987b; Park *et al.*, 2017).

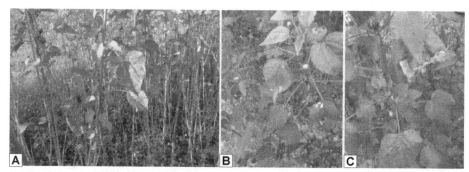

Fig. 8 (A-C). Different stages of Mesta leaf spot/ blight (*Cercospora malayensis*)

Causal organism: *Cercospora malayensis* F. Stevens & Solheim, 1931 [Mycologia 23(5): 394]

Synonyms

= *Cercospora hibisci-esculenti* Sawada (?)

= *Cercospora hibisci-sabdariffae* Sawada (1959)

Morphological/Morphophysiolgical characteristics

Fungal colonies are amphigenous, occasionally in effuse patches. Stromata are composed of a few cells up to 50µm in diameter, dark brown. Conidiophores are mononematous, 5-20 in each fascicle, straight with occasionally geniculate or with small swellings and constrictions, not branched, olivaceous brown to dark brown, often with a slightly reddish tint, 25-260 x 3-5µm, 1-5 (8) septate, with rather thick outer walls. Conidiogenous cells are terminal, on the main axis of the condiogenous cell there are a single thickened terminal scars and less distinct laterally displaced old scars. Conidia are produced in succession from the apex with elongates sympodially after producing each conidium. Conidia are solitary, each with dark thick hilum of1-2.5µm wide, hyaline, acicular, smooth walled, straight to flexuous, 3 to 20-septate, base truncate to sutruncate, tip subacute, 50-270 x 2.5-4 (5.5)µm (Little, 1987b; Park *et al.*, 2017).

After 7 days incubation at 25°C on potato dextrose agar in the dark, raised fungal colonies are formed and showed moderate aerial mycelium with smooth and erose or dentate margins (Park *et al.*, 2017).

Molecular characteristics

The genus *Cercospora* (anamorphs of *Mycosphaerella* Johanson) represents one of the largest genera of hyphomycetes and its species are regarded as major pathogens of a wide variety of plants. Recently, the phylogeny of several *Cercospora* species was evaluated using protein-coding genes, such as translation elongation factor 1-alpha (EF-1α), actin (ACT), calmodulin (CAL), and histone 3 (HIS). Multigene analyses have provided a more robust identification of *Cercospora* species.

Accordingly, the genomic DNA of *Cercospora malayensis* was extracted and five nuclear gene regions were targeted for PCR amplification and subsequent sequencing. The primers ITS4 and ITS5 were used to amplify the internal transcribed spacer areas and 5.8S rRNA gene (ITS)of the nrDNA operon. Part of the elongation factor 1-a gene (EF-1α) gene using the primers EF728F and EF2Rd and part of the actin gene (ACT gene) was amplified using ACT512F and ACT2Rd primers. The CAL228F and CAL737R primers were used to amplify part of the calmodulin gene (CAL) gene and the primers CylH3F and CylH3R to amplify part of the histone H3 gene (HIS gene). The PCR products were separated and visualized under UV light, followed by purification. The amplicons were sequenced in both directions using the same PCR primers. The reactions were monitored using BigDye Terminator Cycle Sequencing Kits and analyzed on an automated DNA sequencer. The possible identity of the isolates was established by comparing their ITS, EF-1α, ACT, CAL, and HIS sequences with those in the GenBank database (Park *et al.*, 2017).

The ITS sequences revealed that the sequences showed over 99-100% identity with several sequences of *Cercospora* species, including *C. cyperina*, *C. zebrina*, *C. kikuchii*, *C. capsici*, and *C. malayensis*. Further, it was revealed that although the ITS, CAL, and HIS sequences of the fungus were same as several *Cercospora* sequences, the ACT and EF-1α sequences showed that the isolate was a distinct species. The phylogenetic tree, created using a five combined sequence ITS, EF-1α, ACT, CAL, and HIS dataset, showed that *C. malayensis* from *H. cannabinus* formed a well-supported clade that was sister to a clade consisting of *Cercospora* spp., as well as revealed a separate clade distinct from other genera, with bootstrap values greater than 98% for each clade. Furthermore, the *C. malayensis* pathogen isolated from leaf spot on *H. cannabinus* was closely related to *C.* cf. *sigesbeckiae* on *Malva verticillata*, *Cercospora* sp. on *H. sabdariffa*, and *C. fagopyri* on *H. syriacus* (Park *et al.*, 2017).

Predisposing factors

The disease favours humid conditions.

Disease cycle

[Like other *Cercospora* spp. described in preceding chapter and following chapters]

Control measures

As regards the control measures of the disease, spraying of Mancozeb (0.25%) or Zineb (0.25%) or Carbendazim (0.1%) is suggested.

ii) Leaf spot (*Coniella hibisci*)*

A leaf spot disease caused by *Coniella hibisci* (= *Coniella musaiaensis* var *hibisci*) has been found to occur in Kenaf (*Hibiscus cannabinus*). The disease (Fig. 9) is known as Coniella leaf spot in Nigeria (Adeoti, 1991), while, as stem rot in Malayesia (Sajili *et al.*, 2017) and as leaf blight in India (Khatua and Maiti, 1977). A field survey in the southern Guinea savanna of Nigeria in 1984, indicated that the leaf spot caused by *C. hibisci* was between 41.5 and 98.2 percent damage to the leaves in kenaf (Adeoti, 1991). Application of NPK fertilizer increased kenaf growth and fibre yield but increased Coniella leaf spot severity on the crop. Supply of N as sole fertilizer also predisposed kenaf plants to attack by the leaf spot (Adeoti, 1991). The morphological and molecular characteristics of the pathogen are similar to that causes disease in Roselle (*H. Sabdariffa* var. *sabdariffa*). The pathogenic strains from kenaf showed 99% similarities with 28S ribosomal RNA of *C. hibisci* (= *C. musaiaensis*) strain AR3534. Referring to previous study on the *C. musaiaensis*, which is found to cause a disease of a serious leaf spot and stem canker on *H. sabdariffa* var. *sabdariffa*, commonly known as Roselle and that is in the same family to *H. cannabinus* (Sajili *et al.*, 2017).

*[N.B. The disease symptoms and fungal characteristics are same with that causes disease in roselle. The disease is described in details with Roselle leaf spot in the following chapter]

Fig. 9. Mesta Leaf spot /blight (*Coniella hibisci*)

6. Root rots/wilts

i) Root rot (*Pythium delicense*)

Occurrences and disease severity

Root rot of kenaf caused by *Pythium delicense* Meurs has been reported both from Africa and Australia (Vawdrey and Peterson, 1990). Generally the pathogen infects potato and beet and occurs in warmer regions only. The fungus is primarily a seedling disease. It causes pre- or post-emergence damping off.

Symptoms

Symptoms of the disease is associated with stunted growth of the plant, brown and dead root tips, wilting of plants at midday, occasionally at night, yellowing of plants and die. Brown tissue on the outer portion of the root is easily peeled off leaving a strand of vascular tissue exposed. The cells of roots contain round, microscopic, thick-walled spores.

Causal organism: *Pythium delicense* Meurs, 1934

[Phytopathologische Zeitschrift, 7: 176]

Synonym

= *Pythium indicum* M.S. Balakr. (1948)

Morphological/morphophysiological characters

Colonies on cornmeal agar medium are with little loose aerial mycelium, while, those on potato-carrot agar with submerged mycelium in a vague radiate pattern. Main mycelium is up to 8 μm wide. Sporangia are mostly terminal, occasionally intercalary, consisting of extended, inflated filamentous structures, often with swollen side branches. Zoospores are formed at 25-30°C; discharge tubes of variable length, mostly terminal; encysted zoospores 8-12 μm diam. Oogonia are smooth, mostly terminal, globose, (19-) 20-24 μm (av. 21.9 μm) diam; oogonial stalks bending towards the antheridium. Antheridia are single, rarely two applied to an oogonium, with a straight stalk, terminal and intercalary, monoclinous, occasionally diclinous, antheridial cells about 8 x 8 μm. Oospores are aplerotic, 16-18 (-19) μm (av. 17.0 μm) diam, wall up to 2 μm thick (Plaats-Niterink, 1981)

Cardinal temperatures are the minimum about 10°C, optimum 30°C and maximum over 40°C. The daily growth rate on potato-carrot agar at 25°C is about 30 mm. (Plaats-Niterink, 1981)

Control

Pythium root rot is difficult to control once rot has begun. Every effort should be directed toward preventing the disease before it begins. The effective fungicides, like, metalaxyl, mefenoxam and propamocarb, are usually systemic fungicides and are recommended to use sparingly and in rotation or mixed with chemicals of different mode of actions, to avoid the risk in development of resistance by their repeated use.

ii) Root Rot (*Pythium (Globisporangium) ultimum*

Occurrence and disease severity

Pythium ultimum is one of the most prevalent and pathogenic fungi associated with damping-off (Watkins, 1981; Minton and Garber, 1983). It is a very widely distributed species that infects the roots of many plants, including Kenaf. Its occurrence was recorded in India in the states, Uttar Pradesh and West Bengal (Krik, 2018). In kenaf, it is observed in South Africa (Tesfaendriasl *et al.*, 2004). The fungus from diseased tissue was grown on malt-extract agar (MEA) and identified the causal fungus as *Pythium* group G, a form of *Pythium ultimum*Trow. Further, it was claimed as the first published report of *Pythium* group G causing stem or root rot of kenaf (Tesfaendrias *et al.*, 2004).

Symptoms

The diseased plants displayed large, black sunken lesions (10-20 cm long) at the base of the stem, and severe root rot. Subsequently, in most cases the affected plants (seedlings & young ones) die. However, in some cases the infected plant is not killed. In that case, the infection leads to poor root development and stunting growth (Hendrix and Campbell, 1973). Plants that survive with infection often are prone to water stress, resulting in lower yield (Seney, 1984).

Causal organism: *Globisporangium ultimum* (Trow) Uzuhashi, Tojo & Kakish, 2010

[N.B. The synonym *Pythium ultimum* var.*ultimum*Trow is mostly used to describe this disease. However, the Kew accepted name (Krik, 2018) is used in this book]

Synonyms

= *Globisporangium ultimum* var. *sporangiferum* (Drechsler) Uzuhashi, Tojo & Kakish, 2010

= *Globisporangium ultimum* var. *ultimum* (Trow) Uzuhashi, Tojo & Kakish, 2010

= *Pythium ultimum* Trow, 1901

= *Pythium ultimum* var. *sporangiferum* Drechsler, 1960

= *Pythium ultimum* var. *ultimum* Trow, 1901

Morphological/ morphophysiological characters

Colonies on cornmeal agar form cottony areal mycelium, on potato-carrot agar these are with a radiate pattern. Main hyphae are up to 11μm wide (Plaats-Niterink, 1981). Sporangia are 18-27 (av. 23) μm diam, terminal and intercalary, globose, non-proliferating; discharge-tubes are 2-4 μm wide and 7-11μm long; number of zoospores per sporangium are 5-10 (Tesfaendrias *et al.*, 2004). Sporangia are not formed mostly and zoospores are very rarely produced through short discharge tubes at 5°C. Hyphal swellings are globose, intercalary, sometimes terminal, (14-) 20-25 (-29) μm diam (Plaats-Niterink, 1981). Encysted zoospores are 9-16 μm diam (Tesfaendrias *et al.*, 2004). Oogonia are terminal, sometimes intercalary, globose, smooth walled, (14-)20-24(-25) (av 21.5)μm diam. Antheridium is 1(-3) per oogonium, sac like, mostly monoclinous originating from immediately below the oogonium, sometimes hypgynous, or 2-3 which are then either monoclinous or diclinous and frequently straight. Oospores are single, aplerotic, globose, (12-)17-20 (-21) (av 18)μm diam., wall often 2μm or more thick (Plaats-Niterink, 1981).

Cardinal temperatures are the minimum 5°C, optimum 25-30°C and maximum 35°C. Daily growth rate on potato carrot medium at 25°C is 30mm. (Plaats-Niterink, 1981).

Molecular characters

The genomes of both *Pythium ultimum* var *ultimum* and *Pythium ultimum* var *sporangiferum* have been sequenced (Levesque *et al.*, 2010; Adhikari *et al.*, 2013). Analysis of the genomes suggests that two species encode 15290 and 14086 proteins, respectively.

Management/control measures

Crop sanitation and use of phosphates are effective to manage the disease to great extent by cultural way of control measure. Biological control agents, such as, *Bacillus subtilis*, *Streptomyces griseoviridis*, *Trichderma harzianum* and *T. virens* are effective to manage the disease up to a satisfactory level (Moorman, ND). The use of fungicides, like, mefenoxam, thiadiazole, etridiazole, propamocarb, dimethomorph, dichlorophen and mancozeb with metalaxyl (Ridumil MZ 78) are very effective to control the disease (Tesfaendrias *et al.*, 2004).

iii) Root rot (*Selenosporella* sp.)

Occurrence and disease severity

The disease is reported to occur in Australia. In Australia, particularly at the area in North Queensland, the root rot and wilt caused by *Selenosporella* sp. has been found as a major disease of Kenaf (*H. Cannabinus*) under water stressed condition (Vawdrey and Peterson, 1990).

Symptoms

The disease causes root rot and wilting in kenaf plants. The wilted mature plant shows dark brown cortical rot of the tap root or main lateral roots. The discolouration of stele extends beyond the crown and bark lesions 10-40cm in length are developed along the stem.

Causal organism: *Selenosporella* sp. (Kew Acc No. IMI 307983)

Morphological characters

The causal fungus can be grown on PDA medium. This fungus is mononematous, hyaline anamorph with sympodially proliferating conidiogenous cell lacking denticles, with dry sickle shaped conidia (Seifert *et al.*, 2000).

iv) Wilt/Root Rot (*Fusarium oxysporium*)

Occurrence and disease severity

The disease is reported to occur in Southern Italy (Corato *et al.*, 1999), Malaysia (Wong *et al.*, 2008) and possibly in almost all mesta (kenaf) growing countries. Further, a study from Malaysia indicates that *Fusarium* species isolated from Kenaf are diverse based on morphology and vegetative compatibility, however, only *Fusarium oxysporium* is pathogenic to kenaf (Nur-Ain-Izzati *et al.*, 2014).

Symptoms

Wilting and necrosis symptoms appear on the above ground parts (Corato *et al.*, 1999). The young seedlings show damping off symptoms, while, the infected older plants show black or brown stem lesions near the ground level, lodging and death of infected plants (Wong *et al.*, 2008). Typical internal symptoms of continuous dark vascular discolouration of the infected root and adjacent stem part are usually evident.

Causal organism: *Fusarium oxysporum* Schltdl., 1824

[Kirk.(2018). Species Fungorum (verson Oct. 2017)]

Synonyms

(Synonyms are many including all form species)

= *Fusarium oxysporum* f. sp. *vasinfectum* G.F. Atk, 1892

Morphological/morphophysiological characters

Cultures on 2% potato sucrose agar pH 6.5 are white, pnkish to violet. Mycelium is sparse to floccose. Stroma is brownish white to violet, plectoparenchymatous. Microconidia are one celled or sparsely one septate, scattered over mycelium, one septate 13-20 x 2.5-3.5µm and non-septate 6.5-10.5 x 2.5-3µm in size. Macroconidia predominately are formed on sporodochia and pionotes, light buff to salmon orange in mass, fusiform-falcate, curved inwards at both ends with a pedicellate base, 3-septate 27-40 x 2.5-4µm, 4-5 septate 32-48 x 3.5-4.5 µm. Chlamydospores are formed both terminal and intercalary, 7-13 µm in diameter (Booth and Waterston, 1964).

7. Tip rot

Occurrence & severity status

The disease was first reported by Ghosh and Mukherjee (1958) from Jute agricultural Research institute (CRIJAF), Barrackpore West Bengal. It appears during the month of July and remain prevalent throughout the months July and August. In Uttar Pradesh the incidence of the disease is 12.01% (at Bahraich) to 13.06% (at Gonda) (Singh *et al.*, 2013).

Symptoms

The disease affects the tips of the growing plants, stipules, young leaves and leaf buds. The tip turns brown and the further growth of the plant is checked. Initially, the disease appears as brown colouration on the tip of stipules of young leaf or leaf buds. Gradually, the entire stipule turns brownish black and withers away. Dark brown or black lesion also develops on the stem adjacent to the base of petiole. The young leaves turn ash-grey and looks somewhat scalded before it completely dries and drops off. Tip of the stem turns brown, becomes shrivelled and eventually, dries up. Numerous black, spherical and erumpent pycnidia appear in scattered way on both stem and leaf near the infected region (Ghosh and Mukherjee, 1958).

Causal organism: *Phoma* spp.[= *Trichospheria* spp.]

Morphological/Morphophysiolgical characteristics

Pinidia are spherical, separate, erumpent, brown, ostiolate with a dark band of cells round the ostiole, glabrous, thin walled, parenchymatous wall ranging

62-131µ. Pycniospores are 1-celled, hyaline with slight bluish-green tinge in mass, held in sticky matrix, thin walled ellipsoid with somewhat narrow ends, size 6.1-13.7µ x 2.7-4.1µ (mostly 10.8 x 2.7µ). Perithecia develop subsequently when the tip of plants are dry. Perithecia are globose, slightly wider at the base, papillate, hairy, hair short, dark-coloured, superficial (not erumpent), 115-249 µm (mostly 230 µm) in diameter. Asci are many, cylindric with narrower tail, wall hyaline, 8-spored, 61.9-65.36 µm x 9.8-10.32 µm in size. Paraphyses are absent. Ascospores are thick walled, lenticular, hyaline, 7.5-12.0 µm x 4.1-6.1µm (mostly 10.0 µm x 6.1 µm) in size (Ghosh and Mukherjee, 1958).

The pathogen can be grown in PDA medium. In this medium, the fungus grows initially as mycelia of whitish sodden and sometimes fluffy. Later it turns greyish due to formation of pycnidia and certain portion (lower half of slant) with dark due to formation of sterile, irregular shaped, dark perithecia (Ghosh and Mukherjee, 1958).

Control measures

Any of the copper fungicides can be used for controlling this disease.

8. White Stem Rot/cottony soft rot/ white mould

Occurrence & severity status

The disease was noticed during December and January of 2012-14, with a disease incidence up to 50% at the CRIJAF Research Farm in Barrackpore, West Bengal, India (Tripathi *et al.,* 2015b). The fungal disease also attacks Kenaf (*H. cannabinus* L.) in Southern Italy (Basilicata), USA (Florida), Australia (New South Wales) and in other countries all over the world (Annonymous, 1955; Averre, 1966; Duke. 1983; Frisullo *et al.,* 1995; Corato, 1996). However, the occurrence of the disease in Kenaf is generally at low scale of incidence, except in few places of seed crops where both kenaf and roselle are grown in certain distance within the vicinity. Since the latter one is very much prone to the disease and the polyphenolic compounds of kenaf stem bark have some resistance against the pathogen.

Symptoms

Initially, the symptoms appear in the form of water-soaked pale to dark brown necrotic lesions on the stem at the soil line. The necrotic lesions are then covered with white cottony mycelia. After a few days, black sclerotia of irregular shaped, 2-5 mm in size are formed. Eventually, particularly during the capsule and boll formation stages, the severely infected plant dies due to dieback (Tripathi *et al.,* 2015b).

Causal organism: *Sclerotinia sclerotiorum* (Lib.) de Bary (1884)

Synonyms

= *Hymenoscyphus sclerotiorum* (Lib.) W. Phillips (1887)

= *Peziza sclerotiorum* Lib. (1837)

= *Sclerotinia libertiana* Fuckel (1870)

= *Sclerotium varium* Pers. (1801)

= *Whetzelinia sclerotiorum* (Lib.) Korf & Dumont (1972)

Morphological/Morphophysiolgical characteristics

The pathogen produces black resting structures known as sclerotia and white fuzzy growths of mycelium on the plant it infects. The sclerotia give rise to a fruiting body (apothecia) in the favourable season that produces spores (ascospores) in a sac (ascus). Apothecia are small; thin stalks ending with a cup-like structure about 5-15mm in diameter. The cup of the apothecium is lined with asci, in which the ascospores are contained.

Fungal colonies on PDA medium at 20°C produce abundant, irregular, large, black-coloured sclerotia (0.5-1.5 cm). Germinated sclerotia produce white-coloured colonies with hyaline, septate, branched hyphae.

Molecular characteristics

Polymerase chain reaction (PCR) amplification of the β-tubulin and calmodulin genes with TU1/TU2/TU3 and SscadF1/SscadR1 PCR primers and the presence of introns and single-nucleotide polymorphisms (SNP) within the ribosomal DNA (rDNA) as detected with NS1/NS8 and internal transcribed spacer (ITS)1/ITS4 PCR primers are effective for rapid and accurate differentiation between the two species of *Sclerotinia* (*Sclerotinia trifoliorum* and *Sclerotinia sclerotiorum*) (Baturo-Ciesniewska *et al.*, 2017). *Sclerotinia sclerotiorum* provides a good model system for studying sclerotial development. The studies on the phosphorylative regulation of sclerotial development in *S. sclerotiorum* indicate that the components such as cAMP-dependent protein kinase, ERK-like mitogen-activated protein kinase and Ser/Thr phosphatases type 2A and 2B, regulate sclerotium development (Erental *et al.*, 2008). In another study, cell wall proteins released from *Sclerotinia sclerotiorum* hyphal and sclerotial cell walls via a trifluoromethanesulfonic acid (TFMS) digestion were determined by proteomic analysis. A comparison of the proteins in the sclerotial cell wall with the proteins in the hyphal cell wall demonstrated that sclerotia formation is not marked by a major shift in the composition of cell wall protein. Further, the *S. sclerotiorum* cell walls contained 11 cell wall proteins that were encoded only in *Sclerotinia* and *Botrytis* genomes (Liu and Free, 2016).

Further, sequncing of a fungicide (fludioxonil)-resistant mutants of *Sclerotinia sclerotiorum* and three wild-type strains shows that the biological differences between the resistant mutants and the wild-type strains are related to the mutation in histidine kinase gene Shk1 (Duan *et al.*, 2014).

Predisposing factors

The disease is favoured by cool temperatures, high soil moisture, high fertility, poor weed management, high plant populations and lack of air circulation under the crop canopy. The optimal temperature for sclerotial germination is in the range of 12 to 18°C at low light intensity regardless of moisture level, while, that is 20°C, at high light intensity and when the soil moisture level is high. Under high light intensity, only a few days are required for sclerotia to develop into apothecia (Sun and Yang, 2000). The apothecia produce ascospores that germinate and cause infection. There is no secondary means of infection (Walker, 1952).

Disease cycle

The lifecycle of *Sclerotinia sclerotiorum* can be described as monocyclic, as there are no secondary inocula produced during the cropping period. Sclerotia survive in soil. Under high light intensity and high soil moisture level they germinate and develop into apothecia. Apothecia develop from sclerotia located either upon or buried in the soil, and eject ascospores which become air borne and land on the host. Irrigation also has been shown to be involved in the spread of the disease. The ascospores germinate and cause infection (Walker, 1952). Infection of susceptible host plants can also occur from mycelium that originates from eruptic germination of small sized sclerotia in soil. Hyphal germination of sclerotia causes infection by first invading non-involving organic matter and forming mycelium which is an intermediate necessity for mycelia infection (Purdy, 1979). The disease can affect the plant at any stage of growth, including seedlings and mature plants. At the end of the growing season, *S. sclerotiorum* will once again produce sclerotia. The sclerotia will then remain on the surface of the ground or in the soil, on either living or dead plant parts until the next season. There is no secondary means of infection (Walker, 1952).

Control measures

Crop rotation for at least 3 years with graminaceous crops is a necessary in case of highly infected zone.

Benomyl is effective to control *Sclerotinia* diseases in any crops. Benomyl may apply at a rate of 50g/100L along with 1% white oil or may apply @ 100g/100L to get effective control measure.

9. Viral diseases

i) Yellow vein disease/ yellow vein mosaic disease

Occurrence & severity status

A yellow vein mosaic disease [Fig. 10 (A-B)] has been observed over the last few years in the southern, northern and eastern part of India. The disease causes complete destruction of the crop and the farmers in many regions are showing reluctance to cultivate this economically important crop as a result of this disease. A survey that conducted on 50 fields over four districts of West Bengal indicated that the incidence of the disease was 89.04% in *H. cannabinus* (Chatterjee *et al.*, 2005a). The disease was also found in Uttar Pradesh (India) to cause significant reduction in plant height and crop yield (Ghosh *et al.*, 2007). The average disease incidence and height reduction were approximately 59% and 22%, respectively. The disease causes marked height reduction and directly affects fibre production. Yellow vein mosaic disease also occurs in Nigeria. There, its incidence is more if the crop grown in late (July) of the season, while, early grown (June) crop can escape the disease with low disease incidence and high plant growth (Kareem *et al.*, 2017).

Fig. 10 (A-B). Yellow vein mosaic disease in Mesta. **A.** Initial stage, **B.** Developed stage

Symptoms

Initial symptom of the disease is veinal chlorosis of leaves. Subsequently, the disease causes yellowing of veins and veinlets followed by complete yellowing of the leaves with upward leaf curling and stunted growth of the affected kenaf plants. Eventually, it causes defoliation and fibre yield reduction at advanced stage (Chatterjee *et al.*, 2005a, 2005b). Biochemically, the disease lowers the activities of catalase, acid phosphatase and peroxidase enzymes in plants, however, it enhances the activities of esterase, polyphenol oxidase and superoxide dismutase. The chlorophyll content, phenolics and soluble protein are decreased in diseased plants, while, free amino acid, proline and disease related proteins are increased on infection. The SDS-PAGE protein profile of total soluble protein from diseased leaves of kenaf shows the absence of one hypersensitive 20kDa, which is pronounced in healthy plants (Chatterjee and Ghosh, 2008).

Causal organism: Begomovirus (Geminiviridae: Begomovirus)

In eastern India Mesta yellow vein mosaic disease is associated with Mesta yellow vein mosaic virus and isolate of *Cotton leaf curl Multan beta satellite* (CLCuMB), while in northern India association of another species namely *Mesta yellow vein mosaic Bahraich virus* (MeYVMBV) and an isolate of *Ludwigia leaf distortion beta satellite* (LuLDB) are reported. Analysis of diseased kenaf showed the presence of at least three different monopartite begomovirus. In eastern and southern India *Mesta yellow vein mosaic virus* (MeYVMV) and *Tomato leaf curl Joydebpur virus* (ToLCJV) while in northern India a new begomovirus that named *Kenaf leaf curl virus* (KLCuV) [Chatterjee *et al.*, 2007a; Das *et al.*, 2008a; Paul *et al.*, 2009; Roy *et al.*, 2009]. The natural association of KLCuB with MeYVMV has also been reported from Pakistan (Anwar, 2017).

Virus

Yellow vein mosaic disease of mesta is whitefly-transmitted and found in endemic form in different parts of India, causing great economic losses. The full-length DNA-A of a begomovirus infecting mesta was cloned and sequenced. The genome was homologous to the DNA-A of monopartite begomoviruses originating from the Old World, with six conserved open reading frames. The complete nucleotide sequence of the DNA-A molecule was 2728 nucleotides in length, having the highest sequence identity (83.5%) with an Indian begomovirus causing cotton leaf curl disease. It thus belongs to a novel Gemini virus species, and the name *Mesta yellow vein mosaic virus* was proposed (Chatterjee and Ghosh, 2007a). In other way, the southern blot analysis and nucleic acid spot hybridisation tests with α-$32P$ radio-labelled probes to the complete cotton leaf curl Rajasthan virus DNA-A sequence and the cotton leaf curl virus β-DNA sequence gave a positive hybridisation signal and, therefore, confirmed the involvement of a Begomovirus with the disease. An expected amplicon of approximately 0.5 kb from infected leaves with Deng Primer confirmed the presence of DNA-A of Begomovirus, and an amplicon of approximately 1.3 kb size with primers specific to β-DNA confirmed the association of a satellite β-DNA with the disease (Chatterjee *et al.*, 2005a; Ghosh *et al.*, 2007). However, no amplification was observed with DNA-A specific primers of cotton leaf curl virus, tomato leaf curl virus and mung bean yellow mosaic virus, which indicated that the causal virus have a difference in the DNA-A component from all these viruses (Chatterjee *et al.*, 2005a).

A begomovirus containing a satellite DNA β was found to be associated with yellow vein mosaic disease (YVMD) in kenaf in the eastern part of India (Chatterjee *et al.*, 2005a). The DNA β satellite molecule and coat protein gene have been found associated with different begomoviruses such as *Ageratum yellow vein virus* (Saunders *et al.*, 2000), *Cotton leaf curl virus* (Mansoor *et al.*, 2003), *Tomato yellow leaf curl virus* (Cui *et al.*, 2004) and *Tomato leaf curl New Delhi virus* (Raj *et al.*, 2007). These viruses are spreading rapidly in a devastating manner throughout different crops and weed species in various parts of the world. Hybridization with the DNA A probe and positive PCR amplification with primers for DNA β and the Bhendi YVMV coat protein gene confirmed the association of begomovirus with this disease (Ghosh *et al.*, 2007).

Transmission

The virus is transmitted both by cleft grafting and white fly vector (*Bemisia tabaci*). The whitefly transmission efficiency is about 78% to *H. cannabinus*. Symptoms appeared on *H cannabinus* within 9 days in case of white fly transmission and 10-12 days in case of cleft grafting (Chatterjee *et al.*, 2005a).

Control

Control of the disease is very difficult, if the disease reaches its high level of infection.

However, certain measures are advocated to prevent the disease attack. In case initial stage of infection, irradication and burning of the diseased plants followed by insecticides spray are suggested. Moreover, in respect of insecticides selection, care should be taken to select insecticides of different mode of actions. Since, the white fly has already developed resistance against several insecticides including dimethoate (Rogor).

ii) Other virus diseases (minor)

Apart from Yellow vein Mosaic disease causing viruses, several other viruses, *Hibiscus chlorotic ringspot virus* (Waterworth *et al.*, 1976), *Alfalfa mosaic virus* (AMV) (Rubies-Autonell and Turina, 1994), *Tobacco streak virus* (TSV) (Bhaskara Reddy *et al.*, 2012) have been isolated from mesta, but their role in causing disease development and economic loss is negligible.

a) Hibiscus chlorotic ringspot virus (HCRSV)

The disease mostly occurs in ornamental *Hibiscuss* (*H. rosa-sinensis*) all over the world. Its occurrence in kenaf (*H. Cannabinus*) was also noticed in various countries including USA (Waterworth *et al.*, 1976), China (Zhou *et al.*, 2006), India (Niu *et al.*, 2014), Iran (Reza *et al.*, 2013), Singapore (Liang *et al.*, 2002a), etc. The virus infects different species of the Malvaceae and other families. The non malvaceous genera with susceptible species are found, *Chenopodium, Gomphrena, Phaseolus, Vigna, Antirrhinum, Digitalis,* and *Torenia*. The best local lesion hosts are *Chenopodium quinoa* and *C. amaranticolor*. The virus causes mosaic or chlorotic rings or spots of various sizes, later they form into distinct necrotic lesions in kenaf (Waterworth *et al.*, 1976). The necrotic and chlorotic leaf spot symptoms are systemic in kenaf (Reza *et al.*, 2013). The virus is systemic and can readily be detected by ELISA in leaves, stems, sepals and petals throughout the year (Raju, 1985). Complete nucleotide sequence and genome organization of *Hibiscus chlorotic ringspot virus* (HCRSV) indicates as a new member of the genus Carmovirus due to presence and expression of two novel open reading frames (Huang *et al.*, 2000). The virus is transmitted by cell sap, while rubbed with infected one. *Hibiscus chlorotic ringspot virus* (HCRSV) coat protein (CP) is required for encapsidation and virus systemic movement. The virus upregulates plant sulphite oxidase transcripts and increase sulphate levels in kenaf. Only P and S domains of CP interact with sulphite oxidase (SO) from kenaf. The SO activity and SO-dependent H_2O_2-generating activity increased in the HCRSV-infected leaves (Zhang and Wong, 2009). In contrast, Niu et al. (2014) observed that HCRSV coat protein is dispensable for viral RNA replication but essential for cell-to-cell movement. Further, a viron is required for virus systemic movement. The proline 63 is crucial for HCRSV viron assembly in kenaf plants and the N-terminal 77 aminoacids including β-annulus domain is required. HCRSV is a member of the genus Carmovirus in the family Tombusviridae. HCRSV possesses a novel open reading frame (ORF) which encodes 23-kDa protein (p23). The expression of p23 can be detected in protein extracts from transfected kenaf (*H. cannabinus* L) protoplasts and HCRSV-infected leaves. The p23 protein is indispensable for host specific reaction. It (p23) does not bind nucleic acids and does not act as a suppressor of post transcriptional gene (Liang *et al.*,

2002b). Zhou *et al.* (2006) reported that *Hibiscus chlorotic ringspot virus* (HCRSV) encodes p27 (27-kDa protein) and two other in-frame isoforms (p25 and p22.5) that are coterminal at the carboxyl end. Only p27, which initiates at the ^{2570}CUG codon, was detected in transfected kenaf (*H. cannabinus* L.) protoplasts through fusion to a Flag tag at either its N or C terminus. Subcellular localization of a p27-green fluorescent fusion protein in kenaf epidermal cells showed that it was localized to membrane structures close to cell walls. Infectivity assays on plants indicated that p27 is a determinant of symptom severity. Without p25, appearance of symptoms on systemically infected kenaf leaves was delayed by 4 to 8 days. In a time course analysis, the delay corresponded to retardation in virus systemic movement which suggested that p25 is probably involved in virus systemic movement. Mutations disrupting expression of p22.5 did not affect symptoms or virus movement.

b) *Alfalfa mosaic virus* (AMV)

This virus is a worldwide distributed phytopathogen. It cases necrosis and yellow mosaic. *Alfalfa mosaic virus* (AMV) is the only Alfamovirus of the family Bromoviridae. Transmission of the virus occurs mainly by some aphids, by seeds or by pollen to the seed. Alfalfa mosaic virus (AMV) has been isolated from kenaf (*H. cannabinus*) in Cardianoin the year 1993 (Rubies-Autonell and Turina, 1994). The thermal inactivation period is between 60-62°C, *in vitro* longevity at room temperature is 24-48h and dilution end point (DEP) is between 10^{-3} and 10^{-4} (Rubies-Autonell and Turina, 1994)

c) *Tobacco streak virus* (TSV)

In India, the TSV is found to cause mosaic and necrotic spotting on leaves and bud necrosis in Kenaf under field conditions in Chittoor district of Andhra Pradesh, India (Bhaskara Reddy *et al.*, 2012). The virus is transmitted by thrips vector. The natural occurrence of *Tobacco streak virus* (TSV) in *H. cannabinus* was detected by enzyme-linked immunosorbent assay using an antiserum raised against TSV and reverse transcription polymerase chain reaction (PCR) using primers specific for the coat protein gene of the virus. Sequence analysis of the PCR products showed 99.6 and 99.5% of maximum identity at nucleotide and amino acid levels, respectively with TSV onion isolate from Kurnool (HM131490). This is the first report of the natural occurrence of TSV on kenaf in India (Bhaskara Reddy *et al.*, 2012).

10. Root knot nematode disease

Occurrence & severity: The root knot nematode disease affects mesta plants mostly in sandy loam soils. Both kenaf (*H. cannabinus*) and roselle (*H. Sabdariffa*) are affected by the nematode disease. A number of plant parasitic nematodes are reported to be associated with roselle in Nigeria (Adeniji, 1970). Among these, root-knot nematode *Meloidogyne incognita* is greatly associated with the crop. This nematode constitutes a major constraint to roselle production. Roselle varies in its susceptibility to *M. incognita*. Minton et al. (1970) have reported that roselle varies in root galling (scale of 1-4) from 1.8-2.9 for *M. incognita*. Vawdrey and Stirling (1992) and Adeniji (1970) have also reported that several roselle accessions are resistant to *M. incognita* and *M. javanica*. Of the two most important species, viz., kenaf and roselle, of fibre crops, kenaf is more susceptible to nematodes than the other one (Wilson and Menzel, 1964). Kenaf is grown throughout the tropics in the world and root knot nematodes (*Meloidogyne* spp.) are a common problem in number of locations (Minton and Adamson, 1979; Minton et al., 1970; Parrado, 1958; Schieber et al., 1961; Tu and Cheng, 1971; Wilson and Summers, 1966). Kenaf is susceptible to plant-parasitic nematodes, especially root-knot nematodes, *Meloidogyne* species, such as *M. incognita*, *M. arenaria* and *M. javanica* (Summers and Seale, 1958; Adeniji, 1970; Minton et al., 1970; McSorley and Parrado, 1986; Adegbite et al., 2005). *Meloidogyne* species have been recognized as an important production constraint wherever kenaf is grown (Adeniji, 1970). In a study to evaluate some commonly grown kenaf cultivars for resistance to *Meloidogyne* spp. it is documented that kenaf varied in their reaction to *Meloidogyne* species (Adeniji, 1970).

Symptoms

Symptoms of infection are the presence of large number of galls on the roots, excessive branching of roots and reduced root system. Poor germination and death of seedlings may be observed in case of heavy infestations and surviving plants are stunted, chlorotic, and yield less dry matter (Summers and Seale, 1958).

Causal organisms: *Meloidogyne incognita, M. javanica* **and** *M. arenarea*

Morphological/ Morphophysiological characters:

Meloidogyne incognita: *Meloidogyne incognita* is widely distributed throughout the world and consists of four pathotypes or races (Taylor and Sasser, 1978). These all four races of *M. incognita* are found pathogenic to kenaf and they can reproduce extensively on each of the several kenaf genotypes (Veech,1992).

[**N.B.** Other characters have been described in the preceding chapter along with jute and ramie diseases]

M. javanica: The pathogenic nematode is mainly occurs in tropical regions. It reproduces by obligatory mitotic parthenogenesis (apomixis). In perennial pattern, double lateral lines are the characteristic feature of the species.

[**N.B.** Other characters have been described in the preceding chapter along with jute diseases.]

M. arenarea: The morphology and cytology of this species are most variable. It has three races (Laura and Sanchez-Puerta, 2012). Race-1 reproduces on peanut, while, Race-2 cannot do so on peanut but can reproduce on tomato plants and Race-3 reproduces on both pepper and tomato (Lopez-Perez, *et al.,* 2011). Adult females are swollen, pear shaped pearly white sedimentary bodies, lacking protuberance in the posterior area. Morphologically this nematode is more or less similar to other *Meloidogyne* species. However in perennial pattern, the dorsal arch is flattened and rounded at the lateral lines and generally form shoulder arch. The stylet is very robust and the cone and shaft are broad.

Management/Control measures

Host Resistance

Roselle, *H. sabdariffa* cv. 4288, shows resistance (Laha and Pradhan, 1987). Further, it is reported that *H. sabdariffa* is immune to *M. incognita* (Laha and Pradhan, 1987). Adamson et al. (1975) reported that, because of the resistance of roselle, it was an effective crop rotation for management of *M. incognita* and *M. javanica* on kenaf. However, it has previously been found to be infested with root-knot nematode (Anon, 1961). A significant reduction in the yield of roselle has also been observed due to direct damage of the root system by the feeding of root-knot nematode *M. incognita* in Nigeria (Adegbite *et al.,* 2008). Heffes et al. (1991) reported that the roselle (sorrel) cv. red was severely galled and supported at least moderate levels of reproduction by race 1 of *M. incognita* population. Kenaf is more susceptible to nematodes than roselle (Wilson and Menzel, 1964). However, the susceptibility of kenaf is varied amongst the varieties (Adeniji, 1970).

Cultural control

Proper crop rotations with paddy have been found to be useful.

Chemical control

The application of nematicides, like carbofuran, fenamiphos and 1,3-dichloropropene, has been found very effective measure for the control of nematodes in mesta. In an investigation, it is observed that the yield of roselle was higher with the application of nematicide-Carbofuran 3G at 2 kg a.i./ha. The percentage increase over control was 48.7 and 40.8% in the years 2004 and 2005, respectively (Adegbite *et al.,* 2008). Further, fenamiphos and 1,3-dichloropropene are found to reduce galling caused by *M. incognita* and to increase yield (Mueller and Lewis, 1993).

B. Roselle diseases

1. Bacterial Wilt

Occurrence & severity status

Bacterial wilt of Roselle (*Hibiscus sabdariffa*) was first reported in Taiwan in the year 2013 (Wu *et al.,* 2013). The disease appeared on seedlings of a cultivar, Chiada 1, at the Chungpu Township of Chiayi County (Taiwan). It affects only at seedling stage and the mature plants remain free from this disease (Wu *et al.,* 2013).

Symptoms

The symptoms appear as weak Leaves that droop while green (epinasty), This is followed by collapse of the whole plant within a few days. Browning of vascular and pith tissues is evident on cutting at the base of the stem. A whitish mass of bacteria oozes out from the cut end of the diseased stems, indicating the involvement of bacteria in causing the disease (Wu *et al.,* 2013).

Causal organism: *Ralstonia solanacearum* Race 1, Biovar-4 (In: Wu *et al.,* 2013)

Synonyms

= *Pseudomonas solanacearum*

Morphological/Morphophysiolgical characteristics

Colonies on tetrazolium chloride medium are round to oval and fluidal, each with a pink or red centre after incubation at 30°C for 48 h (Wu *et al.,* 2013). The bacterium is rod shaped 0.45-1.18 x 1.24-3.15µm in size. It is Gram-negative (G⁻), non-capsulated, non-spore forming bacterium with 1-3 pollar flagella at one end or each end (Zheng *et al.,* 1988). When tobacco leaves are infiltrated with these strains, a hypersensitive reaction (HR) typical of phytopathogenic bacteria is induced (Wu *et al.,* 2013).

Molecular characteristics

According to Wu *et al.* (2013), all strains produced the expected amplicon (282 bp) after PCR with the *Ralstonia solanacearum*-specific primer pair, AU759f and AU760r. Three hexose alcohols (mannitol, sorbitol, and dulcitol), rather than three disaccharides (lactose, maltose, and cellobiose), were utilized, which suggests *R. solanacearum* biovar 4 (Hayward, 1964). *Ralstonia solanacearum* phylotype was determined by phylotype-specific multiplex PCR (Fegan and Prior, 2005).

Predisposing factors

The disease is predisposed by high soil moisture and high temperature. Water logged condition of young plants (seedlings) followed by dry sunny days is the most favoured condition for the disease.

Disease cycle

The bacterium remains in soil and grows saprophytically on the plant debris. It can remain several years in soil. Under favourable conditions it infects the root of the host plant. It may be dispersed by farming tools from diseased plants to healthy plants in the same field or nearby areas. The dead infected plants parts serve as inocula for succeeding years while that fall on soil.

Control measures

Practically, the disease is difficult to control once it established in field. Only crop rotation with non-host plants for at least three years is effective. It is better, to select separate field to grow Roselle crop.

2. Black leg

Occurrence & severity status

The disease is also called as 'stalk base rot and root rot' and 'Root rot and wilt'. Black leg is an important disease of roselle (McClintock and Tahir, 2004). The disease attacks both seedlings and adult plants, causing serious losses in crop productivity and quality in Upper Egypt. There, it is considered as the second most important root infecting wilt disease next to *Fusarium* wilt (Hassan *et al.*, 2014a, 2014b). In Mexico, the disease in roselle was first reported from Guerrero state with its average intensity of 10.6% (Ortega-Acosta, *et al.*, 2015b). In Cuba and El Salvador, the disease is reported to cause stem rot in roselle plants (Wellman, 1977). In Bangladesh, the pathogen of the disease has been isolated from roselle seeds and that considered as a pathogen affecting germination (Islam *et al.*, 2013). In India it causes "collar rot" in coral hibiscus (*Hibiscus schizopetalus*) (Santhakumari *et al.*, 2002).

Symptoms

The symptoms in affected roselle plants appear as cankers and/or soft rot in the roots and on the basal part of their stems adjacent to soil surface (Hassan *et al.*, 2014b). These areas are covered with brown to black coloured necrotic zone which is limited to the base of the stem. Plants are frequently found with detached epidermis and constriction in the area of progress of the disease (Ortega-Acosta, *et al.,* 2015b). There are no secondary roots as they are completely deteriorated and destroyed. In severely infected plants, the adult ones are wilted and dried off from down to up. The leaves of the infected plants lose their turgidity, become flaccid, greenish yellow in colour and later, that wilt, turning yellow to brown in colour and flabby and, eventually, defoliate. At that time, the whole plant becomes brown with dried leaves followed by death of plant (Hassan *et al.,* 2014b). In certain cases, only the young tender shoots are wilted and died. In cross section of infected plant's stem and root, discoloured brown areas appear as complete or incomplete ring representing the discoloured vascular tissue. At Last, the whole plant become brown with dried leaves (Hassan *et al.,* 2014b). Locally, this symptom is known as "black leg" of roselle in Mexico (Ortega-Acosta, *et al.,* 2015b).

Causal organism: *Macrophomina phaseolina* **(Tassi) Goid.**

Synonym

= *Tiarosporella phaseolina* (Tassi) Van der Aa

Morphological/ Morphophysiolgical characteristics

The fungal colonies displayed a gray color in their phase of growth and development, and became darker with age; they developed microsclerotia that varied in size (33-54 µm), round to irregular in shape with a black colour, articulate and hard, septate mycelia (Ortega-Acosta, *et al.,* 2015b).

Molecular characteristics

The amplifications performed with the ITS5/ITS4 initiators amplified a fragment of approximately 600 pb for *M. phaseolina*. The consensus sequences of nucleotides obtained when compared to those available in the GenBank of the National Centre for Biotechnology Information (NCBI), indicated a 99% similarity which confirmed the identity of the isolated organism at the species level (Ortega-Acosta, *et al.,* 2015b). Various recent studies are devoted to understand the genetic and pathogenic variability of *M. phaseolina*. Accordingly, several attempts have been made in molecular detection and differentiation of *M. phaseolina* isolates using Restriction Fragment Length Polymorphism (RFLP), Random Amplified Polymorphic DNA (RAPD), and Amplified Fragment Length Polymorphism (AFLP) analysis (Mayek-Pérez *et al.,* 2001; Su *et al.,* 2001; Purkayastha *et al.,* 2006; Reyes-Franco *et al.,* 2006). However, so far, none of these methods have been able to differentiate isolates of *M. phaseolina* from specific hosts or geographic locations. The lack of a strong correlation between genotype and geographical origin suggests a high diversity level within and among the population of *M. phaseolina* (Jana *et al.,* 2005).

Predisposing factors

The disease favours high soil moisture and soil temperature above 28°C. It also prefers soil pH ranges between 6 and 7 (Ortega-Acosta, *et al.,* 2015b).

Disease cycle

The fungus produces microsclerotia in root and stem tissues of its hosts which enable it to survive adverse environmental conditions (Cook *et al.,* 1973; Meyer *et al.,* 1974; Short *et al.,* 1980). It is a soilborne plant pathogenic fungus that can survive for long periods on plant debris in the soil. The pathogen can be spread on germination of microscleotia. Microsclerotia in soil, infected seeds or host tissues serve as primary inoculum (Bouhot, 1968; Dhingra and Sinclair, 1977; Abawi and Pastor-Corrales, 1990b). Root exudates induce germination of microsclerotia and root infection of hosts. The infective hyphae enter into the plant through root. During the initial stages of pathogenesis, the mycelium penetrates the root epidermis and is restricted primarily to the intercellular spaces of the cortex of the primary roots. As a result, adjacent cells collapse and heavily infected plantlets may die. At latter stage, the fungal hyphae grow intracellularly through the xylem and form microsclerotia that plug the vessels and disrupt host cells (Short *et al.,* 1978; Mayek-Pérez *et al.,* 2002).

Control measures

The disease is very difficult to manage chemically at field level. However, crop rotation with non-host which required stagnant water for its growth, such as, paddy, is the most suitable measure for its management.

As regards the chemical and biological control measures, in a study the in vitro results showed that microelement (copper, 400 ppm), antioxidants (salicylic acid and ascorbic acid. 100 ppm), a fungicide (Dithane M45, 100 ppm.) and biological control agents (*Trichoderma harzianum*, 1×10 cfu/mL) are effective to reduce the linear growth of the causal pathogen (Hassan *et al.,* 2014a). Moreover, under field conditions, additional application of copper, salicylic acid, and *T. harzianum* as seed treatment (by soaking seeds for 24 hours) shows the best results in this respect, suggesting microelements, antioxidants, and biocontrol agents can be used as alternative strategies to fungicides for controlling root rot and wilt diseases (including black leg) in roselle (Hassan *et al.,* 2014a).

3. Foot and stem rot

Occurrence & severity status

The disease is also named as 'root and stem rot', 'collar rot and wilt', 'crown rot', 'black foot' or 'black shank'. The foot and stem rot [Fig. 11 (A-B) disease popularly known as 'collar rot and wilt' is a serious disease in roselle

Fig. 11 (A-D). Stem diseases in Roselle. **A-B.** Stem rot/ Southern blight. **C-D.** Foot and stem rot.

(*H. sabdariffa*). This disease also affects kenaf (*H. cannabinus*) but the disease incidence is relatively low (Majumdarand Mandal, 1996). In India, it causes root and stem rotting in roselle (Kumar and Mandal, 2010), The presence of this disease in roselle has also been reported from Africa, Indonesia, Puerto Rico, Malaysia, Philippines, Ivory Coast, Brazil, Guerrero and other countries (Hernández and Romero, 1990; Erwin and Ribeiro, 1996; Drenth and Guest, 2004; Silva *et al.*, 2014). In Brazil, the disease of roselle is most frequent in the state of Maranhao (Alconero and Stone,1969; Silva, 2001). The disease is known as 'crown rot' or 'black foot' in Mexico. There it is most frequently associated with the "black leg" disease (Escalante-Estrada *et al.*, 2001; Angel *et al.*, 2015). In Mexico, the attack of this disease causes 10 to 30% yield loss (Escalante-Estrada *et al.*, 2001). The disease is also considered as the most important disease of roselle in South-East Asia. The causal agent of the disease has a broad host range.

Symptoms

The disease causes black shank at the lower portion of stem, appearing purplish black or dark brown to black discoloration, encircling the stem about 30 cm above the ground. Sometimes the dark areas are covered with white mould on the surface. The dark lesion normally advances towards the root and above the stem. As a consequence of this rot, the foliage turns yellow and the leaves and flowers drop prematurely (Hernandez and Romero, 1990). Sudden wilting of the plant may occur due to rotting of lower portion of the stem, particularly at collar region and adjacent roots.

Causal organism: *Phytophthora nicotianae* **Breda de Haan, (1896)**
[De bibitziekte in de Deli-tabakveroorzaakt door *Phytophthora nicotianae*. Mededeelingenuit 's Lands Plantentuin Batavia. 15:1-107]

Synonyms

= *Phytophthora nicotianae* var. *parasitica* (Dastur) G.M. Waterhouse 1963

= *Phytophthora parasitica* var *sabdariffae* Mukerjee*

= *Phloeophthora nicotianae* (Breda de Haan) G.W. Wilson (1914)

= *Phytophthora parasitica* var. *nicotianae* (Breda de Haan) Tucker (1931)

= *Phytophthora parasitica* Dastur (1913)

= *Phytophthora parasitica* var. *parasitica* Dastur, 1913

= *Phytophthora melongenae* Sawada (1915)

= *Phytophthora terrestris* Sherb. (1917)

= *Phytophthora parasitica* var. *rhei* G.H. Godfrey (1923)

= *Phytophthora tabaci* Sawada (1927)

[* N.B. *Phytophthora parasitica* var. *sabdariffae* Mukerjee is an invalid name, nom. nud. (Waterhouse 1963)]

Morphological/ Morphophysiolgical characteristics

The fungus grows well on several culture media, such as corn-meal-, oat-meal-, bean-, and potato agar. On PDA medium at 24°C, the fungal colonies are fluffy with cottony mycelium growing in slightly striated pattern (Alvarez *et al.*, 2007). The monosporic culture of *P. nicotianae* grown on V8-agar medium sporulates when that (mycelia disk) transferred in petridishes containing sterile distilled water and incubated at 24 ± 2 °C (Ortega-Acosta *et al.*, 2015b) . This fungus develops into dense cottony colonies in rosette shape. Mycelia are coenocytic, mostly presenting spider-like growth, with predominantly oval shaped papillate sporangia with a prominent papilla and intercalary and terminal chlamydospores (Ortega-Acosta *et al.*, 2015b). Sporangia are produced abundantly in sterile soil extract and on 10% non-sterile soil. The Sporangia are hyaline, papillate, persistent and predominately ovoid or obpyriform, measuring 33.3 -56.0 (-59.1) x 24.5-35.0 (-45.9) μm (average 42 (44.4) x 29 (35.3) μm) with a length-breadth ratio of 1.3:1 to 1.4:1. Chlamydospores are terminal or intercalary in the mycelium, with a diameter of 25.4 to 40.3 μm (average of (28) 33.0 μm). The fungus is heterothallic and isolates from both the A1 and A2 compatibility groups. Oospores measure 23-38 μm in diameter (average: 29 μm). The antheridia are amphygynous (Santos *et al.*, 2005; Alvarez *et al.*, 2007; Ortega-Acosta *et al.*, 2015b).

Cultures of *P. nicotianae* on carrot agar medium (CA) are petaloid, with dense and cottony aerial mycelium; colonies have diffuse edges. The optimum temperature for mycelial growth is between 24 and 32°C, and no growth occurs at 36°C (Santos *et al.*, 2005). The causal fungus grows in a pH of between 5.5 and 6, and their growth is affected when it was higher than 6.5 (Besoain, 2013). In most cases, the fungus is isolated in soils with a pH of between 4.7 and 5.5, whereas in soils with a pH between 6.6 and 6.8 this organism cannot grow (Dasgupta *et al.*, 2012, Besoaín, 2013, Ortega-Acosta *et al.*, 2015b).

Molecular characteristics

The genus *Phytophthora* is confirmed to be monophyletic. Its species are redistributed into 8 clades, providing a more accurate representation of phylogenetic relationships within the genus *Phytophthora*. *Phytophthora nicotianae* is similar to *P. frigida* in various morphological characteristics; however, the latter one is in Clade 2 while *P. nicotianae* is in Clade 1 (Kroon *et al.* 2012). The PCR products as obtained using primers ITS4 and ITS6 are of about 900 bp. Their sequences are compared with ITS sequences of *Phytophthora* species available in GenBank using BLAST searches, and are identical to those of *P. nicotianae* (Alvarez *et al.*, 2007). The draft genome sequence of *P. nicotianae* has been completed (Meng *et al.*, 2014). The genome sequence of *P. nicotianae* includes about 23,121 predicted genes within the 82-Mb genome (Judelson, 2012). The number of predicted genes is more than the two narrow host range species *P. infestans* and *P. sojae*, implicating possible relation to its capability to infect large number of plant species (Tyler *et al.*, 2006).

In recent years, many researchers have focused on effectors, which are key virulence factors of pathogens. Effectors are molecules and typically proteins secreted by the pathogen to manipulate host cell structure and function thereby facilitating infection and colonization (Kamoun, 2006). Recently, a RXLR effector of *P. nicotianae* PSE1, has been identified in a cDNA library for the penetrating stage of *P. nicotianae* (Kebdani *et al.*, 2010). These effectors favour the pathogen infection by modulating the auxin accumulation during the penetration process (Evangelisti *et al.*, 2013).

Predisposing factors
The disease is favoured by waterlogged and badly drained conditions in fields. The fungus is both soil and water borne and the infection starts when there is water stagnation in the field. It grows at the temperature between 6-37°C. Its growth is best at pH 5.8-6.7.

Disease cycle
Foot and stem rot (Black Shank) is a polycyclic soil borne disease, with multiple disease cycles per growing season occurring from May to October. There are important structures this pathogen uses in its disease cycle. Chlamydospores are produced asexually and serve as resting structures, surviving for about four to six years. Chlamydospores are the primary inocula. These spores germinate in warm and moist soil to produce a germ tube that infects plants or produces a sporangium. This is another asexual structure and acts as secondary inoculum. The spores produced in sporangium can either germinate directly or release motile zoospores. Zoospores are kidney shaped with an anterior tinsel flagellum and a posterior whip like flagellum that helps to navigate toward root tips where infection occurs. Black Shank needs water for germination and movement because zoospores swim through soil pores and standing water. Splashing water from rain or irrigation can infect healthy plant leaves leading to more repeating secondary cycles. Zoospores move towards nutrient gradients around root tips and host wounds. Once the root surface is contacted, zoospores encyst and a

germ tube emerges in penetrating the epidermis forming appressoria and germ tubes (Kebdani *et al.,* 2010; Wang *et al.,* 2011). Infection leads to systemic rotting of the root system and wilting and chlorosis in the leaves. Another structure is called hyphae. The hyphae are coenocytic and heterothallic, and require two mating types to produce oospores, the sexual survival structure. Many fields only contain one mating type, so the zoospores rarely germinate and rarely cause epidemics. However, to which extent the sexual reproduction play a role in *P. nicotianae* remains unclear (Meng *et al.,* 2014).

Management/Control measures

Host resistance

The absence of differential interaction between the host and the parasite shows that the resistance of *H. sabdariffa* to *P. nicotianae* is of the general or the horizontal type, polygenetically controlled (Boccas and Pellegrin, 1976). The roselle genotype, HS 4288 is susceptible, while, the genotype 8434-8433-68 is resistant (Majumder and Mandal, 2000). It has been observed that the varieties with pigmented stem are less affected by this disease (Singh, 2010).

Biocontrol

Under greenhouse conditions, plants inoculated with *Trichoderma gamsii* and *T. longibrachiatum* had 30% less incidence of *P. nicotianae* (Eduardo *et al.,* 2014).

Chemical control

Seed treatment with 'Master' (Metalaxyl 8% + Mancozeb 64%) @ 3g/kg + soil application with *Trichoderma viride* 10g/kg of farm yard manure (FYM) or seed treatment with 'Curzate M8' (Cymoxanil 8% + Mancozeb 64%) @ 3g/kg seed + soil application with *Pseudomonas fluorescens* 10g/kg of FYM, recorded lowest foot and stem rot incidence (9.6% and 9.0% respectively) and highest fibre yield (20.7 and 20.9 q/ha respectively) (Manoj *et al.,* 2010).

The chemical metalaxyl (which is the active ingredient of the eumycete-specific fungicides, Subdue and Ridomil) has been the primary fungicide used to control stem and crown rots caused by *Phytophthora* spp. In Guerrero, Mexico, Ridomil MZ 75% is the best fungicide to control the disease of roselle, which is called there 'Jamaica black leg' caused by *P. nicotianae* (Hernandez and Romero, 1990). However, continuous use of metalaxyl as the primary means of control of the disease may develop resistance in *Phytophthora* causing insensitivity to metalaxyl (Ferrin and Rohde, 1992).

4. Foliar blight /Leaf blight

Occurrence & severity status

In the tropical region of South-west Nigeria, 63% disease symptoms in roselle are of leaf blight. The disease occurs during the peak of the rainy season (Amusa *et al.,* 2001). In a study, it was found that more than 40% of leaves of roselle plants in fields under continuous cultivation (endemic plots) were blighted up by the disease. Further, it was estimated that, over 20% leaf yield loss and over 34% edible leaf loss were occurred due to leaf blight. The mean marketable and biological yield loss of calyces in 1999 and 2000 were 35% and 38% respectively, while seed yield loss was 30% and 32% (Amusa, 2004). In West Bengal (India) the disease is now in alarming position (Sarkar and Jana, 2017). The disease mainly occurs during from August to October. During this period over 40% of the foliage of the plants is blighted up while, a few are found to be completely defoliated (Amusa, 2004). Apart from Roselle, it also affects leaves of kenaf (*H. cannabinus*) and jute (*C. capsularis*). The fungus also attacks *Hibiscus mutabilis* wherefrom the pathogen name was first derived (Ellis and Everhot, 1988).

Symptoms

The symptoms at the beginning appear as patches of water-soaked lesions on the foliage of young plants. The lesions mostly start to be appeared at the margin of leaf and progress towards vein and petiole. They eventually develop into greyish blighted irregular necrotic spots of (0.5-) 1.0- (-1.5) 1.6 cm in size on both sides of the leaves. Under dry condition the symptoms restricted but develops black pin head shaped pycnidia on the infected surface (Sarkar and Jana, 2017). Black pin head shaped pycnidia are often formed on the upper side of the spot in a ring around the centre of the necrotic lesion. In advanced stage, the blighted patches spread, causing extensive blight, defoliation and complete wilting (Ellis and Everhot, 1988; Amusa *et al.,* 2001; Amusa, 2004).

Causal organism: *Phyllosticta hibiscina* Ellis & Everh., 1888

[Journal of Mycology 4 (1): 9]

Synonyms

= *Phyllosticta hibiscini* (Ellis and Everhot) [In: Aliyu, 2000]

(unaccepted name mostly used to describe the disease)

Morphological/ Morphophysiolgical characteristics

The fungus develops greyish, scattered, epiphyllous, irregular blighted lesions on leaves. Pycnidia are rather large, dark-brown, subcuticular, erumpted, ostiolate, spherical to globoid, 78 (195-) – 159,6 (-273) μm in diameter. Pycnidiospores are small, hyaline, single celled, two nucleate, smooth, oblong to elliptical, (5.0-) 6.3 - (-6.6) 8.5 x (2.5-) 2.9 - (-3.3) 4.2 μm in size. (Rao, 1962; Ellis and Everhot, 1988).

The pathogen grows on PDA medium, producing highly branched hyaline mycelia with globular black dot like pycnidia. Pycnidium produces small, hyaline, single celled and elliptical- cylindrical conidia. Chlamydospores are also produced in the medium (Amusa, 2004; Sarkar and Jana, 2017).

Predisposing factors

The causal fungus is soil born by nature and harboured at plant debris (Amusa *et al.*, 2001). It favours rainy and humid condition (Amusa, 2004). The temperature range 15°-29°C is the most suitable condition for maximum infection.

Disease cycle

These fungi survive through conidia and stromata on crop debris in soil. During favourable conditions it infects host plant and produces pycnidia and pycnidiospores for reinfection.

Control measures

The disease can be controlled by spraying copper fungicides (Copper oxychloride).

The new fungicides tebuconazole, tebuconazole with trifloxystrobin and triticonazole are effective against *Phyllosticta* leaf spot diseases.

5. Leaf fleck

Occurrence & severity status

Leaf fleck is one of the most important diseases of roselle (McClintock and Tahir, 2004). The disease is also known as 'leaf rot' in India where it is predominantly found (Singh, 2010) and causes widespread damage to the foliage, but seldom leads to mortality (Ghosh, 1983). Chowdhury (1955) observed both leaf rot and stem rot caused by this fungus at Jorhat in Assam (India). The disease appears during early and late stages of the crop. It also causes considerable decrease in fruits number and weight in most of the cultivars of roselle and has disastrous impact on fructification (Lepengue *et al.*, 2013). So far, it is observed that almost all the cultivated plants of roselle are susceptible to this disease (Gomez-Leyva *et al.*, 2008).

Symptoms

The disease induces wet rot on every part of the infected plant's organs (Lepengue et al., 2013). The infection starts at the leaf apex of seedlings which gradually spreads towards the petiole through midrib. In some cases, the infection starts at the base of lamina and moves along the sides of midrib towards the apex and become rotted. The leaf lobes of aged plants are also attacked. In that case, the infected lobes start rotting. First of all, the tip of leaves starts blackening and thereafter whole leaf becomes black. The leaves start falling and ultimately the plant growth is adversely affected (Singh, 2010).

Causal organism: *Phoma sabdariffae* Sacc., 1913

[Annales Mycologici, 11 (6): 554]

Morphological /Morphophysiolgical characteristics

The fungus on roselle produces dense, erumpent, globular pycnidia which are 150 to 180 µm in diameter, with non-elevated ostioles of 14-16µm in diameter. The hyaline pycnospores are 12-14 x 4-4.5µm in size and are borne on short, papilliform conidiophores (Saccaebo, 1884).

The fungus is classified under the imperfect pathogenic fungi. It acts by producing toxic compounds. The fungal culture filtrate stays stable at 0 °C and the filtrate is both thermosensible and photosensible, containing glucidic, proteic and phenolic compounds. Its toxicity is due to the phenolic compounds that possess two active fractions F1 and F4, with physico chemical characteristics close to those of brefeldine A and cytochalasine B. (Lepengue et al., 2009)..

Molecular characteristics

The literature on molecular characters in respect of genomic sequence is not available. Only information that available is that *P. sabdariffae*, like most pathogenic fungi, presents supplementary electrophoretic proteic band of 18 kD (Lepengue et al., 2010).

Predisposing factors

Wounds at the leaf margin may cause incidence of the disease. Increase in percentage of infected leaves and increase in disease intensity may be due to wounds (Perpustakaan ,1991).

Control measures

Three roselle (*H. sabdariffa* var. *sabdariffa*), cultivars, viz., VV1, RV1 and RR1 are found resistant in Gabon (Lepengue et al., 2010).

The spraying of fungicides, Mancozeb (75%) and Copper oxychloride 50%WP @ 0.3% is effective against the disease.

6. Leaf spots

i) Leaf spot (*Cercospora malayensis*)

Occurrence & severity status

The disease occurs in both roselle and kenaf. The presence of *Cercospora malayensis* on roselle in Ghana was reported for the first time by Golato (1970). In Indian sub-continent, the occurrence of the disease in roselle was earlier noticed (Moses, 1988). In India, particularly in states of Uttar Pradesh (Rai, 1980), West Bengal (De, 1986), Bihar and Delhi (Rao, 1973) the disease occurs regularly. The disease appears in the field in May and continues to affect the plants up to the end of rainy season. The disease is most severe during hot, humid and rainy conditions. The late summer rainfall is the major contributor of the disease severity, causing defoliation (Robertson, 2015).

Symptoms

In general, leaf spots are first visible on older leaves at the bottom of the plant then spread upward to the top of the plant. Initial spots are purple and small with a circular shape (Robertson, 2015). Later, the spots become pale brown with a purplish margin and showing greyish patches on the lesion due to heavy fructification (Fig. 12 A).

Fig. 12 (A). Foliar diseases in roselle.
Leaf spot caused by *Cercospora malayensis*.

Causal organism: *Cercospora malayensis* **F. Stevens &Solheim, 1931**
[Mycologia 23(5): 394]

Synonyms

= *Cercospora hibisci-esculenti* Sawada (?)

= *Cercospora hibisci-sabdariffae* Sawada (1959)

Predisposing factors

The disease is favoured by hot, humid and rainy conditions. Rainfall and overhead irrigation are major factors that play a pivotal role in symptom expression and intensity. Late summer rainfall acts as a major contributor for disease severity, causing defoliation and decline (Robertson, 2015).

Control measures

Spraying Mancozeb or Zineb 2 g or Carbendazim 1 g/l can control the disease.

[**N.B.** Other details of leaf spot caused by *Cercospora malayensis* are with Mesta (kenaf)]

ii) Leaf spot (*Coniella hibisci*)

Occurrence & severity status

The disease occurs more frequently in several African countries (Persad and Fortune, 1989; Alegbejo, 2000). In West Bengal (India) its occurrence in Roselle was reported earlier (Jain *et al.*, 1965; Khatua and Maiti, 1977). In Africa, the disease in epidemic form was observed in *H. sabdariffa* var. *sabdariffa* in the first time report from Trinidad and Tobago (Persad and Fortune, 1989). According to Alegbejo (2000), this is the most important disease of roselle plant in Nigeria. It has an overwhelming destructive ability on the plant, which usually in severe cases leads to death of the whole plant (Arowosoge, 2008). However, the incidence and severity of the leaf spot disease varied from farm to farm (Apeyuan *et al.*, 2017) and in many cases, the disease occurs as secondary invaders of plant tissues infected or infested by other organisms.

Symptoms

The disease, at first, appears as greenish water soaked small spots on leaves usually near margin although it may occur at any place of leaf surface. Later, the spot expands rapidly and develops in to dark brown area at the centre with association of numerous pycnidia, as visible by a hand lens. The dark brown zone expands toward petiole. The affected leaf becomes necrotic, distorted, withered and eventually falls off. In severe cases, upper portion of stem shows die back symptom and numerous pycnidia occur on the lesion of stem (Khatua and Maiti, 1977; Adeoti, 1991). Symptoms on stems also include girdling of stem and stem canker. Severe stem infection results in the death of the plant. The disease also attacks pods and flowers (Persad and Fortune, 1989). The pod on infection dehisces prematurely (Apeyuan *et al.*, 2016a).

Causal organism: *Coniella hibisci* **(B. Sutton) Crous, 2017**
[Studies in Mycology 86: 155]

Synonym

= *Coniella musaiaensis* var. *hibisci* B. Sutton, 1980

Morphological/ Morphophysiolgical characteristics

Pycnidia (conidiomata) are separate, immersed or superficial, globose to depressed, initially appearing hyaline, becoming olivaceous to black with age, with plate like structures, up to 350 μm diam. Ostiole is central, 40-80μm diam. Pycnidial wall is consisting of 2-4 layers of medium brown textura angularis. Conidiophores are densely aggregated, slightly thicker, subulate, simple, frequently branched above, reduced to conidiogenous cells, or with 1-2 supporting cells, 25-35 x 3-5 μm. Conidiogenous cells are simple, hyaline, smooth, tapering, 8-15 x 2.5-3 μm wide at apex, surrounded by a gelatinous coating, with visible periclinal thickening. Conidia are hyaline to pale yellowish-brown with age, fusoid to ellipsoidal, inequilateral, apex acutely rounded, widest at middle tapering to slightly truncate base, smooth walled, mono- or multiguttulate, germ slits absent, (10-) 11- 13 (-15) x (3-)3.5-4 (-5)μm (L/W = 3.4), with a mucoid appendage alongside conidium (Martin-Felix *et al.*, 2017).

The fungus grows in PDA medium. The colony in the medium is initially whitish, later dark brown with the production of pycnidia. The fungal culture grows in concentric rings when exposed to alternate light and dark conditons, however, it is uniformly brown when grows in complete darkness (Khatua and Maiti, 1977; Adeoti, 1991).

Molecular characteristics

The genotypic identification on molecular basis is done with the DNA extracted from the fungus. The LROR (5'-ACCCGCTGAACTTAAGC-3') and LR7 (5'- TACTACCACCAACA-TCT-3') primers are used to amplify 28S large subunit ribosomal RNA gene and the product size obtained is 1.5 kb. DNA sequencing is then carried out to further confirm the identity of the test fungi. The pathogenic strain shows 99% similarities with 28S ribosomal RNA of *Coniella hibisci* (=*C. musaiaensis*) strain AR3534; referring to previous study on *C. musaiaensis* collecting from *H. sabdariffa* var. *sabdariffa*, commonly known as Roselle (Sajili *et al.*, 2017).

Predisposing factors

Application of NPK fertilizer increased plant growth and fibre yield but increased Coniella leaf spot severity on the crop. Supply of N as sole fertilizer also predisposed the plants to attack by the leaf spot (Adeoti, 1991).

Control measures

Disease severity can be lowered by using benlate. The fungicide, benlate at the rate of 0.6kg a.i./ha, is very much effective to control of the leaf spot disease caused by *Coniella hibisci* (Nejoa *et al.*, 2007; Apeyuan *et al.*, 2016a). In addition, for better control, a synthetic pyrethroid insecticide (cypermethrin) is to be used in controlling insects that feed on the leaves of roselle plant. In an investigation, it was found that when plants were treated only with insecticides were not infested by insects and consequently escaped infection by the pathogen in Makurdi, Nigeria, indicating that activities of insects help in the initiation and development of this leaf spot disease of roselle (Apeyuan *et al.*, 2016b)..

iii) Leaf spot (*Corynespora cassiicola*)

Occurrence & severity status

The leaf spot disease caused by *Corynespora cassiicola* on roselle (Wei, 1950) is mostly found in South American countries, like, Brazil (Poltronieri *et al.*, 2012) and Mexico (Ortega-Acosta *et al.*, 2015a). In Mexico, the disease causes damage to roselle (*H. Sabdariffa*) in an epidemic form. The disease is currently considered the main limiting factor for the production of this crop, as there has been incidence in the order of 100% (Ortega-Acosta *et al.*, 2015a, 2016). In Brazil, also, the disease causes considerable damage to roselle (Poltronieri *et al.*, 2012). It occurs also on kenaf (*H. Cannabinus*) causing the same disease (Shaw, 1984). The pathogenic causal fungus is an endophyte, and saprophyte. It can be found growing on at least 530 plant species of 380 genera, primarily in the tropics and commonly reported as a plant pathogenic foliar fungus with a wide host range within tropical and subtropical areas (Holliday 1980; Farr *et al.*, 1980).

Symptoms

The symptoms on leaves consist of circular to irregular spots with straw-coloured centre, black borders and purple rings, and these are mostly spread on the leaf veins. The spots coalesce over time into large necrotic areas. Infected calyces are with necrotic, circular to irregular sunken spots. Blight symptom originates at the apex of the calyx and extends toward the base of the calyx (Ortega-Acosta *et al.*, 2015a, 2016).

Causal organism: *Corynespora cassiicola* **(Berk. & M.A. Curtis) C.T. Wei, 1950**

[Mycological Papers 34: 5]

Synonyms

= *Helminthosporium cassiaecola* Berk. & M.A. Curtis (1868)

= *Helminthosporium cassiicola* Berk. & M.A. Curtis (1869)

= *Cercospora melonis* Cooke (1896)

= *Corynespora mazei* Gussow (1906)

= *Helminthosporium warpuriae* Wakef. (1918)

= *Helminthosporium papayae* Syd. (1923)

= *Cercospora vignicola* E. Kawam. (1931)

= *Helminthosporium vignae* L.S. Olive (1945)

Morphological/ Morphophysiolgical characteristics

Conidia and conidiophores are observed on infected leaves incubated for 48–60 h in a moist chamber. Conidiophores are mostly solitary, straight cylindrical, 94.5-162 × 2.7-4 µm, 2-20 septate, and dark brown. The conidia produced singly or in acropetal chains are obclavate, cylindrical, straight or curved, 62-127 × 5-8 µm (mean 100 × 6.8 µm), with 5-20 cells separated by hyaline pseudoseptate and dark brown to pale brown (Chairin *et al.*, 2017).

Molecular characteristics

An isolate (SK01) of the fungus grown on PDA was used for DNA extraction by Cetylmethyl ammonium bromide (CTAB) method. The inter-transcribed spacer (ITS) and large subunit (LSU) gene regions were amplified using BIO-RAD T100™ Thermal Cycler (BioRad, Hercules, CA, USA). Amplification of the gene region used the primer pair PN3/PN16 for ITS and LROR/ LR for LSU. The PCR products were visualised by agarose gel electrophoresis. The ITS and LSU gene regions were sequenced at Macrogen (Seoul, Korea) using the same primers as used in the PCR reaction. Using BLAST analyses, the sequences as obtained were compared with known sequences available in Genbank (The National Centre of Biological Information). The BLAST search revealed that the sequence of isolate SK01 had a 100% match with *C. cassiicola* E9807C (JN541214, *Malvaviscus concinnus*, Ecuador) for the ITS gene, and with *C. cassiicola* C13–1 (KF590123, unknown host, Japan) for the LSU gene. This sequence has been deposited in GenBank with accession number LC177365 for LSU and LC177366 for ITS (Chairin *et al.*, 2017).

Predisposing factors

High temperature and humidity are found to favour the disease incidence. Frequent showers and/or irrigation along with high nitrogen fertility levels may also contribute to increase disease. Tender leaves appear to be the most susceptible.

Control measures

Integration of crop rotation and the sanitary practices are suggested to manage the disease.

The use of water based fungicides is effective for disease control. Mancozeb (Indofil M 45 (2.5 g in 1 litre water), Carbendazim (Bavistin) (1 g/lit water), Copper oxycholride 50WP (Blitox) (2.5 g/lit water) or Bordeaux mixture (1%) can be used for control. However, loss of sensitivity of the casual fungus to carbendazim has been observed in some cases (Avozani *et al.*, 2014). Chlorothalonil and pyraclostrobin have been shown to be more effective than mancozeb, for the control of disease caused by *C. cassiicola* (Vawdrey *et al.*, 2008). Bravo 500EC (Chlorothalonil 500g/L) and Difolatan 80W (Captafol), either alone or in combination with copper oxychloride, provided the best control of *Corynespora* leaf spot (Kingsland and Sitterly, 1986). Thiophanate-methyl has also been found effective in controlling the disease.

7. Boeremia (Phoma) leaf blight*

Occurrence & severity status

Boeremia (*Phoma*) leaf blight of roselle (*H. subdariffa*) frequently appears during the months of July and August in West Bengal, India (Tripathi *et al.*, 2015a). The fungal pathogen is also most frequently isolated from the infecting Roselle seedlings in Malayesia. Further, it is reported that *B. exigua* is highly pathogenic on wounded seedlings of roselle (Eslaminejad and Zakaria, 2011). The disease is also very frequent in kenaf (*H. cannabinus*) under West Bengal climatic conditions (Tripathi *et al.*, 2015a).

Symptoms

Initially the disease symptoms appear as water soaked small circular spots which coalesced to form large, irregular lesions on the leaves. These irregular lesions later appear as greyish brown to dark brown, with light brown margins. Concomitantly, abundant black dot like pycnidia appear on the lesions, especially on the upper surface of the leaves in loose rings. In advanced stage, the disease causes foliage blight and destroys the leaves (Eslaminejad and Zakaria, 2011; Tripathi *et al.*, 2015a).

Causal organism: *Boeremia exigua* (Desm.) Aveskamp, Gruyter & Verkley, 2010

[Krik, P.M. (2018). Species Fungorum (version Oct. 2017), Kew Mycology]

Synonyms

= *Ascochyta nicotianae* Pass., 1881

= *Boermia exigua* var. *coffeae* (Henn.) Aveskamp, Gruyter & Verkley, 2010

= Boermia exigua var. *exigua* (Desm.) Aveskamp, Gruyter & Verkley, 2010

= *Phoma exigua* Desm., 1849

= *Phoma exigua* var *exigua* Desm., 1849

= *Phomopsis perexigua* Traverso, 1906

= *Phyllosticta hydrangea* Ellis & Everth., 1889

Morphological/ Morphophysiolgical characteristics

The pathogen grows on PDA as irregular, velvety white to pale olive grey coloured colony and that looks dark brown to black from the reverse side of the petridishes. The colony grows up to 8.5 cm in diameter on PDA at room temperature within four days of inoculation. The growth of this fungus on oat agar (OA) and malt agar (MA) reaches about 30–67 mm in diameter and 25–62 mm diameter respectively in seven days. The fungus produces unicellular conidia from monophialides and doliform to flask-shaped conidiogenous cells in pycnidial conidiomata. Pycnidia are formed abundant on PDA and these are globose, 70-100 µm in diameter, solitary, glabrous, with one non-papillate ostiole. Conidia are unicellular (aseptae), ellipsoidal to oblong, measuring 2.5 (2.9-) x 9.0 (-9.53) µm in size, with two polar guttules (Eslaminejad and Zakaria, 2011; Tripathi *et al.,* 2015a). However, there are no chlamydospores formed in PDA medium. The fungus produces a colourless diffusible antibiotic metabolite 'E' (named after *B. exigua*). On addition of a drop of concentrated NaOH, this metabolite oxidizes successively into the pigments 'α' and 'β'. Application of NaOH on metabolite E-producing culture appears as a greenish blue spot or ring (pigment α) within 10 min, which changes to browinish red (pigment β) after about 1 h. The pigments are sometimes restricted to cytoplasm or guttules in the hyphae, but usually diffuse into the agar (Eslaminejad and Zakaria, 2011).

Molecular characteristics

Molecular basis for confirmation of *B. exigua* is done by sequencing analysis of the internal transcribed spacer (ITS) region of rDNA. The primers ITS1 and ITS4 are used for PCR and the ITSrDNA sequence is analized by BLAST sequence analysing tool to develop consensus sequence with 100% identity with that of reference sequences of *B. exigua* (of GenBank; here that was *Phoma exigua* var. *exigua* with accessions numbers EU343168; EU343160; EU343139.1 and EU343119.1) (You, *et al.*, 2016).

Predisposing factors

In an investigation wounded seedlings of roselle are found to be more affected (Eslaminejad and Zakaria, 2011). This suggests that wounds caused by insect pests or else predisposed the disease.

Control measures

The spraying of fungicides, Mancozeb (75%) and Copper oxychloride 50%WP @ 0.3% may be effective.

[*N.B. The disease is very similar to Leaf fleck disease of roselle which caused by *Phoma sabdariffae*.]

8. Powdery mildews

i) Powdery mildew (*Fibroidium* (*Oidium*) *abelmoschi*)

Occurrence & severity status

The powdery mildew of roselle plants has been reported to occur in Florida (Morton, 1987), Philippines (Ansari *et al.*, 2013) and several tropical African countries. According to McClintock and Tahir (2004) this is one of the most important diseases of roselle. The Roselle types green leaves are found to be affected more by this powdery mildew than the plants with red leaves (Grubben and Denton, 2004; Brink & Achigan-Dako, 2012).

Symptoms

The disease first appears as white powdery spots on both surfaces of leaves, on shoots and sometimes on flowers and calyces. These powdery spots gradually spread over a large area of the leaves and stems until the whole surface area of leaves and tender shoots are covered. Finally, the infected leaves and twigs become necrotic yellow and dried up.

Causal organism: *Fibroidium abelmoschi* (Thum.) U. Braun & R.T.A. Cook, 2012

[Krik, P.M. (2018). Species Fungorum (version Oct 2017). Kew Myvology, England]

Synonyms

= *Acrosporium abelmoschi* (Thum.) Subram., 1971

= *Euoidium abelmoschi* (Tum.) Y.S. Paul & J.N. Kapoor, 1978

= *Oidium abelmoschi* Thum. 1878

Morphological/ Morphophysiolgical characteristics

Mycelium is amphigenous, effuse, or irregular patches. Conidiospores are erect, foot cells cylindric straight, (30-) 40-60 x 10-12µm in size, followed by 1-2 shorter cells. Conidia are in chains, ellipsoid-ovoid, 24-34 x 16-18µm in size, with fibrosin bodies. Appressoria are indistinct to more or less nipple-shaped. Cleistothecia are unknown (Braun, 1987; Braun and Cook ,2012).

Predisposing factors

The warm and dry weather is favoured for the disease development.

Management/Control measures

Use of Host resistance

Roselle types with green leaves are susceptible to powdery mildew, while, types with red leaves are partially resistant (Grubben and Denton, 2004).

Chemical control

Three sprays of 0.05% tridemorph or 0.2 % sulphur after the appearance of disease symptoms may be recommended for disease control as practiced in other plants affected by the same fungus.

ii) Powdery mildew (*Leveillula taurica*)

Occurrence & severity status

The powdery mildew disease of roselle caused by *Leveillula taurica* occurs in several countries including, Egypt (Khairy and Abd-el-Rehim,, 1971), Iran (Reza *et al.,* 2007), Italy (Frisullo *et al.,* 1995) Pakistan, Thailand (Amano, 1986), USA (Cook and Riggs. 1995) and South Africa (Swart, 2001) in both tropical and subtropical regions (Farr and Rossman 2009). In Iran, near Birjand of southern Khorasan province, about 30% roselle plants of 5-month old were found infected with this powdery mildew (Reza *et al.,* 2007). Roselle plants are also found to be infected by this powdery mildew fungus in Pakistan and Thailand (Amano, 1986). Apart from roselle, the same powdery mildew disease is also occurred in kenaf in South Africa (Swart, 2001), USA, Italy and Egypt (Amano, 1986).

Symptoms

The disease first appears at the lower leaves as whitish floppy mycelia specks both on ventral and dorsal surfaces. The disease progresses in upward direction from bottom leaves. The extensive growth of white superficial mycelium in angular patches is noticed on the lower and upper surfaces of infected leaves. The growth of the mycelia develops yellowish spots on leaves, which are visible from the opposite side of the spots on the bottom leaves (Palti, 1988). The partial defoliation is also occurred due to heavy infection. Newly infected leaves have a sparse covering of powdery mildew (Reza *et al.*, 2007).

Causal organism: *Leveillula taurica* (Lév.) Arnaud, 1921

[Annales des Epiphytes, 7: 92]

Synonyms

= *Acrosporium obductum* (Ellis & Langl.) Sumst. (1913)

= *Erysiphe taurica* Lév. (1851)

= *Oidiopsis taurica* (Lev.) E.S. Salmon, 1906

= *Oidium obductum* Ellis & Langl., 1890

Morphological/ Morphophysiolgical characteristics

In contrast to most powdery mildew species which are epiphytic, *L taurica* is an endophytic fungus colonizing the mesophyll tissues of the leaf. The mycelium of this fungus found on the plant is both extra-cellular and inter-cellular. The extracellular mycelium is epiphytic with highly branched hyphae approximately 5-8 um in diameter, densely woven with conidiophores. The appresoria are inserted at right angles and opposite sides. In this contrast, the intercellular mycelium is endophytic and is found just below the epidermis mostly in the mesophyll cells with some found in the palisade cells. Hyphae are septate and sinuous or straight measuring less than 10 um. Verrucose granulations are in intercellular hyphae. Conidiophores are septate, simple, often branched, vary in length and width depending on whether epiphytic or from intercellular mycelium. Conidiophores are often found emerging from stomata of the host plant. There are two types of conidia each borne singly on conidiophores. There are the primary conidia that are first formed on the conidiophores. The primary conidia are hyaline, lanceolate with distinct apical points, $55.0-62.5 \times 17.5-22.5$ μm (av. 60×18.8 μm) in size. The second type is aptly called the secondary conidia, which is formed after the primary conidia. The secondary conidia are hyaline, cylindrical to oblong, $50.0-67.5 \times 17.5-20$ μm (av. 59.25×18.3 μm) in size. The teleomorph (Chasmothecia) are absent or very rare in Roselle (Palti, 1988; Reza *et al.*, 2007). Chasmothecia are generally scattered, often embedded in dense superficial mycelia, 135-250μm diam., globose or becoming concave at maturity, peridium of polygonal cells, up to 10μm diam. Appendages are numerous, hypha like, simple, densly interwoven, short, indistinctly branched, colourless to olivaceous brown. Asci are usually about 20 but sometimes less or up to 35 in each ascocarp, 2-spored, ovate, distinctly stipitate, 70-110 x 25-40 μm in size. Ascospores are large, cylindrical to pyriform, sometimes slightly curved and variable in size, 25-40 x 12-22μm (Mukerji, 1968a).

Predisposing factors
This disease prefers warm, dry environments.

Disease cycle
The fungus survives as chasmothecia in crop residues above the soil surface. Under favourable climatic conditions, the chasmothecia open and release ascospores, which are dispersed by wind. The ascospores enter the host through its stomata, germinate, and colonize the host's tissues with its mycelia. The pathogen then begins to produce its asexual conidia on branched conidiophores. The conidia exit through the host's stomata and serve as a secondary inoculum to spread disease after initial infection and repeat infection for several times in the crop season. At the end of the season, the pathogen undergoes sexual reproduction and again produces chasmothecia, overwintering structure (Evans *et al.*, 2008).

Control measures
The fungicide Cabrio (BASF), a strobulirin or QOI (Quinone Outside Inhibitor), is effective to control *L. taurica* infections in plants. Nova 40W (myclobutanil; triazole - sterol inhibitor fungicide) is used as a systemic protectant and curative fungicide, and has good activity against powdery mildew fungi. Application of this fungicide, at a rate of 340g in 1500- 3000 L/ha, is needed as soon as possible after initial infection and again 12 days later when disease pressure warrants control. Alternatively, microscopic Sulphur 92% (elemental, inorganic fungicide) at 500-700g a.i./ha is used prior to infection as a protectant fungicide.

iii) Powdery mildew (*Podosphaera* sp.)
A powdery mildew fungus, *Podosphaera* sp, has been reported to cause powdery mildew disease in North East of India (Baiswar *et al.*, 2010). The disease occurs on *Hibiscus sabdariffa* during October 2009.The disease symptoms include whitish circular patches on the upper surfaces of the leaves. Symptoms are present on mature leaves. The presence of ectophytic mycelium has been observed. Conidiophores are mostly curved containing a foot cell

(30–59 x 8–11 µm). Conidia are in chains, ovoid, 27–35 x 14–19 µm in size, with fibrosin bodies and indistinct appressoria. The edge lines of conidial chains are crenate. The basal septum of the conidiophore is just adjacent to the mycelium. Conidia are smooth wrinkles as evident in SEM studies. Germtubes are present on the lateral side of the conidia and are branched. The perfect stage (chasmothecium) has not been found (Baiswar *et al.,* 2010).

9. Seedling blights

Several seed borne fungi are found to cause seedling blight soon after germination of seeds. The cause of this disease is different in different countries based on the occurrences of the pathogens in different agro climatic conditions. In Malaysia, four fungi infecting Roselle seedlings, namely, *Boeremia exigua* (= *Phoma exigua)*, *Gibberella nygamai* (= *Fusarium nygamai)*, *Fusarium camptoceras* and *Thanatephorous cucumeris* (= *Rhizoctonia solani)* in Penang. Amongst them, the fungal pathogen most frequently isolated is *B. exigua* (45% of the samples), followed by *G. nygamai* (25%), *T. cucumeris* (19%) and *F. camptoceras* (11%). *Thanatephorous cucumeris* is the most pathogenic fungus affecting both wounded and unwounded Roselle seedlings, followed by *B. exigua* that is highly pathogenic on wounded seedlings. *G. nygamai* is less pathogenic while the least pathogenic fungus is *F. camptoceras* in Penang, Malaysia (Eslaminejad and Zakaria, 2011).

i) *Boeremia* (*Phoma*) *exigua* seedling blight

Symptoms

The disease symptoms appear as small circular spots which coalesced to form large, irregular lesions on the leaves. These lesions later turn dark brown, with light brown margins. Abundant black dots (pycnidia) appear on the lesions, especially on the upper surface of the leaves in loose rings. Severe infection causes defoliation of the seedlings, resulting in a sparse crown. Nevertheless, the seedlings recover when new leaves appeared (Eslaminejad and Zakaria, 2011).

Causal Fungus: ***Boeremia exigua*** **(Desm.) Aveskamp, Gruyter & Verkley, 2010**

Synonym

= *Phoma exigua* Desm., 1849

Morphological/ Morphophysiological characters

Boeremia (Phoma) species has been isolated from Roselle are identified mainly from their morphological characteristics *in vitro*. Colony growth is reached 8.5

cm in diameter after four day incubation at room temperature. The colony is flat, spreading, and powdery to velvety with scalloped to lobed margins, which are often submerged in the medium. Immersed mycelium is hyaline with abundant pycnidia that covered by a low mat of felt to floccose grey aerial mycelium. The colour of the colony culture is initially white, later turning olive-grey with an occasional tint of pink. The pigmentation of the culture plates is dark brown to black. Later, a reddish-purple to yellowish-brown diffusible pigments is appeared in the colony reverse side. Pycnidia are olive-yellow to dark olive in the older cultures. The biochemical reaction tests indicate the presence of certain metabolites that are common in Phoma. For example, the application of alkaline reagent (sodium hydroxide, NaOH) on fresh cultures causes the change of the colour of pH-dependent metabolites and pigments. The NaOH spot test is performed on the PDA culture by adding a drop of 1 N NaOH solution on the colony margin. The culture shows a change in colour from olive-grey to blue-green initially, and turning brownish-red after a while. Microscopic features include pycnidia, conidia and chlamydospores. Pycnidia are found to be solitary, globose, glabrous slightly papillate and olivaceous buff on the agar surface, with average diameter of 83.28 ± 31.68 µm. Later, they enlarge to black broadly globose or irregular structures with a clear elongated neck around the ostiole, measuring 97.95 ± 34.2 mm. Pycnidia has one non-papillate ostiole from which conidia are released. Conidia of *B. exigua* are mainly aseptate and occasionally uniseptate, and are rather variable in size. The conidia are hyaline, subglobose to ellipsoidal oblong, with two polar guttules and $9.53 \pm 0.64 \times 2.9 \pm 0.39$ µm in size. Chlamydospore are not observed in the culture (Eslaminejad and Zakaria, 2011).

ii) *Gibberella* (*Fusarium*) *nygamai* seedling blight

Symptoms

The disease symptoms appears as small circular to angular brown spots which coalesce to form large lesions on the upper surface of the leaf. These lesions subsequently dry up to form light brown, regular or irregular necrotic blighted spots. Buff coloured, mycelial wefts of the fungus appear on the diseased tissues (Eslaminejad and Zakaria, 2011).

Causal fungus: *Gibberella nygamai* Klaasen & P.E. Nelson, 1997

Synonym

= *Fusarium nygamai* L.W. Burges & Trimboli, 1986

Morphological/ Morphophysiological characters

The fungus grows well on PDA medium at room temperature. The colony is woolly to cottony, flat and spreading. The colour of the colony is violet from top while from reverse, it is colourless to orangish or pinkish. The growth rate of the fungus on PDA is about 1.16 cm per day and the culture starts to sporulate after five days. The fungus produces septate hyphae, conidiophores, phialides, macroconidia, micro-conidia and chlamydospores. The phialides are cylindrical, solitary or sometimes showing complex branching. Only monophialides are observed, usually with falseheads and micro-conidia are formed on long monophialides. Both macro- and micro-conidia are formed, with average measurements of $26.22 \pm 5.29 \times 2.74 \pm 0.62$ µm and $5.1 \pm 0.92 \times 1.48 \pm 0.34$ µm, respectively. Micro-conidia are mostly aseptate or, rarely one or two septate. They are smooth, hyaline, ovoid to cylindrical, and arranged in chains. Macroconidia are two or more celled, thick-walled, smooth, and cylindrical or sickle shaped. They have a distinct basal foot cell and have pointed distal ends. Chlamydospores are sparse, thick-walled, hyaline, intercalary or terminal and they are formed singly or in pairs (Eslaminejad and Zakaria, 2011).

iii) *Fusarium camptoceras* seedling blight

Symptoms

The disease symptoms appear as yellowing between the veins of the leaves, followed by the formation of small brownish-red spots that coalesce to form large, irregular lesion. These lesions turn to light brown with discolouration at the margins. The leaf tips also appear shrivelled and desiccated (Eslaminejad and Zakaria, 2011).

Causal fungus: *Fusarium camptoceras* Wollenw. & Reinking, 1925

Morphological/ Morphophysiological characters

The fungus produces woolly to cottony, flat, spreading colony on PDA. From above, the colour of the colony is ashen with brown pigmentation and from reverse; the colonies are colourless to orange. Macroconidia are produced in cream- to orange-coloured sporodochia. The growth rate of *F. camptoceras* is about 0.35 cm per day. Microscopic features include macroconidia and mesoconidia. The mesoconidia are very similar to macroconidia in appearance, but they are dissimilar in size and produced on polyphialides in the aerial mycelium. The mesoconidia are crescent in shapes, but these are sometimes falcate and conical to pointed, with notched ends. They show 3-5 septa with average measurements of $4.76 \pm 0.59 \times 1.6 \pm 0.35$ µm. Mesoconidia are abundant, but are not clearly distinguishable from macroconidia. At age, the cultures of the fungus produce macroconidia with an average measurement of 18.29 ± 2.66 x 2.18 ± 0.33 μm. Macroconidia have 5-7 septa (Eslaminejad and Zakaria, 2011).

iv) *Thanatephorus cucumeris* (*Rhizoctonia solani*) seedling blight

Symptoms

The disease symptoms appear as small grey spots which coalesce to form large, irregular brown lesions on the leaves. These lesions subsequently turn light brown. A white coloured, mycelial weft of the fungus appears on the lesions especially on the upper surface of the leaves. Some infected leaves appear shrivelled. Severely infected leaves defoliate prematurely and new leaves appear later (Eslaminejad and Zakaria, 2011).

Causal fungus: *Thanatephorus cucumeris* (A.B. Frank) Donk, 1956

Synonym

= *Rhizoctonia solani* J.G. Kuhn, 1858

Morphological/ Morphophysiological characters

The fungal colony on PDA medium is light brown with abundant sclerotia that scattered in the colony, forming more towards the periphery. Sclerotia are irregular in shape, initially white, then turning brown and dark brown after one week. These are formed singly or in clumps. The growth rate of the fungus on PDA is about 2.13 cm per day. Hyphae are varied in size from 4-7 μm, branched at right and acute angles to the main hyphae. There is often a septum close to each branch and the hyphae are usually constricted at its point of origin. Young hyphae are usually possessed more than three nuclei per cell close to their tips. Monilioid cells are observed in older cultures and they are fused together to form a hard structure called asclerotium (Eslaminejad and Zakaria, 2011).

Control measures

The three pathogens, *B. exigua*, *G. nygamai* and *T. cucumeris* are found to be susceptible to the volatile inhibitors produced by *Trichoderma viride*, giving rise to growth inhibition of about 68% in each case. When *T. viride* non-volatile metabolites are applied against the pathogens, maximum inhibition occurred against *T. cucumeris* (73.95% mycelial growth inhibition), followed by *B. exigua* (37.17% inhibition). However, the inhibitory effect of the non-volatile metabolites on *G. nygamai* is very low (Eslaminejad *et al.*, 2012).

10. Stem rot/ Southern blight

Occurrence & severity status

Roselle (*H. sabdariffa*) is susceptible to the attack of Stem rot/ Southern blight disease [Fig. 10 (A-B)]. The disease infects plants at early development stages, when competition from weeds can also be deleterious (Gomez-Leyva *et al.*, 2008). The disease commonly occurs in the tropics, subtropics, and other warm temperate regions, especially the southern United States, Central and South America, the West Indies, southern European countries bordering the Mediterranean, Africa, India, Japan, the Philippines, and Hawaii. In Malaysia cultivation of Roselle is the subject to be affected by stem rot disease (McClintock and Tahir, 2004).

Symptoms

The first visible symptoms are progressive yellowing and wilting of the leaves. Following this, the fungus produces abundant white, fluffy mycelium on the infected tissue at the base of stem near ground level. Later, the fluffy white mycelia become to some extent dull yellowish-white coloured and compressed, with numerous sclerotia on it. Sclerotia are of uniform size, roundish and white when immature then becoming dark brown to black. Mature sclerotia resemble mustard seed. Seedlings are very susceptible and die quickly once they become infected. Older plants that have formed woody tissue are gradually girdled by lesions and eventually die. Invaded tissues are pale brown and soft, but not watery (Taylor and Rodriguez, 1999; Kator *et al.*, 2015).

Causal organism: *Athelia rolfsii* (Curzi) C.C. Tu &Kimbr., 1978

[Krik, P.M. (2018). Species Fungorum (version Oct. 2017), Kew Mycology]

Synonyms

= *Botryobasidium rolfsii* (Cruzi) Ventatar., 1950

= *Corticum rolfsii* Curzi, 1932

= *Pellicularia rolfsii* (Curzi) E. West, 1947

= *Sclerotium rolfsii* Sacc., 1911

Morphological/ Morphophysiolgical characteristics

The fungus is soil borne and produces abundant white mycelium on infected plants and in culture. Advancing mycelium and colonies often grow in a distinctive fan-shaped pattern on lower stems, leaf litter, and soil .After 7 to 14 days, tan-to-brown, mustard-seed-sized (0.5 to 1.5 mm) sclerotia form on the mycelial mat.

Predisposing factors

Athelia rolfsii (=*S. rolfsii*) is a soil borne pathogen. So the infected soil causes stem rot disease on plants. The disease is favoured by the temperature ranging from 25° and 35°C, humid conditions and acidic soil (Punja, 1985; Mullen, 2001; Roberts, 2003).

Disease cycle

The causal fungus overwinters as mycelium in infected tissues or plant debris. It usually persists as sclerotia. Sclerotia are disseminated by cultural practices, irrigated water and wind. The fungus grows and attacks plants at or near the soil line. Before the pathogen penetrates host tissue it grows on the plant surface. Penetration of host tissue occurs when the pathogen produces an enzyme which deteriorates the hosts' outer cell layer (Fery and Dukes, 2002). This results in tissue decay, further production of mycelium and the formation of sclerotia. The latter two rely upon favourable environmental conditions (Kator *et al.*, 2015).

Control measures

- Crop rotation is a common and preferred method of this disease control.
- Antagonistic rhizobacteria, more specifically fluorescent pseudomonads (e.g. *Pseudomonas fluorescens*) exhibit the high activity and produced blue green pigment. This is found as a potential biological control agent against the phytopathogenic fungus (Farrag *et al.*, 2015).
- Plant treatments with fungicides for control of southern blight are generally protective in nature. Good disease control may be achieved by foliar applications of azoxystrobin products, flutolanil, or tebuconazole. The insecticide, chlorpyrifos, which hydrolyzes into a fungicidal product, is also effective to control the disease very well (Kator *et al.*, 2015).

11. Vascular wilt

Occurrence & severity status

Vascular wilt disease of roselle is one of the most important diseases that attack both seedlings and adult plants of roselle, causing serious losses in crop productivity and quality. The disease is called by different names, like, 'Root rot and wilt' and '*Fusarium* wilt'. It occurs all over the world, including the countries, like, Egypt (Hassan *et al.*, 2014b), Malaysia (Ooi and Salleh, 1999), Nigeria (Amusa *et al.*, 2005), United States (Ploetz *et al.*, 2007) and India. The disease is more frequent in young plants and adversely affects the growth in various parts of the world. Boulanger *et al.* (1984) observed that the infection proceeds to stem and, eventually to pod of young plants of roselle.

Symptoms

The disease at first appears as slight vein clearing on the outer portion of the younger leaves, followed by epinasty of the older leaves. At the seedling stage, the infected plants may wilt and die soon after the symptoms appeared. In older plants, the symptoms of the affected roselle are characterized by cankers and/or soft rot in the roots, as well as the basal part of their stems adjacent to soil surface. The leaves turn yellow then brown in colour, die and defoliate, and the whole plant becomes brown with dried leaves followed by death of the whole plant (Ooi and Salleh, 1999; Amusa *et al.*, 2001, 2005; Ploetz *et al.*, 2007; Hassan *et al.*, 2014b). On cutting, browning of the vascular tissue is evident (Agrios, 1988).

Causal organism: *Fusarium oxysporum* Schltdl., 1824

[Flora Berolinensis, Pars secunda: Cryptogamia: 106]

Synonyms

= *Cylindrophora albedinis* Kill & Maire, 1930

= *Fusarium albedinis* (Kill & Maire) Malencon, 1934

= *Fusarium apii* P.E. Nelson & Sherb., 1937

= *Fusarium batatas* Wollenw., 1914

= *Fusarium bulbigenum* Cooke & Massee, 1987

= *Fusarium oxysporum* f *apii* (P.E. Nelson & Sherb.) W.C. Snyder & H.N. Hansen, 1940

Morphological/ Morphophysiolgical characteristics

The fungal colony in Oatmeal Agar medium grows rapidly. Aerial mycelia are white, usually becoming purple, discrete erumpent orange sporodochia may be seen, reverse hyaline to dark blue or durk purple. Conidiophores are short, single, lateral monophialides in the aerial mycelium, later arranged in densely branched clusters. Macroconidia are fusiform, slightly curved, pointed at the tip, 3-5 septate, basal cells are pedicellate, 23-54 x 3.0-4.5 µm in size. Microconidia are abundant, never in chains, mostly non-septate, ellipsoidal to cylindrical, straight or often curved, 5.0-12.0 x 2.3-3.5 µm in size. Chlamydospores are terminal or intercalary, hyaline, smooth or rough walled, 5-13 µm diam. Sclerotial pustules may be present, which are pale to green or deep violet (Hoog, 2000). Colonies on one-half-strength potato dextrose agar (PDA) medium are salmon-coloured. On banana leaf agar medium, single-spored strains produces copious microconidia on monophialides, infrequent falcate macroconidia, and terminal and intercalary chlamydospores (Poletz *et al.*, 2007).

Molecular characteristics

The fungal culture obtained from Roselle was first cultured in Potato Dextrose Broth (PDB) in a rotary shaker (180 rpm) for 4 days at 28±2°C. The mycelia were harvested and ground in liquid nitrogen for DNA extraction. The PCR amplification of DNA at ITS using primers ITS1 and ITS4 fragments was approximately at 500 bp. The molecular identification from DNA sequences was done using BLAST network services against NCBI nucleotide database. GenBank database confirmed that the isolate tested was homology with 100% sequence similarity to *Fusarium oxysporum* isolate TVD-Fungal-Culture 71 with accession number of KF494069.1 (Ng *et al.*, 2017). In another study, partial elongation factor 1-α (EF1-α) sequences were generated for two of the strains, O-2424 and O-2425, and compared with previously reported sequences for the gene (O'Donnell *et al.*, 2004). Strains of *F. oxysporum* recovered from wilted roselle in Egypt, O-647 and O-648 in the Fusarium Research Centre collection were distantly related to the Florida strains. *F. oxysporum*-induced wilt of roselle which had been reported in Nigeria (Amusa *et al*, 2005) and Malaysia (Ooi and Salleh, 1999) where the sub-specific epithet *F.* sp. *roselle* was used for the pathogen (Poletz *et al.*, 2007).

Predisposing factors

As the fungus is soil borne, it can survive in the plant debris thus the presence of plant debris will increase the disease incidence. Plant debris is found to be associated with the fungus (Amusa *et al.*, 2005). Root infections by *M. incognita* increased the colonization of roselle by *F. oxysporum* and subsequently caused higher damage to the roselle seedlings. The high wilt incidence in the presence of *M. incognita* and *F. oxysporum* may be due to the synergistic relationship between these two pathogens (Ooi *et al.*, 1999). Further, the disease prefers humid condition.

Disease cycle

The causal agent overwinters in soil. When the conditions for its development are favourable, i.e., in wet and poorly drained soil, the fungi infects plant roots. On infection it establishes disease in roselle and produces conidia and chlamydospores for reinfection.

Management/Control measures

Cultural control

It is extremely persistent in the soil. The fungal (*F. oxysporum*) strains are ubiquitous soil inhabitants that have the ability to exist as saprophytes, and degrade lignin and complex carbohydrates associated with plant debris. For this reason, it is necessary to remove and discard all infected plants. When a plant dies from root rot, the infected plant should be dug off and disposed as quickly as is practicable. All composts used have potential to suppress the growth of *F. oxysporum*, especially the non-sterilized agro-waste composts have higher potential to be developed as soil suppressive amendment against *F. oxysporum* (Ng *et al.,* 2017).

Biocontrol

The biocontrol agents, *Trichoderma gamsii* and *Trichoderma virens* increase the height and, fresh and dry weight of Roselle plants infected with *F. oxysporum* (Osorio-Hernandez *et al.,* 2014). Further, in greenhouse evaluation, the former one has been found to decrease the disease incidence by 20% along with increase of plant height and fresh and dry weight (Osorio-Hernandez *et al.,* 2014).

Integrated measures

The microelements (copper and manganese), antioxidants (salicylic acid, ascorbic acid and EDTA), a fungicide (Dithane M45) and biological control agents (*Trichoderma harzianum* and *Bacillus subtilis*) are found to reduce the linear growth of the causal pathogens. Additionally, application of the previous microelements, antioxidants, a fungicide and biological control agents significantly reduces disease incidence of root rot and wilt diseases under field conditions. Copper, salicylic acid and *T. harzianum* show the best results in this respect. In conclusion, microelements, antioxidants, and biocontrol agents are suggested to use as alternative strategies to fungicides for controlling vascular wilt disease in roselle (Hassan *et al.,* 2014a).

12. Watery leaf Spot

Occurrence & severity status

The disease has been found to occur in Tabasco, Mexico during 2008-2009 with more than 70% incidence (Correa *et al.,* 2011). The disease may possibly occur in several countries, like, Algeria, Nigeria, Tanzania of Africa; China, India, Japan, Turkey of Asia; Austria, Bulgaria, France, Great Britain, Greece, Hungery, Italy, Rumania, Spain, Switzerland, Russia, Yugoslavia of Europe; Canada, United States of North America and Brazil, Uruguay of South America (CMI Map, 335; Sutton and Waterston, 1966).

Symptoms

The symptoms appear on diseased plants as circular to irregular shaped, light brown coloured necrotic spots on leaves with water-like appearance that extended to the entire leaf blade and petiole. As the necrosis progresses, light brown coloured concentric rings of a soft consistency are seen. Small dark brown coloured pycnidia are found in groups and that protruding the cuticle of the leaf. When the necrosis reaches the petiole, it causes the drying of the leaf. The leaf spots initiate from the lower leaves and advance to upper foliage, resulting in full defoliation. After defoliation, necrosis continues to the branches on which small, dark brown bodies (pycnidia) are produced and concludes with the death of the plants in the production stage, causing yield losses (Correa et al., 2011).

Causal organism: *Coniella diplodiella* **(Speg.) Petr. & Syd., 1927**
[Repertorium Specierum Novarum Regni Vegetabilis Beihefte 42: 460]

Synonyms

= *Phoma diplodiella* Speg., 1878

= *Coniothyrium diplodiella* (Speg.) Sacc., 1884

= *Charrinia diplodiella* (Speg.) Viala & Ravaz, 1894

= *Clisosporium diplodiella* (Speg.) Kuntze, 1898

= *Pilidiella diplodiella* (Speg.) Crous & Van Niekerk, 2004

= *Coniella petrakii* B. Sutton, 1980

Morphological/ Morphophysiolgical characteristics

Colonies are white to greyish coloured on the surface and honey coloured on reverse of petridishes with PDA medium. In plant, pycnidia are immersed but strongly erumpent, honey-yellow to brown (or dark brown), amphigenous but mostly epiphyllous, smooth, globose to subglobose, 85-130µm diam. (differently, 200-300µm in PDA medium), wall up to 4 cells thick, composed of honey-yellow thin walled cells, darker near the protruding portion. Ostiole is central, almost papillate, circular with measures 20µm diam. Conidiophores are hyaline, branched only at the base, non-septate, elongated, 10-14.5 x 2.5-3.5µm, restricted to a convex cushion of small celled, hyaline pseudoparenchyma at the base of pycnidium. Conidiogenous cells are terminal, phialides. Conidia are unicellular, subhyaline then pale brown at maturity, elliptical to obovate, inequilateral, straight to slightly curved, with an obtuse apex and truncate base, 8-16 (-16.2) x (5.0-) 5.5-7.5 µm in size (Sutton and Waterston, 1966; Correa et al., 2011).

Molecular characteristics

The amplification and sequencing of the ITS DNAr gene was performed using the universal primers ITS1-5' TCC GTA GGT GAA CCT GCG G 3', and ITS4-5' TCC TCC GCT TAT TGA TAT GC-3'. The obtained sequences showed

99% similarity with the sequences of *Coniella diplodiella* (= *Pilidiella diplodiella*) with the accession number Ay339331 as deposited in the National Centre for Biotechnology Information's (NCBI) GenBank (Correa *et al.*, 2011).

Predisposing factors

The hailstorms cause damage to the plants that increases disease severity (Bisiach, 1988).

Disease cycle

The pathogen remains viable for 2-3 years on plant debris in soil and the disease is dispersed by soil splash and farm implements. It enters to plants through wounds, sun scorch and through the infection of other pathogens (Sutton and Waterston, 1966).

Control measures

The fungi normally present in stable manure, like, *Trichderma*, *Chaetomium* and *Chaetomella*, exercise a destructive action on *Coniella* spores (Turian, 1954). The use of protective fungicides, like, mancozeb, wettable sulphur or copper oxychloride may be used.

13. Virus diseases

i) Leaf curl

Occurrence & severity status

Leaf Curl disease of Roselle occurs in several countries, including South Africa, China (Arief *et al.*, 2017), India (Ambuja *et al.*, 2017), and possibly other countries where roselle is being cultivated for fibre production. The disease in India, is an emerging threat in production of mesta fibre, generally the fibres from roselle and kenaf together are known as 'mesta'. The incidence of the disease varied from 20 to 80 per cent in southern India with highest disease incidence of 81.66 per cent as observed in Raichur district of Karnataka state (Ambuja *et al.*, 2017). The disease is more prevalent in kharif than rabi season, the latter crop is used mainly for vegetable. The mean disease incidence 47.83 and 32.18 per cent was recorded in Raichur and Bidar districts, respectively, in 2015-2016 (Ambuja *et al.*, 2017). The leaf curl disease occurs in kenaf in both India (West Bengal) and China (Paul *et al.*, 2006; Tang *et al.*, 2015). In West Bengal, the kenaf leaf curl is an emerging major threat for its increasing severity (Paul *et al.*, 2006).

Symptoms

The infected plants exhibit vein clearing, vein swelling, upward leaf curling, chlorosis or yellowing of leaf, enations on the veins of ventral surface, with twisting of petioles, leathery and small sized leaves and stunted growth of the infected plants (Aswathanarayana *et al.*, 2016; Ambuja *et al.*, 2017; Arif *et al.*, 2017). The leaf curl diseased plants produce less or no flowers with poor pod formation (Aswathanarayana *et al.*, 2016; Ambuja *et al.*, 2017).

Causal organisms: Begomovirus [*Cotton leaf curl Multan virus* (CLCuMuV), in conjunction with *Cotton leaf curl Multan betasatellite* (CLCuMuB)]

[**N.B.** In China *Hibiscus sabdariffa* is reported to be affected by *Cotton leaf curl Multan virus* (CLCuMuV), in conjunction with *Cotton leaf curl Multan betasatellite* (CLCuMuB), which causes cotton leaf curl disease in South Asia (Arif *et al.*, 2017). The virus complex, CLCuMuV-CLCuMuB was first found in 2006 to be associated with a leaf curl disease of *Hibiscus rosa-sinensis* in China (Mao *et al.* 2008; Du *et al.* 2015). Recently, CLCuMuV-CLCuMuB has become prevalent in *H. rosa-sinensis* in the southern part of China. The causative organism of leaf curl disease in rosellle (*H. Sabdariffa*) as CLCuMuV-CLCuMuB has been identified in China for the first time, where *H. sabdariffa* is the fourth seed-propagated plant to be affected by this virus complex (Arif *et al.*, 2017).]

Virus

The diseased samples are failed to give amplification for DNA-B component. So the disease is considered as a monopartite begomovirus (Ambuja *et al.*, 2017). In an advanced study, a degenerate PCR was conducted to amplify a conserved region on the genome of the suspected begomovirus infected *H. sabdariffa*. PCR amplicons of the expected size (about 570 bp) were obtained from symptomatic plant samples. A BlastN search of the NCBI non-redundant nucleotide database revealed more than 99% identity of the PCR amplicons to a corresponding fragment of CLCuMuV occurring in China (JX861210). PCR using a primer pair specific to CLCuMuV (CLCuMuVF and CLCuMuVR 5'CAACAGGCATGGACAAACAG3' and 5'- CCAATACGATGGGTCA AACC-3') was used to confirm the presence of CLCuMuV in symptomatic *H. sabdariffa*. PCR fragments of the expected size (197 bp) were obtained from all symptomatic samples. These results indicated that the *H. sabdariffa* showing begomovirus-like symptoms might have been caused by CLCuMuV. To sequence CLCuMuV and CLCuMuB infecting *H. sabdariffa*, rolling circle amplification (RCA) was conducted to enrich circular DNAs. The circular DNAs were amplified using β01/02. The fragments obtained from the PCR were cloned by using a T-vector (pMD18-T, TAKARA, Dalian, China). The colonies each for CLCuMuV and CLCuMuB were used for Sanger sequencing and the sequences were assembled using DNASTAR Lasergene 10 (DNASTAR Inc.). For both CLCuMuV and CLCuMuB a consensus sequence each was deposited in GenBank. The accession numbers KY992859 and KY992858 have been assigned to CLCuMuV and CLCuMuB, respectively. Sequence analysis revealed that the CLCuMuV and CLCuMuB isolates were more than 99% identical to their counterparts occurring in China, but shared a nucleotide sequence identity of less than 94% with those found in South Asia. In phylogenetic trees constructed with the aligned genomic sequences of CLCuMuV and CLCuMuB, CLCuMuV occurring in China, including the new isolate identified in their (Arif *et al.*, 2017) studies, formed a monophyletic clade together with a few of their counterparts occurring in South Asia. However, Chinese CLCuMuV formed a distinct branch within this clade. Similarly, CLCuMuB occurring in China, including the new isolate identified, clustered and formed a monophyletic clade relative to their counterparts from South Asia (Arif *et al.*, 2017).

Virus transmission

The viruses are transmitted by vector (whiteflies; *B. tabaci*), dodder (*Cuscuta subinclusa*), grafting or cell sap (Ambuja *et al.*, 2017).

Vector (white fly) transmission: In case of white fly transmission, the first symptom of vein clearing is observed at 18 days after inoculation, followed by vein thickening, slight curling, chlorosis of leaves, twisting of petioles, leathery leaves, severe upward curling, reduction in leaf size enations and stunted growth of the plants are between 18 to 48 days (Ambuja *et al.*, 2017).

Dodder transmission: Dodder (*Cuscuta subinclusa*) is a natural complete stem parasite of roselle. In case of dodder transmission, the initial symptoms is observed at 27 days after dodder twining (inoculation), and the further symptoms are observed at between 27 to 48 days (Ambuja *et al.*, 2017).

Graft transmission: In case of graft transmission, the initial symptoms appear at 33 days after grafting, and the further symptoms are found between 33 to 48 days of virus inoculation (Ambuja *et al.*, 2017).

Sap transmission: In case of sap transmission, the initial symptoms develop at 30 days after sap inoculation, and the further symptoms are observed from 30 to 48 days of virus inoculation (Ambuja *et al.*, 2017).

Control measures

The disease can be managed by white fly control with the use of effective insecticides as suggested with the yellow vein mosaic disease control.

ii) Yellow vein mosaic disease

Occurrence & severity status

The roselle plant suffers from many virus diseases (Jones and Behncken 1980; Brunt *et al.*, 1996). The virus diseases caused by, *Cotton leaf curl Begomovirus*, *Okra mosaic Tymovirus* and *Malva vein clearing Potyvirus* on roselle were earlier noticed (Burnt *et al.*, 1996), causing slight stunting of plant and reduction in yield, sporadically. Worldwide there are four viruses that affect roselle: *Cotton leaf curl virus* (CLCuV), *Mesta yellow vein mosaic virus* (MeYVMV), *Malva vein clearing virus* (MVCV), and *Okra mosaic virus* (OkMV) (Fauquet *et al.*, 2005; Roy *et al.*, 2009). Amongst them, the yellow vein mosaic disease (Fig. 12B), which has been observed in the southern, northern and eastern part of India, causes considerable damage. A survey was conducted on 50 fields over four districts of West Bengal to record the disease incidence. The survey indicated that the incidence of the disease was 92.46% in *H. Sabdariffa* (Chatterjee *et al.*, 2005a). The disease of roselle is also found in the state of Guerrero (Mexico), where the disease incidence was observed 90-100% in 2014 (Velázquez-Fernández *et al.*, 2016). The disease causes complete destruction of the crop and the farmers in many regions are showing reluctance to cultivate this economically important crop as a result of this disease. The disease also occurs in kenaf (Chatterjee *et al.*, 2005a).

Fig. 12 (B). Foliar diseases in roselle. Yellow vein mosaic virus disease.

Symptoms

The leaves of the affected plant show yellowing of veins, which leads to complete yellowing, sometimes with mosaic and yellowing of leaves, followed by defoliation and yield reduction at an advanced stage (Chatterjee and Ghosh, 2008; Velázquez-Fernández *et al.*, 2016).

The disease lowers the activities of catalase, acid phosphatase and peroxidase enzymes in plants, however, it enhances the activities of esterase, polyphenol oxidase and superoxide dismutase. The chlorophyll content, phenolics and soluble protein are decreased in diseased plants, while, free amino acid, proline and disease related proteins are increased on infection. The SDS-PAGE protein profile of total soluble protein from diseased leaves of roselle shows the absence of 22kDa and 26kDa protein bands, which are pronounced in healthy plants (Chatterjee and Ghosh, 2008).

Causal organism: Begomoviruses (Geminiviridae)

Virus

The association of a novel *Begomovirus*, namely, *Mesta yellow vein mosaic virus* (MeYVMV) with the disease, has been confirmed by electron microscopy and molecular techniques using PCR, sequence information and southern hybridization. In eastern India Mesta yellow vein mosaic disease is associated with Mesta yellow vein mosaic virus and an isolate of *Cotton leaf curl Multan beta satellite* (CLCuMB) (Chatterjee and Ghosh, 2007a; 2007b), while, in northern India association of another species namely *Mesta yellow vein mosaic Bahraich virus* (MeYVMBV) and an isolate of *Ludwigia leaf distortion beta satellite* (LuLDB) are reported (Das *et al.*, 2008a, 2008b). In Southern India the begomovirus complex consisted of MeYVMV and LuLDB (Roy *et al.*, 2009). In Mexico, Yellowing of *Hibiscus sabdariffa* is associated with a begomoviruses complex in which *Okra yellow mosaic Mexico virus* (OkYMMV) is present (Velázquez-Fernández *et al.*, 2016).

Transmission

This virus is transmitted by a white fly (*B. tabaci*) and the transmission efficiency is 85% for roselle (*H. sabdariffa*). The minimum acquisition period is found 12h, while, the minimum inoculation access period is 4h. The minimum incubation period for developing disease symptoms is 9 days. The disease is also transmitted through cleft grafting with 80% efficiency (Chatterjee *et al.*, 2008). Symptomatic plants while grafted onto healthy plants, vein chlorosis and mosaic symptoms are observed in healthy plants at 17 days after grafting (Velázquez-Fernández *et al.*, 2016).

Molecular characteristics

For the DNA-A of the begomoviruses, Velázquez-Fernández et al. (2016) had designed the primers *Sense* DNA-AC (AAAACTCGAGGATGTGAAGG CCCATG) and *Antisense* DNA-AC (AAAAGGGAAGA CGATGTGGGC), which amplified a fragment of approximately 1400 pb corresponding to genes AV1, AC3, AC2, and AC1. For DNA-B, they used the primers proposed by Rojas *et al.* (1993), which amplified a fragment of 500 pb. The reaction mixture consisted of 1 µL of DNA obtained by RCA, PCR reaction buffer (1X), 0.20 mM dNTPs, 2.0 mMMgCl$_2$, 10 µM of each primer, 0.5U of Taq polymerase (PROMEGA®), adjusting to a final volume of 20 µL. The PCR was carried out with the following conditions: initial denaturation 94°C for 4 min, 30 denaturation cycles of 94°C for 1 min, 55°C for 1 min, 72°C for 1.5 min, one final extension cycle of 72°C for 3 min. The amplified products of 1400 pb were sequenced and compared with the GenBank data base. The results showed 95% similarity with *Okra yellow mosaic Mexico virus* (OkYMMV) (Velázquez-Fernández *et al.*, 2016).

Control measures

No specific control measure of the disease has been found, except the control of white fly population. In Soybean, the effectiveness of thiamethoxam (30FS @ 5ml/kg seed) in seed treatment and imidacloprid (17.8 SL @ 500ml/ha) and triazophos (40 EC @ 800 ml/ha) in foliar application has been noticed to minimize the whitefly population (Swathi and Gaur, 2017), that may be used for roselle also.

C. Mesta (kenaf & roselle) Pests

1. Flea Beetle

Occurrence & severity

Flea Beetle is a small jumping bettle that has extraordinary power of jumping. The beetle was first recorded as a pest of kenaf in the year 1961 (Choudhuri, 1962). This is one of the major pests of mesta of early season. The intermittent showers followed by dry spell with high humidity are the conductive conditions for its multiplication (Pandit and Pathak, 2000). The pests attacks mesta (*H. cannabinus*), roselle (*H. sabdariffa*), congo-jute (*Urena lobata*), malachra (*Malachra capitata*) and bhindi (*Abelmoschus esculentus*) in West Bengal. The pest population and infestation levels are related to soil moisture, rainfall and plant growth stage (Pandit 1998).

Types of damage

The beetle feeds on the leaf lamina of mesta. Adult beetle feeds on tender shoots and leaves. Its attack commences when the plants are of a week old and ceases just before harvest. Irregular cut and holes on the leaves are the typical damage symptoms. The larva feeds on roots. A mosaic disease of Kenaf is reported to be transmitted by the Flea beetle (Sardar, 1995).

Scientific name & order: *Nisotra orbiculata* Motschulsky, 1866 (Coleoptera: Chrysomelidae)

Biology

Egg: Eggs are white and oval. They are laid in clusters on the lower surface of leaves or under the cracks and crevices of soil.

Grubs: Grubs are elongated with soft body, about a quarter to a third inch long. Their body is whitish with yellowish or light brown head. They have six short legs.

Pupa: Pupa is rarely seen as it is formed several inches below in the soil.

Adult: Adult beetle has an oval body with shiny black in colour. Length is about 3.5-4 mm. Underside of the body, scutellum and seven apical segments of the antennae are piceous. Legs, prothorax, and head are light brown to dark brown (darker in older specimens). Eyes are black. Anterior coxal cavities are closed behind. Antenna is 11 segmented. Pronotum and elytra are not pubescent; Pronotum is without a distinct antebasal impression. Mid and hind tibiae are not excavated apically. Each side of excavation is with a marginal row of stiff bristles. Rows of elytral punctures are indistinct or confusedly punctuate. Claws are appendiculate. Vertex is evenly convex, disc not elevated, sides not deeply excavated above eye. Sides of pronotum with opposing, short, longitudinal impressions are situated on anterior and posterior margins, though the latter is only visible under high magnification. It moves by quick jumping rather than flying. It has the hind leg femora greatly enlarged.

Life cycle

Egg stage lasts about 7-9 days, grub stage 10-14 days, prepupal stage 1-2 days and pupal stage 7-9 days. Adult's longevity varies from 30-32 days in kenaf to 40-45 days in roselle (Pandit and Chatterji, 1978).

Control measures

Application of *Beauveria bassiana* on the flea beetle may be effective as it was reported pathogen of this beetle (Pandit *et al.*,1979). The entomogenous fungus *B. bassiana* has been found to cause natural mortality of *N. orbiculata* up to 87% (Pandit and Pathak 2000).

Neem oil (5ml/l) is reported as effective plant product against the flea beetle. Carbaryl is very much toxic to the insect (Pandit and Chakraborty, 1986). Seed treatment with Thiamethoxam 75WS @ 5g/ kg seeds is effective as a preventive measure for at least 60 days after sowing. Foliar spray with Profenofos 50EC (0.1%) is also very effective against the flea beetle.

2. Mesta Hairy caterpillar

Occurrence & severity

Mesta Hairy caterpillar (Fig. 13), also known as 'Yellow tail tussock moth', is found in northern India, Sri Lanka, Myanmar and India's Andaman Islands (Annonymous, ND1). Though it is considered as a minor pest, larva can sporadically be a serious pest. It is a polyphagous species. Apple, Ragi, castor, pigeon pea, cowpea, field bean, cucurbits, mango, citrus, hibiscus, rose, ficus, coffee, tea and many more plants are affected by the insect (ICAR, 2013). The pest mostly occurs during July-August. The pest also infests several other crops including sunnhemp and cotton.

Fig. 13. Hairy caterpillar in Mesta/kenaf.

Types of damage

The larvae of Mesta hairy caterpillar are gregarious at the beginning. The young gregarious larvae feed voraciously on foliage, causing defoliation. Later, they disburse and enter the buds and flowers, and feed on sepals and stamens.

Scientific name & order: *Somena scintillans* (Walker, 1856) (Erebidae: Lepidoptera)
(Wang *et al.*, 2011)

Synonyms
= *Artaxa scintillans*

= *Euproctis scintillans*

= *Nygmia scintillans*

= *Prothesia scintillans*

Biology
Egg: Eggs are round, 0.43 ± 0.01 mm in diam., pale yellowish, laid in clusters under the surface of leaves. The female covers them with brown hairs. Prior hatching, the egg colour is changed to brownish. Incubation period is 5-9 (-10) days (Gupta *et al.*, 2013).

Larva: Larva is dark brown with a series of crimson lateral tubercles on a yellow line bearing tufts of grey hair and a mid dorsal red stripe. The third somite banded with yellow. Dorsal tufts of short brown hair are on fourth, fifth and eleventh somites. Fifth to tenth somites are with a broad, dorsal yellow stripe. There is a yellow spot on the anal somite. Head is dark coloured. Larval period is 29-35 days.

Pupa: It pupates in coccon of hairs on the leaves or in leaf folds. Pupal period: 10-12 days

Adult: Adult is yellowish coloured moth with reddish line and spots on edges of wings and tuff of hairs on side. Head is testraceous, somewhat luteceous. Thorax and abdomen is brownish, the later with luteous lip. Forewings are 12-17mm, vinous brown, wholly studded with black tubercules, and adorned with silvery spangles along its exterior side, where the brown hue forms two branches across the yellow marginal area below the apex and to centre of margin, but sometimes not reaching the margin. Costa is often yellowish. Hindwings are yellow to fuscous brown with a broad yellow margin (Hampson, 1892). Male genitalia: Uncus is broad, weakly bifid. Valva is divided in the distal half, the dorsal arm broad, the ventral arm slender, acute apically, saccus triangular, aedeagus simple and vesica unadorned. Female genitalia: Papilla analis is small, ductus short, cylindrical, sclerotized; bursa long, sausage shaped and with a bicornute signum (Wang *et al.*, 2011).

Life cycle

Life cycle consists of six larval instars with a total duration of 24-28 days. Pupal period lasts for 8-10 days.

Control measures

It is suggested to remove mechanically the caterpillars along with damaged pods. Spraying of caster oil-neem oil emulsion with garlic extract diluted in water is also effective.

This should be followed by spraying of *Beauveria bassiana* (20 g or 5 ml/l of water).

Spraying of methyldemeton @ 2ml/litre or phosalane @ 2ml/litre of water and acephate (Asataf/Starthene 75 SP) 0.1 to 0.15% may be applied, if needed.

3. Mealybugs

Mealybugs are the major pests of both Kenaf (*H. cannabinus*) and Roselle (*H. sabdariffa*). Mainly, two types mealy bugs, Pink hibiscus mealybug (*Maconellicoccus hirsutus*) and Cotton mealybug (*Phenacoccus solenopsis*) are reported to cause damage on mesta (kenaf & roselle) (Ghose, 1961, 1971a; Singh and Ghosh, 1970; Satpathy *et al*., 2009, 2013).

i) Pink hibiscus mealybug/ pink mealybug/ hibiscus mealybug

Occurrence & severity

The mealy bug occurs on many plants of 200 genera under 76 families, however, it has some preference for Fabaceae, Malvaceae and Moraceae (Mani, 1989; Garland, 1998). In the fibre crops, *H. sabdariffa*, *H. cannabinus* and *Boehmeria nivea* are found to be affected by the pest frequently (Ghose, 1961; 1971a; Singh andGhosh, 1970; Raju *et al*., 1988).

Types of damage

Both the nymph and adult of this mealy bug attack mesta (kenaf & roselle). The nymphs and female adult feed on the apical parts. The infested growing points become stunted and swollen, appearing stunted plants with bushy-top, popularly known as bunchy top or rosetting (Pena, 2013). The petiole of the fibre crop becomes shortened, the lamina crumples and inter-nodal length reduced. The affected plants show signs of a mouldy substance and a presence of large quantities of wax. The attack of the mealybug causes deterioration of fibre quality and reduction in yield.

Scientific name, Family & order: *Maconellicoccus hirsutus* Green, 1908 (Hemiptera: Pseudococcidae)

Synonym

= *Phenacoccus hirsutus* Green 1908

Biology

Egg: Female lays eggs about 150-600 in ovisacs (egg sacs) of white wax in a week, usually in clusters on twigs and bark of host plant. Eggs are minute, cylindrical, 0.3-0.4mm in length and 0.15-0.21mm in width, initially orange in colour but turn pink on maturity. Egg developmental period is 6-9 days.

Nymph: Nymphs emerge from the ovisacs in batches, corresponding with the sequence of egg-laying. Newly hatched nymphs are called crawlers. Crawlers are very mobile, 0.3 mm long, pink. The nymphal stages appear much like the female in form, but the female nymphs have three instars, while the male nymphs have four instars. The last instar of male is an inactive stage with wing buds within a cocoon of mealy wax. The nymphal stages last for 10-25 days. In certain cases, the nymphal stages may last for as long as 30 days.

Adult: Immature females and newly matured females are greyish-pink, dusted with mealy white wax. Female adults are reddish pink and covered by a waterproof mass of white wax (Miller, 1999), 2.5–4 mm long, soft-bodied, elongate oval and slightly flattened. Entire colony becomes covered by white, waxy ovisac material. Slide-mounted females show the combination of 9-segmented antennae, short caudal filaments, and long dorsal setae anal lobe bars. Numerous dorsal oral rim ducts are on all parts of the body except the limbs and long, flagellate dorsal setae. Females die after depositing the eggs so only mate once in its life time.

Adult males are smaller than the females and live only a few days to mate (Juniora *et al.*, 2013). They are reddish brown, 3mm long, with long antennae, two long tails, lack of waxy coating on body, but white wax filaments are projecting posterior, have a pair of very simple wings, capable of flight, and have non-functional mouthparts.

Life cycle

The life cycle of *M. hirsutus* has been studied in India. Each adult female lays 150–600 eggs over a period of about one week, and these hatch in 6–9 days (Bartlett, 1978; Mani, 1989). A generation is completed in about five weeks under warm conditions. In temperate region during winter, the species survives as eggs (Bartlett, 1978) or other stages, both on the host plant and in the soil (Pollard, 1995). There are as many as 15 generations per year (Pollard, 1995).

Small 'crawlers' are readily transported by water, wind or animal agents. Crawlers settle in cracks and crevices of new host plant and in which densely packed colonies develop. There are three immature instars in the female and four in the male. Development from egg to adult occurs in 23-35 days at 37°C (Pluke et al., 1999). Reproduction is mostly parthenogenetic in Egypt (Hall, 1921) and Bihar (India) (Singh and Ghosh, 1970), but *M. hirsutus* is bi-parental in West Bengal (India) (Ghose, 1971a; 1971b) and probably in the Caribbean (Williams, 1996). The males are attracted to females over a wide distance due to the production of sex pheromones by virgin females (Zhang et al., 2004). Infestations of *M. hirsutus* are often associated with attendant ants (Ghose, 1970; Mani, 1989).

Management/Control measures

Biological control

Although, there are several recommendations to use biocontrol agents for management of pink mealy bugs in different crops, their efficacy in most cases are not promising. For example, *Cryptolaemus montrouzieri* is suggested as a very good candidate for the biological control of M. hirsutus in fibre crops (Mani and Satpathy, 2016). Moreover, the release of lady beetle, C. montrouzieri as an effective measure, used at the vineyards in southern India in the 1970s. But, failure was occurred as *M. hirsutus* could not survive in winter and attacked by several parasites and predators. It is attacked worldwide by 21 parasites and 41 predators (Sagarra and Peterkin, 1999). Moreover, conservation of natural enemies in field has been suggested in many cases. For example, it is suggested to conserve *Scymnus pallidicollis* (Coccinellidae) as it is the most efficient predator that feeds vigorously on the eggs, nymphs and adult females of pink mealy bug. However, its population only increases with the increase of mealy bug population. Thus, it is never able to give satisfactory control of mealy bugs of a given field.

Chemical control

Chemical control is the most effective measure for the mealy bugs. There are several insecticides used for this purpose. Spraying of methyl demeton at 0.2 % is found highly effective in controlling populations of *M. hirsutus* on roselle (Mani and Satpathy, 2016). The insecticide, dimethoate spray @ 1.25 L dimethoate 30EC in 625 L of water per ha gives satisfactory control of Pink hibiscus mealy bug.

ii) Cotton mealybug (*Phenacoccus solenopsis*)*

Occurrence & severity

The occurrence of the pest in mesta is of recent origin. During the cropping season of 2009, a survey report from Bashirhat (North 24 Parganas, India) shows that the mealy bug (*Phenacoccus solenopsis*) infestation in mesta (cv. Local) plants was 60% with damage intensity 4% (Satpathy *et al.* 2013). The morphological features of the mealybug include short to medium sized waxy filaments around the body, anal filaments about 1/4th length of body and the two dark stripes on either sides of the middle "ridge" of the body, long glassy rods are present on the back. They produce an egg mass or ovisac (Satpathy *et al.* 2013). The warm and dry conditions during summer is reported as pre-disposing factor for mealybug outbreak on mesta. For the suppression of this mealy bug attack the predator, *Aenasius bombawalei*, is suggested (Mani and Satpathy, 2016).

4. Spiral borer

Occurrence & severity

Spiral borer is a serious pest of "mesta" (Kenaf) in India (Dutt and Bhattacharjee, 1960). The pest occurs in West Bengal (India) (Dutt and Mitra, 1954), Bangladesh (Bhuiyuan, 1981), Java (Leefmans, 1923), and other south East Asian countries. In West Bengal, the pest infestation is more if kenaf is grown in April (Dutt and Bhattacharjee, 1960) or up to mid May in India (Tripathi and Ram, 1969). The extent of damage by spiral borer had been recorded as 3 to 5% in 1975 and 1976. An alarming higher infestation by the spiral borer to the extent of 89.6 percent was observed at the Dacca central station farm of Bangladesh Jute Research Institute in 1977 (Bhuiyan, 1981). The pest affects a number of different indigenous and exotic types of *Hibiscus cannabinus*. *H. sabdariffa* var. *altissima* is also susceptible but to a much lesser degree. A number of olitorius types of jute are also found to be affected by the pest (Tripathi and Ram, 1969).

Types of damage

The newly hatched larva burrows beneath the cambium layer and starts feeding upon the woody tissues, making a spiral around the main stem beneath the bark. During feeding, the larva travels spirally throughout the entire length of the stem. The larva damages the inner fibre layers and interrupts the flow of nutrient sap of the plant. The spiral ring of stem as made by the larva can be recognised because of callus formation which results in some slight swelling above its path. A portion of the infested region swells up to form an elongated

---

*[N.B. The Cotton mealy bug has been well described under Jute pest in the preceding chapter]

gall (Leefmans, 1923; Dutt and Mitra, 1954; Datta and Dutt, 1968). The lignification in the cell wall is affected so the stem at the level of gall becomes very weak and breaks by a strong gust of wind (Datta and Dutt, 1968). The portion of the stem above the region of the gall dies and dries up after the break. The galls are usually 9-15 cm long, ranging between 3 and 30 cm. It is initiated by late stage larva before it tunnels into the woody tissue to form the pupal chamber (Dutt and Bhattacharjee, 1960; Tripathi and Ram, 1969). On roselle, galls are not usually formed (Dutt and Bhattacharjee, 1960).

Scientific name & order: *Agrilus acutus* **Thunberg, 1787 (Coleoptera: Buprestidae)**

(Jendek, 2004)

Synonym

= *Buprestis acuta* Thunberg, 1787

Biology

Egg: The eggs are ovate and scale like, and are laid singly on the stem, usually near a leaf scar. Egg period 10 -12 (Av. 12.4) days (Dutt, 1969).

larva: The larva bores through the lower surface of the egg directly into the stem and feeds under the bark forming a spiral tunnel. There are normally three moults in the active feeding stages, but rarely four, and a prepupal moult. The larval stage lasts for 26 days. The fully grown larva is about 21 mm. Long.

Pupa: Mature larva bores into the wood to pupate, making a pupal chamber of about 11 mm. long into which it fits itself by adopting an asymmetrical U-shaped posture. After a prepupal stage of 2.5 days it pupates. The pupal period occupies about 11 days and a further seven days are spent by the adult in the pupal chamber.

Adult: Adult beetle is minute, brown or blue metalic in colour with length 5.5 - 8.9 mm. Tarsal claws are bifid, cleft nearly to base, teeth subeaqual in form and length, outer tooth straight, inner one curved inside forming with inner tooth of opposite claw pincette like configuration. Last tergite is acuminate on apex. Ovipositor is short, subquadrate; pronotal sides subarcuate. Elytra is glabrous or with two transversal stripes of white ornamental pubescence. Laterosternites and episternites are with white tomentose patches. Prosternal lobe is subtruncate; elytra concolor in various hues of blue or green, rarely black. Pronotal mediobasal impression is deep. Sexual dichroism is not apparent. Elytral apices are with small or long outer spine. Pronotum is wider (wide/ length = 1.6-1.7) (Thunberg, 1787; Jendek, 2004).

Life cycle: Preoviposition period is about seven days. Incubation period varies within 12 to 14 days at 30°C and 76 per cent relative humidity. Total lifecycle completes in about 60 days at 31°C and 73 per cent relative humidity (Tripathi and Ram, 1969).

Management/ Control measures

Host resistance

The kenaf cultivar, HC-583, shows high degree of resistance to the insect. In a study, it was found that there was no incidence up to two months' age, slight incidence at three months' age, and about 10 per cent infestation at four months age of plants of the same variety (Tripathi and Ram, 1969). Several varieties of *H. sabdariffa* are also found as tolerant to the insect, showing very little incidence even at five months' age of plants (Tripathi and Ram, 1969).

Chemical control

Preventive measure may be taken at the early period of infestation. The pest is easily controlled by spraying with Dichrotophos (Bidrin-8) at 340g (355ml) or Methylparathion at 454g (473ml) per acre. The incidence of pest can be reduced by shifting the period of time of sowing. Spraying of Oxydemeton methyl (Metasystox 25EC) @ 300ml/ha may be effective. Since, it is found that the pest is best controlled when 20 ml per plant of 2.5 per cent Metasystox is applied by cotton swab method on the stem at a height of 37.5 cm from base of the affected plants containing borer inside (Tripathi and Ram, 1971). The pest can also be controlled by Endrin 20EC (0.04%). Carbofuran3G @ 1kg/ha is very effective. Other chemicals, Fipronil 5 % EC @ 1 mllt, Acephate 75 % SP @ 0.75 gm/lit, Triazophos 20 % EC @ 2 ml/lit and Acephate 50 % +Imidachloprid 1.8 % SP @ 2 gm / litre of water, may also be sprayed to control the pest.

5. Mesta stem weevil

Occurrence & severity

Mesta stem weevil is also known as Cotton shoot weevil. It is one of the major pests of mesta (*H. cannabinus*) in India. The younger plants are especially susceptible to this borer. It occurs in Jammu and Kashmir (Beeson, 1919), Punjab (Hussain, 1925), Dehradun (Gardner, 1934); Bengal (including Bangladesh) (Beeson, 1919), South India (Ayyar, 1922), including Coimbatore (Subramanian, 1959) and Karnataka (Thippeswamy *et al.*, 1980), as well as in Ceylon (Hutson, 1930). It causes damage to cotton (*Gossypium* sp.), bhendi (*Hibiscus esculentus*) and mesta/kenaf (*H. cannabinus*) (Aurivillius, 1891; Fletcher ,1919; Ayyar, 1922). In most cases, it is considered as a minor pest but in South India it is the key pest (Deshpande at al., 1994; Regupathy *et al.*, 1997; Halikatti and Patil, 2015). It is a polyphagous pest (Sharma *et al.*, 2012), causing major damage to both kenaf (mesta) and cotton all over the south India including Karnataka (Deshpande *et al.*, 1994; Arora *et al.*, 2010). The rainy season, especially the month of July, is more conducive for breeding of the pest.

Types of damage

Adults feed by scooping the tissue of succulent shoots towards the tip, leaf buds and petioles. During infestation, mature female oviposits by notching out holes with the help of its snout in the tender shoots, petioles or leaf buds. After ovipositing, the mouth of the holes is sealed with a yellow secretion which turns black in a day or two. The feeding of the grub around the site of injury results into a gall like swelling. A frothy secretion oozes out of the holes in the stem due to larval feeding inside the stem. Thus, the grubs cause serious damage to the plant by boring the stems and petioles and feeding on them, resulting in a gall like swelling around the bore hole. Exit holes for the emergence of adults and tunnelled stems are seen after longitudinal splitting. Too many exit holes made by the grubs in heavy damage weakens the stem resulting into its rupture and breakage further reducing yield (Subramanian ,1959; Thimmaiah *et al.*, 1975; Thippeswamy *et al.*, 1992; Sharma *et al.*, 2012).

Scientific name & order: *Alcidodes affaber* **Aurivillius 1891 (Coleoptera: Curculionidae)**

Synonym

= *Alcides affaber* (Aurivillius)

Biology

Egg: Egg is creamy white to light yellowish, oblong with anterior end slightly narrower, having minute pits on the chorion and laid in hollow scoop on the petiole. Egg turns to light brownish on the penultimate day with the two mandibles of the developing grub clearly visible through the chorion. It measures 1.09 ± 0.07 (range: 1.02-1.19) mm in length and 0.60 ± 0.036 (range: 0.56-0.64) mm in width (Devaiah *et al.*, 1981; Sharma *et al.*, 2012).

Grub: Grab is with dark brownish head having a prominent epicranial suture. Body is 'C' shaped at rest with wrinkled and orange brown eyes (Devaiah *et al.*, 1981). There are nine grub instars are recorded, without showing much difference in general characters among the different grub instars but vary in the size of body and head capsule. First instar grub is apodous with soft C-shaped body, pale yellow in colour and best with soft hairs. Head is smooth, freely movable, pale brown with a median dark line on the posterior end of frons. Mandibles are dark brown and bifid. Body is glabrous, wrinkled with posterior end more or less rounded. The terga of thorax and abdomen except that of prothorax and last and second last segments are divided by one or two grooves into two or more folds. The fully grown grub is creamy yellow, apodous with stout, cylindrical, moderately curved and wrinkled body. Head capsule is dark, subcircular with irregular and deeply pitted surface. Frons bears transverse sculpture on the surface along with a dark streak posteriorly which extends forwards to a quarter of its length and bears five pairs of setae. Mouth parts are bitting and chewing type. Prothorax is undivided dorsally with prescutal and scutal areas that roughly indicated by rows of setae. Meso and metathorax are divisible into prescutellum and scutoscutellum. The former one is with two small setae and the latter one with four setae set in a straight line. Abdomen is 10 segmented with three distinct transverse folds viz. prescutellum, scutum and scutellum. Each epiplural lobe of abdomen is with a single setae and each hypoplural lobe with two setae. The last two segments are simple provided with a number of hairs. Spiracles are short, circular tubes and do not project beyond the peritreme. Posterior spiracles are placed more dorsally whereas anterior ones are placed laterally (Sharma *et al.*, 2012).

Pupa: Pupa is exarate with appendages projecting freely on the ventral surface. Colour is creamy yellow, turns darker with age. Head is as long as broad, provided with five pairs of setae that arising from minute tubercles. One pair of setae is situated near the base, two pairs situated behind the eyes and two pairs situated between the eyes. Rostrum is about one fourth of the total length of the body and bears a pair of setae. Prothorax is one and a half times broader than its length having nine pairs of setae. Mesothorax is as long as wide and bears two pairs of setae. Metathorax is one and a half times as long as wide. Abdomen is nine segmented with each of first eight segments bearing six pairs of setae dorsally. Last segment bears a pair of curved pleural processes. Ventral surface is bare. Length of pupa varies from 8 mm to 9.8 mm (Sharma *et al.*, 2012).

Adult: Freshly emerged adult is soft and chocolate brown, turning hard and steel grey. Later greenish yellow patterns develop on the elytra and scales on the body (Devaiah *et al.*, 1981). Male is smaller in both length and width, snout is shorter and stouter than the female. Body is measured as 9.0 -11.3 mm in length and 3.10- 3.40 mm in width. Longivity in case of male is about 28 days, while, for female it is 39 days (Thippeswamy *et al.*, 1992). Geniculate antenna arises one on either side in a groove in the middle of the snout. Scape is long and enlarged at the apex. Funicle is six segmented. Club is five segmented with conical terminal segment. Funicle and club bear setae of varied length. Mouth parts are mandibulate, located at the tip of snout and enclosed in a sheath.

Mandibles, maxilla and labium are clearly visible. Mandible is dark brown, tridentate with the middle denticle being the most prominent. Prothorax is large, narrow at the apex, broad towards the middle with tubercle like elevations all over. Mesothorax bears a pair of elytra. Elytra are dark brown, convex with dark margins and longitudinal rows of punctuations alternating with rows of fine setae. Metathorax bears a pair of membranous hind wings. Ventrally each thoracic segment bears a pair of dissimilar legs. Fore coxa is triangular and broad at the base. Mid coxa is roughly circular and bulged ventrally. Hind coxa is triangular and broad at the apex. Prothoracic femur is better developed than others. Vedge shaped femora is with a prominent spine ventrally. Spine of metathoracic leg is less developed than others. All tibias possess a prominent process postero-ventrally. Tarsi are 4-jointed and end with a pair of bifid claws. Abdominal segments are five with largest last segment (Sharma *et al.*, 2012).

Life cycle: Eggs are laid on tender shoots and petioles in 8-10 days after emergence of adult. Total life cycle lasts for 72-80 days, duration of each stage varying from 6-7 days for eggs, 55-62 days for grub and 10-12 days for pupa. Life of adult is varied from 6-43 days for females and 8-32 days for males (Subramanian, 1959). The weevil has one generation (Subramanian, 1959; Sharma, 2012) to two generations (Thippeswamy *et al.*, 1992) in a year.

Management/Control measures

For management of the weevil followings are recommended:

1. Basal application of FYM 25 t/ha or 250 kg/ha of neem cake.
2. Soil application of Carbofuran 3 G at 30 kg/ ha on 20 DAS and earthed up.

3
Sunnhemp (*Crotalaria juncea*) -Fabales: Fabaceae

A. Diseases

1. Anthracnose

Occurrence & severity status

The Anthracnose [Fig. 14 (A-B)], also called as 'Stem-break', is one of the most serious diseases of sunnhemp (*Crotalaria juncea*) in India. It occurs in all sunnhemp growing areas of India (Mitra 1934; Kempanna *et al*., 1960; Kundu, 1964; Kumar *et al*., 2016). Apart from India, its occurrence in several countries of South Tropical Africa, like, Rhodesia (Whiteside,1955), Zimbabwe, and other tropical countries including Trinidad and Tobago (Damm *et al*., 2009), was also noticed as a major disease (Purseglove, 1974). In many places, the disease is considered as the second most serious disease of sunnhemp, causing wilt and weakening of the stem.

Fig. 14 (A-B). Anthracnose disease caused by *Colletotrichum curvatum* in Sunnhemp. **A.** Showing infection on leaf; **B.** Showing infection on stem.

Symptoms

The disease at first appears on cotyledons of seedling, showing soft and discoloured zone. This is followed by discolouration of stem and growing tip. Later, the whole plant parts except underground portion show brownish spots on them. The affected seedling drops from the point below cotyledons and eventually dies. In case of older plant infection, the infected plants show necrotic symptoms on leaf and stem. The necrotic spots are greyish brown to dark brown, round to irregular in shape and they coalesce to form large spots and cover the entire leaves and larger portion of stem. The spots on older leaves appear at first on one side of leaf blade and later they extend to the opposite side (Kumar *et al.*, 2016).

Causal organism: *Colletotrichum curvatum* Briant & E.B. Martyn, 1929
[Tropical Agriculture 6: 258]

Synonym

= *Colletotrichum truncatum*

[**N.B.** *Colletottrichum curvatum* is an accepted name for this fungus; vide, Kirk, P.M. (2018). Species Fungorum , Kew Mycology]

Morphological/ Morphophysiolgical characteristics

Vegetative hyphae on synthetic nutrient-poor agar medium (SNA) are hyaline, septate, branched, 1-8µm in diam. Chlamydospores are not found. Conidiomata is acervular. Conidiophores and setae are formed directly on hyphae. Setae are hyaline to pale brown, smooth to verruculose, 80-150µm long, 2-5 septate, tapering only little towards the slightly acute to roundish tip, base cylindrical to conical, 4-6µm diam. Conidiophores are hyaline to pale brown, septate, strongly branched, densely clustered, up to 90µm long. Conidiogenous cells are enteroblastic, hyaline to pale brown, cylindrical, 6-20 x 2.5-4µm, opening 1.5-2µm diam., collarette rarely visible, 0.5µm long, periclinal thickening not observed. Conidia are hyaline, smooth walled to verruculose, aseptate, long central part of conidia slightly curved with parallel walls, ending abruptly at the round and truncate base, while tapering towards the acute and more strongly curved apex with granular content, (16.5-) 20-23.5 (-26) x (3-) 3.5-4(-4.5) µm, L/W ratio = 5.7 (described as synonym of *Colletotrichum truncatum* by Damm *et al.*, 2009).

Colonies on SNA are flat with entire margin, no aerial mycelium, filter paper stained slightly saffron, 14 mm in 7 days (23 mm in 10 days). Colonies on oat meal (OA) medium are flat with entire margin, no aerial mycelium, surface buff, covered with olivaceous grey to iron grey acervuli, reverse buff to pale saffron (Damm *et al.*, 2009).

Molecular characteristics

Molecular characterization of *C. curvatum* (= *C. truncatum*) isolated from sunnhemp has been determined (Damm *et al*., 2009). The GenBank accessions for ITS is GU22876, while, ACT: GU22974, Tub2: GU228170, CHS-1: GU228366, GADPH: GU228268 and HIS3: GU228072. The multilocus molecular phylogenetic analysis (ITS, ACT, Tub2, CHS-1, GAPDH, HIS3) reveals that *C. curvatum* along with other *C truncatum* isolates is in Clade-6 of 20 clades formed with *Colletotrichum* species of curved conidia (Damm *et al*., 2009).

Predisposing factors

High temperature (30-32° C), cloudy weather and continuous rain favour the disease. Seedling stage is more vulnerable to be attacked by the disease (Kumar *et al*., 2016).

Disease cycle

The spore spreads by rain splash and infects new plant (Kumar *et al*., 2016).

Control measures

Seed treatment with carbendazim (@ 2g/kg seed) is advocated to control seed borne disease. Whiteside (1955) recommends the use of disease free seed, seed treatment with fungicide, early planting and crop rotations. Kundu (1964) recommends seed treatment with fungicide, early planting and spraying the crop with Bordeaux mixture. But the spraying of Bordeaux mixture in agricultural crop should not be advised due to its phytotoxic effect and that is now not in practice. Instead, copper oxychloride (Blitox) may be sprayed. Further, spraying of carbendazim @ 1.5% in order to control secondary infection is advised (Kumar *et al*., 2016). The efficacy of carbendazim spray may be increased with the addition of mancozeb (0.2%), thus, simultaneous spray may be recommended.

2. Leaf spot

Occurrence & severity status

Leaf spot [Fig. 15(A-B)], also called as *Cercospora* Leaf spot, occurs in Assam, Bangladesh, Guinea, Hong Kong, Madhya Pradesh, New Caledonia, Papua New Guinea, Philippines, Puerto Rico, Sabah, Singapore, Sri Lanka, Taiwan, Uttar Pradesh, Venezuela and West Bengal (Chupp, 1954), and probably in all other tropical countries growing sunnhemp.

Fig. 15 (A-B). Leaf spot disease caused by *Cercospora crotalariae* in Sunnhemp.
A. Dorsal view B. Ventral view

Symptoms

Leaf spots are subcircular (subobicular) to irregular, 4-10 mm. in diameter or width, dingy grey to pale tan, with dark brown very narrow raised margin. Fruitings are amphigenous and chiefly epiphyllous (Chupp, 1954; Hsieh and Goh, 1990).

Causal organism: *Cercospora crotalariae* Sacc., 1913

[Syll. Fung. 22: 129]

Synonyms

= *Cercospora crotalariae-junceae* Sawada, 1942.

= *Mycosphaerella crotalariae* (Petch) Hans., 1942. (teleomorph)

= *Phyllosticta crotalariae* Sacc. & Trotter, (1913)

= *Sphaerella crotalariae* Petch, (1906)

Morphological/ Morphophysiolgical characteristics

Fruiting is amphigenous, chiefly epiphyllous. Stromata are dark brown to black, globular, small, 30-60 µm (mostly up to 40µm) in diameter; Conidiophores are in a dense fascicle of 2-18, pale to medium brown, paler towards the narrow apex, multiseptate, not branched, straight, occasionally once geniculate, subtruncate at the apex, (20-) 45-110 x 4-6 µm; medium conidial scar at the tip, scars conspicuously thickened. Conidia are hyaline, acicular to obclavate, straight to slightly curved, indistinctly multiseptate, base truncate, tip acute or subacute, 40-l00 (-160) x 2.5-4 µm in size (Chupp, 1954; Hsieh and Goh, 1990).

Control measures

Seed treatment with Captan 3g/kg of seed and spraying of copper oxychloride 50 WP (Blitox) @ 3g/l of water may be effective to manage the disease.

Alternatively, spraying carbendazim (50 WP) @ 1g/l or mancozeb (75 WP) @ 3g/l water may be used to manage the disease. Further, the mixture of both carbendazim (0.05% a.i.) and mancozeb (0.2% a.i.) may be sprayed in severe case.

3. Powdery mildews

i) Powdery Mildew (*Golovinomyces cichoracearum*)

Occurrence & severity status

Powdery mildew [Fig. 16] was first observed on sunnhemp in a research field in Hastings (Florida) (Gevens *et al.*, 2009). The pathogen is distributed worldwide and occurs on different host plants. This disease of sunnhemp is important because it has the potential to impact the health and quality of sunnhemp, and this particular powdery mildew can infect cucurbits (Gevens *et al.*, 2009).

Fig. 16. Powdery mildew disease, caused by *Golovinomyces cichoracearum* in Sunnhemp

Symptoms

The disease appears as white, powdery growth on both upper and lower sides of leaves. Initially, the powdery mildew colonies appear as small, round, reddish spots on the adaxial surface of leaf, especially along the mid vein of the infected leaf. Subsequently, those colonies widen and coalesce and move to abaxial surface showing conspicuous hyaline ectophytic mycelia and conidial growth on leaf surfaces, keeping, petiole and floral parts are disease free. Powdery mildew is easily recognized by the conspicuous epiphytic mycelium. The mycelial growth continues throughout the life of the plant. The disease causes slight twisting of the foliage (Koike and Saenz, 1996; Meeus and Wittouck, 1999).

Heavily infected leaves senesce and abscise. Infection is primarily seen on the lower mature leaves of plants and not on the top 0.6 m (2 feet) of the plant (Gevens *et al.*, 2009).

Causal organism: ***Golovinomyces cichoracearum*** **(DC.) V.P. Heluta, 1988**
[*Ukr. bot. Zh.* 45(5): 62]

Synonym

= *Erysiphe cichoracearum* DC. (1805)

Morphological/ Morphophysiolgical characteristics

Mycelia produce white accumulations of conidiophores and conidia. Hyphae are superficial with papillate appresoria, which produce conidiophores with cylindrical foot cells, measuring 48.5 × 10.0 µm, and short chains of conidia. Conidia are hyaline, short-cylindrical to ovoid, lack of fibrosin bodies, borne in chains, have sinuate edge lines with other immature conidia, and measuring 22.5- 40.0 (av. 29.85) × 12.5- 20.0 (av. 15.55) µm. The teleomorph is not observed (Gevens *et al.*, 2009).

Molecular characteristics

In a study of Gevens *et al.* (2009), the nuclear rDNA internal transcribed spacer (ITS) regions were amplified by PCR, using universal primers ITS1 and ITS4, and the amplified products were sequenced (GenBank Accession No. FJ479803). The ITS sequence data indicated 100% homology with *G. cichoracearum* from *Helianthus annus* (GenBank Accession No. AB077679). The causal fungus of sunnhemp powdery mildew was identified as *G. cichoracearum* of the classification *Golovinomyces* Clade III (Gevens *et al.*, 2009).

Predisposing factors

Excessive nitrogen fertilizer promotes the disease.

Disease cycle

The disease is polycyclic. The conidia formed on different hosts and on plant residues are the primary inocula of the disease. The conidia germinate on host surface and establish infection. The conidia produced on the infection of host also serve for secondary infection.

Management/Control measures

Ploughing in or removing plant residues after harvest, crop rotations, and the use of resistant cultivars are recommended for control of the disease (Ivancia *et al.*, 1992).

The powdery mildew can be effectively controlled by common fungicides (e.g. sulphur, myclobutanil, quinoline, strobilurins, *etc.*) and protective compounds (e.g. extract of neem oil, *Reynoutrias achaliensis* extracts) as evident in lettuce affected by same fungus (Lebeda and Mieslerova, 2010).

ii) Powdery mildew (*Erysiphe diffusa*)

Occurrence & severity status

The powdery mildew (*Erysiphe diffusa*) was reported as a serious disease of sunnhemp in USA by Farr *et al.* (1989). It occurs on different host plants including *Crotalaria juncea, C. intermedia* and *C. retusa*. However, its occurrence in Asia is very rare.

Symptoms

The disease appears as small, circular, whitish powdery spots on the dorsal surface. Subsequently, it covers both the dorsal and ventral surfaces of the leaves in patches and proceeds along the vein, stems, petioles and pods as the disease progressed. In severely infected plants, the infected leaves abscise and yield decreases. Sometimes, chlorosis or yellowing, scorching of the leaves, rusty patches on the ventral surface of leaves and premature defoliation may be seen.

Causal organism: *Erysiphe diffusa* (Cooke & Peck) U. Braun & S. Takam., 2000

[Schlechtendalia 4: 7]

Synonyms

= *Microsphaera diffusa* Cooke & Peck, (1872)

= *Trichocladia diffusa* (C. & P.) Jaczewski

Morphological/ Morphophysiolgical characteristics

Mycelium on leaves is amphigenous, thin, effused, occasionally patches, evanescent to persistent. Conidiophores are erect, straight to curved, sometimes flexuous, up to 100 µm, foot-cells cylindric, (24-) 25-38 (-44) x (5-) 7.5-10 (-11) µm, followed by 1-2 shorter cells. Conidia are formed singly, oval to ellipsoid-cylindric (-doliform), 25-35(-40) x (10-)11-17.5 (-20) µm, without fibrosin bodies (Braun, 1987; Fu *et al.*, 2015). Cleistothecia are scattered to gregarious, (65-) 75-135 µm in diam, sometimes larger, cells irregularly polygonal, 10-25(-30) µm diam. Appendages are more or less equatorially arising, 7-30 per ascocarp, stiff to flexuous, horizontally spread, 1-3 times as long as the cleistothecial diam, 6-11 µm wide, hyaline or faintly coloured at the base, aseptate or with 1-3 septa near the base, septa occasionally remote from the base and coloured up to the middle of the stalk, smooth, rarely faintly verrucose below, thin-walled but thicker towards the base, apex 3-6 times branched, branchings rather loose, diffuse, often deeply cleft, divergent, tips straight, not recurved. Branchings of early developed ones are with apexes of mostly richly branched. Asci are 4-10, sessile or shortly stalked, 40-70 x 25-45(-50) µm, 3-6-spored. Ascospores are ellipsoid-ovoid, 16-24 x 9-15 µm (Braun, 1987).

Molecular characteristics

The molecular sequences of *E. diffusa* (EF196675) from soybean has been deposited in Genbank by Almeida *et al.* (2008) from Brazil. The ITS sequence is different from that of *E. glycenes*. In another study, complete internal transcribed spacer (ITS) region of rDNA of the fungus was fully amplified by seminested PCR with the primers ITS5/P3 followed by ITS5/ITS4, then sequenced (Fu *et al.*, 2015). The resulting 559-bp sequences (Accession No. KM 260363) showed 100% similarity to *E. diffusa* (Accession Nos AB078804 and AB078813) (Fu *et al.*, 2015).

Predisposing factors

It affects more during high humidity and moderate temperatures.

Disease cycle

Same as other powdery mildew as described earlier.

Control measures

Several fungicides can control powdery mildew effectively. However, due to cost effectiveness, decision should be made based on more economically prudent on field grown crop. The fungicides, hexaconazole, myclobutanil, penconazole, tridimefon and propiconazole, are effective against the fungus.

4. Rust

Occurrence & severity status

The rust is said to cause considerable damage in India and an outbreak has also been recorded in the Ivory Coast (Africa). The occurrence of this disease has been recorded in Africa (Ivory Coast, Ghana, Guinea) and Assia (Ceylon, Japan, India, Pakistan) on different species of *Crotalaria* (*Crotalaria albida, C. juncea, C. mascilloseis, C. maxillaris, C. medicaginea, C. medicapinea, C. orixensis, C. pumila, C. retusa, C. shrica, C. verrucosa* and *C. vitatoni*) (Joshi, 1960; Punithalingam, 1968b).

Symptoms

The disease symptoms appear as pale to dark brown erumpent rust pustules on both stem and leaves. The pustules are of uredinia, which are mostly hypophyllous, subepidermal and scattered.

Causal organism: *Uromyces decoratus* Syd., 1907 [Annls mycol., 5:491]

Synonyms

= *Nigredo decorata* (Syd. & P. Syd.) Arthur, 1926

Morphological/ Morphophysiolgical characteristics

Uredia are hypophyllous, rarely epiphyllous, numerous, scattered, pulverulent and yellowish-brown. Uredia are also formed on other green parts of the plant. Urediospores are globose to subglobose or ellipsoid, 21-26 μm diam.; wall 1.5-2 μm thick, light brown, echinulate, with more or less equatorial 4-6 germpores. Telia are hypophyllous, smaller than uredia, dark brown to black, erumpent, scattered or circinate, pulverulent. Teliospores are globose to oblong, single celled, 20-32 x 14(16 -) -20 (-26) μm; wall uniformly 1.5-2 μm thick, verrucose, apex hyaline, 3-5 μm thick. Pedicel is hyaline, slender, easily detached. Spermogonia and aecia are unknown (Butler, 1918; Punithalingam, 1968b).

Physiological specialization: Joshi (1960) inoculated 8 species of *Crotalaria* with urediospores of *Uromyces decoratus* which orginially collected from *C. juncea* and found 5 species were susceptible.

Disease cycle

It has been suggested that in India the rust could survive the summer in the lower hills either on regular crops or on intermediate hosts such as *C. medicaginea*, reported to be a perennial in the hills (Joshi, 1960).

Control measures:

Sharma (1985) demonstrated the antagonistic ability of some phylloplane fungi against sunnhemp rust (*Uromyces decoratus*).

5. Root and stem rot

Occurrence & severity status

A root and stem rot disease was reported by Farr *et al.* (1989). The disease is also called as 'Foot rot' or 'Southern blight'. It occurs in almost all tropical countries, including Thailand (Piriyaprin *et al.*, 2007). Sunnhemp is moderately susceptible to the disease (Chaurasia *et al.*, 2014).

Symptoms

The disease is characterized by a conspicuous web of white ropy or fluffy, fan shaped mycelial growth on stem near the soil level as well as on the soil around the plant. Later, the white mycelia disappear and masses of small, hard, round to ovoid, light to dark-brown sclerotia, about the size of mustard seeds (1-2mm dia.) are formed at the collar region. The fungus attacks the root zone. The plant starts wilting and finally collapses.

Causal organism: *Athelia rolfsii* (Curzi) C.C. Tu and Kimbr. 1978

[Kirk, P.M. (2018). Species Fungorum, Oct 2017, Kew Mycology]

Synonyms

= *Botryobasidium rolfsii* (Curzi) Venkatar, 1950

= *Corticum rolfsii* Curzi 1932

= *Pellicularia rolfsii* (Curzi) E. West, 1947

= *Sclerotium rolfsii* Sacc., 1911

Morphological/ Morphophysiolgical characteristics

Colonies on PDA medium are white, usually with many narrow mycelia strands in the aerial mycelium. Cells of primary hyphae at advancing zone of colony usually are with one or more clamp connections at septa. Secondary hyphae are arisen immediately below the distal septum of cell and often grow adpressed to the primary hyphe. Tertiary and subsequent branches are narrow, having comparatively short cells and a wide angle of branching not closely associated with septation and are usually without clamps. Sclerotia are developed on colony surface, near spherical, slightly flatten below, mostly 1-2mm across when fresh, shrinking slightly when dry, with smooth or shallow pitted shiny surface. On agar media white to ochraceous stromata are sparse or aggregated in irregular masses. Fructifications are dense, forming crustose hymenial layer. Basidia are clavate. Basidiospores are smooth, hyaline, globose-pyriform, 4.5-6.5 x 3.5-4.5µm (Mordue, 1974).

The fungal culture on PDA medium is appeared as with abundant, radial growth of white, septate, branched mycelia in about 4 days. The fungus grows rapidly, covering the entire petriplates in about 9-10 days and later, small, dense, white, pin-head like structures are formed, which modify gradually in colour and size and finally become brown or black coloured sclerotia within 10-12 days.

Molecular characteristics

Parsimony analysis of LSU and ITS sequences both suggests that there are close relationships among *Sclerotium rolfsii*, *S. rolfsii* var. *delphini* and *S. coffeicola*. They are grouped with *Athelia* species with 99% bootstrap support (Xu *et al.*, 2010).

Predisposing factors

The temperature range 25-35°C is optimum for the growth of the fungus. The sclerotia germinate only near 100% relative humidity. Light, except some blue-green and ultra-violet wavelengths, favours the mycelia growth of the fungus (Mordue, 1974).

Disease cycle

The pathogen survives as sclerota on plant debris in soil. The sclerotia germinate on soil and grow there for a certain period before parasitization. After establishment on soil, the mycelia infect stem of plant near soil line. Sclerotia are produced again on the infected plants and serve as inocula for next year.

Control measures

The biological agent, *Gliocladium virens* is effective to minimize root and stem rot disease in sunnhemp by the mechanisms, like, antibiosis, competition and mycoparasitism (Piriyaprin *et al.*, 2007). Antagonism of *Trichoderma harzianum* against *S. rolfsii* has also been observed (Khattabi *et al.*, 2001). Hexaconazole is the best fungicide to control the disease along with *T. harzianum*, since; it has negligible effect on the antagonist (Khattabi *et al.*, 2001).

6. Twig blight

Occurrence & severity status

The disease is also called as '*Choanephora* leaf blight' or 'Tip rot'. It occurs sporadically and its sporadic incidence has been reported by Tripathi (1973) from Varanasi (U.P., India). In Sunnhemp Research Station, Pratapgarh, U.P., the occurrence of the disease is in regular feature during high humid conditions. About 20% incidence of this disease is recorded in the month of August (Sarkar *et al.*, 2015). The disease also occurs on several wild species of *Crotalaria*, like *C. spectabillis*, *C. sericea* and *C. retusa* (Patel *et al.*, 1983, Bandopadhyay and Mitra, 1994, Sarkar *et al.*, 2015). It causes flower and stemblight on *C. spectabillis* in Brazil (Alfens *et al.*, 2018).

Symptoms

This disease is characterized by rotting of tender shoots. Brown discolouration occurs just below the infection point. Affected portions decay, break and droop. White mycelial growth along with black coloured sexual bodies of the fungus is seen on the affected parts from the leaf tip towards petioles. A characteristic whitish brown discolouration appears in leaves on advancement of disease development. Infected leaves loss their chlorophyll and droop from the stem. The pathogen affects the epidermal and outer cortical layers, and the affected cells are disintegrated, (Tripathi *et al.*, 1975; Sarkar *et al.*, 2015);

Causal organism: *Choanephora cucurbitarum* (Berk & Ravenel) Thaxt., 1903

Synonyms

= *Choanephora americana* A. Moller, (1901)

= *Choanephora heterospora* B.S. Mehrotra & M.D. Mehrotra, 1962

= *Mucor cucurbitarum* Berk & Ravenel, (1875)

= *Rhopalomyces elegans* var *cucurbitarium* (Berk & Ravenel) Marchal, (1893)

Morphological/ Morphophysiolgical characteristics

The fungus grows very fast on potato dextrose agar (PDA) medium at 28°C (Bandopadhyay and Som, 1984). It also grows very rapidly on Potato-Carrot Agar (PCA) medium (Krik, 1984). Colonies on PCA are white at first but later turns pale yellow. Aerial mycelia are abundant, nonseptate. Sporangiophores with sporangia arise from substrate mycelium of aerial hyphae, erect or ascending, nonseptate, unbranched, 1-10mm high, hyaline and smooth at the base of 6-12μm wide. Sporangia are with few or many spores, often numerous in old cultures, larger sporangia circinately borne, smaller sporangia borne on straight sporangiophores, spherical, (25-) 40-160 (-200) μm in diameter, initially white but turning through yellow and pale brown to very dark brown (appearing black by reflected light) at maturity. Sporangial wall is persistent, encrusted with more or less capitates spines up to 1μm high, or appearing smooth, dehisces at a single perforated meridional suture into 2 (rarely 3 or 4) segments. Sporangiospores are ellipsoid to broadly ellipsoid, brown or reddish brown to pale brown, distinctly or indistinctly longitudinally striate, with or apparently without a group of fine hyaline appendages at each pole often up to twice the length of spore, spore content granular, (12-) 16-26 (-48) x (7-) 8-12 (-20)μm in size. Zygospores are formed from tongs like often slightly unequal, apposed suspensors, brown to reddish brown, translucent, with smooth thin outer layer and a striate thick inner layer, 30-105μm in diameter (Kirk, 1984).

Molecular characteristics

The molecular characterization of the fungus isolating from *Crotalaria spectabillis* was done in Brazil by Alfenas *et al.* (2018). In their studies, the internal transcribed spacer of ribosomal DNA (rDNA) sequences was generated using ITS1 and ITS4 primars and the sequences were deposited in GenBank under assession numbers MF942131 and MF942130. The generated sequences had 100% nucleotide identity with corresponding *Choanephora infundibulifera* f. *cucurbitarum* (KJ461159, KM20034, NJ943006). The Neighbor-joining phylogenetic analysis was also performed using Tamura-Nei model in MEGA7 software and that revealed that the tested isolates from *C. spectabillis* were clustered in a clade of *Chanephora cucurbitarum* supported by a high bootstrap value (Alfenas *et al.*, 2018).

Predisposing factors

The disease favours high temperature (30°C) and high humidity (Naito and Sugimoto, 1989).

Disease cycle

The pathogen is soil borne, where it survives as saprophytes. Under favourable conditions numerous spores are produced and disseminated by wind, insects and water. The spores infect the host tissue, which subsequently, repeat in host canopy with the spores formed in the host tissue

Control measures

Mancozeb, dinocarp and thiabendazole may be effective to control the disease (Hammouda, 1988). Panja (1999) has reported that Captan 50% (2.0g/litre), Ziram 27% (1.5ml/litre) and copper oxychloride (50%) (3.0g/litre) can inhibit the mycelial growth and sporulation of *C. cucurbitarum* completely. Thus, these fungicides may also be used in sunnhemp disease control.

7. Wilts

i) *Ceratocystis* wilt

Occurrence & severity status

In Brazil, the only disease reported on the sunnhemp crop is *Ceratocystis* wilt (National Research Council, 1979; Wang *et al.*, 2002). The disease also occurs on diverse species of annual and perennial plants.

Symptoms

The wilt symptoms in sunnhemp are associated with necrotic and elongated lesions on the stem and sporulation near ground level (Holliday, 1980).

Causal organism: *Ceratocystis fimbriata* Ellis & Halst., 1890

[Bulletin of the New York Agricultural Experimental Station: 14]

Synonyms

= *Sphaeronaema fimbriatum* (Ellis & Halst.) Sacc., (1892)

= *Ceratocystis fumbriata* f. *fimbriata* Ellis & Halst. (1890)

= *Ceratocystis moniliformis* f. *coffeae* (Zimm) C. Moreau (1954)

= *Ceratostomella fimbriata* (Ellis & Halst.) J.A. Elliott, (1923)

= *Ophiostoma fimbriatum* (Ellis & Halst.)Nannf., (1934)

= *Endoconidiophora fimbriata* (Ellis & Halst.) R.W. Davidson, (1935)

= *Rostrella coffeae* Zimm., (1900)

Morphological/ Morphophysiolgical characteristics

Perithecia are superficial or partly to completely embedded in the substrate, dark brown to black, globose, (110-) 140-220 (-250) µm diam., unornamented, necks long, straight, dark brown to black, pale brown to subhyaline towards the tip, tapering slightly, 440 to770 (-900) µm long, 28-40 µm wide at the base and 16-24 µm wide at the tip. Ostiolar hyphae are light brown to hyaline, erect or moderately divergent, non septate, smooth walled, 20-120 µm long. Asci are not seen. Ascospores are elliptical, with a gelatinous sheath, forming a brim, giving a hat-shaped appearance, hyaline, non-septate, smooth, 4.5-8.0 x 2.5-5.5 µm. Conidiophores (endoconidiophores) are slender, scattered or arising laterally in clusters from the vegetative hyphae, 1-8 septate, phialidic, hyaline to very pale brown, up to 160 µm long, usually tapering towards the tip and producing a succession of conidia through the open end. Conidia (endoconidia) are cylindrical, unicellular, straight, biguttulate, truncate at the ends, hyaline, smooth-walled, (9-) 11-25 (-33) x (3.0-) 4.0-5.5 µm. Doliform endoconidia are absent. Aleurioconidiophores are arising laterally from the mycelium, with 0-5 septa, 9-98 x 4.5-6.5 µm. Aleurioconidia are brown, globose to pyriform, thick walled, 11-16 x 6.5-12 µm, occurring singly or in short chains (Morgan-Jones, 1967; Engelbrecht and Harrington, 2005).

Isolates from sunnhemp differs in pathogencity with those causing diseases in sweet potato and coffeae (Holliday, 1980).

Molecular characteristics

The isolates of *C. fimbriata* from different plant species differ in ITSrDNA sequences (Baker *et al.*, 2003). There are two closely related sublineages exist within this species, one centered in western Equador and the other containing isolates from Brazil, Colombia and Costa Rica. The two sublineages differ very little in morphology but they are intersterile and have unique microsatellite markers (Steimel *et al.*, 2004) and ITSrDNA sequences (Baker *et al.*, 2003). Their ITS sequences differ from all other isolates from the *C. fimbriata* complex. Representative ITS-rDNA sequences of the Ecuadorian and Brazilian sublineages have been deposited in GenBank with Acc. Nos. AY157950 and AY157951-157953, respectively (Engelbrecht and Harrington, 2005). The draft nuclear genome of *C. fimbriata* is comprised of 29 410 862 bp. *De novo* gene prediction produces 7266 genes. The genome of *C. fimbriata* is relatively small (29.4Mb) and harbours fewer genes for an ascomycete fungus than other fungal species (Wilken *et al.*, 2013).

Predisposing factors

Frequent rains at the temperature range 18-30°C favours the pathogen to cause disease (Huang *et al.*, 2003).

Disease cycle

The pathogen is a soilborne fungus. It survives in soil, in water, and on plant debris in the field as mycelia or as thick walled aleurioconidia (Accordi, 1989). It can survive for several years in the soil. The fungus is spread by wind, water, soil, on harvesting baskets, on farm machinery, by some insects, by humans (clothing), by contaminated tools. It invades in plants through wounds which are important entry points for infection by the fungus. The ascospores of the sporulating fungus accumulate in a sticky drop at the tip of their perithecial necks. The combination of the fruity aroma and sticky nature of the ascospores promotes the fungal dispersal by insect vectors.

Management/Control measures

It is suggested to use disease-free seeds in management of the disease. Therefore, sowing of healthy seeds or seed treatment with carbendazim (0.1% a.i.) may be effective in controlling the disease in field. Destruction of all diseased plants and removal of debris after harvest are essential for taking up any management programme for this disease. Since, they harbour the disease causing organisms. Disinfection of farm equipments and crop rotation are the most important practice for controlling the disease. Soil drenching of carbendazim or propiconazole (0.2%) along with chloropyriphos (0.2%) may also be effective to control the disease in sunnhemp as it is found as effective to control wilt in Pomegranate caused by the same fungus (Sharma *et al.*, 2010).

ii) *Fusarium* wilt

Occurrence & severity status

Fusarium wilt [Fig. 17(A-B), also called as 'Vascular wilt' is an important disease of sunnhemp reported for the first time from Mumbai (India) by Chibber (1914). It is an important soil borne disease in tropical regions. It occurs in different sunnhemp growing areas of India (Mitra, 1934; Mundkur, 1935; Padwick, 1937; Uppal and Kulkarni, 1937; Armstrong and Armstrong, 1950, 1951) as well as in other sunnhemp growing countries, like, China (Zhang *et al.* 1986), Rhodesia (Booth, 1978), Taiwan (Wang and Dai, 2018), Trinidad (Briant and Martyn, 1929), USA (Wang and McSorley, 2009), Zimbabwe (Farr and Rossman 2017), and probably in others. In UP (India) the incidence of wilt was reported as about 13.4% during the year 1997-1998 (Sarkar *et al.*, 2000). In Taiwan also about 13% of plants were found symptomatic at the most affected farm (Wang and Dai, 2017).

Fig. 17 (A-B). Fusarium wilt in sunnhemp caused by *Fusarium udum*.
A. Early stage. B. Late stage of infection

Symptoms

Plants with several weeks old become affected by showing yellowish symptoms and death of plant within 3-4 days after the first obvious symptoms (Holliday, 1980). However, external symptoms vary with severity, ranging from rapid yellowing of the lower leaves to wilting of the plant. Discoloration of vascular bundles first appear at the crown and continues upward on one side of the stem. Often, white to pinkish fungal spore masses are formed on the stem surface at a later stage (Wang and Dai, 2018). Sometimes brown streaks are observed on the stem surface near the ground level (Holliday, 1980). In case of young plants, which are mostly affected, the disease appears as by causing wilt and necrosis. The emergent seedlings, while infected, show poor and slow growth and the seedlings remained stunted and weak. Infection is mainly confined to the roots and base of the stem. This disease is characterized by browning of stem at the collar region adjacent to the soil lines and watersoaked cream to brown-coloured lesions on cotyledon and hypocotyl. Invaded hypocotyls become thin and soft. The lower parts of tap roots and lateral root systems are completely destroyed resulting into drooping and wilting of infected seedling.

Causal organism: *Fusarium udum* **E.J. Butler, 1910**

[Mem. Dept. Agric. India, 2(9): 54]

Synonyms

= *Fusarium udum* (Berk) Wollenw, 1913

= *Fusarium udum* var *pusillum* Wollenw, 1913

= *Fusarium udum* var *solani* Sherb, 1915

= *Fusidium udum* Berk, 1931

= *Fusisporium udum* Berk, 1841

= *Gibberella indica* B. Rai & R.S. Upadhyay, 1982

= *Pionnotes uda* (Berk) Sacc. 1886

Morphological/ Morphophysiolgical characteristics

The fungus grows well in PDA medium. The cultures are pale sulphureous to rose buff, becoming salmon-orange or somewhat purple; growth rate medium, covering the surface of a petridish after 10-14 days at 25°C. Aerial mycelium is felled. Conidia are initially produced on simple or verticillately branched conidiophores and later from pionnotal or small sporodochia on monophialide conidiogenous cells when they form a salmon-coloured mass. Macroconidia are falcate, 21-41 x 3-4.5µm, straight to slightly curved, 1-4 septa (mostly 3 septa), with indistinct foot cells, and with strongly curved to hooked apex. Microconidia are oval to reniform, 6-10 x 2-4µm, and 0-1 septa (mostly 0-septa). In some isolates, there is no clear distinction between microconidia and macroconidia. Chlamydospores are sparse, produced singly, in pairs, or in short chains, oval to globose, 8-11 x 8-12 µm. Perithecial state is unknown (Booth 1978; Wang and Dai, 2018).

Molecular characteristics

The partial TEF1α sequences (GenBank Acc. No. KY706083 and KY706084) and β-tubulin sequences (GenBank Acc. No. MF893323 and MF893324) of the two isolates of suspected *Fusarium udum* from sunhemp wilt were amplified and sequenced. Phylogenetic analysis inferred from concatenate sequences of TEF1α and β-tubulin was performed to reveal the relationships between the two isolates and six reference isolates referred to species closely related to *F. udum* and three distant species. The result indicated that *F. udum* (NRRL22949) was the closest species to the two isolates in a highly reliable clade (Wang and Dai, 2018).

Predisposing factors

The occurrence of wilt is positively correlated with crop age and negatively correlated with temperature. Relative humidity has no significant effects on the occurrence of wilt on the crop (Sarkar, 2005a).

Disease cycle

This is an important soil borne disease, but also transmitted by roots (Subramanian, 1971). On infection through penetration at roolets, the fungus establishes infection inside the host tissues and produces conidia (microconidia and macroconidia) and chlamydospore on reproductive and vegetative hyphae, respectively. These spores germinate in soil and grow on plant debris to survive as saprophytes or remain as it is (in case of chlamydospores) in soil to overcome dormant period. During favourable conditions the dormant spores germinate and the saprophytic mycelia grow to infect new hosts.

Management/Control measures

Use of host resistance

It is reported that the sunnhemp variety 'D-IX' is resistant to wilt and the variety 'K-12 yellow' shows certain degree of resistance (Chaudhury *et al*, ND.).

Cultural methods

Crop rotation and hygienic condition are the useful measures to keep the disease at low level. Well drained soil with good moisture retention capacity helps to escape infection. As regards the crop rotation, it is suggested that the rotation of fields every 3 years is effective to avoid outbreak of the disease (Cook *et al.*, 2005). It is also stated that the wilt is caused by continuous planting of sunnhemp on the same land and, thus, recommended crop rotation (Medina, 1959; Chaudhury *et al*, ND).

Chemical control

Application of combined of micronutrients, like, zinc, boron and iron, may decrease wilt incidence in field (Sarkar *et al.*, 2000). Desai *et al.* (1984) have conducted in vitro tests with twenty five fungicides against a *Fusarium* sp. pathogenic to *Crotalaria juncea* and found largest growth inhibition zones with Carbendazim, Thiram, Hexaferb and Aureofungin. Seed treatment with 0.15% Bavistin is also suggested as an effective and feasible remedy, along with, spraying of 0.15% Bavistin on the first appearance of disease (Tripathi *et al.*, 1978).

8. Seedling diseases (blight, damping off, rot & wilt)

Seedling diseases caused by several soil borne fungi viz., *Fusarium* spp., *Pythium butleri, Rhizoctonia solani, R. bataticola, Sclerotium rolfsii, Sclerotinia sclerotiarum* and *Colletotrichum curvatum* are mostly responsible for poor stand in the field (Mitra, 1937; Tripathi *et al.*, 1978, Pal and Basuchudhury, 1980; Choudhury and Pal, 1981, 1982; Bandhopadhyay, 1981). Amongst them, *Colletotrichum* spp. cause seedling blight, while, *Pythium butleri* causes damping off of seedlings and different fungi viz., *Fusrium, Rhizoctonia, Sclerotium* and *Macrophomina* are responsible for seedling wilt and root rot. Different pathogenic fungi produce different types of characteristic symptoms on seedlings. Amongst them, seedling wilt caused by *Fusarium udum* has been described earlier along with *Fusarium* wilt in this chapter. The rest of the diseases as caused by other pathogens are discussed hereunder.

i) Seedling blight (*Colletotrichum* spp.)

The fungus, *Colletotrichum coccodes* (Wallr.) S. Huges, 1958 (= *Colletotrichum crotalariae* Petch) has been reported in causing seedling blight on *Crotolaria striata* in Srilanka (Petch, 1917). This fungus has also been recorded from Southern parts of USA by Mckee and Enlow (1931) on *C. striata* and *C. spectabilis*. A different species, *Colletotrichum curvatum* Briant & E.B. Martyn, 1929, has been found to cause seedling blight of *Crotolaria juncea* in Trindad (Briant and Martyn, 1929). In India, also, Mitra (1937) has reported the same seedling blight caused by *C. curvatum* in *C. juncea* at Pusa. He has mentioned that it causes severe seedling mortality in thickly sown crop in the month of August when the weather is cloudy and accompanied by rain.

ii) Damping off (*Pythium* sp)

The fungus, *Pythium butleri* Subraman.1919, which has more copious robust aerial mycelium and larger all sporangia, zosporees, oogonia and oospores than

P. aphanidermatum, has been reported to cause damping off disease in sunnhemp sporadically in fields (Tripathi *et al.*, 1975). They also observed damping off and wilt symptoms caused by *P. butleri* during the month of August at Varanasi (India). Seedling shows typical damping off in patches in the morning hours. Water soaked lesions appear on the hypocotyls, which gradually encircle it completely, thus causing decay of tender tissues of seedlings. Young plants show symptom of wilting in healthy dense population. On closer examination of the affected plant, discoloration and necrosis of the stem near collar region are also noticed. Bark of the affected region becomes water soaked and soft, which gradually move in both upward and downward directions.

iii) Thanatephorus (Rhizoctonia) wilt

The disease is caused by *Thanatephorus cucumeris* (A.B. Frank) Donk, 1956 (= *Rhizoctonia solani* J.G. Kuhn, 1858) was first reported from Pusa (Mitra, 1934). The disease also occurs in Varanasi (Tripathi, 1973) and other places where sunnhemp is grown (Pal, 1978; Pal and Basuchoudhury, 1980). It causes pre- and post-emergence damping-off of seedling. Seeds may be rotted completely prior to germination. Damping off is mainly due to colonization of the fungus in hypocotyls and radicle just after emergence. A reddish brown coloured decay of outer cortical layer appears on stem base, crown and roots. This decay develops into shrunken reddish brown cankers, which sometimes girdle the stem at or just above the soil line. The cankerous areas may be extended in any direction on stem, resulting wilt or death of seedlings.

iv) Athelia (Sclerotium) rot

The disease is caused by *Athelia rolfsii* (Curzi) C.C. Tu & Kimbr. 1978 (= *Sclerotium rolfsii* Sacc. 1911), showing both pre- and post-emergence rotting of seedling and non-germinated seeds. The affected seeds and seedlings are covered with white mycelial growth of the pathogen. In some cases, particularly while post-emergence infection occurs, rotting at collar region appears, resulting in seedling blight. The collar portion and seeds are colonized by white cottony mycelial growth of the fungus, sometimes with small, round and brown sclerotia. Rotting of roots is also evident on uprooting the seedlings (Pal, 1978; Pal and Basuchoudhury, 1980).

v) Macrophomina rot

The soil born pathogen, *Macrophomina phaseolina* (Tassi) Goid. 1947, causes seed rot and infects the seedlings that may show blackish brown discoloration at emerging portion of the hypocotyls at soil line or above. The discoloured area turns dark brown to black on death of seedlings. In older seedlings the fungus causes a reddish brown discolouration in the vascular tissues of tap root and that progress up to the stem. Numerous small dark brown or black sclerotia may be appeared on removing the epidermis of tap root and lower stem.

Management/Control of seedling diseases

Hygienic cultivation is the most important aspect in management of seedling diseases caused by soil borne pathogens. The removal of plant debris and ploughing after each crop to expose inner soils to sun are prerequisite for sunnhemp cultivation. As regards the chemical measure, the most effect measure to control seedling diseases of sunnhemp is seed treatment by any suitable chemical fungicides, like, Carbendazim (0.1% a.i.), Thirum (0.2% a.i.) or Mancozeb (0.2% a.i.). However, amongst them the latter two may be better to reduce pre- and post-emergence rotting of seedlings (Sarkar, 2003).

9. Virus diseases

i) Sunnhemp mosaic

Occurrence & severity status

The disease was first reported from Delhi in India by Raychoudhury (1947). Later it is found to occur in almost all sunnhemp growing tracts in India (Ghosh *et al.*, 1977; Sarkar, 2005b) and in other countries including, Australia (Bowswell and Gibbs, 1983), Hawaii (Jenson, 1950), Nigeria (Ladipo, 2008) and United States (Varma, 1986). Severe incidence of mosaic disease was observed in south India, which caused appreciable loss in fibre yield, seed and green matter (Capoor, 1950, 1962). The sunnhemp mosaic disease is widespread in nature. The occurrence of the disease is more in monsoon months when temperature ranged between 22-41° C and relative humidity remained near 80 % (Sarkar *et al.*, 2015). Verma and Awasthi (1976) reported around 40-60 % incidence of this disease at Lucknow, Uttar Pradesh.

Symptoms

The disease first appears as mottling on the tender leaves. Subsequently, thin elongated enations running more or less parallel to each other are developed on the under surface of the leaves.. Patches of light yellow and dark green areas become prominent and finally, it causes yellowing and crinkling of the leaves and weak stems that lodge easily. The infected plant, also, becomes shorter in height with reduced leaves and scanty flush of flowers. Thus, the disease causes low fibre and less seed yields (Raychoudhury, 1947; Capoor, 1962; Chandra *et al.*, 1975; Sarkar, 2005b; Sarkar *et al.*, 2015). Some variants are causing severe leaf distortion and rosette of leaves as disease symptoms in sunnhemp (Verma and Awasthi, 1976). The virus causes cellular mutations, stunted growth, and damages plants, photosynthesis ability. Cellular mutations usually manifests as disclouration and misshapen leaves. Discolouration usually manifests as yellow or grey mottling that can form a spotted, mosaic, or streak pattern. Misshappen leaves can be the result of the damage to the plants at a cellular level, making them appear contorted and/or twisted (Wikipedia).

Causal organism: *Sunn-hemp mosaic virus* **(SHMV)**
(Kassanis and Varma, 1975)

Synonyms

= *Bean strain of tobacco mosaic virus*

= *Dulichos enation mosaic virus* (Capoor and Varma, 1948)

= *Southern sunn-hemp mosaic virus* (Capoor, 1950, 1962)

= *Sunn-hemp rosette virus* (Verma and Awasthi, 1976; 1978)

= *Crotalaria mucronata mosaic virus* (Raychoudhuri and Pathanian, 1950)

= *Cowpea strain of tobacco mosaic virus*

= *Cowpea chlorotic spot virus* (Sharma and Varma, 1975)

= *Cowpea Yellow mosaic virus*

= *Cowpea mosaic virus* (Lister and Thresh, 1955)

= *Hemp mosaic virus*

Virus

The disease is caused by a Tobamovirus. The structure of the virus was determined under electron microscope (Dasgupta *et al.*, 1951). It is a positive-sense single stranded RNA virus. The virus is rod shaped particles, 300 nm in length and 17 (-18) nm in width (Anand, 1968; Nariani *et al.*, 1970;.Kasanis ana Verma, 1975). Axial canal is obscure, 4nm in diameter. Basic helix is obvious, pitch of basic helix is 2.3nm (in Gibbs, 1983). The nucleotide sequence of the 5'-untraslated region, the 129 kD protein gene, and a portion of 186 kD protein gene of SHMV have been determined (Silver *et al.*, 1996). The 4,683 nucleotides (nts) are the complete sequence of SHMV genome (Silver *et al.*, 1996). As regards the physical properties the virus has three sedimenting components. The fastest one is 187 s of the others, 35 s and 75 s. Density is 1.318 cm^{-3} in CsCl (Boswell and Gibbs, 1983).

The virus has wide host range among the legumes, including cowpea (Jensen, 1950), sunnhemp, *Crotolaria mucronata*, *Cyamopsis tetragonoloba*, *Dolichos lablab*, *Mucruna aterrima* and *Pisum sativum*. The hosts of other families are included, *Nicotiana tabacum*, *Datura strumonium* and *Lycopersicon esculenta* (Capoor, 1962; Das and Raychoudhury, 1963; Frison *et al.*, 1989).

Transmission

Vector: The virus is readily sap transmissible. It does not transmit by vector or via seed of cowpea (Kassanis and Varma, 1975; Kulthe and Mali, 1979). In sunnhemp, there is little or no seed transmission (Capoor *et al.*, 1947; Capoor, 1962; Nagaich and Vashish, 1963) but, in contrast, it is reported to be seed transmitted (5-10%) in nature if the plants are infected before flowering (Verma and Awasthi, 1976). The virus is transmitted by mechanical inoculation or by contacts. The viral movement inside the host is systemic (Sastry and Vasudeva, 1963).

Predisposing factors: It is observed that the severity of the virus increases with the onset of monsoon (Sarkar *et al.*, 2015). As the virus occurs in different hosts, so, the disease may be initiated from the other weed host (s). The application of higher dose of nitrogen increases the disease, while, potassium application reduces that (Sastry and Vasudeva, 1963).

Management/ Control measures

Host resistance

Amongst the different varieties, K-12 yellow shows less disease incidence (Ghosh *et al.*, 1977). In addition, the recently released varieties, SH-4 and SUN 053, from ICAR-Central Research Institute for Jute and Allied Fibres (Barrackpore, India) are reported as resistant to sunnhemp mosaic (Sarkar *et al.*, 2015).

Induction of systemic resistance by leaf extracts has been found promising to manage the disease in field. In this context, a number of plant extracts like *Cleondendrum aculeatum*, *Chenopodium murale*, *Boerhaavia diffusa* (root extract), *Cuscuta reflexa*, *Solanum melongena* and *Pseuderanthemum bicolour* are found to reduce the disease under laboratory conditions (Verma and Khan, 1984; Srivastava and Verma, 1995; Verma and Versha, 1995).

Cultural methods

Drastic reduction of the disease has been noticed in the early sown (mid-April) crop (Sarkar and Tripathi, 2003) than the monsoon crop. Thus, it is suggested to sow the crop early in order to escape the disease. There, is no known method to cure the sunn-hemp mosaic virus (SHMV). Once a plant become infected with the virus, the plant will never be free from infection. The virus is needed to destroy by uprooting the infected plant and burning it.

ii) Phyllody and witches' broom

Occurrences and severities

This disease was first reported by Solomon and Sulochana (1973) at Chennai (India) after rainy season. It occurs sporadically in different parts of India. Plant once affected with mosaic virus was never affected by this disease. This disease was also reported from Brazil (Bianco *et al.*, 2014a), Myanmar (Win *et al.*, 2011), Oman (on *Crotolaria aegyptiaca*; Al-Subhi *et al.*, 2017) and Srilanka (Shivanathan *et al.*, 1983). It is rarely noticed in the field of Sunnhemp Research Station, Pratapgarh. In a study, it was observed that during the recent four years span the disease incidence was varied from 1.0-3.5%. The disease spreads more rapidly in seed crop than fibre crop (Sarkar *et al.*, 2015).

Symptoms

It is characterized by yellowing of apical leaf, followed by big bud formation at the terminal raceme, conversion of floral meristem in the lateral raceme to the vegetative state, leading to buds becoming phylloid and forming dwarf shoot. The sequence of symptoms follows a definite pattern, such as, at first yellowing of apical leaves, subsequently, big bud formation in the terminal raceme, convertion of floral meristem in the lateral racemes to the vegetative state and the buds become phylloid. Almost all the floral parts are transformed into leafy structure and virtually no seed setting took place. Similar kinds of symptoms are also noticed in Barrackpore, West Bengal. Sharma (1990) reported Witches' Broom like symptoms, such as, chlorotic leaves, internodes shortening, reduced sized leaves and shoot proliferation, which appeared after 30-60 days of sowing in Pratapgarh (U.P, India).

Causal organism(s): Sunn-hemp phyllody phytoplasma (Schneider *et al.*, 1995)

and/or

Sunnhemp witches' broom (ShWB) phytoplasma (Win *et al.*, 2011).

Phytoplasma

Phytopsama is wall-less prokaryotes that lives as an obligate parasite. The phytopsma of sunnhemp is belonged to faba bean phyllody strain cluster (Schneider *et al.*, 1995). The Sunnhemp phyllody and stem fascination phytoplasma of India is reported as under the subgroup 16SrIX-E (Candidatus Phytoplasma phoenicium – Related) phytoplasma (Biswas *et al.*, 2018). The phytoplasma which associated with witches broom symptom of sunnhemp in Myanmar described as sunnhemp witches' broom (ShWB) phytoplasma (Win *et al.*, 2011). The ShWB phytoplasma is in 16Srll (Peanut WB group). The phylogenetic analysis and nucleotide sequence of 16S rDNA, the ShWB phytoplasma is closely related to strains SUNHP from diseased sunnhemp in Thailand which is classified in 16Srll group. The putative restriction fragment length polymorphism maps indicate that the ShWB phytoplasma belongs to subgroup of 16Srll-A (Win *et al.*, 2011). In Brazil, the phytoplasmas of sunnhemp belong to 16Sr I, III, VII, IX and XV, of which, the most abundant is 16Sr IX. The 16Sr IX phytoplasma shows 100% similarity to the citrus phytoplasma and develops witches broom symptoms in sunnhemp (Bianco *et al.*, 2014b).

Transmission: The phytoplasma transmits mechanically by sap inoculation (Sharma, 1990) and by vector, leaf hopper.

Control

The mycoplasma can be killed by the application of tetracycline @ 500ppm (Sharma, 1990). However, the spraying of tetracycline cannot be recommended to control the disease in any field crops, since, the antibiotic is used for human disease control and there is ample chance to develop resistance in human disease causing organisms which may lead to fatal for human health.

iii) Leaf curl

Occurrence and disease severity

A severe leaf curl disease was observed on sunnhemp in and around Lucknow during the rainy season of 2001 (Khan *et al.*, 2002). The severity of the disease was observed as about 50-60%. However, the recent studies indicate that the disease incidence in and around Lucknow is about 20-40% (Kumar *et al.*, 2010). A survey works conducted during July, 2007 at farmers' fields in Pratapgarh district (UP) and observed that the disease severity was more (almost cent percent) in the crops which were sown later (May-June) than the early (April) sown crop (severity 10-20%). The disease affected plant height and fibre yield that reached up to 3.4 q/ha fibre yield loss (Sarkar, 2010).

Symptoms

Infected plants show stunted young leaves and shoots. The plant grows slowly and become stunted. The leaf margins curl mostly upward and a few inward. The leaf curling is associated with darkening of veins, vein swelling and enations that frequently develop into cup shaped structures on the ventral surface. The leaf becomes stiff with yellow margin. The infected leaf is thicker with leathery texture. The young leaves of infected plant become yellowish.

Causal organism: *Indian tomato leaf curl virus* (**IToLCV**) (Ahmad *et al.*, 2011)

[**N.B.** The name of the virus is used in this book as mentioned by the respective authors in their publication. Probably, the name, '*Tomato leaf curl virus* (ToLCV)' is more appropriate for this disease]

Virus

The virus genus *Begomovirus* (family: Geminiviridae) is typically a bipartite, circular single-stranded DNA (ssDNA) genomes (Khan *et al.*, 2002; Raj *et al.*, 2003). It has all functions as required for virus replication, control of gene expression and encapsidation encoded DNA-A and genes involved in intra and inter-cellular movement encoded on DNA-B (Bowdoin *et al.*, 1999).

Molecular characterisation

In a study of Kumar *et al.* (2010) DNA templates isolated from symptomatic leaf tissues were used in polymerase chain reaction (PCR) using specific primers to amplify coat protein (CP) gene of DNA-A as well as betasatelite DNA associated with leaf curl disease. CP gene showed 97% sequence identity with that of Cotton Leaf Curl Burewala virus (CLCuBwV). Further studies revealed betasatellite DNA molecule sequence similarity with betasatellite DNA of begomoviruses affecting malvaceous crops of different regions. Maximum similarity (Ã 90%) was observed with *Cotton leaf curl Multan betasatelite* (CLCuMB) and other betasatellite DNA from Pakistan. Their studies confirmed the possible infection of begomovirus in sunnhemp leaf curl disease (Kumar *et al.*, 2010). In another study of Ahmad *et al.* (2011) sequence alignment of the 897bp amplicon obtained from sunnhemp leaf curl disease. The DNA of the diseased plant revealed a complete 771bp coat protein (CP) gene flanked by 3´ regions of AV2 and AC3 genes. Southern hybridization using (α-32P)dCTP labelled (CP) gene probe of Indian tomato leaf curl virus (IToLCV) demonstrated the association of begomovirus with the leaf curl disease of sunnhemp (Ahmad *et al.*, 2011).

Phylogenetic analysis revealed that the AV2, CP and AV3 genes had closest genetic relationship with begomovirus isolates from India, China and Bangladesh, respectively (Ahmad *et al.*, 2011). Further, in silico recombination analysis elucidated 297 nucleotides hot spot (346 to 643 nucleotides) within AV2 overlapping region of CP gene, amenable to genetic rearrangements with lineage from *Tomato leaf curl virus Bangalore* (ToLCuVB) and *Indian cassava mosaic virus-Ind* (ICMV) as major and minor parents, respectively. Thus, they (Ahmad *et al.*, 2011) had concluded that the recombinant CP genes related to begomoviruses are evolved from Indian isolates, causing broad host specificity and molecular diversity among the related begomoviruses across the geographical limits of Southeast Asia.

Transmission: The disease is transmitted by a vector white fly (*B. tabaci*).

Control

The disease is very difficult to control as the vector pest (white fly) is very much resistant to many commonly used insecticides. However, proper monitoring of the field and uprooting and burning the diseased plant, followed by insecticides spray may give some better result. The mixture of both organic oil, like, canola (*Brassica napus*) oil, 25% solution, i.e. 25ml oil mixing with 75ml vinegar and insecticide (Bifenthrin 10% EC) while sprayed at recommended doses gave better control of cotton leaf curl (Humza *et al.*, 2016). The same spray mixture may be applied to control sunnhemp leaf curl disease.

B. Pests

1. Crotalaria pod borer

Occurrence & severity

Crotalaria pod borer [Fig. 18(A-B)] is a tiger moth, very much present in Africa (eastern countries), Australia, Burma, China, India, Pacific Islands, Philippines and Sri Lanka. The moth becomes active in the evening. It is a serious pest of sunnhemp at its caterpillar stage (Lefroy, 1909), causing extensive damage in South Indian states. About three broods are noted during the total duration of

Fig. 18 (A-B). Crotalaria pod borer (*Argina astrea*) attack on leaves and pods of sunnhemp. **A.** Showing damages on leaf. **B.** Showing damages on both leaves and pods.

the crop from June to November. The larvae feed on Crotalaria species. The pod borer species prefers secondary habitats ranging from the lowlands to the montane region (Cerny, 2011).

Types of damage

The pest at its larval stage feeds on the leaves of *Crotalaria juncea* (Lefroy, 1909) and causes defoliation. Particularly, the young larvae feed on leaves, while, the older ones on the seed pods. In severe cases, the pest attacks leaves, pods and flowers by eating those completely (Sarkar *et al.*, 2015)

Scientific name & order: *Argina astrea* Drurv, 1773 (Lepidoptera: Erebidae)

[Illustrations of Natural History. 2: 1–92.]

Synonyms

Many including the followings:

= *Argina cribraria* Clerck. 1759 (?)

= *Phalaena astrea* Drury, 1773

= *Phalaena cribraria* Clerck, 1759.

= *Phalaena (Noctua) astrea* Drury, 1773

= *Phalaena cribraria* Clerck, 1764

= *Bombyx pylotis* Fabricius, 1775

= *Deiopea dulcis* Walker, 1854.

Biology

Egg: The eggs are spherical, pale green (yellowish), laid in clusters on the under surface of leaves.

larva: The mature larva is about 4 cm long with typical caterpillar type body and reddish brown head. The body is black with white intersegmental rings that contain broken black transverse lines. These rings are preceded by transverse white bars dorsally on A1-6, with dots in front of these on A3-6. The prothorax has a dorsal, longitudinal white streak. The spiracles are set in orange patches. There are secondary setae on blue-black verrucae (Robinson, 1975).

Pupa: The caterpillars pupate under litter on the plant or surface of soil.

Adult: The adult moth is active in the evening. The wingspan is about 40 mm. It is an orange coloured moth with black spots on all wings, some of the spots are encircled by white ring-like coloured patches. The species is extremely variable in wing pattern as well as ground colour. In this species, the head, thorax and forewing are orange-yellowish or whitish in color. The abdomen and hind wings are bright orange. Forewing spots are quite large, with light edges (Arora andChaudhury, 1982). Palpi are upturned, reaching the vertex of head with short third joint. Antennae are ciliated in both sexes. Mid and hind tibia are with minute terminal spur pairs. Hind wing of male is with a fold on inner margin containing a glandular patch near the base and a tuft of long hair beyond it. The anal angle produced to a point (Hampson, 1892).

Life cycle: Eggs are laid in small clusters on the under surface of the leaves. The larva passes through five instars and pupates at its sixth mould. The larval period lasts 18-21 days. Pupation takes place on the plant or on the surface of the soil. The total life cycle is about 26-31 days.

Management/ Control measures

Protection of natural enemies

The larval parasite, *Bracon brevicornis* (wasp) suppresses the population of the insect pest under natural condition. So care should be taken for not harming the wasp if it remains in field.

Mechanical control

The pest can be suppressed by hand picking and killing the caterpillars at its initial stage of infestation.

Chemical control

Foliar spray with methyl parathion @ 2 ml / lit. water or dusting methyl parathion 2 % D @ 10-12 kgs / acre are the effective measures for the control of the pest. The pest can also be controlled by dusting of 5% BHC or spraying of 0.1% dieldrin (Sarkar *et al.*, 2015).

2. Eriophid mite

Occurrence & severity

This is an obligate plant-feeding erithrophid mite. The eriophyoid mites are highly host-specific, with nearly 80% of them known from a single host species, 95% from one host genus, and 99% from one host family. Furthermore, non-monophagous species are often found on closely related host species (Skoracka *et al.* 2010). In sunnhemp, the mite is a gall mite, causing erineum galls. It is found as active from May to July in the field in West Bengal (India) (Pandit and Pradhan, 1982). It also occurs in different states of India. Its occurrence was first reported from Karnataka (Channabasavanna, 1966), Later, it is reported to occur in Tamil Nadu during the year 2003 (Annonymous, ND2), and in Tripura (Ghosh and Chakraborti, 1988). The erithrophid mite of sunnhemp is polyphagous, infesting *Crotalaria juncea, C. biftora, C. retusa, C. verrucosa* and *Desmodicum triflorum* (Gupta, 1985).

Scientific name & order: *Aceria crotalariae* Chnnabasavanna (1966) (Trombidiformes: Eriophyidae)

[Null. Univ. Sci. Agric. Hebbal, Bangalore: 154]

Types of damage

Mite infestation causes twisting of apical leaves which become shortened, discoloured and ultimately yellowish in the first phase of attack. Later on, the infested portion of the stem becomes brownish in colour. The mites induce slight erineal growth on tender shoots which show a slight stunting (Channa basavanna, 1966). However, in some cases profuse erineal growth as hairs on the infested leaves and stems (Pandit and Pradhan, 1982) may occurs. The attacked portion of the stem turns into deep brown in colour, due to presence of thick growth of brown erineum mites and that is distinctly visible from to some extent distant place. The length of internodes of infested stem becomes shortened and the plants remain stunted (Pandit and Pradhan, 1982). During heavy infestations, large number of galls is developed on leaf surface, causing a considerable degree of distortion and curling of leaves (Ghosh and Chakraborti, 1988).

Biology

Egg: Eggs are soft, round and white, deposited on the leaf or inside the gall.

Larva: larva is immature mite, similar in appearance as adult but smaller in size. This is white to yellowish. There is no sexual dimorphism among immature ones.

Nymph: The larva moults into nymph. There are no differential nymphal stages. The nymph also looks like adult. The last juvenile stage moults into adult mite.

Adult: Adult female is worm like, whitish or creamy, 175µm long, 45 µm thick, with two pairs of legs near the anterior of the body. Rostrum is 20µm long projecting obliquely forward. Antapical seta is 3µm long. Shield is distinctly designed, 30 µm wide, 25µm long, lacking longitudinal wrinkles, narrowly truncated at the apex; median line straight, evident only on the basal half of shield. Abdomen lines are complete, close together and parallel to each other on apical portion, diverging gradually to the rear shield margin. Median line on rear half shield is without a dart shaped mark. Dorsal tubercles are at rear shield margin. Dorsal seta, tibial seta and basal tarsus all are about 6µm long each. Claws are gently tapering towards apex. Feather claw is 5 rayed. Hindleg is long. Forecoxae are narrowly connate, heavily studded with microtubercles; hind coxae with sparser microtubercles. Microtubercles are more or less evenly developed dorsally and ventrally. Abdomen is with about 65 rings, uniformly microtuberculate. Second ventral seta on abdomen is shorter than third ventral seta. Female genittalia 19µm wide, 11µm long; coverflap with about 12 longitudinal ribs, genital setae are shorter than the length of genitalia (Anonymous, ND2; Gupta, 1985).

Adult male is 125µm long, 35µm thick, with two pairs of legs near the anterior of the body, genitalia 16µm wide, gential seta long as female (Annonymous, ND2)

Life cycle

Colonies establish in stem buds, transforming them into rounded pubescent galls. It has four life stages: egg, larva, nymph and adult. At maturity, the male deposits a spermatophore which is collected by a female via her genital flap. Fertilised eggs are produced by the female, one at a time. The female lays eggs either inside the gall or on leaves to colonise new buds on the same plant or be carried by wind to a new host plant. The egg incubation period lasts for 4-7 days under natural conditions. A generation period (from egg to egg) lasts 15-19 days. Several overlapping generations develop in galls during the crop season.

Control measures

Sulphur (e.g., wettable suphur @ 0.4% a.i.) sprays are useful to manage the mites. Apart from this, several chemical acaricides, like, bifenthrin, carbaryl, dicotofol, fenbutatin oxide and lambda-cyalothrin, are effective at their recommended doses. So, anyone can be used.

3. Sunnhemp flea beetle

Occurrence & severity

Sunnhemp flea beetle is a minor pest of sunnhemp (Reddy, 1956; Ayyar, 1963; Tripathi and Ram, 1971). It becomes active during June-July and October-December in India. It occurs in the states, like, Bihar, Kamataka, Kerala, Maharashtra, Manipur, Meghalaya, Sikkim, Tamil Nadu, Tripura and Uttar Pradesh, on different hosts, like, *Crotalaria juncea, C. grahamiana, C. pallida* and *C. striata* var. *acutifolia*. Amongst the hosts, former two are highly preferred (Sarkar *et al.*, 2015).

Types of damage

The adult beetle feeds on tender leaves at apical parts of plants and as a result several small elongated holes of about 2-4 mm in length appear on the leaves. The larva after hatching enters into the root and cotyledon of germinated seed and thereby feeds making a tunnel (Sarkar *et al.*, 2015).

Scientific name & order: *Longitarsus belgaumensis* Jacboy, 1896 (Coleoptera: Chrysomelidae)

[Ann. Soc. Ent. Belg., 40 : 260.]

Biology

Egg: Eggs are laid in clusters, orange in colour, cylindrical in shape and measuring 0.50 mm x 0.22 mm. With development, the colour of eggs is changed to dusky dull with a prominent black spot at the interior region. Incubation period varies from 7-8 days (Pandit and Pradhan, 1991).

Larvae: The first instar larva looks creamy white in colour with black head and caudal region, measures about 1.3 mm X 0.20 mm, and lasts for about 4 - 5 days. The second instar larva is about 2.16 mm x 0.29 mm, similar in colour with that of first one and lasts for about 4-5 days. The third instar larva is about 3.78 mm x 0.56 mm in size, lasts for 5-6 days. The full-grown third instar larva stops feeding and prepares a small circular cell in the soil for pupation (Pandit and Pradhan, 1991). The pre-pupal larva is creamy white in colour, crescent shaped and lasting for 4-5 days (Pandit and Pradhan, 1991).

Pupa: The freshly formed pupa is creamy white in colour, exarate type and body including head is covered with setae, measuring 2.5 mm x 1.0 mm. Pupal period varies from 6-7 days (Pandit and Pradhan, 1991).

Adult: The adult is small oblong in shape, almost parallel-sided, light yellowish brown in colour, measures 2.45 to 2.68 mm long and 1.22 to 1.40 mm breadth. Dorsal surface is pale brown with the sutural margin blackish. Head is impunctate. Antennae are long and slender. Prothorax is slightly broader than length, pronotum with a few fine punctures and its basal part with fine longitudinal wrinkles. Elytra are subcylindrical, granulate and confusedly punctured. Apical segments are dark brown. Hind femora are enlarged. Fore- and middle legs, hind tibia and tarsi are light brown (Basu *et al.*, 1981; Pandit and Pradhan, 1991).

Life cycle: The incubation period is 7-8 says. The first instar creamy white larva mines into tender growing roots and cotyledons of germinated seeds of sunnhemp plants. Fully grown larva is 3.8 mm long. It pupates in an earthen cocoon in soil. The duration of 1st, 2nd and 3rd instars are 4-5, 4-5 and 5-6 days respectively. The prepupal stage requires 4-5 days. The pupa is creamy white and exarate type, 6-7 days. The life cycle completes within 31-34 days (Pandit and Pradhan, 1991)

Control measures

There are many insecticides, like, pyrethrin, carbaryl, malathion, spinosad, permethrin, bifenthrin and esfenvalerate. Anyone of these may be applied as foliar spray to protect the foliage against the feeding of the adult beetle.

4. Hairy caterpillar

Occurrence & severity

The pest is also called as 'crimson-speckled flunkey' or 'crimson-speckled moth'. This is a polyphagous moth; its larvae feed on a range of herbaceous plants. The moth is a common widespread species of Africa, Asia, and Europe. It is also widely distributed in Indo-Australia and the southern Oceanic islands (Hampson, 1894). It inhabits at dry open places, meadows, shrub lands, grasslands and parks. Its caterpillar is one of the serious pests of sunnhemp, occurring all over India, most commonly, in the states of Andhra Pradesh, Bihar, Maharashtra, Orissa, Tamil Nadu, Uttar Pradesh and West Bengal (Ayyer, 1963; Mallick, 1981b; Atwal, 1991). Generally it attacks in sporadic form but sometimes in hot summer it becomes a serious problem in North India (Singh, 1977; Sarkar *et al.*, 2015). The insect is active from March to October (Pande, 1972). In North Africa and in tropical areas, it occurs throughout the year.

Types of damage

The young caterpillars are gregarious in nature and feed on upper foliage of the plants (Sarkar *et al.*, 2015). The caterpillars cause two types of damage. At the beginning, they feed voraciously on the foliage, skeletonise that and defoliate the affected plants. However, at later stage when pod formation starts, they migrate upwards and bore into pods and eat away the unripe seeds, causing decrease in seed production.

Scientific name & order: *Utetheisa pulchella* (Linnaeus, 1758) (Lepidoptera: Arctiidae)

Synonyms

= *Tinea pulchella* Linnaeus, 1758

= *Utethesia callima dilutior* Rothschild, 1910

= *Deiopeia pulchella* var *candida* Butler, 1877

= *Noctua pulchra* Denis & Schiffermuller, 1775

= *Utethesia thytea* Rothschild, 1914

= *Utethesia shyama* Bhattacherjee et Gupta, 1969

Biology

Egg: Eggs are small, whitish (dirty white) and round, laid singly on the tender leaves and shoots but often 3 to 7 eggs are found in groups. The female moths lay 80-100, differently 258-397 eggs (Pande, 1972). Incubation period is 3-4(-5) days.

Larva: Larva is warty, dark brown or greyish caterpillar, with brown head and tufts of greyish hairs. An orange cross-line is on each segment, a wide whitish line along the back and two other lateral white lines (Pande, 1972). In some cases the hairy caterpillar (Fig. 19) is white with red colour band and black spots (Bhatt, 2016). There are five larval instars which took 13 to 18 (-21) days after which it pupates (Pande, 1972).

Pupa: It pupates on soil surface or in dry fallen leaves for 6-7 (-8) days (Pande, 1972)

Adult: Adult is a medium sized, short lived (Pande, 1972) colourful tiger moth with wingspan 29-42 mm. It has pale white or cream coloured narrow forewings with a variable pattern of numerous small black spots located between the larger-sized bright red spots. Sometimes the red spots are merged to transversal bands. The hind wings are wide, white, with an irregular black border along the outer edge and two black markings in the middle of the cell. The head and thorax are cream colour to buff yellow in colour, with the same pattern as the wings. The antennae are long and monofiliform. The abdomen is smooth, with a white background.

Fig. 19. Hairy caterpillar (larva) on Sunnhemp

Life cycle

The moth has several generations per year. The incubation period lasts for 3 to 5 days. There are 5 larval instars which take about 13 to 18 days after which it pupate on soil surface or in dry fallen leaves for about 6 to 7 days. The adults are short-lived. One complete life-cycle takes 22 to 30 days. The insect is active from March to October. During November to February the insect is found overwintering in the pupal stage (Pande, 1972). This species in southern Europe overwinters as a caterpillar. During mild winters in temperate and typically Mediterranean climates this species hibernates as pupae and completes three generations in a year, but in the tropics, they develop continuously. The adults fly both day and night and come to light.

Management/Control measure

Cultural control: In this method the adult moths are collected by the nets and destroyed. Hand picking of larvae at early stage is also effective in reducing the population builds up (Ayyar, 1963).

Biological control: Several insects have been found to parasitize on sunnhemp hairy caterpillar. Ayyar and Margabandhu (1934) have noticed *Phanurus* sp. as egg parasite of the pest, *Utetheisa pulchella*, at Coimbatore. A tachinid insect, *Drino inconspicua*, has been found to parasitize on its caterpillar (David and Kukaraswami, 1960). Ayyar (1963) reported a braconid, *Bracon brevicornis* and a tachinid parasite, *Padomyia setosa* from South India as hyperparasites. Protection of these hyperparasites may be beneficial in the pest management programme.

Chemical control: It is found that the dusting of stomach poisoning insecticide, Geigy Kutra dust @ 33.5 kg /ha as the best measure to control the pest (Srivastava, 1964). Further, Atwal (1991) has suggested spraying 500 g of BHC/DDT 50WP in 250 l of water for the control of this pest. Mallick (1981b) has recommended 3-4 spraying of Diazinon 0.04% at 15 days interval starting at 20 days of crop age.

5. Stem grinder

Occurrence & severity

Stem girdler is a minor pest of sunnhemp. It was first reported on sunnhemp in 1972 at Sunnhemp Research Station, Pratapgarh, Uttar Pradesh. The incidence was noticed in August-October (Sarkar *et al.*, 2015). It also affects jute (*Corchorus olitorius*) (Dutt, 1952b). The insect is distributed in India, China and Taiwan.

Types of damage

The ovipositing female while preparing for laying eggs, causes loss of stem length by girdling the stem above and below a slit in which the egg is laid. The slit is made by two successive indentations of the mandibles. It is composed of a central pit and two lateral punctures. The depth of the slit reaches down to the pith level where a single egg is deposited. By this time real damage to the stem has already been done because the portion of the stem above the girdles withers to die.

Scientific name & order: *Nupserha bicolor* (Thomson, 1857) (Coleoptera: Cerambycidae)

Synonym

= *Stibara bicolor* Thomson, 1857

Biology

Egg: About 30-50 (-60) (Av.35) yellowish eggs are laid singly over a month's period. The egg hatches after 3-4 days to larva.

Larva: Larva is round headed borer. It is 1.4 cm long when grown up. The total larval period is 20-25 (differently, 30-50) days. The larva remains dormant in the soil for over a year. Seventy percent (70%) of the larvae emerge as adults when RH is about 97%.

Pupa: The larvae pupate in a chamber made in the hollow of the stem.

Adult: Adults are long horn beetle. The beetles have long antennae, as long as the beetle's body, dark coloured, jointed with different segments. Head is rusty brown. Wing covers are bicoloured with rusty brown and black. Three pairs of legs are light yellowish, with dark thorny claws.

Life cycle

After hatching from their eggs the longhorn larvae proceed to feed upon their chosen food source- stem wood. With the advent of winter, the larvae cut out small portions of the stem in which they encase themselves and diapauses. The larval stage continues up to the next spring and pupation takes place only after rains have started. After pupation, they burrow out of the wood and emerge as adults and proceed to feed on nectar from flowers. They then find a mate using their antennae to smell each other out. After breeding, the female lays eggs inside the stem.

Management/Control measures

It is suggested to remove and destroy the drooping stem portions and stem casings containing the larvae in diapauses. Different chemical insecticides have been suggested to control the pest of stem grinder. Most commonly, it is suggested to mix 25kg of phorate 10G per hactre in the top soil followed by light irrigation. Further, spraying of phosalone 0.07% or endosulfan 0.07 % at fortnight interval may also be effective. Control involves spraying the apical portion with Endosulfan 0.075% (@ 21.4ml/10l water) or Methyldemetos 0.04% (16ml/10L of water) or Dimethoate 600ml/500l of water/ha. Encouraging results were also noticed with Folidol (0.01%), Diazinon (0.15%), Metasistox (0.05%) and Endrin (0.3%).

6. Sunnhemp mirid

Occurrence & severity

Sunnhemp mirid is also called as 'Sunnhemp capsid'. This is a small, active, green coloured plant bug, which sucks plant sap and when present in large swarms, causes appreciable damage. This plant bug is of Paleotropical (Kerzhner and Josifov, 1999) or Indo-Pacific distribution (Schuh, 1984). Its occurrence in Reunion Island has recently been reported by Ratnadass *et al*. (2018). In India, particularly in South India the pest is widely distributed (Reddy, 1956).

Types of damage

Nymphs and adults suck the cell sap from tender leaves and shoots. Their feeding causes minute chlorotic spots on leaves. Later, these spots coalesce to cause yellowing of leaves.. Leaves become distorted and desiccated, and may drop prematurely. The plants become stunted and produce lesser seeds. In severe infestations, due to foliar chlorosis the entire field looks like whitish. Sometimes, the plants may die due to severe infestation (Reddy, 1956; Gopalan and Basheer, 1966; Gopalan, 1976b; Agarwal and Gupta, 1983). Several salivary enzymes, like, amylase, cellulose, polyphenol oxidase and protease, as produced on feeding, affect respiration, transpiration, moisture content and oxidative enzyme activity in sunnhemp plants (Gopalan, 1976a; Gopalan and Subramanian, 1978).

Scientific name & order: *Moissonia importunitas* (Distant, 1910)

Synonyms

= *Ragmus importunitas* Distant 1910

= *Marshalliella unicolor* Poppius 1914

= *Atomoscelis hyalinus* Lindberg 1958

= *Psallus impictus* Odhiambo 1960

= *Elleniaim portunitas* Linnavuori 1993

= *Moissoniaim portunitas* Duwal, Yasunaga and Lee 2010

Biology

Egg: Fecundity is 38-42 eggs. The bugs lay white cylindrical eggs singly in the plant tissue, generally under the surface layer of leaves. Incubation period is about 7-8 days (Gopalan and Basheer, 1966).

Nymph: Nymphal developmental period is about 18 days (Gopalan and Basheer, 1966).

Adult: Adult is macropterous, light green coloured bugs measuring 3.3 mm in length, Head is yellow, weakly produced in front, posterior margin weakly concave with row of brown setae. Eyes are small, brown. Antennal segments are brown, short and stout, gradually thickening towards apex, light brown or yellow. Labium is reaching mesocoxae. Pronotum is yellow or greenish, with recumbent brown spine-like setae and shining yellow simple setae. Calli are not prominent. Lateral margins are linear, rounded, posterior margin medially concave. Metathoracic scent gland evaporatory area is pale yellow with prominent opening. Mesoscutum is yellow, flat. Scutellum is yellow with green patch posteriorly. Hemelytra are hyaline, shiny, parallel sided, extending beyond abdomen, with brown spine-like setae whose bases are light brown. Cuneus is broad, membrane hyaline with outer small triangular and inner elongate cells. Legs are elongated, spines on femora with basal black spots, meta femora swollen, tibiae with prominent black spines, last segment of tarsi and claws brown or black. Claws are long, smoothly curved, base broad, pulvilli small, parapodia setiform, fleshy. Male genitalia are with broad smooth pygophore. Eendosoma is S-shaped, stout, with series of notches reaching secondary gonopore, apical region thin, with spines at apex. Phallotheca is L-shaped. Right paramere is lanceolate. Left paramere is larger with sensory setae, posterior process elevated and strongly curved (Yeshwanth, 2013).

Life cycle

The bugs lay 38-42 eggs singly in the plant tissue, generally under the surface layer of leaves. Incubation period of egg is about 7-8 days. The young nymphs on emergence begin to feed on the plants. The nymph takes about 18 days to develop into adult. Longevity of adult is about 21 days for males and 32 days for females (Gopalan and Basheer, 1966). All the stages of the insect may be found simultaneously.

Management/Control measures

The population of the pest can be suppressed by collecting the bugs with nets or sticky traps and killing them. Spraying the crop with malathion 50EC (0.1%) or dimethoate 50 EC (0.1%) @ 625 lit. insecticide solution per ha. may be effective. Dusting of 5% Heptachlor is also suggested to manage the pest (Annonymous ND3).

7. Top shoot borer

Occurrence & severity

The shoot borer moth is found to occur in India, Indonesia (Java), Sri Lanka, and probably, in China. In India, it is the most serious pest of sunnhemp, causing 11.5-

20.6% fibre yield loss (Kundu, 1964; Anonymous, 1969, 1970; Ram and Tripathi, 1969; Prakash, 1990). The crop sown in late (June and July) is most suffered by the pest (Ram, 1968a). Generally, this insect becomes active on sunnhemp in the second week of July and continues to breed till the first week of November, with a peak of occurrence in the month of September and October. The pest is mostly confined on sunnhemp. Ayyar (1963) reported that this pest is monophagous. However, in this contrast, it is found that the larvae of this pest feed on several crops, like, *Crotalaria juncea, Phaseolus mungo, Dolichos lablab, Dolichos biflorus* and *Tephrosia purpurea*. Further, Teotia and Pathak (1956) have observed that dhaincha (*Sesbania aculata*) is affected by this pest.

Types of damage

The larva of the moth initially attacks the sunnhemp plant while it is still young, about five or six inches high. This attack does not stop the growth of the plant. In later stages of the plant, the attack takes place at node on stem near leaf axil of young shoot, where a swelling is formed as a gall due to feeding of larva inside the stem. The fibre obtained from such plant is short, coarse and specky, thus, affecting fibre quality. Finally when the pod formation takes place in seed crop, the larvae bore into the pod and feed upon the seeds (Ram, 1968a, Tripathi and Ram, 1971, Mallick, 1986).

Scientific name & order: *Fulcrifera tricentra* Meyrick, 1907 (Tortricidae: Lepidoptera)

[Journal of the Bombay natural History Society 18:137–160]

Synonyms

= *Laspeyresia tricentra* Meyrick, 1907

= *Laspeyresia crocopa* Meyrick, 1907

= *Laspeyresia pseudonectis* Meyrick, 1907

Biology

Egg: Egg is small, broadly oval and whitish when laid. Later, the colour changes to orange yellow on maturation. Before hatching, a big spot of the impression of inside larval head appears at the anterior end.

Larva: The newly hatched larva is greenish yellow with brown head and a brown pronotum. It measures about 2-3 x 0.20-0.25 mm. There are five larval instars. As it matures, the colour of larva changes and the full grown larva is bright red in colour with well developed chocolate brown head and a black prothoracic shield. It measures about 9 mm X 2.5-3.0 mm.

Pupa: Pupa is brown, which remains inside the gall or in a rolled leaf.

Adult

Male: Adult male is about 4.0 mm long (Sarkar *et al.*, 2015), with 10-13 mm wingspan. Head and thorax are rather dark fuscous, closely irrorated with ochreous-grey-whitish. Forewing is elongated, not or only slightly dilated, costa gently arced, apex obtuse, termen slightly sinuate, little oblique. General colouration of forewing is dark fuscous, finely irrorated with ochreous-grey-whitish, with indistinct darker transverse striae. Interspaces between dark fuscous costal strigulae are whitish, producing short bluish-leaden metallic striae, becoming longer posteriorly. A very faint, hardly paler, slightly leaden tinged, subtriangular blotch is on the middle of dorsum. Ocellus is laterally margined with leaden-metallic, marked with three black dashes. Cilia are fuscous irrorated with whitish. Hindwing is with dorsal fold, dark fuscous with blackish dorsal suffusion, lighter towards base. Cilia are fuscous with darker sub-basal line. Genitalia are with narrow tegument, with tufts of long scales. Valva is broad, both ventral and dorsal margin with slight notch before cucullus, slightly variable in shape. Aedeagus is slender, armed with long dentate process, originating from the anellus above the base of the coecum penis (Meyrick, 1906, 1907).

Female: Female adult is about 6.0 mm long (Sarker *et al.*, 2015), with 10-13 mm wingspan. Forewing colouration and markings are similar to male. Hindwing is without dorsal fold and without blackish suffusion dorsally. Genitalia are with sterigma of weakly sclerotized postvaginal plate. Surrounding of ostium bursae is sclerotic, edges of sclerite extending posteriorly. Ductus bursa is slender. Antrum is not sclerotized. Iinception of ductus seminalis is situated medially. Corpus bursa is with two small signa (Meyrick, 1906, 1907).

Life cycle

The adult female lays 35-175 eggs, singly, on both sides of the leaves, which turns orange in colour before hatching. Egg hatches within 3-4 days and newly hatched larva feeds on the tender leaves of upper portion of plant and after 1-2 days enters into the upper most portion of the stem. Larva remains inside the stem for 12-20 days and the mature larva pupates in the soil for 4-8 days and then adult comes out. Life cycle completes within 23-39 days (Gupta, 1955; Teotia and Pathak, 1956). Seeds are also attacked, particularly late in the season, and occasionally leaves are rolled. The moth hibernates as larva or pupa from November to February, and then aestivates from March to June. It may be in the stem if the plant remained in the field or, if the pods are collected, the larva forms cocoon amongst the debris and remains there (Fletcher, 1920; Perrin and Ezeuh, 1978; Singh *et al.*, 1990).

Management/ Control measures

Use of host resistance: Sowing of K-12 yellow sunnhemp variety may be effective to minimize the pest attack. Ghosh *et al.* (1977) have reported K-12 yellow as resistant. However, its resistance nature is now minimized due to cross pollination amongst the varieties as suggested by Sarker *et al.* (2015).

Cultural control

Crop sanitation by removing debris and clean cultivation with proper digging are effective to minimize the dormant pest from the field. Collection and destruction of galls are also suggested to reduce the population build up, which partially reduces the infestation (Sarkar and Tripathi, 2003). It is observed that the late sown crop suffers most while their incidence in early sown crop during April and May is significantly less resulting in higher fibre yield (Ram, 1968a). So, early sowing may be advocated.

Biological control: In India, a number of parasites of top shoot borer were reported by Reddy (1956) and Ram (1968b). Five parasites, namely *Apanteles tetragammae*, *Cremastus* sp, *Goniozus* sp, *Elasmus homonae* and *Sphyracephala hearciana* were reported from Uttar Pradesh (Ram, 1968b). The extent of parasitization was varied from 3.0 to 57.5 % in different months.

Chemical control: Spraying of Diazinon @ 0.04% for five times at ten days interval starting from 20 days after sowing has been found effective to control the pest (Ram and Prakash, 1968). In Madhya Pradesh, Ghumary and Bisen (1967) have reported that the spraying of BHC-50% or DDT or both in combination are effective measure to control the pest in sunnhemp. Soil application of Carbofuran 3G @ 0.5kg a.i./ha followed foliar spray (twice) with Carbaryl 50WP @ 0.1% are found effective measures to control the borer pest (Prakash, 1989).

4
Ramie (*Boehmeria nivea*)- Rosales: Urticaceae

A. Diseases

1. *Alternaria* leaf spot

Occurrence & severity status

In the years 2012 and 2013, black leaf spot disease was observed on ramie plants in Hunan and Hubei Provinces, China (Yu *et al.*, 2016). Yu *et al.* (2016) claimed that their report was the first report of *Alternaria* leaf spot of ramie in China.

Symptoms

In the field, the symptoms of this disease included dark green to black big spots on leaves (Fig. 20A), often resulting in upwardly curled leaf margins (Yu *et al.*, 2016).

Causal organism: *Alternaria alternata* **(Fr.) Keissl.1912**
[Beihefte zum Botanischen Centralbalt, 29: 433]

Synonyms

= *Alternaria alternata var alternata* (Fr.) Keissl., 1912

= *Alternaria alternata var rosicola* V.G. Rao, 1956

= *Alternaria tenuis* Nees, 1816

= *Alternaria tenuis f. tenuis* Nees, 1816

Fig. 20 (A). Leaf spot/blight in ramie. Altenaria leaf spot/blight

Morphological/ Morphophysiolgical characteristics

Conidiophores are mostly unbranched, with one or a few conidial scars, up to 50μm long and 3-6μm wide. Conidial chains are profusely branched. Conidia are obclavate to ellipsoidal, with a short cylindrical beak, 23-56 x 8-17μm, medium brown, rugulose with muriform separation, with a single scar at tip, arising mostly unbranched chains of ten or more (Hoog et al., 2000)

Colonies on Plate Count Agar (PCA) is expanding, grey to olivaceous powdery or felty (Hoog et al., 2000). Surface sporulation in light-exposed rings is a dense turf of multiple branched chains of conidia. The aerial portion of the colony (light-deprived rings) consists of a less dense growth of abundant subarborescent hyphae that produce open, entangled heads of branched chains of conidia. Sporulation on cut PCA surface is an abundant but not dense turf of simple and branched chains of 4-6 conidia on short conidiophores. However, sporulation of cut V-8 Juice Agar (V-8) is dense. All surface and aerial elements are conidiogenous (Simmons, 2007).

Molecular characteristics

The pathogen isolates were identified as *Alternaria alternata* (Fr.) Keissler on the basis of morphology and sequence similarity of 99–100% to the published data for internal transcribed spacer (ITS), and glyceraldehyde-3-phosphate dehydrogenase (gdp) (Yu et al., 2016). The ITS rDNA sequence of *A. alternata* strains HBxn2-1 from China causing ramie leaf spot is HQ645083 (Kwon et al., 2016).

Predisposing factors

The temperature 20-30°C is most favoured by the pathogen. However, it can grow at temperature range 0-35°C and remain pathogenic even at 0°C and onwards (Zhu and Xiao, 2015). The infection is triggered by rainfall, or even just a sudden drop in humidity which helps to release conidia from the source (Dewdney, 2015) and by night-time dew required for conidial germination (Timmer et al., 2003).

Disease cycle

The conidia produced in lesions on mature or dying leaves are dispersed by air currents and their release is triggered by rainfall or sudden drop of humidity. When the conidium lands on leaf surface, it germinates mostly during night in the presence of dew. The germ tube enters through the stomata or directly penetrates through the top of the leaf, using appressorium. The infection establishes later on and produces conidia for repeated infection.

Control measures

In a study to determine the fungicide sensitivity of *A. alternata*, it is reported that Bordeaux mixture (0.6%) and tricyclazole (0.1%) are the most effective to check the fungal growth and sporulation (Nagrale and Gawande, 2016). In addition, other effective fungicides in inhibiting sporulation and fungal growth as reported are difenoconazole (0.1%), propiconazole (0.1%), hexaconazole (0.1%), iprodione + carbendazim (0.1%), captan (0.1%) and mancozeb (0.1%) (Nagrale and Gawande, 2016). So, any of the fungicides, giving preference to former two, may be applied to *Alternaria* leaf spot disease control in ramie.

2. Angular Leaf spot

Occurrence & severity status

The disease was earlier named as '*Cercospora* leaf spot' based on earlier name of the causal fungus which is at present is a synonym. It has been recorded on ramie throughout the subtropical, tropical and temperate regions (Bodeijn, 1962; Chupp, 1953; Dennis, 1970; Fukui, 1918; Katsuki, 1965; Tai, 1979; Watanabe, 1948; Yamamoto, (1934). In Indonesia, Boedijn (1962) reported that the disease was first found at Bogor, West Java Province in 1949. The disease has been observed only on ramie in Indonesia (Alan *et al.*, 1994). In Phillipine, the leaf spot fungus of ramie was found to cause severe attack with 40% to 60% of infected leaves (Clara and Castillo, 1950). In India, the disease was first reported from Assam by Chowdhury (1957). He reported that in October 1953, the leaf spot disease was noticed for the first time at Barbheta (Assam). In India, the disease was earlier considered as a minor disease of Ramie (Chowdhury, 1957; Sarma, 1981; Gawande et.al., 2011). But nowadays, due to change in climatic conditions, this disease occurs in alarming situation and the incidence of this disease as observed is in the range of 5 to 30 % on different genotypes (Gawande *et al.*, 2016). The disease, in severe infection, affects more than 60 % of the leaves and on an average 35-60 % of the leaf area is destroyed by spotting. In Assam (India), the disease occurs during the months March to October, with its peak of occurrence in May and June, when frequent rainfall is followed by warm weather condition (Gawande *et al.*, 2016).

Symptoms

The disease appears as numerous small, 1-2 mm in size, irregular to angular, and pale brown to brown or dark brown to nearly black amphigenous spots on the upper leaf surface [Fig. 19 (A-B). Then the spots enlarge to (2-) 4-5 (-13) mm in size limited by leaf veins, and are often surrounded by the yellow-greenish halo. The centre of older spots, however, turns paler and becomes greyish brown; adjacent spots may coalesce. On the upper leaf surface of the spots, numerous fruit bodies of the causal fungus are formed and later they cover the whole surface of the spot as greyish green powdery masses consisting of conidia and conidiophores. In the field, the occurrence of disease starts from the lower part of ramie plants and rapidly progresses upwards. Heavily attacked plants are markedly retarded their growth. In the dry season, the disease symptom usually does not appear on the new sprouts or ramie plants after harvested. No symptoms have been seen on the stem; the petioles, however, often show brown elongated spots. Leaves severely spotted turn yellow and fall prematurely (Chowdhury, 1957; Alan *et al.*, 1994; Silva *et al.*, 2016).

Fig. 21 (A-B). Angular leaf spot caused by *Pseudocercospora boehmeriigena* in ramie. **A.** Showing spots on green leaves **B.** Leaves showing yellowing after infection.

Causal organism

Pseudocercospora boehmeriigena **U. Braun, 1997**

[Trudy Botanicheskogo Instituta im. V.L. Komarova 20: 42]

Synonyms

= *Pseudocercospora boehmeriae* (Peck) Y.L. Guo& X.J. Liu, 1989

= *Cercospora boehmeriae* Peck, 1883 (1881?).

= *Cercospora boehmeriae* Fukui

= *Cercospora fukuii* W. Yamamoto

= *Cercospora boehmeriana* Woron

Morphological/ Morphophysiolgical characteristics

Stroma are small, but distinct, epiphyllous, 17-40 x 15-55µm, brown to olive brown. Internal mycelium is indistinct. External mycelium is absent. Conidiophores are epiphyllous, arise on the surface of stroma, simple, straight or flexuous, rarely geniculate, aggregated in loose fascicles, pale brown to pale olive brown, cylindrical, 10-15 (-26.5) x 2.5-3 µm. Conidiogenous cells are terminal, subcylindrical, proliferating sympodially, 6–20 × 2.5–3 µm, brown, smooth. Conidiogenous loci are inconspicuous, unthickened, not darkened. Conidia are subhyaline to pale brown, solitary, guttulate, straight or curved, narrow obclavate to cylindric, tapered to the apical end, truncate at basal end without thickening or hilum, 35-120 x 2.5-4 (-4.5) µm in size, (0-) 3-10 (-12) - septate, smooth (Chowdhury, 1957; Alan *et al.*, 1994; Silva *et al.*, 2016).

The fungus grows well in a number of culture media but fails to produce spores under varied conditions of temperature, hydrogen-ion concentration, light and humidity (Chowdhury, 1957). Alan *et al.* (1994) observed that the fungus grew well on PDA medium at 20-24°C with best at 20°C. Differently, Watanabe (1948) recorded the optimum temperature of his isolate as 25°C. He also noted that the fungus grew on PDA and diluted soy agar at pH 4.2. It grows very slow in culture, growing 12–14 mm diam in 20 days as corrugated, compressing the medium, raised, erumpent, aerial mycelium sparse, irregularly lobate margins, white and grey, reverse iron-grey, sterile (Silva *et al.*, 2016). The fungal conidia germinate and extend the germ tubes at 20-24°C, though 24°C seems to be optimum for the germination (Alan *et al.*, 1994).

Molecular characteristics

In a study of Silva *et al.* (2016) a total of 27 *Pseudocercospora* spp, including *P. boehmeriigena*, were used to a multigene analysis. Four genomic regions (LSU, ITS, tef1 and actA) were amplified and sequenced. A multigene Bayesian analysis was performed on the combined ITS, actA and tef1 sequence alignment. Their results based on DNA phylogeny revealed rich diversity amongst the *Pseudocercospora* spp. of Brazil. Further, in Bayesian phylologenetic tree, *P. boehmeriigena* was grouped separately with the nearest fungi, *P. pouzolziae* and *P. nephrolepidis*. From the study, it indicates that phylogenetically, *P. boehmeriigena* is distinct from other species and it has a position basal to a clade containing *P. nephrolepidis* and *P. pouzolziae*. It is not possible to distinguish *P. boehmeriigena* from *P. nephrolepidis* and *P. pouzolziae* based solely on ITS data. In the actA and tef1 phylogenies it is distinct from all other species (Silva *et al.*, 2016).

Predisposing factors

Moderate temperature (20-24°C), frequent rain, and high humidity (moist weather) are the most important factors for the development of this disease (Chowdhury, 1957; Alan *et al.*, 1994; Gawande *et al.*, 2016). The fungal conidia germinate well at 24°C which seems to be optimum (Alan *et al.*, 1994). The relative humidity of 90-100% favours the disease. Diseased seeds and planting materials also play an important role in disease spread (Gawande *et al.*, 2016).

Disease cycle

Conidia and mycelium remain in diseased leaves in the winter, which may occur in plant debris or in active plants. In the beginning of the following year the mycelia and conidia become the source of infection. During the growth period of ramie, the conidia disperse by air and rain splashes. Thus, the disease is spread by multiple re-infections during favourable season. This cycle repeats again for survival of the pathogen in the following years.

Management/Control measures

Cultural control

Crop sanitation is an important measure to minimize the sources of inocula of the disease in ramie. Proper digging and cleaning of plant debris after winter is necessary. Further, importance should be given to select the field for ramie. The field should have well drainage system. Application of increased phosphatic (P) and potassic (K) fertilizers may increase tolerance capability of the plant against the disease.

Chemical control

At the early onset of heavy infection season, spraying of 50% benomyl wp diluted to 1500 times or 36% thiophanate-methyl sc diluted to 600 times while sprayed twice at 10 to 15 days interval @ 75-80l diluted solution per $667m^2$ give satisfactory control. Further, application of propiconazole (0.1%) or difenconazole (0.1%) or mancozeb (0.25%) at 15 days interval after the previous crop harvest may be effective to minimize the disease (Gawande *et al.*, 2016).

3. Anthracnose

i) Anthracnose (*Colletotrichum higginsianum*)

Occurrence & severity status

The anthracnose is regarded as one of the most widespread and devastating disease of ramie. This disease is most severe during warm and humid conditions. In China, ramie anthracnose is found in almost all ramie fields, causing yield losses averaging 20% and ranging as high as 55% in some fields. In September 2010, a typical anthracnose symptom was observed in cultivated fields near Xianning, HuBei Province, China (Wang *et al.*, 2011). The anthracnose disease is reported to cause economic damage in a wide range of cruciferous crops (Higgins, 1917; Narusaka *et al.*, 2004; Wei *et al.*, 2016). The ramie anthracnose caused by the fungal pathogen (*Colletotrichum higginsianum*) was reported for the first time from China by Wang *et al.* (2011).

Symptoms

The disease lesions on leaves are initially small, scattered, bluish white and water soaked. As the disease progresses, the lesions on the leaves become irregular spots and the spots turn grey in the centre with a brown margin. The diameter of the spots is approximately 1 to 3 mm. Initially, the spots on the stems are fusiform and then expand, causing the stem to break (Wang *et al.*, 2011).

Causal organism: *Colletotrichum higginsianum* Sacc., 1977

[Journal of Agricultural Research 10: 161]

Morphological/ Morphophysiolgical characteristics

On PDA, the fungus initially produces grey colonies with an orange conidial mass and then the colonies turn black after 5 days. Spores are single celled, colourless, straight, oval, obtuse at both ends, $10.0\text{-}20.0 \times 3.0\text{-}5.0\mu m$ with an average size of $15.8 \times 4.6\mu m$, and a length/width ratio of 3.47. The setae are dark brown, 1 to 3 septa (Wang *et al.*, 2011)

Wang *et al.* (2011) isolated the pathogen from the Leaf and stem tissue adjacent to and including lesions, which were surface disinfected in 0.1% sodium hypochlorite and then planted on potato dextrose agar (PDA) plus oxalic acid to inhibit bacterial growth. The plates were incubated at 25°C for 3 to 5 days until the appearance of pink spore masses with numerous dense clusters of black setae.

Molecular characteristics

The ITS1, 5.8S, and ITS2 sequences with primers ITS1 and ITS4 of this fungus have been determined and obtained the GenBank Accession No JF830783, which are 99% similar to the sequences of multiple isolates of *C. higginsianum* (GenBank Accession Nos, GU935872 and AB042303) (Wang *et al.*, 2011).

The role of RNA silencing machinery in the fungus *C. higginsianum*, by generating deletions in genes encoding RNA silencing components has been investigated (Campo *et al.*, 2016). Severe defects are observed in both conidiation and conidia morphology in certain strains. Analysis of transcripts and small RNAs reveals an uncharacterized double-stranded RNA (dsRNA) virus [termed *Colletotrichum higginsianum non-segmented dsRNA virus1* (ChNRV1)] persistently infecting *C. higginsianum*. The virus has been shown to be de-repressed in those strains. Further, phylogenetic analyses clearly show a close relationship between ChNRV1 and the members of the segmented Partitiviridae family, despite the non-segmented nature of the genome. Immuno-precipitation of small RNAs associated with AGO1 (agronaute protein) shows abundant loading of 5'U-containing viral siRNA. The fungal (*C. higginsianum*) strains are cured of ChNRV1. It reveals that the conidiation and spore morphology defects are the primarily caused by ChNRV1. Based on these results, RNA silencing involving ChDCL1 (dicer protein) and ChAGO1 in *C. higginsianum* is proposed to function as an antiviral mechanism and cause the conidiation and spore mutant phenotypes. This indicates that *C. higginsianum* employs RNA silencing as an antiviral mechanism to suppress viruses and their debilitating effects (Campo *et al.*, 2016).

Predisposing factors

The pathogen favours the temperature range 25-28°C, and pH 5.8-6.5. It is usually inactive in dry season but during moist condition it causes anthracnose disease

Disease cycle

The pathogen establishes interaction with host by producing melanised appressorium and then penetrates the host cuticle. After penetration, infection vesicles and primary hyphae are formed. Later, secondary hyphae are developed and spread to kill the host cell and established infection. It produces conidia on acervulum which serve for re-infection.

Management/Control measures

Use of plant extract: The methanol extract of a medicinal plant, viz., *Melaphis chinensis*, is found effective to control the anthracnose disease caused by *C. higginsianum*, suggesting the possibility anthracnose disease management by plant derived pesticides (Kuo *et al.*, 2015).

Fungicidal control: The mixed formulation with azoxystrobin 200g/l + difenoconazole 125g/l (trade name 'Amistar Top', Syngenta) @ 800-100 ml per hectare, diluted in 250-1000l of water, is suggested for tractor application (Syngenta SA (Pty) Ltd.).

ii) Anthracnose (*Colletotrichum gloeosporioides*)

Occurrence & severity status

The fungus (*Colletotrichum gloeosporioides*), which is one of the most important plant pathogens, occurs in both tropical and subtropical regions of the world. Its occurrence in ramie was noticed from June 2007 to September 2010, when typical anthracnose symptoms were observed in cultivated ramie fields in HuBei, HuNan, JiangXi, and SiChuan provinces, in China. There, the diseased area was estimated to more than 10,000 ha. Ramie yield was reduced by 20% on average with up to 55% yield losses in some fields (Wang *et al.*, 2010). In India, also, the disease has been reported to occur on 129 ramie germplasm accessions in Assam, with varying degree of resistance (Gawande and Sharma, 2016).

[**N.B.** The pathogen was earlier named as *Colletotrichum boehmeriae* based on the host specificity by some Chinese researchers (Li and Ma, 1993). However, in the revision of *Colletotrichum* by von Arx (1957) and Sutton (1980b), the name *C. boehmeriae* based on the host specificity was cancelled. Thus, Wang *et al.* (2010) claimed that their report is the first report of *C. gloeosporioides* causing anthracnose of ramie in China.]

Symptoms

The disease [Fig. 22(A-B)] appears as minute necrotic lesions on leaves. The lesions are initially small, scattered, round, and grey with brown margins on leaves. As the disease progress, irregular spots are developed and expanded until the leaves wither. Initial lesions on stems are fusiform and expanded, causing the stem to break. Finally, the fibres rupture (Wang *et al.*, 2010).

Fig. 22 (A-B). Anthracnose (*Colletotrichum gloeosporioides*) in ramie.
A. Disease on leaf B. Disease on stem

Causal organism: *Colletotrichum gloeosporioides* **(Penz.) Penz. & Sacc., 1884**

[Attidell'Istituto Veneto Scienze 2: 670]

Synonyms

= *Vermicularia gloeosporioides* Penz., (1882)

= *Phyllosticta asclepiadearum* Westend., (1851)

= *Gloeosporium fructigenum* Berk., (1856)

= *Gloeosporium affine* Sacc., (1878)

= *Phyllosticta araliae* Ellis & Everh., (1894)

= *Ramularia arisaematis* Ellis & Dearn., (1897)

= *Colletotrichum tabacum* Böning (1932)

= *Glomerella cingulata* (Stoneman) Spauld. & H. Schrenk, (1903)

= *Plectosphaera atractina* (Syd. & P. Syd.) Arx & E. Müll., (1954)

Morphological/ Morphophysiolgical characteristics

Conidiomata is acervular, formed only on host leaves abaxially on laminae and veins, setose. Setae are straight, cylindrical, tapering towards pointed apex, 62.5-120µm long, 4 to 5µm width, 2-septate, brown, smooth. Conidia are straight, cylindrical, apices obtuse, 9-22.5 x 3-4.5 µm, aseptate becoming 1-septate at germination, guttulate, hyaline, smooth. Appresoria are variable in shape, 8-25 x 5-17.5µm, aseptate or sometimes septate, thick walled, dark-brown, smooth (Seixas et al., 2007; Wang et al, 2010). Ascocarp is perithecia, solitary or aggregated, spherical to subspherical, 150-309 µm diam, walls composed of brown textura angularis, mostly 4-5 cells thick, 8.5-25 µm, smooth. Dehiscence is ostiolate. Ostiole is single, central, circular, 8.5-25.5 µm diam, lined with periphyses. Asci are unitunicate, cylindrical to clavate, 44-66 x 9-11.5 µm, rounded or slightly flattened at apex, paraphysate, 8-spored. Ascospores are straight to fusiform to slightly curved, allantoid, 12.5-17.5 x 4.5-6 µm, aseptate, guttulate, hyaline, smooth (Seixas et al., 2007).

On potato dextrose agar (PDA), the fungus initially develops white colonies with orange conidial mass and the colonies turn to gray or brown after 5 days of incubation (Wang et al., 2010). The colonies are relatively fast growing (54-86 mm diam/11 d on PDA). Mycelial growth is mostly within medium, having a central area of white to greyish, sparse, woolly aerial mycelium, where sporulation is concentrated in orange mucilaginous masses of conidia. Reverse view of colonies is greyish-white with no pigmentation of medium or evidence of diurnal zonation (Seixas et al., 2007).

Molecular characteristics

Genomic DNA was extracted from the five isolates and sequences of rDNA-ITS with primers ITS1 and ITS4 were obtained and that deposited in GenBank with Accession Nos. GQ120479–GQ120483. Comparison with the sequences in GenBank showed 99 to 100% similarity with *C. gloeosporioides* (Accession Nos, FJ515005, FJ459930, and HM016798) (Wang et al., 2010).

Wang et al. (2012) reported that for the purpose of screening putative anthracnose resistance-related genes of ramie (*Boehmeria nivea* L. Gaud), a cDNA library was constructed by suppression subtractive hybridization using anthracnose-resistant cultivar Huazhu no. 4. The cDNAs from Huazhu no. 4, which were infected with *C. gloeosporioides*, were used as the tester and cDNAs from uninfected Huazhu no. 4 as the driver. Sequencing analysis and homology searching showed that these clones represented 132 single genes, which were assigned to functional categories, including 14 putative cellular functions, according to categories established for Arabidopsis. These 132 genes included 35 disease resistance and stress tolerance-related genes including putative heat-shock protein 90, metallothionein, PR-1.2 protein, catalase gene,

WRKY family genes, and proteinase inhibitor-like protein. Partial disease-related genes were further analyzed by reverse transcription PCR and RNA gel blot. These expressed sequence tags are the first anthracnose resistance-related expressed sequence tags reported in ramie (Wang *et al.*, 2012).

Predisposing factors

The optimum temperature for growth of this pathogen is 25-28°C and pH 5.8-6.5. It is usually inactive in dry season but during favourable conditions it causes anthracnose disease (Sharma and Kulshrestha, 2015).

Disease cycle

The pathogen establishes interaction with host by producing melanised appressorium and then penetrates the host cuticle. After penetration, infection vesicles and primary hyphae are formed. Later, secondary hyphae develop and spread to kill the host cell and establish infection (Munch *et al.*, 2008). It produces conidia on acervulum which serve for re-infection. *C. gloeosporioides* follows the hemibiotrophic mode of infection where, biotrophic and necrotrophic phases are sequentially occur (Sharma and Kulshrestha, 2015).

Management/Control measures

Use of host Resistance

Anthracnose-resistant cultivar, Huazhu no. 4 (Wang *et al.*, 2012), may be cultivated to minimize the disease attack and yield loss.

Fungicidal control

Pageant (pyraclostrobin + boscalid) 38% at 12.0 oz (340.2g) /100 gal (378.54L) and Insignia (pyraclostrobin) 20% at 8.0 oz (226.8g) /100 gal (378.54L) provide excellent control of anthracnose caused by *C. gloeosporioides* (McMillan, 2011).

4. Brown root rot

Occurrence & severity status

In the year 2013, brown root rot disease was observed on ramie plants in Yuangjiang, Hunan Province, China (Yu *et al.*, 2016). The authors claimed that their work was the first report of brown root rot disease on ramie.

Symptoms

The disease symptoms appear as stunted plants, reduced number of ramets per plant, and brown to purple-black discolouration of the root (Yu *et al.*, 2016). The disease causes extensive necrosis of the root system. The infected plants lack feeder roots and exhibit reduced growth, leaf chlorosis, premature defoliation and death.

Causal organism: *Phytopythium vexans* **(de Bary) Abad, de Cock, Bala, Robideau, Lodhi & Levesque, 2014**

[Persoonia 34: 37 (2014)]

Synonyms

= *Pythium vexans* de Bary, (1876)

= *Pythium allantocladon* Sideris, (1932)

= *Pythium complectens* Hans Braun, (1924)

= *Pythium piperinum* Dastur, (1935)

= *Ovatisporangium vexans* (de Bary) Uzuhashi, Tojo & Kakish., 2010

= *Pythium vexans* var. *minutum* G.S. Mer & Khulbe, 1983

= *Pythium vexans* var. *vexans* de Bary, 1876

Morphological/ Morphophysiolgical characteristics

The main hyphal branches are 1.5-6.6 µm wide (average 3.3 µm) with finer side branches of approximately 2µm wide and the tips are curled. Sporangia are formed in large numbers in water at 20°C, terminal, subglobose, broad ovoid or obpyriform on short sporangiophores, measuring 12.9-27.9 µm long x 12.3-19.3 µm wide (average 18.6 x 15.9 µm), with mean length∆ breadth ratio of 1.2 (0.9-2.0). Oogonia are spherical and terminal, usually on short side branches, measuring 18-22µm (average 20µm) in diameter. Oospores are spherical, single and aplerotic in most cases. Antheridia are bell-shaped with a large contact surface area with the oogonia, usually one per oogonium, monoclinous or diclinous on long stalks. Encysted zoospores are 8-12µm in diameter. The cardinal growth temperatures on V8A (V8 Agar) are 9, 26 and 36°C as minimum, optimum and maximum, respectively. Average daily growth rate at 26°C is 13.4 mm. Hyphal swellings are not observed (Tao *et al.*, 2011).

Molecular characteristics

The pathogen shows sequence identity (99-100%) to the published data for internal transcribed spacer (ITS), 18S and 28S rDNA sequences of *Phytopythium vexans* (= *Pythium vexans*) (Yu *et al.*, 2016). In another study with the same pathogen collecting from Dendrobium, Tao *et al.* (2011) diagnosed the pathogen based on ITS and β tubulin gene sequences. In their studies, a single band was amplified based on the internal transcribed spacer-1, 5.8S and large parts of the ITS2 region (884 bp) and the β tubulin gene (989 bp) of the *Phytopythium* (= *Pythium*) isolate collected in Simao City, Yunnan Province. Alignment of the ITS and b tubulin gene sequences were submitted to Genbank with the Accession no. GU931701 and Accession no. GU931700, respectively. The sequences were homologous to the published Genbank database of *Phytopythium vexans* (=*Pythium vexans*) (Accession no.:AM701804, Isolate: IFAPA-CH538, from feeder avocado roots and orchard soils); Accession no.: EU080484, Isolate: PD00391[P3980], from the World *Phytophthora* Genetic Resource Collection) with sequence identities of 99% and 98% for both regions, confirming that the pathogen was *P. vexans*.(Tao *et al.*, 2011).

Predisposing factors

The causal fungus needs high humidity for sporangia and zoospores production those are important infective propagules (Tao *et al.*, 2011)

Disease cycle

The pathogen may live saprophytically on plant debris in soil. It infects the young saplings or root lets by the germinated zoospores. On infection, the pathogen with entire mycelium remains in host tissues. The hyphae as developed are both intracellular and intercellular within the host tissues. They do not produce haustoria. When the infection is established the fungus produces zoospores in sporangium for reinfection. During that time hyphae are also occurred externally of the host. These external hyphae infect the neighbourhood plants. The fungus may form sexual organs like oogonia and antheridia and may perform sexual reproduction producing fertilized oospores, the overwintering structures, and later, they germinate to zoosporangia from which zoospores are formed for reinfection.

Control measures

Drip-irrigation may be used instead of sprinkling or flood irrigation to avoid the spread of zoospores. Since, the pathogenic fungus needs high humidity for the production of sporangia and zoospores which are important infective propagules (Tao *et al.*, 2011).

Etridiazole, fosetyl-aluminium and metalaxyl are effective and commonly used fungicides for controlling *Pythium* root rot (Chase, 1987). These fungicides may be effective against the brown root rot disease, while applied as soil drenching at the base of the ramie plant.

5. *Curvularia* leaf blight

Occurrence & severity status

The disease has been found to occur in a wide variety of plant hosts. In Ramie, the disease occurs in Assam during the months June-November with10-18% infection (Gawande *et al.*, 2016). Its occurrence is more prevalent during the month of August, particularly in ramie genotype R-1452 and in the seedlings generated from seed population (Gawande *et al.*, 2016).

Symptoms

The disease appears on leaves as rusty or reddish brown irregular lesions, which form ring shaped reddish brown spots, with central fade zone and surrounded by chlorotic tissue (Fig. 23A). Then, the spots rapidly expand and appear as oblong shaped which coalesced with each other on leaf surface. The infected leaf becomes completely covered with the spots within 6-8 days after its first appearance, which ultimately disrupt the photosynthetic activity of the plant that, eventually, leads to reduction in the quality and quantity of fibre yield (Gawande *et al.*, 2016).

Fig. 23 (A). Leaf spot/blight in ramie.

Causal organism: *Curvularia eragrostidis* (Henn.) J.A. Mey., 1959

[Publications de l'Institut Agronomique du Congo Belge 75: 183]

Synonyms

= *Brachysporium eragrostidis* Henn., (1908)

= *Cochliobolus eragrostidis* (Tsuda & Ueyama) Sivan., (1987)

= *Pseudocochliobolus eragrostidis* Tsuda & Ueyama, (1985)

= *Spondylocladium maculans* C.K. Bancr. (1913)

Morphological/ Morphophysiolgical

The fungal ascocarp is superficial, globose to subglobose, black, 375-750µm wide, with a protruding ostiolar beak, developing from columnar or flat stromata firmly adhering to the substratum at the base. Ostiolar beak is 250-1125 x 85-190 µm, with a hyaline apex. Asci are vestigially bitunicate, almost cylindrical with a short stalk 1-8-spored, 150-240 x 12.5-22 µm, along with filamentous, hyaline pseudoparaphyses. Ascospores are hyaline, filiform or flagelliform, 12-22-septate, parallel to loosely coiled in the ascus or rarely coiled in a helix,

175-240 x 4-6.5 µm, with or without a thin mucilaginous sheath. Conidiophores are solitary or in groups, simple or rarely branched, straight to flexuous, sometimes geniculate above, multiseptate, light brown to brown, variable in length, up to 5 µm wide. Conidia are 3-distoseptate, ellipsoidal or barrel-shaped, the middle septum almost median, appearing as a thick black band, central cells brown to dark brown, end cells paler, smooth, on natural substrata 22-33 (27.4) x 10-18 (14.5) µm, in culture 18-37 (28) x 11-20 (14) µm. Stromata are formed on the substrates (Ellis, 1966; Matsushima, 1975; Tsuda and Ueyama, 1985; Siyanesan, 1987)

Molecular characteristics

As per the NCBI records, the 18S ribosomal RNA gene, partial sequence; internal transcribed spacer 1, 5.8S ribosomal RNA gene, and internal transcribed spacer 2, complete sequence; and 28S ribosomal RNA gene, partial sequence of *Curvularia ergostidis* strains (UPM 1299, UPM 902, UPM 1188, UPM 1199, UPM 1281, UPM 1128) have the sequence length 580bp linear DNA (Nor Azizah *et al.*, 2016). The GenBank accessions of the sequences are KP340050.1, KP340049.1, KP340048.1, KP340047.1, KP340046.1 and KP340045.1, respectively. The large subunit ribosoml RNA gene partial sequence is 794bp liner DNA (Goh and Hyde, 1999).

Predisposing factors

The fungus is able to germinate and grow in a very wide range of temperature (10-40°C) or pH (2-11) conditions, although 28°C and pH 6 are optimal (Zhu and Qiang, 2011).

Disease cycle

The pathogen survives on infected plants for next season and can remain in soil for about 3 years. The conidia are dispersed by wind and infect new host tissues on germination. The pathogen establishes infection within 2-5 days and completes its life cycle within a week in warm rainy weather (Kenneth Horst, 2001). The conidia produced on the infection during the favourable season serve as secondary inoculums.

Control measures

Carbendazim 0.25% and hexaconazole 0.1% are found effective with 67.03 and 65.75% disease reduction over control, respectively (Gawande, *et al*, 2015). Further, fungicides, like, thiram and organo mercury formulations, proved to be useful in inhibiting the disease

6. White cane rot

Occurrence & severity status

This disease is also named as 'Collar rot' or '*Sclerotium* rot' or 'Stem rot'. The pathogen of the disease affects over 500 plant species all over the world (Flores-Moctezuma *et al*., 2006). The disease was first observed in ramie on 24 of 36 germplasm lines during the rainy season of 1985 at the Ramie Research Station, Sorbhog, Assam (India) and reported as 'White cane rot' for the first time from India by Singh in 1987. The disease of ramie also occurs in Philippines and was reported for the first time in 1988 (Dizon *et al*., 1988). In India, the disease occurs during the months April to October with 5-20% infection (Gawande *et al*., 2016).

Symptoms

The disease attacks ramie stem primarily, although it may infect any part of a plant under favourable environmental condition (Gawande *et al*., 2016). Symptoms of the disease appear as water soaked and brown lesions on the basal portion of the stem, followed by yellowing, wilting and defoliation of the plant. Severely infected plant turns brown and ultimately died. Profuse growth of the white mycelia of the fungus covers the infected stem (Dizon *et al*., 1988). Later, mustard like sclerotia which initially white later deep brown, are formed on white mycelia at the infected region of the base. Ramie seedlings are very susceptible and die quickly once they become infected (Gawande *et al*., 2016).

Causal organism: *Athelia rolfsii* **(Curzi) C.C. Tu & Kimbr., 1978**

Synonyms

= *Botryobasidium rolfsii* (Curzi) Venkatar., 1950

= *Corticium rolfsii* Curzi, 1932

= *Pellicularia rolfsii* (Curzi) E. West, 1947

= *Sclerotium rolfsii* Sacc., 1911

[N.B. *Sclerotium rolfsii* is the anamorphic stage of the pathogen which was earlier under the group of Incertae sedis. For a long time, its classification was controversial as the teliomorphic stage i.e. the sexual stage was rarely observed. The teleomorph of *Athelia rolfsii* (Curzi) C.C. Tu & Kimbr. is a basiomycete and this name is now accepted by IMI, Kew {Kirk, P.M (2018). Species Fungorum (version Oct 2017), Kew Mycology}]

Morphological/ Morphophysiolgical characteristics

The fungus produces effused basidiocarps (fruit bodies) that are smooth and white. Microscopically, the fruit bodies consist of ribbon-like hyphae with clamp connections. Basidia are club-shaped, bearing four smooth, ellipsoid basidiospores, measuring 4-7 x 3-5 µm. Small, brownish sclerotia are also formed, arising from the hyphae (Tu and Kimbrough, 1978).

The fungus produces abundant mycelia on various agar media except on pechay (cabbage) agar, glycerine agar, sweet potato agar, and water agar. Good mycelial growth occurs at 25, 30, and 35° C with optimum growth at 30° C. At 40° C, the mycelial growth is completely inhibited. Continuous light, continuous darkness and alternate light and dark enhance mycelial growth of the fungus. Longer exposures to ultraviolet light (16-24 hrs.) suppress growth. The fungus does not produce any fruiting structures on agar media but forms basidiocarps on the surface of inoculated ramie stem tissues (Dizon *et al.*, 1988).

Molecular characteristics

The molecular study for identification of the same pathogen from Cucurbit fruit rot was done by Mahadevakumar *et al.* (2016). In their studies the genomic DNA was isolated and internal transcribed region of ribosomal DNA of *Athelia rolfsii* (=*Sclerotium rolfsii*) was amplified using ITS1-ITS4 universal primers. The rDNA sequence results showed 99–100 % similarity with reference sequence AB075298.1 and JF966208.1 confirming pathogen identity (Mahadevakumar *et al.*, 2016).

In another study, a total of 13 isolates were sequenced, and all sequences of ITS and LSU gene showed 99–100% similarity with *A. rolfsii* (anamorph: *S. rolfsii*) sequences from GenBank by BLAST search analysis. Parsimony analysis of ITS sequences showed three species of *Sclerotium* (*S. rolfsii, S. delphinii,* and *S. coffeicola*) produced a single group with a high (99%) bootstrap value. For construction of a phylogenetic tree, Paul *et al.* (2017) used sequences of *S. rolfsii* collected worldwide, along with *S. delphinii* and *S. coffeicola*. The fungi, *S. delphinii* and *S. coffeicola* produced a separate subgroup, supporting that the isolates identified in this study were different from these two species. The maximum parsimony (MP) analysis of the large subunit (LSU) sequences resulted in two equally most parsimonious trees. A tree was selected to illustrate the phylogenetic location of the present isolates. The maximum parsimony analysis (MP) of all sequences produced three major clusters, designated as S1, S2, and S3. *S. denigrans, S. cepivorum, S. sclerotiorum,* and *S. perniciosum* were included in the S1 cluster together with *Monilinia fructicola* and *Botryotinia fuckeliana*, which was under sclerotiniaceae (Helotiales, Ascomycota). The fungi, *S. hydrophilum, S.*

rhizodes, *Thanatephorus cucumeris*, and *Ceratorhiza oryzae-sativae* clustered together as the S2 group within Cantharellales (Basidiomycota). *Athelia rolfsii* (*S. rolfsii*), *S. delphinii*, and *S. coffeicola* clustered together with a maximum bootstrap value (100%) in the S3 group (Paul *et al*., 2017).

Predisposing factors

Mustard-like sclerotial bodies germinate very rapidly as soon as the favourable condition occurs in the field. Frequent rain followed by warm temperature 20-30°C are the most important factors for the development of this disease (Gawande *et al*., 2016). In more precisely, it can be mentioned that temperature and moisture are very important factors in the spread and development of this pathogen. Hyphal growth occurs over a temperature range of 8-40°C, but optimal growth and sclerotia production occurs between 27-35°C. In addition to temperature effects, the hyphal growth and sclerotia germination require a water-saturated soil. High humidity also favours fungal development. *Athelia rolfsii* survives under adverse conditions as sclerotia or as mycelium in diseased plants or plant debris. The sclerotia formation is favoured by temperature of 30°C, ample moisture, soil pH below 7, and well-aerated, light soil. Sclerotia survive best when those present at or near the soil surface in well-drained soil (Mullen, 2001).

Disease cycle

The pathogen is soil-borne and exists only as mycelium and sclerotia. It overwinters as sclerotia (Agrios, 2005). The sclerotium is a survival structure and considered the primary inoculum (Mersha, 2017). Sclerotia germinate to mycelium very rapidly as soon as the favourable condition prevails in field. The fungus attacks the host crown and stem tissues at the soil line as mycelium by producing a number of compounds such as oxalic acid and several pectinolytic and cellulytic enzymes. These compounds effectively kill plant tissues and allow the fungus to enter the plant. After gaining entry, the pathogen uses the plant tissues to produce mycelium (often forming mycelial mats), as well as additional sclerotia. Sclerotia formation occurs when conditions are especially warm and humid (Flores-Moctezuma *et al*., 2006; Mersha, 2017).

Management/Control measures

Cultural methods

The management of the disease is critical, although, there are several practical ways to reduce disease pressure. Avoidance of infected fields is the most straight forward management practice; however it is not possible in case of ramie. Thus, the other methods, like, practicing proper sanitation, deep tillage which is effective to reduce the disease occurrence by burying infected plant tissues and creating an anaerobic environment that hinders pathogen growth (Mersha, 2017), may be adopted. Certain organic amendments (e.g. composted chicken manure and rye-vetch green manure), as well as introducing certain *Trichoderma* spp. have also been shown to reduce plant death in various crops and number of sclerotia produced in the field (Latunde-Dada, 1993; Flores-Moctezuma *et al.*, 2006;Liu *et al.*, 2007). Similar way the disease in ramie can be minimized.

Chemical control

In addition to these cultural methods, chemical methods (e.g. fungicides) can also be employed (Keinath and DuBose, 2017; Mersha, 2017). Several systemic fungicides, such as, Propiconazole 25% EC and Tricyclazole 75% WP, and among the non-systemic fungicides, Mancozeb 75% WP and Thiram are found most effective ones in inhibiting 100 % mycelial growth of *A. rolfsii* at 500ppm (Rangarani *et al.*, 2017). In a recent trial with peanut, azoxystrobin has been found to control southern blight significantly (**Mullen, 2001**). Hence, these fungicides may be effective to control white cane rot disease in ramie.

7. White fungus disease

Occurrence & severity status

The most serious disease of ramie is "white fungus disease", also known as 'white root rot'. It occurs in India, Philippines, Vietnam and Japan (Sarma, 2009; Brink, 2011). The disease affects a wide range of plant species. In China, the white root rot disease in ramie occurs most predominantly in Xianning City of Hubei Province and Yichun City of Jiangxi Province (Zhang, 2014).

Symptoms

The disease generally appears on young roots which initially covered with white mycelium. Later, the mycelium changes to brown or almost black on older roots. Superficial black sclerotia are formed and the hyphae at septa region become swollen. The infection is mostly confined to root only. The root system is destroyed and the dead roots are covered with a white veil of thread-like forms. The above ground parts show the symptoms of wilting leaves, either in slow or fairly rapid wilting. According to the age of the plant, the disease development may be slow (death occurs after several years of decline) or very rapid (sudden wilting occurs following a period of drought).

Causal organism: *Rosellinia necatrix* Berl. ex Prill. 1904
[Bull. Soc. mycol. Fr. 20: 34]

Synonyms

= *Dematophora necatrix* Hartig, 1883

= *Hypoxylon necatrix* (Berl. ex Prill.) P.M.D. Martin, 1968

Morphological/ Morphophysiolgical characteristics

The pathogen is a soilborne ascomycetous fungus. Its perithecia are densely aggregated, globose, black, shortly pedicellate at the base, 1-2 mm diam., embedded in a ropey subiculum of brown, septate hyphae which form a thin crust on the host substratum at the base of the perithecium. Hyphae of subiculum are of two types; some are uniform, 5-8 µm wide, and others are exhibiting characteristic pyriform swellings 2-3 times the width of the hypha and are formed immediately above a septum. Wall is of 3-layered, the thicker outer layer is composed of dark brown, thick-walled, rounded to polygonal cells, the middle layer is of thin-walled, flattened, elongated, brown, polygonal cells and the innermost layer is of cells similar to those of the outermost layer, these cells are becoming gradually less coloured as they progress towards the interior. The pedicel is composed of fused, intertwined filaments with thickened walls. Ostiole is papillate. Asci are cylindrical, long-stalked, unitunicate, 8-spored, 250-380 x 8-12µm, with an apical apparatus blued by iodine. Ascospores are monostichous, cymbiform, straight or curved, dark brown, 30-50 x 5-8µm, with a longitudinal germ slit running parallel to the long axis of the spore for about one-third of its length. Paraphyses are numerous and filiform. Conidiophores are produced independently or in association with perithecia on brown ropey synnemata which project straight outwards as rigid columns. Synnemata are up to 1.5mm high. Stipe is 40-300 µm wide, composed of flexuous, intertwined, repeatedly branched threads, 2-3.5µm wide, often branched dichotomously towards the apex and splaying out to form a pale to brown head. Conidiogenous cells are polyblastic and integrated and terminal on branches or discrete, sympodial, geniculate, denticulate, with short thin-walled separating cells which break across the middle leaving a minute frill at each geniculation which corresponds to a frill at the base of each conidium. Conidia are solitary, acropleurogenous, simple, ellipsoid or obovoid, hyaline to pale brown, non-septate, smooth, 3-4.5 x 2-2.5µm (Sivanesan and Holiday, 1972).

Molecular characteristics

The use of double-stranded RNA (dsRNA) for attenuating virulence of *R. necatrix* is considered a beneficial approach in plant disease management. Biological control with a double stranded fungal virus that reduces virulence should be beneficial. In this regard, a mycovirus, viz. *Rosellinia necatrix partiti virus* 1-W8 (RnPV1-W8), of the fungal pathogen was characterised by Sasaki *et al.*(2005). *Rosellinia necatrix* (W-8 isolate) that causes white root rot contained three segments of double-stranded (ds) RNA, namely L1, L2 and M. This mycovirus with particles of about 25 nm in diameter contains an RNA segment with almost the same mobility as M-dsRNA, but the band is sensitive to S1 nuclease (Sasaki *et al.*, 2005). In another study, Yaegashi *et al.*(2013) observed that novel mycoviruses, from an unknown source, were infecting *R. necatrix* strains W563 (virus-free) and NW10 [carrying a mycovirus-like double-stranded (ds) RNA element (N10)] in the soil, and that N10 dsRNA was being transmitted between incompatible strains, NW10 to W563.

Predisposing factors

This is a low temperature pathogen. It can grow well within the temperature range 3°C to 27°C, with an optimum in soil of 21-24°C (Vegheli *et al.*, 1995). There appears to be no evidence for spread by spores. Mycelium of plant roots and debris in soil is the main source of inoculums. Diseases caused by infection from the soil have occurred in widely separated areas. The pathogen spreads easily in loose soil characterized by high sand content and average quantity of water, being soil at field capacity the best for fungal growth (Anselmi and Giorcelli, 1990).

Disease cycle

Long distance dispersal of *R. necatrix* mainly occurs with infected propagating materials. In field, the infested soils and infected plant debris can be distributed by cultural practices or by water (Anselmi and Giorcelli, 1990). Pathogen diffusion in the soil may occur through direct root contact between host plants and by diffuse mycelium or by mycelial strands which grow through soil cavities from infected plants to healthy ones. Due to this mechanism of diffusion, *Rosellinia* root rots are often characterized by their occurrence in patches that extend in a circular pattern. Sexual and asexual spores have been historically considered not important for the conservation and dissemination of the pathogen (Thomas *et al.*, 1953; Khan, 1959; Teixeira de Sousa and Whalley, 1991; Nakamura *et al.*, 2000).When the fungal mycelia make contact with the root surface a condensation of the hyphae occurs. On young tissues, penetration occurs with the formation of mycelial aggregates, defined as "cone of penetration", which penetrates deeply into the cortical parenchyma, without destroying cellular walls (Perez-Jimenez, 2006). This mycelium penetrates root tissues, destroying the suberized layers and afterwards invades the phelloderm (Tourvieille de Labrouhe, 1982). In addition to the mechanical action in penetration, enzymatic and toxigenic activities also play an important role during the infection process. *Rosellinia necatrix* possesses high activity of cellulolitic enzymes (Araki, 1967; Sztejnberg *et al.*, 1989; Melo and Ferraz, 1990) and produces different metabolites with phytotoxic effects (Abe and Kono, 1957). The fungus establishes disease in the host tissues and extends some hyphae in soils. The secondary infection occurs to the nearby plants through hyphal strands that extended in soil.

Management/Control measures

The infected areas should be dug out and burned or disinfected by formalin (0.5%) spray or by chloropicrin solution. However, enough care should be taken for the use of last one, since it is hazardous chemical with high toxicity.

The weeds, viz., *Prosopis farcta*, *Amaranthus gracilis* and *Conyza bonariensis*, promote the disease spreading. So, the field sanitation by proper weeding is highly beneficial to check the disease spread (Sztejnberg and Madar, 1980).

The fungicide, Fluazinam (50 EC) @ 0.5ml/l is suggested to apply as soil drenching at the infected areas, before and after 12 months of plantation. This is very effective measure to control white root rot disease, caused by *R. necatrix* in field (Sugimoto, 2002; Stephens, 2003).

8. Ramie mosaic

Occurrence & severity status

The disease occurs in Assam (India) during the months October to March with 30-70% infection (Gawande *et al.*, 2016). In China, the infection is found as highest during September to December and infection is in the range of 30 to 70% on different genotypes of ramie where the temperature remains in the range of 15 to 26°C (Liang et. al., 1994; Li et.al., 2010). The incidence of the disease is found greater in the first and second harvests than in the third harvest in India. In a field, most of the plants may be diseased and the infection may start at any stage of plant growth. The disease never kills plants but lowers the quality and quantity of the crop, particularly when the young plants are infected (Gawande *et al.*, 2016).

Symptoms

The disease appears with yellow mosaic symptoms (Li *et al.*, 2010).The mosaic symptoms are with intermingled patches of normal and light green or yellowish colours on the leaves of infected plants and yellowing of the entire network of veins in the leaf blade [Fig. 24(A-B)]. In severe infections the younger leaves turn yellow, become reduced in size and the plant is highly stunted (Gawande *et al.*, 2016).

Fig. 24 (A-B). Ramie mosaic disease caused by *Ramie mosaic virus* (RamMV). **A.** Showing pluckering, deformation, yellowing of leaves with stunted growth of ramie plant. **B.** Showing yellowing of ramie leaves.

Causal organism: *Ramie mosaic virus* (RamMV)

Synonyms

= *Ramie mosaic begomovirus*

= *Tomato leaf curl Hsinchu virus* (ToLCHsV)

Virus

The virus is bipartite linear Begomovirus with lineage ssDNA. The size is 1326-1411 x 13.04-13.18 nm (Liang *et al.*, 1994). The complete nucleotide sequences of both DNA-A and DNA-B of the virus have been determined (Feng, 2008; Feng and Guo, 2008, direct submission to the INSDC). The reference sequences NC_010791 and NC_010792 for DNA-A and DNA-B, respectively, have been deposited in GenomeNet. The sequences are identical to EU596959 for DNA-A and EU596960 for DNA-B in NCBI review. *Ramie mosaic virus* is now described as *Ramie MV Begomovirus* NC_010791 EU596959 NC_010792 EU596960 bipartite China ow Ramie.

In a study of Li *et al.* (2010) the virus isolates were obtained from three ramie samples showing yellow mosaic symptoms collected in Jiangsu and Zhejiang provinces, China. Comparison of partial DNA-A fragments amplified with begomovirus universal primers PA/PB revealed that these viral isolates shared a high sequence identity. The complete DNA-A sequences of two isolates J4 and Z1 were determined to be 2736 and 2737 nts, respectively, sharing 94.7% nucleotide sequence identity with each other. Also, the DNA-B components were identified for J4 and Z1 isolates and comprised 2717 and 2719 nts, respectively, sharing 88.6% nucleotide sequence identity with each other. Furthermore, sequence alignment and phylogenetic analysis showed that J4 and Z1 isolates had the highest sequence identities (93.6-94.7%) with isolates of *Ramie mosaic virus* (RamMV) for DNA-A. These molecular data suggested that J4 and Z1 were two different isolates of RamMV (Li *et al.*, 2010).

Transmission: The virus is transmitted by white fly (*B. tabaci*). The vector allows rapid and efficient propagation of the virus because it is an indiscriminate feeder. The disease is also found to spread by propagating materials (Liang *et al.*, 1994).

Predisposing factors

The temperature range of 15 to 26 °C is favourable for the spread of the disease (Liang et. al., 1994; Li et.al., 2010; Gawande *et al.*, 2016).

Management/Control measures

Use of disease resistant variety and addition of potash are suggested to manage the disease (Liang *et al.*, 1994). The disease may also be managed by the prophylactic spray of Folidol (Ebos @ 1 ml/5 l of water) or Endrin (20 EC; @ 7.5 ml/5 l of water) for killing the vector. Recently, Rogor (0.5%), Metasystox (0.0 2%) or Dimecron (0.05%) is being suggested as prophylactic spray. However, it should be remembered that the white fly has enormous ability to develop resistance against systemic fungicides. So, the fungicides of different mode of actions should be applied simultaneously or alternatively.

[N.B. Recently, *Corchorus golden mosaic virus* (CoGMV) has been reported to cause yellow mosaic disease of ramie, with disease incidence 2 to 10%., during summer (May) in the experimental fields of CRIJAF research farm, Barrackpore, India (Biswas *et al.*, 2016). The disease appears as vein-clearing in young leaves and shows complete yellowing in older leaves. Further, the infected plants become stunted and bushy (Biswas *et al.*, 2016). *Corchorus golden mosaic virus* (CoGMV) is a bipartite Begomovirus and transmitted by whitefly, *B. tabaci*. This virus is found to affect ramie in such an area that located about 500 km away from the commercial or traditional ramie growing areas, so, the disease is not considered to describe separately in this book]

9. Nematode diseases

Ramie is a perennial bush, which mostly grown in India in the North Eastern states and to some extent in Western Ghats of Maharastra. Several nematode species, viz., *Meloidogyne incognita, M. thamesi, Pratylenchus elachistus, P. brachyurus, P. cofleae, Scutellonema brachyurum, Helicotylenchus mucronatus, Hoplolaimus indicus, Xiphinema* sp. and *Paralongidorus* sp. are recorded to be associated with this crop. Among these, root knot nematodes, particularly *M. arenaria* (= *M. thamesi*), cause considerable crop losses in ramie in India (Mishra, 1995). Root knot nematode is considered as a major pest on ramie crop in the Philippines also. In 1967, a severe decline of ramie in Mindanao was attributed to root-knot nematodes (Madamba *et al.*, 1968; Davide, 1988).

i) Root-knot nematode (*Meloidogyne incognita*)

Occurrence & severity

The root-knot nematode, *Meloidogyne incognita*, is commonly called the "southern root-knot nematode" or the "cotton root-knot nematode". This parasitic roundworm has numerous hosts. It is an important plant parasite prefers to attack the root of its host plant. It is widespread in Asia, Southeast Asia and usually occurs in warmer areas. It occurs regularly in ramie. It is considered major pest on ramie crop in the Philippines (Madamba *et al.*, 1968; Davide, 1988).

Types of damage

The root knot nematode, *M. incognita*, is an obligate parasite that causes significant damage to the host plant. When it attacks the roots of host plants, it sets up a feeding location, where it destroys the normal root cells and establishes giant cells with secretion of proteins. The roots become gnarled or nodulated, forming large galls or "knots" throughout the root system of infected plants. Severely infected plants become stunted with yellow leaves, which, eventually, reduce fibre yields.

Scientific name & order: *Meloidogyne incognita* **(Kofoid & White, 1919) Chitwood 1949 (Tylenchida: Heteroderidae/ Meloidogynidae)**

Biology

Egg: Eggs are inside the white females, ellipsoid, with size about 94.37 µm in length and 41.24 µm width (Calderon *et al.*, 2016), with rounded ends. Ist instar larva remains inside and moults in the egg.

Larva (Juvenile): In juvenile the head is not offset, truncate cone shaped in lateral view, sub-hemispherical in dorso-ventral view. Head-cap is wide followed by 2 clear annules on sublateral head sectors, 3 annules on lateral head sectors. Stylet knobs are prominent, rounded. Hemizonid is 3 annules long just anterior to excretory pore. Lateral field is with 4 incisures, outer bands cross striated.

Rectum is inflated. Tail is tapering to subacute terminus, striae coarsening posteriorly. Length is about 371 μm (Whitehead, 1968).

Adult

Female: The endoparasitic body of female is pear shaped (spherical with projecting neck). Head is with 2 or occasionally 3 annules behind head-cap. Cuticle is thickening abruptly at base of relaxed stylet. Stylet knobs are rounded or drawn out laterally. Excretory pore is at level of or posterior to stylet knobs, 10-20 annules behind head. Posterior cuticular pattern displays bewildering variation. Typical pattern is 'incognita type', with striae closely spaced, very wavy to zig-zag especially dorsally and laterally. Dorsal arch is high, rounded. Lateral field is not clear, sometimes marked by breaks in striae, broken ends often forked, pattern merging into body striae. 'Acrita type' is with striae smoother, more widely spaced. Dorsal arch is variable, may be flattened at top or trapezoid. Striae are often forked along a 'lateral line'. Limits of pattern are more or less well-defined. Aberrant patterns occur. Size: 609 x 415 μm (Whitehead, 1968).

Male: Body is vermiform. Head is not offset, a high truncate cone shape, clearly annulated. Head-cap is with stepped outline in lateral view. Aannule number behind head-cap is very variable, usually 1-3 on sublateral head sectors and 1-5 on lateral head sectors. Conus of stylet is longer than shaft, stylet knobs prominent, usually of greater width than length, with flat, concave or 'toothed' anterior margins. Excretory pore is at level of posterior end of isthmus with hemizonid usually 0-5 annules anterior. Lateral field is with 4 incisures, outer bands areolated, inner band rarely cross-striated except at posterior end. Testes are 1 or 2. Tail is bluntly rounded, terminus unstriated. Phasmids are at cloaca level or just anterior. Spicules are slightly curved, gubernaculum crescentic. Length is 1583 μm, width at stylet base 5.8 μm, width at median bulb 11.2 μm (Whitehead, 1968).

Life cycle

All nematodes pass through an embryonic stage, four juvenile stages (J1–J4) and an adult stage. Juveniles hatch from eggs as vermiform second-stage juveniles (J2). The first moult occurs within the egg. Newly hatched juveniles have a short free-living stage in the soil. They may reinvade the host plants of their parent or migrate through the soil to find a new host root. J2 larvae do not feed during the free-living stage, but use lipids stored in the gut. Once the juvenile nematode begins feeding on tissue of a favourable host, the second, third and fourth moults occur giving rise to the third, fourth and fifth or adult stages, respectively. Between moults, there is further growth and development of the nematode, with concurrent development of the reproductive systems in the two sexes. Upon maturity, the female deposits eggs and the life cycle is repeated. The length of the life cycle is temperature-dependent. Its life cycle completes in about 4-8 weeks of time.

Management/Control measures

A biocontrol agent *Paecilomyces lilacinus* while used as soil drench (200 x 10⁶ power spores/ml) shows the most reduction of root knot nematode infection. The nematicide, Phenamiphos (100 ppm) is also effective to kill the eggs of nematode (Tandingan and Asuncion, 1990). Nematicide treatment of soil planted to ramie improved growth and yield of the crop (Madamba *et al.*, 1968). Several nematicide, such as, Nemagon, Mocap, Dasanit, Nemacur, Furadan, Temik, Vydate, etc., have been found very effective against the root-knot nematode

ii) Root-knot nematode (*Meloidogyne arenaria*)

Occurrence & severity

The nematode (*M. arenaria*) is widely distributed around the world in tropical, subtropical and temperate climates where the average temperature in the warmest month is 36°C or lower and the average temperature in the coldest month is at lowest 3°C (Taylor *et al.*, 1982). This species is most common in climates with average annual temperatures of 18-27°C, with annual precipitation averages 1000-2000 mm (Taylor *et al.*, 1982). In India ramie is grown mostly in North Eastern states, where root knot nematode, *M. arenaria* (= *M. thamesi*), causes considerable crop loss. It is so severe, that after 3 to 4 years, no roots or rhizomes remain free of galls and hence plantations are rendered uneconomic after 5th year. Even fresh plantations do not survive in a freshly reclaimed ramie field (Mishra and Mandal, 1988; Mishra, 1995). The nematode more frequently infects tea.

Types of damage

The root knot nematode, *M. arenaria*, is also an obligate parasite that causes significant damage to the host plant. The roots become gnarled or nodulated, forming round gall (2-5 mm dia. in size) or fusiform swellings of the roots, often produce many small bead-like galls, i.e., "knots" throughout the root system of infected plants. In severe cases, no roots or rhizomes remain free of galls (Mishra and Mandal, 1988). Severely infected plants also become stunted with chlorotic leaves, which, eventually, reduce fibre yields. New plantations die in a freshly reclaimed ramie field (Mishra and Mandal, 1988; Mishra, 1995).

Scientific name & order: *Meloidogyne arenaria* (Neal, 1889) Chitwood, 1949 (Tylenchida: Heteroderidae/ Meloidogynidae)

Synonyms

= *Anguillula arenaria* Neal, 1889

= *Heterodera arenaria* (Neal) Marcinowski, 1909

= *Meloidogyne arenaria arenaria* (Neal) Chitwood, 1949

= *Meloidogyne arenaria thamesi* Chitwood in Chitwood *et al.*, 1952

= *Meloidogyne thamesi* (Chitwood *et al.*) Goodey, 1963

= *Tylenchus arenarius* (Neal) Cobb, 1890

[**N.B.** The morphological variant that in many literatures described as either subspecies *M. arenaria thamesi* or elevated to species *M. thamesi* does not correlate with other taxonomic characters and the two rankings are considered as synonyms of *M. arenaria* (Eisenback and Triantaphyllou, 1991; CABI, 2018)]

Biology

Egg: Eggs are inside the shining-white females. The egg is elliptic cylindrical, size is 77-105μm long and 32-44μm wide (Chitwood, 1949).

Larva: Second stage larva (juvenile) of this nematode is short, (400-) 450-490 (–600) μm, with very short stylets (10-15m μm). The stylet is lightly sclerotized with knobs 2μm across by 1μm long, whch taper into stylet. The cephalic framework is also weakly sclerotized. The dorsal gland orifice is 3μm posterior to base of stylet (Chitwood, 1949). The tail is moderately long (44-69 (56) μm), and the poorly defined hyaline tail terminus is moderately long (6-13 (9) μm), tapers to a finely rounded to pointed tip with a clear terminus.

Adult

Female: The body of the female is swollen, pearly white and pear-shaped, sedentary, 500-1000 μm long by 400-600 μm wide. The conical neck of the female is in line with the spherical portion of the body. The stylet is dorsally curved, robust, 13-17 (16) μm long and characteristically shaped with large, posteriorly sloped, tear-drop-shaped knobs. The distance of the dorsal oesophageal gland orifice to the base of the stylet is comparatively long (3-7 (5) μm). The excretory pore is located closer to the base of the stylet than to the median bulb.

Perineal pattern

The perineal pattern may be characteristic for the species, but some populations may contain individual variants that restrict the usefulness of this character. This is very similar to that of *M. incognita* and other *Meloidogyne* species. Patterns that contain short, lateral incisures resemble that of *M. javanica*, and patterns that are rounded to hexagonal, often containing wings, are like that of *M. hapla*. The perineal pattern of *M. arenaria* has a low and rounded dorsal arch, but in some individuals it may be high and squarish. The dorsal arch of *M. arenaria* race 1 is flattened and rounded. The striae in the arch are slightly indented at the lateral lines and generally form a shoulder on the arch (Eisenback *et al.*, 1981). The perineal patterns of *M. arenaria*, race 2 show moderate morphological variations. The overall shape is rounded in most cases, but it is oval-shaped in a small proportion. The dorsal arch varies from low to high, with a few intermediate. The lines in the post-anal region are smooth or wavy, continuous or broken, occasionally forming shoulders in some females. Wings are generally not found. Phasmids are visible in some patterns and the mean distance between them is about 24 mm. Only in few cases perineal patterns are with lateral lines weakly demarcated by forked striae. However, in general, the striae are coarse and smooth to wavy, and some striae may bend toward the vulva. The most useful character of the perineal pattern is the lines in the lateral areas of the dorsal arch that sharply curve toward the tail terminus and meet the ventral striae at an angle. These striae become forked and the distance between them increases near the lateral areas which are often demarcated, but not delineated by distinct lateral incisures. Very short lateral incisures may be present very near the tail terminus. Some perineal patterns of *M. arenaria* form one or two 'wings' that extend laterally and are marked by fusion of the striae in the dorsal and ventral arches.

Male: Males of *M. arenaria* are long (0.9-2.3 mm; avg.1.6mm) and narrow (27-48 µm) (Garcia and Sanchez-Puerta, 2012). The head is flat to concave. The labial disc and medial lips form a smooth, posteriorly sloping head cap. The head annule is smooth and usually not marked with additional head annulations. Both the head annule and the body annulations are in the same contour. The stylet is long (20-28 µm), robust with distinct knobs and a bluntly pointed tip. The dorsal oesophageal gland orifice to the base of the stylet is long to very long (4-8) 6 µm, overlaps the intestine ventrally. The tail is short and rounded and has no bursa. Spicules open a short distance from the tail tip.

[**N.B.** A DNA probe that is specific for *M. arenaria* has been developed and may be useful for diagnosis of this species (Baurn *et al.*, 1994). Cytological and biochemical characterization provide additional characters for identification of *M. arenaria* (Triantaphyllou, 1979; Esbenshade and Triantaphyllou, 1990).]

Life Cycle

The egg of a root-knot nematode develops into a vermiform first-stage juvenile that undergoes first moult into a second-stage juvenile. The second-stage juvenile hatches from the egg. It moves freely in the soil, penetrates the root just behind the root cap, migrates intercellularly in the root and establishes a feeding site within the developing vascular cylinder. As it feeds on the nematode-induced giant cell system, the second-stage juvenile loses its mobility and begins to increase in girth. After it has imbibed a sufficient quantity of sustenance, the flask-shaped second-stage juvenile moults three times without feeding and matures into a saccate adult female. Females of *M. arenaria* reproduce by mitotic parthenogenesis; as soon as they are mature adults they begin producing eggs (Triantaphyllou and Hirschmann, 1960). Male second-stage juveniles undergo a metamorphosis during the third moult into elongate vermiform fourth-stage juveniles. The fourth-stage juvenile male remains enclosed in the cuticle of the second and third stages where it moults again to form an adult vermiform male. The male escapes from the cuticles and the root system. It moves freely in the soil, not feeding, only mating with mature adult females. As populations of *M. arenaria* reproduce by mitotic parthenogenesis, males serve no reproductive function (Triantaphyllou and Hirschmann, 1960). The length of one generation of *M. arenaria* is greatly affected by temperature. At very high temperatures (>29°C), the life cycle takes approximately 3 weeks, but at very cool temperatures it can be extended to 2-3 months.

Management/Control measures

Experimental trials incorporating granular pesticides, saw dust, mustard oil cake and tagetes did not show any effect as the crop is grown in high rainfall areas along slopes that leaches all the active principles very soon. Hence, while planting ramie, it is suggested to take quarantine measures and the plantation should be frequently checked for the presence of galls and larvae so that control measures can be taken at the earliest (Mishra, 1995).

The biocontrol agent *Bacillus cereus* S2 as found effective against other *Meloidogyne* species, may be effective in to control *M. arenaria* in Ramie. The bacterium strain produces sphingosine ($C_{16}H_{35}NO_2$) which has nematicidal activity (Gao *et al.*, 2016).

The application of fumigants SN556 (40% methyl isothiocynate [MIT]), 1,3D at 84litres/ha, and CO562 (30% methyl isothiocyanate, 70% 1,3-D) as found effective to control the nematode attacks in peanut (Dickson and Hewlett, 1988) may be effective in ramie also. The broadcast application before 7-8 days of plantations of 1, 3-D is superior to row applications (Dickson and Hewlett, 1988). Both DBCP (1,2-Dibromo-3-chloropropane) and EDB (ethylene dibromide) had longer residual activity in soil than 1,3-D because of their vapour pressure and higher water solubility (Hague and Gowen, 1987). However, the latter two chemicals are banned in many countries. Addition of nonfumigant nematicide, aldicarb, in 1,3-D treated plots gives better control (Dickson and Hewlett, 1988).

iii) Root lesion nematode/ lesion nematode

Occurrence & severity

Root-lesion nematode (RLN, *Pratylenchus coffeae*), which is a migratory root endoparasite, is one of the three most devastating plant parasitic nematodes. The root lesion nematode is a major ramie pest, causing plant mortality and large fibre yield losses in China annually (Yu *et al.*, 2012; Zhu *et al.*, 2012, 2014). The yield damage is so high that there use of nematicides can give increased yield ranging from 11 to 60% more (China, 1990). The nematode occurs worldwide both in tropical and temperate areas and has a broad host range, including several important tropical and temperate crop species (Edwards and Wehunt, 1973; O'Bannon and Esser, 1975; Das and Das, 1986). The lesion nematode can survive in moist soil for about eight months in the absence of host (Colbran, 1954).

Types of damage

The nematode causes root rot, which inhibits the growth of plants. It also causes remarkable changes in stem shape, resulting in a significant decrease in bast fibre yield (China, 1990; Zhu *et al.*, 2012). The nematode is a migratory endoparasitic and develops mainly in the cortical parenchyma where it causes severe root damage and hinders the absorption of water and nutrients from the soil, resulting in the inhibition of plant growth (Zhu *et al.*, 2014).

Scientific name & order: ***Pratylenchus coffeae*** **(Zimmermann 1898) Filipjev & Schuurmans Steckhoven 1941 (Tylenchina: Pratylenchidae)**

Synonyms

= *Pratylenchus musicola*

= *Tylenchus coffeae* Zimmermann, 1898

= *Tylenchus mahogani* Cobb, 1920

= *Anguillulina mahogani* (Cobb) Goodey, 1932

= *Pratylenchus mahogani* (Cobb) Filipjev, 1936

= *Tylenchus musicola* Cobb, 1919

[**N.B.** A different species of this genus, viz., *Paratylenchus elachistus*, which is known as 'Ramie pin nematode' has been reported to affect ramie. However, the literature on this nematode in ramie is scanty (Ravichandra, 2013).]

Biology

Egg: Eggs are laid singly or in small groups within the tissues of the root. The eggs hatch in 6-8 days at 28-30°C.

Larva: The larva hatches within egg. The first moult takes place within the egg and three moults occur outside.

Adult

Female: Female body is slender in young and fatter in old, 0.45-0.7 mm long, spear 15-18 μm, Neotype female 0.59 mm long, spear 18μm (Sher and Allen, 1953). Lateral fields normally with four, sometimes five or six incisures. The head (lip) region is low and flattened, with two distinct annules, occasionally three annules, on one side of the lip region. It has distinct head skeleton, continuous with the body contour. Basal knob of spear (stylet) is round to oblong. The esophagus has a well developed median bulb, and the posterior gland lobes overlap the intestine ventrally. Post-uterine branch 1.0-1.5 times body-width long, but may be upto 90μm long, with a terminal rudimentary ovary which sometimes have distinct oocytes. Spermathecae is large, broadly oval to nearly rounded, often with sperms. Intra-uterine eggs may contain embryos. Tail 2.0-2.5 times anal body width. Terminus indented, sometimes appearing smoothly rounded, truncate or irregularly crenate.

Male: Male adult is abundant. Body is 0.45-0.70mm long, spear 15-17μm (Loof, 1960). Spicules is slender with well marked manubria and ventrally arcuate shaft, 16-20μm long. Gubernaculum is 4-7μm in length. Hypotygma is prominent. The male tail is conical with a distinct bursa that reaches the tail tip. Bursal margins are faintly crenate.

Life cycle

All life stages and both sexes invade and feed in the root tissues where the eggs are laid. The life cycle of *P. coffeae* completes within 21-28 (27) days at 25°-30°C (Siddiqi, 1972). This species has the typical nematode life cycle with four juvenile stages, and the adults. The juveniles of *P. coffeae* mature and differentiate within the root, and adult females deposit eggs singly or in small groups within the tissues of the root.

Molecular characteristics

The lesion nematode (*P. coffeae*) is a widespread and variable species which may represent a species complex. Recent molecular and morphological studies on isolates of *P. coffeae* and closely related species have demonstrated difficulties involved in identification (Duncan *et al.*, 1999). The complex has yet to be satisfactorily resolved. The *P. Coffeae* nematode genome encodes fewer than half the number of genes found in the genomes of root knot nematodes, comparative analysis to determine genes *P. coffeae* does not carry, that may help to define development of more sophisticated forms of nematode-plant interactions. The *P. coffeae* genome sequence may help to define timelines related to evolution of parasitism amongst nematodes. The genome of *P. coffeae* is a significant new tool to understand not only nematode evolution but animal biology in general (Burke *et al.*, 2015).

Management/Control Measures

Use of resistance

Ramie cultivar Qingdaye is resistant to the root lesion nematode, while, the cultivars, 'Heipidou' and 'Zhongzhu 1' are susceptible in China (Zheng *et al.*, 1992; Zhu *et al.*, 2014).

Molecular studies on host resistance

In order to determine resistance mechanism in ramie against the root-lesion nematode (RLN), Zhu *et al.* (2014) identified genes that are potentially involved in the RLN resistance in ramie using Illumina pair-end sequencing in two RLN-infected plants (Inf1 and Inf2) and two control plants (CO1 and CO2). Approximately 56.3, 51.7, 43.4, and 45.0 million sequencing reads were generated from the libraries of CO1, CO2, Inf1, and Inf2, respectively. De novo assembly for these 196 million reads yielded 50,486 unigenes with an average length of 853.3 bp. A total of 24,820 (49.2%) genes annotated for their function. Comparison of gene expression levels between CO and Inf ramie revealed 777 differentially expressed genes (DEGs). The expression levels of 12 DEGs were further phenylpropanoid biosynthesis were strongly influenced by RLN infection. A series of candidate genes and pathways that may contribute to the defense response against RLN in ramie will be helpful for further improving resistance to RLN infection (Zhu *et al.*, 2014). More recently, a ramie cystatin gene *BnCPI* (Cysteine protease inhibitor, Unigene 11292) has been found to be regulated in *P. coffeae*-infected resistant ramie but not in a susceptible cultivar, which suggests that *BnCPI* may be involved in pest resistance (Yu *et al.*, 2015).

Organic amendments

Addition of organic amendments, like, chicken manure is very effective in reducing the nematode.

Chemical control

One common management option for lesion nematodes is application of nematicides, like, Aldicarb (Temik) (Saxena and Mukerji, 2007). However, the use of aldicarb and methyl bromide, also effective against lesion nematode, is banned because of their negative environmental impact. A number of organophosphate, oxime carbamate and carbamate nematicides are used to control this nematode as granular or emulsifiable concentrate formulations. The nematicides have been successfully used to control *P. coffeae* in Central America. Effective control of *P. coffeae* can also be achieved with organocarbamate and organophosphate nematicides under field conditions.

iv) Stunt nematode

Occurrence & severity

Stunt nematode has been reported in various literatures that it affects ramie however literature, pertaining its impact in ramie cultivation is scanty (Anderson and Potter, 1991). The nematode is a soil inhabiting plant parasite. It is widely distributed and can infect wide host range. In ramie, it occurs at Yuanjiang (Hunan) in China (Song *et al.*, 2017). The nematode is also reported to affect ramie in the Philippines. However, it does not cause much damage there, though; it is found in high population densities (Davide 1988).

Types of damage

The nematode damages by feeding ectoparasitically on roots. It produces brownish lesions at the feeding site, and can cause severe stunting and chlorosis. Symptoms of stunting and yellowing are most severe when plants are under stress (Radewald *et al.*, 1972).

Scientific name & order: *Tylenchorhynchus* sp. (Cobb, 1913) (Tylenchida: Belonolaimidae)

Biology

Egg: Eggs are cylindrical with round end and are laid as unsegmented. The protoplasm in the egg is granular with irregular margin. Egg size is about 62-68 μm long and 20-23 μm width as in *T. claytoni* (Wang, 1971).

Larva: Three larval stages occur between hatching and the adult stage. The length of genital primordium of second stage larva is about 8-10 μm and is located at the right of the ventral chord. In third stage female gonad differed markedly from the male gonad. The gonad is measured 20-26 μm long and at the time of moult it extended in length to 62-86 μm. In third stage the male larva possesses same number of cells in gonad as in 2nd stage larva. The gonad measured 17-

24μm in length. The gonad increased to 50-55μm. In fourth stage female larva each germ cell divides into two to three cells of equal size and the vaginal primodium cells increased in number. The gonad is 96-105μm long that distinctly more developed at the time of moult. The anterior part of the gonad is increased to 115-162μm in length, while, the posterior one is measured to 126-173μm in length. In fourth stage male larva the gonad is measured 71-97μm in length. The clocal primordium is increased in size. During the moult gonad is joined with the cloacal primordium. By this time spermacytes and sperms are observed. Spicule, gubernaculums and caudal alae developed well at the same time as the stylet. The gonad increased to 290-330μm in *T. claytoni* (Wang, 1971).

Adult: There is no sexual dimorphism in adults. Adult is vermiform. Esophageal glands are bound by a membrane into a large basal bulb. Sometimes a lobe of the basal bulb slightly overlaps intestine. Lateral fields marked by 4, 5 or 6 incisures. Stylet well developed, 15 to 30 μm long, thin to slender, has strog basal knobs, with cone about as long as shaft, sometimes needle-like.. Deirids are usually inconspicuous. Phasmids are conspicuous, located near middle of tail. Vulva is near middle of body. In female ovaries are two, amphidelphic, outstretched. Female tail is cylindrical, conoid, with terminus usually bluntly rounded, not acute. Male is monorchic, tail is slightly arcuate, enveloped by bursa; phasmids about middle of tail. Spicula is with well-developed velum (gubernaculum). Cephalic framework is lightly to heavily sclerotized (Allen, 1955; Fortuner and Luc, 1987).

Life cycle

Life cycle is similar to other parasitic nematodes with four moults and four larval stages. Sex can be differentiated in the third stage larva. Male is required for reproduction. The lifecycle from egg to egg requires 31-38 days. The second stage 7-8 days, third stage 6-7 days, the fourth stage 7-9 days and preovipostion period of adult is 8-10 days in case of *T. claytoni* (Wang, 1971).

Management/Control measures

Organic Amendments

Addition of organic amendments, like, chicken manure is very effective in reducing the nematode.

Chemical Control

The application of nematicides, like, Aldicarb (Temik), may be effective to control stunt nematode in ramie (Saxena and Mukerji, 2007). A number of organophosphate, oxime carbamate and carbamate nematicides may be used to control this nematode as granular or emulsifiable concentrate formulations.

v) Reniform nematode

Occurrence & severity

Reniform nematodes in the genus *Rotylenchulus* are semiendoparasitic (partially inside roots) species in which the females penetrate the root cortex, establish a permanent-feeding site in the stele region of the root and become sedentary or immobile. The anterior portion (head region) of the body remains embedded in the root whereas the posterior portion (tail region) protrudes from the root surface and swells during maturation. The term reniform refers to the kidney-shaped body of the mature female. The reniform nematode is largely distributed in tropical, subtropical and warm temperate zones in the Americas, Africa, Europe, Asia, and Australia (Ayala and Ramirez 1964). Mostly affects pineapple and some legume crops in Philippines (Davide, 1988). It is rapidly becoming the most economically important pest associated with cotton in the southeastern United States (Lawrence *et al.*, 2005). In Ramie sporadic reports indicated the occurrence of this nematode as a minor pest (Brink and Achigan-Dako, 2012).

Types of damage

Only females infect plant roots. After infection, a feeding site composed of syncytial cells is formed. A syncytial cell is a multinucleated cell resulting from cell wall dissolution of several surrounding cells. The nematode infestation can cause symptoms in the plant that resemble those of moisture and nutrient deficiencies. It causes hypertrophy in the pericycle cells of seedling roots and in the periderm cells of the roots of older plants (Oteifa, 1970). Root growth slows and secondary root development is limited. Root necrosis has been observed. Infested plants can become stunted and chlorotic.

Scientific name & order: *Rotylenchulus* sp. (Tylenchida; Hoplolaimidae)

Biology

Egg: Egg is elongated oval but a little curved at one side. It hatches one to two weeks after laid. About 60 to 200 eggs are deposited in soil surrounded by a gelatinous matrix (egg sac).

Larva: The first stage juvenile moults within the egg, producing second-stage juvenile that emerges from the egg. The second stage larva goes through three more moults to become adult. The average body length is about 0.34 to 0.42 mm for juveniles. The juvenile tail tapers to a narrow, rounded terminus with about 20 to 24 annules. The lateral field of juveniles has four incisures that are not areolated (Mai and Mullin, 1996).

Adult

The average body length is about 0.34 to 0.42 mm for males and 0.38 to 0.52 mm for mature female nematodes. Mature female is kidney shaped, while, mature male is vermiform. The lip region of the young female is not offset, and the cephalic framework is conspicuous. The stylet is 16 to 21 µm long, of moderate strength with a small rounded knob. The dorsal gland orifice is more than one-half the stylet length and is posterior to the base of the stylet knobs. The basal glands overlap the intestine laterally or, less often, ventrally. The vulva is post-median (V > 63%). The female reproductive system is amphidelphic with two flexures in immature females and highly convoluted in mature females. The female tail is usually more than twice the length of the anal body diameter. Phasmids are pore like, about the body width or less, behind the anus. Males have weak stylets and stylet knobs, a reduced esophagus, and an indistinct median bulb and valve. Caudal alae are adanal. The lateral field of males and young females has four incisures that are not areolated (Mai and Mullin 1996).

Life cycle

Egg hatches to the second-stage juvenile (J2) that emerges from the egg. The infective stage is reached one to two weeks after hatching. Once root penetration occurs, one or two more weeks are required for a female to reach maturity. The male remains outside of the root. It can inseminate the female before female gonad maturation. Sperms are stored in the spermatheca. Soon after female gonad maturation, the eggs are fertilized with sperm, and about 60 to 200 eggs are deposited into the soil covered by gelatinous matrix (ovisac). The life cycle of this nematode depends on soil temperature. It completes egg to egg life cycle within (22-) 27-29 (-32) days at 30-32°C (Patel, 1982). However, it can survive at least two years in the absence of a host in dry soil through anhydrobiosis, which is a survival mechanism that allows the nematode to enter an ametabolic state and live without water for extended periods of time (Radewald and Takeshita, 1964).

Management/Control measures

Cultural control

Use of trap and antagonistic crops: *Tagetes erecta*, *T. patula* (marigold), *Crotalaria juncea* (sunhemp) *and C. spectabilis* in nematode infested soil has been found effective against nematode (Sasser and Carter, 1985).

Biological control

Paecilomyces lilacinus, a fungal egg parasite is found effective against reniform nematode (Galano *et al.*, 1996).

Chemical control

Several nematicides, such as, Nemagon, Mocap, Dasnit, Nemacur, Furadan, Temik, Vydate, have been reported to be effective against reniform nematode (Birchfield and Martin, 1968; Sasser and Carter, 1985). The soil fumigant, 1,3-Dichloropropene (1,3-D) (Telone) while applied prior planting @ 8 gal/acre is very much effective to control reniform nematodes in cotton and pineapple (Robinson *et al.*, 1987). The effectivity may further increase when 1,3-D (Telone) and aldicarb are applied simultaneously. The nematicide, Temik 150G (6lb/acre) may also be applied in soil to control the nematode.

B. Pests

1. Indian red admiral caterpillar

Occurrence & severity

Indian red admiral, also called as 'Asian admiral', is a butterfly found in the higher altitude regions of India and its neighbouring countries (Bingham, 1905; Kunte, 2013; Van Der Poorten and Van Der Poorten, 2016). It is mostly found in the higher altitude regions (above 2,000 feet (610 m) of Himalayas in India and of the Nilgiri Hills in southern India. It also occurs on smaller hill ranges in Peninsular India such as the Nandi Hills near Bangalore. The red admiral occurs over a wide range of habitat types and altitudes, from coastal lowlands to montane areas, and woodland to urban centres. The red admiral can often be abundant in gardens and orchards where rich sources of nectar are present, and are not restricted to areas with available larval food plants. The occurrence of the insect pest in ramie has been found in germplasms under Assam agroclimatic conditions in India. The infestation was in the range of 10-30% and it was highest in the month of December and January later that decreased in the month of February. The infestation was found more on R-67-34 (kanai) germplasm (Gawande *et al.*, 2012, 2014).

Types of damage

The red admiral causes considerable damage by feeding on tender leaves which affect the growth of the plant. The adults lay their eggs on tender leaves in which larvae emerge and feeds on young tender leaves, preferably at neck region of the leaves. As a result, these leaves droop down and remain in hanging position and latter get dry. The mature caterpillar feeds on young leaves and folds the leaves in such a way that both the margins are attached by a silky web , and pupates inside the folded leaf which appearing as a ball. This ball-like folded leaves lead to branching of ramie cane (Gawande *et al.*, 2016).

Scientific name & order: *Vanessa indica* **Herbst, 1794 (Lepidoptera: Nymphalidae)**

Synonyms

= *Vanessa calliroe* (Hübner, 1808)

= *Pyrameis asakurae* Matsumura, 1908

= *Pyrameis atalanta* Cramer, 1779

= *Pyrameis buana* Fruhstorfer, 1898

= *Pyrameis callirhoe* Miller, 1868

= *Pyrameis horishanus* Nire, 1917

= *Pyrameis occidentalis* Felder, 1862

Biology

Egg: Eggs are bright green with vertical, lighter green ridge, measuring about 0.7-0.9 mm in size. The eggs are usually laid singly on the upper side of stinging nettle leaves, on stinging spines, but may be laid anywhere on the plant. Sometimes these are laid in small batches of 2-4 (Kunte and Sengupta, 2018).

larva: Newly hatched larva (caterpillar) is more hairy than spiny, brown, with black head and no markings. The spines grow larger in successive instars. From second instar onward, the caterpillar has a series of large white markings on the sides, which grow more prominent in successive instars. The fourth and fifth instar caterpillars have white and yellow dots all over the body. The final instar caterpillar is deep maroon-brown with creamy greenish yellow spines, and with a white-yellow dorsal band. The head remains black, but in later instars there develop prominent white tubercles, each bearing a long shiny hair. There are no cephalic spines (Kunte and Sengupta, 2018).

Pupa: Pupa is formed inside shelter similar but larger to the larval shelter. It usually hangs from near the margin rather than from the roof of the shelter. Pupa is either shiny marble-white or pale brown, with two rows of small, light orange tubercles on the dorsal side of the abdomen. These tubercles are small near the tip of the abdomen and increase in size towards the thorax. The last three pairs of tubercles, just behind the pointed projection on the dorsal side of the thorax, have much larger, shiny silvery-golden bases. There is an additional pair of small tubercle on the thorax (Kunte and Sengupta, 2018).

Adult: The adult is a medium sized rich dark gray brown and fast flying butterfly; with wing span about 55-60 mm (Fig. 23). Both male and female are similar in appearance. The underside has a blotched appearance with many shades and patterns of brown and gray. Its ground colour is dark both on the upperside and underside, and the orange markings are deep and rich in tint. The underside of its forewing has the ochraceous orange-red on disc, and across cell proportionately of less extent and uniform, not getting paler towards the apex of the cell. The upper four spots of the preapical transverse series on the black apical area are minute. Hindwing has narrow postdiscal transverse band and short, not extending below vein 1, margined inwardly by a series of broad black subcrescentic marks. The tornal angle is with a small patch of violet scales bordered inwardly by a short black transverse line. Underside is very much dark, the orange red on the disc and in the cell on the forewing are restricted as on the upperside. Three small transversely placed blue spots are beyond the cell. Hindwing mottling is very dark, purplish blade, with slender white margins, shaded on disc with rich dark olive-brown. The postdiscal series of ocelli are dark and somewhat obscure. An inner subterminal transverse series is of blue and an outer very much slenderer transverse series of black lunules. Cilia of both forewings and hindwings are white, alternated with brown. Antenna is black, tipped with pale ochraceous. Head, thorax and abdomen are with dark olive-brown pubescence. Beneath, the palpi, thorax and abdomen are pale ochraceous brown (Bingham, 1905).

Fig. 25. Indian Red Admiral caterpillar (*Vanessa indica*) fed ramie leaf along with just emerged adult butterfly

Life cycle

Indian male red admiral is a territorial butterfly that patrols its area in order to find female mate. The male typically perches upon a sunlit spot, in the mid-afternoon, to wait for female to fly by. Once fertilized, the female red admiral will lay its eggs on the upper surface of host plant leaves. Young larvae (caterpillars) hatch after about a week in good summer weather, and the caterpillars pass through five instars. Caterpillars create loosely closed shelters for themselves. During the first and second instars, when they are too small to bend over leaf margins, they weave a thick mat of silk over their resting place. In later instars, they start to make loose cells by turning over leaf margins, or by drawing together margins of adjacent leaves. They lie inside the shelters curled sideways, and come out to feed when hungry. They change their shelters when a large part of the leaf is eaten and the shelters no longer adequately conceal them (Kunte and Sengupta, 2018). Larvae pupate within the 'leaf tent' created from a folded nettle leaf. Adult emerges from the pupae within 2-3 weeks. In some cases, the butterfly may overwinter as eggs or caterpillars

Control measure

The adult red admiral butterfly shows a colour preference to yellow and blue for its visit. It even prefers odourless yellow colour than floral scents (Omura and Honda, 2005). So, yellow sticky trap may be useful to control the caterpillar damage if it causes economic damage.

2. Six-spotted zigzag ladybird

Occurrence & severity

The six-spotted zigzag ladybird beetle is usually known as a friend insect which feeds on harmful aphids and scale insects in various crops (Singh, 2007). In ramie crop, however, this is a harmful pest that feeds on tender leaves, causing considerable damage, particularly, during July to September in Assam (India) as reported by Gawande *et al.* (2016), considering it as an Epilachna beetle.

Types of damage

The adult beetle insect scraps the tender leaves and feed on chlorophyll, resulting initially in whitish longitudinal broad streaks on leaves and later in membranous leaves giving scorched appearance which affect the photosynthetic activity of the plant (Gawande *et al.*, 2016).

Scientific name & order: *Cheilomenes sexmaculata* **Fabricius , 1781 (Coleoptera: Coccinellidae)**

Synonyms

= *Chilomenes sexmaculata* (Fabricius),

= *Menochilus sexmaculatus* (Fabricius),

= *Menochilus quadriplagiatus* (Swartz)

Biology

Egg: The freshly laid eggs are small, cigar shaped, shiny yellow in colour, turn light grey just before hatching. Incubation period is about 2.85 days (Singh *et al.*, 2008).

Grub: The young grubs are black with long legs, body tapering to the hind end as it grows older, white spots appear and full grown grubs (larvae) are black with yellow and white blotches (Shanmugapriya *et al.*, 2017). The duration of first, second, third and fourth grub instars are 2.45, 2.25, 2.60 and 3.55 days, respectively. Total grub period is about 11 (10.83-11.08) days (Bhaduria *et al.*, 2001; Singh *et al.*, 2008).

Pupa: Pupation takes place on the leaf; the grub (larva) undergoes pupation by fixing itself by the tail. When the grubs are about to pupate, they turn dark brown colour, attaches itself either on upper or lower surface or sometimes even on stems. Black spots are established symmetrically on the segments of fully formed pupae (Shanmugapriya *et al.*, 2017). The pupal period is 4.48 days (Singh *et al.*, 2008). The pupal period varies from 3.60 to 5.5 days depending upon the types of foods they fed (Bhadauria *et al.*, 2001; Reddy *et al.*, 2001).

Adult: The adult beetles are broadly oval to sub-rounded in shape, dorsum moderately convex and shiny, with body length 3.3-6.2 mm and width 3.0-5.3 mm. Ground colour is orange, light red, yellow or pinkish with the following markings in the typical form: head with a black marking in posterior half ; pronotum with a T-shaped median marking connected to a broad black band along posterior margin; elytra with six black maculae including two zigzag lines and a posterior black spot, sutural line with a narrow to moderately broad black stripe. Ventral side is uniformly yellow. Antenna is short and compact. Prosternal process is with a pair of sub parallel carinae reaching up to middle (Anonymous, ND4). Male and female adult beetles can be differentiated on the basis body size and structure of external genitalia. The last abdominal segment is entire or notched to some degree in males and it is narrow or evenly rounded or medially divided or rounded in case of females (Shanmugapriya *et al.*, 2017). Longevity of adult is about 89 to 96 days (Singh *et al.*, 2008).

Life cycle

The adult female lay eggs a total of about 277-300 eggs during its ovipostion period of 24-30 days (Bhadauria *et al.*, 2001; Singh *et al.*, 2008). Egg incubation period is about 3 (2.85) days; total grub period is about 11 (10.83-11.08) days ; the pupal period is 4-5 (4.48) days. Total duration of life cycle is about 18-21 days (Debraj *et al.*, 1997; Singh *et al.*, 2008).

Control measures

No control measure has been suggested. It may be sensitive to any insecticides, like, dimethoate 30EC at 0.1-0.2% concentration.

[**N.B.** *Micraspis discolor* Fabricius, 1798 (Coleoptera: Coccinellidae) (Syn: *Verania discolor*; *Coccinella discolor*) is also another ladybird beetle affecting ramie, causing same type of damage (Gawande *et al.*, 2016). It is distributed in different areas of India, Bangladesh, Taiwan, Malaysia, Thailand, Indonesia, Philippines, China, Japan (including Ryukyu Islands), Myanmar, Pakistan and

Sri- Lanka. In India, the adults of *M. discolor* occur throughout the crop year in aphid infested fields and then disperse to weeds. *Micraspis discolour* is the most abundant species of coccinellid in rice ecosystems and touted as a biocontrol option for brown plant hopper (BPH), *Nilaparvata lugens* (Stal), a key pest of rice and a series of laboratory-based influence of host species *Aphis gossypii*. *Micraspis discolor* has been well known as both carnivorous and pollinivorous (Shanker *et al*., 2013; Hllaing, *et al*., 2017). Biology of *M. discolor* was studied under laboratory condition (26±2°C, 65±3% RH) at Department of Biotechnology, Mandalay Technological University. It has been found that incubation period and total larval period are 4-5 days and 8-9 days, respectively. The larvae pass through four instars within 8 to 9 days. The female and male longevity are 44-45 and 34-35, respectively (Hllaing, *et al.*, 2017).]

3. Leaf beetle

Occurrence & severity

The leaf beetle-attack in ramie was first reported in the year 2013 (Annonymous, 2012-13; Gawande et.al., 2013). This beetle is nocturnal in habit and it is attracted to lights at night. During the day time it remains under leaves, loose bark or as shallowly buried in the soil, and emerges at dusk to feed. Nowadays, this is considered as one of the most important insect pests of ramie crop. It causes highest damage with 5-10% infestation during May to August at temperature range 25-32°C (Gawande et.al., 2013, 2016; Annonymous, 2013-14). The pest also attacks jute. However, in jute it is considered as a minor pest in Bangladesh (Das, 1948; Rahman and Khan, 2006).

Types of damage

Adults feed on plant foliage at night, creating a lace like or shot with holes appearance on leaves by feeding on plant tissue between leaf veins. In severe cases most leaves are skeletonised (Gawande *et al*., 2016). Its larvae also live in the soil at the roots of the plants and probably do some damage (Fletcher, 1917).

Scientific name & order: *Pachnephorus bretinghami* **Baly, 1878 (Coleoptera: Chrysomelidae).**

[Zoological Journal of the Linnean Society, 14(75): 246-265 (256)]

Description: Adults are sturdy, piceous with aeneous gloss or pale golden-brown, very small in size with length about 2 mm. Its body is clothed with whitish scales (fine white hairs) that can give the beetle a greyish appearance (Baly, 1878; Selvaraj *et al*., 2016). Antennae and legs are obscure rufo-piceous.

Head is strongly punctured, excavated between the eyes. Clypeus is transverse, thickened at the base. Antennae are more than a third the length of the body, terminal five joints stained with fuscous. Thorax is scarcely broader than long, sides straight and diverging from the base to far beyond the middle, thence rather abruptly rounded and converging to the apex, above sub-cylindrical, convex on the disk, deeply punctured, clothed with deeply bifid, narrowly oblong, adpressed scales. Elytra are broader than the thorax, obovate, convex, each elytron deeply excavated below the basilar space, deeply and coarsely punctate-striate, the punctuation rather finer towards apex, surface clothed with similar scales to those on the thorax; here and there they are more densely congregated, and form ill-defined patches (Baly, 1878).

Control

No specific control measures are suggested. The insect may be controlled by spraying any insecticides, like, dimethoate 30EC at 0.1-0.2% concentration.

4. Leaf roller

Leaf rollers are small caterpillars, reaching about an inch in length. They feed inside nests made from leaves of their host plants, rolled together and tied with silk. Once inside their leaf nests, leaf rollers chew holes through the tissue, sometimes adding more leaves to the nest to keep themselves protected from predators. Leaf roller damage is usually minor, but some years it may be quite severe. When there are lots of nests in a plant, defoliation may occur. In Ramie mainly three different types of leaf roller species, viz. *Pilocrocis ramentalis* (Litzenberger, 1976), *Sylepta derogata* (Ghosh and Ghosh, 1971) and *Patania spp.* (Bendicho-Lopez, 1998; Gawande *et al.*, 2015) have been reported as pests.

i) Leaf roller (*Pilocrocis ramentalis*)

Occurrence & severity

The pest occurs in America (north), Costa Rica, Cuba, Galapagos Islands, Honduras, Mexico and Puerto Rico. The habitat consists of open woods, clearings and damp areas. Adults have been recorded on wing from February to December (Heppner, 2003). However, the main flight period is August to October. The leaf roller is the most serious and widespread pest of ramie (Litzenberger, 1976). The larvae of this species also feed on *Boehmeria cylindrica, Odontonema strictum* (Heppner, 2003), *Pachystachys spicata* and *P. coccinea* (BugGuide; bugguide.net).

Types of damage

The larvae feed on ramie leaves and pupate in the leaves that they have caused to roll up. Heavy infestations have caused complete defoliation and cessation of growth (Litzenberger, 1976).

Scientific name & order: *Pilocrocis ramentalis* Lederer, 1863 (Lepidoptera: Crambidae)

[Wien. Ent. Monats., 7(12): 430]

Synonym:

= *Zinckenia perfuscalis* Hulst, 1886

Biology

Larva: The developing larvae are whitish-green in color, up to 10 mm in length, (Litzenberger, 1976)

Pupa: The larva pupate in the leaves (Litzenberger, 1976)

Adult: The wingspan is 24–29 mm. The forewings are greyish-brown with a slightly irregular, white antemedial line and a sinuous white postmedial line edged in black. Postmedial line has squarish convex lobe near middle. The reniform spot has the form of a small white arc, edged in black basally. The hindwings are greyish-brown with a single white postmedial line. Adults have been recorded on wing from February to December (Heppner, 2003). Males harbour distinctive secondary sexual characters in the modified, broadly bent antenna basally and the 'flap' of modified scales from the base of the forewing costa covering part of the wing's base (Landry, 2016).

Life stages

There are four life stages, such as, embyro (ova or egg), larva (in this case, caterpillar), pupa (chrysalis), and imago, (or adult/ butterfly).

Control measures

It is suggested to cut the damaged leaves and toss the caterpillars into a bucket of soapy water. *Bacillus thuringiensis* works as a stomach poison to feeding caterpillars, and is extremely effective if applied to these pests and their food source while they are young. Further, good control has been obtained by the use of 5 percent impregnated DDT dust @ 12 kg per ha (Litzenberger, 1976).

ii) Leaf roller (*Syllepte derogata*)

Occurrence & severity

This leaf roller is found in almost every country where ramie is grown (PROSEA, 2016). In India, this ramie leaf roller is found from time to time (Ghosh and Ghosh,1971). The species also occurs in cotton throughout the South East Asia and considered as a minor pest.

Types of damage

The larvae feed inside nests made from leaves of the ramie, rolled together and tied with silk. Once inside the rolled leaf, the larvae chew holes through the tissue. When there are lots of nests in a plant, defoliation may occur.

Scientific name & order: *Syllepte derogata* **(Fabricius, 1775) (Lepidoptera: Crambidae)**

(NBAIR, 2013)

Synonyms

= *Sylepta derogata* (F.)

= *Haritalodes derogata* F.

Biology

Egg: Eggs are laid both singly and in batches on both sides of the leaf, but more on the abaxial surface. Newly laid eggs are colourless but later changes to dirty white towards the time of eclosion. Size: 0.76 x 0.51mm (Anioke, 1989).

Larva: In general, the larva is glistering green with black head. The number of larval instars varies from 5 to 6. The larva possesses three pairs of thoracic legs and four abdominal prolegs. The 1st Instar larva is light green and the body is slender and slimy. It has small head and enlarged prothorax. Each thoracic segment bears a pair of legs. Four abdominal prolegs are located on the 3rd to the 6th segments. The last abdominal segment bears the 5th abdominal prolegs or claspers. The body is covered with numerous setae. The 2nd instar larva is green. Two small brown spots are located on the dorsal aspect of the prothorax. The 3rd instar larva is with distinct epicranial and frontal sutures. The head capsule is slightly bilobed. The posterior portion is brownish while the anterior portion is light brown. The two brown sopts on the prothorax start to develop into thoracic shields. The 4rth Instar larva is deep green. The bilobed head capsule and thoracic shields are black. The prothoracic segment becomes slightly flattened on ventral surface. The thoracic legs are black with white claws. The abdominal prolegs are dirty white. The stematae become visible by the sides of the head near the antennae. The 5th instar larva is green but towards pupation turns light green. The black thoracic shields start to degenerate from posterior portion. The mandibles are stout with 5 teeth, 3 of which are more prominent. The 6th Instar larva is similar to that of 5th instar but the prothoracic shilds are continued to degenerate, thereby exposing the greater part of the prothoracic segment (Anioke, 1989).

Pre-Pupa: The larva shortens and losses characteristic green colour and becomes pinkish. It becomes inactive (Anioke, 1989).

Pupa: The pupa is obtect in shape with more or less convex head. The thoracic segments are not visible on the ventral aspect having been concealed by the appendages. A dorsomedial ridge runs from the mesothorax and extends to the last abdominal segment. Single pair thoracic spiracles are located between the pro and metathoracic segments, towards the dorsal aspect. The abdominal spiracles are only visible on the 3rd to the 8th segments, the one on the 4th segment being the most prominent. All the spiracles have thick round to oval and dark-brown peritremes. The abdomen is conical in shape. The 9th and 10th segments are not distinctly separated. Pairs of proleg scars are conspicuous on the ventral surface of the 5th and 6th segments. On the last segment is attached a black cremaster which is conical in shape. It bears six curved hooks. The genital pore occurs ventrally below the 8th abdominal segment in the female and in the male on the 9th segment. The abdomen of the male is more slender than that of the female. The pupa is dark brown towards the time of adult emergence (Anioke, 1989).

Adult: The adult is a small and delicate moth which is silvery white to gold in colour with a network of brown markings on the wings. Both the wings and the body are covered by scales. The abdomen of male is slender and tapers gently towards the tip. In the female the abdomen is stouter and blunter at the tip. Female moths are darker in colour than the males (Anioke, 1989).

Lifecycle: The fecundity is about 443. Pre-oviposition, oviposition and post-oviposition periods are 2.5, 6.5 and 3.0 days, respectively. The number of larval instars varies from 5 to 6. The life cycle completes in about 33.9 ± 0.5 days (Anioke, 1989).

Management/Control measures

It is suggested to collect and destroy rolled leaves. Hand picking and destruction of grown on caterpillars. As regards the chemical measure, the spraying of carbaryl 50 WP 1.0 kg or phosalone 50 EC 1.0 l in 500 l water/ha is effective. Several other insecticides, such as, chloropyriphos 20 EC @ 2.0 l/ha and dichlorovos 76WSC @ 1.0 l/ha, are also found effective to control the pest by any (TNAU, 2016). The attack of this insect can easily be controlled by spraying 0.04% of endosulfan or sumithion (Singh, 1998).

iii) Leaf roller (*Patania* sp.)

Occurrence & severity

Surveys conducted in Assam, India, during the year 2014-15 cropping season revealed the presence of this insect on ramie. The pest was identified based on various morphological analyses. This is thought to be the first report of this insect infesting ramie in Assam, India. Maximum plant damage was observed during the month of October and November (Gawande *et al.*, 2015, 2016).

Types of damage

The larva feeds on the leaves of ramie, which it rolls up as a shelter with strands of strong silk. At first, only the edge of the leaf, but from about its third instars it starts rolling up a whole leaf. After having consumed the inner layers of the tube, the larva moves (mostly) up and rolls another leaf. In this way, as it develops, it produces a trail of rolled leaves on the plant stem. As a result, photosynthesis of the plant gets affected, which eventually affects the plant growth. During severe infestation, the leaf margins dry completely (Gawande *et al.*, 2015, 2016).

Scientific name & order: *Patania sp.* Moore, 1888 (Lepidoptera: Crambidae)

Synonym

= *Pleuroptya* sp. Meyrick, 1890

[N.B. Several *Patania* spp. are found to affect ramie leaf. For example, *Patania sabinusalis* (Walker, 1859), which has been found in Borneo, Cameroon, Congo, Fiji, India, Japan, Java, Kenya, Sechelles, Somalia, Solomon islands, Srilanka, Taiwan, Uganda and Zambia, affects ramie. It is reported that the larvae of this moth feed on ramie (*B. nivea*) (Website: *Patania sabinusalis*). Further, *Patania silicalis,* which was described by Guenée in 1854, is found to affect ramie. Its larva feeds on ramie leaves (Bendicho-Lopez, 1998). The moth is found in America (north), Brazil, Costa Rica, Cuba, Ecuador, French Guiana, Guatemala, Guyana, Hispaniola, Jamaica, Mexico, Panama, Puerto Rico and Venezuela. The adult moth is with wingspan of 24-26 millimetres. The forewings range from yellowish grey to orangish with dark grey antemedial and postmedial lines, as well as a discal spot in a dark grey crescent. The hindwings have similar colour and markings, but lack an antemedial line (Website: *Patania silicalis*)]

Biology

Egg: Members of this genus lay their eggs singly in rows.

Larva: The first instar larva is greenish yellow, very small and delicate. As the caterpillar grows, it turns greener with dark spots along the body (Gawande *et al.*, 2015).

Pupa: The larva pupates inside the leaf fold of the rolled up leaf and the pupal exuviae remains attached to the leaf after emergence (Gawande *et al.*, 2015).

Adult: The adults are about 25 to 40 mm in size, pearly-cream coloured with beige mottling. Females have their abdomens slightly shorter and broader than males (Gawande *et al.*, 2015).

Control measures

No control measure has been suggested. It may be sensitive to several organophosphate insecticides, like, dimethoate at 0.1% concentration.

5. Mealy bug/ Pink hibiscus mealybug

Occurrence & severity

The pink hibiscus mealy bug-attack in ramie (*Boehmeria nivea*) was first recorded by Dutt and Kundu (1960). The mealy bug causes considerable damages in several fibre crops (*Hibiscus cannabinus, H. sabdariffa* and *B. nivea*) including ramie (Ghose, 1961; 1970; Singh and Ghosh, 1970; Raju *et al.*, 1988). It is a serious pest of this crop (Tripathi and Ram, 1971).

Types of damage

The mealy bug causes stunting and swelling of the stem, deformation of leaves and decreases both the quantity and quality of fibre (Tripathi and Ram, 1971).

Scientific name & order: *Maconellicoccus hirsutus* **(Green) (Hemipetra: Pseudococcidae)**

Synonym

= *Phenacoccus hirsutus* (Green)

(**N.B.** Further details of the mealy bug have been described in Chapter II along with Mesta and Roselle pests)

Management/Control measures

Natural Enemies

Protection of natural enemies may be helpful. Ghose (1970) has reported the activity of natural enemies in relation to *M. hirsutus*. He has observed that *Brumus suturalis* Fab.and *Nephus* sp. are in association with the mealy bug on ramie (*B. nivea*) in April.

Chemical control

An emulsion spray of 0.01 per cent Parathion at 35 gallons/acre is reported to give effective control (Tripathi and Ram, 1971). Spraying of Dimethoate 30EC @ 1.7ml/l alone or with fish oil rosin soap may be very effective to control the mealy bug attacks in ramie. Since, in an investigation with other crop it was found that the treatment of Dimethoate 30 EC at 1.7 ml + Fish oil rosin soap at 5g/l gave highest protection (85.1%) against the same mealy bug species on grape bunches followed by Dimethoate 30 EC at 1.7 ml/l (82.0%) (Katke, 2008).

6. Ramie moth

Occurrence & severity

The Ramie Moth is a member of the Noctuidae family. The insect is an economically important pest that seriously impairs the yield of ramie fibre crop (Zeng *et al.*, 2016). This pest is widely distributed in south-east Asia, including China, Fiji, India, Japan, Korea, Myanmar, New Guinea, Sri Lanka, Taiwan, Norfolk Island and in the Malay peninsula, as well as in Australia. In central and south China, up to four generations of the moth per year can occur during the vegetative growth phase of ramie (Zeng *et al.*, 2013). The larvae of ramie moth feed on *Boehmeria nivea* (ramie) and *Boehmeria australis*.

Types of damage

The moth causes severe ramie fibre yield reductions, as it feeds on ramie leaves and new shoots (Ide, 2006). Female moth lays eggs on the abaxial surface of ramie leaf, which causes the leaves to yellow. After the eggs' hatching, the larvae begin to feed on the leaves and that results in a net-like pattern of damage, which seriously impairs the plant's photosynthetic capacity and eventually, causes fibre yield reduction.

Scientific name & order: *Arcte coerula* (Guenée, 1852) (Lepidoptera: Noctuidae)

Synonyms

= *Cocytodes coerula* Guenée, 1852

= *Arcte coerulea*

= *Cocytodes coerulea*

Biology

Egg: Female adult lays about 400 eggs.

Larva: Larva is black with brown or black head. Somites are with transverse dorsal white bars each enclosing a black line. Stigmata are ochreous, black ridges and with some red color around, situated on white patches from near the top of each of which springs a white hair. There is an inter-spiracular disconnected white line and a broader spiracular line with a black spot from which springs a white hair below each spiracle. A broad ventral white band is present. The 11th somite is humped and black above. Extremity is orange, above spotted with black. Head and somites are covered with long white hairs (Guenee, 1852). Larva passes through six instars. The caterpillars are frequently gregarious and change their colour according population density. Under crowded condition, the black transverse stripes on the thorax and abdomen increase in size. Solitary caterpillar has black head however the head is brown under crowded condition.

Pupa: Mature larva pupates in a slight silken cocoon between leaves drawn together with silk.

Adult: The adult moth has heavy body with wingspan of about 84mm. Mid and hind tibia spiny. Hind wings of male are with long hairy inner margin. Head is black and thorax is vinous reddish brown. Pectus is white. Abdomen is bluish fuscous, with a white tuft in male below claspers. Forewings are with brown suffused with black, except costal area as far as postmedial line and the apical area, and irrorated with a few bluish-white scales. A short almost basal line, two black sub-basal patches, an oblique waved antemedial line are present. A black spot is in cell and two lunules are at end of cell. Traces of a pale waved sub-marginal line can be seen. Hindwings are black with a bright blue patch on disk and a maculate with post-medial band and patch near anal angle (Hampson, 1892).

Life cycle: The lifecycle from egg-laying to eclosion of adult completes within 55 to 68 days.

Molecular studies (on host resistance)

In order to understand defence mechanism in ramie, transcriptome analysis of ramie (*Boehmeria nivea* L. Gaud.) in response to ramie moth (*Arcte coerula*) infestation was studied by Zeng *et al.*(2016). In their studies, two cDNA libraries derived from RM-challenged (CH) and unchallenged (CK) ramie leaves were constructed to understand the ramie defence mechanisms against the moth. The subsequent sequencing of the CH and CK libraries yielded 40.2 and 62.8 million reads, respectively. Furthermore, de novo assembling of these reads generated 26,759 and 29,988 unigenes, respectively. An integrated assembly of data from these two libraries resulted in 46,533 unigenes, with an average length of 845bp per unigene. Among these genes, 24,327 (52.28%) were functionally annotated by predicted protein function. A comparative analysis of the CK and

CH transcriptome profiles revealed 1,980 differentially expressed genes (DEGs), of which 750 were upregulated and 1,230 were downregulated. A quantitative real-time PCR (qRT-PCR) analysis of 13 random selected genes confirmed the gene expression patterns as determined by Illumina sequencing. Among the DEGs, the expression patterns of transcription factors, protease inhibitors, and antioxidant enzymes were studied. Overall, these results provide certain useful insights into the defense mechanism of ramie against RM (Zeng et al., 2016).

Management/Control measures

Use of host resistance

The ramie variety, Chuanzhu No.8, is found tolerant to Ramie moth attack and suggested to plant in China. Further, for using induced host resistance, it is suggested to spray elicitor (1.2 mM Salicylic Acid or 80 mg/ mL Chitosan oligosaccharide 300-450L/ha) when the plants are 15-20 cm in height. This spray improves resistance in ramie and reduces oviposition by ramie moth (Plantwise, 2015).

Cultural methods

Plantation of trap crops: The plantation of susceptible variety, like, 'Yuanyeqing' (*B. nivea*), around the field is suggested to kill the larvae and pupae of ramie moth. This is done by cleaning and destroying residues after harvest (Plantwise, 2015).

Use of light traps: It is suggested to use black light traps (e.g. TDB-2218 frequency trembler grid lamp) to kill the adult moths between 18:00-22:00h, from April to September. One light trap is required per ha (Plantwise, 2015).

Biocontrol

Two different measures are suggested for early control of young larvae (before 3rd instar larvae) that have not spread. In case of first one, spraying of 0.3% nimbin EC (azadiractin) at 750-800mL/ha is required. While, in second method, placing of egg cards of *Trichogramma* wasps, 1 card per 10m, 3 days after the adult full flight period is suggested (Plantwise, 2015). Further, spraying of *Bacillus thuringiensis* products against the larvae is effective to control ramie moth. This organism is biological multisite stomach poison for insects. Usually suspension of 4000 IU @ 450-600 L/ha is suggested to use as soon as possible after preparation (Plantwise, 2015).

Chemical control

Chemicals such as dichlorovos, Trinox and pyrethrum ester insecticides are effective in controlling ramie moth larvae (Zheng et al., 2016). Several other insecticides are also suggested to control the ramie moth as follows:

Spraying with emamectin-benzoate –based products is suggested. This is Avermectin group of pesticide. Usually 1% EC product is used at 150-200 mL product per hectare, which is 10 mL product per 30 litres. This is needed to use against larvae in cloudy or afternoon (Plantwise, 2015).

Spraying with beta-cypermethrin is another chemical advocated to use against ramie moth. This is in Pyrethroid group of pesticide. Usually 4.5% EC product is used at 300-375 mL product per hectare, which equals to 20 mL product per 30 litres. This is better to use against larvae in cloudy weather or late afternoons (Plantwise, 2015).

Spraying with chlorantraniliprole –based products is also suggested against the pest. This is the pesticide of Diamides group. Usually 20% EC product is used @ 150-225 mL product per hectare, which is equal to 10 mL product per 30 litres. This is recommended to use against larvae from 10 to 16 O'clock (Plantwise, 2015).

7. Termite

Occurrence & severity

Termite is an ancient social insect in the world for about 0.13 billion years. Termites are commonly known as "white ants'. They are soil inhabiting and build nests both underground and huge termite mounds over ground. In Assam (India), termite is a major problem in ramie cultivation particularly when the ramie is planted by stem cuttings and the soil remains dry for a considerable period after planting (Mitra *et al.*, 2013; Gawande *et al.*, 2016).

Types of damage

The young plants grown from the hard wood stem cuttings are affected by termites. The termites feed on the cambium as well as pith inside the stem cutting, thus making hollow, which leads to drying and rotting of the whole plant. Sometimes they also damage the rhizomes in established plants (Mitra *et al.*, 2013; Gawande *et al.*, 2016).

Scientific name & order: *Microtermes* sp. (Termitidae: Blattodea/ Isopteraa)

Biology

The termites are soft bodied insects, occurring in three forms or castes, e.g., reproductives (sexuals), workers and soldiers. In general, the termites are 3-4 mm in size, but, their queen is about 4 inch long giant that lies in the royal chamber motionless. Hence workers have to take care of all its daily chores. Termite queen is an egg-laying machine that lays up to 60,000 eggs per day.

Generally termites can lay at the rate of 6000-7000 egg per day. If a queen is lost by accident or death, nymphs in various stages of development are quickly matured to neotenic female forms which can lay eggs parthenogenetically. The other castes, workers and soldiers are highly devoted to the colony, working incessantly and tirelessly. Soldiers have long dagger-like mandibles with which they defend their nest and workers chew the wood to feed to the queen and larvae and grow fungus gardens for lean periods. Nasutes are specialized soldiers which specialize in chemical warfare. In breeding season which usually coincides with the rains, newly produced males and females in large numbers grow with wings (alates). Alates have nuptial flight, disperse to long distances, make pairs and find a new place to start a colony. The phenomenon is known as swarming. After nuptial flight, the males and females pair off, shed their wings and move away. They both then dig a pit and create a nest site with a nuptial chamber in which eggs are laid. Once the nuptial chamber is constructed the king and queen mate. Initially the queen lays only a few eggs of 100 to 300 eggs 3-10 days after swarming. The eggs have an incubation period of 20-40 days and the hatched nymphs all develop into workers which make new galleries and enlarge the nest. Queen settles down and undergoes physogastry, a phenomenon in which abdomen of the queen enlarges enormously by the expansion of the intersegmental membranes, the sclerites appearing only as brown spots. The worker's life span is one to two years. After a few years, sexuals (alates) are released from the colony and they will in turn form new colonies.

Management/Control measures

Biocontrol

Application of *Bacillus thuringiensis* as a biocontrol agent may be effective against termite. Since, it is reported that *B. thuringiensis* strains caused mortality of above 80% in the termite species (*Microtermes obesi*) and the *B. thuringiensis* sub sp. *israelensis* is very virulent (Singha *et al.*, 2010).

Chemical control

Soil drenching near the base of cuttings/saplings with chloropyriphos (20EC) at 0.1% (1ml/L) solution is very much effective to control termites in field (unpublished personal observation based on other crops). Further, the control measures as practiced for jute and other crops may also be used in ramie. In Jute the termite is controlled by soil application with aldrin or dieldrin (5%) dust or heptachlor dust 5% @ 15 kg (or at 1lb a.i.) per acre before planting (Bigger, 1966; Chandra, 2017). On the standing crop parathion 0.01% gives good control on the stems (Chandra, 2017).

8. White grubs

White grubs are the slug-like larval stage of many insects. They often live just below the surface in the topsoil. With white grub infestations, plants can be damaged and will likely die.

i) White grub (*Lepidiota* sp.)

Recently the species *Lepidiota* sp. indet (Coleoptera: Scarabaeidae) has been reported as a serious root feeder pest of ramie crop at Sorbhog, Assam (Gawande *et al.*, 2016). The beetle is also reported to feed on leaves of ramie under field conditions (Bhattacharya *et al.*, 2014). The outbreak of *Lepidiota* sp. is being experienced in Ramie Research Station, Sorbhog, Assam since 2012. The emergence of beetles starts from the 1st week of April and continues up to the 1st fortnight of May (Gawande *et al.*, 2016). Beetles are found to emerge from soil at 6-6.30 PM onwards and are phototactic in nature. Freshly emerged beetles are brown to chocolate brown in colour and body length varies from 2.5 to 3.5 cm (Gawande *et al.*, 2016). The *Lepidiota* sp. is credited as the second species of white grub belonging to the genus "*Lepidiota*" reported from Assam after *Lepidiota mansueta* which had appeared as an extremely serious key pest of many field crops in Majuli, Assam the largest midriver deltaic island of the world (Bhattacharyya *et al.*, 2013). Endemism of white grub species belonging to the genera "Lepidiota" is chiefly governed by the presence of river/rivulet/large water bodies and light soil with high organic carbon content (Bhattacharya *et al.*, 2014)

ii) White grub (*Rhinyptia meridionalis*)

Another species, *Rhinyptia meridionalis* Arrow, 1911 (Syn: *Rhinyptia puncticollis* Arrow, 1917) (Scarabaeidae: Coleoptera), has also been reported (as *Rhinyptia meridionous* Arrow) to affect ramie in Debijhora, India (Pandit, 1995). The grubs of *R. meridionalis* damage young rhizomes and roots of ramie. *Rhinyptia* is a genus of scarab beetles. The genus is characterised by the clypeus narrowing into a long and pointed tip (a "snout"). The male has the outer claw of the mid leg unforked. The larvae are white and curled (also known as white grubs) living under the ground and feeding on roots of various plants. Adults feed on flowers and their emergence is associated with the flowering of various grasses including cultivated millet and cereal crops. Adults are attracted to lights (Jameson *et al.*, 2007).

Management and Control measures

Cultural and mechanical methods

Deep ploughing may be effective to kill the grubs directly those exist in the upper part of the soil. This type of ploughing may expose some grubs to predator birds for their feeding, that follow behind the ploughing, and for high levels of parasitization. Pal (1977) has carried out light trap studies in Jodhpur area of Rajasthan during 1971-72. *Rhinyptia meridionalis* is the most common species attracted in light. Hence, light trap may be used in minimizing the damage of white grubs (*R. meridionalis*).

Chemical control

As regards the insecticidal control, Lindane is most effective but that is banned for use in many agricultural crops. Alternatively, Furadon 3G or Phorate 10 G @ 3kg per acre may be applied.

5
Hemp (*Cannabis sativa*)- Rosales Cannabaceae

A. Diseases

1. Anthracnose

Hemp anthracnose is caused by two different fungi, viz., *Colletotrichum coccodes* (=*C. atramentarium*) and *Colletotrichum dematium* (Saccardo, 1882; Cavara, 1889).

i) Anthracnose (*Colletotrichum coccodes*)

Occurrence & severity status

The anthracnose disease occurs on many plants, including hemp (*Cannabis sativa*). The disease causing fungus is well-studied and it is an important pathogen responsible for black dot disease on potato. The fungus is an unspecialized pathogen, infecting a wide range of host families. It is most common in Canada and commonly regarded as a minor pathogen or secondary invader, but the fungus may seriously affect yield where intensive cropping without rotation occurs. It primarily affects the growth rate of plants and diminishes harvest. Moreover, it becomes fatal occasionally. Hoffman (1958, 1959) reported heavy hemp losses in central Europe. Conversely, Gitman (1968b) considered the pathogen was of little importance in the USSR.

Symptoms

The disease first appears on leaves as light green, water soaked, sunken spots. The spots later enlarge to circular or irregular shapes with greyish centres and brownish-black borders. Larger spots may become zonate. Affected leaves soon wrinkle then wilt (Hoffman 1959). Then black dot-like acervuli arise in the lesions, lending a salt-and-pepper appearance (McPartland *et al.*, 2000). The disease also affects stems. The lesions on stem initially appear as white. The affected stems slightly swell up and develop cankers. The periderm is easily peeled off. Stems sometimes snap at cankers. Distal plant parts become stunted and often wilt. The affected young plants die.

Causal organism: *Colletotrichum coccodes* **(Wallr.) S. Hughes 1958**
[Canadian Journal of Botany 36 (6): 754]

Synonyms

= *Colletotrichum atramentarium* (Berk. & Broome) Taubenh. 1916

= *Vermicularia atramentaria* Berkeley & Broome 1850.

Morphological /Morphophysiolgical characteristics

Acervuli on stems are round or elongated, reaching 300 μm in diameter, acervular tissue intra- and sub-epidermal, disrupting the outer epidermal cell walls of host irregularly and exuding slimy conidia and bristling setae. Setae are smooth, stiff, septate, slightly swollen and dark brown at the base and tapering to sharpened and paler apices, up to 100 μm long. Conidiophores are hyaline, cylindrical, and occasionally septate and branched at their base. Conidiogenous cells are phialidic, smooth, hyaline, with a minute channel and periclinal thickening of the collarette. Conidia are hyaline, honey to salmon-orange coloured in mass, aseptate, smooth, thin walled, guttulate, cylindrical or fusiform with obtuse ends, straight, often with a slight median constriction, averaging 14.7 x 3.5 μm on Cannabis stems but ranging 16-22 (-24) x (2.5-) 3-4 (-4.5) μm in culture (Mordue, 1967; McPartland *et al.*, 2000). Setose sclerotia are common, usually abundant evenly distributed over the agar surface; when young greyish and glabrous, rapidly becoming dark and setose (Mordue, 1967). Appressoria are club-shaped, medium brown, edge irregular to almost crenate, 11-16.5 x 6-9.5 μm, rarely becoming complex (McPartland *et al.*, 2000).

Colonies on potato dextrose agar have scanty white aerial mycelium usually with a profusion of evenly distributed sclerotia. Sclerotia are smooth and greyish when young, quickly becoming black and producing setae. Acervuli develop in conjunction with sclerotia or from separate aggregates of setose mycelium. Phialides are often found as single on mycelium. Conidiophores are occasionally septate and branched. Conidial masses are typically small, colourless to orange, with conidia commonly shorter in proportion to their width than on host. Reverse of colonies is grey, darkening with age due to formation of appressoria. Appressoria very readily are formed in slide cultures, cinnamon buff or amber coloured, ovate or obclavate to elliptical, occasionally irregularly lobed (5-) 6.5-11.5 (-14) x (3.5-) 4-8 (-10.5) μm, borne on hyaline, thin-walled sigmoid supporting hyphae, variable in shape (Mordue, 1967). Colonies on oatmeal agar (OA) medium are dark brown, consisting of numerous black sclerotia, occasionally with sparse, white aerial mycelium. Conidial masses are honey-coloured; reverse dark brown (Hoog *et al.*, 2000).

Molecular characteristics

In an investigation of Zornitsa *et al.* (2013), two PCR primer sets were used to sequence the ribosomal internal transcribed spacer (ITS1 and ITS2) regions. PCR amplification with genus-specific primers (Cc1F1/Cc2R1) gave one single band of ~450 bp in all isolates analyzed (*C. coccodes*, *C. acutatum* and *C. gloeosporioides*) confirming that they belong to the genus *Colletotrichum*. The nested primer set Cc1NF1/Cc2NR1 amplified one single PCR band of ~350 bp only in the reactions containing DNA from *C. coccodes* isolates as a template (Cullen *et al.*, 2002). Molecular identification of *C. coccodes* with species-specific primer was a successful method for the confirmation of species. Some morphological differences were observed between the isolates on potato dextrose agar (PDA) and predominantly on sucrose soy protein agar (SSPA). ITS region was successfully amplified for the isolates of *C. coccodes* from pepper, but additional methods of identification were required for genetic diversity (Zornitsa *et al.*, 2013). In another study, the correlation between morphological and molecular-based clustering demonstrated the genetic relationships among the isolates and species of *Colletotrichum* and indicated that ITS rDNA sequence data were potentially useful in taxonomic species determination (Photita *et al.*, 2005).

Predisposing factors

The pathogen is an opportunist that incites problems only under suboptimal growing conditions or in the presence of other pathogens. It prefers a comparatively cooler temperature to infect.

Disease cycle

The pathogen is soil borne. Transmission of this disease is through soil, but no rapid or extensive growth of hyphae occurs through soil. Survival is probably only in decaying roots and other trash and on weed hosts. Hyphae are growing primarily through decaying roots and other organic matter. Living roots are infected when they contact the contaminated organic material. Sclerotia overwinter on trash and develop into acervuli in spring. New plants are infected from dead material by conidia. Persistence on weeds may be implicated in carry-over from crop to other crops (Mordue, 1967).

Management/Control measures

Avoidance: The pathogen can be avoided by creating optimal growing conditions and using disease-free seeds.

Prevention: Sanitation is the best prevention. Removal of plant residues, use of clean and sharp farm appliances or sterile tools to keep wounds clean and cultivation in new fields, avoiding underwatering or overwatering are the best measures to keep the disease away from the field. Further, it is suggested to avoid sowing hemp in clay soils, or amend clay soils to improve drainage. Further, we should not allow to wet foliage and stems when watering.

Control: Use of *Trichoderma harzianum* (Trichodex) is suggested to control anthracnose. Sulfur or copper powders or sprays may be applied weekly as soon as the disease is noticed in the field. This will help to keep spores from spreading, but they will not kill the disease. The fungicide, Chlorothalonil and many systemic fungicides are able to kill the pathogen of this anthracnose.

ii) Anthracnose (*Colletotrichum dematium*)

Occurrence & severity status

The disease is also known as Leaf spot. Its occurrence on hemp stem was reported earlier from Italy by Saccardo (1882) and Cavara (1889). Hoffman (1958, 1959) noted high losses of hemp due to anthracnose in "bog" soils, where the disease was escalated in plants under stress from drought or frost damage. The heaviest infection was observed after the plants had flowered, with males succumbing before females (Hoffman, 1958, 1959; Cook, 1981). Concurrent infection by the nematode *Heterodera schachtii* or the fungus *Rhizoctonia solani* may increase plant's susceptibility to the anthracnose (Smith *et al.* 1988).

Symptoms

The disease initially appears on leaves and stems as small, light brown spots (1 mm in diameter) with red-to-purple margins. As the disease progresses, the lesions expand and turn brown in the centre, with 3 to 5mm diameters. Later, the affected leaves wither and die, while the stems break at affected region. The surface of the lesions also covers with black acervuli in irregular groups.

Causal organism: *Colletotrichum dematium* (Pers.) Grove, 1918
[J. Bot. 56: 341]

Synonyms

= *Colletotrichum bakeri* (Syd. & P. Syd.) Mundk., (1938)

= *Colletotrichum brassicae* Schulzer & Sacc., (1884)

= *Colletotrichum lysimachiae* Duke, (1928)

= *Colletotrichum pucciniophilum* Togashi, (1936)

= *Colletotrichum sanguisorbae* Bres., (1894)

= *Colletotrichum volutella* Sacc. & Malbr., (1882)

Morphological/ Morphophysiolgical characteristics

Acervuli are round to elongated, 50-400 µm diameter, black, abundant, gregarious, frequently confluent, at first covered by the cuticle and epidermis but later strongly erumpent with considerable stromatic development, exuding pale smoke grey conidial masses with divergent setae. Setae are stiff, 60-200 µm long, 4-7.5 µm wide at the base, abundant, erect, rarely curved but occasionally irregular at the apex, divergent, smooth-walled, 0-7-septate, usually 3 or 4 septate, tapering to an acute apex, thick-walled, vandyke brown throughout development, basal cells, apart from the colour, are undifferentiated from the stromatic tissue from which they arise (Chupp, 1964; Matsushima, 1975; McPartland *et al.*, 2000). Conidiophores are simple, hyaline, swollen cylindrical. Conidiogenous cells are phialidic. Conidia are 18-26 (-30) x (2-) 3-3.5 (-5.49) µm, hyaline, aseptate, becoming two-celled during germination, smooth-walled, falcate, central part almost straight with parallel walls, bent abruptly to a roundish, forming an angular shape, apex acute, base truncate, guttulation irregular. Appressoria are clavate to circular, brown, edge usually entire, 8-11.5 x 6.5-8 µm, often becoming complex (Chupp, 1964, Matsushima, 1975; Hoog *et al.*, 2000; McPartland *et al.*, 2000; Guan *et al.*, 2016).

Colonies on oatmeal agar (OA) are mouse grey, with woolly aerial mycelium, becoming felt-like; reverse dark brown. Microscopically these are found as with abundant sclerotia, setae, conidia and appressoria (Hoog *et al.*, 2000). The colonies on PDA display changes from initially whitish to grey after 4 days, and become olivaceous to black when mature, with abundant setose sclerotia. Conidia are rare on PDA. When cultured on synthetic nutrient-poor agar, conidia are produced (Guan *et al.*, 2016). The fungus is a weak parasite on many hosts (von Arx, 1957).

Molecular characteristics

A multilocus molecular phylogenetic analysis (ITS, ACT, Tub2, CHS-1, GAPDH, HIS3) of 97 isolates of *C. lineola, C. dematium* and other *Colletotrichum* species with curved conidia from herbaceous hosts resulted in 20 clades, with 12 clades containing strains that had previously been identified as *C. dematium*. The epitype strains of *C. lineola* and *C. dematium* reside in two closely related clades (Damm *et al.*, 2009). In the multigene analyses of 98 isolates of *C. dematium* and other *Colletotrichum* species with curved conidia 2333 characters including the alignment gaps were processed, of which 740 characters were parsimony-informative, 157 parsomony-uninformative and 1436 constant. However some clades, e.g. *C. circinans* and *C. spinaciae* were very short-branched in the ITS phylogeny and some, e.g. *C. dematium* and *C. lineola* were only distinguished in three (Actin, HIS3 and GAPDH) of the six phylogenies (Damm *et al.*, 2009).

Predisposing factors

The disease favours cool damp weather, especially in heavy soils. More infection occurs in "bog" soils (Hoffman, 1958, 1959). Continuous wetness (more than 95 % humidity) for more than 10 hours is required for infection by the fungus (Dillard 1992). In an investigation with cowpea it was found that a dew period of 12 hours of high humidity was required by *C. dematium* to initiate disease and that extended period of high humidity promoted infection. The suitable temperature for the pathogen attack is ranging 25-30°C (Pakela *et al.*, 2002). The stress condition from drought or frost damage and mono-cropped system favours the disease to occur in hemp (Smith el al. 1988). The rain water and wind-driven rain help to spread of conidia of the pathogen. Hoffman (1958,1959) and Cook (1981) described heaviest infections after plants had flowered, with males succumbing before females. Concurrent infection by the nematode *Heterodera schachtii* or the fungus *Rhizoctonia solani* increases plant susceptibility to anthracnose (Smith *et al.* 1988).

Disease cycle

The causal fungus overwinters as sclerotia in plant debris or soil. In the spring, sclerotia sporulate and conidia splash onto seedlings. Conidia form appressoria may directly penetrate epidermal tissue or enter via stomata and wound. As soon as the infection established, conidia are produced on acervulus. Conidia as formed are spread via splashing water and wind-driven rain and causes reinfection.

Management/Control measures

Sanitation by eliminating crop residues, deep ploughing and eliminating weeds is the best preventive measure for the disease. Avoidance of heavy soils and drought or waterlogged condition by proper site and irrigation is necessary to manage the disease by cultural way (McPartland *et al.*, 2000).

Application of biocontrol agent, like, *Trichoderna harzianum* (Trichodex), is suggested to manage the disease (Samuels, 1996). Further, prophylactic spray of lime sulphur on plants during favourable weather of the disease may be effective to minimize the occurrence of disease in the crop (Yepsen, 1976).

2. Bacterial blight /leaf spot of hemp

Occurrence & severity status

The disease in hemp was first reported from Yugoslavia (Sutic and Dowson, 1959). It is a common disease of hemp in Europe.

Symptoms

The disease appears on leaves as yellowing of foliage along with small (2-5mm), angular, water-soaked lesions. The lesions on leaves are surrounded by chlorotichalos and with tan or brown centres. Eventually, the lesions become necrotic spots.

Causal organism: *Pseudomonas cannabina* (ex Sutic[1] and Dowson) Garden *et al.* 1999

[International Journal of Systematic Bacteriology, 49(2): 469-478]

Synonyms

= *Pseudomonas cannabina* Sutic and Dowson 1959

= *Pseudomonas syringae* pv. *cannabina* Dye *et al.* 1980

Morphological/ Morphophysiolgical characteristics

The pathogen is a Gram negative, fluorescent, motile, flagellated and aerobic bacterium. The cells are rod shaped, 3.0-4.0 x 1.1-3.0 mm, motile by means of one to four polar flagella (Bull *et al.*, 2010).

Colonies are grey, slightly convex and produces a fluorescent pigment on King's B medium. Metabolism is aerobic. Nitrate is not reduced. Hydrolysis of starch is negative. Arginine test (Thornley), indole production, DNase activity, and gelatin hydrolysis are negative. The species hydrolyses tween 80, nucleates ice and assimilates D-glucose, D-fructose, D-galactose, D-mannose, D-ribose, glycerol, D-saccharate, mucate, L(-)malate, citrate, D-glucuronate, D-gluconate, L-aspartate, L-glutamate, L-proline, L-alanine and L-serine. The organism contains genes for the production of coronatine (clf) and ethylene (eth). The DNA G+C content is 60.2 mol%. This pathogen produces necrotic halos on *Cannabis sativa* but, does not cause disease on *Broccoli rabe* (cv Sorento) or on oats (*Avena sativa* cv. Montezuma). All strains produce a brown pigment on King B medium and produce levan from sucrose (Bull *et al.*, 2010).

Molecular characteristics

In a molecular study Bull *et al.* (2010) observed that there is sequence similarity in the 16S rDNA of gene of *Pseudomonas cannabina* with the crucifer pathogen of *P. syringae* pv. *alisalensis*. In reciprocal DNA/DNA hybridization experiments, DNA relatedness was high (69–100%) between these two species. Sequence similarity was also noticed in five gene fragments used in multilocus sequence typing, as well as similar rep-PCR patterns when using the BOXA1R primers (Bull *et al.*, 2010).

Predisposing factors

It prefers wet, cool conditions. Optimum temperatures for disease tend to be around 12-25 °C.

Disease cycle

The bacterium is spread with the seeds and dispersed between plants by rain. It enters into the host either through natural openings or injured cells.

Control measures

In Canada it is a Quarantine pest since 2000.

Spraying of Copper hydroxide (@ 0.2% a.i.) and seed treatment with Streptomycin sulphate solution (0.05% a.i.) may be effective to manage the disease. The chemical, Acibenzolar-S-methyl (ASM, BTM, Bion or Actigard), which is an inducer of systemic resistance, is often used with Copper to manage bacterial diseases in various crops (Louws *et al.*, 2001). This combination may be used in hemp disease management. Further, the mixture of copper bactericides (e.g., Copper oxychloride) and ethylene bis-dithiocarbamate (e.g., Mancozeb) fungicides as effective in other crops (Conlin and McCarter, 1983) may be used to control the disease in hemp.

3. *Xanthomonas* leaf spot

Occurrence & severity status

The disease occurs on over 350 plant species, including cannabis (*Cannabis sativa* L.). Initally, this bacterial disease of hemp was first reported in Japan by Watanabe (1947). It was observed in Tochigi Prefecture, Japan in 1982 to cause bacterial leaf spot of hemp (Takikawa *et al.*, 1984). In Japan the disease appears in May and continues until harvest season of August (Netsu *et al.*, 2014). The hemp disease is also found in Romania (Severin 1978).

Symptoms

The disease appears as pin-head sized water-soaked lesions on leaves. Subsequently, the lesions enlarge to 1–2 mm diameter, and eventually the central part of lesions becomes necrotic with a brownish tinge. The lesion is often accompanied by a yellow halo of 2–3 mm wide. The lesions are generally circular, sometimes become polygonal and are delineated by leaf veins and scattered on the leaf blade. In some cases, the lesions are developed extensively along the veins. In case of shoot apex infection, bud blight is developed. No symptoms are formed on stems. Finally, the lesions become necrotic and dry spots, measuring 1-2 mm, accompanied by a yellow halo of 2-3 mm wide (Severin, 1978; Netsu *et al.*, 2014).

Causal organism: *Xanthomonas cannabis* pv. *cannabis* Jacobs *et al.*, **2015** [Front. Plant Sci., 6: 431]

Synonyms

= *Xanthomonas campestris* pv. *cannabis* Severin, 1978

Morphological/ Morphophysiolgical characteristics

The pathogen is a gram-negative, short rod shaped aerobic bacterium that motile with one polar flagellum. The bacterium develops yellow bacterial colonies in medium. These are circular, entire, raised, translucent and smooth colonies with shiny surfaces and are yellow on nutrient agar plates. The bacterum produces abundant extracellular polysaccharides on various media containing sugars. It also produces a faint, water-soluble, brown pigment on various media. Fluorescent pigment is not produced on King's B medium. The bacterium gives positive reactions in the following tests, such as, catalase, production of ammonia, production of hydrogen sulfide, mucoid growth on the 5 % sucrose containing medium, production of reducing substance from sucrose, hydrolysis of esculin, hydrolysis of starch, production of lecithinase, hydrolysis of cotton-seed-oil, hydrolysis of Tween 80, growth on 3 % NaCl, growth at 35 °C, and potato rot test. It gives negative reactions in the following tests, e,g., nitrate respiration, nitrate reduction, production of indole, arginine dihydrolase, phenylalanine deaminase, urease, production of acetoin, methyl red test and oxidation of gluconic acid. In purple milk culture, alkalization and digestion of casein is observed. The bacterium produces acids from carbohydrates, such as, ribose, raffinose, mannitol, etc. The casual bacterium of hemp develops water-soaked lesions on inoculation to the leaves of tomato and geranium after 2–3 days (Netsu *et al.*, 2014).

Molecular characteristics

In a study of Netsu *et al.*(2014) for phylogenetic classification of the causal bacteria isolated from hemp in Japan, gyrase B gene (gyrB) was amplified by polymerase chain reaction from the extracted genomic DNA.The gyrB sequence of SUPP546, one of the hemp isolates in Japan was analysed. The phylogenetic tree was drawn and modified using MEGA5.2 and Illustrator CS5.1 (Adobe Systems, San Jose, CA, USA). The obtained gyrB sequence for SUPP546 was completely identical to the retrieved sequence of *X. campestris* pv. *cannabis* (EU285190). The sequence data for the gyrB gene of SUPP546 was deposited in DDBJ (accession AB778813) (Netsu *et al.*, 2014). In further study of Jacobs *et al.* (2015), to gain insight into the evolution of *Xanthomonas* strains pathogenic to cannabis, they sequenced the genomes of two geographically distinct *Xanthomonas* strains, NCPPB 3753 and NCPPB 2877, which were previously isolated from symptomatic plant tissue in Japan and Romania. Comparative multilocus sequence analysis of housekeeping genes revealed that they belong to Group 2, which comprises most of the described species of *Xanthomonas*. Interestingly, both strains lack the Hrp Type III secretion system and do not contain any of the known Type III effectors. Yet their genomes notably encode two key Hrp pathogenicity regulators HrpG and HrpX, and hrpG and hrpX are in the same genetic organization as in the other Group 2 xanthomonads. Promoter prediction of HrpX-regulated genes suggested the induction of an aminopeptidase, a lipase and two polygalacturonases upon plant colonization, similar to other plant-pathogenic xanthomonads (Jacobs *et al.*, 2015).

Predisposing factors

The pathogen grows well at temperature 25-30°C.

Disease cycle

The pathogen can live in a soil for over a year and spreads through any movement of water including rain, irrigation and surface water.

Control measures

There are some limited chemical control methods that can be used. Spraying healthy plants with copper fungicides may reduce the spread of the bacteria in the field.

4. Downy mildew

Occurrence & severity status

The disease was earlier believed to occur in the countries of Asia (China, India, Japan, Pakistan, Azerbaijan, Kirghizia, Uzbekistan) and Europe (France, Hungary, Italy, Poland, Portugal, Rumania) (Hall, 1991). However, at present time, it occurs on every continent except Antarctica (McPartland and Cubeta, 1997). The economic impact of the disease is greatest in warm temperate regions, such as southern Europe, Italy and southern France. The disease is particularly destructive in regions where hemp is continuously cropped without rotation. Reports of disease intensity range from 7 to 10% of the leaves per plant (Juneja *et al.*, 1976) to serious (Barloy and Pelhate, 1962) to severe (McCain and Noviello, 1985) and to complete devastation (Zabrin, 1981). It is also reported that a single infected plant introduced into Colombia or Jamaica during a wet season can cause complete devastation (Zabrin, 1981).The disease occurs on *Cannabis ruderalis* and *C. saliva* (Hall, 1991).

Symptoms

Downy mildew infects half-grown to mature plants. Its symptoms begin as yellow leaf spots of irregular size and angular shape, limited by leaf veins. Opposite to the spots, on the underside of leaves, the fungus emerges from stomata to sporulate. Mycelial growth on the underside of leaves is best seen in early morning when dew turns the mycelium a lustrous violet-grey colour. Lesions enlarge quickly (McPartland *et al.*, 2000) and form brown to black spots on lower leaf surfaces that covered with a light violet to grey felt, with corresponding yellow necrotic patches on upper surfaces (Hall, 1991) . The affected leaves become contorted. Leaves soon necrose and fall off. Whole plants and entire fields may follow this course (McPartland *et al.*, 2000).

Causal organism: *Pseudoperonospora cannabina* (Otth) Curzi, 1926
[Riv. Pat. Veg.Pavia 16:234]

Synonyms

= *Peronospora cannabina* Otth 1869

= *Peronoplasmopara cannabina* (Otth) Peglion 1917

= *Pseudoperonospora cannabina* (Otth) Hoerner 1940

[N.B. Another fungus, *Pseudoperonospora humuli*, had been reported to cause downy mildew disease, but, so far, only three reports had indicated as a causal fungus of hemp (Hoerner, 1940; Ceapoiu, 1958; Glazewska, 1971). The pathogen normally infests hops. *Pseudoperonospora cannabina* and *P. humuli* can be distinguished as different species, based on sporangiophore branching patterns and sporangia papillae. In *P. humuli*, the sporangiophore branching is abundant dichotomous, occasionally up to sixth order and sporangia papillae are blunt apical (McPartland *et al.*, 2000)]

Morphological/ Morphophysiolgical characteristics

The fungus is an obligately biotrophic plant parasite. Its mycelium is intercellular, colourless, aseptate, thin-walled, hyphae 8-12 µm wide, bearing small, vesiculate haustoria, sometimes having finger-like projections. Sporangiophores are arborescent, hyaline, thin-walled, arising several per leaf stomate, dichotomous branching sparsely up to the third order, often with swellings on the main axis and branches,(100-) 160-350 x (4-) 5-8 µm. Branching in main branches are alternate and straight, dichotomous or irregular. Branchlets are predominately dichotomous, with a slight curvature, 40-100 µm long, arising at an acute angle to the main axis. Tips are straight, 10-18 µm long, diverging at an acute angle (Hall, 1991). Sporangia are ovoid to ellipsoid, usually (23-) 26-30 (-36) x (12-) 16-19 (-24) µm, grey-violet (turning brown with age), sometimes with a very short (1-2 µm long) pedicel, thin-walled, brown and poroid, with a protruding apical papilla 3-6 µm wide, 1-3 µm deep, germinating into zoospores or hyphae (Hall, 1991; McPartland *et al.*, 2000). Zoospores are reniform, laterally biflagillate, rounding up to producing hyphae (Hall, 1991). Oogonia are with oospores rare. Oospores were reported once in cotyledons, but no description or measurements were given (Peglion 1917).

Molecular characteristics

In an investigation of Choi *et al.* (2005), phylogenetic analysis of the ITS rDNA region was carried out with *Pseudoperonospora cannabina*, *P. celtidis*, *P. cubensis* and *P. humuli* to reveal taxonomic relationships. They found that all four species formed a well-resolved clade when compared with the ITS sequences of other downy mildew genera, using Bayesian inference and maximum parsimony.

Predisposing factors

The pathogen survives in infected buds and crowns. Downy mildew causes more damage during wet conditions and moderate temperatures 16- 20°C than in dry, hot conditions. Since leaf wetness duration is critical for leaf and shoot infection.

Disease cycle

Sporangia are spread by wind or water. Germination of sporangia requires a wet period (heavy dew will do). Sporangia germinate into hyphae or cleave into zoospores. Hyphae penetrate plant epidermis directly. Downy mildew fungus may persist in a field, and become progressively worse over the years (McPartland *et al.*, 2000).

Management/Control measures

Use of host resistance

Two Italian hemp cultivars, 'Superfibra' and 'Carmagnola Selezionata,' are reported as resistant to *P. cannabina* (McCain & Noviello, 1985).

Cultural and Mechanical Control

The disease can be minimized by certain cultural methods, like, crop sanitation, crop rotation with other crops for a minimum three years, optimizing soil structure and nutrition, and rouging the infected plants.

Biological and Chemical Control

A unique strain of *Bacillus subtilis* is sold as a foliar spray for controlling downy mildew. Ferraris (1935) controlled *P. cannabina* epidemics in Italy with copper sulphate. Hewitt (1998) treated *P. humuli* with Bordeaux mixture or copper oxychloride. The undersides of leaves must be sprayed.

5. *Fusarium* wilt

Occurrence & severity status

Fusarium wilt has been reported to cause serious loss in the cultivated hemp crop in Italy (Noviello and Snyder, 1962). It is also reported to occur in other countries of eastern Europe. In southern France its prevalence is very high, infecting plants up to 80% (Tiourebaev *et al.*, 1998). At present, the disease has been flourished and is found throughout the Northern hemisphere. The disease in the Chu River Valley, Djambul region, Kazakhstan, causes severe wilting and plant death of *Cannabis sativa*. In an experiment, the disease was found to affect about 35% of *Cannabis* plants (Tiourebaev, 1999). Nowadays, *Fusarium* wilt is a menace to hemp cultivation worldwide because its causal fungi cannot be constrained by known organic control measures. The disease threat is compounded by misguided attempts to stop marijuana cultivation around the world (McPartland and Hillig, 2008).

Symptoms

The pathogen attacks from the ground and colonises the cannabis plant through its xylem conduits. The fungus colonizes in the xylem tissues of plants, blocking the vessels and the flow of sap. The affected seedlings do not emerge (pre-emergent damping off), or wilt and fall over (post-emergent damping off). Symptoms in older plants begin as small, dark, irregular spots, initially appearing on lower leaves. The affected leaves turn yellow-tan and dry but remain hanging on the plant. The disease progresses rapidly upwards, usually unilaterally. Later, cortical tissues of the stem may loosen and become covered with mycelium and the plant generally dies (Noviello and Snyder, 1962). Stems also turn yellow-tan. Cutting into wilted stems reveals a reddish-brown discolouration of xylem tissue. Pulled-up roots show no external symptoms. Eventually, the disease may cause whole plants wilting (Barloy and Pelhate, 1962) or one side wilting of plants (Noviello and Snyder, 1962). The surviving plants become stunted.

Causal organism: *Fusarium oxysporum* f. sp. *cannabis* Noviello & W.C. Snyder, (1962)

[Phytopathology 52: 1317]

Morphological/ Morphophysiolgical characteristics

The mycelium of the pathogen in culture is floccose or felty, white to pink to purple, growing abundantly. Conidiogenous cells are hyaline, short, barrel-shaped, phialidic, in tufts of one to four atop metulae, 10-12 µm long.

Macroconidia are hyaline, three to five septa, sickle-shaped, ends curved inward with a hooked apex and pedicellate base, 45-55 x 3.5-4.5 μm. Microconidia are hyaline, aseptate (rarely one septum), oval to cylindrical, 5-16 x 2.2-3.4 μm. Chlamydospores are hyaline, thick walled, with a rough or smooth surface, spherical, borne singularly or in pairs, formed atop conldiophores or intercalary within hyphae or macroconidia, 7-13 μm in diameter (McPartland *et al.*, 2000).

Predisposing factors

There are various factors that can promote both the infection and reproduction of the fungus. Using seeds, growing spaces, soil or tools which are already infected will undoubtedly lead to the growth of this fungus if propagation conditions are favourable. *Fusarium* wilt is a warm-weather disease. Optimal temperature for fungal growth is 26°C (Noviello & Snyder 1962). Barloy & Pelhate (1962) reported epidemics in sandy, alluvial soils. Soils deficient in calcium and potash predispose plants to wilt disease Excess nitrogen increases wilt disease. Zhalnina (1969) reported that "acid fertilizers" increased wilt in hemp. Soils deficient in calcium and potash predispose plants to wilt disease (McPartland *et al.*, 2000).

Disease cycle

The fungus overwinters as chlamydospores in soil or crop debris. In the spring, chlamydospores produce hyphae and colonization takes place at soil level. The hyphae directly penetrate roots of seedlings. In older "root-hardened" plants, the hyphae enter via wounds. After hyphae penetrate roots, *F. oxysporum* invades water-conducting xylem tissues. Microconidia arise in these vessels and flow upstream to establish a systemic infection. Disease symptoms may not become evident until the advent of hot summer temperatures. There is no windborne transmission. Conidia arising on dead plants may be rain-splashed onto neighbours. Chlamydospores also arise on dead plants. *F. oxysporum* invades Cannabis seeds. Seed borne infections lay dormant until seedlings sprout the following spring (Pietkiewicz 1958; Zelenay 1960).

Management/Control measures

Use of Host resistance: The Romanian hemp cultivar 'Fibramulta 151' is resistant to *Fusarium* wilt while the Italian hemp cultivar 'Super Elite' is susceptible (Noviello *et al.* 1990).

Cultural methods: Selection of proper site for raising hemp, crop sanitation, periodical crop rotation and avoidance of infected seeds are helpful to check the disease occurrence in field. Noviello *et al.* (1990) reported different soil types affect plant resistance. Barloy and Pelhate (1962) reported epidemics in sandy alluvial soils. So, that kind of soils should be avoided. Soils deficient in calcium and potash and excess nitrogen increases wilt disease and that must be corrected by adding proper fertilizers, avoiding acid forming ones. Continuous cropping of Cannabis causes a build-up of inoculum and the creation of "wilt-sick soil" (Czyzewska and Zarzycka 1961). So, they suggested rotating hemp after wheat. Ceapoiu (1958) laid soil fallow for five years to eliminate fusaria.

Biocontrol : Application of biocontrol agent to manage the disease may be effective. The application of any commercially available biocontrol agents, such as, *Burkholderia cepacia, Streptomyces griseoviridis, Gliocladium* sp., *Trichoderma harzianzim* and *Glomus intraradices*, is suggested to manage the wilt disease (McPartland *et al.*, 2000). However, amongst them, the latter two work well in combination against *F. oxysporum* (Datnoff *et al.* 1995). Czyzewska and Zarzycka (1961) mixed *Trichoderma lignorum* into soil to protect their hemp crop.

Chemical control: It is suggested to disinfect the seeds with chemicals, either by fungicide (like, carbendazim or thirum) or by formalin solution (Booth, 1971). Spraying wilted plants with fungicides is not useful. Some farmers fumigate soil with nematicides to reduce root wounding in *Fusarium* infested fields.

Above all, it is to mention that, till now, no effective remedy has been developed to fight the disease in field, so the best thing we can do is to remove the infected plants (Kushka , 2017).

6. Grey mould

Occurrence & severity status

The disease, 'Grey mould', also called as 'Graymold', has become the most common disease of *Cannabis* (McPartland *et al.*, 2000). It causes serious economic losses in a wide variety of crops, including protein crops, fibre crops, oil crops, and horticultural crops. It is reported that the pathogen of grey mould affects more than 200 dicotyledonous plant species and a few monocotyledonous plants in both temperate and subtropical regions (Williamson *et al.*, 2007). In hemp, the disease affects mature plants prior to harvest, or at seedling stage. The disease is so serious in hemp that it can destroy the crop within a week. The disease peaks in drizzly, maritime climates (e.g., the Netherlands, the Pacific Northwest). In these climates grey mould reaches epidemic proportions and can destroy a *Cannabis* crop in a week (Barloy and Pelhate, 1962; Frank, 1988). Further, in another way it can be said that, the disease is more common for the regions with high humidity and cool to moderate temperatures. Damage from grey mould in *Cannabis* is compounded by humid climates, above 50%. It afflicts fibre and drug cultivars, outdoors and indoors (glasshouses). The grey mould fungus tends to attack *Cannabis* in two places-flowering tops and stalks.

Flower infestations tend to arise in drug cultivars and seed cultivars with large, moisture-retaining female buds. Scheifele (1998) reported 30-40% incidence of "head blight" in fields of early-maturing seed hemp. Stalk rot seems more common in fibre varieties (Patschke *et al.* 1997). The grey mould fungus can also infect seeds, destroy seedlings and attack plants after harvest (McPartland *et al.*, 2000).

Symptoms

The disease appears in both mature plants prior to harvest, or in seedlings. In case of seedling infection, the disease causes damping off. Seedlings are also died. In mature cannabis plants, it mostly affects stem, which appears as grey-brown mat of mycelium covered with fungal spores. Stems become chlorotic and often snap at canker sites. *Botrytis* cankers develop to the point of limb breakage or stem splitting, especially if plant canopies are dense and heavy. Small, black sclerotia can be developed within stem tissues. The cankers produce gray-brown masses of spores. In "drug type" or in seed yielding cannabis plants, grey-mould infects flowering tops, especially during flowering and near harvest time. Infections start within buds, so initial symptoms is not visible. It causes brown, water-soaked spots on buds. Fan leaflets turn yellow and wilt and pistils begin to brown. In high humidity, whole inflorescences become enveloped in grey fuzz, and then degrade into grey-brown slime. The grey fuzz is a mass of microscopic conidia. In low humidity, the grey fuzz does not emerge; infested flowers turn brown, wither, and die. Stalk rot begins as a chlorotic discolouration of infected tissues. Chlorotic sections turn into soft shredded cankers. Stalks may snap at cankers. Cankers may encircle and girdle stalks, wilting everything above them. In high humidity, cankers become covered by conidia. Conidia are liberated in a grey cloud by the slightest breeze. Small, black sclerotia may form within stalks (McPartland *et al.*, 2000). Eventually, the buds are rotted, producing a grey-brown mass of spores.

Causal organism: *Botrytis cinerea* Pers. 1801

[ITIS Species 2000 Catalogue of Life; Species Fungorum, Kew Mycology]

Synonyms

= *Botrytis cinerea* sub sp. *cinerea* Pers. 1794

= *Botrytis cinerea* var. *cinerea* Pers. 1794

= *Botrytis cinerea* f. *cinerea* Pers. 1794

= *Botrytis cinerea* f. *coffeae* Hendr. 1939

= *Botrytis cinerea* var. *dianthi* Voglino 1909

Morphological/ Morphophysiolgical characteristics

Conidiophores are upright, septate, grey-brown, pale brown toward apex, straight to slightly sinuous, up to 25 µm wide, thick-walled, branching near the apex, simple or dichotomously branched, branching again alternately, forming at each end a globose, swollen conidiogenous cell. Conidia are borne on pedicels of globose conidiogenous cell at the apex of conidiophore in botryose clusters, synchronously formed, with a slightly protuberant hilum, obovoid to ellipsoid, subhyaline to pale brown, smooth, 8-16 x 6-12 µm (Matsushima, 1975; McPartland *et al.*, 2000; Soares *et al.*, 2009). Microconidia are rare, *Myrioconium*-like state, arising from phialides, are hyaline, oval, aseptate, 2.0-2.5 µm in diameter. Apothecia are rare, arising from sclerotia, on a 3-10 mm stalk, topped by a yellow-brown disc. Disc is flat to slightly convex, 1.5-5mm in diameter, capped by a single layer of asci and paraphyses. Asci are cylindrical, with long tapered stalks, eight-spored, 120-140 x 8µm. Paraphyses are hyaline, septate, filiform. Ascospores are aseptate, unlguttulate, uniseriate, hyaline, ovoid to ellipsoid, 8.5-12 x 3.5-6 µm (McPartland *et al.*, 2000). Sclerotia are hard, black, rough, planoconvex, irregularly round to elongate, 1-15mm long (av.5mm) on *Cannabis* (Flachs 1936), sometimes in chains; cross-section of sclerotium reveals a thin black rind with a hyaline interior (McPartland *et al.*, 2000).

Molecular characteristics

In a study, *Botrytis cinerea*, collected from infected bottle gourd was used to characterize the pathogen on molecular basis in China. The species specific primers, Bc-f /Bc-r [Bc-f (5'- GGAAACACTTTTGGGGATA3')/Bc-r (5'-GAGGGACAAGAAAATCGACTAA-3')] and Bc108+/Bc563- [Bc108 + (5'-ACCCGCACCTAATTCGTCAAC3')/Bc563 - (5'-GGGTCTTCGATACG GGAGAA-3')] were found to produce 354 and 450 bp fragments, respectively. Sequence of PCR amplification of the rDNA- ITS region produced a 547 bp fragment using the universal primer pair ITS1/ITS4. In the phylogenetic tree based on ITS nucleotide sequences, the representative isolate was located within a clade comprising reference isolates of B. cinerea (Kamaruzzaman *et al.*, 2017). In another study, a set of 30 isolates from rose of different regions were identified with a single band of 0.7 kb that developed by using specific primers as designed by Rigotti *et al.*, (2002) for *B. cinerea* detection (Khazaeli *et al.*, 2010).

In order to compare the differences in ploidy level by determining their DNA content per nucleus, the strain SAS56, an ascospore line of *B. cinerea* was used for genetic analyses. Molecular analyses (RAPD) of the haploidized strains indicate a very limited degree of heterozygosis of the parent strain SAS56. However, analysis of field isolates of *B. cinerea* showed that their DNA content per nucleus varied considerably, indicating that aneuploidy/polyploidy is a widespread phenomenon in this species (Büttner et al., 1994).

Since one or several allelic mutations in the histidine kinase (Bos1) and β-tubulin genes generally confer the resistance to fungicides, the sequences of these target genes were investigated in the selected isolates of *B. cinerea*, which allowed the identification of two different haplotypes. Mating types were also determined by PCR assays using primer specific for MAT1-1 alpha gene (MAT1-1-1) and MAT1-2 HMG (MAT1-2-1) of *B. cinerea*. Twenty-two out of 50 isolates (44%) were MAT1-2, while 38% were MAT1-1. Interestingly, out of whole studied samples, 9 isolates (18%) were heterokaryotic or mixed colonies. In addition, cluster and population structure analyses identified five main groups and two genetic pools, respectively, underlining a good level of variability in the analysed panel. The results highlighted the presence of remarkable genetic diversity in *B. cinerea* (Polat et al., 2018).

Predisposing factors

The fungus thrives in high humidity and in temperatures of cool to moderate (McPartland et al., 2000). Although the fungus can survive between 10°C and 27°C, it grows well in temperature range 18.3-23.9°C with optimum temperature for its growth 20°C. The high humidity and free water on plant surfaces favour the disease, as do close plant spacing and irrigation practices that keep plants wet for a longer time. Temperature, relative humidity, and wetness duration produce a conducive environment that is favourable for inoculation of mycelium or conidia (Ciliberti et al., 2015). Standing water on plant leaf surfaces provides a place for spores to germinate (Elad and Evensen, 1995). The threat of Grey mould increases as flowers mature. This is because when flowers swell, during the last weeks of flowering; their burgeoning size hinders intrafloral ventilation. Large moister-retaining female buds are more vulnerable. A low pH is preferred by the grey mould to perform well. *Botrytis cinerea* can acidify its environment by secreting organic acids, like oxalic acid (Amselem et al., 2011). By acidifying its surroundings, cell wall degrading enzymes (CWDEs) are enhanced, plant-protection enzymes are inhibited, stomatal closure is deregulated, and pH signaling is mediated to facilitate its pathogenesis (Amselem et al., 2011).

Disease cycle

The fungus is an opportunistic pathogen that easily invades weak, damaged, or senescing tissue. The pathogen overwinters in infected seeds. Infected seeds may mould in storage or give rise to seedborne infections of seedlings the following spring (Pietkiewicz 1958, Noble & Richardson 1968). The pathogen also overwinters within stalk debris as sclerotia or dormant hyphae. Melanized sclerotium allows the pathogen to survive for years in the soil. Sclerotia and the asexual conidia spores contribute to the widespread infection of the pathogen (Amselem *et al.*, 2011). Spring rains induce sclerotial germination, producing conidia or rarely producing apothecia with ascospores. Hyphae from sclerotia may also penetrate plants directly. Conidia (rarely ascospores) are blown or splashed onto seedlings. After infection, the fungus grows within the host. In humid conditions, the fungus forms new conidia, which spread to sites of secondary infections. Cycles of secondary infections eventually build to epidemics in mature plants (McPartland *et al.*, 2000).

Management/Control measures

Host Resistance

In general fibre cultivars are more resistant than drug cultivars (Van der Werf *et al.*, 1994; Mediavilla *et al.*, 1997). In a study, the fibre cultivar 'Livoniae' was found most resistant with 25% infection, while, the drug cultivar, 'Swihtco' was most susceptible, with 87% infected plants (Mediavilla *et al.*, 1997).

Cultural control

Sanitation by removing dead or dying plant tissue in the fall will decrease inoculum levels as there is no debris for the sclerotium or mycelia to overwinter. Removing debris in the spring will remove inoculum from the site. It is suggested to limit irrigation during and after bloom. Irrigation should be done in the morning so plants do not stay wet more than 12 hours. In greenhouses, it is better to maintain the relative humidity below 50%, temperatures warm and high light intensity. Filtering out UV light may prevent epidemics since sporulation requires UV light. Avoidance of over-fertilization with nitrogen is always beneficial.

Biocontrol

Foliage, flowers, and stems can be sprayed with *Gliocladium roseum* and *Trichoderma* sp. Bees can successfully deliver these fungi to flowers of other crops (Yu and Sutton, 1997). Damping off caused by *B. cinerea* can also be prevented by mixing *Gliocladium* and *Trichoderma* species into the soil (McPartland *et al.*, 2000).

Chemical control

Grey mould can be chemically controlled with well-timed fungicide applications starting during the first bloom. Timing can reduce the chance of resistance and will save on costs. Cyprodinil + fludioxonil mixture is the most effective in inhibiting mycelial growth of this pathogen, both *in vitro* and *in vivo* conditions. Cyprodinil+fludioxonil at dose of 3 µg ml-1 inhibit spore germination of it (Selvi *et al.*, 2016).

7. Powdery mildew

Powdery mildew disease in hemp is caused by two different fungi, *Leveillula taurica* and *Podosphaera macularis*. The disease occurs both in temperate and subtropical regions, infecting outdoor hemp (Transhel *et al.*, 1933) and indoor drug cultivars (Stevens, 1975).

i) Powdery Mildew (*Leveillula taurica*)

Occurrence & severity status

Powdery mildew caused by *Leveillula taurica* is distributed worldwide, including France (Hirata, 1966), other eastern European countries (Jaczewski, 1927; Transhel *et al.*, 1933; Gitman ,1935) and Turkistan (Gitman and Boytchenko 1934). This disease prefers warm, dry environments. Globally, it is a minor problem with limited occurrences in the Middle East, Europe, and South America. It parasitizes a wide range of hosts and causes disease on several other crop plants, most predominantly on onion (Zheng *et al.*, 2013). Different races of *L. taurica* can only infect certain crops, and even specific cultivars within the same crop. An accurate way to describe its host specificity is that this disease is, a composite species consisting of many host-specific races (du Toit *et al.*, 2004).

Symptoms

The disease appears as white powdery patches with cotton like texture (formed by the mycelia and conidia) on the surface of the largest leaves. These patches are small at first, later they cover most of the leaves, and even the small sugar leaves of the buds.

Causal organism: *Leveillula taurica* (Lév.) G. Arnaud 1921

[Annales des Épiphyties 7: 92]

Synonyms

= *Leveillula taurica* f. *cannabis* Jacz. 1927 (?)

= *Erysiphe taurica* Lev. (1851)

= *Tigria taurica* (Lev.) Trevis. (1853)

= *Oidiopsis taurica* (Lev.)

Morphological/ Morphophysiolgical characteristics

The fungus, *Leveillula taurica* is an obligate fungal pathogen. The superficial hyphae of the fungus often cover whole plants. The hyphae are pale buff to white, persistent, densely compacted, tomentose. Conidiophores are often two-celled, upright, straight or occasionally branched, hyaline, emerging from host stomata, up to 250 μm long. Conidia are born singly atop conidiophores, either cylindrical or navicular in shape, single-celled, hyaline, 50-79 x 14-20 μm. Cleistothecia are rarely formed, globose, black, smooth, 135-250 μm in diameter, with numerous hyphal appendages, scattered, embedded in mycelium. Appendages are hyaline to light brown, indistinctly branched, less than 100 μm in length. Asci are usually 20 per cleistothecium, ovate, distinctly stalked, two-spored, 70-110 x 25-40μm. Ascospores are cylindrical to pyriform, sometimes slightly curved, 24-40 x 12-22 μm (McPartland et al., 2000).

Molecular characteristics

In a study to identify powdery mildew, caused by *Leveillula taurica*, on molecular basis, DNA was extracted from fresh conidia washed off leaf lesions by using a standard rapid boil extraction procedure and were amplified with primers ITS1 and ITS4. The ITS fragment was sequenced and a 444-bp sequence matched with the ITS sequence of *Leveillula taurica* (gb|GQ167201.1) with 100% identity (Koike et al., 2015). In another study of Zheng et al. (2013), a pair of *L. taurica*-specific primers was designed based on the internal transcribed spacer sequence of *L. taurica* and used in real-time polymerase chain reaction (PCR) assay to quantify the fungal DNA during infection. The specific primers were designed as the forward primer LV-F (52 AGCCGACTAGGCTTGGTCTT32) and the reverse primer LVR (52 GCGGGTATCCCTACCTGATT32), which amplified a 208-bp fragment.

Predisposing factors

The powdery mildew species is most favoured by the temperature of 25°C and low relative humidity. It even adapts well to xerophytic conditions; conidia can germinate in 0% RH. Films of water actually inhibit *L. taurica* germination. The disease generally increases as rainfall decreases. Poorly ventilated areas, in greenhouses, are perfect for its propagation. Low light intensity (indoors) or shaded areas (outdoors) increase disease severity (McPartland et al., 2000).

Disease cycle

Leveillula taurica causes polycyclic disease. It overwinters as chasmothecia in plant debris above the soil surface. During favourable climatic conditions, the chasmothecia open and release ascospores which are dispersed by wind. The ascospores enter the host through its stomata, germinate, and colonize the host tissues with mycelia. Thereafter, the pathogen begins to produce its asexual conidia on conidiophores. The conidia exit through the host's stomata and serve as secondary inoculum to spread the disease. In the fall, the pathogen undergoes sexual reproduction and again produces chasmothecia, its dormant, overwintering structure (Evans et al., 2008).

Management/Control measures

In order to prevent the powdery mildew attack during growth stages, is useful to spray potassium soap or sulphur as preventives and also to treat the first symptoms of an infection. We can also use systemic fungicides, which remain in the plant tissues and protect it for several weeks. The fungicidal spray should be done right before switching to bloom, so that plants will be protected during the first weeks of this crucial stage. Spraying of high pH water (8-8.5) and repeating every 3-4 days is also effective to manage the disease. Several other ways are also in practice. Such as spraying water and milk solution (90% water and 10% milk) as well as spraying a solution of water with a dash of hydrogene peroxide are found effective against the disease. The superficial nature of *L. taurica* renders it susceptible to fungicide sprays. A simple solution of sodium bicarbonate or potassium bicarbonate is sufficient. Some botanical insecticides and miticides, such as Neemguard and Cinnamite, also kill powdery mildews. Horticultural oil (Sun spray) when mixed with baking soda works well. Sulphur fungicide controls powdery mildew when applied at two week intervals. Copper oxychloride may also be effective, Physcion extract (Milsana), a new botanical, protects plants by inducing plant resistance (McPartland *et al.*, 2000).

ii) Powdery mildew (*Podosphaera macularis*)

Occurrence & severity status

The disease caused by *Podosphaera macularis* occurs worldwide. The pathogen commonly parasitizes hops (Miller *et al.* 1960). The fungus usually occurs in conidial state on leaves and stems, sometimes also on sepals and fruits (Mukerji, 1968b). Only the conidial (anamorphic) state has been found on drug plants in the USA (McPartland, 1983a; McPartland and Cubeta, 1997), on Cannabis in South Africa (Doidge *et al.*, 1953), and on fibre varieties in Russia and Italy (Hirata, 1966). The pathogen is most commonly occurs on *Humulus lupulus* and many other plants, e.g. *Fragaria vesca*, especially in the members of Rosaceae (Junell, 1967; Blumer, 1967; Dennis, 1960) in Europe, Asia (Central Asia, Siberia, China, Far East of the USSR, Japan), North America (USA, Canada), South America and South Africa (Mukerji, 1968b).

Symptoms

The disease may occur at all stages of growth, affecting leaves and buds. The lower leaves are the most affected site of the disease, but it can appear on any part of the plant that is above the ground (Madden and Darby, 2011). Infections on susceptible leaves appear at first as whitish, powdery spots on either the upper or lower leaf surface. Later, the spots cover the entire leaf surfaces and buds with whitish powdery spores (conidia) and mycelia of the fungus. Finally, the diseased leaves are covered by numerous limited white patches which are persistent and characteristic in appearance and maculate. The leaf petioles can also be infected showing same symptoms as well as flower bracts. In case of older powdery mildew specks, ascocarps (chasmothecia), which are small nearly spherical fruiting bodies, are formed. Later, the chasmothecia turn from yellow to dark brown-black in colour (Rodriguez *et al.*, 2015; Ward-Gauthier *et al.*, 2015; Pscheidt and Ocamb, 2018).

Causal organism: *Podosphaera macularis* **(Wallr.) U. Braun & S. Takam. 2000**

[Schlechtendalia 4: 30 (2000)]

Synonyms

= *Desetangsia humuli* (DC.) Nieuwl. 1916

= *Sphaerotheca macularis* (Wallr.) Magnus 1899

= *Albigo humuli* (DC.) Kuntze 1898

= *Sphaerotheca humuli* (DC.) Burrill 1887

= *Leucothallia macularis* (Wallr.) Trevis. 1853

= *Erysiphe macularis* (Wallr.) Fr. 1829

= *Alphitomorpha humuli* (DC.) Wallr. 1819

= *Alphitomorpha macularis* Wallr. 1819

= *Erysiphe humuli* DC. 1815

Morphological/ Morphophysiolgical characteristics:

Mycelium is hyaline, septate, amphigenous. Haustoria are intraepidermal, globoid, 12.5-17.5 µm diam (Parmelee, 1975). Conidiophores are hyaline, short, simple branches, erect, foot-cells cylindric, 40-70 x 9-11 µm, followed by 1-3 shorter cells. Conidia are narrowly to broadly ellipsoidal, hyaline, smooth, nonseptate, basipetally catenate, about 20-33 x 13-20 µm, with fibrosin bodies. Appressoria are indistinct, sometimes nipple-shaped. Chasmothecia are black globoid bodies, with appendages, mostly hypophyllous, gregarious to scattered or caespitose, groups of ascocarps (chasmothecia) and coloured hyphae often forming dark patches on the leaves, (60-) 70-110 (-120) µm in diam., cells irregularly polygonal,. Appendages are few to many, numerous, in the lower half and partly in the upper half, long, 3-5 (-6) times the chasmothecial diam, brown, often brittle, flexuous, fairly straight, sometimes mycelioid, interlaced with each other and with the mycelium, septate, thin-walled, smooth, simple, coloured throughout when mature, brown or yellowish, width variable, (2.5-) 4-8 (-10)µm, rarely exceeding. Ascus is 60-90 x 50-80 µm in size with spores 4-8, seldom 2. Ascospores are ellipsoid-ovoid, 16-24 x 11-18 µm, formation of ascospores often irregular, spores of 2- or 4-spored asci are larger, up to 30 x 21 µm (Mukerji, 1968b; Parmelee, 1975; Braun, 1987). This species is well distinguished by the appearance of the maculate spots on infections and the long appendages of chasmothecia.

Molecular characteristics

In a study of Patzak (2005), a molecular device was developed to detect early infection of the powdery mildew *(P. macularis)* fungus in plants. He reported that the molecular analysis of internal transcribed spacer (ITS) regions of rDNA is a nobel and very effective method of this species determination. In doing so, he used species specific PCR assay. The specific primers combination S1 + S2 (S1 = CCCGAACTCATGTAGTTAGTGC; S2 = GAGCACATCGGTACCGCC ACTA) amplified specific fragment of 282bp nuclear rDNA (ITS1, 5.8S, ITS2) region from the fungus. He further used the PCR primer combinations R1 + R2 (R1 = GTGAACCWGCGGARGGATCATT; R2 = TTYGCTRCGTTCTT CATCGATG) and R3 + R4 (R3 = GCATCGATGAAGAACGYAGCRA; R4 = TATCTTAARTTCAGCGGGT) in multiplex PCR detection of this fungus. The PCR primers R1 and R2 amplified a fragment of 248bp nuclear rDNA region, while, R3 and R4 amplified a fragment of 312bp nuclear rDNA region. Phylogenetic relationships from nuclear rDNA (ITS1, 5.8S, ITS2) were inferred for 42 species including this causal fungus. The causal fungus, *P. macularis*, collecting from hop plant, showed close relationship with *P. filipendule* and *P. ferrugunea* (Patzak, 2005).

Predisposing factors

Favourable conditions include low sun exposure, excess fertility, and high soil moisture (Mahaffee *et al.*, 2009; Marks and Gevens, 2014). Additionally, optimal infection is observed when the temperature is between 18 and 25°C (Peetz *et al.*, 2009; Madden and Darby, 2011). In addition, periods with small temperature differences between night and day, with a minimum of 10°C at night and a daily high of 20°C increase the risk of infection (Engelhard, 2005). During the period from mid-July to August, conidia infectivity and germination is highest around 18°C (Peetz *et al.*, 2009). Conidia germinate best at 100% RH, although they can tolerate RH down to 10-30% (McPartland *et al.*, 2000).

Disease cycle

The fungus overwinters on the soil surface in plant debris as fungal survival structures (chasmothecia) or as mycelia in plant buds (Marks and Gevens, 2014). During favourable conditions, these chasmothecia rupture and ascospores are discharged. The ascospores act as the primary inoculum and are dispersed passively by wind. Upon encountering a susceptible host plant, the ascospores germinate and cause infection. Following infection, masses of conidia are produced during the season (Marks and Gevens, 2014). These conidia are dispersed through wind during the growing season and can further infect additional host plants. Under optimal conditions, this polycyclic disease can potentially grow 20 generations in a growing season (Peetz *et al.*, 2009).

Management/Control measures

Cultural control: Cultural practices that can help to prevent the disease include carefully monitoring water and nutrient, reducing initial inoculum, and removing basal growth. It is suggested to avoid drought stress that increases disease, overcrowding and overhead irrigation and to remove the infected leaves early in disease development which can aid in delaying epidemics (McPartland *et al.*, 2000). It is also suggested to maintain adequate nitrogen levels and not to over-apply because more succulent tissue is more susceptible (Rodriguez *et al.*, 2015; Ward-Gauthier *et al.*, 2015; Pscheidt and Ocamb, 2018).

Biocontrol : Different biocontrol agents have seen identified as effective to control powdery mildew disease in different crops. For example, The hyperparasitic fungi such as *Ampelomyces quisqualis*, a strain of *Verticillium lecanii* and *Sporothrix flocculosa* (Kendrick 1985) are found effective against powdery mildews. However, their use in practical field is so far negligible due to certain limitation in use.

Chemical control: Chemical control is the only measure use extensively for the control of powdery mildew disease. Chemical control primarily consists of spraying fungicides in hopes of preventing the disease through the use of early, continuous spraying during the growing season. Thus, prophylactic fungicide programs can be a very effective way in preventing the disease (Marks and Gevens, 2014). In this method thorough spray coverage is essential to protect leaves. Application of protective fungicides, like, sulphur, copper, before appearance of disease is effective to check the initial inoculum. Chemical control can help to keep disease levels low, both in terms of disease incidence and severity before plants begin flowering (Rodriguez *et al.*, 2015; Ward-Gauthier *et al.*, 2015; Pscheidt and Ocamb, 2018).

B. Pests

1. Aphids

Aphids also known as "plant lice" are tiny, soft-bodied, pear shaped insect with longer legs and antennae. Some of them have bumps on the head between the antennae. A pair of tube-like cornicles (siphons) project backwards. The rear end tapers to a pointed, tail-like caudum. Most adult aphids do not have wings, but some do. The wings are much longer than the body. Several aphid species are found to attack hemp of which six species are more common (McPartland *et al.*, 2000). The aphids affect plants in warm, moist weather, with gentle rain and little wind. They suck sap from a plant's vascular system, using long narrow stylets. Most of them are phloem-feeders, but some also suck on xylem (Hill, 1994). Besides sucking sap, aphids damage plants by vectoring pathogens, especially viruses. Viruses and aphids are reported to have a symbiotic relationship (Kennedy *et al.*, 1959).

Types of damage

Aphids congregate on the undersides of leaflets and cause yellowing and wilting. Some aphid species prefer older, lower leaves (e.g., *Myzus persicae*), and some prefer younger, upper leaves (e.g., *Aphis fabae*). Some species even infest flowering tops (e.g., *Phorodon humuli, P. cannabis*). At first, the damage symptoms develop on the undersides of leaflets as light-coloured spots, especially near leaflet veins. Later, the leaves and flowers become puckered and distorted (Kirchner, 1906). Heavily infested plants may completely wilt and die. Surviving plants remain stunted. Honeydew exudes from the anus of feeding aphids. The honeydew causes secondary problems-ants eat it, and sooty mould grows on it. Sooty mould reduces plant photosynthesis and leaf transpiration, and hinders the movement of aphid predators and parasites (McPartland *et al.*, 2000).

i) Green Peach Aphid

Occurrence & severity

The Green Peach Aphid (*Myzus persicae*) attacks dozens of plant species, and now lives worldwide (Spaar *et al.*, 1990). The pest is exceptionally restless. The alatae repeatedly land on plants, probe briefly and then take off for other plants. This behaviour makes the aphid as a notorious vector of plant viruses (Kennedy *et al.*, 1959). It infests feral hemp in Illinois and marijuana in India (Sekhon *et al.*, 1979).

Scientific name & order: *Myzus persicae* **(Sulzer) 1776 (Hemiptera: Aphididae)**

Synonyms

= *Phorodon persicae* Sulzer

Biology

Eggs: Eggs are elliptical in shape, measuring about 0.6 mm long and 0.3 mm wide, initially yellow or green, but soon turn black. Eggs are deposited on *Prunus* spp. trees (Capinera, 2017a)

Nymphs: Nymphs initially are greenish, but soon turn yellowish, greatly resembling viviparous adults. There are four instars in this aphid, with the duration of each averaging 2.0, 2.1, 2.3, and 2.0 days, respectively. Females give birth to offspring six to 17 days after birth, with an average age of 10.8 days at first birth. The length of reproduction varies considerably, but averages 14.8 days. The average length of life is about 23 days. The daily rate of reproduction averaged 1.6 nymphs per female. The maximum numbers of generations as observed are 20 to 21 annually, depending on the year (Horsfall, 1924). In contrast, MacGillivray and Anderson (1958) have reported five instars with a mean development time of 2.4, 1.8, 2.0, 2.1, and 0.7 days, respectively. Further, there are a mean reproductive period of 20 days, mean total longevity of 41 days, and mean fecundity of 75 offspring (Capinera, 2017a)

Adults: Apterae are green, sometimes yellow-green or pink, oval in outline, averaging 2.0-3.4 mm in length. Antennal tubercles are mammary-like bumps of less than half as long as the first antennal segment, and point inward, converging towards each other. Cornicles are of same colour as the abdomen, except for darkened tips. Cornicles are long, thin, slightly swollen near the midpoint, and grow twice as long as cauda. Cauda are lightly bristled and constricted slightly at the midpoint. Alatae nymphs are often pink or red, and develop wing-pads. Alatae (winged adults) have black-brown heads, a black spot in the middle of their abdomens and hold their wings in a vertical plane when at rest (McPartland *et al.*, 2000).

Life cycle

Green peach aphid (*M. persicae*) is heteroecious (or holocyclic). It migrates between two hosts. The overwintering host is called the primary host. The aphid overwinters on *Prunus*, as shiny black eggs. Eggs hatch into rotund "stem mothers" or fundatrices. Fundatrices multiply parthenogenically. They are born fertile and within days begin giving birth. They give birth viviparously to nymphs. Fundatrices bear 60-100 fundatrigeniae. Soon the fundatrigeniae begin to give birth to more live females (apterous viviparae or apterae). In late spring the first alatae (winged aphids) develop. They are called spring migrants and fly off to secondary hosts, such as cannabis. Aphids terminate migration by actively flying downward and settling on plants. Once settled on Cannabis the alatae give birth to apterae; apterae undergo four moults in about ten days to reach sexual maturity. Each aptera gives birth to 30-70 young. Crowded conditions or a lack of food induce new alatae, which fly off seeking unexploited Cannabis. They are called summer migrants. At the end of the summer, special alatae called sexuparae fly back to the primary host and give birth to ten sexuales. Sexuales are either females or males, alatous or apterous. They mate and the females become oviparae and lay five to ten eggs, which overwinter. Hill

(1994) claims that a single springtime fundatrix can give rise to up to 12 generations of aphids in one year-and theoretically 600,000 million offspring. In tropical regions, aphids do not migrate between hosts, nor they lay overwintering eggs. They reproduce parthenogenically all year long, and remain on their secondary host. Thus, some aphids that are normally heteroecious may become autoecious in warmer climates. *M. persicae*, for instance, overwinters as adult females on secondary hosts in the south or in warm glasshouses (Howard *et al.*, 1994; McPartland *et al.*, 2000).

Management/Control measures

Biological control

The natural enemy wasps are very beneficial to kill the aphids at the beginning of season when the temperature is comparatively high or where temperature can be controlled, as in some greenhouses (Mackauer, 1968). Thus, protection of natural enemies may be helpful in hemp also. However, in the field, biological control agents may be differentially affected by the cropping system. For example, Tamaki *et al.* (1981) have found that the wasp *Diaeretiella rapae* (Hymenoptera: Braconidae) is more effective in broccoli, whereas lady beetles (Coleoptera: Coccinellidae) and big eyed bug (Hemiptera: Lygaeidae) predators are more effective on radish. In addition, a few parasitoids, like, the entomopathogenic fungus *Verticillium lecanii*, and the predatory midge *Aphidoletes aphidimyza* (Diptera: Cecidomyiidae), are found effective in greenhouse-grown vegetables in Europe (Gilkeson and Hill, 1987, Milner and Lutton, 1986). These biological control agents including the parasitoids may be effective in hemp.

Chemical control

Green peach aphid has evolved high levels of resistance to almost all classes of insecticide. At present, the application of neonicotinoids (such as, imidacloprid) insecticides is mostly considered as an effective control measure, however, recent resistance monitoring work in southern Europe represents a threat to the long-term efficacy of this chemical class. The aphid, *M. persicae*, can resist a broader range of pesticides, including neonicotinoids than any other insect pest. The resistance to neonicotinoids is probably due to the over-expression of P450 gene and cuticular proteins, and reduced penetration of insecticide through the cuticle (Puinean *et al.*, 2010). In spite of the resistance problem, many producers are dependent on insecticides for suppression of green peach aphid abundance. Systemic insecticide applications are especially popular at planting time, most of which provide long-lasting protection against aphid population build up during the critical and susceptible early stages of plant growth (Powell, 1980) and some of which provide protection for 3 months (Palumbo and Kerns, 1994). The granular pesticides, Aldicarb granules at 1 g/plant and disulfoton at 2 g/plant as found most effective in controlling the aphid in tobacco, may also be effective in hemp (Prasadrao *et al.*, 1982)

ii) Black Bean Aphid

Occurrence & severity

The Black Bean Aphid migrates between two hosts: *Euonymus* and *Viburnum* as primary hosts then to *Cannabis* as secondary host. Blackman and Eastop (1985) listed *Cannabis* as a host for black bean aphid. It is the second most common aphid in Dutch glasshouses, and attacks outdoor crops of hemp in China and South Africa (Dippenaar *et al.*, 1996; McPartland *et al.*, 2000). The black bean aphid may have been infesting a seed-oil cultivar ('FIN-315') in Finland (Callaway and Laakkonen 1996). The Black bean aphids are found in all temperate regions except Australia. This aphid infests many crops and vectors over thirty viruses (McPartland *et al.*, 2000).

Scientific name & order: *Aphis fabae* Scopoli, 1763 (Hemiptera: Aphididae)

Biology

Egg: The eggs overwinter on the aphid's primary hosts, *Euonymus* spp., *Viburnum* spp., and *Philadelphius* spp. Aphid eggs are elliptical (0.02 in. long and 0.01 in. wide), yellow or green in color turning black as they develop. Once developed, they are shiny-black. The eggs are usually laid in the crevices of bud, stems, and barks of the plant. Aphids usually do not lay eggs in warm parts of the world.

Nymph: Nymphs look like apterae (wingless adults) but are smaller. They have black bodies and yellow legs with black tarsi. The cornicles of the nymphs are short. Alatae nymphs have prominent white markings on their abdomen (McPartland *et al.*, 2000). They become adults within 7 to 10 days.

Adult: Apterae are oval to pear-shaped, with a small head and bulbous abdomen, olive green to dull black, averaging 1.5-3.1 mm in length (Chinery, 1993; McPartland *et al.*, 2000). Legs are light green to white or yellow, tarsi brown-black. Antennae are light yellow. Antennae tubercles are not prominent. Cornicles are relatively short (0.3-0.6 mm) and cylindrical. Cauda are heavily bristled. Alatae are slightly smaller than apterae (1.3-2.6 mm long). Their bodies are dark green-black with variable white stripes. Wings are spotted with white wax. Alatae hold their wings vertically over their abdomens when at rest. The aphid colonies are regularly ant-attended and they may congregate in great numbers (McPartland *et al.*, 2000).

Life cycle

The aphid (*A. fabae*) is heteroecious. It migrates between two hosts. The overwintering hosts are the spindle tree (*Euonymus europaeus*), or on *Viburnum* or *Philadelphius* plants. The aphids that hatch from these eggs in spring are all special wingless females (fundatrices), known as 'stem mothers'. These stem mothers are able to reproduce asexually through a process known as 'parthenogenesis' that does not involve mating. Furthermore, they do not lay eggs but give birth to live offspring, fundatrigeniae. Soon the fundatrigeniae begin to give birth to more live females (apterae), which are able to reproduce without mating. The next generation to be produced are typically winged forms, called 'first alatae' (spring migrants) and these undertake migrations to Cannabis and other secondary hosts. At the end of the summer, autumn migrants (special alatae called sexuparae) fly back to the overwintering host and give birth to sexuales (males & females). Sexual females and males mate and the females become oviparae and lay eggs, which overwinter (McPartland *et al.*, 2000; Natwick, 2010).

Management/Control measures

Biological control

The black bean aphid is attacked by predatory Chrysopidae, Coccinellidae and Syrphidae, as well as by endoparasitoids, especially *Lysiphlebus fabarum* (Marchall). Several entomopathogenic fungi also infect the pest (Sengonca *et al.*, 1995; Volkl and Stechmann, 1998; Kluth *et al.*, 2002).

Chemical control: It is found that the pest is susceptible to organophosphates and to acetamiprid, whereas its reproduction seems to be enhanced by pyrethroids (Hutt *et al.*, 1994). In an investigation, it is observed that parathion (0.02% w/v) spray gives excellent control of *A. fabae*. The systemic insecticides Isopestox and Systox at 0.05% (w/v) active ingredient and nicotine at 0.1% (w/v) are also equally effective to control the aphid (Way *et al.*, 1954).

iii) Bhang Aphid

Occurrence & severity

The Bhang Aphid or Hemp Louse, which is also known as cannabis aphid or hemp aphid, is nearly colourless and about twenty five percent smaller than the Green Peach Aphid. It never alternates hosts. Cannabis is the only known host of this aphid, although, recently *Rullia prostrata* Poir has been recorded as host plant of the aphid (Mall *et al.*, 2010). The species is native to Eurasia, ranging from Britain to Japan. Now it is well established in Europe, Asia, North Africa, and North America (McPartland *et al.*, 2000; Annonymous, 2017a).

According to Muller and Karl (1976) the bhang aphid infests *C. sativa*, *C. indica* and *C. ruderalis*. The pest also attacks hops (Blackman and Eastop, 1984). It is a potential vector of viruses and other plant pathogens (Annonymous, 2017a).

Types of damage

The aphid sucks fluids from the phloem of hemp plant. In high populations this can cause reduced growth, yellowing and wilting (McPartland *et al.*, 2000), although plants that are well hydrated can well tolerate this type of injury. This aphid damages female buds. It sits between female flowers and seeds, and sucks plant sap (Kirchner, 1906). It is a potential vector of viruses. It vectors hemp streak virus (Goidanich, 1955), hemp mosaic virus, hemp leaf chlorosis virus (Ceapoiu 1958), cucumber mosaic, hemp mottle virus, and alfalfa mosaic virus (Schmidt and Karl, 1970).

Scientific name & order: *Paraphorodon cannabis* **(Hemiptera: Aphididae)** (Bisby *et al.*, 2011).

Synonyms

= *Diphorodon cannabis* (Heinze. 1960)

= *Paraphorodon omeishanensis* Tseng & Tao, 1938

= *Phorodon asacola* Matsumura, 1917

= *Phorodon cannabis* Passerini, 1860

= *Myzus cannabis* (Passerini) [McPartland *et al.*, 2000]

= *Aphis cannabis* Passerini [McPartland *et al.*, 2000]

Biology

Egg: Oviparous females lay eggs in the flowering tops of *Cannabis*. Most eggs are destroyed when the hemp crop is harvested. Overwintering eggs are ovate, shiny green-black, 0.7 mm long (McPartland *et al.*, 2000).

Nymph: Nymph is like apterae but smaller in size and colourless at first; later becomes light green with three longitudinal dark green coloured stripes.

Adult: Apterae have flattened elongate-oval bodies. They closely resemble *M. persicae* but are about 25% smaller, averaging 1.9-2.7 mm in length. They are yellow to light green or bright green with darker green longitudinal stripes (Kirchner, 1906; Annonymous, 2017a). Their heads are covered with tiny bristles. The bristles have knobbed apices (Muller and Karl, 1976). Antennae are 1.1-2.2 mm long; antennal tubercles are prominent, converge slightly, and several bristles sprout from each tubercle. Between the tubercles arises a smaller midline knob, also bristled. Cornicles are white, up to 0.8 mm long, nearly a third of the body length, cylindrical, and taper towards their tips. Cauda tapers evenly to its tip. Alatae are slightly smaller than apterae, and develop black-brown patches on heads and abdomens. They hold their wings vertically over their abdomens when at rest. Males are smaller and darker than females, averaging 1.6- 1.8 mm long. Fundatrices are more oval than other apterae but slightly shorter (1.8-2.4 mm), with long antennae (McPartland *et al.*, 2000).

[N.B. Cannabis aphid is very much similar to hop aphid (*Phorodon humuli*). Differentiating cannabis aphid from hop aphid is difficult. Microscopic examination is required to do so. The hairs (bristles) on the head, thorax, and basal antennal segments of cannabis aphid are swollen at the tip (knobbed apices), while those in hop aphids are not as such (Muller and Karl 1976).]

Life cycle

The life history of bhang aphid (*P. cannabis*) is autoecious (monoecious). The pest never alternates hosts (Balachowski and Mesnil, 1936; Muller and Karl 1976). Sexuparae of bhang aphid never flies away. So, oviparous female lays eggs in the flowering tops of *Cannabis*. Most of the eggs are destroyed when the hemp crop is harvested (McPartland *et al.*, 2000).

Management/Control measures

Aphid control is difficult due to its reproductive rate and ability to develop resistance to most of the systemic insecticides. However, certain level of control can be done with the use of Azadirachtin at early life stages. The biocontrol agents, *Beauvaria bassiana* and *Chromobacterium subtsugae* strain PRAA4-1 can be used at low population levels. Application of potassium salts of fatty acids are also an option (Annonymous , 2017a). Control for established populations may also be accomplished through the use of predators. Effective and commonly available species are green lacewing, *Chrysoperla carnea*, aphid predator midge, *Aphidoletes aphidimyza*, and some ladybugs (larvae), such as *Hippodamia convergens* (Annonymous, 2017a).

Use of insecticides: A few of the insecticides are allowed for uses on cannabis. These include sprays of products that contain the active ingredients pyrethrins which have potential for control of aphids.

iv) Hop Aphid

Occurrence & severity

Hop Aphid (*Phorodon humuli*) lives in Europe, central Asia, North Africa, and North America, but not in Australia. In California spring migration peaks in early June. The autumn migration peaks in late September. This aphid is aided by a large black ant, *Formica subsericea* Say (Parker, 1913a). Blunck (1920), Flachs (1936) and Eppler (1986) reported *P. humuli* infesting Cannabis. This pest normally attacks hops. It vectors many plant viruses and *Pseudoperonospora humuli*, a fungus causing downy mildew of hops and hemp (Sorauer 1958). Hop aphid hosts various plants in summer and winter seasons. In winter season, they are known to hosts in *Prunus* spp. such as, *Prunus domestica*, *P. spinosa*, *P. padus*, and *P. cerasifera*. During summer, it hosts within *Humulus lupulus*, *H. japonicus*, and *Urtica dioica* (Annonymous, ND5).

Types of damage

Aphids on hemp gather on young shoots. They can reduce growth of plants and cause yellowing and wilting. Sooty moulds may develop on the honeydew, which appears as a shiny and sticky layer on the upper side of leaves; photosynthesis is thus reduced (Annonymous, ND5).

Scientific name & order: *Phorodon humuli* (Schrank) 1801(Hemiptera: Aphididae)

Synonym

= *Myzus humuli*

Biology

Egg: Gynopara lays 16-12 eggs. Eggs are laid in October (after mating) in protected areas, on axils of the buds in autumn. Overwintering eggs are ovate, shiny green-black, 0.7 mm long (similar to bhang aphid).

Nymph: Nymph is like apterae but smaller in size and colourless (transparent) at first. later becomes light green with yellowish tinge or spots.

Adult: Apterae are small to medium sized, whitish to pale yellowish green and relatively shiny. The abdomen is marked with three dark green longitudinal stripes. The head has a characteristic pair of elongate projections on the inside of the antennal tubercles. Hop aphids closely resemble bhang aphids, except the head of *P.humuli* possesses few bristles (Muller and Karl 1976). The bristles on antennal tubercles have blunt or pointed apices, but, never knobbed. The midline knob is scarcely apparent. The cornicles are pale with slightly dusky tips and more than twice as long as the pale cauda. Apterae are larger on Prunus in the spring (2.0-2.6 mm) than apterae on Humulus in the summer (1.1-1.8 mm). Alatae are 1.4-2.1 mm long (McPartland *et al.*, 2000).

Life cycle

Hop aphid (*P. humuli*) is heteroecious. The overwintering host is *Prunus*. Eggs hatch when the temperature reaches 22-26°C. These eggs hatch in spring to give wingless fundatrices. These fundatrices produce fundatrigeniae, then apterae on the young *Prunus* leaves. The apterae give birth toalatae- spring migrants. The spring migration mostly takes place in June (Dixon, 1985; Parker, 1913a). *Phorodon humuli* migrants fly straight up into a moving air mass, and can travel 150 kmor more. They spend the summer on secondary hosts. Then fly back to overwintering hosts as winged gynoparae. For *P. humuli* this autumn migration peaks in late September (Dixon 1985).

Management/Control measures

Biological control

The insect predators, particularly *Anthocoris nemoralis* and *A. nemorum*, can usually contribute greatly to control the aphid (Cranham, 1982). A biological control programme has been initiated in northern France, with the release of *Chrysoperla carnea* and *Harmonia axyridis* against *P. humuli* (Trouve *et al.*, 1996). Studies of biological control, using the introduced parasitoid *Aphidius colemani*, have shown promise in Poland (Puszkar and Jastrzebski, 1999). Traps releasing synthetic sex pheromone of *P. humuli* also have potential for catching the males (Hartfield *et al.*, 2001). The biological control alone is insufficient for aphid control, however, it provides an important component within the integrated pest management approaches where insecticide use is minimized.

Chemical control

Chemical control measure is same as bhang aphid. However, the hop aphid, *P. humuli*, has developed marked resistance to many organophosphates and lowered the susceptibility to *Aphidius colemani* which used as foliar sprays. However, the hop aphid control mostly depends on the use of systemic organophosphate mephosfolan, which is applied as a soil drench (Cranham, 1982). More recently, imidacloprid, cyfluthrin, pymetrozine and amitraz, are registered for the control of sucking pests in German hop gardens and are useful against *P. humuli* (Lars and Ralf, 2003). Further, in a study on insecticide resistance in *P. humuli*, it was observed that the resistance to diagnostic concentrations (LC_{95} for reference strains) of imidacloprid, amitraz and pymetrozine was not detected. However, moderate resistance to pyrethroids was observed in May and August using the diagnostic concentrations of 10 mg litre^{-1} cyfluthrin (LC_{95} of the susceptible reference strain H5) (Lars and Ralf, 2003).

2. European Corn Borer

Occurrence & severity

The European corn borer (ECB) is a major pest of hemp (*Cannabis sativa* L.). During World War II, when the hemp was last produced in major quantities, there only this insect was mentioned as significant to production (Willsie *et al.* 1942). The damage by the insect to fibre hemp crops mainly occurs in eastern and southeastern Europe. The last such damages worth noting were registered in the late 1950s. The insect is particularly known as a pest of corn, in which it is one of the most important pests. Thus, the insect pest is very severe in the hemp crop which grown in the areas where there is a significant acreage of corn. The insect is native to Europe, originally infesting varieties of millet, including broom corn. Now it is found in Europe, North America and some areas of northern Africa. European corn borer is quite polyphagous (Caffrey and Worthley, 1927). Mugwort (*Artemisia vulgaris*), hemp (*C. sativa*) and hop (*Humulus lupulus*) are among the native hosts, but the host record includes also apple, cotton, sorghum, and many vegetables (Straub *et al.* 1986, Eckenrode and Webb 1989). The females of this borer exhibit upwind flight to three main host plants, corn (*Zea mays*), hemp (*C. sativa*) and hop (*H. lupulus*) (Bengtsson *et al.*, 2006). This confirms the role of plant volatiles in host finding in European corn borer (Lupoli *et al.* 1990). There are at least two, and possibly more, strains present of this borer (Capinera, 2017b). In Europe, the E-race feeds on native hosts, using the E-isomer of 11-tetradecenyl acetate as the main pheromone compound as found on the ancestral (native) hosts, mugwort, hemp and hop, whereas the Z-race feeds on corn, using Z-11- tetradecenyl acetate (Langenbruch *et al.*, 1985; Huang *et al.*, 2002; Pelozuelo *et al.*, 2004; Bengtsson *et al.*, 2006).

Types of damage

The moth lays eggs on hemp plant. The eggs hatch to a larva which does the damage to the host plant. Borer larva enters the hemp stalk through a pin hole and feed in the centre of the stalk. This weakens the hemp stalk reducing the yield potential and leading to stalks breaking. At the site of invasion the main stem was typically destroyed, and the plant became strongly branched. The fibre quality, are detrimentally affected by the ECB. The insect showed a preference for larger plants (Small *et al.*, 2007). Boring by corn borers also allows several fungi to affect the host plants (Capinera, 2017b).

Scientific name & order: *Ostrinia nubilalis* Hubner, 1976 (Lepidoptera: Crambidae)

Synonyms

= *Pyralis nubilalis* Hubner, 1796

= *Pyrausta nubilalis* Meyrick, 1890

= *Botis nubilalis* var. *paulalis* Fuchs, 1900

= *Pyralis glabralis* Haworth, 1803

= *Botys appositalis* Lederer, 1858

= *Pyrausta nubilalis* f. *fanalis* Costantini, 1923

Biology

Egg: Eggs are deposited in irregular clusters of about 15 to 20. The eggs are oval, flattened, and creamy white in colour which become pale yellow and finally translucent before hatching. Eggs normally are deposited on the underside of leaves, and laid in an overlapping configuration. Eggs measure about 1.0 mm in length and 0.75 m in width. Eggs hatch in four to nine (differently, three to seven) days (Phelan *et al.*, 1996; Capinera, 2017b).

Larva: The fully grown larva is 0.75 to 1 inch (1.9-2.5 cm) in length. Larva varies in colour from light brown to pinkish grey dorsally, with a brown to black head capsule and a yellowish brown thoracic plate. The body is marked with conspicuous small, round, brown spots on each segment. The larva feeds on the stalk of the host plant. Mortality tends to be high during the first few days of life, but once larva establishes a feeding site within the plant survival rates improve. Duration of the instars varies with temperature. Under field conditions development time was estimated at 9.0, 7.8, 6.0, 8.8, 8.5, and 12.3 days for instars 1 through 6, respectively, for a mean total development period of about 50 days, but this varies considerable from year to year according to weather conditions (Anonymous, 2015b; Capinera, 2017b).

Pupa: The pupa is normally yellowish brown in colour. It measures 13 to 14 mm in length and 2 to 2.5 mm in width in male and 16 to 17 mm in length and 3.5 to 4 mm in width in female. The tip of the abdomen bears five to eight re-curved spines that are used to anchor the pupa to its cocoon. The pupa is ordinarily, but not always, enveloped in a thin cocoon formed within the larval tunnel. Duration of the pupal stage under field conditions is usually about 12 days (Capinera, 2017b). In European corn borers larval diapause is induced by temperature and changes in daylight length. At higher temperatures (18° to 29°C), shorter photoperiods are sufficient to induce diapause. Moreover, at high temperatures and long photoperiods, fewer larvae enter diapause (Beck and Hanec, 1960).

Adult: The adult (moth) of European corn borer is about 1 inch (2.5 cm) long with a 0.75 to 1inch (1.9-2.5 cm) wingspan. The female is pale yellow to light brown with dark, irregular, wavy (or, zigzag) bands across the wings and bearing pale, often yellowish, patches. The male is slightly smaller and darker; usually pale brown or greyish brown, but also with dark zigzag lines and yellowish patches. The tip of its abdomen protrudes beyond its closed wings. They are most active before dawn. The adults spend most of their time feeding and mating. Males and females of different strains have been found to produce differing sex pheromones. The female often deposits 400 to 600 eggs during her life span. Total adult longevity is normally 18 to 24 days (Capinera, 2017b).

Life cycle

The corn borer overwinters as a caterpillar and changes to a pupa in the soil in the spring when conditions are ideal. The adult emerges as a moth in mid- May to June. The moth lays eggs on a suitable plant host which can include hemp. A female during its adult life of 18 to 24 days can lay a total of 400 to 600 eggs (Capinera, 2017b). It first lays eggs in June. The eggs are laid on the underside of the leaves near the midvein. Brood size ranges from 15 to 30 eggs and egg masses are about 6 mm in diameter (Anonymous, 2017b). The period of egg laying is about 14 days with an average of 20 to 50 eggs per day (Capinera, 2017b). The egg hatches to a larva which does the damage to the host plant. The young larvae disperse over several plants; during this phase, mortality is high, especially in cool and wet weather. After 3-4 weeks, the caterpillars either pupate, to produce a second generation of moths about three weeks later, or they start diapause. In both cases, they stay inside the plants. Overwintering caterpillars are very cold-resistant and have often bored so far down within the plant that they are able to survive the winter in the stubble. They pupate in the following spring. In the northern parts of its distribution range, it produces one generation per season; in the southern parts two, or occasionally more, depending on the climatic conditions in spring and summer.

Management/Control measures

Use of host plant resistance

The use of host resistance is the most effective way to control the pest attacks. However, no such resistant cultivar has been developed in hemp. Thus, extensive breeding research is needed.

Cultural control: It is necessary to destroy the crop residues. If possible by ploughing them in carefully to a depth of 20 cm, significantly reduces the number of surviving caterpillars. Mowing of stalks close to the soil surface eliminates greater than 75% of larvae, and is especially effective when combined with ploughing (Capinera, 2017b).

Biological control: Biological control has been attempted repeatedly in several crops susceptible to European corn borer attack. *Bacillus thuringiensis* products can be as effective as many chemical insecticides, but often prove to be less effective than some insecticides. *Bacillus thuringiensis*, have proven to be erratic. Release of native *Trichogramma* spp. provide variable and moderate levels of suppression (Capinera, 2017b). In an investigation, one egg parasitoid species has been identified as *Trichogramma evanescens* (Westwood). It is found that this natural enemy occurred constantly in fields but its number greatly fluctuated from year to year (Tancik, 2017).

Chemical control

Liquid formulations of insecticide are commonly applied to protect against the corn borer attacks. Liquid applications are usually made to coincide with egg hatch in an effort to prevent infestation. A popular alternative to liquid insecticides is the use of granular formulations, which can be dropped into the whorl for effective control of first generation larvae because this is where young larvae tend to congregate. Insecticide is more persistent when applied in a granular formulation. The insecticides, like, pyrethroids (lambdacyhalothrin and bifenthrin), carbamate (methomyl) and spinosyn (spinosad) are mostly used to control European corn borer attacks. However, except methomyl, effectiveness of these insecticides against the pest varies with the environmental conditions (Musser and Shelton, 2005). Insecticides from the same class often have similar temperature responses. Organophosphate and carbamate insecticides generally have stable toxicities at all temperatures. Pyrethroid insecticides often have reduced efficacy at high temperatures, but several studies have found pyrethroids to have a positive temperature coefficient against some species (Scott, 1995). Spinosad now represents an alternative class of insecticides available for control this pest (Thompson *et al.*, 1995). Temperature effects on the toxicity of spinosad was found unaffected with the changing in temperature (Fang and Subramanyam, 2003), although, this finding does not corroborate by the later study of Musser and Shelton (2005).

3. Hemp Flea Beetle

Occurrence & severity

Hemp flea (hop flea) beetle is the most common beetle on Cannabis. It, as adult and larva (grub), damages plants. Adult feeds on foliage, flowers, and unripe seeds. Grub feeds on roots or occasionally acts as leaf miner. This beetle specializes in infesting only one or two plant species of Cannabaceae, mainly hops, also hemp and sometimes nettles (Silantyev 1897). Rataj (1957) considered the hemp flea beetle as insignificant, but several other reports indicated it as a major pest (Ragazzi, 1954; Dempsey, 1975; Bosca and Karus, 1997). The flea beetle is considered as the number one problem in the North Caucasus (Durnovo, 1933). It also occurs in black soil zone of Russia (Aandkeeva, 1930), California in USA (DPR, 2017), Slovenia (Cizej and Zolnir, 2003), Great Britain and France to eastern Siberia, northern China, and Japan (McPartland *et al.*, 2000). Hemp Flea is considered as the most dangerous insect pest to industrial hemp.

Types of damage

Damages are noticed only when the temperature of the top layer of soil and the air rises to 10-15°C (50-60°F) and the weather is dry. The hibernating fleas then crawl from the soil and feed on the cotyledons and on the hypocotyl (the seedling stem below the cotyledons). If the plants have already developed, the fleas will even chew small holes in the leaves. In the leaves, damage consists of many small, round to irregular holes between leaf veins. Leaves of heavily infested plants may be completely skeletonized and that is considered as the chief injury (Aandkeeva, 1930). Young plants are killed. Grubs feed on roots, usually at depths of 4-8 cm underground. They feed on cambium, making longitudinal or tortuous galleries in roots. Grubs sometimes act as leafminers, mining beneath the upper surface of leaves, tracking in a tight spiral that ends in a brown blotch (McPartland *et al.*, 2000).

Scientific name & order: *Psylliodes attenuata* (Koch, 1803) (Coleoptera: Chrysomelidae)

Synonyms

= *Haltica attenuata* Koch, 1803

= *Psylliodes attenuatus* Koch, 1803

= *Psylliodes japonica* Jacoby 1885

Biology

Egg: Egg is sub-oval, pale yellow, deposited singly around plant roots near the soil line (Flachs, 1936).

Grub: Grub is cylindrical, white with tiny spots and bristles, with six minute legs, a brownish head, and reach 4.5 mm in length (Flachs, 1936). It has 3 instars and develops in 21-42 days (David¹yan, 2003).

Pupa: Pupa is initially pearly white but slowly darken, beginning with the eyes, followed by other parts of the head, legs, and finally the elytra (Flachs, 1936).

Pupation occurs inside the soil cradle. Its development lasts 6-34 days (Davidyan, 2003).

Adult: Adult is oval in outline, metallic green or bronze or black with minute bronze-grey hairs, and average 1.3-2.6 (-2.8) mm long. Elytra (wing covers) are striated and punctuated. Tips of elytra are usually reddish-brown, but may be entirely as the basic colour. Head exhibits sharp outgrowths crossed in an "X" at the eyes, and sprout antennae reaching half of their body length. Leg colour is red-brown, hind femora darkened (Flachs 1936; McPartland et al., 2000).

Life cycle

Adults overwinter in soil and emerge at the end of March to feed on young seedlings. Adults mate in April and begin laying eggs ten days later. Each female may lay about 300 eggs, which are deposited in the upper layer of the soil (Aandkeeva, 1930; David'yan, 2003). Egg laying lasts from late April to the end of July. Larvae hatch in 5 to 16 days (differently in 8-10 days, Aandkeeva, 1930), and grubs feed within roots when young. By June grubs exit roots and live in soil. Pupation occurs from mid-June to August, 4-15 cm underground. Adults emerge from pupae from late June to September. Adults feed on plant tops until autumn, then burrow into soil to depths of 20 cm (Angelova 1968). The total life-cycle from egg to adult is completed in 52-70 days (Aandkeeva, 1930). Only one generation arises per year.

Management/Control measures

Fighting the flea is a challenge because it is not a threat every year. However, the use of white sticky traps is effective to catch the flea beetles in hemp field. In case of seedling raising, protecting the seedlings with fine mesh or screens is effective mechanical method to minimize the flea attacks, since flea beetles prefer bright sunlight (Parker, 1913b). Application of *Bacillus thuringiensis* may also be effective as the adults are susceptible to this bioagent.

4. Hemp borer

Occurrence & severity

Hemp borer, also called 'Eurasian hemp moth', 'hemp moth', 'hemp leaf rollers' and 'hemp seed eaters', is a species of tortricid moth in the family Tortricidae (Walker, 1863). It is a native of East Asia that has spread into almost all hemp growing countries all over the world. Hemp borer arrived in North America around 1943 (Miller, 1982). In Europe it was first detected in 1960, in Ukraine (Kryachko et al., 1965) and Russia (Danilevski and Kuznetsov, 1968). Subsequently, the hemp borer appeared in Romania by 1963 (Manolache et al.,

1966), Hungary by 1964 (Nagy, 1967), and Bosnia-Herzegovina by 1967 (Bes, 1967). After that it was reported in Armenia (Shutova and Strygina, 1969), Moldavia (Shutova and Strygina, 1969), Serbia and Montenegro (Lekic and Mihajlovic, 1971), Bulgaria (Gerginov, 1974), Greece (Vassilaina-Alexopoulou and Mourikis, 1976), Slovakia (Bako and Nitri, 1977) and Slovenia (Camprag *et al.* 1996).This pest, besides boring into stems, destroys flowering tops. In a report it is indicated that it had destroyed 80% of flowering tops in Russia (Kryachko *et al.*, 1965). Bes (1978) reported 41% seed losses in unprotected Yugoslavian hemp. Each borer at larval stage consumes an average of 16 Cannabis seeds (Smith and Haney 1973). Hemp Borer has two or three generations per year. The first generation larvae damage fibre hemp. Apart from fibre hemp, they also infest marijuana, feral hemp (*Cannabis ruderalis*), and hops (Mushtaque *et al.*, 1973, Baloch *et al.*, 1974, Scheibelreiter, 1976).

Types of damage

Hemp borer feeds on leaves in the spring and early summer. After hatching, the young larvae skeletonize leaves (or leaf mine, according to Mushtaque *et al.*, 1973) for several days before they bore into petioles, branches and stalks. Feeding galleries within the branches and stalks cause fusiform-shaped galls and splitting (Manolache *et al.*, 1966). Branches and stalks may break at galls, although the length of tunnels within galls averages only 1 cm (Miller, 1982), or at most 2 cm (Nagy, 1967). Boring near the terminal shoot may kill the shoot and cause branching at that point (Manolache *et al.*, 1966). In warmer regions where two or more generations occur per year, late-season larvae that hatch in the autumn feed on leaves, flowers and seeds, hence the common names 'hemp leaf roller' and 'hemp seed eater' are given. The seed eaters destroy young seeds. In some cases, plants fail to form flowers and seeds (Vassilaina-Alexopoulou and Mourikis, 1976). The larvae spin loose webs around terminal buds, especially the seed clusters of female plants and they pupate in curled leaves within buds, bound together by strands of silk (Kryachko *et al.*, 1965; Smith and Haney, 1973).

Scientific name & order: *Gropholita delineana* **Walker, 1863 (Lepidoptera: Tortricidae)**

Synonyms

= *Cydia delineana* (Walker)

= *Laspeyresia delineana* (Walker)

= *Gropholita sinana* Felder 1874

= *Cydia sinana* (Felder)

Biology

Egg: Egg is white to pale yellow (light yellow), transparent, oval, 0.6 mm in length, 0.4 in width, with thin and wrinkled shell. Egg is laid singly on stems and undersides of leaves. Egg hatches to caterpillar in 8-10 days at temperature 20-22° C (Ovsyannikova and Grichanov, 2003).

Larva: Larva (caterpillar) is orange (or pinkish-white to pale brown), small, size 6-8 mm (differently 9-10 mm; Vassilaina-Alexopoulou and Mourikis, 1976) long. Several pale bristles per segment are barely visible. Its head is dark yellow-brown, with black ocelli, averaging 0.91 mm wide. Five developmental stages occurred in the larva's lifetime (Vassilaina-Alexopoulou and Mourikis, 1976). Duration of its feeding is 21-33 days (Ovsyannikova and Grichanov, 2003). Caterpillar of the last (usually 5th) instar overwinters in the cocoons covered with soil particles and located at roots, in a surface soil layer to a depth of 5-10 cm, within plant residues in fields, in places of thrashing, retting, storing, and also in thickets of wild growing hemp (Ovsyannikova and Grichanov, 2003).

Pupa: Pupa is light brown, 5-7 mm in length, with 2 rows of spinules directed backward on dorsal side of 2nd-8th segments. Pupation of caterpillars of summer generations occurs mainly in stalks, seeds, and braided leaves (curled leaves within buds, bound together by strands of silk). Development of pupae lasts 16-22 days (Ovsyannikova and Grichanov, 2003).

Adult: Adults are tiny moths, with greyish- to rusty-brown bodies and brown, fringed wings. Body length and wingspan average 5 mm and 9-13 mm respectively in males, and 6-7 and 10-15 mm respectively in females. Forewings exhibit three or four white stripes along the anterior edge with four chevron-like stripes near the centre. Male genitalia are characterized by an aedeagus that narrows abruptly. Female genitalia are characterized by two signa in the corpus bursae. Fertility of females of the wintered generation is 100-200 eggs on average and 350-500 eggs in the subsequent generations (Ovsyannikova and Grichanov, 2003). Adults live less than two weeks.

Life cycle

The species has two or three generations per year. The first generation larvae damage fibre hemp. During the winter, they hibernate in the stalk and stubble. After mid-May the following year when temperatures reach 15°C (60°F), the larvae spin cocoons around themselves. A week or two later, the moths are fully developed and emerge from the cocoons. Moths of 1st generation fly from mid-May to the end of June, second generation from 1st or 2nd third of July to the last third of August, the third generation from the end of August to the last third of September. Flight of moths of the 2nd generation coincides with the growth stages of flowering and formation of seeds. The generations overlap, and all stages of the Hemp Moth development meet during the vegetation period. Caterpillars of the 3rd generation eat mainly seeds. Having no time to finish their development before harvesting, they usually perish. The majority of caterpillars diapause by the end of August under the influence of seasonal change of daylight hours (Ovsyannikova and Grichanov, 2003).

Management/Control measures

Use of host resistance

It is reported that Vermont feral hemp is more resistant to *G. delineana* than feral hemp growing in the Midwestern USA. The Vermont germplasm may have descended from plants imported in the 1830s, called "Smyrna" hemp, a western European landrace devoid of Chinese ancestry (McPartland., 2002).

Quarantine measure

Careful phytosanitary measures are able to prevent the spread of *G. delineana* into quarantine areas, such as western Europe, Canada, and the entire southern hemisphere (McPartland, 2002).

Cultural Control and Sanitary Methods

Manolache *et al.* (1966) have reported that good control can be achieved by destroying all hemp crop debris, and deep ploughing in the autumn. Deep ploughing buries overwintering larvae and pupae into too deep in soil that they are not able to emerge (Sandru, 1972). Early harvest of hemp may decrease the population of overwintering pests, because this would destroy a high percentage of the larval population (Nagy, 1979). Camprag *et al.* (1996) noted that mono-cultured cannabis is attacked more by *G. delineana*, so, the crop should be rotated. The moth is nocturnal. Hence, light traps are useful to catch adults. Nagy (1980) used female sex hormones to attract and trap male moths, preventing reproduction. He noted that female sex hormones of *G. delineana* can also attract males of *Grapholita compositella*.

Biological Control

Native organisms heavily parasitize *G. delineana* larvae, but none of these species has been commercially developed. *Trichogramma* and *Macrocentrus* species are commercially available. These two species may be used to minimize the pest population (McPartland*et al.*, 2000).

Chemical Control

Chemical measures are fumigation of hemp seeds, insecticide treatments during leafing-out against 1st generation caterpillars, and later against 2nd generation caterpillars. Any organophosphate insecticide may be applied against the moth.

5. Mites

i) Two-Spotted Spider Mite

Occurrence & severity

This two-spotted spider mite, also known as 'Red spider mite', was originally native only to Eurasia, but has acquired a cosmopolitan distribution. It is the most destructive pest of *Cannabis* of glasshouse and growing room. Outdoor crops become more infested in warm climates. Cherian (1932) has reported 50% of field crop losses near Madras, India. The spider mite is common in North American and European glasshouses. It also attacks outdoor crops in temperate climates in Canada (Frank & Rosenthal, 1978; McPartland *et al.*, 2000). This spider mite is extremely polyphagous. It can feed on hundreds of plants, including most vegetables, food crops and ornamental plants such as roses (Thomas and Denmark, 2009). It is the most prevalent pest of ashwagandha (*Withania somnifera*) in India (Sharma and Pati, 2012).

Types of damage

Spider mites bite into leaves and suck up exuded sap. The mites tend to infest crops in a patchy distribution, so early infestation is difficult to notice. Each mite puncture produces a tiny, light-coloured leaf spot ("stipple"), which appears on both sides of the leaf. The stipples are grey-white to yellow in colour. These stipples begin as the size of pinpricks and then enlarge. Many stipples arise in lines parallel to leaf veins, which significantly reduce the photosynthetic capability of plants (Thomas and Denmark, 2009). Ultimately, whole leaves turn a parched yellow colour, droop as if wilting, then turn brown and die (Kirchner 1906). Inspecting the underside of leaf surfaces, particularly along the main veins, reveals silvery webbing, eggs ("nits"), faecal deposits ("frass"), and the mites themselves. Leaves near the bottom of the plant are usually infested first. Symptoms are the worst during flowering, when whole plants dry up and become webbed together (McPartland *et al.*, 2000).

Scientific name & order: *Tetranychus urticae* **Koch 1886, (Acari: Tetranychidae)**

Synonyms

= *Tetranychus bimaculatus* Harvey 1898,

= *Tetranychus telarius*

= *Epitrtranycus athaea* von Hanstein 1901

Biology

Egg: Eggs are laid on leaves, spherical and 0.14 mm in diameter. Initially the egg is translucent to white, later, turns a straw colour just before hatching (McPartland *et al.*, 2000).

Nymph: Hatched larvae have six legs and two tiny red eye spots. Protonymphs become eight-legged and moult into deutonymphs. Deutonymphs moult into yellow-green coloured adults (McPartland *et al.*, 2000).

Adult: Adult mite is extremely small, barely visible with the naked eye as reddish or greenish spots on leaves and stems. The mite has two brown-black spots enlarge across their dorsum. Female mite is 0.4 to 0.5 mm in length. Male is slightly smaller, with a less-rounded posterior. Its dorsal spots may not be as evident. The male's knobbed aedeagi are at right angles to the neck and symmetrical. As winter approaches, the two-spotted spider mite turns bright orange-red (McPartland *et al.*, 2000). The adult mite spins a fine web on and under leaves (Thomas and Denmark, 2009).

Life cycle: The mite overwinters as an adult and emerges in the spring. Female lays eggs on underside of leaf or in a small web, one at a time. It lays as many as 200 eggs. Eggs hatch into larvae, which moult three times before they are capable of reproduction. Under optimum conditions for development (30°C with low humidity), the life cycle of two-spotted spider mites repeats every eight days. Shortened photoperiods in autumn usually induce a reproductive diapause, where females stop feeding, turn orange-red, migrate into clusters, and then hibernate in underground litter. Warm temperatures inhibit diapause (McPartland *et al.*, 2000).

Management/Control measures

Biocontrol

The mite's natural predator, *Phytoseiulus persimilis,* preys mainly or exclusively on spider mites (Thomas and Denmark, 2009).

Chemical control

The acaricide, Fenazaquin (Magister 10 EC), at 0.0025% concentration is found to be most effective in control of the two spotted spider mite (Singh *et al.*, 2017). Further, in another study it is observed that under the polyhouse conditions Diafenthiuron 50 WP at 0.055% concentration is found most effective in reducing the eggs and mobile stages of *T. urticae* (Pokle and Shukla, 2015). Shah and Shukla (2014) also recorded diafenthiuron and fenazaquin as very effective chemicals against mobile stages of *T. urticae* on gerbera under polyhouse condition.

ii) Hemp Russet Mite

Occurrence & severity

Hemp russet mite was first observed on hemp in 1958 in Hungary (Farkas, 1960). The hemp (*Cannabis sativa* L.) was the only host plant for this species; the relation of the given russet mite to the host was defined as vagrant. The mite later noticed to infest hemp in other parts of central Europe (Farkas 1965) and feral hemp in Kansas (Hartowicz *et al.*, 1971). At Indiana University, the pest thrives in glasshouses and feeds on all kinds of Cannabis, including European fibre cultivars, Southeast Asian drug landraces, Afghan landraces, and ruderals. It was assumed that the mite population at Indiana University was possibly imported on seeds from Nepal or northern India (McPartland *et al.*, 2000). The hemp russet mite is becoming a serious pest. The mite is striking Washington cannabis growers hard during the recent years. Numerous growers are fighting off these tiny pests, and these are spreading like a plague of locusts. Russet mites are very small and they can actually be distributed to neighbouring areas via wind- which is a big concern in Eastern Washington.

Types of damage

The hemp russet mite is a sap sucker working at the cellular level. It is extremely harmful pest which primarily feeds on leaves, petioles and meristems. Initial signs of infestation are subtle and easily missed. Mites cause curling of leaf edges, followed by leaf russeting. Petioles become brittle and leaves break off easily. In severe infestations the mite crowds the plants by thousands, giving leaves a beige appearance. The mites feed on inflorescences of both sexes, and on glandular trichomes, severely reducing resin production. The mites also infest flowering tops; they selectively feed on pistils, rendering female flowers sterile (McPartland *et al.*, 2000; McPortland and Hillig, 2003).

Scientific name & order: *Aculops cannabicola* (Farkas) 1965, (Acari: Eriophyidae)

Synonym

= *Vasates cannabicola* (Farkas, 1960).

Biology

The biology of hemp russet mite is very little studied but is likely similar to related tomato russet mite (*Aculops lycopersici*). It reproduces by production of eggs that, upon hatch, is followed by two immature stages (protonymph, deutonymph) and an adult. The entire life cycle (egg to egg) of the tomato russet mite is completed in about a month at temperatures of 25°C (77°F). Each adult produces 10-50 eggs. The egg and nymph morphologies are given below based on the characters of tomato russet mite (*Aculops lycopersici*).

Egg: Eggs are round and colourless to white (Kay, 1986). The eggs are laid on leaves and stems of plants. Both males and females hatch in 2-3 days.

Nymph: The protonymphal stage lasts about 11 hours for females and about 7 hours for males. The protonympha chrysalis (quiescent moulting period between stages) lasts about 13 hours for both males and females. Protonymphs are white in colour and look similar to the adults, but they are smaller and less active. The second nymphal (deutonymph) stage lasts a little over a day for females and about 19 hours for males. The deutonymphal chrysalis lasts 18 hours for females and 16 hours for males.

Adults: Hemp mites are tapered, translucent, wedge-shaped cylinders that take on a yellow tint, especially in groups.

Female adult: Body is fusiform, light orange in colour. Gnathosoma is downcurved with long coxal setae, dorsal genual setae, apical setae, cheliceralstylets. Prodorsal shield is with acuminate frontal lobe. Dorsal tubercles are well developed, on rear margin of frontal lobe and prodorsal shield. Scapular setae are widely separated, directed to the rear divergently. Shield pattern consists of one median line over one half of the base and two admedian lines longer than the base of the frontal lobe. Median and admedian lines are connected by two transverse lines. Legs are only two pairs with all usual segments and setae. Coxae are almost smooth, with some granulae around the forecoxal tubercles. Sternal line is unforked. Coxisternal area is with five microtuberculated annuli. Genitalia is wide, with eight longitudinal ridges. Opisthosomal setae are long apart on annulus (Petanovic *et al.*, 2007).

Male adult: Body is smaller than female. Gnathosoma is downcurved with coxal, dorsal genual and subapical tarsal setae, and cheliceral stylets. Prodorsal shield is wide. Dorsal tubercles are well developed, on rear margin of prodorsal shield. Scapular setae are apart, directed to the rear divergently. Legs are only two pairs with all usual segments and setae. Coxae are with 6 long sternal lines, unforked. Coxal setae are apart. Coxisternal area is with five (5-7) microtuberculated annuli. Genitalia is wide. Opisthosomal setae are long apart on annulus (Petanovic *et al.*, 2007).

Life cycle

Very little is known about *A. cannabicola*'s life cycle (McPartland *et al.*, 2000). Outdoor populations probably overwinter in contaminated seed. Indoor populations remain on plants year round. The mites move towards the top of dying plants, where they spread to other plants by wind or splashing water. A turn of the life cycle takes about 30 days under optimum conditions of 27°C and 70% RH. Related eriophyid mites lay ten to 50 eggs during a life-span of 20-40 days (McPartland *et al.*, 2000).

Management/Control measures

Cultural control

It is suggested to keep clean seed in order to keep mites out. Once mites infest in an area, a grower's best efforts would be only to decrease populations, but not to eliminate them (McPartland *et al.*, 2000). Regular and close observation is required, especially around the leaves nearest the soil line if outdoors, or anywhere a plant is flowering, is crucial to early detection. The mite infestation may appear as an iron or magnesium deficiency symptoms. This symptomatic leaves should be examined by magnifying glass for infestation of russet mites, otherwise, if in doubt, treat for mites by spraying neem oil which may repel and kill mites.

Biocontrol

No effective biocontrol of *A. cannabicola* is known. *Phytoseiulus persimilis* does not feed on *A. cannabicola* (McPartland *et al.*, 2000).

Chemical control

Application of synthetic chemical acaricides, except the use of sulphur, is mostly prohibited in cannabis cultivation. However, based on similar type russet mite control in other crop plants, the following effective chemicals may be applied in fibre producing cannabis. Sulphur is usually recommended for the control of mites. However, in northern Queensland, a study by Kay and Shepherd (1988) reveals that sulphur is ineffective. They have found that the most effective acaricides against an established infestation of tomato russet mite are dicofol (Kelthane), cyhexatin (Plictran), azocyclotin (Peropal), sulprofos (Bolstar), and monocrotophos (Azodrin). Royalty and Perring (1987) have evaluated five acaricides on tomato russet mite. According to them, for tomato russet mite, avermectin B1 (Avid, Agrimer) is the most toxic, followed by dicofol, cyhexatin, sulfur, and thuringiensin. Selective doses of Avermectin B1 may provide good control of hemp russet mite.

6
Flax (*Linum usitatissimum*)- Malpighiales: Linaceae

A. Diseases

1. Anthracnose

Occurrence & severity status

Anthracnose has a serious impact on yield and fibre quality of flax (*Linum usitatissimum*) and is well-known in Europe, Asia and America. Flax anthracnose increased in Germany when flax production was expanding in the 1930s (Rost 1938). The anthracnose pathogen is seed- and soilborne, causes damping off of flax seedlings (Rost 1938), and is one of the causal organisms of so-called flax-sick soils (Bolley & Manns 1932). It has been recorded in nearly every country where the crop is grown either for fibre or oil. It is known to occur in Belgium, the British Isles, Canada, Formosa, France, Germany, Holland, Japan, Latvia, Lithuania, New Zealand, Poland, U.S.A. and the U.S.S.R. (including Siberia), in most of which countries its incidence has resulted in appreciable crop losses (Muskett and Colhoun, 1947). The disease is most common in cool and humid flax-growing areas (Nyvall, 1989). The disease is destructive to the crop, especially when the seed source is contaminated by the fungus. It can spread rapidly under favourable conditions causing severe local epidemics and heavy losses in yield and quality of both fiber and seed (Mercer, 1992c; Rashid, 2003b). The losses of 50% in fibre yield in an epidemic year were reported by Muskett and Colhoun (1947). Ondrej (1985) indicated a considerable loss in fibre strength on attack by anthracnose.

Symptoms

The disease first appears on young seedlings. One or both seed leaves may be attacked and the symptoms are most easily noticeable when the plants are 5-7 cm tall. Initially typical symptoms on cotyledons are small, circular, water soaked light dull green to red-brown zonate spots. The whole cotyledon quickly becomes brown and shrivelled. Sometimes the seedling may be killed outright if the growing point is infected. A reddish canker develops on the stem at or below the soil line and may girdle the stem, thereby killing the plant. Seedlings which contract slight stem canker and are not killed may grow on to reach maturity, the plants thus produced being frequently thin, stunted and sub-normal. In the affected plants leaf spots, stem cankers and acervuli can be observed throughout the growing season. Acervuli may be observed in older lesions during moist weather as tiny coral pink bee-hive shaped spore masses with dark hair-like objects (setae) projecting up from them. Sepals of flowers become infected and form a substrate on which the fungus grows to infect the bolls and eventually the seed (Musket and Colhoun, 1947; Nyvall, 1989). Under favourable conditions, the disease develops rapidly through the crop canopy giving the plants a burning appearance and destroying the crop (Rashid, 2003b). Therefore, apart from causing the death of the young plant, the disease may affect a proportion of the grown crop and by the production of sub-normal plants adversely affect the fibre yield. The thinning out of the plants resultant upon the deaths of seedlings also results in unevenness of crop which encourages the production of sterns of unequal thickness thereby affecting the quantity and quality of the fibre produced (Musket and Colhoun, 1947).

Causal organism: *Colletotrichum linicola* **Pethybr. & Laff. 1918**

[Scientific Proceedings of the Royal Dublin Society 15: 368]

Synonyms

= *Colletotrichum lini* (Westerd.) Tochinai 1926

= *Gloeosporium lini* Westerd. 1916

Morphological/ Morphophysiolgical characteristics

The fungus is described based on characters of anamorphs in different culture media. It forms light green to dark olive-green colonies on potato dextrose agar (PDA). The upper side of the colony is creamy white with neat margins, and the reverse side is creamy orange or pink. Mycelium is hyaline, septate, 1.5-6 μm diam, smooth-walled, branched. Chlamydospore is not observed. On potato dextrose agar (PDA), the fungus develops black acervuli (conidiomata) around the centre of the colony. Acervuli are formed on pale brown, angular cells, 3-8.5 μm. Conidiophores are hyaline, smooth-walled, septate, branched, up to 40 μm long. They may form directly on hyphae or on pale brown, angular cells, 3-8.5 μm diam from which condiomata are developed. Conidiogenous cells are hyaline, smooth-walled, cylindrical to elongate ampulliform, (8-) 9-32 × 2.5-4.5 μm, opening 1-1.5 (-2) μm diam, collarette 0.5-1 μm long, periclinal thickening may be observed. Conidia are hyaline, produced in acervuli with setae, smooth-walled, aseptate, slightly curved to straight, fusiform, tapering slightly, with one end pointed and the other slightly rounded, with dimensions (12-) 12.5 to 25.0 × 2.5 to 7.5 μm (mean 19.83×4.42 μm). Setae are numerous, formed on pale brown, angular cells, 3-8.5 μm diam. Setae are medium brown, slightly darker at the bottom and lighter at the top, smooth-walled to verruculose, 1-3 septate, with knee-bent-shaped base, base cylindrical to conical, with 3.5-6.5 (-7) μm diam. Setae dimensions are (15-) 100 -185.5 × 2.5- 5 (-5.5) μm (average 160.9 × 3.12 μm). Appressoria are single or in loose groups, pale brown, smooth-walled, ellipsoidal to subglobose outline, with an entire or undulate margin. Teliomorph is not found (Damm *et al.*, 2014; Vasic *et al.*, 2014, 2016; Wang *et al.*, 2018).

Culture characteristics: Colonies on SNA are flat with entire margin, hyaline to pale luteous, filter paper partly pale luteous, agar medium partly covered with floccose white aerial mycelium, reverse same colours; growth 27.5-30 mm in 7 days. Colonies on OA are flat with entire margin; buff to rosy-buff, aerial mycelium lacking, reverse buff, growth 26-29 mm in 7 days (Damm *et al.*, 2014). Colonies on PDA are light green to dark olive-green. At 25°C in the dark the upper side of the colony is creamy white with neat margins, and the reverse side is creamy orange or pink (Wang *et al.*, 2018).

Physiological races

Several races have been identified based on virulence on flax genotypes (Zarzycka, 1973).

Zarzycka (1976) has identified 11 races of *C. linicola* (= *C. lini*), two of which are predominant.

Recently, 25 races of this fungus are identified using a set of flax genotypes (Kudriavtseva, 1998).

Molecular characteristics

Molecular techniques are being employed to overcome the limitations of morphological approaches for determination of *Colletotrichum* species, (Guerber *et al.*, 2003; Johnston and Jones, 1997; Lubbe *et al.*, 2004; Talhinhas *et al.*, 2002). Although ITS regions are known to be potentially informative regions for phylogenetic studies at species level, they are not informative enough to distinguish several species. Moriwaki *et al.* (2002) showed that *C. destructivum* was not separated from *C. higginsianum*, *C. linicola* and *C. fuscum* based on the sequences of ITS1 region. In addition, Sreenivasaprasad *et al.* (1996) reported that *C. destructivum* and *C. linicola* were not differentiated based on ITS1 sequences. However *C. linicola* (= *C. lini*) is distinguishable by CHS-1, HIS3, ACT and TUB2. The ITS and GAPDH sequences are the same as those of *C. americae-borealis* (Vasic *et al.* 2014, 2016). The PCR amplification of the fungal Genomic DNA, using ITS1-ITS4, GSF1-GSR1, GDF1-GDR1 and T1-Bt2b primer pairs yielded fragments of approximately 495 base-pairs (bp), 900 bp, 200 bp and 750 bp, respectively (Vasic *et al.*, 2016). The molecular identification of total DNA of an isolate (viz., JL01) of *A. linicola* was done by extracting genomic DNA from mycelia using a Rapid Fungi Genomic DNA Isolation Kit. Sequences of internal transcribed spacer (ITS), actin (ACT), glyceraldehyde 3-phosphate dehydrogenase (GAPDH), and chitin synthase 1 (CHS-1) regions were amplified using primers ITS1/ITS4, ACT-512F/ACT-783R, GDF/GDR, and CHS-79F/CHS-345R, separately. Then they were sequenced and deposited in NCBI with accession numbers KX364055, MF563540, MF563539, and MF563541, respectively. BLAST analysis showed that these sequnces were 100, 98, 100, and 99% identical to the sequences JQ005765, JQ005828, KM105581, and JQ005786, respectively of *C. linicola* ex-type culture CBS172.51 (Wang *et al.*, 2018).

Predisposing factors

The fungus is normally seed-borne and is carried by the seed as mycelium or spawn located in the outer layers of the seed coat. Spores (conidia) are also found adhering to the seed but the mycelium is largely responsible for carrying over the parasite. So, contaminated seed is one of the factors for this disease occurrence. When a contaminated seed is sown and commences to germinate, warm moist soil conditions also stimulate the growth of the fungus which develops in the outer layers of the seed coat and produces large numbers of spores, anyone of which is capable of germinating and infecting, the young seedling. The extent to which infection occurs will depend largely upon weather conditions, the onset of the disease in epidemic form being most favoured by continuous wet warm weather. Thick sowing favours the rapid spread of the disease in the crop and the first outbreaks of infection are most readily observed where the seedlings tend to occur in clumps due to the slight but inevitable unevenness of sowing (Musket and Colhoun, 1947).

Disease cycle

The causal fungus survives most commonly as mycelium in seed and conidia on the seed coat. It also overwinters as mycelium, acervuli, and possibly conidia on infected residue. Seedlings are infected during cool, wet weather. Acervuli are eventually formed subepidermally in lesions produced from primary inoculum and rupture the epidermis to release conidia (Nyvall, 1989). Conidia are normally produced in abundance on the infected seed leaves and are washed down the stem of the seedling by rain or heavy dew. This is one way in which the stem may become attacked, the point of infection most frequently occurring at the collar of the plant at or just below soil level (Musket and Colhoun, 1947). The released conidia are also windborne that serve as secondary inoculum. Sporulation occurs throughout the growing season on plants that have died or on stem cankers. Seed becomes infected by conidia originating from stem lesions. Lodged plants are likely to have infected seed (Nyvall, 1989).

Management/Control measures

Use of host resistant

It is suggested to manage the disease by sowing clean seeds of resistant varieties (Nyvall, 1989; Rashid, 2003b). However, till date resistant variety of high yielding nature is yet to develop, although, there is considerable varietal diversity (Popisil, 1976). The varietal differences in resistance to this disease have not been observed in flax varieties cultivated for fibre (Musket and Colhoun, 1947). The major constraint, in developing resistance against this disease is the presence of different physiological races of *C. linicola* (= *C. lini*) (Zarzycka, 1976).

Cultural control

Crop sanitation, by cleaning plant debris, digging and sun drying the field before sowing, cleaning weed hosts are the general preventive measures effective for almost all diseases, including this anthracnose. However, in case of anthracnose of flax, crop rotation with other crops has major impact in disease management (Nyvall, 1989; Rashid, 2003b). Adequate rotation and removal of crop debris should also minimise the risk of the pathogen being picked up from the field (Mercer, 1992). Early sowing may also be effective in restricting the spread of the disease (Muskett and Colhoun, 1947; Ambrosov and Neofitova, 1978). Disease reduction may also be achieved by uniform sowing and avoidance of clumps of plants (Mercer, 1992).

Chemical control

Seed treatment

The causal fungus of anthracnose is seed borne. So, seed treatment with suitable protectant fungicide is suggested to eradicate the disease before sowing (Nyvall, 1989; Rashid, 2003b). Seed treatment is the most effective means of control. In general, seed treatment with the mixture of carbathiin and thiram is suggested. In addition, seed treatment with Sportak 45 EC (prochloraz 450 g/l) at the dose 0.8 l/t and Maxim Star 025 FS (fludioxonil 18.75 g/l, cyproconazole 6.25 g/l) at the dose 1.5 l/t have been found to give the best control of seedling blight causal agents (Gruzdeviene and Dabkevicius, 2003). These fungicides may also be effective against the anthracnose disease, which also in many cases consider as a seedling blight disease (Musket and Colhoun, 1947). Anthracnose, as a seed-borne disease, may be completely prevented by disinfecting the seed with dressings in powder form such as "Nomersan" and "Arasan" which contain tetramethylthiuram disulphide as their active constituent; for this purpose they are applied at the rate of 12 oz. per cwt. of seed (Musket and Colhoun, 1947).

2. Basal stem blight

Occurrence & severity status

Basal stem blight disease, also known as "Foot rot disease", is common in most flax-growing areas in Europe (Muskett and Colhoun, 1947). The disease of flax was first recorded in Ireland by Pethybridge *et al.* in 1921. Since then it has been noted to occur consistently in the Northern Ireland crop. However, a review of the literature indicates that the disease has been attacking the flax crop in the Continent of Europe since the 1890's, and, more recently, in Australia (Millikan, 1944; Muskett and Colhoun, 1945). It is widespread and nearly always present in the soil where flax is grown. The disease at seedling stage is of less importance in comparison to the damage caused by this disease at the flowering stage. Occasionally, severe epidemics occur, resulting in a total loss of the crop. Under warm and humid conditions in the spring, the disease develops rapidly, resulting in severe losses in yield and fibre quality. In fibre flax the disease is particularly serious as infected stems are not decomposed during the retting process (de Tempe, 1963a; Ondrej, 1983a). There are no described cultivars with resistance to this disease but there are some indications that linseed flax cultivars are more resistant than fibre flax cultivars (Muskett and Colhoun, 1947). The disease is caused by a noxious seed-borne pathogen of cultivated flax (*Linum usitatissimum*), causing damping-off, foot rot and dead stalks.

Symptoms

Seedlings grown from infected seed or growing in infested soil are attacked in the pre-emergence stage or shortly after emergence. The hypocotyls are attacked at ground-level. Elongated brown lesions appear on the stem just above the ground. The infected seedlings turn yellow and soon fall over and die. On mature plants, the first symptom is wilting and shrivelling of the lower leaves. On diseased plants, an area at the base of the stem becomes brownish and discoloured. Root tips also become brownish and discoloured as evident on uprooting. Pycnidia can be observed in this discoloured area. Often cortical tissues bearing pycnidia peel away from the woody tissues. Pycnidia are usually confined to stem lesions but may be found higher on the stem or on the sepals and walls of the capsules (Kerr, 1953). Eventually, the infected plants turn yellow, wilt and die. Microscopic examination of the dead tissue reveals the presence of numerous small pycnidia releasing masses of tiny unicellular spores in a yellowish tendril (Muskett and Colhoun, 1947).

Causal organism: *Boeremia exigua* var. *linicola* (Naumov & Vassiljevsky) Aveskamp, Gruyter & Verkley, 2010

[Studies in Mycology 65: 39]

Synonyms

= *Ascochyta linicola* Naumov & Vassiljevsky, (1926)

= *Phoma exigua* var. *linicola* (Naumov & Vassiljevsky) P.W.T. Maas, (1965)

= *Phoma exigua* f.sp. *linicola* Malc. & E.G. Gray, (1968)

= *Phoma linicola* Naumov, (1926)

Morphological/ Morphophysiolgical characteristics

In potato dextrose agar (PDA) medium the fungus after 7 days develops black pycnidia with 140 to 225 µm in diameter The pycnidia release hyaline, elliptical, and aseptate conidia measuring 4.8 to 9.2 × 1.6 to 4.3 (avg. 6.9 × 3.1) µm in size (Garibaldi *et al.*, 2016).

Identifying characters

NaOH oxidation reaction is positive, blue-green, later brownish red (E+ reaction). Growth-rate is relatively slow on oatmeal agar (OA) and malt agar (MA), 20-45 mm. Colonies are compact, olivaceous grey to olivaceous black. The pathogen is seed-borne of *Linum usitatissimum* (van der Aa *et al.*, 2000).

[**N.B.** In combination with other characteristics, the presence (+) or absence (-) of ca colourless metabolite known as 'antibiotic E' is a useful diagnostic criterium for species of *Boeremia* (=*Phoma*). The production of 'E' was first described from ubiquitous strains of *Boeremia* (*Phoma*) exigua (E is derived from exigua) (Boerema and Howeler, 1967). It can be demonstrated by adding a drop of concentrated NaOH to the margin of colonies on MA which produces an initial blue-green colour reaction, turning to brownish-red (oxidation reaction) (van der Aa *et al.*, 2000)]

Molecular characteristics

The molecular characters of *Boeremia exigua* var. *linicola* causing disease in Autumn Sage (*Salvia greggii*) in Italy were determined by Garibaldi *et al.* (2016). In their studies, the Internal Transcribed Spacer (ITS) region of rDNA was amplified using the primers ITS4/ITS6 (White *et al.* 1990) and sequenced. BLASTn analysis (Altschul *et al.* 1997) of the 463-bp amplicon (GenBank Accession No.KU512286) showed 100% homology with *B. exigua* var. *linicola* (syn. = *Phoma exigua* var. *linicola*) (EU573009.1) (Garibaldi *et al.*, 2016).

Management/Control measures

Use of host resistance

At present there is no resistant fibre yielding flax cultivar against the disease. Ondrej (1983b) did not observe any significant resistance to *B. exigua* var. *linicola* in fibre-flax cultivars.

Cultural methods

This disease is seed borne, and as such, the primary means of control should be the production of disease-free seed. Sanitation is an important approach to protect the crop from this disease. The use of clean and disease-free seed, following crop rotation of at least three years between flax crops and destroying crop residue after harvest is very useful measures in sanitation programme (Rashid, 2003b).

Chemical control

Seed treatments is failed to protect the crop from this disease, while foliar applications of fungicides may be effective in reducing the disease (Turner, 1987; Decognet *et al.*, 1994). However, seed treatment with suitable fungicides is effective to kill the seed borne pathogen in seeds. Maddens (1987) described a number of suitable seed treatments. The fungicides, difenoconazole, and flusilazol plus tridemorph, have been tested and proved very effective while applied at early stage of infection (Decognet *et al.*, 1994). In the later stages of the disease, control is more difficult.

3. Pasmo

Occurrence & severity status

Pasmo disease, also known as 'spasm' or 'septoriosis', is a widespread disease that affects leaves, stems and bolls. The disease starts to infect at the seedling stage, however, the severity of pasmo is not generally recognized until after boll setting and the ripening stage. Epidemics can occur early in the season when favourable conditions of high humidity with frequent rain showers prevail. Splashing rain and wind spread the infection to flax leaves and stems vertically and horizontally to adjacent plants (Rashid, 2015). Pasmo can cause defoliation; premature ripening and can weaken the infected pedicels resulting in heavy boll-drop under rain and wind conditions. Depending on the earliness and severity of the infection, pasmo reduces seed yield as well as seed and fibre quality (Flax Council of Canada). It occurs in North and South America (Garassini, 1935; Sackston and Gordon, 1945; Rashid *et al.*, 1998b), in Europe (Colhoun and Muskett, 1943; Muskett and Colhoun, 1947; Martelli, 1961; Holmes, 1976), in Africa (Nattrass, 1943), and in Australasia (Newhook, 1942; Millikan, 1951).

In recent years, the occurrence of pasmo disease has increased in Lithuania (Jovaišiene and Taluntyte 2000; Markevicius and Treigiene 2003). Pasmo is more common in western Canada and occurs in about 79-100% of the fields of Manitoba and Saskatchewan (Rashid *et al.* 2010a, 2011, 2012, 2013). The disease was reported to cause 20% losses in Minnesota in 1928 (Rodenhiser, 1930). Severe losses were reported in Argentina in 1930 and 1931 (Bolley, 1931). The disease also caused yield losses in South Dakota (Ferguson *et al.*, 1987) and England on winter linseed (Perryman and Fitt, 1999). Stem incidence and severities exceeded 15% in 2002 and 20% in 2003 in North Dakota (Halley *et al.*, 2004). The pasmo disease is on the rise. These days, this fungal disease is the most prevalent disease of flax on the Canadian Prairies. It affects flax seed yield and quality, with severe inpections reducing yields by 50 percent (Rashid 2003a; King, 2013). In susceptible cultivars, pasmo disease can reduce oil content by 9% and total seed and fibre yield by 40-70% (Plonka and Anselme, 1956; Frederiksen and Culbertson, 1962). In addition, the quality of both oil and fibre is diminished, the latter being short and breaking easily on stripping. Subsequent industrial losses are considerable, as was reported in New Zealand in 1961-1962 (Sanderson, 1963).

Symptoms

Initially the symptoms of pasmo are small circular to oblong shaped lesions that change from green to yellow and eventually dark brown in the centre (Rashid, 2003b) or dark brown to black lesions (spots) on the lower leaves of flax plants. Symptoms may appear on flax leaves from the seedling stage to maturity. Dark brown to black bands of pasmo infection start showing up on flax stems shortly after flowering and this weakens the stems (Rashid, 2015). Symptoms of this disease are striking and easily recognized, especially during the latter part of the growing season. At that time, the disease is characterized by circular, brown lesions on the leaves and by brown to black infected bands that encircle the stem. These bands alternate with green, healthy bands, making pasmo easy to identify. The alternate brown and green banding of the stem occur until the infection becomes severe and the lesions coalesce (de Tempe, 1963a; Martens *et al.*, 1984). Pasmo causes defoliation and weakening of the stems that resulting in stem breakage and lodging (Rashid, 2015). Infected flax tissue contains tiny black pycnidia which are the fruiting bodies of the fungus (Bedlan, 1984). Lesions are also developed on the sepals and flower buds. Seed infected by the organism may exhibit a bluish black tinge. White eruption of fungal growth on seed has also been reported (Halley, 2007). Flowers and young bolls are also blighted. Older bolls are discoloured and contain shrivelled or non-viable seed.

Causal organism: *Mycosphaerella linicola* Naumov 1926
[Mater.Mikol.Fitopatol.Rossii: 2]

Synonyms

= *Mycosphaerella linicola* var. *linicola* Naumov 1926

= *Phlyctema linicola* Speg. 1911

= *Septogloeum linicola* Speg. 1911

= *Septoria linicola* (Speg.) Garass. 1938

= *Sphaerella linicola* (Naumov) Wollenw. 1938

Morphological/ Morphophysiolgical characteristics

Pycnidia develop in the lesions. Pycnidia are immersed, up to 120 µm wide, ostiolate, globose to subglobose, sometimes flattened at the base, thin-walled. Conidiogenous cells are hyaline, obpyriform, non-septate, formed from the cells lining the pycnidial cavity. Conidia are hyaline, filiform, 1-3-septate, 17-40 x 1,5-3 µm, straight or curved (Sivanesan and Holliday, 1981). Conidia (pycnidiospores) are exuded in slime and are transported by rain splash, animals, and insects. Because of their large size, pycnidiospores are usually not transported by the wind (Christensen *et al.*, 1953). Sporulation can occur in 8 days after infection. In the teleomorph, Perithecia (pseudothecia) develop on diseased flax stems and resemble black spherical dots. Pseudothecia scattered, immersed, globose to subglobose, 75-120 µm wide, ostiolate. The pseudothecial wall is up to 13 µm thick, composed of 3-5 layers of pseudoparenchymatous cells which are thick-walled and dark brown towards the outside and hyaline and thin-walled towards the inside. Asci are fasciculate, bitunicate, 8-spored, 30-50 x 8-9 µm, thick-walled. Ascospores are hyaline, 1- septate, constricted at the septum, 13-17 x 2.5-4 µm (Sivanesan and Holliday, 1981).

Molecular characteristics

The structure of two *Mycosphaerella linicola* (=*Septoria linicola*) populations in Manitoba was studied using Random Amplified Polymorphic DNA (RAPD), a PCR based molecular marker system. Plants were collected from two commercial fields and used to generate single spore isolates for use in that study. The level of polymorphism detected using RAPD suggests that it is plausible that there is sexual recombination occurring between the two populations, or that there is extensive movement of individual isolates throughout the province. Limited grouping based on site was seen in the dendrograms. Analysis of Molecular Variance (AMOVA) revealed that 88% of the total genetic variation was due to within population variation and 12% to between population variations. This suggests a mix of clonal and sexual reproduction during the growing season (Grant, 2008). In another study on the phylogenetic relationships inferred from ITS and D2-LSU nrDNA sequences revealed that intraspecific variation in ITS was limited in *Septoria*. The pathogens *M. linicola* (*S. linicola*), *S. apiicola* and *S. populicola* cluster in a major clade containing *Phloeospora ulmi*. Short branch lengths in this clade suggest a very recent evolution (Verkley et al., 2004).

Predisposing factors
The fungus can infect plants over a range of temperatures, from about 15 to 30°C with an optimal temperature of 20-21°C (Brentzel, 1926). The causal organism can persist in the seed for over a year and in crop residue for up to four years. Pasmo infections increase with prolonged wet periods in late summer and autumn. Although flax is most susceptible to pasmo in the ripening stage, it can occur early in the season when warm, moist conditions prevail. The disease is favoured by wet conditions, including splashing rain and high humidity. Lodging favours the development of pasmo, because of increased humidity within the crop canopy (Rashid 2000a; Vera *et al.* 2012).This leads to patches of dead plants completely covered with the fungus. A dense, weedy or lodged canopy that traps moisture around the flax plants provides ideal conditions for the disease, especially under regimes of high soil moisture and high soil N fertility (Franzen 2004). Nitrogen applications frequently increased leaf and stem pasmo (Grant et al., 2016).

Disease cycle
The pasmo pathogen is seed-borne. It also overwinters in the soil on infected flax stubble. The main source for the spread of the disease is infected seed, but the disease causal agent remains viable in the soil for 9–12 years even where flax is no longer grown (Kornejeva and Loshakova 1976; Neofitova *et al.* 1984; Andruszewska and Korbas 1989). Inoculum survives the winter on flax stubble. With rain and warmer weather of spring, fungus spores ooze out of the pycnidia on the flax stubble and are splashed by rain droplets to flax seedlings. Spores, dispersed by wind and rain, cause the initial infections on leaves and stems. It attacks above-ground parts of the plant. The infection on the leaves produces more pycnidia and more conidia. The spores are splashed by raindrops, spreading the infection up higher on the plant and to adjacent plants. Infected seed can also be a source of pasmo. In that case, the fungus spreads from the seed to the seedlings, and the infected seedlings become foci for disease spread and development. Pasmo may complete multiple life cycles within a flax field (Rashid, 2015).

Physiological races

Two races, designated A and B, have been identified by differences in virulence (Rodenhiser, 1930) and geographic distribution of the fungus. Race A, a less aggressive race that grows well on artificial media, is widespread in Germany, Hungary, the former Yugoslavia and the Americas. Race B is widespread in Argentina and Czechoslovakia, grows with difficulty on artificial media and forms no perithecia on agar but is characterized by a very short incubation period. The races can be distinguished easily in culture by their colour, surface consistency and topography (Halley, 2007).

Management/Control measures

Use of host resistance

Flax cultivars differ in the amount of disease shown in Argentina, Australia, Canada, New Zealand, UK, Uruguay, USA and USSR. However, most commercial cultivars lack resistance to this pathogen; although, marked differences in susceptibility among varieties do exist (Kudryavtseva, 1988; Bauers and Paul, 1991). Practically, there are no registered flax varieties with resistance to pasmo (Rashid, 2015).

Cultural methods

In the absence of genetic resistance to this pathogen, the best disease management is achieved by early seeding to avoid high moisture conditions in late summer and fall, using clean seed with recommended seed treatment to protect the crop at the seedling stage. Seed treatment is suggested to do with thiram, however, internal seed infection cannot be controlled by thiram alone. Since, the pathogen tends to be deep-seated making eradication difficult (Turner, 1987). Thus, the seed with the fungicide Thiram and a hot water treatment of 38 °C for ten minutes are suggested. A minimum of three years of crop rotation can reduce overall inoculums in field. It is suggested to avoid adjacent fields of flax in avoiding disease spread from field with flax stubble to new seedlings. The variety which possesses lodging character should be avoided. Plant lodge resistant varieties are recommended. It is suggested to use recommended seeding rate, since, very high seeding rates increase the probability of lodging. Control weeds, which cause thick canopy of crop and favour wet microclimate for disease development, is essential to manage the disease in cultural way (Sivanesan,and Holliday, 1981; Flax Council of Canada, ND2).

Chemical control

Seed treatment with thiram and four additional sprays (when first symptoms occur) with mancozeb, benomyl or thiophanate-methyl have been found to reduce infection and disease development. (Batalova and Kumacheva, 1983). Pasmo can be reduced by preventative use of fungicides. Quadris, Priaxor, and Headline are all labeled for Pasmo. If environmental conditions are favorable for infection, spray timing is at early to mid-flowering according to label instructions (NDSU, 2017).

The application of pyraclostrobin fungicide reduced disease severity and increased seed yield of flax. The increase in disease severity due to increased N application occurred in the absence of fungicide, but when fungicide was applied the effect of N on fungicide treated field was less, only low levels of disease severity (near 20%) were observed as compared to without fungicide sprayed field with extreme disease severity (near 100%). The occurrence of lodging took place late in the development of flax (Vera *et al.*, 2014).

4. Powdery mildews

Different fungi have been reported to cause powdery mildews in flax. The species *Erysiphe polygoni* DC was reported from Japan (Homma, 1928), *E. lini, E. Cichracearum* DC from Siberia (Badayeva, 1930) and Minnesota (Allison, 1934), *Sphacelotheca lini* from Russia (Tvelkov, 1970), *Leveillula taurica* from India (Saharan and Saharan, 1994b) and *Golovinomyces orontii* (= *Oidium lini* Skoric) from UK, India (Saharan and Saharan, 1994a) and Canada (Rashid *et al.*, 1998a). However, amongst them, the powdery mildews caused by *Golovinomyces orontii* and *Erysiphe polygoni* are described hereunder.

i) Powdery Mildew (*Golovinomyces orontii*/ *Oidium lini*)

Occurrence & severity status

Powdery mildew (*Golovinomyces orontii*) is common on flax in the countries, like, Canada (Rashid *et al.*, 1998a), China(Li *et al.*, 2007), Egypt (Mansour, 1998), India (Pandey and Misra, 1992), Lithuania (Grigaliūnaitė 1997) and North America (Allison, 1934). In China the field occurrence of the pathogen of flax powdery mildew was initially studied in Xinjiang (China). There, the flax powdery mildew was a major disease of flax and widely occurred in flax planting area in Xinjiang (Li *et al.*, 2007). In North America, the powdery mildew has been reported only from Minnesota (Allison, 1934). In Egypt powdery mildew occurs annually in all flax-production areas and all commercially grown flax cultivars are susceptible to the disease (Mansour, 1998). In western Canada, this disease was first reported in 1997 and has been noted to spread widely in a short period of time (Rashid, 1998; Rashid *et al.*, 1998a). The incidence and severity have been on the rise in most flax growing areas on the Canadian Prairie causing an estimated 10-20% yield loss (Rashid, 2017). Yield losses of up to 38 percent have been reported in susceptible cultivars in India (Pandey and Misra, 1992).

Symptoms

The symptoms are characterized by a white powdery mass of mycelia that start as small spots and rapidly spread to cover the entire leaf surface. Heavily infected leaves wither and die. In flax the mycelia of powdery mildew appears on leaves, stems and sepals. A close examination of the colonies reveals the abundance of mycelia and microscopic conidia which are borne singly or in chains on short conidiophores. Early infections may cause complete defoliation of flax plants and reduce the yield and quality of seed (Flax Council of Canada, ND2).

Causal organism: *Golovinomyces orontii* (Castagne) V.P. Heluta 1988

[N.B. Mostly used name is *Oidium lini* Bondartsev 1913, which is now considered as an ambiguous synonym]

Synonyms

= *Acrosporium lini* (Skoric) Subram. 1971

= *Erysiphe cichoracearum* f. *apocyni* Jacz. 1927

= *Erysiphe orontii* Castagne 1851

= *Erysiphe tabaci* Sawada 1928

= *Euoidium lini* (Bondartsev) Y.S. Paul & J.N. Kapoor 1987

= *Oidium lini* Bondartsev 1913

= *Oidium lini* Skoric 1926

Morphological/ Morphophysiolgical characteristics

Mycelium is amphigenous, thinly effuse or patches, initially forming indistinct white patches, finally covering whole leaf surfaces and stems, persistent, hyphae slightly flexous, branched at right angles, 6–8 μm wide. Appressoria on the mycelium are nipple-shaped, well developed or occasionally moderately lobed. Conidiophores are 100 to 260 × 10 to 12 μm in size, mostly erect, containing a foot cell (36–65 × 7–12 μm) that has a distinctly curved base, followed by 2 or 3 shorter cells (8– 12 × 8–10 μm), produces 2 to 6 immature conidia in chains with a sinuate outline. The basal septum of the conidiophore is just adjacent to the mycelium. Conidia are produced abundantly presenting a white powdery appearance; hyaline, ellipsoid to ovate or oblong or cylindrical-oblong or dolliform, (26–) 30 -34 (-43) × (12-) 14–18 μm, with a length/width ratio of 1.7 to 3.2, formed in chains of two or three and in basipetal succession and lacked distinct fibrosin bodies (differently, containing the fibrosin bodies of well-developed special forms; Homma, 1928). Germ tubes are produced on the perihilar or lateral position of conidia. Chasmothecia are not generally found (Homma, 1928; Cho et al., 2016; Mulpuri et al., 2016).

Predisposing factors

The pathogenic fungus of flax powdery mildew conidia can germinate in the temperature range of 15-30°C, however, the optimum is 20-25°C. The occurrence of flax powdery mildew has a familiar relation with air temperature and rainfall. It occurs at relatively low air temperature and much rainfall season (Li et al., 2007). The rainy and high humid season accelerates the development of powdery mildew of flax. The more seed density and the later time of seed period in spring the development of powdery mildew of flax is more severe (Yang et al., 2007).

Disease cycle

The powdery mildew is an obligate polycyclic disease. The conidia produced on host plants repeatedly infect. Chasmothecia are overwintering structures may remain in plant debris for several years. They germinate to ascospores which may infect during favourable season. On infection conidia are produced in host plant and repeat the infection.

Physiological races

Although the indications from the reactions of flax cultivars in different flax-growing areas under natural powdery mildew epidemics in Europe and North America point to the presence of different races of powdery mildew (Beale, 1991; Saharan and Saharan, 1992), there is no scientific confirmation for the existence of races in powdery mildew.

Management/Control measures

Use of host resistance & molecular characterization of resistance

Investigations on the genetics of resistance to powdery mildew in flax resulted in the identification of several resistant Canadian and introduced cultivars. A single dominant gene for resistance to powdery mildew, designated PM1, was identified in Canadian flax cultivars ('AC Watson', 'AC McDuff', and 'AC Emerson') and introduced flax cultivars ('Atalante' and 'Linda'). In addition, two putative dominant genes for resistance to powdery mildew were identified in 'Linda' (Rashid and Duguid, 2005). Canadian flax cultivars are moderately resistant (MR) to PM. The results from the 4-years field testing showed that AC Emerson and WestLin 70 flax cultivars are resistant, while, the cultivars, viz., AAC Bravo, AAC Prairie Sunshine, CDC Bethune, CDC Buryu, CDC Glass, CDC Neela, CDC Plava, CDC Sanctuary, CDC Sorrel, Lightning, Prairie Blue, Prairie Grande, Prairie Sapphire, Prairie Thunder, Taurus and VT50, are moderately resistant (Rashid, 2017). The genetics of resistance against powdery mildew in flax was analyzed. The F1 plants from reciprocal cross of resistant materials 9801-1 and three susceptible cultivars ILONA, VENUS and DIANE were resistant to powdery mildew. The ratio of resistant and susceptible plants in F2 generation fitted the excepted 3 to 1. It was postulated that 9801-1 carried a single dominant and resistant gene (Yang et al., 2008). F2 populations were obtained from the cross between 9801-1 and DIANE. Bulked segregate and RAPD analyses were employed to identify molecules linked to the resistance to powdery mildew. OPP02 amplified about 792 bp polymorphic band in all individuals from 9801-1 and resistant bulk, but absent in all individuals from DIANE and susceptible bulk. By further analysis in F2 segregating population, the polymorphic band was found to be co segregated with the resistant gene possibly. The fragment was sequenced, and settled a foundation for transforming it into SCAR marker (Yang et al., 2011). In Egypt, the powdery mildew resistant flax cultivars are viz., Dakota, Wilden and Williston Brown, while the susceptible and less suscetible flax cultivars are Giza 8, Cortland, Linore, C.I. 2008 and Gzia 7 (Hussein et al., 2011). SDSPAGE, RAPD analysis by using four primers and double diffusion test (DDT) were used to differentiate between resistance and susceptible genotypes. RAPD analysis by using primer no. 6 and DDT were able, at least partially, to differentiate between resistant and susceptible genotypes, while SDS-PAGE was unable to differentiate between cultivars of the two groups (Hussein et al., 2011).

Cultural control

The most economical control is through the use of any of the resistant varieties as mentioned above. Early seeding will reduce the impact of this disease on yield loss by avoiding early infections and build-up of epidemics.

Biocontrol

Essential oils of jojoba and coriander when applied as foliar sprays were highly effective as they reduced powdery mildew of flax disease severity by 66.24 and 68.64%, respectively (Aly et al., 2013).

Chemical control

Foliar application of recommended fungicides around flowering time may protect the crop from severe powdery mildew epidemics and reduce losses in yield and seed quality (Flax Council of Canada, ND2). An experimental results show that 40% Flusilazole EC, 43% Tebuconazole SC and 40% Myclobutanil WP have better effect on flax powdery mildew (Wang et al., 2006).

ii) Powdery mildew (*Erysiphe polygoni*)

Occurrence and severity

Powdery mildew of flax, caused by *Erysiphe polygoni*, was first observed in the year 1922 to cause disease in flax in Japan (Homma, 1928). Later, it was described by Tanda and Hirose (2003) giving a new name, *Erysiphe lini* Tanda, sp. nov. and considered *Erysiphe polygoni* which described by Homma (1927, 1928, 1937) as a synonym. However, the new name is not prevailed now a day.

Symptoms

The disease appears as circular or irregular, thin, greyish white mycelial patches with obscure margin, covering the whole surface of leaves and stems, persistent on the upper surface, evanescent on the under surface. The white mycelium develops not only on both surfaces of the leaf, but also on the stem and peduncle (Homma, 1928; Tanda and Hirose, 2003).

Causal fungus: *Erysiphe polygoni* DC. 1821 [Kirk, 2018]

Morphological Characters

Mycelia are amphigenous, also cauligenous, from which small ellipsoidal or globular haustoria are sent into the epidermal cells of the host. The conidia are produced in a chain, but look as if they are borne singly at the tip of the conidiophore. They are ellipsoidal or oblong-ellipsoidal in shape. Conidia are solitary, ellipsoidal, broadly ellipsoidal or rarely subglobose, vacuolate, fibrosin body absent. The size of the conidia is (28) 30-38 (38.4) X (16) 19.4-23 (24.0) µm. The conidiophores are erect, branching from the hyphae creeping on the surfaces of leaves and stems, straight or rarely curved, 1 or 2 septate, 49-47 (62.4 µm) x 8-10 µm. Foot-cells are cylindric, 49-70 x 8-10 µm. Cleistothecia are scattered on the under surface of the leaves, dark brown, subglobose or depressed globose, (68) 91-105 (107) µm ; wall cells irregularly polygonal, 11-18 x 7-18 µm. The white mycelium is evanescent, disappearing before the perithecia reach their maturity. Cleistothecial appendages are (2-) 3-4 (-6) (differently about 8) in number measuring (47) 168-217 (186) µm in length, produced on the lower half part of the cleistothecium, mycelioid, simple, thin-walled throughout, brownish or hyaline upwards, curved or flaccid, 1-3-septate (differently, 5-7 septate), brown-coloured toward the base. Usually each perithecium contains six asci. The asci are (2-) 3-4 (-5) in number, ovate in shape, with short stalk, colourless, (43) 48.0-55.2 (75) x (24) 25-31.2 (50) µm. Each ascus contains (2) 3-4 ascospores. The ascospores are ellipsoidal, hyaline, (18.0) 21-27 (20.4) x 9.8 (12)-12.0 (14) µm in size (Homma, 1928; Tanda and Hirose, 2003).

Disease cycle

The fungus remains dormant as cleistothecium in plant debris in soil. The cleistothecium releases ascospores in favourable conditions. The ascospores germinate in the middle of June in Japan and form thin, greyish white mycelial patch on a few leaves of flax, and the patches expand. The conidia form on the mycelia in the patches. Later, the conidia germinate on the leaves of flax. The mycelial patches again develop in five to six days after infection (Tanda and Hirose, 2003).

Control measures

Same as described earlier with other powdery mildew disease, caused by *Golovinomyces orontii* in flax

5. Rust

Occurrence & severity status

The disease occurs in all flax-growing regions of the world. This is potentially the most explosive disease affecting flax. Early infections and rapid disease development favoured by weather conditions may completely defoliate flax plants and cause major losses in yield and quality of both seed and fibre (Flor, 1944; Hora *et al.*, 1962; Hoes and Dorrell, 1979; Acosta, 1986; Shukla, 1992). The major damage to commercial flax (*L. usitatissimum*) is through weakening and disfiguring of the fibres (Laundon and Waterston, 1965). In Canada, rust is the most destructive disease affecting flax. The last major rust epidemic occurred in the 1970's (Flax Council of Canada, ND2). Thirteen flax cultivars licensed in Canada had to be replaced in the 1970s because of susceptibility to the new races of rust (Hoes, 1975; Anonymous, 1996). In North America, the build-up of rust during the 1930s and continuing until the early 1950s resulted in severe epidemics and heavy losses in yield (Flor, 1946; Vanterpool, 1949). The latest outbreaks of flax rust in North America were in the 1960s and 1970s when a new group of races were identified (Hoes and Tyson, 1963; Zimmer and Hoes, 1974; Hoes and Zimmer, 1976). Presently, this disease is under control in North America where all commercial flax cultivars are immune to local races of rust (Rashid and Kenaschuk, 1992; 1994). Rust remains a constant threat to flax production worldwide because of the ability of the fungus to produce new races that can attack resistant cultivars. The reasonable levels of control of this disease worldwide have been attributed to the availability of the sources of genetic resistance to most flax-breeding programs all over the world and the wise deployment of these genes in commercial flax cultivars (Rashid, 2003b). The rust has quite a wide host range in the genus *Linum*. For example, it grows on numerous European species of *Linum* (Gaumann, 1959), on several of North

American species (Anon., 1960), on the sole Australian species, *L. marginale* (McAlpine, 1906), and on the sole New Zealand species, *L. monogynum* (Gaumann, 1959). The pathogen has also been found on 12 of the 13 species in the genus *Hesperolinon* in the field (Springer, 2006) and an isolate from cultivated flax in the Netherlands was found to be virulent on an accession of *Radiola linoides* (Kowalska and Niks, 1998)

Symptoms: The leaves are the first to show the symptoms and gradually all the aerial parts of the plant get infected. Large, bright orange or reddish yellow coloured pustules generally appear on the leaves, on stems and bolls, and all aerial plant parts (Laundon and Waterston, 1965; Rashid, 2003b). These pustules are usually circular on leaves and bolls but elongated on the stems. Small pustules are initially surrounded by chlorotic areas. Little necrosis of the leaves is at first observed but it grows, becomes more general and the leaves prematurely die. Severe epidemics cause the leaves to dry and wither resulting in heavy defoliation of the plants (Rashid, 2003b). The pustule (uredopustules) is the site of production of innumerable summer spores (uredospores) each of which is orange coloured and borne on a separate stalk. When ripe the spores become detached and are readily dispersed by wind and air currents over long distances. Later in the growing season, as the plants mature, elongated slightly raised, black incrustations (telial pustules) appear on the main stem, branches, leaves and seed bolls on the sites occupied by the pustules producing summer spores. These incrustations are made up of winter spores (teleutospores) of the fungus which are arranged as a compact palisade like layer encrusting the affected portion of the plant. It is the occurrence of these spectacular black streaks resembling burnt patches on the stems and boughs that led to the popular name of "firing" for this phase of the disease (Muskett and Coehoun, 1947). The contents of telial pustules do not break the epidermis of the host and remain buried subepidermally appearing glossy.

Causal organism: *Melampsora lini* var. *lini* (Ehrenb.) Lév. (1847)

Synonyms:

= *Lecythea lini* Berk. (1860)

= *Melampsora lini* (Ehrenb.) Lév. (1847)

= *Uredo lini* DC. (1805)

= *Xyloma lini* Ehrenb. (1818)

Morphological/ Morphophysiolgical characteristics

Flax rust is autoecious and the fungus can complete the four stages of its life cycle, namely pycnia, aecia, uredia, and telia, on the flax plant. The pycnia and aecia are usually formed on cotyledons and lower leaves early in the season and are rarely noticed. Pycnia are varied in size, color, and fertility. Fertile types are classed as normal and large. Sterile types are classed as necrotic, chlorotic, and small white (Statler, 1983). In general, pycnia are small, rounded or oval yellow on both the surfaces of the leaf. The pycnia contain paraphyses and pycniosporophores producing pycniospores at their apical ends. The aecia are small, orange coloured and scattered and contain polygonal, binucleate aeciospores. But the aecia do not contain paraphyses and peridia (Prasada, 1948). Pycnia and aecia are the most important stages for completing the sexual life cycle of the fungus and for the development of new races. The third stage of the development of this fungus is the uredia. The uredial stage is the most destructive stage to the crop since uredia produce cycles of urediospores that can create new infections with each cycle throughout the growing season (Rashid, 2003b). The mycelium is intercellular, septate, branched, dikaryotic, subepidermal and usually found within the parenchymatous tissues of the host. The urediospores are echinulate, with spines of 1μ long over their surface. The spines are electron-transparent, conical projections, with their basal portion embedded in the electron-dense spore wall. The entire spore, including the spines, is covered by a wrinkled pellicle. The spore wall consists of three recognizable layers in addition to the pellicle. Spines form initially as small deposits at the inner surface of the spore wall adjacent to the plasma membrane. Endoplasmic reticulum occurs close to the plasma membrane in localized areas near the base of spines. During development, the spore wall thickens, and the spines increase in size. Centripetal growth of the wall encases the spines in the wall material. The spines progressively assume a more external position in the spore wall and finally reside at the outer surface of the wall (Littlefield and Bracker, 1971). At the end of the season, the fourth stage produces the hardy overwintering teliospores which can survive the adverse weather conditions (Rashid, 2003b). Telia are black or reddish brown crust formed on the stems. These telia are consisted of palisade layers of black to brown coloured, sessile, cylindrical, unicellular and produce binucleate teleutospores.

Molecular characteristics

Melampsora lini has chromosome number of n= 8 (Boehm and Bushnell, 1992), and an earlier estimation genome size is of 170 Mb (Eilam *et al.*, 1992; Leonard and Szabo, 2005). However, presently, that is estimated nearly to 200Mb (Peter Dodds in Melampsora lini-Science Direct). The 189.52 Mb genome of *M. lini* is dramatically expanded and riddled with transposable elements (Nemri *et al.*, 2014). During infection, *M. lini* secretes virulence effectors to promote disease. To assess the number of these effectors, their function and their degree of conservation across rust fungal species, Nemri *et al.* (2014) sequenced and assembled de novo the genome of *M. lini* isolate CH5 into 21,130 scaffolds spanning 189 Mbp (scaffold N50 of 31 kbp). Global analysis of the DNA sequence revealed that repetitive elements, primarily retrotransposons, make up at least 45% of the genome. Using ab initio predictions, transcriptome data and homology searches, they identified 16,271 putative protein-coding genes. An analysis pipeline was then implemented to predict the effector complement of *M. lini* and compare it to that of the poplar rust, wheat stem rust and wheat stripe rust pathogens to identify conserved and species-specific effector candidates. Previous knowledge of four cloned *M. lini* a virulence effector proteins and two basidiomycete effectors was used to optimize parameters of the effector prediction pipeline. Markov clustering based on sequence similarity was performed to group effector candidates from all four rust pathogens. Clusters containing at least one member from *M. lini* were further analyzed and prioritized based on features including expression in isolated haustoria and infected leaf tissue and conservation across rust species. They described 200 of 940 clusters that ranked highest on their priority list, representing 725 flax rust candidate effectors. Their findings on this important model rust species provide insight into how effectors of rust fungi are conserved across species and how they may act to promote infection on their hosts (Nemri *et al.*, 2014).

Predisposing factors

The disease is favoured by cool, moist weather. The dikaryotic urediniospores germinate on the surface of a leaf under conditions of high humidity or free water. Temperatures above 23–27 °C inhibit germination (Littlefield and Bracker, 1972). It is spread by spores produced in the pustules. The rust organism overwinters in flax straw, from which spores are produced that cause primary infection.

Disease cycle

In temperate countries, primary infection takes place through basidiospores, which are produced as a result of germination of teleutospores, perennating in the soil. Early infections produce the aecial stage with aeciospores on volunteer flax seedlings which subsequently produce the uredial stage. But in tropical countries, the teleutospores, produced at the end of growing season i.e. April-May; lose their viability due to excessive heat of summers (Prasada, 1948). Uredospores too are killed due to excessive temperatures. It is presumed that the uredospores produced on flax at hills come down to plains to cause infection. Thus the primary inoculum, windblown, fall on the host, germinate and cause infection at later stages of plant growth. Uredospores produced as a result of primary infection can cycle through several generations during the growing season resulting in completely defoliated flax plants and reduction of seed yield and fibre quality. Flax rust completes its life cycle on the flax plant (Flax Council of Canada, ND2).

Physiological races

Physiological specialization is a common phenomenon among the rust pathogens. This phenomenon was observed and studied in flax rust early in the twentieth century (Hart, 1926; Flor, 1931). Interactions between flax genotypes and different rust races and the phenotypic expression of these interactions in resistant or susceptible infection types led to the identification of distinct races (Flor, 1954; 1955), and the development of the gene-for-gene theory of pathogenesis (Flor, 1956; 1971). In North America, 239 races of *M. lini* were identified (Flor, 1954) using a set of 29 single-gene flax genotypes. The number of races identified was up to 400 by the 1970s, and an additional flax differential genotype was added to the differential set (Hoes and Tyson, 1966; Hoes, 1975; Hoes and Zimmer, 1976). Additional races have also been identified in India, Australia and Europe (Misra, 1963; Misra and Lele, 1963; Saharan and Singh, 1978; Statler *et al.*, 1981).

Management/Control measures

Use of host resistance

The use of host resistance is the most effective measure to manage rust in flax. The rust may be completely controlled by the use of rust-resistant varieties. Some varieties of flax have the ability to retard disease development in the field, suggesting tolerance to rust (Rashid, 1997). It is suggested the existence of multiple resistance genes as observed in some flax varieties (Fincham and Day 1965). Rust resistance genes in flax are dominant traits and are characterized by race-specific interaction. However, various studies have demonstrated the effectiveness of field resistance or slow-rusting and partial resistance in flax cultivars (Hoes and Kenaschuk, 1980; Parish and Statler, 1988; Rashid, 1991; Hoes and Kenaschuk, 1992; Rashid, 1997). In general, the most effective control method for *M. lini* appears to be resistance breeding (Paul *et al.*, 1991). The molecular basis of resistance has also been investigated (Sutton and Shaw, 1986; Timmis and Whisson, 1987; De Wit, 1992; Lawrence *et al.*, 1994, 1995; Roberts and Pryor, 1995). Breeders have applied the gene-for-gene concept to produce many resistant cultivars (Kenaschuk and Rashid, 1994; Rashid and Kenaschuk, 1994; Kenaschuk *et al.*, 1996). To date, a total of 31 resistance genes have been identified in *Linum*. These genes occur in five series of closely linked or allelic genes at loci designated K, L, M, N and P, which contain two, 13, seven, three and six of the 31 resistance genes, respectively (Lawrence, 1988). The five resistance loci show independent inheritance except for N and

P, which are linked with about 10% recombination (Kerr, 1960; Shepherd, 1963). Many of the resistance genes have now been cloned and sequenced. The current list includes 11 genes at the L locus, one at the M locus, three at the N locus and two at the P locus (Anderson *et al.*, 1997; Dodds *et al.*, 2001a, b; Ellis *et al.*, 1999; Lawrence *et al.*, 1995). These genes all encode resistance proteins of the Toll Interleukin 1 Receptor–Nucleotide Binding Site–Leucine Rich Repeat (TIR-NBS-LRR) class, although the P locus proteins have an additional C-terminal domain of 150 amino acids downstream of the LRR region (Lawrence *et al.*, 2007).

In Canada, the following varieties are resistant to rust. These are CDC Arras, CDC Bethune, CDC Gold, AC Carnduff, AC Emerson, Flanders, Hanley, Lightning, AC Linora, Macbeth, AC McDuff, McGregor, CDC Mons, NorLin, NorMan, CDC Normandy, CDC Sanctuary, Prairie Blue, Prairie Grande, Prairie Thunder, Sapphire, Shape, Somme, CDC Sorrel, Tauras, CDC Valour, Vimy, AC Watson, 1084, 2047, 2090, 2126 and 2149 (Rashid and Kenaschuk, 1992; Rashid and Kenaschuk, 1994; Rashid, 1997; Rashid and Kenaschuk; 1999).

Cultural control

As additional safeguards, several cultural practices may be applied to avoid or minimize rust attacks. These include destroying of plant debris, using certified and disease-free seed of a recommended variety, crop rotation and planting the flax crop in a field distant from that of the previous year (Flax Council of Canada, ND2). Early planting may help the crop mature before rust becomes serious. Flax should not be planted on flax stubble. Avoidance of excessive nitrogen fertilization, plough in infected stems and stubble after harvest and removal wild or volunteer plants are also useful in this respect.

Chemical control

Seed treatment: Seed treatment with fungicide, like, thirum to kill the teleutospores in hills has been suggested.

Foliar spray: Several strobilurin fungicides (Group 11) are labelled for use as spray in field. However, it should not make more than one (1) application of a Group 11 fungicide before alternating to a labelled fungicide with a different mode of action. In such condition, fungicides of sulphur formulations are labelled for use (Laundon and Waterston, 1965).

6. *Sclerotiana* Stem Rot

Occurrence & severity status

Sclerotinia stem rot, also known as white mould, watery soft rot, drop or blossom blight, is a limiting factor in lodged flax. This disease has been reported to affect a few fields in England (de Tempe, 1963b; Mitchell *et al.*, 1986) and Pakistan (Mirza and Ilyas, 1984). In Canada, sclerotinia infections were reported in irrigated flax in Alberta (Mederick and Piening, 1982) and from heavily lodged flax in Manitoba and Saskatchewan (Rashid, 2000b). The severity of the disease depends on the level of *Sclerotinia* inoculum in the soil from previous crops, the soil water saturation and the severity of lodging. The disease is widespread affecting sunflower, dry bean, soybean, canola, potato, alfalfa, mustard, safflower, lentil, flax, field peas and many garden vegetables (Flax Council of Canada, ND2).

Symptoms
Initial symptoms of sclerotinia rot appear as soft, water-soaked white to grey longitudinal lesions girdling on the stems. Plant parts above the affected area often turn pale green or yellow, wilt and eventually die. Mature lesions are bleached and are easily shredded resulting in premature ripening, stem collapse and lodging in heavily infested fields. White mycelia grow on the stem surface and cylindrical shaped sclerotia are formed inside the stem (Flax Council of Canada, ND2). When heavily lodged flax remains lodged, small sclerotia can also be formed on the surface of the stem (Rashid, 2003b)

Causal organism: *Sclerotinia sclerotiorum* (Lib.) de Bary, 1884
[Vergh.Morph. Biol. des Pilze, Mycet. Bact., p. 236]

Synonyms

= *Hymenoscyphus sclerotiorum* (Lib.) W.Phillips (1887)

= *Peziza sclerotiorum* Lib. (1837)

= *Sclerotinia libertiana* Fuckel (1870)

= *Sclerotium varium* Pers. (1801)

= *Whetzelinia sclerotiorum* (Lib.) Korf & Dumont (1972

Morphological/ Morphophysiolgical characteristics
Hypha is hyaline, septate, branched and multinucleate (Bolton *et al.*, 2006). Mycelium may appear white to tan on the host surface while spread over and ramified both inter-cellular and intra-cellularly. No asexual conidia are produced. Sclerotia, the pigmented, multi-hyphal structures, develop on external and internal surfaces and are sometimes numerous in pith cavities. They frequently do not form until after death of host. Sclerotia undergo a period of dormancy after maturation, so that apothecium production is rarely seen in association with diseased plants.

Colonies on PDA are white or faintly grey, usually with aerial mycelium but sometimes showing tufting or stranding. Sclerotia are developing on colony surface mainly near edge of Petri-dish, black, rounded or elongated, up to 1 cm across, broadly reniform in vertical section, with smooth or shallowly pitted surface. Cells of primary hyphae at advancing colony edge are thin-walled, with dense granular contents, usually 9-14 (-18) µm wide. Cells of secondary and subsequent branch hyphae are narrower than those of primary hyphae. Sclerotium initials are developed by repeated branching of long aerial primary hyphae, usually several merging to form a single sclerotium. Mature sclerotia show a sharply differentiated rind with evenly thickened strongly pigmented walls, a narrow cortex (2-3 cells thick) of almost iso-diametric thin-walled cells, and a medulla of intertwined branched hyphae of approximately the same diameter as the primary hyphae and with colourless unevenly thickened walls. Cells of cortex and medulla show granular contents, rind cells do not. Interhyphal spaces in the medulla do not contain gelatinous material. Phialidic spermatial state is usually present, and occasional small aggregates of cells occur superficially similar to those of the sclerotium rind (Mordue and Holliday, 1976).

Molecular characteristics

Molecular markers become useful criteria for taxonomic discrimination of *Sclerotinia* species. These markers include proteins (Wong and Willetts 1975; Petersen *et al.* 1982; Cruickshank 1983; Tariq *et al.* 1985), random ampliûed polymorphic DNA (RAPD) (Ekins, 2000) and restriction fragment length polymorphisms (RFLP) (Kohn *et al.* 1988). The RFLP probe, pMF2, which contains a Pst-1 fragment from rDNA of *Neurospora crassa*, was earlier used for separation of *S. sclerotiorum*, S. minor and *S. trifoliorum* (Kohn *et al.* 1988). Later, Ekins *et al.* (2005) have observed that the restriction fragment length polymorphism (RFLP) probes generated from cloned genomic DNA fragments of *S. sclerotiorum* are effective for accurate species designation and to compare against other markers. Several RFLP probes, either singly or in combination, enabled clear separation of the *Sclerotinia* species (Etkins *et al.*, 2005). Liu and Free (2016) used a proteomic analysis to identify cell wall proteins released from *Sclerotinia sclerotiorum* hyphal and sclerotial cell walls via a trifluoromethane sulfonic acid (TFMS) digestion. The analysis identified 24 glycosylphosphatidylinositol (GPI)-anchored cell wall membrane proteins and 30 non-GPI-anchored cell wall proteins. A comparison of the proteins in the sclerotial cell wall with the proteins in the hyphal cell wall has demonstrated that sclerotia formation is not marked by a major shift in the composition of cell wall protein. It is found that the *S. sclerotiorum* cell walls contain 11 cell wall proteins that are encoded only in *Sclerotinia* and *Botrytis* genomes (Liu and Free, 2016). In an another study, β-1,3-glucanases, which are known to be involved in different morphogenetic processes in fungi, were characterized on molecular basis and analysed their amino acid sequences in this fungus (Ezzine et al., 2016). The genomic DNA and cDNA of the *S. sclerotiorum* β-1,3glucanase were amplified using the corresponding genome and total RNA, respectively. Ezzine et al. (2016) succeeded to amplify the fragments of 1090 bp and 909 bp corresponding to the expected sizes of the genomic DNA and cDNA of the Glucan β-1,3glucosidase sequences, respectively. Nucleotide sequences of the obtained PCR products were subsequently cloned and sequenced (vide, NCBI: XM_001595467.1.). One extracellular enzyme with laminarinase activity was also identified by performing a rapid proteomic analysis of the induced secretome of *S. sclerotiorum* and using Label free mass spectrometry analysis. The genomic DNA and cDNA sequences were amplified by PCR and RT-PCR respectively, cloned and sequenced to reveal a gene with two intron sequences. The open reading frame of 909 bp encoded a polypeptide of 302 aminoacids having a calculated molecular mass of 32312.06 Da. The authors claimed that the molecular modeling and comparative investigation of different resolved structures showed that this laminarinase belongs to the family 17 of glycoside hydrolases (Ezzine et al., 2016).

Predisposing factors

The disease favoured by temperatures between 20 and 25°C, moist conditions and dense crop canopies. *Sclerotinia* infection is observed in moist patches in the field such as in hollows or nitrogen-rich areas. Lodged crops are more susceptible to sclerotinia infection (Government of Saskatchewan, ND). Differences in flax cultivars to *Sclerotinia* stem rot have also been reported (Pope and Sweet, 1991).

Disease cycle

The *Sclerotinia* fungus spends most of its life cycle in the soil as a hard-walled resting structure called a sclerotium (plural: sclerotia). Sclerotia can germinate to produce mycelia or apothecia depending on environmental conditions. The most common way is called carpogenic germination, resulting in the formation of mushroom-like structures called apothecia (singular: apothecium). Apothecia produce ascospores, which are the primary means of infection in most host plants (Bolton et al., 2006). The second type of germination is called myceliogenic germination. Sclerotia germinate by producing small hyphae that grow through the soil and directly infect host roots, resulting basal stem infection. Apparently, stems of heavily lodged flax pick up the infection by direct contact with the fungus from the soil (Rashid, 2003b). Regardless of the type of infection, the fungus invades and advances within its host using cell-wall degrading enzymes and toxins that kill plant tissues. Lesion and sclerotia are develop within or on diseased plant tissue and the sclerotia are returned to the soil with crop residue or are harvested with the seed, thus completing the disease cycle (Government of Saskatchewan, ND).

Management/Control measures

Cultural control

Sclerotinia stem rot is a tough disease to prevent. Cultural control measures, like crop rotation, are ineffective because sclerotia can survive in the soil for many years and ascospores can travel by wind up to several kilometres. Further, there are over 400 host plant species, including most broadleaf weeds, so even in the absence of a flax crop, the disease lives on. However, certain measures can be taken to minimize the attacks. First of all, the removal of crop residues from field after harvest is needed to minimize sclerotia loads in field. Sowing of certified or disease free seed is another important aspect to limit the amount of disease inoculums in the field. Management of moisture levels, especially in irrigated crops is an important aspect to prevent the disease attack. Flax crops are needed to cultivate into well-drained soil. Wide, uniform row spacing and lower seeding rates may increase air movement within the crop canopy and reduce the moist soil microclimate required for sclerotia to germinate. It is suggested to avoid irrigation where extended periods of high humidity occur. Irrigation should be done early in the day, allowing plants to dry before evening. It is also necessary to keep the top of the soil as dry as possible as the crop matures. High nitrogen in soil favours the disease by creating a lush, humid plant canopy and delaying crop maturity. It is valuable to test for soil nutrients and add only the necessary amount of nitrogen fertilizer (Government of Saskatchewan, ND).

Use of host resistance

It is recommended to use lodging resistant varieties (Flax Council of Canada, ND2).

Fungicidal control

The best *Sclerotinia* stem rot management strategy is applying a fungicide. Correct timing, proper water volume and the right rate of a fungicide are critical for success. Carboxamide (FRAC Group 7) formulations are registered for use. However, the fungicide should not use more than two (2) sequential applications before alternating to a labeled fungicide with a different mode of action. Chemical fungicides, such as, Benlate or Rovral while applied as soil drenching or spray is effective to control this disease as evident from other crops. The phenylpyrrole fungicide fludioxonil has been reported to have high activity against *S. sclerotiorum* (Duan *et al.*, 2014).

7. Seedling blight

i) Seedling blight (*Alternaria linicola*)

Occurrence & severity status

This seedling blight (*Alternaria linicola*) disease is also known as 'Brown stem blight' and 'Alternaria blight' at different places. It occurs in flax all over the world (Groves and Skolko, 1944; Moore, 1946; Muskett and Colhoun, 1947; de Tempe, 1963a; Fitt *et al.*, 1991d). However, it is more commonly reported from wherever the crop is grown in Europe (Fitt and Vloutoglou, 1992). Earlier, *A. linicola* was not considered as an aggressive pathogen of either flax or linseed when it was first recorded in Canada (Groves and Skolko, 1944), Denmark (Neergaard, 1945) and UK (Moore, 1946; Muskett and Colhoun, 1947). However, it is now considered as a major disease in certain areas. In the United Kingdom the yield loss due to this disease goes up to 35 percent (Mercer *et al.*, 1989). On contrary, in North America, this disease has been observed occasionally but no major epidemics have been reported. This disease reduces the quality and viability of seed harvested from diseased plants (Rashid, 2003b). The pathogen, *A. linicola*, has been found on only two species of the genus *Linum*, namely *Linum grandiflorum* and *L. usitatissimum* (Neergaard, 1945). On the latter species it seems that the fungus occurs more frequently on linseed than on fibre flax (Beaudoin, 1989a).

Symptoms

The fungus attacks the young seedlings as they emerge, causing damping-off symptoms with development of a brown moist rot (Mercer *et al.*, 1991a) or the seedlings may be blighted up. Short red streaks and water soaked areas may also be visible on the hypocotyls and cotyledons of some infected seedlings (Sheppard, 2005). Cotyledons become brown to dark brown, wither and fall off at a touch. Closer examination reveals the presence of minute dark spots on the first pair of true leaves. The spots spread irregularly to give necrotic areas of 1-2mm across. Sometimes, the first pair of infected true leaves of seedling may be discoloured and shrivelled. A pale brown streak is often present on the stem just above the cotyledons and more frequently a canker-like area with reddish brown rim develops on the hypocotyl or at the base of the main root (Moore, 1946). Brick-red lesions can also appear on the stems and lower leaves. Infected seedlings can be seriously weakened, stunted or killed. The upper leaves of linseed crops seem to be free of disease for most of the growing season (Mercer *et al.*, 1991a). The disease reduces the root numbers and the root may be swollen on infection. On older plants, the fungus causes spots on the tip or the base of the leaves. In this case, an elongated spot often appears on the stem above or below the point of attachment of the leaf (Neergaard,

1945). Symptoms in grown up plants generally appear on the upper leaves, sepals and capsule cases late in the season just before harvest (Mercer et al., 1991a). On infected leaves, the symptoms are dark brown lesions that usually coalesce to cover the entire leaf which turns chlorotic and dies.

Causal organism: *Alternaria linicola* J.W. Groves & Skolko 1944
[Canadian Journal of Research 22 (5): 223]

[**N.B.** In India, the fungus, *Alternaria lini* Dey, 1933, has been reported to affect on aerial parts of flax, particularly on Linseed (Dey, 1933). The disease is more severe on the flowers. Dark brown spots are common on leaves, calyx and pedicels. Infected buds may produce shrivelled seed or fail to produce any seed. Under favourable conditions of the disease, these lesions develop on stem and the whole plant may be killed. Infected seeds become dark black or dark brown, and some normal appearing seeds may have enough inocula to cause seedling blight. The disease is favoured by 26-33°C and humid conditions (Gill, 1987b). In India, a yield loss in Linseed due to this disease was recorded 28-60% (Chauhan and Srivastava, 1975). Well drained fields are recommended to avoid this disease due to the severe epidemics in poorly drained fields (Gill, 1987b). Further, control measures may be the same as presented with *A. linicola*. The disease, mainly reported from Linseed, the oil producing variety of flax, and its occurrence in fibre yielding flax crop is negligible or nil, so, further details of this disease is not discussed separately in this book.]

Morphological/ Morphophysiolgical characteristics
Hyphae are septate, branched, hyaline to olive-brown, (2-) 3-6 (-7) μm wide. Conidiophores are simple (unbranched), occurring single or in bundles, pale olive-brown, septate, erect, often geniculate with 1-2 or more scars, variable in length, (5-) 6-8 (-9) μm in diameter. Conidia form singly, are smooth-walled, olive-brown, paler on the host than in culture, elongated conical to ellipsoid or obclavate, muriform, with (4-) 7-11 (-16) transverse septa and occasionally 1-4 longitudinal septa, not or only slightly constricted at the septa, gradually tapering towards the beak, 20-130 x (7-) 17-24 (-30) μm. Secondary conidiophores may be produced from cells of the conidium body. Beak of conidium is long, filiform, often branched (more often in culture), septate, occasionally swollen at the tip when producing a secondary conidiophore (more often in culture), 16-230 x 3-4.5 μm (Groves and Skolko, 1944; David, 1991)

The appearance of the fungus in culture depends on the medium. On malt agar (MA) the aerial mycelium is cottony, white, "deep olive-grey" to "dark olive-grey" paler at the centre to "pale olive-grey". The submerged mycelium is dense, mostly radiating, "olive-brown". The colony reaches about 6-7 cm in diameter after 10 days. On potato dextrose agar (PDA) the aerial mycelium is dense, cottony, rough, and white to "leaf-green". The submerged mycelium is radiating, white to "dark greyish olive" or "olivaceous black". On most media the areas where the sporulation is concentrated appear darker and under alternating light and darkness conditions that induce sporulation, forming light and dark concentric zones on cultures (Neergaard, 1945).

Molecular characteristics

Nucleotide sequences of the ribosomal DNA (rDNA) internal transcribed spacers (ITS) 1 and 2 and a 1068 bp section of the beta-tubulin gene divided seven designated species of *Alternaria* into five taxa. *Stemphylium botryosum* formed a sixth closely related taxon. Isolates of A. linicola possessed an identical ITS sequence to one group of *A. solani* isolates, and two clusters of A. linicola isolates, revealed from beta-tubulin gene data to show minor variation, were as genetically similar to isolates of *A. solani* as they were to each other. It is suggested that *A. linicola* falls within the species *A. solani*. Similar results suggest that *A. lini* falls within the species *A. alternata*. RAPD analysis of the total genomic DNA from the *Alternaria* spp. concurred with the nucleotide sequence analyses. An oligonucleotide primer (ALP) was selected from the rDNA ITS1 region of *A. linicola/A. solani*. PCR with primers ALP and ITS4 (from a conserved region of the rDNA) amplified a c. 536 bp fragment from isolates of *A. linicola* and *A. solani* but not from other *Alternaria* spp. nor from other fungi which may be associated with flax or linseed. These primers amplified an identical fragment, confirmed by Southern hybridization, from DNA released from infected linseed seed and leaf tissues. These primers have the potential to be used also for the detection of *A. solani* in host tissues (McKay et al., 1999).

Predisposing factors

The temperatures from 10°C to 25°C are favoured by the fungus. It is found that the temperatures ranging from 10°C to 20°C are suitable for conidial production, with increased production at higher temperature. Conidial germination of the fungus is favoured by temperatures between 10°C and 25°C. Penetration of the leaf tissues starts 12 h after inoculation at 15°C. Light (Sun light) favous sporulation of *A. linicola*. The greatest numbers of conidia are produced under continuous leaf wetness and alternating dark/light periods (12 h each). Conidia are mainly dispersed by the wind (air-borne conidia) and their dispersal follow seasonal and diurnal periodicities, which are influenced by the weather conditions and the incidence of the disease in the crop at first in dry day following periods in rain (Irene, 1994).

Disease cycle

Alternaria linicola is a seed-borne pathogen which can survive for long periods, up to 5 years (Mercer *et al.*, 1991a). Although seed is considered to be the main source of primary inoculum, the possibility that the pathogen survives on infected plant debris, volunteer plants or alternate weed hosts. The pathogen is carried in the seed coat as resting hyphae (Mercer *et al.*, 1991a). When the infected seed takes up water on sowing, the hyphae are activated and later, depending on the environmental conditions, symptoms appear on the cotyledons and the lower leaves. It seems that the upper parts of the plants grow free of symptoms for most of the growing season. When the crop matures and the capsules begin to change colour, they may become infected, especially if the period between flowering and harvest is wet (Mercer *et al.*, 1991a). Conidia do not appear to be an important means of dissemination of *A. linicola* between seasons (Mercer and Hardwick, 1991; Mercer, 1992). However, within the season, conidia produced on volunteer plants or weed hosts or at early season on flax host may significantly contribute in disease occurrence and in secondary infection. Conidia are mainly dispersed by the wind (air-borne conidia) and that their dispersal follows a seasonal periodicity related to rainfall (Fitt and Vloutoglou, 1992).

Management/Control measures

Cultural control

Crop rotation may likely be an effective method for decreasing the amounts of primary inoculums of this fungus for the infection of flax crops (Irene, 1994).

Seed Treatment

The most effective means of control is the use of seed treatment which can effectively control the pathogen even at high levels of infection (Mercer and Hardwick, 1991). The most effective seed treating chemical for the control of *A. linicola* appears to be prochloraz. However, iprodione has also been occasionally used. In the UK iprodione has been replaced by prochloraz since 1986, as there have been indications of an increase in the proportion of iprodione-resistant strains of *A. linicola* (Mercer *et al.*, 1988). Although prochloraz, as a seed treatment, is highly effective against most of the seed-borne pathogens of flax, it is considered to have a fungistatic rather than a fungitoxic effect on *A. linicola*, as the fungus can be isolated from seedlings grown from prochloraz-treated seed (Mercer *et al.*, 1989).

Biological agents used as seed treatments have been reported to be alternatives to fungicides for control of other *Alternaria* species, although they can control superficial but not internal infections (Vannacci and Harman, 1987).

Biocontrol

Some control of *A. linicola* with sprays of spore suspensions from *Trichoderma* spp and *Epiccocum nigrum* may be achieved (Mercer and Hardwick, 1991; Mercer *et al.*, 1992a). In some investigations (Mercer *et al.*, 1991b; 1992a) it was found that control of *A. linicola* with sprays of spore suspensions of *Trichoderma viride* and *Epicoccum nigrum*; with some isolates the level of control was equivalent to that achieved by prochloraz sprays although less than that achieved by iprodione sprays.

Fungicidal spray

The use of fungicidal sprays has given mixed results. In Northern Ireland, Hardwick and Mercer (1989) and Mercer *et al.* (1990) found little evidence of control with single fungicide sprays. Multiple, uneconomic sprays resulted in some reduction of levels of *A. linicola* (Mercer and Hardwick, 1991). In this contrast, Fitt and Ferguson (1990) found greater level of control in the south east of England in 1988 and 1989. In general, sprays appear to be more effective in drier production areas. The chemicals, prochloraz and iprodione, are generally used for spray. Between the two, the latter one is more effective with a problem of developing fungicidal resistance in *A. linicola*.

ii) Seedling blight (*Thanatephorus cucumeris*)

Occurrence and severity status

Seedling blight, also known as 'Seedling blight and root rot', caused by *Thanatephorus cucumeris* is the principal causal agent and can be particularly destructive in soils that are loose, warm and moist. The fungus, *T. cucumeris* (= *Rhizoctonia solani*), is a plant pathogenic fungus with a wide host range and worldwide distribution. It frequently exists in its vegetative form in nearly all agricultural soils. In this non-spore-producing phase, the fungus lives saprophytically on dead plant remains, but it can become vigorously parasitic when roots or other parts of a susceptible host penetrate the infested zone (Watkins 1981). The *Thanatephorus* seedling blight is commonly present in most flax-growing areas. Severe epidemics reduce the stand and result in heavy yield losses. In spite of seed treatment, seedling blight and root rot can develop, and cause reductions in yield. Yellow-seeded varieties are more affected by blight and root rot than brown-seeded varieties due perhaps to the thin seed coat (Vest and Comstock, 1968; Groth *et al.*, 1970). Different strains of *T. cucumeris* attack flax. Seed coats of yellow-seeded flax varieties are more prone to cracking and splits, which renders them more susceptible to infections causing seedling blight and root rot than brown-seeded varieties (Flax Council of Canada, ND2).

Symptoms

The pathogen attacks flax at an early stage of development, destroying the root and causing thinning or, in severe infection, death of seedlings by blighting up (Krylova 1981). The blighted seedlings initially show red to brown lesions on the roots below the soil surface. Later, the lesions turn dark and the roots become shrivelled. While the above ground symptoms appear as yellowing of seedlings that later shrivel, wilt and die. The infected seedlings may occur singly or in patches (Rashid, 2003b). Seedling blight may be inconspicuous and gaps in the row may be the principal sign of disease occurrence (Flax Council of Canada, ND2). The pathogen also causes root rot symptoms, which appear in plants after the flowering stage (Hartman 1996).

Causal organism: *Thanatephorus cucumeris* (A.B. Frank) Donk 1956
[Reinwardtia 3: 367 (1956)]

Synonyms

= *Botryobasidium solani* (Prill.& Delacr.) Donk 1931

= *Ceratobasidium solani* (Prill.& Delacr.) Pilát 1957

= *Corticium sasakii* (Shirai) H. Matsumoto 1934

= *Rhizoctonia solani* J.G. Kühn 1858

= *Thanatephorus corchori* C.C. Tu, Y.H. Cheng & Kimbr. 1977

= *Thanatephorus sasakii* (Shirai) C.C. Tu & Kimbr. 1978

[N.B. *Thanatephorus cucumeris* is most widely recognized as *Rhizoctonia solani* J.G. Kühn 1858 and the latter name is mostly used to describe this disease. It is the most occurring species of *Rhizoctonia* that originally described by Julius Gotthelf Kühn on potato in 1858 (Kuhn, 1858). The fungus is a member of basidiomycete, which does not produce any asexual spores (i.e. conidia) and occasionally produces sexual spores (basidiospores). In nature, it reproduces asexually and exists primarily as vegetative mycelium and/or sclerotium. At teleomorphic stage, the basidiospores are not enclosed in a fleshy fruiting body. The sexual fruiting structures and basidiospores were first observed and described in detail by Prillieux and Delacroiz in 1891. The sexual stage of this fungus has undergone several name changes since 1891 and is now known as *Thanatephorus cucumeris* (A.B. Frank) Donk 1956. In this book, the latest IMI, Kew accepted name of the fungus is used to describe the disease]

Morphological/ Morphophysiolgical characteristics

Anamorph: The vegetative mycelium is colourless when young but becomes brown coloured as it grows and matures. The hyphae are monilioid with variable, ellipsoid to subglobose compartments (Butler and Bracker, 1970). The hyphal system is monomitic. Hyphae are without clamps, basal ones hyaline or some with a light brown colour, thin or usually with walls thickened or more rarely distinctly thick-walled, 10-12 μm wide, other hyphae thin-walled and forming a very thin subhymenial tissue (Hjortstam et al., 1988). The hypha is partitioned into individual cells by a septum containing a dough-nut shaped pore. This septal pore allows for the movement of cytoplasm, mitochondria, and nuclei from cell to cell. The hyphae are often branched at a 90° angles and usually possess more than three nuclei per hyphal cell. The anatomy of the septal pore and the cellular nuclear number (CNN) are used extensively by researchers to differentiate *R. solani* from other *Rhizoctonia* fungi. *R. solani* is characterized by: CNN close to the tips in young hyphae is greater than two, main runner hyphae usually wider than 7mm, mycelium buff-colored to dark brown, sclerotia (if present) irregular shaped, light to dark brown, not differentiated into rind and medula (Parmeter, 1970).

Teleomorph: Basidiome is effused, thin, hypochnoid, smooth, ochraceous. Hymenium comprises one or more layers of basidia on vertically branching, cymose, thin-walled hyphae arising from a subicular layer of wider, thick-walled, basal hyphae. Hyphae are multinucleate, subhymenial hyphae thin-walled, hyaline, with short hyphal compartments, somewhat swollen, 5-9 μm diam. Basal hyphae are ochraceous to brown, with long hyphal compartments, straight, 6-13 μm diam., with walls up to 2.5 μm thick, often double-laminate. Clamp connexions are absent. Septal pores are dolipores with discontinuous parenthesomes (Andersen, 1996). Basidia are ellipsoid to oblong or cylindrical (12-23 x 8-11 μm. Sterigmata are (2-) 4 μm, occasionally producing subsidiary sterigmata and then appearing furcate. Basidiospores are ellipsoid to oblong (5.5-) 7-10 (-12) x (3.5-) 4-5.5 (-6.5) μm, hyaline to ochraceous, thin-to slightly thick-walled, producing secondary spores by replication (Roberts, 1999).

Molecular characteristics

Thanatephorus cucumeris is a species complex composed of divergent populations. The genetics of *T. cucumeris* is multifaceted and is currently not well understood (Lubeck, 2004). From hyphal anastomosis reaction of different isolates, the fungal isolates are divided into different anastomosis groups (AGs). However, currently, a growing number of other means for characterization and grouping of isolates are reported, in particular RELPs, isozymes, PCR fingerprinting, DNA/DNA hybridization and sequence analysis. In general, these methods have supported the AG concept and revealed that the AGs consist of distinct phylogenetic entities. The methods have also revealed that some of the AGs is further genetically isolated in subgroups. Recently, a universally primed PCR (UP-PCR) cross hybridization assay has been developed for rapid identification of isolates into correct AG subgroup. The assay is based on the fact that UP-PCR products of isolates within AG subgroups cross hybridize strongly; where as there is little or no cross hybridization of UP-PCR products from isolates belonging to different AG subgroups. The UP-PCR product cross hybridization assay represents a micro-array technique and has potential for routine diagonistics of *T. cucumeris*. At present, about 400 DNA sequences derived from *T. cucumeris* are present in GeneBank. However, most of them are sequences of the nuclear encoded rRNA genes, which are used for taxonomic purposes (Lubeck, 2004). In a separate study, *T. cucumeris* of AG4 group, which affects flax and several other crops severely, was characterized. In the subgrouping of AG4 isolates, PCR-RFLP patterns in the rDNA-ITS were used. After obtaining the genomic DNA belonging to *T. cucumeris* AG4, an approximately 700bp amplification product of ITS1-5.8S-ITS2 region was obtained with PCR, using ITS1 and ITS4 universal primers (Melike and Ibrahim, 2010).

Predisposing factors

The disease favours by warm soils (20-25°C) and high soil moisture (50-75%).

Disease cycle

The fungus can survive for many years by producing small irregular-shaped, brown to black structures (called sclerotia) in soil and on plant tissue. The fungus also survives as mycelium by colonizing soil organic matter as a saprophyte, particularly as a result of plant pathogenic activity. Sclerotia and/or mycelium present in soil and/or on plant tissue germinate to produce vegetative threads (hyphae) of the fungus that can attack the flax plants. After attachment with the host surface, the fungus continues to grow on the external surface of the plant and will causes disease by producing a specialized infection structure (either an appresorium or infection cushion) that penetrates the plant cell and releases nutrients for continued fungal growth and development. As the fungus kills the plant cells, the hyphae continue to grow and colonize dead tissue, often forming sclerotia. New inoculum is produced on or in host tissue, and a new cycle is repeated when new substrates become available.

Management/Control measures

Use of host resistance

Pathogenicity of the causal fungal isolates to flax hypocotyls is not host specific and is controlled by several dominant factors. Further, separate genetic systems in the fungus control its ability to cause seed rot and hypocotyl infection (Anderson and Stretton 1978). There are various reports on the differences in susceptibility among flax genotypes to *T. cucumeris*, but flax cultivars with resistance or immunity to *T. cucumeris* are not yet known (Omran *et al.* 1968, Anderson, 1977, Islam 1992, Bos & Parlevliet 1995), except the variety Linott, which is reported as resistant to *Thanatephorus* blight (Anderson, 1973). Yellow-seeded varieties are more prone to cracking, which renders them more susceptible to seedling blight and root rot than brown-seeded varieties (Hartman 1996). Differences in susceptibility of flax cultivars to *T. cucumeris* and differences in aggressiveness among fungal isolates have been demonstrated in greenhouse experiments with 40 varieties of flax, while inoculated with the most virulent AG-4 isolates. The result shows that three fibre and four linseed varieties are with some degree of resistance (Anderson 1977). Kangatharalingam (1987) has found significant cultivar x isolate interaction in each of two greenhouse experiments, while, conducted to study the aggressiveness of *T. cucumeris* isolates on flax cultivars. These also imply that the resistance of the tested cultivars is only horizontal (polygenic), and there are significant differences among the cultivars in this type of resistance (Omran, *et al*, 1968; Vest and Comstock, 1968; Aly *et al.*, 2013). Similarly, pathogenicity of the tested isolates is only aggressiveness, and the isolates significantly differ in this type of pathogenicity (Aly *et al.*, 2013).

Cultural control

Seedling blight and root rot can be controlled by a combination of farm practices. Use of certified seed of a recommended variety and reduction of cracking of seed by adjusting combine settings during harvest are essential to manage the disease primarily. Use of fungicide treated seed to kill the seed borne pathogen and to check the early infection is always beneficial. It is also suggested to practice a crop rotation of at least three years between flax crops and to grow the crop in a field that is distant from fields sown to flax in the previous year. It is also needed to avoid legumes and sugar beets in the rotation. Use of recommended fertilizer and seeding practices to promote vigorous stands are essential to develop inbuilt resistance in the host (Flax Council of Canada, ND2).

Biocontrol

The studies on the antagonism between flax rhizobacteria and *T. cucumeris*, the causal of flax seedling blight indicate that *Pseudomonas flurorescens*, *P. cepacia* and *Pseudomonas* sp. are the most effective antagonists of the pathogen. Linear growth of *T. cucumeris* is inversely proportional to the concentration of the culture filtrates of Pseudomonas strains. Individual strains as well as their mixtures significantly increase the percentage of surviving seedlings. However, the mixtures are much more effective than the individual strains. Moreover, the mixtures are more effective in increasing the seed and straw yields in field trials (Ashour and Afify, 2000).

8. Stem break

Occurrence & severity status

Stem break disease, also known as 'Stem break and browning', is caused by a seed- and soil-borne fungus. The disease was first recorded in Ireland in 1921 (Sanderson, 1965). Now, it occurs in Europe, North America, Africa, and Asia. Under heavy infestations in Europe, fibre flax can suffer a considerable loss in yield and fibre quality (Muskett and Colhoun, 1947). This disease is of minor importance in western Canada, although, some damage has been reported in the parkland regions of Saskatchewan and Alberta (Henry, 1934; Henry and Ellis, 1971). It also occurs on wild *Linum* spp. (Conners, 1967). The pathogen of this disease is well known as a naturally occurring epiphyte or endophyte of a wide range of plant species (e.g. apple, grape, cucumber, green beans, cabbage) without causing any symptoms of disease.

Symptoms

The disease appears on mature stems, leaves and sepals as diffuse and irregular light brown lesions. In case of early infection which occurs by the pathogen from seed, it causes severe canker at the stem just above ground level, resulting stem break. The stem break is the first conspicuous disease symptom. Development of a canker at the stem base weakens the plant. The stem may break at this point when the plants are still young, or at a later stage (Flax Council of Canada, ND2). Initially, dark coloured circular lesions are produced on cotyledons and lower stem parts originating from infected seeds during seedling emergence. These lesions develop slowly and may result in seedling blight or stem break. Cankers on the stem base weaken the plant, and the stem may break early at the seedling stage or later on during the season depending on the severity of infection (Rashid, 2003b). Stem break is a distinguishing sign of this disease. Plants may remain alive after stem-breakage, but produce few seeds which will be lost at harvest. The browning phase of this disease starts late after the flowering initiated by infections on the upper part of the plant. The lesions on the branches, leaves and bolls are oval or elongated in shape, dark brown in colour, and surrounded by narrow purplish margins. These lesions often coalesce giving the leaves and stem a brownish appearance. Patches of heavily infected plants appear brown (Rashid, 2003b). In the late infection, the disease causes leaf fall and early maturity. Sepal infection leads to the seed borne carry over the disease. Conidia are produced freely on the surface of diseased tissue forming tiny creamy white pustules of large numbers of both radulospores and bud spores (Sunderson, 1965).

Causal organism: *Kabatiella lini* **(Laff.) Karak. 1957**
[Fungi imperfecti Parasitici: Pars II. Melanconiales, : 1-680 (209), 1950; Corus, P.W. *et al.* (2018). Dothideomycetes, Oct 2017, Kew Mycology]

Synonyms

= *Polyspora lini* Laff. (1921)

= *Pullularia pullulans* var. *lini* (Laff.) N.H. White (1945)

= *Microstroma lini* (Laff.) Krenner (1954)

= *Aureobasidium pullulans* var. *lini* (Laff.) W.B. Cooke (1962)

= *Aureobasidium lini* (Laff.) Herm.-Nijh. (1977)

Morphological/ Morphophysiolgical characteristics

Anamorph: On flax, mostly conidial state of the pathogen occurs. The conidia are smooth, hyaline, thin walled, borne as radulospores on stalks of 1-4μ long, formed on any part of the superficial mycelia, 9-15 x 3-5μ. Secondary conidia are produced as yeast-like buds from primary conidia, globose or oval, hyaline, 7-9μ, thin walled. Chlamydospores are globose, darkly pigmented, with thick walls up to 10μ wide, formed within host tissue (Sunderson, 1965).

Teleomorph: Pseudothecium (ascocarp) forms in single or in groups of two or three, usually borne on a poorly developed stroma which is formed in host tissue immediately below the epidermis, subglobose, sometimes laterally flattened, 75-135 x 75-100μ, dark brown and black. Pseudothecial wall is about 15-20μ thick, composed of 2-4 layers pseudoparenchymatous cells. There is no ostiole. Asci are bitunicate, clavate, eight spored, arising from basal tissue or pseudothecia, 60-65 x 10-12μ. Ascospores are biseriately or irregularly arranged in ascus, nonseptate, hyaline, obovate or elliptical, thin walled, smooth, 12-16 x 4-5μ in size. Ascospores are liberated from the ascocarp by the rupturing of the wall as a thin tagential irregular crack. (Sunderson, 1965).

Colonies grown on potato dextrose agar are white and yeast-like in appearance, soon becoming black with the production of radiating, frequently branching mycelium. Concentric rings develop at the centre of the older colonies, while the outer edge soon becomes very irregular owing to persistent sectoring. The surface may become grossly convoluted, matt, or shiny, with only the occasional production of tufts of short white aerial mycelium (Sanderson, 1965). Colonies grown on malt agar at 24°C attaining a diameter of about 24 mm in 7 days. Colonies are velvety with smooth margin, whitish to grey. Hyphae are hyaline, smooth, thin walled, up to 16 μm wide. In older culture dark brown, thick walled hyphae may occur. Conidiogenous cells are undifferentiated, internally and terminal forming blastic conidia, simultaneously in scattered, usually dense groups. Conidia are one celled, hyaline, smooth, ellipsoidal, rather variable in size and shape, 11-16 x 4-5.5 µm. Secondary conidia are sometimes formed, being smaller than the primary conidia. Endoconidia occasionally are present, ellipsoida, 8-13.5 x 3.5-4.5 µm (Hoog and de Hermanides-Nijhof, 1977).

[N.B. The fungus is very similar to *Aureobasidium pullulans* in morphology, but, differs by having larger conidia, by chlamydospores chains with only inconspicuous constrictions and by slightly larger endoconidia (Hoog and de Hermanides-Nijhof, 1977)]

Molecular characteristics

The molecular sequences of ITS 1 and 2 of the nuclear rDNA gene of *Kabatiella lini* and allied fungi have been determined. The molecular diversity of ITS2 of the different strains is larger than that of ITS 1 (Yurlova *et al.*, 1999). In other investigation (Loncaric *et al.*, 2009), while studied the intra-specific diversity of different strains of *K. lini* and their relatives, the sodium dodecyl sulfate-polyacrylamide gel electrophoresis technique (SDS-PAGE) of whole-cell proteins as well as enterobacterial repetitive intergenic consensus (ERIC)-, repetitive extragenic palindromic (REP)- and BOX-PCR techniques (collectively known as rep-PCR) were used. The result indicated that Rep-PCR was an efficient procedure for discrimination of strains in terms of simplicity and rapidity. However, when the RFLP-PCR technique was applied for the identification of isolates and distinction from related species, that technique was insufficient for investigation of intra-specific diversity. Only, by using 18S rDNA gene sequence analysis of selected isolates, three strains of genera *Aureobasidium* and *Kabatiella* were identified (Loncaric *et al.*, 2009). The sequences of different genes of *K. lini* strain CBS 125.21 were determined by Zalar *et al.* (2008) as follows. For example, the partial sequence (GenBank: FJ150946.1) of 28S ribosomal RNA gene was 587 bp linear DNA, while, partial cds (GenBank: FJ039840.1) of putative fatty acid elongase ELO1 (ELO) gene was 700 bp linear DNA; partial sequence (GenBank: FJ150897.1) of internal transcribed spacer 2 was 486 bp linear DNA; partial sequence (GenBank: FJ157908.1) of EF1 gene was 267 bp linear DNA; partial sequence (GenBank: FJ157873.1) of 21 beta-tubulin (TUB) gene was 400 bp linear DNA and the partial sequence (GenBank: EU707925.1) of 18S ribosomal RNA gene was 2,977 bp linear DNA (Loncaric *et al.*, 2009).

Predisposing factors

Contaminated seed is one of the major sources of the disease. It is found that using naturally contaminated seed the incidence of the seedling blight, stem-break and browning phases of attack by the fungus is more. Heavy attacks of the disease do not occur when the number of contaminated seeds is less than 5% (Colhoun, 1946). Development of stem-break is, however, greatly influenced by conditions during growth of the plants, the highest levels of attack being associated with high air humidity (Colhoun, 1959). The development of the disease is favoured by warm and wet conditions in a thick crop canopy (Rashid, 2003b).

Disease cycle

The pathogen is primarily seedborne, overwintering as conidia and mycelium in or on the seed coat, or in plant debris (Muskett and Colhoun, 1947; Nyvall, 1989). The primary infections may start during seedling emergence, or from spores produced on diseased stubble that are spread by wind and rain to infect emerging seedlings (Muskett and Colhoun, 1947). The fungus can also survive and reproduce on secondary hosts and weeds (MacNish, 1963). The fungus produces masses of unicellular spores freely on the surface of infected tissue (Rashid, 2003b). These spores can reinfect and complete lifecycles in repeated manner before overwintering.

Physiological races

Differences in pathogenicity among the isolates of this causal fungus based on resistance and/or levels of infection on a set of host differentials were reported (Henry, 1934; Colhoun, 1960).

Management/Control measures

Cultural control

Use of disease-free seed produced by healthy plants is the most important measure to control the disease. Crop rotation of two years between flax crops would reduce spread of infection from diseased stubble. Currently recommended varieties may differ in susceptibility to stem break and browning (Rashid, 2003b).

Chemical control

Seed treatment: Fungicidal seed treatment controls surface-borne inoculum, but is unlikely to be effective when the fungus has penetrated coat-deep into the seed (Muskett and Colhoun, 1947; Turner, 1987). Rotating crops and planting flax in a field distant from that of the previous year reduces spread of infection from diseased stubble (Flax Council of Canada, ND2).

Fungicidal spray: In the conditions of the Republic of Belarus, Derozal, CS (carbendazim, 500 g/l) while applied twice in dose 1.0 l/ha at 5-10 cm stage + 1.0 l/ha at the phase of flower-bud formation gave better yield by managing the disease along with other diseases (Belov and Prudnikov, 2011).

9. Wilt

Occurrence & severity status

Flax wilt [Fig. 26 (A-B)], also known as 'Fusarium Wilt of flax', is one of the most widespread diseases affecting flax wherever it has been grown on the same land for a lengthy period of time. The wilt disease has been associated with flax since early cultivation. It was first noted by Pliny in the first century, as scorching of the ground where flax was grown for successive seasons, leading to deterioration of soil (Boyle, 1934). Early work on fusarium wilt of flax was conducted by Otto Lugger in 1889, but Hiratsuka (1897) and Bolley (1901) were the first to identify the causal organism to be a *Fusarium* species. Fusarium wilt has been identified as one of the main diseases affecting the plant growth and yield of flax in Canada. It can occur in flax at any growth stage and may result in 100% disease incidence in certain cultivars (Kommedahl *et al.*, 1970). Flax wilt has been found as the most destructive in the United States and Japan and less so in Argentina and Canada. In most of Europe, where long term rotations have been practised for centuries, the disease has not become a major problem. In other parts of the world, such as the UK where soil temperatures are relatively low during the early growth period, flax wilt has never been destructive (Rawlinson and Dover, 1986). In the former Soviet Union, Krylova and Voronova (1981) indicated that 66% of the crops were affected by wilt annually. Fusarium wilt was observed in 86-93% of flax

Fig. 26 (A-B). Wilt in flax caused by *Fusarium oxysporum* f.sp. *lini*. **A.** Early stage of infection showing yellowing and wilting of leaves. **B.** Wilted flax plants.

crops surveyed in Manitoba and eastern Saskatchewan, Canada, in 1995 (Rashid *et al.*, 1998a, 2000). The amount of damage may vary from a very little when only a few plants are attacked, to catastrophic with almost the entire crop killed. The severe epidemics can result in 80–100 percent yield reduction (Kommedahl *et al.*, 1970; Sharma and Mathur, 1971; Kroes *et al.*, 1999). In North America, flax wilt became a destructive disease by 1890 (Lugger, 1890). The development of resistant cultivars worldwide has reduced the negative impact of this disease. In North America, all commercial flax cultivars are characterized by resistance or moderate resistance to fusarium wilt (Kenaschuk and Rashid, 1993; Kenaschuk *et al.*, 1996). Resistance has also been reported in European cultivars of oil and fibre flax (Kroes *et al.*, 1998c; Kroes *et al.*, 1999), and from China (Liu *et al.*, 1993).

Symptoms

The disease appears on the plant at any growth stage, with the seedling stage as the most susceptible one (Kommedahl *et al.*, 1970). Infection occurs through the roots and the pathogen colonizes the xylem, ultimately obstructing water movement, resulting in wilting. Early infections may kill flax seedlings shortly after emergence or seedlings may die even before emergence, while delayed infections cause yellowing and wilting of leaves, followed by browning and death of plants. The tops of wilted plants often turn downward, forming a "shepherd's crook". Later, wilt can also result in premature ripening and reduced number of seeds per boll or plant, leading to yield loss. At the base of the stem of infected plants, white mycelial growth can be observed, especially under humid conditions. Roots of dead plants turn ashy grey. Affected plants occur more commonly in patches but may also be scattered throughout the field (Muskett and Colhoun, 1947; Kommedahl *et al.*, 1970; Rashid and Kenaschuk, 1993; Flax Council of Canada, ND2).

Causal organism: *Fusarium oxysporum* f. sp. *lini* W.C. Snyder & H.N. Hansen, (1940)

[American Journal of Botany 27: 66]

Synonyms

= *Fusarium lini* Bolley, (1901)

= *Fusarium oxysporum* f. *lini* (Bolley) W.C. Snyder & H.N. Hansen, (1940)

Morphological/ Morphophysiolgical characteristics

The fungus produces white to white with pinkish tinch or white to pale violet or violet white or white with brown tinch like mycelial growth on artificial medium.

Mycelium is sparse or abundant producing small pale brown, blue or violet sclerotia, sometimes abundantly. Mycelium produces terminal or intercalary, smooth or rough walled "chlamydospores" abundantly and quickly, usually singly, in pairs, in clusters or in short chains, 5–10 μm in diameter. Macroconidia sparse in some strains, but usually abundant in sporodochia, straight to slightly curved, sickle shaped, hyaline, thin walled, (25-) 27- 40 (-60) × 3-5 μm (differently 16.90-31.60 × 3.89-5.53 μm; Ashwathi *et al*., 2017), usually (2-) 3- to 5-septate, 3-septate being most common, each with a tapering and curved apical cell and foot shaped basal cell. Microconidia are abundant, oval shaped to elliptical, straight to curved, hyaline, usually 0-septate, small, (5-) 8.70- 12 (-14.05) × (2-) 3.30-3.5 (-5.0) μm in size, produced in false heads on short monophialidic conidiogenous cells (Kommedahl *et al.*, 1970; Brayford, 1996; Ashwathi *et al*., 2017).

This species usually produces pale violet or magenta pigmentation on Potato Dextrose Agar (PDA) medium. It grows on Glycerol Nitrate Agar (G25N) with white or pale yellow aerial mycelium and yellow pigmentation. It grows on mannitol sucrose medium as white or reddish white or red aerial mycelium and with red or brownish red or grayish red pigmentation. Growth on Czapek-Dox Iprodione Dichloran Agar (CZID) is as white or pinkish white or pink aerial mycelium and brownish grey or purplish grey pigmentation. The species produces urease and phosphatase enzymes, but does not produce acid on creatine sucrose agar and acetylmethylcarbinol compound. Different isolates vary in production of peroxidase and pyrocatechol oxidase enzymes.

Molecular characteristics

The genetic diversity in *F. oxysporum* f .sp. *lini* from three natural populations (from Hebei, Ningxia and Inner Mongolia) was investigated by inter simple sequence repeat (ISSR) by Liu and Yuan (2011). In their studies, the unweighted pair-group mean analysis (UPGMA) indicated that clustering of *F. oxysporum* f .sp. *lini* individuals mainly relates to their populations and geographic distances separating those populations. Genetic differentiation was also indicated by ISSR analysis. Yuan *et al*. (2013), while studying the genetic diversity in *F. oxysporum* f .sp. *lini* from six provinces in China with the use of molecular markers, inter simple sequence repeat (ISSR), observed that the number of bands amplified by each of 12 ISSR primers ranged from 43 to 142, with sizes ranging from 250 to 4,500 bp. Most of the bands were found polymorphic (99.62%). The percentage of polymorphic loci varied from 17.25% in Gansu and Inner Mongolia to 33.75% in Sinkiang. Nei's gene diversity index (h) ranged from 0.0428 in Gansu to 0.0666 in Sinkiang, and Shannon's information index (I) ranged from 0.0675 in Gansu to 0.1117 in Sinkiang. The genetic identity using the Nei's genetic identity varied from 0.9643 between the populations from Hebei and

Gansu to 0.9844 between the populations from Sinkiang and Shanxi. Unweighted pair group mean analysis (UPGMA) cluster analysis, as indicated by the Nei's genetic distance, showed the distances ranging from 0.0158 between the populations from Sinkiang and Shanxi to 0.0364 between the populations from Hebei and Gansu. The six populations were clustered into three subgroups. The Gansu population was clustered into one subgroup, the same as the Inner Mongolia population. The four other populations were clustered into the third subgroup. The Nei's GST (0.2972) and gene flow among populations (Nm =1.1825) revealed large gene exchanges among populations (Yuan *et al.*, 2013)

Predisposing factors

Warm weather favours the disease. The disease severity is correlated with temperature (Wilson, 1946). The environment, particularly temperature, is the most important factor in flax wilt disease development. Even the most resistant cultivars wilt at high temperature, while even the most susceptible ones do not wilt at low temperature (Sherbakoff, 1949). The optimum temperature for the pathogen to grow in the soil is considered to be 24-28°C, although it can survive over a wide range, 14-38°C (Muskett and Colhoun, 1947; Saharan *et al.*, 2005).

The incidence of wilt increases at low soil moisture (dry soil), which enhances the effect of the pathogen (Kommedahl *et al.*, 1970). Further, it is found that disease incidence in peat soil is low and the addition of nitrogen, potash or phosphate reduces disease incidence in susceptible plants (Kommedahl *et al.*, 1970). Sandy soils promote flax wilt, while acidic pH and availability of Ca, Mg and Fe iron to the pathogen all provide a conducive environment (Hoper *et al.*, 1995).

Disease cycle

The fungus persists in soil, while the mycelia and spores survive for many years in flax debris and other organic tissue. More importantly, the chlamydospores have a high germination rate and can survive for about 50 years in a field without a host (Houston and Knowles, 1949). Spores are dispersed by wind and rain from one field to another (Flax Council of Canada, ND2). The fungus invades the plants through roots. Infection takes place at three sites: 1) near the root apex, at the site of cell differentiation, 2) at the cell elongation zone, and 3) through root hairs and the epidermis in the older part of the root (Turlier *et al.*, 1994). After infection, it continues growth inside the water-conducting tissue. The fungus interferes with water uptake. The fungus is a hemibiotroph, initially survives on live hosts and eventually on the dead infected cells (Ma *et al.*, 2013). The mycelia of this fungus are observed to grow within and between undifferentiated cells (intercellular and intracellular) in the roots such as the apical cells in root tips. The pathogen may grows intercellularly through the cortex, then colonizing cortical cells and developing radially through the endodermis to reach the xylem; or it may enters through the root apex into undifferentiated xylem precursor cells and then into the xylem (Turlier *et al*., 1994; Kroes *et al*., 1997). This is followed by cell death or production of multiple cell layers (Kroes *et al*., 1997). The root tip is colonized and the cortical region is degraded within 8-16 days after infection. Roots become hollowed out after severe colonization the pathogen. Spores produced at the infected region serve as inocula for further infection of the next season crops.

Physiological races

The pathogenic variability in *F. oxysporum lini* (Fol) was first reported in 1926 (Broadfoot and Stakman, 1926). This fungus comprises an indefinite number of biotypes with differences in cultural characteristics and pathogenic types (Armstrong and Armstrong, 1968; Kommedahl *et al*., 1970). Isolates of this fungus have been identified to vary in: 1) morphology, with respect to the amount and type of sporulation, production of different types of conidia, size and number of septa and pigment production on growth media; 2) physiology, rate and type of growth on substrates and in host; 3) environmental preferences, antibiotic capabilities; and 4) pathogenicity. Therefore, it is considered an ideal pathogen to demonstrate diversity within a species with numerous biotypes and pathotypes (Kommedahl *et al*., 1970; Saharan, *et al*.,2005). Different pathogenic races (pathotypes) have been reported from different regions of the world: Argentina, United States, Canada, Australia, India, and Japan (Borlaug, 1945; Millikan, 1945; Millikan, 1948; Houston and Knowles, 1949; Tochinai and Takee, 1950; Sharma and Mathur, 1971; Kroes *et al*., 1999; Kroes *et al*., 2002). High pathogenic variability has also been reported in the population of this fungus in western Canada (Rashid and Kroes, 1999; Mpofu and Rashid, 2000a; 2000b). In India, ten races of the pathogen were identified based on the pathogenicity on eight differents flax cultivars (Saharan *et al*., 2005). Vegetative compatibility determines the ability of different strains of a fungus to form heterokaryons, from which the resulting strain might differ in pathogenicity and host range compared with the original individuals (Leslie, 1993). Mpofu and Rashid (2001) examined genetic variation of this fungus using nitrate non-utilizing mutants. They reported 12 vegetative compatibility groups (VCGs) among 74 isolates, while 22 were not assigned to any VCG, which suggested that there is a considerable amount of genetic diversity within this fungal species, giving rise to minor differences in pathogenicity. It is also reported that Fol has multiple independent origins, suggesting the existence of multiple races and VCGs supporting this theory (Baayen *et al*., 2000).

Host resistance & its molecular basis

Resistance to flax wilt is considered to be polygenic, because of the failure to observe any major resistance genes and the lack of evidence supporting race specific or vertical (single gene) resistance (Kommedahl *et al*, 1970). Knowles and Houston (1955) reported that two major genes conferred wilt resistance in flax. These genes, FuA and FuB, were identified using phenotypic ratios with two specific clones (isolates) of the pathogen. In a later study, the high wilt resistance of Dakota 48-94 to a third clone, was attributed to another gene FuC, inherited independently of the other two, proposing three major genes along with several minor genes that confer wilt resistance in flax (Knowles *et al.*, 1956). Flax wilt resistance was proposed to be due to major genes by Spielmeyer *et al.* (1998b). In their investigation, a significant number of recombinant doubled haploid population was derived from the haploid component of polyembryonic F_2 seeds originating from a cross between a wilt resistant, twinning Linola™ line CRZY8/RA91 (Linola is a registered trademark of CSIRO Plant Industry, Canberra, Australia) and the wilt susceptible Australian flax cultivar Glenelg. The segregation of resistance was studied in 143 doubled haploid lines under glasshouse and field conditions. Most of the phenotypic variation was attributable to the segregation of two independent genes with additive effects. Minor resistance genes may have also contributed by modifying the resistance response (Spielmeyer *et al.*, 1998b). In recent studies, it is suggested that the differences in plant resistance can be due to the fact that the identified genes in flax pre-constitute closely-linked units, which provide the effect of a monogenic control (Rozhmina , 2015). However, some genes of these oligogenes may be influenced by the temperature leading to a decrease in expression like in the Fu 7 gene. Thus, an increased aggressiveness of some races of the pathogen and a decreased expression of some plant R-genes at air temperature above 25 °C have been found. Gene Fu 7 which controls *Fusarium* wilt resistance in the line Siciliana 285 was mostly temperature influenced as at 26-28 °C the line became susceptible and moderately resistant to some isolates (Rozhmina and Loshakova, 2016). Although, the effectiveness of Fu 7 gene was significantly influenced by the raised temperatures, the effect of R-genes in the k-5657 (Minnesota, the USA) did not depend on the temperature (Rozhmina and Loshakova, 2016).

The resistance in flax varies, depending on the interactions between specific cultivars and isolates of the pathogen. This interaction has a strong molecular basis. To understand the genomic information on how the plant responds to attempted infection, the most resistant cultivar (CDC Bethune) was used for a full RNA-seq transcriptome study through a time course at 2, 4, 8, and 18 days post-inoculation. Several key genes, such as, an induced RPMI-induced protein kinase; transcription factors WRKY3, WRKY70, WRKY75, MYB113, and MYB108; the ethylene response factors ERF1 and ERF14; two genes involved in auxin/glucosinolate precursor synthesis (CYP79B2 and CYP79B3); the flavonoid-related enzymes chalcone synthase, dihydroflavonol reductase and multiple anthocyanidin synthases; and a peroxidase implicated in lignin formation (PRX52) have been identified (Galindo-González and Deyholos, 2016). In another study, the transcriptomes of two resistant and two susceptible flax cultivars with respect to *Fusarium* wilt, as well as two resistant BC2F5 populations, which were grown under control conditions or inoculated with *F. oxysporum*, were sequenced using the Illumina platform. Genes showing changes in expression under *F. oxysporum* infection were identified in both resistant and susceptible flax genotypes. It is observed that the predominant over expression of numerous genes are involved in defense response. This was more pronounced in resistant cultivars. In susceptible cultivars, significant down regulation of genes involved in cell wall organization or biogenesis was observed in response to *F. oxysporum*. In the resistant genotypes, upregulation of genes related to NAD(P)H oxidase activity was detected. Upregulation of a number of genes, including that encoding beta-1, 3-glucanase, was significantly greater in the cultivars and BC2F5 populations resistant to *Fusarium* wilt than in susceptible cultivars in response to *F. oxysporum* infection (Dmitriev *et al.*, 2017).

Management/Control measures

Use of host resistance

Use of host resistance is the most effective measure to control wilt in flax. Several resistant cultivars have been successfully identified, developed and released (Kommedahl *et al.*, 1970; Ondrej, 1977; Kenaschuk and Rashid, 1993; Kenaschuk *et al.*, 1996). In Canada, the fax cultivars, Bison and Aurore are moderately resistant, while, the cultivars, Novelty and Oliver are susceptible (Edirisinghe, 2016). In USA, Bison, was released in 1925 (Thompson and Zimmer, 1943). This cultivar has maintained resistance since release and has been a progenitor of many wilt resistant cultivars (Ausemus, 1943). In Russia, it was observed that the accessions from North and South America and from East Asia had above average wilt resistance, while accessions from Europe and the Indian subcontinent were below average (Diederichsen *et al.*, 2008). This suggests that resistant cultivars from one location becoming susceptible at another location (Kommedahl, *et al.*, 1970). Canadian cultivars AC McDuff, Dufferin, Noralta and AC Emerson are highly resistant to flax wilt and displayed no symptoms.

The resistant/moderately resistant varieties which are mentioned above may use, since; this is the most effective measure to control the flax wilt. The most popular cultivars grown, for example Ariane, Natasja, Marina, Saskia, Viking and Vimy, are resistant to wilt (Beaudoin, 1989b). Later releases flax varieties with varying degrees of resistance include Day and Prompt, moderately resistant varieties from South Dakota, USA (Grady and Lay, 1994b); Geria, from Romania (Popescu *et al.*, 1994); and AC Emerson a moderately resistant variety developed in Morden, Manitoba, Canada (Kenaschuk *et al.*, 1996). North American flax cultivars Bison, AC Linora, AC Emerson, Hanley, Lightning, Macbeth, Prairie Thunder and Shape are reported to exhibit a high level of resistance to flax wilt (Kommedahl *et al.*, 1970; The Western Committee on Plant Diseases, 2012). However, the resistance among varieties may differ due to the variability of pathogen races in different geographical regions with varying temperatures and environmental conditions (Kommedahl *et al.*, 1970; Rashid and Kenaschuk, 1993).

Cultural measure

Crop rotation is another important control measure for flax wilt control. The crop rotation of at least three years between flax crops helps to maintain low levels of inoculum in the soil (Flax Council of Canada, ND2). Avoiding early seeding in acidic soils is suggested to reduce disease development (Saharan *et al.*, 2005). Acidic, sandy soils provide a conducive environment for flax wilt, resulting in high disease severity. Increasing soil pH by liming (addition of $CaCO_3$) has been observed to reduce disease severity along with the addition of clay minerals montmorillonite and illite, which changes soil texture. Iron availability and sand content are negatively correlated with soil suppressiveness of the pathogen (Hoper *et al.*, 1995). Flooding of infected fields may also reduce pathogen inoculum and may reduce disease severity (Saharan *et al.*, 2005).

Fungicidal control

Seed treatment: Seed treatment with recommended fungicides may protect the crop from early infection at the seedling stage and helps in maintain good stands and seedling vigor (Rashid and Kenaschuk, 1996b; Flax Council of Canada, ND2). The fungicides which are most commonly used for seed treatment are Thirum and Carbendazim.

10. Other minor diseases

i) Damping-off

Occurrence & severity status

Damping-off is very common disease affecting flax worldwide (Millikan, 1951). The disease is more common under cold soil temperatures and in the fields where water logging delayed germination and seedling emergence, which facilitated prolonged period of exposure for the seedlings to be infectious (Rashid, 2003b).

Symptoms

The affected seedlings shows reddish brown spots on the cotyledons, stems and roots. Less severely affected seedlings may produce secondary roots and partially recover. However, most of the seedlings are killed before emergence or shortly after emergence. Microscopically, the presence of spherical oospores in the infected root tissue may be observed.

Causal agent: *Pythium* **spp**. These are soil-borne pathogens which can survive on soil organic matter for a long time.

Control

It is suggested to adopt seed treatment to protect the seedlings during the early growing stage when these are most susceptible to this disease. Further, avoidance of early seeding when the soils are wet and cold may help the crop to escape the disease in temperate countries (Rashid, 2003b).

ii) Dieback

Occurrence & severity status

This disease was reported to occur at Saskatchewan in Canada (Vanterpool, 1947), with minor importance at that time. Since then, the disease has not been reported to cause any major damage of flax in Canada or other flax-growing areas (Rashid, 2003b).

Symptoms

The disease appears on stems and branches of early maturing flax plants. The upper parts of the plants die prematurely. Numerous pycnidia are also formed on infected tissue.

Causal agent: *Selenophoma linicola* **Vanterp., 1947**. The fungus overwinters on flax stubble. There is no evidence of a seed-borne phase of this disease (Vanterpool, 1947). This type of fungus possesses mycelium as immersed, branched, septate, pale brown. Condiomata is pycnidial, separate or aggregated, immersed, globose, dark brown. Ostiole is absent, dehiscence by irregular rapture of upper wall. Conidiphores are absent. Conidiogenuous cells enteroblastic, philidic, hyaline to pale. Conidia are aseptate, falcate, fusiform, eguttulate, hyaline, later pale brown (Sutton, 1980b).

Control

It is suggested to follow an adequate crop rotation and proper sanitation to reduce the level of fungal inoculum in the soil (Rashid, 2003b).

iii) Gray mould

Occurrence & severity status

Gray mould is a common disease affecting a wide range of host plants. It reduces both yield and quality of seeds of the disease affected plants. This disease is capable of producing local epidemics and major yield losses in flax fields with heavy lodging or under stress from abiotic factors (Rashid, 2003b).

Symptoms

The disease at first appears in seedlings. It develops as brownish spots on the base of the stem of seedlings at the soil surface and damping-off. Symptoms on mature plants are light brown patches on the stem and that become soft and decayed, resulting in the death of the upper part of the plant.

Causal agent: *Botrytis cinerea* Pers. 1801. This fungus produces conidiophores as erect, subhyaline to brown, septate, branched towards the apex, branches lateral, alternate at a wide angle to the axis, successively developed from the base to the apex, branching again alternately, forming at the end a globose, swollen conidiogenous cell bearing simultaneous conidia on pedicels (Matsushima, 1975). The fungus can survive on decaying organic matter for a long time. It is seed borne in flax, and the seed-borne phase acts as a source of primary inoculum. This fungus produces small, hard black sclerotia which can survive adverse weather conditions (Rashid, 2003b).

Control

It is suggested to adopt seed treatment with fungicides to protect the seedlings from early infections. Further, the disease is difficult to control at the mature stage of the plant growth, since, the fungus is widespread in nature and has an abundance of inocula (Rashid, 2003b).

iv) Scorch

Occurrence & severity status

This disease mainly occurs in Europe and may be present in other parts of the world (Muskett and Colhoun, 1947; Wiersema, 1955). It affects plants in patches of various sizes in the first part of the growing season. The most severely affected plants in the centre of the patch will soon die, while the less affected plants at the periphery of the patch may partially recover and produce some seeds.

Symptoms

The disease symptoms appear both on above ground parts and on roots. The symptoms on the above ground plants reveal as stunted growth of infected plants along with short inter-nodes and shrivelled brown leaves. Initially, the leaves turn yellow and the yellowing starts from the lower leaves to upper ones. Later, the infected leaves turn brown with dry tips. The leaves of the tender shoots grow closer and that rapidly dry and wither. The infected roots appear as glossy and brittle.

Causal agent: *Globisporangium megalacanthum* **(de Bary) Uzuhashi, Tojo & Kakish, 2010** (= *Pythium megalacanthum* De Bary, 1881). This pathogen is a soil-borne fungus which infects the roots of young plants. The fungus produces sporangia which are globose, elongate or oval, terminal or intercalary, proliferating internally. Encysted zoospores are 18-20 µm in diam. Oogonia are globose, 36-45 µm diam, covered with conical, pointed protuberances; spines 6-9 µm long and 5-6 µm diam at the base, sometimes forming proliferations terminating in swellings, which can act as zoosporangia. Antheridia are 0-5 per oogonium, mono- or diclinous. Antheridial cells are 13.5-17.5 x 11.0-13.5 µm, applied broadly to the oogonium. Oospores are aplerotic, 27-31 µm diam. (Plaats-Niterink, 1981). This fungus is primarily a saprophyte which can survive in the soil for many years (Muskett and Colhoun, 1947; Wiersema, 1955).

Control

Late seeding may be helpful in reducing the incidence since the wet and cold spring conditions in temperate countries favour the disease development. Long term crop rotation may also help to reduce the inoculum in soil. Seed treatment is also suggested to protect the seedlings from early infections especially to cracked or damaged seed (Turner, 1987; Rashid, 2003b)

v) *Verticillium* blight

Occurrence & severity status

This disease has been reported to occur in different countries, such as, United Kingdom, Germany and the Netherlands of Europe (de Tempe, 1963a; Turner, 1987; Fitt *et al.*, 1991a) and USA (de Tempe, 1963a). It can cause severe epidemics under prolonged wet weather conditions when the crop approaches maturity (Fitt *et al.* 1991a). The severity of this disease is compounded when the nematode *Pratylenchus penetrans* is present in the soil (Coosemans, 1977).

Symptoms

The disease appears on stem as dark brown with purplish tinge lesions starting from the base and extending upward. All plant parts are affected and progressive desiccation results in the death of the plants. Dark black microsclerotia may form on the infected stems (Fitt *et al.*, 1991a).

Causal agents: *Verticillium dahliae* Kleb., 1913 and ***Verticillium albo-atrum* Reinke & Berthold, 1879**. These two fungi affect the flax plants and can survive in plant debris, in soil. They can spread as hyphae through the plant and also as spores. These fungi produce conidia on conidiophores and once the conidia are released in the xylem, they can quickly colonize the plant. The pathogen *V. dahliae* survives in plant debris by forming microsclerotia, while, *V. albo-atrum* survives as mycelium (Rashid, 2003b).

Control

It is suggested to use resistant cultivars if available. Crop rotation for long term and avoidance of any susceptible crops, such as, sunflowers and potatoes, in the rotation are suggested to minimize the disease occurrences in the field. Practice of strict sanitation by using disease-free seed and destroying the plant debris after harvesting are very effective measures to manage the disease (Rashid, 2003b).

11. Virus/ virus Like Diseases

i) Aster yellows

Occurrence & severity status

Aster yellows is a minor disease in flax, caused by a phytoplasma, a bacteria-like pathogen that requires living plant and insect hosts to survive, spread and reproduce. In North America, the disease occurs annually at low levels of infection. However, the epidemics in 1957 and 2012 caused widespread severe yield losses in flax and other crops (Conners, 1967; Flax Council of Canada, ND2). In Canada, the disease is common in Saskatchewan (Western Canada), but usually at very low incidence levels (Aster Yellows, ND; Flax Council of Canada, ND2). Aster yellows can affect a number of crop species including canola, camelina, flax, cereals, herbs and spices, pea, chickpea, sunflower, alfalfa, bromegrass, The six-spotted leafhopper (*Macrosteles quadrilineatus*; *Macrosteles fascifrons*) is the insect vector that transmits the phytoplasma (Frederiksen, 1964). The phytoplasma overwinters in perennial broad-leafed weeds and crops, but most infections are carried by leafhoppers that migrate from the southern United States. The severity of the disease depends on the stage at which plants become infected, the prevailing temperature, and the number of insect vectors that carry the organism (Frederiksen, 1964; Rashid, 2003b).

Symptoms

The disease appears as yellowing of the top part and stunted growth of the plant, with high numbers of malformed, sterile flowers that fail to form bolls or seeds (Bailey *et al.*, 2003). The disease also causes excessive branching (witch's broom), smaller leaves, proliferation, alteration of tissue pigments (red, orange, yellow, and purple) and phyllody (Frederiksen, 1961). All flower parts including the petals are converted into small, yellowish green leaves (Bailey *et al.*, 2003). Symptoms of this disease are often mimicking other diseases (Philip, 2007).

Causal organism: Aster Yellows Phytoplasma (AYP)

Morphological/ Morphophysiolgical characteristics

The aster yellows phytoplasma (formerly a mycoplasma like organism or MLO), is a single celled prokaryotic microorganism, lacking a cell wall (Philip, 2007). These are small (0.5-1 μm in diameter) that reproduce by division or budding in the phloem sieve cells of the host plants, as well as the bodies of their leafhopper vectors (Davis and Raid, 2002).

Molecular characteristics

The aster yellow phytoplasma is characterized by polymerase chain reaction (PCR) using the specific primer pair R16 (1) F1/R1. In an investigation with false flax, Aster Yellow infection was detected based on a nested-PCR assay using the Aster Yellow specific primer pair, R16 (1) F1/R1, which gave a PCR product of 1.1 kb from each aster yellow phytoplasma-infected false flax. The DNA amplification with the specific primer pair R16 (1) F1/R1 and restriction fragment length polymorphism indicate the presence of AY phytoplasma in the infected plant sample (Khadhaira *et al.*, 2001).

Transmission

The Aster Yellow Phytoplasma is transmitted by the six-spotted leafhopper or Aster leafhopper (*Macrosteles quadrilineatus* Forbes, 1885). The leafhopper is wedge-shaped, small, about 2-3 (-4.3) mm long. The forewings are greyish green while the abdomen is yellowish-green. There are six distinctive dark coloured spots on its head and a black mustache. Adults will readily fly when disturbed (Gavloski and Derksen, 2015; BugGuide, ND). Aster yellows phytoplasma can overwinter in plant roots. Infected biennial and perennial crops and weeds may be the source of aster yellows phytoplasma in the spring. When a leafhopper feeds on an infected plant, the insect can pick up the phytoplasma. An incubation period of 10 to 18 days is necessary for the phytoplasma to circulate and reproduce within the insect before it becomes infective. The infective leafhopper then feed on a healthy plant for a substantial period to pass on the phytoplasma. The leafhopper species *Macrosteles fascifrons* (Stal, 1858), is also reffered as Aster Leafhopper in certain literature (Lee and Robinson, 1958). The adult of this species migrates from the south into Manitoba each year, and produce one generation before autumn. Lettuce was shown to be the preferred host over aster, parsley, carrot and flax (Lee and Robinson, 1958). The aster leafhopper will not fly at temperatures below 15°C. The migrants are attracted to grasses and forages, such as winter wheat and alfalfa, for breeding purposes. The egg takes two weeks and five nymphal stages to reach the first generation, which appear in late June to early July. These cause most of the crop damage. The feeding itself is not economically damaging, but in the feeding process the plants are infected with aster yellows (Westdal *et al.*, 1961).

Predisposing factors

Cool and wet conditions are conducive to the spread of aster yellows. Abundant rainfall will make plants more succulent and attractive to leafhoppers.

Disease cycle

Phytoplasmas can only survive inside their plant hosts and insect vectors and are not considered to be soil, air or wind borne. Most aster yellows infections come from infected leafhoppers carried north from the United States on wind currents. The aster yellows phytoplasma is transmitted when the leafhopper feeds on a plant's sap. The infective leafhopper must then feed on a healthy plant to pass on the phytoplasma. Phytoplasma can over-winter in plant roots. Therefore, locally infected biennial and perennial crops and weeds can serve as a source of aster yellows phytoplasma in spring (Aster Yellows, ND).

Management/ Control measures

Use of host resistance

The variety, Abyssinian (C.I. 302) from the world variety of flax is found resistant to the disease and suggested to use in breeding programme (Martin *et al.*, 1961)

Cultural control

Early seeding may help to avoid the migrating leafhoppers in mid- and late-season, and to reduce the incidence and severity of aster yellows (Rashid, 2003b). However, no cure is known for plants infected with aster yellows. Infected plants should be removed immediately to limit the continued spread of the phytoplasma to other susceptible plants.

Control of Vector pest

In order to monitor and control of Leafhoppers, sticky traps and sweep nets can be utilized for early detection of leafhoppers. This method will give producers an early warning of potential problems.

Chemical control

Although it is suggested to control leafhoppers by pesticide application, depending on the leafhopper population carrying the inoculums (Rashid, 2003b), pesticide (insecticide) applications may also kill the beneficial organisms and can lead to secondary pest problems. There are no commercially available chemicals that will kill phytoplasma present inside the plant sap (Aster Yellows, ND).

[**N.B.:** In India, phyllody and stem fasciation in flax has recently been reported from West Bengal (Biswas *et al.*, 2014). The infected plants shows floral virescence, phyllody, and stem fasciation (flattened stem). Floral malformation is very conspicuous with abnormal structures replacing normal flowers. All the floral parts, including petals, turn into green leaves. The disease is caused by phytoplasma, whose amplified DNA sequence shared 99% similarity with the 16Sr DNA sequence of the '*Candidatus* Phytoplasma asteris' reference strain (GenBank HQ828108), which belongs to 16SrI group. The *Candidatus* Phytoplasma asteris is a member of the aster yellows phytoplasma (Lee *et al.*, 2004). Further analyses showed that the studied strain had 16SrDNA sequences in the 16SrI-B group with a similarity coefficient of 1.00 and the authors claimed as the first report of 16SrI-B phytoplasma associated with flax in India (Biswas *et al.*, 2014).]

ii) Crinkle

Occurrence & severity status

Crinkle is a minor disease of flax occurring only in traces. Its occurrence was reported from Western Canada (Hoes, 1975; Rashid *et al.*, 2000; Growing Flax, ND; Flax Council of Canada, ND2). The disease is caused by a virus called *Oat blue dwarf virus* that causes disease in oats, wheat and barley. The occurrence of oat dwarf disease was reported from the North American Great Plains from Kansas (Sill *et al.*, 1954) to Manitoba (Creelman, 1965). The virus has many host plants of 7 families of both monocotyledonous and dicotyledonous plants (Banttari and Moore, 1962; Westdal, 1968). Some of the hosts are infected symptomlessly (Westdal, 1968).

Symptoms

The disease is characterized by a conspicuous puckering of leaves, stunted growth, reduced branching and reduced flower-setting in flax. The swelling of lateral veins on the margins of leaves, small indentations along adaxial surfaces of the veins, and enations on abaxial surfaces of leaves are common symptoms of this disease (Banttari and Zeyen, 1971). Enations along the lateral veins of leaves causing 'crinkle' (Fredericksen and Goth, 1959; Banttari and Fredericksen, 1959). Flowering may appear normal but seed production is reduced (Flax Council of Canada, ND2).

Causal organism: *Oat blue dwarf virus* (OBDV) (Marafivirus: Tymoviridae)

Synonyms

= *Blue dwarf virus* (Rev. appl. Mycol. 31: 379)

= *Flax crinkle virus* (Rev. appl. Mycol. 39: 173)

Morphological/ Morphophysiolgical characteristics

Oat blue dwarf virus (OBDV) is a single-stranded RNA containing virus, apparently homogeneous with isometric particles 28-30 nm in diameter, when mounted in sodium phosphotungstate. The particles exhibit subunits in the 20-40 Å range when shadow-cast or negatively stained (Zeyen and Banttari, 1972). Sedimentation coefficients, established by linear-log sucrose density gradients, are 31.9 S before formaldehyde treatment and 21.1 S following treatments (Pring *et al.*, 1973). Substitution of the latter value into the formula of Brakke and Van Pelt (1970) gives an estimate of 2.13 x 10⁶ daltons for the M. Wt of the RNA. The RNA is easily liberated from the particles at pH 9.0 in a variety of buffer systems. The A_{260}/A_{280} ratio is 1.63 (Banttari and Zeyen, 1969; Pring *et al.*, 1973).

Molecular characteristics

The complete nucleotide sequence and genome organization of *Oat blue dwarf marafivirus* (OBDV) were determined (Edwards *et al.*, 1997). The 6509 nucleotide RNA genome encodes a putative 227-kDa polyprotein (p227) with sequence motifs similar to the methyl transferase, papain-like protease, helicase, and polymerase motifs present in the nonstructural proteins of other positive strand RNA viruses. The 32 end of the open reading frame (ORF) that encodes p227 (ORF 227) also encodes the two capsid proteins: a 24-kDa capsid protein is presumably cleaved from the p227 polyprotein, whereas the 21-kDa capsid protein appears to be translated from a subgenomic RNA (sgRNA). Encoded amino acid and nucleotide sequence comparisons, as well as the OBDV genome expression strategy, show that OBDV closely resembles the tymoviruses. OBDV differs from the tymoviruses in its general biology, in its lack of a putative movement gene that overlaps the replication-associated genes, and in its fusion of the capsid gene sequences to the major ORF. OBDV also possesses a 32 poly (A) tail, as compared to the tRNA-like structures found in most tymoviral genomes. Due to the strong similarities in genome sequence and expression strategy, OBDV, and presumably the other marafiviruses, it was suggested to consider this as a member of the tymovirus lineage of the alpha-like plant viruses (Edwards *et al.*, 1997). In order to study the capsid proteins (CP) expression strategy, Edwards and Weiland (2014) produced a series of point mutants in the OBDV CP encoding gene and examined expression in protoplasts. Their results support a model in which the 21 kDa major CP is the product of direct translation of a sgRNA, while the 24 kDa minor CP is a cleavage product derived from both the polyprotein and a larger ~26 kDa precursor translated directly from the sgRNA (Edwards and Weiland, 2014).

Transmission

By Vectors

Like aster yellows, crinkle is a disease of flax that depends for infection via transmission by the six-spotted leafhopper (Flax Council of Canada, ND2). In North America the only known vector is the aster leafhopper, *Macrosteles fascifrons*. Approximately 25-30% of individuals in natural populations of the leafhopper are capable of transmitting (Banttari and Zeyen, 1970). The ability of the vector to transmit is genetically controlled and may be altered by in-breeding (Timian and Alm, 1973). Leafhoppers become viruliferous after feeding on plant extracts through membranes (Long and Timian, 1971). The virus multiplies in the leafhopper vectors and, when these vectors feed, the virus is transmitted to the plant (Banttari and Zeyen, 1972a; Edwards and Weiland, 2010)

By Dodder

Transmission by *Cuscuta* sp. from flax to flax has also been reported (Fredericksen and Goth, 1959).

Predisposing factors

Lower air temperatures in the field are associated with more severe disease for oat blue dwarf virus in crops (Johnson *et al.*, 1977).

Disease cycle

The virus overwinters either on living host plants or in the vector leaf hopper body. The virus is fed by six-spotted leafhopper. Inside the vector body it multiplies and may be transmitted on feeding of host plants, including flax. It may also transmit by dodder, but not by seed (Banttari and Moore, 1962). Inside the fax it multiplies again. Crinkle symptoms develop sooner than aster yellows in the field, probably due to a shorter virus incubation period at cooler temperatures. The vector insect again feeds on the infected plants and transmits the virus to any living host. By this way the disease cycle repeats.

Management/ Control measures

Cultural way

Early seeding is effective to avoid migrating leafhoppers in mid to late season in Canada (Flax Council of Canada, ND2).

Chemical control

It is found that the insecticides (phorate and Disyston) are effective in eliminating or depleting the leafhopper population and to reduce the incidence of virus.

iii) Curly top

Occurrence & severity status

This disease is of minor importance in flax but the virus infects sugar beet. It was first reported from California in 1944 (Muskett and Colhoun, 1947).

Symptoms

The distinguishing symptoms are stunted growth of the plants with thickened, yellowed, and bunched or curled leaves that frequently die early. Young plants often die quickly.

Causal agent: *Beet curly top virus* (BCTV) (Curtovirus: Geminiviridae). The virus contains a single-stranded circular DNA that is encapsulated in a twinned icosahedral capsid (Strausbaugh *et al.*, 2008; Horn *et al.*, 2011). The virus contains monopartite genome that is made up of four complementary open reading frames (ORF). The ORF Complementary 1(C1) is responsible for initiating replication with a host cell, while, C2 is involved in causing disease, C3 plays an important role in replication process and C4 play an important role in developing major symptoms (Strausbaugh *et al.*, 2008; Chen *et al.*, 2010; Horn *et al.*, 2011). The virus is transmitted by the beet leafhopper (*Eutettix tenellus* Baker), which is native to the Western United States.

Control measures

Late seeded crops can escape the disease due to early migration of the insect vector. Control of the insect vector by pesticide application depends on insect population and the presence of the virus (Rashid, 2003b).

B. Pests

1. Aphid

Occurrence & severity

Aphid in flax is 'Potato aphid', also known as 'Pink and green potato aphid'. The Potato aphid is the most serious insect pest of flax grown in the plain regions of western Canada and the north-western United States. It originated in North America and now it has a worldwide distribution (Blackman and Eastop, 1984). It is widespread across the United States and Canada, and the species has spread from the Nearctic region to the Palearctic, Ethiopian, and Neotropical regions. Its range has increased to an almost worldwide distribution, and the aphid is a significant crop pest. Populations can be found throughout Europe, Asia, Africa, South America, and Australia (Finlayson, *et al.*, 2009; Le Guigo, *et al.*, 2012; Raboudi, *et al.*, 2011; Stary, *et al.*, 1993; Valenzuela, *et al.*, 2009), except for the Indian subcontinent. In Hawaii, it was first reported in 1910, and is now present on Hawaii, Kauai, Maui and Oahu. Outbreaks of the potato aphid on flax were first reported in western Canada in the 1980s (Lamb, 1989), and field studies have since shown that this aphid can cause serious yield losses (Wise *et al.*, 1995). Yield losses of 20 percent or more in oilseed flax may occur when aphid densities exceed 50 or more per plant (Wise *et al.*, 1995). Wise *et al.* (1995) demonstrated the pest status of the potato aphid in flax and determined the yield loss in flax is 0.021 t/ha per aphid per plant for crops sampled at full bloom and 0.008 t/ha per aphid per plant for crops sampled at the green boll stage. Most of this yield loss was due to a reduction in the number of seeds (Gavloski *et al.*, 2011). The potato aphid is now considered to be the most serious insect pest of flax in western Canada. The aphid can vector other plant diseases also (Boyce and McKeen, 1967; Halbert *et al.*, 1980), although, it is not known to transmit any diseases of flax. The Potato aphid is highly polyphagous, feeding on over two hundred species in more than twenty plant families, but their preference is for plants in the family Solanaceae (Le Guigo, *et al.*, 2012; Petrovic-Obradovic, 2010; van Emden and Harrington, 2007). It can be found on many commercial vegetable and field crops but mostly considered as a serious pest of tomatoes and potatoes (Hodgson *et al.*, 1974; Lange and Bronson, 1981; Walker *et al.*, 1984a).

Types of damage

The potato aphid, at its adult and nymphal stages, damages flax by extracting plant fluids from the stems, leaves and developing bolls. The potato aphid is an efficient herbivore of flax, with a conversion rate of plant tissue to aphid tissue of near 1 mg/mg (Lamb and Grenkow, 2008). Aphid feeding causes yield losses mainly by lessening the plant's ability to set healthy seed. The weight of individual seeds is only slightly reduced and oil content and quality is not affected by the aphids. Plants under drought stress may be prematurely desiccated by high aphid densities. Aphid densities generally do not get high enough to kill plants because of the impact of natural enemies on aphid populations.

Scientific name & order: *Macrosiphum euphorbiae* Thomas, 1878 (Hemiptera: Aphididae)

Synonyms

= *Macrosiphon solanifolii* Ashmead

= *Macrosiphum amygdaloides*

= *Macrosiphum cyprissiae* var. *cucurbitae* del Guercio

= *Nectarophora ascepiadis* Cowen ex Gillette & Baker

= *Siphonophora euphorbiae* Thomas

Biology

Egg: Egg is pale green when first laid and turn shiny black in a few hours. In temperate regions, these aphids overwinter during the egg stage, but, in Hawaii, eggs are not produced by females.

Nymph: Nymphs resemble smaller adults and and go through several moults in the course of about ten days. Nymph is elongated and paler than adult with a light covering of white-gray wax and a dark stripe running down its back. It grows through four nymphal instars, each lasting from 1.5 to 3 days, though development time varies with temperature. Total development time from birth to reproductive maturity ranges from about 6 to 12 days. Development times in sexually reproductive and parthenogenetic populations are similar (Macgillivray and Anderson, 1964; Alyokhin, *et al.*, 2011; Boquel, *et al.*, 2011; De Conti, *et al.*, 2011).

Adult: Adult is of medium size. It has winged and wingless forms. Apterous (wingless) forms are 1.7 to 3.6 mm long and alate (winged) forms are 1.7 to 3.4 mm long. The adult is spindle or pear-shaped. It has a soft body; long antennae dark at the joints between the segments; and a pair of cornicles at the end of its abdomen. Its color can vary among shades of green, pink, or magenta, often with a darker dorsal stripe, while its eyes are reddish. The legs are long, pale green but dark at the apices. The siphunculi are pale coloured, cylindrical with dark tips and operculi are about one third the length of the body. The tail is sword-shaped and bears 6 to 12 hairs and is much shorter than the siphunculi. Alatae has a uniform darker coloured body and appendages and has a green abdomen. Apterous adults usually appear shinier than nymphs (Wightman, 1972; Stoetzel, 1994; van Emden and Harrington, 2007; Petrovic-Obradovic, 2010; Boquel, *et al.*, 2011; Kaplan and Thaler, 2012).

Life cycle

Female potato aphids overwinter as eggs on weeds. They usually emerge in April and begin feeding on perennial weeds. In May or early June, they migrate to different crops where they feed on shoots, the lower side of leaves, buds and flowers, often on the lower parts of the plant. Each unmated female may give birth to 50-70 nymphs within three to six weeks. A generation develops on every 2 or 3 weeks and there may be ten generations over the summer. The optimum temperature for population increase is 20°C (68 °F) (Barlow, 1962). When populations build up, alate (winged) individuals are produced and fly off to infest new host plants. The production of alate is also dependent on the day length, the temperature, the parent type (winged or wingless) and the generation (MacGillivray and Anderson, 1964). The aphids migrate back to primary hosts in August and overwinter as eggs on weeds. Holocyclic populations of Potato aphid occur only in North America, while, anholocyclic populations occur throughout the rest of its global range. Egg-laying females (oviparae) produce a pheromone to attract male mates. The pheromone is produced by a gland on the hind tibia, and the female lifts her legs to release it (Goldansaz and McNeil, 2006; Alyokhin, *et al.*, 2011; Boquel, *et al.*, 2011).

Management/Control measures

Biocontrol

Various factors influence aphid populations. High temperatures or heavy rainfall may reduce infestations. High temperatures increase mortality (Walker, 1982). Heavy rainfall washes aphids off plants (Hughes, 1963; Maelzer, 1977), however, this mortality factor is small because aphids usually gather on the protected under surface of leaves where they are less likely to be washed off (Walker et. al., 1984b). Aphid numbers are naturally controlled by predators, parasites and pathogens (Walker *et al.*, 1984b). In North America they are heavily parasitized by the braconid wasp *Aphidius nigripes*, which lays its eggs in the aphid nymphs and these, are eventually killed by the wasp larvae developing inside them (Brodeur and Mcneil, 1994).The potato aphid is highly susceptible to attack by fungi in the Order Entomophthorales (Shands *et al.*, 1962). Aphids are attacked by fungal diseases, thus, populations are often decimated before they reach economic thresholds. However, the fungi are effective only when the aphid population is less. A number of predators such as ladybird beetles, lacewings, hoverfly larvae and parasitic wasps attack the potato aphid, but they are largely ineffective in controlling aphid populations during the years of high levels of adult emigration and rapid colonization on flax. The aphid populations are suppressed by predators and parasitoids only during the years of slow colonization (Walker *et al.*, 1984a).

Chemical control

The most effective method to control potato aphids in flax is with a single insecticide application at full bloom or at the early green boll stage (Wise *et al.*, 1995).Treatments applied at these crop stages provide season-long control because there is insufficient time left for aphid populations to recover. To determine if an insecticide treatment is needed, a minimum of 25 plants at full bloom and 20 plants at early green boll should be collected randomly in the field to provide an accurate estimate of aphid densities (Wise and Lamb, 1995). If the aphid populations exceed three aphids per plant at full bloom or eight aphids per plant at early green boll crop stages (Wise and Lamb, 1995) chemicals should be applied. Insecticidal soaps offer some control against aphids. Applications should be done at regular intervals for maximum efficacy (Koehler *et al.*, 1983). However, users should carefully consider the use of soaps. Excessive use can cause a drop in yield of the crop.

2. Army Cutworm

Occurrence & severity

Army cutworm, also known as 'Miller moth', is the most common 'cutworm' in the High Plains Rocky Mountain region. It is commonly found in the Western section and prairies of the United States. It is an important pest in southern Alberta, to a lesser extent in southern Saskatchewan, and rarely in Manitoba (Flax Council of Canada, ND1). The army cutworms are known to travel to alpine climate regions in late June and early July where they feed at night on the nectar of wildflowers. Army cutworms are seasonal migrants and considered as a minor pest of flax. Eggs overwinter and larvae hatch and feed on flax crops and other plants in the spring. Economic thresholds for cutworms in flax have not been developed. A nominal threshold of 4-5 larvae/m^2 in flax has been suggested. The loss of some flax plants may be partially compensated by a small increase in yield of remaining plants (Flax Council of Canada, ND1). Army cutworm larvae eat the foliage of wheat, oats, barley, mustard, flax, alfalfa, sweet clover, peas, cabbage, sugar beet, various weeds (notably stinkweed) and grasses (Anonymous, 2006).

Types of damage

The adult moths lay eggs in late-summer. Eggs overwinter and larvae hatch and feed on the crops in the spring. The gray-brown larvae feed at night, most often around the soil surface. The first signs of damage are holes in leaves and semi-circular notches eaten from the edges of leaves (Anonymous, 2006). Young tender plants may be cut or girdled and killed. Larvae are also fair climbers and may destroy buds. Damage can be of any severity up to complete defoliation. In severe infestations, the area defoliated has ranged from an individual field to thousands of acres with larval densities of up to 200 per square metre. (Anonymous, 2006).

Scientific name & order: *Chorizagrotis auxiliaris* **Grote, 1873 (Lepidoptera: Noctuidae)** (Guala and Doring, 2018)

Synonyms

= *Euxoa auxiliaris* (Grote, 1873)

[**N.B.** In North America the army cutworm, *C. auxiliaris* (= *E. auxiliaris*) and the bertha armyworm, *Mamestra configurata* Walker (King, 1928) are two noctuids that feed sporadically on flax, whereas, in northern China the cabbage armyworm *Barathra brassicae* L. (Li, 1980; Chen *et al.*, 1990; Anonymous, 1998) feeds on flax. The red backed, *Euxoa ochrogaster*, and the pale western, *Agrotis orthogonia*, are two of the more common species of cutworms that feed on flax in Canada (Flax Council of Canada, ND1).]

Biology

Egg: Eggs are laid amongst lush vegetation during September and October at lower elevation/plains areas.

Larva: The hatched larvae feed for a short time, and then overwinter underground as partially grown larvae. Once mature, they are up to 4-5 cm long. It has a light brownish-gray head with pale brown spots. The body is pale greyish with white splotches and a brown-tinged top line. The lower portion of the body has darker, top-lateral stripes and an indistinct band of white splotches.

The caterpillars curl up into a tight "C" when disturbed. Mature larvae then burrow back down into the soil to pupate, emerging as moths in late spring/early summer.

Pupa: Pupation occurs in the soil.

Adult: The adult moth, known as the grey miller moth, is a grey-brown moth with a wingspan of 4-5 cm. Forewing length is about 1.7-2.2 cm (Powell and Opler, 2009). Adults are very capable fliers (Koerwitz and Pruess, 1964). The adults that emerge have a peculiar habit among insects. Instead of laying eggs, they spend the summer in a state of semi-development. Eggs are not matured in them. The moths seek sources of carbohydrates and cool moist conditions for a period of increasing fat reserves. This causes them to migrate to high elevations and can involve migrations of hundreds of miles. All summer is spent at those locations where they feed on mountain flowers and rest under rocks or other cover. In early fall there is a reverse migration to lower elevations and at that time eggs are laid.

Life cycle

The army cutworm is univoltine, having one generation a year. The female moth lays about 1,000 eggs in soft soil in late August through October. The egg hatches in a few days to two weeks. The larva feeds on plant foliage at night and remains below ground during the day. There is some feeding and development during fall, the amount depending on temperature. Development stops when the ground freezes; larva is usually about half-grown by this time. It remains inactive throughout the winter just beneath the surface in loose soil. Larva begins to feed in April and continues to feed until pupation in May to early June. Moth appears in June for a brief flight period, aestivates (summer hibernation) in building and under trash and clod during June and July, and then becomes active again for the egg-laying period (Anonymous, 2006).

Management/Control measures

Natural enemies

Five or possibly six species of wasp parasites have been recorded for army cutworm in Canada, mostly from Alberta (Anonymous, 2006). The most effective one is Copidosoma, which is a tiny parasitic wasp (about 2 mm in length), that lays a single egg in cutworm larvae. The egg produces multiple embryos from which over a thousand offspring may be produced. Sixty per cent or more of army cutworm larvae were found to be parasitized by this organism during an outbreak in Alberta in 1990. Copidosoma also prolongs the larval stage of this cutworm into June; an unusual circumstance since it normally pupates by that time (Anonymous, 2006). The army cutworms are one of the richest foods for predators, such as brown bears, in this ecosystem, where up to 72 per cent of the moth's body weight is fat, thus making it more calorie-rich than elk or deer.

Monitoring & Chemical treatment

Army cutworms are considered nearly impossible to control through normal pest extermination techniques because a new batch shows up every day as they migrate. However, some measures can be taken during their active growth period. During the day, larvae remain within the top 5-7 cm of soil surface. It is suggested to count the larvae within each 50 cm of row in the sample area. Plants that have adequate moisture and are vigorously growing with 12 to 15 cm of top growth can withstand four larvae per 30 cm of row without loss of yield. If plants are under 10 cm in height and two or more larvae per 30 cm of row are present, chemical treatment is required (Anonymous, 2006). Insecticide applications, if needed, should be made late in the afternoon or evening. It may be most economical to just treat infested patches and not entire fields (Flax Council of Canada). Pyrethroids (Astro, Talstar), chlorpyrifos (Dursban), and spinosad (Conserve) are among the insecticides effective against army cutworm.

3. Linseed Blossom Midge

Occurrence & severity

The linseed blossom midge or linseed bud fly is found in all flax-growing regions of the Indian subcontinent, and seriously limits flax production in central and northern India. The pest is serious in Andhra Pradesh, Madhya Pradesh, Bihar, Uttar Pradesh, Delhi and Punjab. The incidence of the pest goes up to 20 per cent (Awaneesh, 2009). In Raipur, Chhattisgarh, India, the infestation of bud fly as recorded during 1995-96 is about 0.05-2.88 per bud (Patel and Thakur, 2005). The linseed bud fly is the major limiting factor in the production of flax for seed in India (Jakhmola and Yadev, 1983). Yield losses by the midge can be expected in most areas each year, and in the central and northern plains of India losses in some years make the growing of flax uneconomical. During years of severe bud infestation, yield losses by the bud fly can exceed over 90 percent (Malik *et al.*, 1996b). The minimum and maximum temperatures 12°C and 28.1°C, respectively, are reported as favourable for its multiplication (Patel and Thakur, 2005). Infestation levels can remain high throughout the period from January to mid-March and are maximized by mean temperatures of 16–20°C and relative humidity of 60–75 percent (Malik *et al.*, 1998). Rainfall in January and March has little effect on midge populations (Shrivastava *et al.*, 1994). It also attacks pigeon pea, *Cajanus cajan* (Pruthi and Bhatia, 1937), and sesame, *Sesamum orientale* (Narayanan, 1962), but has not been found on any noncrop hosts.

Types of damage

Damage to the crop is caused by the larvae (maggots) feeding on buds which prevent the flower from opening. Consequently the seed dose not set properly. Due to their feeding, galls are produced and there is no pod formation (Awaneesh, 2009).

Scientific name & order: *Dasineura lini* (Barnes, 1936) (Diptera: Cecidomyiidae)

[**N.B**. One other midge, *Dasineura sampaina* Tav., produces galls on the terminal leaf buds of flax in Europe and Algeria, but it is of no consequence as a pest (Barnes, 1949).]

Biology

Egg: Eggs are smooth, transparent, laid in the folds of 8-17 flowers or in tender green buds, either singly or in clusters of 3-5. The eggs hatch in (1-) 2 -5 days, depending on temperature (Prasad, 1967).

Larva: Just after emergence, the larvae are transparent, with a yellow patch on the abdomen. They pass through four instars in 4-10 days and when full-grown become deep pink or orange yellow and measure about 2-2.3 mm in length.

Pupa: The full-grown maggots (larvae) of fourth instar drop from the flower as deep pink or orange-red, of about 2.3 mm long, and immediately begin to prepare a cocoon and pupate 5–7 cm below the soil surface (Prasad, 1967).

Adult: Adult midge is small narrow-bodied orange fly, 1–1.5 mm long, with long legs and hairs on the back edge of the wings. The adult emerges early in the morning during late December or early January and can be seen hovering from plant to plant in calm weather (Pruthi, 1936). During the day female lays its eggs singly or in clusters of three to five within the calyx of young flower buds (Pruthi and Bhatia, 1937). A single female can lay up to 100 eggs (Narayanan, 1962).

Life cycle: Eggs hatch into larvae in 2 -5 days. The hatched larvae enter the buds and feed on the internal parts of the flower (Prasad, 1967). Typically two to four larvae develop in infested buds, although as many as ten may be found. They pass through four instars in 4-10 days. The light orange coloured larvae feed for about seven days, causing greatest damage in the second and third instar stages. The full-grown larvae (maggots) become deep pink or orange-red and measure about 2-2.3 mm in length. The full-grown maggots drop to the ground, prepare a cocoon and pupate in the soil. The pupal period lasts 4-9 days. A generation is completed in 10-24 days. There are four overlapping generations during the season (Awaneesh, 2009).

Management/Control measures

Use of host resistance

There is no flax cultivar has been found to be completely resistant to the midge (Jakhmola and Yadev, 1983; Kumar *et al.*, 1992). However, the flax varieties with short flowering periods, thin buds and thin sepals are less susceptible to the midge (Sood and Pathak, 1990; Malik *et al.*, 1991; Mishra *et al.*, 1996), as are varieties with higher polyphenol contents (Malik *et al.*, 1996c).

Cultural control

This is very effective measure to minimize gall midge attack in flax. Early seeding, no later than the first week of November, for most flax cultivars, or the growing of early flowering cultivars, can reduce the severity of damage by the fly (Jakhmola *et al.*, 1973; Pal *et al.*, 1978). Early maturing plants not only are less susceptible to attack by first-generation larvae but are also more likely to escape damage by the later generations. However, the effectiveness of either method at reducing bud fly infestation should be weighed against their yield disadvantages.

Use of traps: The adult flies can be killed by using light traps. The flies are also attracted in day-time to molasses or juggery (gur) added to water.

Biological control

Protection of natural enemies: It is suggested to conserve larval parasitoids viz., *Systasis dasyneurae* Mani (Miscogasteridae), *Elasmus* sp. (Elasmidae), *Eurytoma* sp. (Eurytomidae), *Torymus* sp. (Torymidae) and *Tetrastichus* sp. (Eulophidae). Eight species of parasitoids of larval bud fly have been recorded (Narayanan, 1962), and play a role in fly population reduction. Application of pesticide treatments late in the season should be avoided because of potential damaging effects on parasites such as the chalcid wasp *Systasis dasyneurae* Mani. The larva of this wasp parasitizes midge larvae in the buds, attaching to the midge and extracting body fluids from its host. Each wasp larva can destroy three to four midge larvae (Pruthi, 1937). Levels of parasitism of late instar midge larvae can exceed 50 percent (Ahmad and Mani, 1939).

Chemical control

Two or three insecticide applications at biweekly intervals can effectively control later generations of the midge (Jakhmola, 1974; Singh and Pandey, 1980). In most years insecticides should be used only in areas where bud fly infestation is likely to exceed 7 percent, in order for growers to realize an economic return (Malik *et al.*, 1996a). The crop may be sprayed with monocrotophs 36 SL 750 ml or endosulfan 35EC 1.5 litre/ha (Awaneesh, 2009). However, recently, the insecticide, Fenvalerate 20 EC, has been highly effective for the control of gall midge, with highest grain yield and maximum cost:benefit ratio. Its half dose has been recorded lower incidence with comparatively higher grain yield and cost:benefit which is at par with Monocroptophos, Chlorpyriphos and Endosulfan (Singh *et al.*, 2008).

4. Flax Bollworm

Occurrence & severity

Flax bollworm is a grub (larva) of a moth that is closely related to the cotton bollworm but feeds chiefly in the seedpods of flax. The flax bollworm, unlike cutworm, feeds on buds and flowers at its larval stage for a short time before it enters a capsule and feed until the contents are exhausted. The larva then exits the capsule and either feeds on foliage or enters another capsule (Wise and Soroka, 2003). The flax bollworm, *Heliothis ononis*, though has other host plants, but, prefers flax over other crops (Twinn, 1944; Twinn, 1945; Putnam, 1975). It is widespread species that found in China, Kazakhstan, central Asia, northern Mongolia (Khangai), the Russian Far East the Korean Peninsula, southern European part of Russia, southern and central Europe, southern and eastern Siberia (Transbaikalia, Yakutia) and Turkey. In North America it is found from south-central Manitoba west to British Columbia, north to the Northwest Territories and Yukon and Alaska and south to Colorado. Economic infestations of this pest have been limited so far to west central Saskatchewan in Canada. The adults fly from May to July and population of this pest is usually kept low by parasites and disease (Growing Flax, ND).

Types of damage

The moths of flax bollworm deposit their eggs in the open flowers, and the young larvae eat the developing seed within the boll in the case of flax leaving an empty husk. The older green and white-striped worms leave the bolls and complete development by feeding on other bolls from the outside (Growing Flax, ND; Flax Council of Canada, ND1).

Scientific name & order: *Heliothis ononis* (Denis & Schiffermüller, 1775) (Lepidoptera: Noctuidae)

Synonyms

= *Noctua ononis* (Denis &Schiffermuller, 1775)

= *Melicleptria septentrionalis* H. Edwards, 1775

= *Heliothis ononidis* Guenee, 1775

= *Heliothis intensiva* (Warren, 1911)

= *Heliothis lugubris* Klemensiewicz, 1912

Biology

Egg: Eggs are spherical with flattened base, deposited in open flowers of food plant.

Larva: The larva is a green and white striped worm with thin greenish microspines and dark minute spots on body that dense on head. Young larva eats the developing seeds within the flax boll, and older larva leaves the boll from which it emerged and feeds on the seeds in surrounding bolls. The larvae can undergo 5 to 6 instars to reach pupal stage.

Pupa: Full grown larvae pupate in soil forming earthen cocoon.

Adult: Adult is a small moth with wingspan is 24–26 mm. Forewing is light brown with darker reddish-brown scaling at base and on shoulders (tegulae); dark brown median band incorporates large black reniform spot; subterminal band dark, narrower in the middle; fringe reddish-brown. Hindwing is white with large black discal crescent and broad black terminal band containing pale double spot about midway along outer margin; fringe white. Adult is on the wing from May to July in North America.

Life cycle: Females of the flax bollworm lay their eggs in open flax flowers. Young larvae eat the developing seeds within the flax boll, and older larvae leave the boll from which they emerged and feed on the seeds in surrounding bolls. High populations are sporadic, and populations are usually kept low by parasites and diseases (Wise and Soroka, 2003).

Management/Control measures

Cultural methods

It is suggested to sow the flax seed in time to avoid having plants in flower during peak moth flights. This is effective to reduce feeding injury greatly. Fields should be inspected at flowering stage for the presence of adult moths.

Biocontrol

Predation and parasitism should be encouraged. In India, preying on the larvae by the black drongo bird reduces flax capsule damage by more than 50 percent (Malik, 1998). Damage can also be further reduced by the placement of bird perches near or in flax fields (Wise and Soroka, 2003).

Chemical control

An insecticide treatment may be applied ten days after peak flowering to minimize feeding damage and to prevent reinfestation by late emerging larvae (Passlow *et al.*, 1960).

5. Cutworms

The larvae of many cutworm species (Lepidoptera: Noctuidae) attack flax crops in nearly all areas of the world where flax is grown. Localized outbreaks generally are composed of only one species, but several species often occur in a flax-growing area.

i) Red-backed cutworm (*Euxoa ochrogaster*)

Occurrence & severity

The red-backed cutworm, *Euxoa ochrogaster* (Guenee), is the most common in North America (King, 1926; Philip, 1977). This species is widely distributed in dry, open habitats across the northern regions of both Eurasia and North America. It is found from Iceland and northern Europe, through the Baltic to the Amur region. In North America, it is found from Alaska south to central California and east to Newfoundland and Labrador, south into the northern part of the United States, south in Rocky Mountains to Arizona and New Mexico (Grote, 1882; Lafontaine, 1987). In the Pacific Northwest, it is moderately common on open sagebrush steppe at low elevations east of the Cascades during wet years, but is often less common during drought years.

Types of damage

All the cutworm species infesting flax have similar feeding habits. The cutworms damage flax plants by totally or partially severing the stems of the seedlings at the soil surface. The damaged plants are either completely destroyed or are severely weakened and made susceptible to further damage by wind or disease. Plant injury usually occurs too late in the season to reseed, and results in partial or complete yield loss in the affected areas. A population of 12 red-backed cutworm larvae per m^2 can cause a 10 percent reduction in flaxseed yields (Anonymous, 1996). A density of 32 red-backed cutworm larvae per m^2 can destroy all flax plants seeded at a rate of 45 kg/ha (Ayre, 1990)

Scientific name & order: *Euxoa ochrogaster* (Guenee, 1852) (Lepidoptera: Noctuidae)

Synonyms

= *Noctua ochrogaster* Guenee, 1852

= *Agrotis insignata* Walker, 1857

= *Agrotis cinereomacula* Morrison, 1875

= *Agrotis gularis* Grote, 1875

= *Agrotis islandica* Staudinger, 1857

Biology

Egg: Egg is whitish initially, becomes bluish as the embryo completes its development. Eggs are usually laid on loose sandy soils in autumn. Often co-occurs with other cutworm species with similar oviposition patterns. Eggs do not hatch upon completion of embryonic development, the embryos remaining in diapauses until spring.

Larva: Mature larva is 35-40mm long, often with brick-red or reddish-brown stripes along the back, usually extending the entire length of the body. The top stripe is divided by a dark line and bordered with darker bands. The head and prothoracic shield are yellowish-brown (Berry, 1998), though, the head bears dark brown submedial arcs. Normally there are six instars, but when reared at 30°C only five instars develop. Total larval duration is about 72, 39, 32 and 22 days at 15°, 20°, 25° and 30°C, respectively (Jacobson, 1970). Larvae tend to live below-ground during daylight hours, but come to the surface, and even climb plants at night. This species is a soil-surface feeding cutworm that feeds on general herbaceous vegetation (Grote, 1882; Lafontaine, 1987).

Pupa: Larva pupates in soil at a depth of 2.5-5.0 cm. The pupa is reddish brown and measures about 2cm long. Pupal developmental times are about 33-42 (av. 36.8), 18-23 (av. 21.5), 13-16 (av. 14.3) and 12-14 (av. 12.6) days at 15°, 20°, 25° and 30°C, respectively (Jacobson, 1970).

Adult; Adults are moths; vary in colour, pale clay-yellow, yellow-brown, orange-brown, red-brown medium or dark-red. The wingspan is about 40 mm (Berry, 1998). Forewing is 16–19 mm in length, with several distinct forms that usually have pale ochre lining and dark gray filling of the reniform spot and dark gray terminal area. One common wing pattern is uniformly dark reddish-brown with bean-shaped and round spots on the front wings (forewing) bearing light border. Another common wing pattern is much lighter in colour, often bearing a greyish bar along the leading edge of the wing. This latter colour form also tends to have a black bar connecting the two spots. The spots similarly have a light margin. The hindwing in both forms is greyish basally, with brown distally and along the veins (Hardwick, 1965a). Females produce sex pheromone (Struble, 1981a; Palaniswamy *et al.*, 1983). Males respond to pheromone soon after sunset (Struble and Jacobson, 1970). Adults normally survive for about 20 days (Jacobson, 1970).

Lifecycle: The moths mate in dark. Multiple matings are common, ranging from one to seven with an average of two or three. Egg development accelerates with the age after emergence. The preoviposition period is about 8 days. The mean fecundity per female is 411, ranging from 251-705. Maximum longevity of both sexes' adults is approximately 20 days (Jacobson, 1970). The redbacked cutworm overwinters as a first instar within the egg. Eggs hatch in the spring usually in the late March and April. Larvae feed beneath the soil surface and on foliage for six to eight weeks. After that, the larvae pupate in an earthen cell in the soil. Adults begin emerging in late June (or end of July) and continue emerging until late August and early September (Grote, 1882; Lafontaine, 1987; Berry, 1998). There is one generation per year.

ii) Pale Western Cutworm (*Agrotis orthogonia*)

Occurrence & severity

The pale western cutworm (*Agrotis orthogonia*), like the red-backed cutworm, is also common in North America (Parker *et al.*, 1921; Philip, 1977). This moth is more common in dry and semi-desert areas of Western North America from southern Canada, such as, Alberta east to southwestern Saskatchewan, south to southern California, central Arizona, New Mexico and western Texas (Markku, 2008). The species is occasionally of economic importance. It has also been reported from wheat, small grains, corn and sugar beets.

Types of damage

The type of damage is same as described with red bucked cutworm. It damages flax plants by totally or partially severing the stems of the seedlings at the soil surface. The damaged plants are either completely destroyed or are severely weakened and made susceptible to further damage by wind or disease. The plant injury usually occurs too late in the season to reseed, and results in partial or complete yield loss in the affected areas.

Scientific name & order: *Agrotis orthogonia* Morrison, 1876 (Lepidoptera: Noctuidae)

Synonyms

= *Agrotis orthogonoides* McDunnough, 1946

Biology

Egg: Egg is extremely small, spherical, white or pale yellow when laid, changing to brown before hatching. It is laid in soil between August 15 and September 15. Egg commences development as soon as it is laid and in most years the embryo is fully developed by winter. Hatching occurs in the spring, usually soon after all snow has melted and the soil has thawed (Jacobson, 1962).

Larva: New larva is almost colourless or greyish white and about 3 mm (0.12 inch) long. Young larva is small and very difficult to find. Until it is about 0.5 inch long, it is greyish white. As it gets bigger it becomes a greyish green or pale yellow gray colour, darker on the sides than dorsally. The larva is pale with no distinct markings on its body. The pale middorsal line is bounded on both sides by diffuse darker lines. There are also diffuse thin dark subdorsal lines. The head capsule is usually yellow-brown with or without dark shading. When fully grown, the larva is about 30 to 36 mm (1.2-1.4 inch) long. The larva curls into 'C' shape when disturbed and during the day (Hein et al., 1993; Powel and Opler, 2009).

Pupa: Larva pupates in earthen cell, constructed several inches below the soil surface, in early August shortly before it emerges as adult. Pupa is dark brown to orange in colour with two spines on one end. Size is about 1 inch long. Pupal duration is about 6-7 days.

Adult: Adult is a medium sized moth; very light dusty brown-gray or mottled greenish grey with dark grey terminal area, lines and spots. Its body length is just under 19 mm and wingspan about 34-35 mm. The distinctive characteristic of these moths is the white underside of the wings (Lafontaine, 2004). The fore wing length is 13-15(-17) mm (Powel and Opler, 2009), with pale putty to light gray brown ground colour, usually pale, with scattered individual dark scales that produces a dusty appearance. The terminal area is darker gray. The transverse lines are black, double and filled with the ground colour. The antimedial line is irregular with a long lateral loop at the trailing margin. The claviform spot is small to medium in size, black and filled with darker gray. The postmedial line is dentate, slightly convex toward the margin near the reniform spot and then slightly oblique toward the base and trailing margin. The subterminal line is pale and forms a broad W mark toward the margin on veins M3 and CuA1. The hindwing is dirty light grey, slightly darker toward the outer margin and veins, and with an entirely whitish fringe in males or slightly darker gray with pale fringe in females. The head and thorax are uniformly coloured. The male antennae are bipectinate (Lafontaine, 2004). Most of the female mate once. Multiple matings is with a maximum of 3 times, occurred in less than 20% females. Maximum ovipostion by one female is 564 with means 250 -300 (Jacobson, 1965).

Lifecycle: Moths begin to emerge in late August and quickly increase in numbers by mid September. They are attracted to areas with loose soil to deposit their eggs. Each female lays about 250-300 eggs in upper 0.5 inch (12.7 mm) of soil. Moth activity decreases by early October. Some eggs may hatch during warm spell in the fall or winter, but most of the eggs hatch early in the spring when temperature rises to 21°C. This occurs from February through March. The hatched larvae start ground feeding. Larvae pass through six to eight stages before they cease feeding and pupate. The larvae are capable to survive a month or more without food. They then burrow into soil and form earthen cells where they pass most of the summer, pupating in early august shortly before they emerge as adults (myFields, ND)

iii) Black cutworm (*Agrotis ipsilon*)

Occurrence & severity

The black cutworm (*Agrotis ipsilon*), also known as 'dark sword-grass', 'greasy cutworm,' or 'floodplain cutworm', is a small noctuid moth found worldwide (Showers, 1997). The cutworm is a pest on many crops in tropical and subtropical regions all around the world, causing significant losses, in Brazil (Secchi, 2001), Canada (Archer *et al.*, 1980), Chile (Carrillo *et al.*, 2001), China (Zhao *et al.*, 2007), Egypt (Amin and Abdin, 1997), India (Verma and Verma, 2002), Myanmar (Morris and Waterhouse, 2001), Poland (Walczak, 2002), Spain (Amate *et al.*, 1998) and USA (Showers *et al.*, 1993). The moth gets its scientific name from black markings on its forewings shaped like a letter "Y" and resembles the Greek letter upsilon (McLeod, 2005). This species is a seasonal migrant that travels north in the spring and south in the fall to escape extreme temperatures in the summer and winter (Showers, 1997). The black cutworm typically causes 10 to 30 percent damage to flax in northern China (Anonymous, 1998).

Types of damage

This cutworm damages flax plants by totally or partially severing the stems of the seedlings at the soil surface. The damage from the black cutworm is usually observed as a cutting of young seedlings, often causing death of the cut seedlings. Sometimes wilting is observed because of partial cutting. Larvae are destructive out of proportion to the actual plant material they consume because several plants may be cut by a single larva. A larva often cuts one plant, quickly moves on to other plants and continue cutting. The plant injury usually occurs too late in the season to reseed, and results in partial or complete yield loss in the affected areas.

Scientific name & order: *Agrotis ipsilon* (Hufnagel, 1766) [Lepidoptera: Noctuidae]

Synonyms

= *Phalaena ipsilon* Hufnagel, 1766

= *Noctua suffusa* Denis & Schiffermüller, 1775

= *Noctua ypsilon* Rottemburg, 1777

= *Phalaena idonea* Cramer, 1780

= *Bombyx spinula* Esper, 1786

Biology

Egg: Egg is nearly spherical, initially white or whitish yellow but turn brown as hatching approaches. It measures 0.43 to 0.53 mm high and 0.51 to 0.58 mm wide with a slightly flattened base. The surface of the egg possesses 35–40 ribs that radiate from one apex. The ribs are alternately long and short. The egg stage lasts 3–6 days (Capinera, 2006, 2015a; CABI, ND)

Larva: Larval body is light grey or grey-brown to black in colour, without any distinct stripe or marking, and usually uniform above the spiracles. The subventral and ventral areas are lighter in colour, with numerous pale flecks. The abdominal segments are equal in width. The head is pale-brownish with black coronal stripes and reticulation. The skin bears convex, rounded, distinctly isolated, coarse granules with smaller granules interspersed between the larger granules. The spiracles are black. The setigerous tubercles on the abdomen are large, heavily pigmented with black. There are six to seven instars (usually six; occasionally up to nine, Capinera, 2006, 2015a). The third instar larva is about 7 mm in size, while, the fourth instars 10-12 mm, fifth instars 20-30 mm and sixth instars 35-50 mm in length. By the 4th instar, the larva becomes light sensitive and spends most of the daylight underground. The larval stage lasts 20–40 days (Capinera, 2006; CABI, ND)

Pupa: Pupa is brown to dark brown and approximately 17-25 mm in length and 5-6 mm in width. It appears almost black in colour just before the moth emerges. This species pupates under the soil approximately 3–12 mm below the surface. The pupal stage lasts 12–20 days (Capinera, 2006; CABI, ND).

Adult: The adult is a moth with long, narrow and dark brown forewings, darker than the hindwings and marked with black dashes or 'daggers'. The basal two-thirds of the forewing are dark, with the outer third pale grey to brown. Orbicular is tear-shaped black marking on its outer margin. Claviform is small, dark, oblong, and filled with dark scales. There is a zigzag line of pale scales on a dark background in the subterminal area (CABI, ND). The hindwings are whitish to gray and have darker colored veins (Capinera, 2006). The male antennae are plumose (feathered), and the female antennae are filiform. The wingspread is approximately 35-50 mm (CABI, ND). A Female can lay eggs, singly or in groups, up to 1200 to 1900.

Life cycle

Female black cutworm moth usually deposit eggs singly or a few together. Female can deposit 1200 to 1900 eggs. The eggs are firmly attached to the substrate. Densely growing plants of low height and fine textured plant debris in untiled fields are the preferred substrates for egg laying. In some parts of India, female black cutworm moths prefer to lay their eggs on wet or muddy soil just after flood waters have receded (Narayanan, 1962). The egg hatches in 3-6 days and the larva moves to the soil where they remain during day time. Larva moves to the surface at night and feeds on young plants. There are six to seven instars in 25-35 days to pupate. Pupation occurs in soil at a depth of 2.0 cm to 10.0 cm and pupal stage lasts about 12-15 days. There are several generations in a year. One complete generation from egg to adult lasts 35–60 days. The female preoviposition period lasts 7–10 days.

Management/Control measures

All the cutworm species infesting flax have similar feeding habits, they are considered as one pest when determining their potential for economic damage to the crop. Hence, the control measures are same for all the species.

Cultural control

It is suggested to avoid seeding in field with a known history of cutworm problems. Leaving a protective crust on fallowed fields near the end of the growing season reduces the attractiveness of these fields as oviposition sites. Cultivation of fallow fields harbouring cutworm pupae buries them or exposes them to predators. Also, fallowed fields kept weed free early in the season prevents the fields from being a source of future infestations. Delaying seeding effectively prevents feeding injury by some early cutworm species, but this method is not effective against some of the cutworm species.

Further, monitoring of larvae with larval cutworm bait traps (Story and Keaster, 1982; Munson *et al.*, 1986), adults with both blacklight and sex pheromone traps (Hachler and Brunetti, 2002; Capinera, 2015a) and weather (Bhagat and Sharma, 2000; Zhou and Chen, 2004) is very helpful to predict black cutworm attacks in order to determine the need of suitable management tactics at correct times.

Biocontrol

Bacillus thurigiensis is the most effective against bio control agent for first and second instar larvae of *A. ipsilon* (Yang *et al.*, 2000). Calcium oxide potentiated *B. thuringiensis* has been found effective against *A. ipsilon* in Egypt (Salama *et al.*, 1999). *Agrotis ipsilon multiple nucleopolyhedrovirus* (AgipMNPV), a naturally occurring baculovirus, infects black cutworm on central Kentucky golf courses. In a study, it is found that the virus has potential as a preventive bioinsecticide, targeting early instar black cutworms (Prater *et al.*, 2006). Many species of birds can help to reduce cutworm feeding injury by predation on larvae and pupae. Early season cultivation can encourage predation by bringing these stages to the soil surface. In India, damage to flax by the black cutworm *A. ipsilon* has been reduced by about 70 percent from predation by the pond heron (Malik, 1998).

Chemical control

Use of natural Insecticides

Pills made of powdered *Nerium oleander* leaves, wheat bran, and cotton seed meal are effective for traping and killing *A. ipsilon* on cotton in China (Ma *et al.*, 2003). Neem based products are effective for young seedlings (Viji and Bhagat, 2001a). A methanol extract of *Melia azadirachta* fruits has been found as toxic to *A. ipsilon* (Schmidt *et al.*, 1997). Further, leaf extracts of Lantana, Parthenium, Hyptis and *Ipomea carnea* are toxic to *A. ipsilon* and other pests (Ramesh-Chandra, 2004).

Use of synthetic insecticides

Clothianidin, a new synthetic chloronicotinyl insecticide, has been found effective as a seed treatment against *A. ipsilon* (Andresch and Schwarz, 2003). In addition, several other insecticides, viz. Diazinon 20EC, Quinalphos 25 EC, Chlorpyrifos 20EC, Fenitrothion 50 EC, Deltamethrin 2.8 EC and Malathion 5% dust, are found effective against *A. ipsilon* in India. However, amongst them, Chloropyriphos 20 EC is the most effective (Tripathi *et al.*, 2003).

6. Flea Beetles

Occurrence & severity

Several flea beetle species are found to attack flax (Yaroslavtzev, 1931; Lakhmanov, 1970), but except the two, such as, the large flax flea beetle, *Aphthona euphorbiae* (Schrank), and the flax flea beetle, *Longitarsus parvulus* (Paykull), their damage is either much localized or not significant. The large flea beetle (*A. euphorbiae*) occurs in Europe, Northern Africa and Anterior Asia. It inhabits the European part of the former USSR northward to Arkhangelsk Region, the southern part of West and Mid Siberia, the Caucasus, Near East, Asia Minor, Middle Asia and Kazakhstan (Burkejs, 2009). The flax flea beetle (*L. parvulus*) is widespread and fairly common in Great Britain and Ireland. This beetle is reported from all Baltic and Fennoscandian territories (Silfverberg, 2004) and also from Belarus (Alexandrovitch *et al.*, 1996). It has also been reported from Kaliningrad, Pravdinsk and Chernyakhovsk (Bercio and Folwaczny, 1979). This species is not uncommon in suitable habitats at Kaliningrad region in western Russia (Alekseev and Bukejs, 2011). Cultivated flax is a favoured host of the large flax flea beetle (Kurdiumov, 1917) and the flax flea beetle (Popov and Firsova, 1936). These two flea beetle species are serious pests of flax cultivars grown for fibre and seed in the British Isles (Rhynehart, 1922; Ferguson *et al.*, 1997) and throughout mainland Europe (Grandori, 1946; Palij, 1958; Fritzsche and Lehmann, 1975; Lewartowski and Piekarczyk, 1978;Cate, 1984; Sultana, 1984; Beaudoin, 1989a; Voicu *et al.*, 1997), Russia (Yaroslavtzev, 1931), and Turkey (Lodos *et al.*, 1984). The two species are usually found together (Jourdheuil, 1960), except in the southern areas of Russia where the flax flea beetle is largely absent (Popov, 1941). A widespread feeding damage by flea beetles to flax seedlings was reported for the first time in the United Kingdom in 1994 (Haydock and Pooley, 1997). Since then, flea beetles have become the most serious pest of flax in England (Ferguson *et al.*, 1997). The economic impact of flea beetles is influenced by the time and type of their attacks. For example, feeding by the overwintered adults on seedlings can result in complete crop losses (Carpenter, 1920). Plants with severe damage to the cotyledons and to the first true leaves produce less and lower quality fibre and less seed than plants defoliated even more severely later in their growth (Sokolov and Bezrukova, 1939). Later attack can slow down vegetation growth (Sultana, 1984). Adults that emerge in the summer are not known to cause significant damage to oilseed flax but their feeding on upper and middle stems of fibre flax, the most important parts in the production of fibre, can cause fibre cells to become corky and decrease the value of the crop (Durnovo, 1935).

Types of damage

The most serious damage of flea beetles is caused to the flax seedlings. Their infestation on the young crop may kill all the sprouts. The larvae of flea beetles feed on the roots of flax plants, causing weakened plants and growth retardation that eventually reduce seed yield and develop poor quality of fibre. The flea beetles may cause damage by reducing seedling emergence, particularly along the field borders. This reduction of seedling emergence is mainly caused by the adults. The feeding of adults kills the seedlings before emergence from soil and can damage the young plants by notching. The leaves and cotyledons on young plants may be notched on the edges and have a "shot hole" appearance due to feeding of upper and lower surfaces of the plant tissue. The stems may also be fed upon, causing the plants to wilt or die on sunny days in severe infestation, or those become stunted and develop an increased number of basal tillers. Damage by new generation adult beetles occurs by their semicircular notching on upper and middle stems of flax plant that causes fibre cells as corky (Durnovo, 1935). Adults are also vectors of antracnose and fusariosis.

i) Large flea beetle

Scientific name & order: *Aphthona euphorbiae* **Schrank, 1781 (Coleoptera: Chrysomelidae)**

Synonyms

= *Aphthona aeneomicans* Allard, 1875

= *Aphthona virescens* Foudras, 1860

Biology

Egg: Eggs are light-yellow, 0.62 mm in length. Development of egg lasts 11-25 days.

Larva: Larva is milk-white, with light-yellow head. Thorax is with 3 pairs of legs. Last abdominal segment is the narrowest, without apical spine. Larva eats flax roots; its development lasts 26-29 days.

Pupa: Pupation occurs in soil, in a cradle, at a depth of 14 cm.

Adult: Body is slightly convex, metallic green in colour, upper side dark-green, blue or bronze. Frons and vertex are smooth, with thin punctation, elytra exhibit fine and dense punctation. Spur of hind tibia is situated on the outside of apical margin. Body length reaches 1.5-2 mm. Female lays 1-3 eggs in the surface layer of soil on roots of flax seedlings or near host plants. Fecundity is about 300 eggs.

Lifecycle: Large flea beetle (*A. euphorbiae*) is moderately mesophilous species. Adults appear early in spring during bud blossoming of birch, under the soil moisture with levels of 25% and less. Beetles fly actively at air temperatures of about 20°C. Adults stay first on grains, crucifers, beets, and weeds. They migrate to flax after appearance of seedlings, and here their mating and oviposition occur. Adults of new generation emerge in 2.5-3 weeks and appear in southern regions at the end of June and the beginning of July. After additional feeding on flax, the beetles migrate for hibernation.

ii) Flax flea beetle

Scientific name & order: *Longitarsus parvulus* **(Paykull, 1799) [Coleoptera: Chrysomelidae]**

Description

The flax flea beetle *L. parvulus*, also known as the springing black beetle, is 1.0-1.2 (-1.5) mm long, shiny reddish-brown, dark brown or black, with orange-brown leg, hind femora darkened, and has long first tarsal segments on its hind legs.

Life cycle: Flax flea beetle adult begins to emerge from winter hibernaculum when air temperature is 9–12°C (Popov and Firsova, 1936) and the temperature in the top soil reaches 11°C (Fritzsche, 1958). It prefers dry soil in spring to emerge (Popov and Firsova, 1936). Adult starts feeding as spring temperature near the soil surface reaches 15–20°C (Fritzsche and Hoffmann, 1959; Jourdheuil, 1960). Its peak flight activity occurs in May at the time seedling of spring-sown cultivar begins to emerge from the soil and air temperature exceeds 20°C (Fritzsche, 1958). Adult feeds on the cotyledon, vegetative bud and stem of newly emerged seedling. Female lays eggs in soil surface crack on or near the lateral root of the plant in June to early July. The larva emerges within about three weeks. It feeds on the lateral roots and tunnel into the tap roots of young plants, or feed on the root cortex and lateral roots of older plants (Popov and Firsova, 1936; Jourdheiul, 1963). New generation beetle emerges from August to early September, although summer adult has been found as early as mid-July in Ireland (Lafferty *et al.*, 1922). Adult feeds on the leaves, stems and seed capsules of flax until the plants mature before seeking dry sites in wooded or grassy areas, in crevices of walls, or other protected places in late September and October to overwinter.

Management/ Control measures

Cultural Control

Generally, cultural control methods have been less effective than the use of insecticides in areas of high flea beetle populations. Most part of population of Large flea beetle hibernates in the debris of deciduous forests; therefore, crop rotation has no influence on pest population density. However, in areas of lower infestations, cultural practices such as good seedbed preparation, the use of faster establishing cultivars, and very early (late April) seeding can allow the seedlings to outgrow flea beetle damage. Thus the cultural control measures, as suggested, include the use of fertilizers to accelerate seedling development, fast sowing of flax as early as possible, autumn ploughing immediately after harvesting and pre-sowing pesticide treatment of seeds. Early sown crops that develop under cool conditions are less susceptible to damage in areas where the large flax flea beetle dominates, while later sowings are more effective where the flax flea beetle is more common (Popov, 1941). Historical reports showed that fields with increased seeding rates or narrower row spacings are often less attacked (Kurdiumov, 1917).

Natural enemies

Adults of Flax flea beete are parasitised by a braconid wasp (probably Perilitus sp.), a nematode (*Howardula phyllotretae*) and a mite (*Trombidium* sp.). Eggs are parasitised by a mymarid wasp (*Anaphes regulus*).

Chemical control

Seed treatment: The use of seed dressing insecticides is the most common means of preventing fleabeetle injury to the emerging seedlings (Gheorghe and Doucet, 1987; Gheorge*et al.*, 1990; Horak, 1991; Trotus *et al.*, 1994), particularly to seedlings that may be attacked before they are fully emerged. Jankauskiene *et al.* (2005) observed in Lithuania.that 49.0 to 90.5% of flax plants were affected by flea beetles (*Aphthona euphorbiae* Schr., *Longitarsus parvulus* Payk.) annually in the plots sown with untreated seed. Significantly fewer affected plants were identified in the plots sown with the seed treated with phurathiocarb 900g + metalaxil 60g + fludioxonil 6g 100 kg^{-1}seed (Rapcol, 3 1100 kg flaxseed^{-1}) and thiamethoxam 28 g + metalaxil-M 3.33 g + fludioxonil 0.8 g 100 kg^{-1} seed or thiamethoxam 70 g + metalaxil-M 8.325 g + fludioxonil 2 g 100 kg^{-1} seed (Cruiser,100 ml or 250 ml 100 kg^{-1}). Thiamethoxam + metalaxil-M + fludioxonil, applied at higher dose rates (Cruiser, 500–1000 ml 100 kg^{-1}) gave a higher efficacy. However, Seed treatment is found as less effective than the standard post emergence treatment with pyrethroid insecticides. For example, in an investigation seed treatments were compared with untreated seeds and standard post emergence sprays with deltamethrin or parathion-methyl. Application of thiamethoxam 280 g/l at a rate of 1.1g a.i./kg seed showed no decrease of attack in comparison with the standard spray treatment. Seed treatment with thiamethoxam 600 g/l at 0.6g a.i./kg also showed insufficient protection (Huiting and Ester, 2007).

Foliar spray: Foliar treatments will also protect young seedlings (Oakley *et al.*, 1996; Haydock and Pooley, 1997) if applied when flea beetles invade fields after seedlings have emerged. It is suggested to apply an approved pyrethroid insecticide (deltamethrin or parathion-methyl) if new damage is seen, when the crops under 5 cm tall. The insecticide application is to be repeated (or needed in case of first application) if three or more Flax Flea Beetles are found on a 5cm row of flax cotyledons.

7. Foliar caterpillars (climbing cutworms)

Climbing cutworms are a group of noctuid moth larvae or caterpillars that occasionally attack flax, often causing extensive damage within a short period. The caterpillars feed almost exclusively above ground parts of the plant. Damage to flax capsules is generally done by later instar caterpillars after they fed on flowers, buds or leaves. The most serious climbing cutworm in flax is the flax caterpillar *Rachiplusia nu* (Guenee). It is frequently found in many of the flax-growing areas of Brazil and Argentina (Griot, 1944; da Silva, 1987), and can produce as many as five generations per year. Larvae are mostly green with longitudinal dorsal markings, and can reach a length of 4 cm. They walk in a typical inchworm manner by arching their back upwards to bring the posterior part of their abdomen forward. Larval populations are highest in October, when seeds in the capsules are developing, and can cause extensive defoliation (da Silva, 1987). The adults are mainly dark coloured, stout-bodied moths with mottled forewings and wing spans of up to 40 mm. The females are prolific egg producers, laying hundreds of eggs over a period of about two weeks. In India, the larvae of the semi-looper, *Plusiaori chalcea* F., has been found widely to cause damage in linseed (Narayanan, 1962). Several other noctuids are also reported to feed sporadically on flax. For example, the army cutworm, *Euxoa auxiliaries* (Grote), and the bertha armyworm, *Mamestra configurata* Walker are found as pest of flax in North America (King, 1928). The flax worm or silver Y moth, *Autographa gamma* L., is found in Europe and Russia (Boldirev, 1923; Ruszkowski, 1928), and the cabbage armyworm, *Barathra brassicae* L.is in northern China (Li, 1980; Chen *et al.*, 1990; Anonymous, 1998). The army cutworm and bertha armyworm are univoltine, having one generation a year, while the flax worm and cabbage armyworm have two generations, of which the second one is of greatest concern in flax. These species overwinter either as partially developed larvae or as pupae. The larvae of army cutworm and flax worm become active in early spring and feed on various weeds. In addition, the larvae of a pyralid moth, the beet webworm *Loxostege sticticalis* L., occasionally found to cause damage flax in some areas of western Canada (Strickland and Criddle, 1920), Russia (Esterberg, 1932; Berezhkov, 1936), and northern China (Lei, 1981). Early instar larva of this species is dark green, becoming black near maturity, with two white lines along the length of its back.

Management/Control measure

Cultural control: Controlling weeds in flax largely eliminates problems with these insect pests, except during years when larval populations may move into flax from nearby infested fields.

Biocontrol: Application of nuclear polyhedrosis virus (NPV) solution may be effective to control the climbing cutworms. Since, the populations of all climbing cutworm species are held in check by parasites and diseases, and both the bertha armyworm and the flax worm are highly susceptible to nuclear polyhedrosis viruses during outbreaks (Vago and Cayrol, 1955; Erlandson, 1990).

8. Thrips

Several thrips species are found to attack flax (Franssen and Mantel, 1961; Walters and Lane, 1991; Abrol and Kotwal, 1996), however, not more than two or three (or four) species breed on flax (Morison, 1943). Most species, except *Thrips linarius* (= *Thrips lini*) and *T. angusticeps*, are relatively rare and are not known to cause noticeable crop damage. In India, a different types of thrips, known as 'Linseed thrips', *Caliothrips indicus* is recognised as the key pest of linseed (flax), causing considerable losses in seed production (Rawat and Kaushik, 1983; Deshmukh *et al.*, 1992). Distribution of thrips tends to be general across a field, with little congregation at the edge (Walters and Lane, 1991).

i) Flax thrips (*Thrips linarius*)

Occurrence & severity

The species is distributed in Europe and Northern Africa. In the former USSR the area of Flax Thrips coincides with the territory of flax growing from western state borders eastward to Krasnoyarsk Territory. The adult and larvae of *Thrips linarius* can seriously damage flax grown in western Europe (Bonnemaison and Bournier, 1964; Czencz, 1985; Brudea and Gheorghe, 1989) and Russia (Uvarov and Glazunov, 1916). The thrips (*T. linarius*) feeds primarily on flax (Czencz, 1985) and is commonly known as flax thrips. Thrips are the most serious insect pest of flax in many flax-growing areas of Europe.

Types of damage

Feeding by larvae and adults of *T. linarius* on the growing points and young leaves of mature flax plants either kills the growing points or causes abnormal cell division which distorts growth. This can develop profuse branching on the main shoot or produce flowering tillers which delay crop maturation and make difficulties to harvest. Feeding by thrips causes flax foliage to assume a red and spotted appearance. Heavily infested plants are often short, thus, reducing the length of fibre. Feeding of thrips also increases self sterility in fibre flax by transferring pollen from flower to flower and reduces seed yields (Rataj, 1974). Enzymes excreted during feeding and the extraction of sap from buds and flowers can cause withering of flowers and leaves at the top. The damaged flowers often drop prematurely or produce drastically reduced numbers of seeds. Capsules on infested plants frequently burst open before ripening, causing a reduction in their weight and in number of seeds.

Scientific name & order: *Thrips linarius* **Uzel, 1895 (Thysanoptera: Thripidae)**

Synonyms

= *Thrips lini* Ladureau, 1877 (?)

= *Thrips armeniacus* Pelikan, 1973

= *Thrips ponticus* Knechtel, 1965

= *Thrips tenuisetosus* Knechtel, 1923

Biology

Egg: Egg is transparent-whitish and about 0.3 mm in size. Both fertilized and unfertilized eggs are laid by Thrips. The eggs laid by fertilized females produce mostly female offspring. Males develop parthenogenetically from unfertilized eggs (Zawirska, 1963).

Larva: Larval stage completes in two instars. Larva is similar in appearance to adult, transparent after hatching, then yellowish in color; lateral surface of pronotum darkened; meso- and metanotum with black spots. Antennae are short, 3rd and 4th segments thickened. Length of mature larva is about 0.9 mm and lack of wings. larva actively moves about on the host when disturbed and is most active on calm, clear days (Zawirska, 1963).

Prepupa: Prepupa is yellowish; its wing pads transparent; antennae directed anteriad.

Pupa: Pupa is the same in colour of prepupa. Its wing pads are longer; antennae 7-segmented, directed backward.

Adult: Adult thrips are very small insects, with narrow bodies and prominent legs and antennae. Body length of female is 0.9-1.0 mm, dark-brown, almost black. Head is with ocellar triangle between compound eyes; piercing and suctorial type of mouthpart. Antennae are 7-segmented, mostly black, apical part of the 2nd and 3rd segments yellowish, 4th segment as long as 3rd and longer than 5th, 7th segment narrow. Pronotum is with 3 pairs of posteromarginal setae. Forewings are narrow, acute apically, slightly darkened, without transversal veins, the 1st vein with 3 setae in distal part. Only macroptera form is known. Abdomen is widened. Legs are black, only fore-tibia yellowish, with somewhat darkened anterior surface. Femur is toothless, tarsus 2-segmented, with tarsal bladder at apex. Ovipositor is curved downward. Male is smaller than female and lighter in color. Ventral surface of 2nd-6th segments of abdomen are with transparent ellipsoidal spots.

Life cycle

Adults hibernate in fields, in soil at a depth of 20-40 cm. Mass appearance of adult occurs in spring, at temperature about 20°C at a depth 20 cm, and then mating begins. In spring adults appear on weeds for the first time. Generally, only females fly in search of flax fields. Males of *T. linarius* are few in number and usually remain on weeds at the overwintering site, while females move to flax after mating. Imagoes concentrate near stem growing-point and under sepals of buds and ovaries. Females lay eggs in slits cut with the ovipositor into the apical part of host, around base of leafstalk and into buds. Facundity is near 80 eggs. One generation develops in 42-46 days; egg develops in 5 days and larva in 23-25 days; After feeding on the flax larva drops to the ground and pupates in the soil at a depth 10-25 cm into the small chamber (Ermolaev, 1940). After emergence, adults stay in the ground before the next spring. Population density of thrips in nature depends on the activity of predators, e.g.: larvae of *Aeolothrips fasciatus* L. There are several generations in a year. In most areas of Europe, *T. linarius* is univoltine, having only one complete generation per year, while, it is thought to have a partial second generation in Romania (Brudea, 1990) and at least two generations in Poland (Zawirska, 1960).

Management/Control measures

Cultural methods

Control measures include deep autumn ploughing, crop rotation of flax, increase of seeding rate, using of fertilizations for the development acceleration of seedlings and sowing the flax at the earliest time. Sowing flax early in the growing season can reduce thrips injury, particularly where *T. linarius* is common. Thrips are very susceptible to drowning by heavy rains when they are burrowing into the soil to overwinter, when they emerge in the spring, or if they are dislodged from plants by rains in the summer. Cool and rainy summers suppress thrips reproduction and can result in over ten-fold differences in overwintering populations from year to year (Franssen and Mantel, 1961; Bonnemaison and Bournier, 1964).

Chemical control

Foliar application of insecticides, especially those with systemic activity, can effectively prevent feeding injury by thrips. Application timing should coincide with the most susceptible stage of the thrips and before significant feeding damage has occurred. In most flax fields insecticides are applied in early June or at the onset of flowering when larvae begin to hatch or are of early instars (Brudea and Gheorghe, 1989).

ii) Field thrips (*Thrips angusticeps*)

Occurrence & severity

Field thrips (*Thrips angusticeps*), also known as 'flax thrips' or 'cabbage thrips', is the most serious insect pest of flax in many flax-growing areas of Europe. It is the main damaging pest of flax and linseed in continental Europe, where it has been one of the main pests for decades (Franssen and Huisman, 1958; Bonnemaison and Bournier, 1964; Czencz, 1985; Beaudoin, 1989a). The adult and larvae of field thrips can seriously damage flax grown in Western Europe (Bonnemaison and Bournier, 1964; Czencz, 1985) and Russia (Uvarov and Glazunov, 1916). However, the thrips is only of economic importance in crops on clay soils (Franssen and Mantel, 1963). Two stages of the crop are vulnerable: seedlings can be attacked in spring by brachypterous adults, and flowering plants can be attacked in summer by macropterous adults and their larvae (Franssen and Mantel, 1963). Flax seedlings or young plants sown in or near fields infested by thrips the previous year can be severely damaged or killed by the first generation of *T. angusticeps*. Seedlings whose growth is retarded by cool, dry weather are particularly susceptible (Franssen and Huisman, 1958; Czencz, 1985; Beaudoin, 1989a). As regards the yield loss, in the Netherlands, the yield of flax was increased by 30% following application of insecticide for *T. angusticeps* at the start of flowering (Franssen and Kerssen, 1962). The field thrips was reported to cause as much as 14% yield loss in the flax and linseed grown in the UK in 1990s (Ferguson *et al.*, 1997). The thrips prefers flax but is common on many other crops, particularly peas (Doeksen, 1938), beets, onions, radish, wheat and barley (Franssen, 1955; Franssen and Mantel, 1961).

Types of damage

Seedlings or young plants of flax can be severely damaged or killed by the first generation of *T. angusticeps*. Seedlings whose growth is retarded by cool, dry weather are particularly susceptible (Franssen and Huisman, 1958; Czencz, 1985; Beaudoin, 1989a). The affected flax plants grow more slowly, turn yellowish-grey with twisted leaves and the tops of the stems are erect instead of hanging down. Leaves assume a more horizontal position and may fall off in the top 5-10 cm of the plant. Feeding by larvae and adults of second generation *T. angusticeps* on the growing points and young leaves of older plants either kills the growing points or causes abnormal cell division which distorts growth. This can also result in profuse branching on the main shoot or the production of flowering tillers which could delay crop maturation and make harvesting difficult. Feeding by thrips can cause flax foliage to assume a red, spotted appearance. Leaves and stems may show the typical thrips 'silvering' produced by feeding.

These areas may then turn yellow or brown or necrotic. Heavily infested plants are often short, which can reduce the length of their fibre but usually does not affect its quality. The thrips transfer pollen from flower to flower on flax plants and this is the main and often sole cause of self-sterility in fibre flax, leading to reduced seed yields (Rataj, 1974). Enzymes that excreted during feeding and the extraction of sap from buds and flowers can cause withering of flowers and leaves at the top. The damaged flowers often drop prematurely or produce drastically reduced numbers of seeds. The capsules on infested plants frequently burst open before ripening that cause reduction in seeds (Bonnemaison and Bournier, 1964; Nakahara, 1984; Kirk, 1996).

Scientific name & order: ***Thrips augusticeps*** **Uzel, 1895 (Thysanoptera: Thripidae)**

Biology

Egg: Egg is transparent-whitish and about 0.3 mm in size. Eggs are laid on host plants in the spring. The eggs are laid as both fertilized and unfertilized (Zawirska, 1963).

Nymph: There are five instars. Nymphs are almost colourless feed on seedling leaves for up to 4 weeks before developing dark-coloured long-winged adults (Fransen and Huisman, 1958). Larva actively moves about on the host when disturbed and is most active on calm, clear days.

Adult: Adults are dark-brown or black coloured insects, about 1-1.3mm in length, with short legs. As wingless (short winged) individuals, they live in the soil over winter before laying eggs on host plants in the spring. All adults of the summer generation of *T. angusticeps* have four long narrow wings that are fringed with long hairs, and are strong fliers, whereas the adults of the overwintering generation of *T. angusticeps* have short wings and are flightless. The eggs laid by fertilized females produce mostly female offspring. Males develop parthenogenetically from unfertilized eggs (Zawirska, 1963). Adults of thrips actively move about on the host when disturbed and are most active on calm, clear days. Adults which are produced in the summer have long narrow wings, fringed with setae and can fly away from the crop.

Life cycle: Adults of field thrips overwinter in the soil at depths of 20 cm or more and emerge in the spring when soils begin to warm (Ermolaev, 1940; Franssen and Huisman, 1958) and daytime temperatures average 8 to 10°C (Gheorghe, 1987), or remain to hibernate another year in the soil (Franssen and Mantel, 1961). The short-winged adults of *T. angusticeps* often emerge well before flax is seeded, and seek out other plants on which to feed and lay eggs. The larvae remain on these plants until they become long-winged adults. Then the adults of both sexes either remain or move to flax where they feed on young leaves or growing points at the top of the plant. Females lay eggs in June and July near the growing points on young plants or on the inside of flower buds on more mature plants. Larvae feed for about four weeks. The young adults remain in the soil and move to greater depths with the onset of cold weather. In most areas of Europe, *T. angusticeps* produces two generations: the short-winged colourless females of the overwintering generation lay eggs which develop into the long-winged summer adults (Franssen and Huisman, 1958). However, in Finland, there is only one generation of *T. angusticeps*, which is entirely short winged (Hukkinen and Syrjanen, 1940),

Management/ control measures

Cultural Control

Crop rotation may also be helpful to manage the thrips attack. In case of severely affected areas, it is suggested not to grow flax or other susceptible crops, like, linseed, peas, wheat, barley, rye or brassicas for at least two years. Instead, the non-susceptible crops, winter cereals, caraway, winter oilseed rape and oats, can be grown in the field where the field was infested with the field thrips. Franssen and Huisman (1958) have recommended the following rotations for growing flax: red clover-oats-flax; potatoes-oats-flax or beetroot-oats-flax. Damage to flax by the overwintering generation of *T. angusticeps* can be prevented by sowing flax in fields not sown with a susceptible crop in the previous year. Flax should not be grown after peas, mustard, or cole crops, but by following a rotation that includes cereal crops, red clover or potatoes (Franssen and Huisman, 1958).

Chemical control

Foliar application of insecticides, especially those with systemic activity, can effectively prevent feeding injury by thrips. Application timing should coincide with the most susceptible stage of the thrips and before significant feeding damage has occurred. In most flax fields insecticides should be applied in early June or at the onset of flowering when larvae begin to hatch or are early instars (Brudea and Gheorghe, 1989). Application of an insecticide when the flax is about 2.5 cm high followed by a second treatment about two weeks later may be needed in fields being attacked by overwintering populations of *T. angusticeps* and their offspring (Franssen and Kerssen, 1962). Several pyrethroid insecticides are effective against the field thrips. The systemic insecticide, Fentrothion, gives effective control of this field thrips. Ferguson *et al.*, (1997) have found Deltamethrin and Dimethoate as effective against *T. angusticeps* in trials. Further, the insecticide Chlorpyrifos has been found effective against the field thrips in UK (Bateman *et al.*, 1997).

iii) Linseed thrips (*Caliothrips indicus*)

Occurrence & severity

Linseed thrips (*Caliothrips indicus*), also known as 'black thrips' or 'sesbania thrips' is known to be widespread in India. Amongst the insect pests attacking linseed (flax) in India, the linseed thrips is recognised as the key pest, causing considerable losses in seed production (Rawat and Kaushik, 1983; Deshmukh *et al.*, 1992). At Raipur (Chhattisgarh) in India the incidence of the black thrips is very high as observed in a study during the year 1995-1996. The incidence as recorded was 0.04-12.94 per bud (Patel and Thakur, 2005). The thrips attack is more prevalent if the crop sown in light dry soil in dry season as compared to the crop sown in heavy moist soil and in humid weather. The thrips (*C. indicus*) under dry weather conditions multiplies faster and its population devastates the crop voraciously. The Linseed thrips, *C. indicus* is one of the serious polyphagous species of cultivated and wild plants, aubergine, groundnut, soybean, peas, clover and flax. It occurs most frequently on leguminous plants.

Types of damage

Both the nymph and adults suck the sap from the leaves and cause the damage to the plants. The surface of the leaf tissue is lacerated deeply and the thrips suck up the ooze out saps which result in the formation of white patches (Saxena, 1971). Usually plants attacked by *C. indicus* survive but the infestation slowly moves towards the base and drying of leaves start from the tip and turn brown. The symptoms develop more rapidly, if the crop sown in light dry soil in dry season as compared to the crop sown in heavy moist soil and in humid weather. The linseed thrips under dry weather conditions multiplies faster and its population devastates the crop voraciously.

Scientific name & order: *Caliothrips indicus* (Bagnall, 1913)

Synonym

= *Heliothrips indicus* Bagnall, 1913

(Bagnall, 1913)

Biology

Egg: Eggs are bean shaped, laid singly in leaf tissues. A female lays 19-96 eggs during the oviposition period of 17-23 days. Egg incubation period is 6-15 days.

Larva: Larval period ranges from 4-10 days. There are two larval instars.

Prepupa: The prepupal period varies from 1 to 5 days.

Pupa: The pupal period varies 1 to 6 days. Pupation normally takes place in the soil.

Adult: The adult female is dark brown or black coloured, while, the male is light black. The adults contain heavy net-like sculpture with internal markings. The wings are fringed with brown bands, more than half long as abdomen. Antanne are 8 segmented. Mid and hind tarsi in legs are with one segment. Longevity of adults varies from 21-30 days.

Life cycle

Eggs hatch to larvae in 6-15 days. Larvae live for 4-14 days and adults live for 7-15(differently 21-30) days both dependent on temperature (Lewis, 1973).

Control measures

Lambda-cyhalotherin (0.005 %) treated plots proclain significantly low thrips population (0.96 thrips/ plant), followed by monocrotophos (0.05%) (Sahu, 1999). Spraying of dimethoate 0.03% or malathion 0.05% or ethion 0.05% or quinalphos 0.025% is also effective to control the thrips.

7
Sisal (*Agave sisalana*)
Asparagales: Asparagaceae

A. Diseases

1. Anthracnose

Occurrence & severity status

Anthracnose of sisal is also called as leaf spot or ring spot disease. The disease is frequently accompanied with sun scorch in Kenya and Tanganyika (Morstatt, 1930). This is considered as a leaf disease of minor importance, occurring on numerous species of *Agave* as well as the closely related genus *Furcraea*, including *F. macrophylla* Baker (wild sisal). The disease is distributed in warm temperate and tropical regions including southern Europe, British Guiana (Bancroft, 1914), the Neotropics, Cyprus (Georghiou and Papadopoulos, 1957), Kenya (Nattrass, 1961) and USA (Parris, 1959). It is also been reported to occur in various countries of Asia, including China (Tai, 1979), Korea (Cho and Shin, 2004), the Philippines (Teodoro, 1937) and Taiwan (Anonymous, 1979). In India, the disease was first reported as leaf spot caused by *Colletotrichum agaves* Cav. (Butler, 1905). Its occurrence was also reported from Antigua (Bancroft, 1910) and Puerto Rico (Macedo, 1943). Apart from sisal, the disease also occurs on tomato (*Lycopersicon esculentum*), potato (*Solanum tuberosum*) and over 35 other hosts, representing 13 families, chiefly in the Cucurbitaceae, Leguminosae and Solanaceae (Chesters and Hornby, 1965). The disease also occurs on onion and strawberry (Mordue, 1967).

Symptoms

The disease appears as small, circular, darker in colour than the adjacent parts or chlorotic or pale yellowy soft which eventually turn brown or grey, with or without a brown margin, scattered irregularly, slightly raised spots on the upper surface. Sometimes the spots are arranged in rows near the leaf margin. The spots are usually 2-4 cm diam, up to 6 cm diam or more, several on one leaf, occasionally becoming confluent, and often exuding resin droplets, eventually causing death of plant tissue that becomes purplish to dark brown (Bancroft, 1910; Morstatt, 1930; Farr *et al*, 2006).

Causal organism: ***Colletotrichum coccodes*** **(Wallr.) S. Hughes 1958**
[Canadian Journal of Botany 36 (6): 754]

Synonyms

= *Colletotrichum agaves* Cavara, 1892

Morphological/ Morphophysiolgical characteristics

Acervuli are epiphyllous, scattered or forming concentric rings, distinct to confluent, developing beneath epidermis, becoming erumpent with age, often surrounded by remnants of epidermis, globose to oblong, attaining approximately 300 µm diam, in longitudinal section of hyaline to dark brown cells forming textura angularis, brown to black, producing masses of pale orange conidia. Setae are numerous, projecting beyond hymenial surface, brown to black, thick-walled, septate, straight to irregularly crooked, tapering to acute or rounded apices. Conidiophores are dark brown to hyaline arising from upper cells of acervuli. Conidiogenous cells are hyaline, enteroblastic, 18.5–24 x 3.5–7 µm, cylindric, periclinal thickenings moderate, channel narrow to broad. Conidia are hyaline, (17.5–) 19.0–30.5 (–33) x 5–8 (–9.5) µm, aseptate, cylindrical with obtuse ends, straight or slightly curved, with one or two guttules (Farr *et al*, 2006).

Colonies on potato dextrose agar show sparse whitish aerial mycelium. Sclerotia are usually abundant evenly distributed over the agar surface; when young, greyish and glabrous, rapidly becoming dark and setose. Acervuli are formed in association with sclerotia or as separate aggregates of setose mycelium; solitary phialides often found on mycelium. Conidiophores are sometimes septate and branched. Spore masses are normally small, colourless to salmon orange. Conidia are frequently shorter in proportion to their breadth than on host, many irregular forms occur. Reverse of colony is grey, darker with age because of formation of appressoria. Appressoria are very readily formed in slide cultures, cinnamon buff, ovate or obclavate to elliptical, occasionally irregularly lobed, borne on hyaline thin-walled sigmoid supporting hyphae (Mordue, 1967). Colonies are 3.7–4.2 cm diam, rosy buff to sepia, peach at margin, margin smooth, broadly wavy, reverse rosy buff to peach, profuse sporulation on agar. Appressoria are not seen on agar. All other characters are as on natural substrata (Farr *et al*, 2006).

Molecular characteristics

The molecular characterization of *Colletotrichum coccodes* (=*C. agaves*) with culture voucher/specimen numbers CBS 318.79 (as *C. crassipes*)/BPI 871832 has been done. The pathogen was collected from the host Agave in Netherlands. The Gene bank Accession Numbers for LSU and ITS are DQ286220 and DQ286219, respectively. The molecular characterization of *C. coccodes* collecting from the host Agave in Mexico was also done for the culture voucher/specimen numbers AR 3920, CBS 118190/BPI 871831.The Gene Bank accession numbers for LSU and ITS are DQ286222 and DQ286221, respectively (Farr *et al*, 2006).

[N.B. The abbreviations denoted as: AR, Amy Rossman; BPI, US National Fungus Collections; CBS, Central Bureau voor Schimmel cultures]

Predisposing factors

The optimum temperature for this fungus is between 25 and 30°C, which accounts for it's more rapid development in the summer periods, although it is quite capable of developing at much lower temperatures. It is often associated with light sandy soils, inappropriate fertilisation and short rotations. Microsclerotia that are present in the soil can be infectious for more than 2 years or up to 8 years (Dillard and Cobb, 1998). The spores are distributed by rain splashes or windborne in rain. Wet weather promotes disease development, and splashing water in the form of rain or irrigation favours the spread of the disease (Dillard, 1987).

Disease cycle

The pathogenic fungus can survive the winter as hard, melatinized structures called sclerotia. It may also survive in debris as thread like strands called hyphae in plant debris in soil. In late spring the lower leaves and fruit may become infected by germinating sclerotia and spores in the soil debris. Infections of the lower leaves are important sources of spores for secondary infections throughout the summer and rainy seasons. Leaves with injury are especially important spore sources because the fungus can colonize and produce new spores in these wounded areas. The pathogen also produces acervuli which are full of conidia that help to spread the infection (APPS, 2009)

Management/Control measures

Good cultural and sanitation practices can reduce inoculum and disease dispersal. For example, elimination of weeds or wild hosts from field can reduce levels of inoculum and survival of the pathogen in the soil. Further, removal of leaves with active lesions of the plant can minimize the spread of the disease. Overhead watering where disease has occurred should be avoided. As regards the chemical control meausure, spraying the fungicide, like, thiophanate-methyl, during wet weather is effective and advocated (Ryczkowski, 2011; Rana, 2017)

2. Bole rot

Occurrence & severity status

Bole rot, also known as red rot, is the most serious disease of sisal caused by the fungus *Aspergillus niger* entering through the leaf bases after leaves are cut. The disease was first reported from Tanzania (Central east Africa) in the year 1952 (Wallace and Diekmahns, 1952). Later it was reported from Brazil in the year 2006 (Coutinho *et al.*, 2006). In Brazil the disease is much prevalent in Monteiro and Pocinhos Counties (Paraíba State) and Santa Luz County (Bahia State), where the disease can vary from 5% to 40% in the production areas (Coutinho *et al.*, 2006). The sisal cultivation in Brazil is mainly concentrated in the northeast region, where Bahia State is the largest producer, with 94% of the national production (Silva *et al.*, 2008). This pathogenic agent is contributing to the continuous decrease in sisal plantation in semi-arid region of Bahia in Brazil, and in many areas, it already witnessed the crop abandonment due to high disease incidence (Gama *et al.*, 2015). Similarly, the disease is also an increasing concern in most production areas of Tanganyika.

Symptoms

Soon after infection, the bole of sisal plant starts to rot. The internal tissues become brown, surrounded by a reddish border. After two to three months the rot can spread into the entire region below the meristem. The plant meristem is completely affected, interrupting the communication between the bole and leaves and making it as water soaked in appearance. When the bole is completely rotten, the leaves turn yellow and collapse and the plant dies (Coutinho *et al.*, 2006; Gama *et al.*, 2015). The infected plants exhibit wilting and yellowing of leaves and reddening of the stem or bole and base of the leaves followed by death (Sa, 2009). The causal fungus, entering through the leaf bases after leaves are cut, causes a wet rot which becomes yellowish-brown and soft, with a pinkish margin, and it may lead to plant collapse and death, but, when it enters the base of the bole through an injury it causes a basal dry rot (Brink and Escobin, 2003; Oyen, 2011). The wet type rot almost always originates in cut leaf bases, spreading inwards and to adjacent leaf stumps. In longitudinal section there is no sharp differentiation between rotted and healthy tissues. The dry type of rot affects uncut as well as cut plants. Both forms of the disease are fatal and caused by the same pathogen, *A. niger*. When the affected boles are split and left for two days in an enclosed space almost the whole of the rotted surface become covered with fructifications *of A. niger*. It appears that both types are different manifestations of the same disease and that whether the rot is to be wet or dry possibly depends upon environmental factors or the chemical status of the plants (Wallace and Diekmahns, 1952).

Causal organism: *Aspergillus niger* Tiegh., 1867

[Annales des Sciences NaturellesBotanique 8: 240]

Synonyms

= *Sterigmatocystis nigra* (Tiegh.)T iegh., 1877

= *Aspergillopsis nigra* (Tiegh.) Speg., 1910

= *Rhopalocystis nigra* (Tiegh.) Grove, 1911

Morphological/ Morphophysiolgical characteristics

On PDA medium, the colonies are initially white, but turn black with the development of conidia after 5 days of incubation at 25°C. The fungal hyphae are septate and hyaline. Conidial heads are initially radiate, splitting into columns at maturity. Conidiophores arise from long, broad, thick-walled, mostly brownish, sometimes branched foot cells. Conidiophores are long, 250-360µm in length, smooth, hyaline, becoming darker at the apex with a globose vesicle of 20-40µm in dia., metulate and phialides cover the entire vesicle; Conidia are in large, radiating heads, mostly globose, brown to black, irregularly roughened, (3-)4.0-5.0 µm diam,, uninucleate (Coutinho *et al.*, 2006). Colonies are reaching 2.5-3 cm diam in ten days at 24-26°C on Czapek Agar, on Malt Extract Agar 5-6 cm, typically black powdery.

Scanning Electron Microscopy (SEM) and Transmission Electron Microscopy (TEM) studies of young conidiophores show the synchronous development of metulae and phialides. The wall of the first-formed conidium is continuous with that of the phialide while subsequent conidia are formed in a typical phialidic manner. During germination, a new inner wall layer is formed which gives rise to the germ tube. The rodlet pattern of the conidium surface as seen in freeze-etch replicas is prominent and interlacing with prominent localized raised areas. Heterokaryotic conidial heads are observed after recombination with an albino strain. Some isolates produce sclerotia, particularly at 30°C and to a lesser extent at 20°C. Sclerotium production is inversely correlated to conidium production. DNA analysis gives a GC content of 50.1-52% with nearly equal portions of adenine, guanine, cytosine and thymine (Domsch *et al.*, 2007)

Molecular characteristics

The DNA of the 23 *Aspergillus* isolates collecting from sisal fields (one was from a diseased stem, eight from leaves, 10 from soil, and four from the air) were used for amplified polymorphic DNA (RAPD) analysis with primers A1, A6 and OPA4 (Santos *et al.*, 2014). The programme FreeTree was used to construct the dendrogram using the distances of Jaccard and the UPGMA method. Seven isolates of different groups were selected from the RAPD analysis for molecular identification by sequencing a fragment of 520 pb of the β-tubulin gene. Primers Bt2a and Bt2b were used for amplification and sequencing, which were done according to standard protocols using an ABI 310 sequencer. Alignment of the obtained sequences and reference sequences from public databases was performed using the programme MAFFT version 6.0. The phylogenetic tree was constructed with the maximum likelihood method implemented in the programme MEGA version 5.0, with the Kimura-2 parameter nucleotide substitution model and bootstrap analysis with 1000 resamplings. RAPD analysis showed that the isolates were grouped into 16 different genetic groups with 100% similarity. From these 16 groups, only *Aspergillus alabamensis* (ANS 181) and *A. aculeatus* (ANS 113) could be distinguished from the others in terms of morphology. Sequence analysis showed the species, *A. niger*, *A. aculeatus*, *A. alabamensis*, *A. tubingensis* and *A. brasiliensis*, were in association with sisal. The identity of the sequences varied from 99 to 100% when they were compared with other sequences from the databases. Pathogenicity tests showed that only *A. niger*, *A. tubingensis* and *A. brasiliensis* were pathogenic to sisal (Santos *et al.*, 2014). In another study the putative pathogenicity genes of *A. niger* in sisal and their expression *in vitro* were studied to investigate the involvement of hydrolytic enzymes (cellulase, cutinase, protease and lipase) and ochratoxin A in an *in vitro* simulation of the interaction between *A. niger* and sisal to understand the mechanisms of fungal pathogenicity. Primers for cutinase lipase, protease, ochratoxin A, and elongation factor 1-α (EF), with this last one used as endogenous control were designed and optimized for real time quantitative PCR (qPCR) with the SYBR Green methodology. The genes potentially involved in the pathogenicity of *A. niger* showed higher expression in medium supplemented with sisal as compared to the minimal medium. Primers for cutinases, proteases, cellulases and ochratoxin A, genes putatively involved in the pathogenicity of *A. niger* to sisal were designed and optimized for qPCR. The qPCR analyses revealed that both hydrolytic enzymes and ochratoxin A were expressed at some moment during fungal growth in minimal medium and in the medium supplemented with sisal extract (de Souza *et al.*, 2017).

Predisposing factors

In Tanzania, it has been found that the bole rot infected parts of sisal hemp while eaten by the weevil (*Scyphophorus interstitialis*) ingests *A. niger* spores that remained viable at its excreta, suggesting the weevil attack in sisal as one of the possible factors for spreading the disease (Wienk, 1967).

Disease cycle

This fungus is spread via the air, soil, and water. It is generally a saprophyte, living on dead and decaying matter. The pathogen enters through the cut end leaf base or infested site of insect pests. The conidia after landing on the above places start to germinate producing foot cells. The mycelia enter into the host cells through the raptures. More branching occurs and elongation of hyphae creating a mass of hyphae or mycelium. Soon conidiophores grow from the foot cells and then the conidial head.

Control measures

Bole rot of sisal can be reduced through removal of infested material and harvesting under dry conditions (Brink and Escobin, 2003; Oyen, 2011). It can also be managed by providing proper drainage and spraying with Mancozeb (WP) @ 0.2% (Annonymous, 2013). Further, the use of homeopathic drugs, like Ferrum metallicum 9CH and Natrum muriaticum 5CH, are reported as a good strategic control measure of *A. niger* incidence in sisal plants causing bole rot (Gama *et al.*, 2015).

3. Korogwe leaf spot

Occurrence & severity status

Korogwe leaf spot disease of sisal is characterised by chocolate brown concentric scab like eruption that was first observed in 1951 in Korogwe District, Northern Tanzania by the Mlingano Agricultural Research Institute (ARI) - one of the renowned centres in Africa. Later, it gradually spread to neighboring plantations, and has now covered most of the major sisal growing regions in the country (Keswani and Mwenkalley, 1982). At Mingaro (Tanga region) in Africa the disease has been found to cause havoc damage of sisal fibre, causing losses estimated 30% in yield and a sharp quality drop, resulting substandard under grade. The disease mostly affects sisal hybrid 11648, the main commercial high yielding variety (Mohamed, 2016). Suckers arising from rhizomes of plants are prone to get the disease than those obtained from bulbils (Keswani and Mwenkalley, 1982). Korogwe leaf spot is normally severe on sisal raised on poor soils, particularly those deficient in potassium. Based on this observation, it was earlier considered as a deficiency disorder (Lock, 1969). However, now the disease is thought to be caused by a virus. The disease is now spread in Asia, particularly, in China. In a study a total of 64 affected leaf samples from 5 sisal farms located in Guangxi and Hainan provinces in the Southern China, and 13 farms located in the North Eastern Tanzania were collected in July to September 2013 for pathogen isolation and identification. Result showed that the disease is wide spread to all locations visited in a varying severity. Incidence up to 100% was recorded in some fields (Mtunge *et al.*, 2014). The disease causes difficulty in fibre extraction leading to darkened, low quality fibre (Mtunge *et al.*, 2014)

Symptoms

Korogwe leaf spot disease at first appears as tiny spots, which develop into corky, grey brown or chocolate brown circular scab-like spots or eruptions on both surfaces of mature leaves (Mpunami, 1986). Spots range from 1-30 mm diameter, and in severe cases, coalesce to cover a large portion of the leaf surface (Lock, 1969). The disease causes internal disorder with necrosis of the cell at the base of the stoma. The necrosis later spreads to the guard cells and adjoining tissues of the leaf. A cambium forms in response of infection, and cells in the infected area become suberized. When suberization reaches the leaf surface, the epidermis cracks and a tiny spot, the size of a pinhead, appears. The spot continues to expand with time as more cells become infected and suberized, eventually producing the characteristic leaf spot, with concentric, brown eruptions of scab appearance (Anon, 1955; Lock, 1969; Keswani and Mwenkalley, 1982; Mpunami, 1986). The disease causes brown staining of fibre and reduces tensile strength, subsequently lowering quality. Severely infected leaves are unsuitable for fibre extraction (Mwenkalley, 1978; Keswaniand Mwenkalley, 1982; Mpunami, 1986).

Causal organism: Virus (?)

The cause of the disease is thought to be a virus; however, different opinion still exists, making the cause as unclear (Mohamed, 2016). A school of scientists believe that the disease is caused by a Virus (Kimaro *et al.*, 1994), while, the others consider as combined effects of different fungi and other factors (Mtunge *et al.*, 2014). According to first group, a long, very flexuous, rod-shaped virus has been isolated from Korogwe leaf spot (KLS)-infected sisal leaves at Oregon State University (Mpunami, 1986). The flexuous virus particles are the infectious agent of KLS. Presence of these flexuous rods in leaf dip extracts and purified preparations from symptomatic sisal leaves and sap-inoculated leaves, as well as particles in phloem cells of KLS-infected tissue, is strong evidence for relating the virus to the disease. KLS-associated virus is very sensitive to shear under stress during mechanical extraction and when sap is clarified with organic solvents. The tendency for shearing during extraction is probably the main reason for inability to band the virus in sucrose rate-zonal gradients. This property is the main obstacle in purification of CTV in sucrose gradients (Garnsey *et al.*, 1981). Adsorption of virus particle fragments onto light weight, open-structure, tubular bodies during purification is a unique property of KLS-associated virus (Mpunami, 1986). This virus is partially characterized by mechanical transmission, host range, cytopathological effects, electron microscopy, serology and determination of some biophysical properties. The virus is sap-transmissible with difficulty to young sisal and common bean (Mpunami, 1986). However, it is unable to produce symptom in sisal during pathogenicity test in growth chamber (Mpunami, 1986). Since, Korogwe leaf spot disease is known to have a long incubation period (Lock, 1969; Keswani and Mwenkalley, 1982). KLS-associated virus is distinct from the virus-like organisms associated with parallel streak, a disorder reported to occur naturally in sisal in Kenya (Pinkerton and Bock, 1969). The second group reported through a combined fungal morphological identification and ITS sequence analysis that a total of 11 fungal species have been identified in 110 isolates from both China and Tanzania. Among these, *Fusarium equiseti* has the highest frequency of occurrence covering all locations, followed by *Alternaria alternata, A. tenuissima* and *Phoma herbarum*. The findings suggest the possibility of association of a disease with several fungal pathogens in combination with other factors (Mtung'e *et al.*, 2014).

Predisposing factors
The disease favours the sisal grown on poor soils, particularly those deficient in potassium (Lock, 1969).

Control measures
The disease is mainly managed by the cultural control measures. It is suggested to maintain clean fields by effective weed control measures, providing adequate soil fertility in field and to use of clean planting material. The production of disease-free plants is important to maintain areas that are free of Korogwe Leaf Spot disease, especially in the coastal plantations of East Africa (Lock, 1985).

4. Zebra disease

Occurrence & severity status
Zebra disease [Fig. 27(A-D)], also known as 'Zebra leaf rot' or 'Zebra leaf spot', is a serious threat to the main cultivar Agave hybrid No.11648 (H.11648) worldwide. The disease affects sisal in Tanzania (Wienk, 1968a), China (Gao, *et al.*, 2012) and India (Roy *et al.*, 2011). Most sisal farms in Tanzania grow Hybrid 11648 which is very much prone to the disease (Wienk, 1968a, 1968b; Peregrine,, 1969a; Kimaro *et al.*, 1994). In Odisha (India) the zebra disease is the most common and dreaded disease of sisal (Roy *et al.*, 2011). The main economic fibre yielding species *Agave sisalana*, Hybrid 1 (Leela) and *Agave* hybrid (Bamra 1) are mostly affected by zebra disease in the state where sisal

is mostly cultivated in India (Roy *et al.*, 2011; Jha *et al.*, 2014). The disease is endemic in nature and losses caused by the disease in Odisha are about 10-20% or even more. The hybrid sisal is more susceptible to zebra disease (15.4 to 33.1%) as compared to *A. sisalana* (Jha *et al.*, 2014; Anonymous, 2015a). Recent survey on the disease incidence in Sambalpur, Sundargarh and Jharsugda districts of Odhisha (India) indicates the disease incidence 13.3 to 34.7% in *Agave sisalana* and 17.0 to 48.3% disease incidence in Barma Hybrid-1 (Anonymous, 2018).

Fig. 27 (A-D). Zebra disease caused by *Phytophthora nicotianae* in sisal. **A-C.** Different kind of spots on leaves. **D.** Boll rot caused by *P. nicotianae*.

Symptoms

The disease appears in three distinct phases, namely bole-rot (always fatal), leaf-spot, and spike-rot (often fatal in that the rot extends into the bole). Each phase can occur independently or with another phase, and anyone could lead to another or all three phases (Clinton and Peregrine, 1963; Peregrine, 1963). In *Agave sisalana* only leaf spot is detected. In hybrid sisal leaf spot, bole and spike rot phases of the disease are prevalent (Roy *et al.*, 2011). Zebra disease nearly always starts with the leaf rot which subsequently grows into the stem or bole causing the striped lesions on the leaves and rotting of the bole and spike.

The zebra leaf spot, begins as small lesions on the lamina which rapidly enlarge, developing alternate concentric rings of dark purple and green, with a light greenish-yellow margin. Subsequently the centre darkens, often with a sticky exudate, dead tissues become rugose, and the rings dark grey and yellowish white. The whole leaf may be covered by lesions. The 2nd syndrome is bole rot. The spike rot syndrome appear as a wet rotting of the unexpanded leaves which may occur if advanced bole rot reaches the base of the spike.

Causal organism: *Phytophthora nicotianae* Breda de Haan, (1896)
[Mededeelingen uit's Lands Plantentium Batavia, 15: 57.]

Synonyms

= *Phytophthora nicotianae* var. *parasitica* (Dastur) G.M. Waterh.1963

= *Phytopthora lycopersici* Sawada, 1942

= *Phytopthora parasitica* var. *nicotianae* Tucker 1931

= *Phytopthora tabaci* Sawada 1927

= *Nozemia nicotianae* (Breda de Haan) Pethybr. 1913

= *Phytophthora nicotianae* var *nicotianae* Breda de Haan, 1896

Morphological/ Morphophysiolgical characteristics

Fungal hyphae are colourless, transparent, and coenocytic and fairly uniform in width, up to 5µm (differently, 7-10µm; Hall, 1994), hyphal swellings frequent especially in water culture and on sporangiophores (Laudon and Waterston, 1964), but colonies may yellow with age. Also, there is much morphological variation in colony type with different isolates of *P. nicotianae* and the growth may differ when grown on different media. In certain media mycelium may be dense or loose rosette, or with no pattern and dome shaped or low, spreading aerial mycelium, radial growth on V8 agar at 25°C (3-) 6 (-10) mm/day (Hall,

1994). Sporangiophores are thin (2 µm wide), closely or sparsely branched. Sporangia are appears ovoid, pear, or spherical in shape, terminal or, less frequently, intercalary, broadly turbinate, the basal part near spherical and the apical third or quarter narrowed or prolonged into a beak, sometimes skewly oriented to the pedicel, papilla conspicuous, often two present, apical thickening hemispherical (Laudon and Waterston, 1964). Inside sporangium zoospores are produced. Zoospores are kidney shaped with an anterior tinsel flagellum and a posterior whip like flagellum. Chlamydospores are usually produced asexually in abundance. The hyphae are heterothallic and require two mating types to produce oospores, the sexual survival structure (Gallup *et al.*, 2006).

Molecular characteristics

In a study, 12 diseased samples of Zebra leaf spot of sisal were collected from the main sisal production areas in Hainan, Guangdong and Guangxi of China. The genomic DNA from these samples was extracted and amplified the rDNA-ITS region. The sequence analysis of the results using morphological identification was consistent (comply) with the results, indicating that the strains tested were *P. nicotianae*. The neighbour-joining (NJ) phylogenetic tree was constructed and the results showed that 12 tested isolates could be divided into two groups and have some genetic differences (Zheng *et al.*, 2011-2012).

Predisposing factors

Poorly drained soils favour the occurrence of the disease. Hybrid 11648 is especially vulnerable in Tanzania .Weather variables like minimum temperature, mean temperature and minimum relative humidity have effect on the occurrence and severity of zebra disease in sisal (Anonymous, 2015a). The spores of the pathogenic fungus germinate in warm and moist soil.

Disease cycle

The Pathogen is soil borne and it infect host primarily by germination of chlamydospores which are the primary survival structure, produced asexually in abundance and serve as long lived resting structures, surviving from four to six years (Gallup *et al.*, 2006). These spores germinate in warm and moist soil to produce germ tubes that infect plants or produces sporangia. Another asexual structure and secondary inoculum is sporangium. The spores produced in it and can either germinate directly or release motile zoospores. Zoospore may cause secondary infection by entering into the host through wounds, or through host surface where the zoospore encysts and a germ tube emerges in penetrating the epidermis. The infection leads to systemic rotting of the plant tissues. Another is the sexual survival structure oospore that produced from hyphae. The hyphae are heterothallic and require two mating types to produce oospores. Many fields only contain one mating type, so the oospores rarely germinate to cause epidemics (Gallup et al., 2006).

Management/Control measures
Use of host resistance
In Odisha (India), in all 11 species of Agave were tested under natural epiphytotic condition for reaction to zebra disease. Amongst them, only one species (*A. miradorensis*) showed moderately resistant reaction and 3 species (*A. cantala, A. angustifolia* and *A. amanuensis*) showed moderately susceptible and the rest 7 species showed susceptible to highly susceptible reaction (Jha et al., 2014; Anonymous, 2015a). A zebra disease resistance factor has been incorporated into the high-yielding Agave hybrid no. 11648 (H. 11648) by crossing this hybrid with *A. lespinassei* (Allen, 1971). To select germplasm materials with zebra disease resistance for breeding, the fluorescent amplified fragment length polymorphism (AFLP) technique was used in China. Among the 40 genotypes tested, *Agave attenuata* var. *marginata*, Dong 109, Nan ya 1 and *A. attenuata*, were identified as resistant and suitable for hybridization with H.11648 to breed a new disease-resistant variety (Gao et al., 2012). Zebra disease requires a long period to obtain anti-*P. nicotianae* variants by crossbreeding, given that 10 years are needed for this plant to bloom. Therefore, transgenic technology is useful for the production of new agave lines that develop significant tolerance to *P. nicotianae* within a relatively short period. In a study, the integration stability and expression level of the transgene in the H.11648 genome were analyzed by using Southern blot and reverse transcription polymerase chain reaction. The mycelial growth rate of *P. nicotianae* was significantly inhibited by the crude leaf extracts from the transgenic H.11648 plants. After further study *in vivo*, the resistance of transgenic H.11648 plants to *P. nicotianae* infection was enhanced (Gao et al., 2014).

Cultural method
It is suggested to prevent any part of the plant from contact with surface drainage water.

Chemical control
Fungicidal control: Both mancozeb and copper compounds controlled the leaf-spot phase of zebra disease in Tanzania's high-yielding Agave hybrid no. 11648. As the copper compounds were phytotoxic, Dithane M-45 (Mancozeb) was recommended as a protectant during the rainy season (Peregrine, 1969b). The fungicides mancozeb (0.2%) and ridomyl MZ 72 (mixture of metalaxyl-8% and mancozeb-64%) @ 0.2 (%) may be applied for the management of the disease (Roy et al., 2011)

B. Pests

1. *Agave weevil*
Occurrence & severity

Sisal is relatively free from pests, though 'Agave Weevil' (*Scyphophorus acupunctatus*), also known as 'Mexican sisal weevil', '*Agave Snout Weevil*' or 'Sisal weevil', is a major pest of economic importance and has a wide distribution in the world (Vaurie, 1971; CABI/EPPO, 2006). The insect pest was first recorded in Tanzania in 1914. There, it was recorded as far west as Lambeni, but was most common in the coastal belt from Moa to the Pangani River (Harris, 1936). It subsequently appeared near Kediri in Java in 1916 (Kalshoven, 1951, 1981). From Java it was introduced to Sumatra's east coast and Aceh, where it was found in 1925 (Kalshoven, 1981). In Honolulu (Hawaii, USA), the agave weevil was first found by Muir in 1918. The weevil was recorded for the first time in South Africa during 1975, where a severe outbreak was occurred in the Eastern Transvaal near Komatipoort on *A. sisalina* (Verbeek, 1976; Annecke and Moran, 1982). The agave weevil was intercepted on Yucca originating from Guatemala in the Netherlands in 1980 (Rossem *et al.*, 1981). At present, it is the most important pest of cultivated agaves. According to Kalshoven (1981), the agave weevil is the destructive pest in agave nurseries in Indonesia. After World War I, the pest was very serious on Sumatra's east coast because of inadequate maintenance of plantations. It is also considered as a pest creating conditions that cause cultivated and ornamental agaves to die before they bloom or can be harvested (Waring and Smith, 1986). The weevil has been a major problem in the tequila and henequen (*A. fourcroydes*) industries of Mexico (Woodruff and Pierce, 1973), the sisal (*A. sisalana*) industry of Africa and Indonesia (Clinton and Peregrine, 1963; Lock, 1969) and in the nursery businesses and landscapes of the south-western United States, where the plant is cultured as an ornamental (Pott, 1976). Besides causing mechanical damage and consuming stored resources, larvae may be involved in symbiotic relationships with microorganisms that break down plant tissues (Waring and Smith, 1986). The fungal pathogen, *Aspergillus niger*, induces rotting in agaves that have been attacked by the weevil (Wallace and Diekmahns, 1952; Clinton and Peregrine, 1963). Waring and Smith (1986) showed that the pest is the primary colonizer species and has a role in initiating stem and leaf rots. In Tanzania, sisal plants suffering from stem or bole rot, which is normally caused by *A. niger*, are often heavily infested with the agave weevil. Wienik (1967) showed that the fungal (*A. niger*) spores are ingested by adult weevils and that they are still viable and pathogenic after passing through the alimentary canal of the insect. However, the weevil is not considered as primarily responsible for spreading bole rot, but it may aid its spread.

Types of damage

The larvae damage the subterranean parts of young plants and may cause substantial losses. They also feed on leaves in the central bud. The hatched larvae fed the hearts of sisal plants so that the leaves, when unfolded, appeared as if riddled by bullets (shot-hole effect). They may form innumerable fine holes in the outer heart leaves on the outer surface of the leaf-edge approximately 5 cm from the base. The injury often becomes noticeable only after 1-2 years, when the large percentage of discoloured fibres attracts attention (Schwencke, 1934). Adults bore into leaves of young plants or plants with weak fibres, and if perforated sisal hearts are infected by fungi, the central shoot becomes red and soft, producing conditions suitable for development of the larvae, which die quickly in the absence of moisture (Harris, 1934). The adult weevil damages the crop by feeding on the youngest leaves before and shortly after unfurling (Brink and Escobin, 2003). The weevil attack on healthy produces a mottled area of dead epidermis approximately 20 cm from the leaf base; this damage is only distinguishable from that caused by friction because it occurs before the leaf has unfolded. Suckers that have been too deeply planted may begin to rot at the base and weevils are then attracted to them as secondary pests (Harris, 1934).

Scientific name & order: *Scyphophorus acupunctatus* Gyllenhal, 1838 (Coleoptera: Curculionidae)

Synonyms

= *Rhynchophorus asperulus* Le Conte, 1857

= *Scyphophorus interstitialis* Gyllenhal, 1838

= *Scyphophorus robustior* Horn, 1873

Biology

Egg: Egg is more or less regular ovoid, with only a slight difference in curvature at both ends; 1.6-1.75 mm in length and 0.7 mm in width; creamy-white with a thin, smooth chorion (Harris, 1936).

Larva: Larva is creamy-white with yellow shiny pronotal plate. Body is moderately large, robust, strongly thickened through abdominal segments 4 and 5. Body length is up to 18 mm, maximum width 9 mm. Head is dark or chestnut brown with convergent, non-pigmented stripes dorsally; free, slightly longer than broad, oval posteriorly. Endocarina is absent; head width 4.0-4.5 mm; mandibles dark brown or black. Legs are absent. Typical abdominal segments are with 3 dorsal folds. Asperities are inconspicuous. Spiracles are on abdominal segments distinct. Posterior margin of abdominal is segment 9 with a pair of projections which are longer than broad; each projection bearing 3 elongate setae (Cotton, 1924; Harris, 1936; Anderson, 1948).

Pupa: Pupa is formed in length in the cocoon of approximately 2.5 cm in length that sometimes covered with clay-like material (Kalshoven, 1981). The pupa is 15-19 mm in length, at first pale yellow, but darkens in colour as the black pigment of the developing weevil becomes visible through the skin (Duges, 1886; Harris, 1936).

Adult: Adult is black and/or reddish, fully winged, without scales or dorsal setae, rather flattened dorsally with body length 9-19 mm. Antennae are inserted at base of rostrum, funicle with segment 2 subequal in length to 3, terminal segment twice as wide as long, club corneous, basal segment with apex much wider than base, spongy apical part retracted, concave, not visible in lateral view. Rostrum is almost straight, sinuate ventrally at base, in males broadly sulcate and bicarinate ventrally. Eyes are very large, elongate, touching below. Pronotum is rather oblong, but shape somewhat variable, subquadrate (then elytral interstices flat) or distinctly longer than wide (then elytral interstices convex), usually finely punctate, surface opaque or shining. Scutellum is small, scarcely wider than base of sutural interval. Elytra are with bases broadly emarginate, elytral interstices very finely punctate. Procoxae are narrowly to moderately separated, separation narrower than rostral apex. Mesocoxae are separated by their diameter. Femora are clavate with inner edge emarginate subapically. Tibiae are with inner edge straight, outer apices strongly bidentate, males with double rows of tibial setae, longer and denser than those of females, in latter, protibiae with much longer and more abundant hairs. Tarsi are with third segment dilated, bilobed, glabrous ventrally except for uniform, dense fringe of yellow, erect setae along apical border. Prosternal process is overlapping mesosternum, mesepimera angulate anteriorly, often with irregular border. Metasternum is flat or gently tumid anteriorly. Metepisterna at apical third is distinctly narrower than greatest width of mesofemora, scarcely narrowed posteriorly. Abdomen is with basal sternite in males with median, basal, sparsely pubescent depression.

Life cycle

Agave weevil (*S. acupunctatus*) can live at elevations as high as 3,000 feet in Tanzania and breeds throughout the year, multiplying rapidly during the rains (Harris, 1934). Eggs are laid in batches of two to six at a time on the bracts of leaves (Custodio, 1944) or in bases of young bulbils or suckers, or in very young or senescing sisal tissues, or in tissues weakened by disease, or in the hole made by the adult weevil in the central shoot of a larger plant. The stems of plants that have flower poles are the only important breeding locations. Usually only one or two larvae develop, but plants in which the flower pole has been cut or dead stumps generally support more larvae. Eggs survive only if there is a certain amount of moisture, and larvae also die if exposed to dry conditions. Larvae emerging from eggs make small channels by eating out the soft epidermal layers of two closely pressed leaves, and thus move away from the hole made by the adult itself. This channel increases in diameter with larval growth and follows an erratic course. Where general infection of the shoot has not taken place, larvae do not appear to wander far from the original hole, but return later to complete the work of the adult in boring right through the shoot, and pupate in the hole thus made. The larvae bore through the central shoot or make irregular tunnels through the tissues until full-grown, and pupate in cocoons made from fibre and leaf debris, cemented together lightly on the inside. In Tanzania, egg, larval and pupal stages last 3-4, 28-55 and 19-36 days, respectively. Larvae and pupae both develop most rapidly during the rainy season. The adult female takes a minimum period of 25 days after emerging to reach sexual maturity, so that 11 weeks (about 3 months) are needed to complete the life cycle resulting in the possibility of 4 generations a year. Three females produced on average 62 eggs each over a period of 3 months (Harris, 1936), while, in Indonesia it takes 2 to 3 months in the lowlands (Kalshoven, 1981). In Kenya, egg, larval, prepupal and pupal stages last 3-5, 21-58, 4-10 and 7-23 days, respectively, and complete development requires 50-90 days. Females lay 25-50 eggs each in a moist environment over 6 months, at a rate of approximately two a week (Lock, 1958).

Management/Control measures
Cultural control

The insect is commonly introduced into new areas with planting material, so such material should only be obtained from localities where *S. acupunctatus* does not occur (Ballou, 1920). Planting before or in the early rains and the application of insecticides in the soil around young plants can control the pest (Brink and Escobin, 2003). It is suggested to burn all residues on land recently cleared of old sisal before new plantings. Since, plants that have finished flowering are potential sources of infestation, and old stumps provide a breeding ground for at least 8 months (Harris, 1934). Collection of weevils, particularly just before and during the rainy season, by trapping and destroying potential breeding spots are useful in minimize the pest attacks (Harris, 1934).

Use of Pheromone trap: Pheromone traps, while buried about 5 cm into the soil, filled with soapy water (3%) for capturing weevils attracted to the lure, and keeping distance between two traps 30m in field, are found effective to capture both male and female weevil (*S. acupunctatus*) at day time (Lopez-

Martínez *et al.*, 2011). Further, it is known that the males of agave weevil (*S. acupunctatus*0 produce a pheromone attractive to both sexes and older males release more pheromone than younger males (Ruíz-Montiel *et al.* 2003, 2009).

Biocontrol

Protection of natural enemies: There is no definite observation on natural enemies of *S. acupunctatus* larvae in Tanzania. Although, in Mexico, the carabid beetle *Morion georgiae* and histerid beetle *Hololepta yucateca* are reported as predators of its larvae, but their role as biological control agents is yet to be established (Halffter, 1957). So far, the sisal weevil is not a suitable target for biological control (Sellers, 1951).

Chemical Control

Application of insecticides is the main suppression method available for *S. acupunctatus* (González *et al.* 2007). There are certain insecticides malathion, endosulfan, methomyl, and fipronil showed high biological efficacy. The insecticides of different mode of actions are suggested to apply in rotation (Terán-Vargas *et al.*, 2012). In Tanzania, the insecticide isobenzan, while applied to the centre of each plant individually or as a spray along the row, gave the best results (Hopkinson and Materu, 1970a, 1970b). However, sometimes, these insecticides may not be effective, because larvae and adults are frequently found feeding in the interior of the "ball" of agave plants, far from the reach of the insecticide. In case of new plantation, planting material should be dipped in dimethoate solution before planting that retained concentrations toxic to weevils for a period of 10 weeks.

8
Cotton (*Gossypium hirsutum G. barbadense, G. arboretum G. herbaceum*)- Malvales: Malvaceae

A. Diseases

1. Bacterial Disease: Angular leaf Spot

Occurrence & severity status

Angular leaf spot disease, also known as 'Black arm disease' or 'Bacterial blight' is a disease of cotton potentially destructive in all cotton-growing areas of the world. The disease became prevalent in the USA during the 1950s (Schnathorst *et al.*, 1960) and in India in the 1970s (Verma, 1986). In the USA, losses of between 34 and 59% were reported (Leyendecker, 1950; Bird, 1959). In India, losses of 5-20% were common in crops (Verma and Singh, 1971). Late and very late sowing of crops resulted in more yield reductions (Meshram and Raj, 1992). In general, losses due to bacterial blight in India are ranged between 1% and 27% depending on the cultivar and crop age (Mishra and Krishna, 2001). In Africa, data on crop losses is available for only a few countries. In Sudan, losses of 20% were common from severe infection (Last, 1960). However, nowadays, due to the wide availability of resistant varieties, its importance has been declined, with certain deviations where proliferation of new races has occurred as in Australia (Anonymous, 1980).

Symptoms

The disease appears both in seedling and mature plant during all growth stages, infecting stems, leaves, bracts and bolls. It causes seedling blight, leaf spot, blackarm (on stem and petioles), black vein and boll rot. On cotyledons small, green, water-soaked rounded (or irregular) spots form which turn brown. Cotyledons can be distorted if the infection is intense. The disease manifests on leaves as angular leaf spot which begins with scattered, dark-green, water-soaked spots, initially more clearly visible on the underside of the leaf lamina. The spots are angular in shape, being delimited by the smaller veins. Older spots become dark-brown or black and are visible on the upper surface of the leaves. The spots often coalesce and the leaf gradually becomes yellow and drop-off. Elongated, sunken and dark brown to black lesions appear on stem, petioles and branches. The young stems may be girdled and killed in the black arm phase. Sunken black lesions may be seen on the bolls. Young boll may falloff. Bacterial slime exudes on the brown lesions. Discolouration of lint may take place.

Causal organism: *Xanthomonas citri* subsp. *malvacearum* (ex Smith 1901) Schaad *et al.* 2007

[Validation List-no. 115. Int. J. Syst. Evol. Microbiol., 2007, 57, 893-897.]

Synonyms

= *Pseudomonas malvacearum* Smith, 1901

= *Xanthomonas malvacearum* (Smith) Dowson 1939

= *Xanthomonas campestris* pv. *malvacearum* (Smith) Dye 1978

= *Xanthomonas axonopodis* pv. *malvacearum* (Smith) Vauterin *et al.* 1995

Morphological/ Morphophysiolgical characteristics

The bacterium is rod-shaped with a single polar flagellum, capsulated but forming no spores, motile and aerobic. It occurs singly or in pairs. Stain reaction is gram negative.

Molecular characteristics

The molecular characteristics of *Xanthomonas citri* subsp. *malvacearum* (Xcm) have been determined and the sequence accession numbers of type strain DSM 3849 are FR749942 (clone 1), FR749943 (clone 2) and FR749944 (clone 3) (Anonymous, ND6). The PCR amplifications of different Xcm isolates of Brazil with REP, ERIC and BOX primers produce genomic fingerprints with 22 to 28 reproducible bands ranging from ca. 230 to 3000 bp (Oliveira *et al.*, 2011), while, isolates of Iran with ERIC primer produce 21 reproducible bands, ranging from ca. 250 to 2500 bp (Madani *et al.*, 2010). Rep-PCR analysis using the primers REP, ERIC, and BOX disclosed a high similarity among strains from diseased cotton plants, indicating limited diversity of the Brazilian Xcm population (Oliveira *et al.*, 2011). The cluster analysis generated from the combined data from the REP, ERIC, and BOX amplifications reveals the existence of two large clusters and three smaller size clusters and some strains with isolated profiles. The strains from different cotton cultivars and diverse origins are within the two major clusters (Oliveira *et al.*, 2011).

Predisposing factors

Disease severity can be influenced by many factors. There is a general tendency for stem infections to be closely associated with early leaf infection (Wickens and Logan, 1957). High humidity and moderate temperature (28°C) favours the development of the disease. Primary infection is favoured by 30°C and secondary infection is better at 35°C. Presence of moisture is very important for the first 48 hours. In Uganda, disease incidence is found to increase when rainfall in the third quarter of the year is above average, particularly, when 12 cm or more rainfall in August (Jameson, 1950). Planting regimes can also influence severity. The disease incidence becomes low if the seeds are sown singly than the more seeds per pit (Arnold and Arnold, 1960).

Disease cycle

The pathogen enters the mature seed through the basal end of the chalaza. It can remain as slimy mass inside the seed or as contaminants on the surface of the seeds or in the lint attacked to it. Volunteer seedlings are also the source of primary inoculum when cotton is planted after cotton. The disease may be carried over through infected leaves, bolls and twigs on the soil surface. The secondary infection is through windblown soil, rain and irrigation water.

Physiological races/strains

In all, thirty two pathogenic races of *X. citri* subsp. *malvacearum* (Xcm) have been described in the world (Verma and Singh, 1974), although, based on disease reactions of 10 host differentials, El-Zik *et al.* (2001) have reported 19 races. In India, Xcm is highly variable and as many as 26 races have been reported. The virulent races can overcome 3-4 major bacterial blight resistant genes (B-genes) in the host and are most widely distributed in India (Verma, 1986). The race 32 is considered as the most virulent because it can neutralise 4 different bacterial blight resistance genes namely B7, B2, BIn and BN, whereas the races 1-6, are considered the least virulent because they can neutralize either only polygenes or one major B gene (Verma, 1986). In Tamilnadu (India), eight bacterial isolates of Xcm have been found to be pathogenic to the susceptible cotton variety TCB-209. The eight pathogenic isolates are grouped into clusters A and B based on protein profiles. The isolates are morphologically similar, but varied in their cultural and biochemical characteristics (Khabbaz *et al.*, 2017). In Brazil, the population of Xcm has a low phenotypic and genotypic variability (Oliveira *et al.*, 2011). There, the occurrence of Xcm races 3, 7, 8, 10, 18, and 19 is reported (Moresco *et al.*, 2001; Juliatti and Polizel, 2003). In the United States, as well as other countries, race 18 is the most frequently encountered race (Verma and Singh, 1975; Hussian, 1984; Allen and West, 1991; Thaxton *et al.*, 2001). In Texas, more than 75% of the cultivars are susceptible to race 18 of Xcm (Thaxton *et al.*, 2001).

Management/Control measures

Use of host resistance

Cotton plants show considerable genetic variability for resistance to bacterial blight. The variety *Gossypium hirsutum* var. *punctatum* shows highest degree of resistance, while, *G. barbadense* is with little natural resistance. Knight (1945, 1946, 1954) have identified 10 major genes (B1 to B10) for blight resistance. Eight of these genes are dominant or partially dominant in their expression. Most of these genes are used to transfer to the long-staple cotton cultivars to produce cottons with a combination of B-genes which confers immunity to all known races of the blight pathogen (Siddig, 1973; Hillocks, 1992a). In Sudan, most of the blight-resistant varieties contain the genes B2 and B6. However, a new race of the pathogen has appeared which is virulent to that gene combination. There, a line (S2950) has been developed which is resistant to the new race (Wallace and El Zik, 1990). Varieties with resistance based on the genes B2 and B3 continue to provide protection against blight in many African and South American countries (Hillocks and Chinodya, 1988; Ruano *et al.*, 1988). However, the gene combination B2B3 does not completely prevent disease occurrence, particularly during the most favourable condition of the disease (Hillocks and Chinodya, 1988). In the USA, immunity to blight has been stable for over 20 years in some of the Tamcot varieties (Wallace and El Zik, 1990). The Tamcot varieties carry B2, B3 and B7 genes and are immune to the disease in India also (Meshram and Raj, 1989).

Cultural control

Field sanitation: The bacterial pathogen cannot survive in the soil without crop residues. Hence, it is suggested to remove crop residues from field soon after cotton harvest. In addition, crop rotation for one year is usually sufficient to minimize the disease.

Seed treatment: It is suggested to eradicate externally seed borne infection by treating the seed with Conc. H_2SO_4 for 5 minutes and then washing with lime solution to neutralise the effect and finally washing with running water to remove the residue and drying seeds. The internally seed borne infection can be eradicated by soaking seeds overnight in 100 ppm streptomycin sulphate or Agrimycin.

Biocontrol

A mixture of *P. fluorescens* (Pf32, Pf93) and *B. subtilis* (B49) as talc-based powder formulations is effective to increase yield and to reduce incidence of BLB disease in cotton (Salaheddin *et al.*, 2010). It is considered that the biocontrol agents induced systemic resistance against bacterial blight of cotton (Raghavendra *et al.*, 2013).

Chemical control

Copper oxychloride is mostly used to control angular leaf spot in cotton plants (Suassuna *et al.*, 2006). In certain cases, foliar spray of combination of Paushamycin/Plantomycin 100 mg + 3 gms of Copper Oxychloride per lit of water for 3 or 4 rounds at 15 days interval from the time of disease appearance is recommended. However, copper and antibiotics are not much effective for controlling angular leaf spot, thus the use of healthy seeds and of resistant varieties is crucial for protecting the crop from infection (Metha *et al.*, 2005).

2. *Alternaria* Leaf spot

Occurrence & severity status

Alternaria leaf spot, also called *Alternaria* blight, is one of the most common cotton diseases. Different *Alternaria* species, such as, *Alternaria alternata A. macrospora, A. gossypina, A. gossypii* and *A. malvacearum* have been found to affect cotton in different countries all over the world (Hopkins, 1931; Hillocks, 1991a; Mehta, 1998; Simmons, 2007; Yu, 2015). However, amongst them leaf spot caused by *Alternaria macrospora* and *A. alternata* are common in cotton crops around the world. Both the species are usually considered as minor pathogens of Upland cotton (*Gossypium hirsutum*) but *A. macrospora* causes premature defoliation with resulting yield loss in cultivars of *G. barbadense* in Israel and in the USA (Hillocks, 1991a). In southern Brazil, *A. macrospora* has been reported to cause leaf spot disease on cultivars belonging to *G. barbadense* and *G. hirsutum* (Cotty, 1987; Hillocks and Chinodya, 1989). In India, *A. macrospora* is the most commonly occurring leaf spot disease of cotton in Andhra Pradesh (south India). Under favourable conditions, losses to the tune of 26.59 % (Monga *et al.*, 2013) and 38.23% (Bhattiprolu and PrasadaRao, 2009a) were recorded. The leaf spot disease caused by *A. alternata* is more predominant on *G. hirsutum* under natural field conditions in southern Brazil (Bashi *et al.*, 1983; Bashan, 1986; Ephrath *et al.*, 1989; Mehta, 1998). There, the disease caused by *A. alternata* is considered to be a late-season disease of cotton (Mehta, 1998). The fungus, *A. gossypina*, also causes boll rot of cotton (Hopkins, 1931; Yu, 2015).

Symptoms

Alternaria leaf spot is primarily a leaf disease but symptoms may also develop on stems, cotyledons and bolls. Infection on leaves first appears as small, pale to brown, round or irregular spots which vary in size from 0.5-10 mm. They often develop concentric ridges with a target board appearance on the upper surface. Mature spots have dry, greyish centres which may crack and fall out. The spots may coalesce and produce large irregular dry spots on the leaf

(Annonymous, 2012a). Stem lesions first appear as small sunken spots which develop into cankers. The tissue splitting and cracking may cause a break at stem. The glandular areas on the receptacle are also affected, causing failure of the boll to develop. Flowers and bolls are fall out (Annonymous, 2012b). The bolls become mummified and the fibre attacked. Boll rots can also be caused by *A. gossypina* (Hopkins, 1931; Ellis and Holliday, 1970; Yu, 2015). The seeds may get infected and carry the infection (Padaganur, 1979). Spots occur on the cotyledons when severely affected.

Causal organisms

1. ***Alternaria macrospora* Zimm. 1904**

    [Ber. über Land. und Forstwirth. Deutsch-Ostafrica 24]

    [**Synonyms:** *Alternaria longipedicellata* Snowden 1927;

    Macrosporium macrosporum (Zimm.) Nishikado & Oshima 1944]

2. ***Alternaria alternata* (Fr.) Keissler, 1912**, [Beih. Bot. Zbl., 29: 433]

    [**Synonyms:** *Alternaria tenuis* Nees, 1816; *Torula alternata* Fr., 1877]

3. ***Alternaria gossypina* (Thüm.) J.C.F. Hopkins, 1931** [Trans. Br. Mycol. Soc., 16(2-3): 136]

    [**Synonym:** *Macrosporium gossypium* Thüm., Heb., 1877]

Morphological/ Morphophysiolgical characteristics

Alternaria macrospora

The fungus forms amphigenous colonies on leaves. Conidiophores are arising solitary or in small groups, erect, unbranched, straight or flexuous, almost cylindrical or tapering slightly towards the apex, septate, pale brown to mid brown, smooth, with one to several conidial scars, up to 80 (-90) μm long, 4-9 μm wide. Conidia are solitary or uncommonly in chains of 2, straight or curved, obclavate to broadly obclavate, or broadly ellipsoid, straight or slightly curved, mid to dark brown, verruculose when juvenile, becoming smooth at maturity, rostrate, tapering rather abruptly to a very narrow beak, with (4-) 6-8 (-9) transverse and several longitudinal or oblique septa, often slightly constricted at the septa, (35-)40- 80(-) 90×20-30 μm in size, beak filiform, pale, septate, 80-280 μmlong, 1.5-2 μm wide (Ellis and Holliday, 1970; Yu, 2015).

Colonies on PDA are woolly, pale gray to dark gray with abundant aerial mycelium and profuse sporulation in a 12/12 hr NUV light/darkness cycle, usually a yellow or reddish yellow pigment produced into the medium, some cells of the older hyphae may swell, forming structures resembling a string of beads after 7 days (Yu, 2015).

Alternaria alternata

Colonies on leaves are amphigenous, Conidiophores and conidia are solitary or arising in small groups of two or more, mostly unbranched, straight to flexuous, occasionally extending geniculately, smooth, 1-3 septate, slightly swollen, golden brown and almost hyaline at the apex with a single conidial scar or geniculate with 1 to several conidial scars, variable length, but commonly short, 20-50 (-100) μm x 3-6μ min size, commonly develop a cluster of conidia in branching chains. Conidiogenous cells polytretic, terminal or intercalary, integrated, cylindrical, with conspicuous scars, proliferating sympodially. Conidia are pyriform to ovoid-obclavate, formed singly or in long to long chains of 5-10 or more, commonly branched (differently unbranched), pale brown to golden brown, or smoky brown, with (2-)3- 6(-8) transversal septa, (1-) 2-4 longitudinal septa, base rounded, with a thickened, darkened, 2-3 μm wide scar, (15-) 19-35(-40) × (7-) 8-12 (-16) μm in size, surface usually granulate or densely verrucose, mostly beakless, or with a short cylindrical beak not exceeding one third the length of conidium, or having secondary conidiophores (Srinivasan, 1994; Soares et al., 2009; Yu, 2015).

Colonies on PDA are effuse, olivaceous black to gray black, abundant sporulation with little aerial mycelium, no pigment is realeased into the medium, 60-80 mm in diam. after 7 days. Mycelium immersed or partly superficial; hyphae branched, septate, hyaline at first, later pale olivaceous brown, smooth-walled, 3-5 μm wide. In culture the conidiophores are arising directly from the substrate, or produced terminally and laterally on the hyphae, up to 150 μm long (Yu, 2015).

Alternaria gossypina

The fungus produces amphigenous spots on leaves. Condiophores are solitary or arising in small groups, mostly unbranched, smooth-walled, usually only one pigmented conidiogenous site at the apex, sometimes proliferating sympodially, 80-120 μm long, 3-5 μm wide. Conidia are mostly in short or moderately long chains of 3-10, usually unbranched, sometimes branched; dark brown, 30-45 (-55)× (12-)15-25 μm in size, mostly conspicuously verruculose, occasionally minutely roughened, obclavate to narrowly ovoid or ovoid to ellipsoid, with 3-9 transverse septa and 1-2 (-3) longitudinal or oblique septa, slightly constricted at the transverse septa, occasionally having a darker median transverse septum that is more constricted than other septa; occasionally have a short beak, mostly beakless or having a secondary conidiophore (David, 1988; Yu, 2015).

Colonies on PDA are cottony or velvety, olive green or light brown, darker in the centre with sporulation, with varying amounts white aerial mycelium, no pigment is released into the medium, 60-70 mm in diam after 7 days (Yu, 2015).

[**N.B.** *Alternaria gossypina* was reported to cause leaf spot and boll rot of cotton in Southern Rhodesia by Hopkins (1931). He distinguished the pathogen from *Alternaria macrospora* on the basis of spore size. The fungus, *A. macrospora* differs from *A. gossypina* in producing larger conidia with a very long beak, usually solitary or rarely in chains of two, whereas *A. gossypina* produces small conidia in moderately long chains of 3-9 (Nishikado *et al.*, 1940). However, still there is some confusion about the naming of the causal fungus (Brown, 1976).]

Predisposing factors

The presence of residues of previous cotton crops in field favours the disease, since; the pathogens survive on it (Anonymous, 2012a). The disease infection is favoured by wet weather, intermittent rains and temperatures of 25-28°C. Further, it is reported that 95% mean morning relative humidity is the promising predisposing factor for the disease development (Venkatesh, 2014). The pathogen primarily spreads through irrigation water. The secondary spread is mainly by air-borne conidia (Anonymous, 2012b). Seedling stage and the plants of late stage are more vulnerable to be attacked by the disease. Physiological and nutritional stresses, such as, heavy fruit load and potassium deficiency, favour the disease (Hillocks, 1991a). Pima cotton varieties (Pima S6 and Pima S5) are more susceptible to the disease caused by *A. macrospora* (Cotty, 1987; Anonymous, 2012a)

Disease cycle

The undecomposed crop residues and infected seeds provide the primary source of inoculum. The primary inoculum develops early infection on cotyledons. The conidia splashed from the crop residues or the infection sites of cotyledons to the lower leaves of cotton plants and causes disease in lower side of canopy. The spores are multiplied there and spread to other parts either by rain splash or by wind to cause infection. The causal fungi also attack the bolls and grow on exposed lint if bolls open in wet weather, giving rise to contaminated seed. The infected leaves fall to the ground and serve for primary inoculum of the next year (Annonymous, 2012b).

Management/Control measures

Use of host resistance

Chattannavar *et al.* (2004) have reported 11 Bt-cotton genotypes as resistant to the disease. The genotype, JKDCH-501, is moderately resistant (Hosagoudar *et al.*, 2008) while the other genotypes, viz., DCH 32, RAMSHH 7, GSHB 895, CCHB 2628, CCCHB 702, DHB 782, NSPL 414, HAGHB 12 and Ajeet 999, are found resistant to *Alternaria* leaf spot (Chattannavar *et al.*, 2009).

Cultural control

Reducing plant stress and insuring proper soil fertility, especially with potassium, can minimize disease severity (Hillocks, 1991a).

Biocontrol

It is reported that *T. viride* and *P. fluorescens* isolate Pf1 significantly reduce *Alternaria* leaf spot (Chidambaram *et al.*, 2004). Gholve *et al.* (2012) have recorded the highest inhibition (63.64%) of *A. macrospora* of cotton with *T. viride*. Seed treatment with *P. fluorescens* @ 10 g/kg followed by foliar spray @ 0.2% significantly reduces *Alternaria* leaf spot disease in cotton and increases yield (Bhattiprolu and Prasad Rao, 2009b; Bhattiprolu, 2010).

Chemical control

It is suggested to spray Mancozeb 75WP @ 2 g/lit or Copperoxychloride @ 3g/litre of water to minimize the disease caused by *Alternaria* spp. in cotton (Bhaskaran and Shanmugam, 1973; Chattannavar *et al.*, 2006; Annonymous, 2012b). In addition, zineb, brestan, duter and hexaferb are also very effective fungicides against *A. macrospora* in cotton (Padaganur and Siddaramaiah, 1979). Spraying 0.1% propiconazole also protects cotton crop from *Alternaria* leaf spot and increases seed cotton yields by 67.34% (Bhattiprolu and PrasadaRao, 2009a). Finally, it is suggested to manage the disease by seed treatment with mancozeb 3 g/kg and foliar spray with 0.1% propiconazole at 50, 70 and 90 days after sowing (Venkatesh, 2014).

3. Boll rot

Boll rots occur all over the world and are caused by a number of pathogens, including fungi and bacteria (Bagga, 1970; Silva *et al.*, 1995; Belot and Zambiasi, 2007). Boll rots cause damage to bolls, lint and seed. The term seed rot is also used in certain cases to describe a boll rot which begins in the seed (Roughley *et al.*, 2015).The bacteria and fungi each have characteristic lifecycles and mechanisms of destruction (Guthrie *et al.*, 1994). The boll rots develop when the bolls are maturing and opening are exposed to wet weather (Smith *et al.*, 2011-'12). Cotton bolls rot can cause 20-30% losses in productivity (Iamamoto, 2007). In Bangladesh, Shamsi and Naher (2014) estimated 75% of total fruit loss due to boll rot symptom of cotton. In Georgia, Ranney *et al.* (1971) observed yield losses in the order of 1.5% caused by cotton bolls rot, in a particularly dry year, while under higher humidity and temperature conditions in the next year, the losses increased to 14%. The losses of cotton due to boll rot vary from year to year due to weather conditions, insect vector pressure, pathogen presence, and geographic location (Jones, 1928; Hillocks, 1992b; Mitchell, 2004; Schumann and D'Arcy, 2010). The excessive vegetative growth, high planting density and unbalanced fertilization are also factors that can increase the incidence of cotton bolls rot (Zancan *et al.*, 2013).

i) Phytophthora boll rot

Phytophthora boll rot, caused by *Phytophthora nicotianae* Breda de Haan, 1896, is the most common form of boll rot in Australia (Allen and West, 1986), while, in Europe, *Phytophthora boehmeriae* Sawada, 1927, is found to cause cotton boll rot and considered as a new threat in the mediterranean region of Greece (Elena and Paplomatas, 1998). Phytophthora boll rot is most severe in crops that are lodged or have bare soil exposed between the plants and then have the coincidence of heavy rain onto wet soil when bolls are about to open. Boll rotting is most prevalent on the lower bolls. The infected bolls quickly turn brown, become blackened and open prematurely with light brown tight locks, and eventually falling out of the boll and onto the ground (Allen and West, 1986; Allen, 2007; Roughley *et al.*, 2015). In certain cases, the bolls become decayed, with the subtending dried carpels being dark-brown to black and white mould with numerous sporangia over rotten bolls (Elena and Paplomatas, 1998).

ii) Nigrospora boll (lint) rot

Nigrospora boll (lint) rot, caused by *Nigrospora oryzae* (Berk.& Broome) Petch, 1924, is found to occur in Alabama state which located in the south-eastern region of the United States (Palmateer *et al.*, 2003). During the years 2000 and 2001, the disease was found in the coastal region of Alabama when the precipitation was 55% lower than the 5-year average. About 48% of developing bolls at full bloom were affected. The affected bolls at an early stage of opening contain grey mycelium within the locules. At maturity, the lint within the infected locules becomes discoloured, and the fibres remain compact resulting in the characteristic "gray lock" sign (Palmateer *et al.*, 2003). The mite, *Siteroptes reniformis*, serves as a vector of this fungus. Therefore, the occurrence of the disease may be dependent on the mite association (Laemmlen and Hall 1973; Mirzaee *et al*, 2013).

iii) Sclerotinia boll rot

Sclerotinia boll rot, caused by *Sclerotinia sclerotiorum* (Lib.) de Bary, 1884, usually occurs within the humid canopy of tall rank crops. The disease characteristically has large black globular sclerotia (2 to 10mm diameter) within and on the affected bolls, and sometimes also on adjacent branches and stems (Allen, 2007). A white cottony fungal growth may be present inside the boll and adjacent areas (Charchar *et al.*, 1999; Roughley *et al.*, 2015).

iv) *Fusarium* boll rot

Fusarium boll rots are caused by different species of *Fusarium* to those causing wilt. Several *Fusarium* species, viz., *Fusarium roseum, F. oxysporum, F. solani, F. lateritium, F. moniliforme* and *F. semitectum* have been reported as the causal agents of cotton boll rot disease (Kirkpatrick and Rothrock 2001). The *Fusarium* boll rots are not common and usually occur within the humid canopy of tall rank crops that have been exposed to extended periods of wet weather late in the season. The disease has been reported to cause significant damage to cotton by inciting in India and Columbia under hot and humid conditions (Costa *et al.* 2005). The disease affected bolls never open properly and are covered with light pink spores. It is observed after October rainfall as dried boll rot with cream to pinkish discoloration. The fungus typically colonises the surface of the carpel and boll and produces orange masses (Mirzaee *et al*, 2013)

v) *Lasiodiplodia* (*Diplodia*) boll rot

Lasiodiplodia (*Diplodia*) boll rot is caused by *Lasiodiplodia theobromae* (Pat.) Griffon & Maubl., 1909. The disease was first reported from India in 1879 (Cooke, 1879). The disease is more prevalent under conditions of high rainfall and high temperature during the late summer. It is also more prevalent when plants grow tall, thus shading the lower parts and keeping them moist. It causes damages and destroys cotton bolls and fibre in many areas of the south-eastern part of the United States. Neal and Gilbert (1955) showed that *Lasiodiplodia* in the Mississippi valley causes the greatest damage to the lower bolls and bolls in contact with the soil. Recently, its occurrence has been noticed in crops growing in the Central Highlands of Queensland (Smith *et al.*, 2011-'12). The disease appears at first as small brown lesions on the bolls and cotton bracts, and in high humidity conditions, the spots may expand, affecting the whole cotton boll. At later stage, the bolls become black covered with sooty black spores (Paiva *et al.*, 2001; Smith *et al.*, 2011-'12). Under these conditions, the disease affects the lint and seed. The boll dries and opens prematurely, exposing blackened fibres and seeds (Edgerton, 1912). However, when boll before half grown becomes diseased, it will not open and the fibres remain nonfluffed. This condition is referred to as "tight lock" bolls (Brown and Ware, 1958).

vi) Anthracnose boll rot

Anthracnose boll rot [Fig. 28 (A-B) is caused by *Colletotrichum gossypii* Southw., 1891.The disease occurs more frequently in most cotton fields of developing countries. In a survey in major cotton producing areas in China during 1984, it was observed that boll rot disease (including anthracnose boll rot) caused yield loss of 5.05-24.94% (Smith *et al.*, 1986). However, the boll rot symptom does occur always in anthracnose affected cotton plant in Australia. It occurs occasionally in Queensland cotton crops (Roughley *et al.*, 2015). The disease is characterized by depressed greyish or brown spots with reddish borders on bolls that expand and darken over time. In the central part of the lesion, a spore mass may be observed with a pink coloration (Suassuna and Coutinho, 2007). This is followed by production of small black setose fruit bodies as the boll dries out (Hillocks, 1992b). Badly diseased bolls become mummified (darkened and hardened) and never open. In certain cases, the bolls open partially, leaving the fibre darkened and difficult to remove (Suassuna and Coutinho, 2007). In partially affected bolls, the fungus grows through and infects the seed. Lint from diseased bolls becomes tinted pink and of inferior quality.

Fig. 28 (A-B). Cotton boll rot. **A.** Anthracnose disease spots on bracts. **B.** Tight boll rot symptom.

vii) Bacterial boll rot

The Bacterial boll rot, caused by *Pantoea agglomerans*, has been reported from America (Medrano and Bell, 2007; Medrano *et al.*, 2007) and China (Ren *et al.*, 2008). In China, it causes as much as 20% yield loss in several fields of Xinjiang Province. The disease does not produce any symptom on the outer surface of carpel. In the infected cotton bolls, fibres do not mature completely and seed tissue exhibits brown necrotic coloration (Ren *et al.*, 2008). The cotton boll rotting bacteria (*Pantoea agglomerans* strain Sc 1-R) is found to be vectored by the southern green stink bug (*Nezara viridula* L.) and brown stink bug, *Euschistus servus* (Say), into cotton bolls (Medrano *et al.*, 2007, 2016). All Australian cultivars of *G. hirsutum* are immune to the bacterial blight pathogen. However, the 'barbadense' cultivars have been very susceptible (Allen, 2007).

Control measures

Chemical control with fungicides is a rapid and effective tactic in the management of cotton boll rot diseases. Both through seed treatment and foliar sprays are effective to manage the diseases effectively (Zancan *et al.*, 2013). The fungicide mixture, Azoxystrobin (strobilurin) + cyproconazole (triazole), is suggested to control cotton bolls rot infection (Zancan *et al.*, 2013). Further, spraying of copper oxychloride 2.5kg/h along with an insecticide for bollworm from 45th days of sowing twice or thrice at 15 days interval is also suggested to manage the boll rot disease. Copper oxychloride in combination with azoxystrobin may also be used against boll rot disease (Zancan *et al.*, 2011). The efficacy of fungicides will be further augmented if optimum row spacing along with recommended doses of fertilization is maintained (Zancan *et al.*, 2013).

4. *Corynespora* leaf blight

Occurrence & severity status

Corynespora leaf blight, also referred as "Target spot", of cotton was reported for the first time in 1959 from Alabama, USA (Jones, 1961). Later on, the disease has been reported from Brazil (Mehta *et al.*, 2005), India (Lakshmanan *et al.*, 1990), China (Wei *et al.*, 2014) and several other countries. In India, this disease causes severe boll rots in cotton of TamilNadu fields (Lakshmanan *et al.*, 1990). In USA, also, during the recent years the disease is a serious problem (Conner, 2004; Flumer *et al.*, 2012).The disease causes 70% premature defoliation of cotton in Georgia (Flumer *et al.*, 2012). A seed cotton yield loss of 336 kg/ha or more has been estimated in Alabama (Conner *et al.*, 2013). Apart from cotton, the disease attacks several other crops and is most abundant in tropical countries (Dixon *et al.*, 2009).

Symptoms: The disease first appears in the lower canopy and spreads upward through the canopy toward the shoot tips (Conner *et al.*, 2013). The initial symptoms first develop as brick red dots that lead to the formation of irregular to circular lesions with tan-to-light brown centres. Lesions further enlarge, varying between 2 mm and 10 mm, and often demonstrate a target like appearance formed from irregular concentric rings of alternating light and dark brown bands within the spot (Mehta *et al.*, 2005; Flumer *et al.*, 2012; Wei *et al.*, 2014). Leaves typically retain their original green colour. Leaves with multiple lesions become senesced prematurely. Lesions are also formed on the boll bracts and bolls (Lakshmanan *et al.*, 1990). Severely infected bolls lose quality and produce infected seed (Galbieri *et al.*, 2014).

Causal organism: *Corynespora cassiicola* **(Berk. & M.A. Curtis) C.T. Wei 1950**

[Mycological Papers 34: 5]

Synonyms

= *Cercospora melonis* Cooke 1896

= *Cercospora vignicola* E. Kawam. 1931

= *Corynespora melonis* Cooke ex Lindau 1910

=*Corynespora vignicola* (E. Kawam.) Goto 1950

= *Helminthosporium cassiicola* Berk. & M.A. Curtis 1868

= *Helminthosporium papayae* Syd. 1923

= *Helminthosporium vignae* L.S. Olive 1945

Morphological/ Morphophysiolgical characteristics

Colonies are effuse, grey or brown, thinly hairy. Mycelium is mostly immersed. There is no stroma. Conidiophores are scattered or clustered, erect, simple or occasionally branched, straight or slightly flexuous, pale to mid brown, smooth, septate, often nodose, monotretic, percurrent, with several cylindrical proliferations, (100-) 110-850 μm long, 4-11 μm thick. Conidia are solitary or catenate, very variable in shape, obclavate to cylindrical, straight or curved, subhyaline to rather paleolivaceous brown or brown, smooth, 4-20-pseudoseptate, 40-220 (-230) μm long, 9-22 μm wide in the broadest part, 4-8 μm wide at the truncate base; scar often dark with a slight rim, apex obtuse (Ellis and Holliday, 1971; Hoog *et al*, 2000; Wei *et al.*, 2014).

On potato dextrose agar (PDA) at 28°C, the initial colour of the colonies is olivaceous, turning dark brown after 5 days (Wei *et al.*, 2014). Colonies on Plate Count Agar (PCA) are with rapid growth, hairy or velvety, olive-green to greyish-black (Hoog *et al.*, 2000).

Molecular characteristics

Molecular studies indicated that the *C. cassiicola* isolates attacking cotton and soybean belong to the same strain of the pathogen in Brazil (Galbieri *et al.*, 2014). In order to confirm the species, DNA was extracted from a week-old culture and amplified with specific primers for loci "ga4" and "rDNA ITS"

(Flumer *et al.*, 2012). The DNA sequences obtained with the Applied Biosystems 3730 x 1 96-capillary DNA Analyzer showed 99% identity to *C. cassiicola* from BLAST analysis in GenBank (Flumer *et al.*, 2012). In another study, the internal transcribed spacer region (ITS) of one isolate was amplified using primers ITS1 and ITS4 and sequenced. BLAST search in GenBank revealed 100% homology with sequences of *C. cassiicola* (Wei *et al.*, 2014). Further, the internal transcribed spacer region (ITS) of two isolates, one representing each location, was amplified using primers 2234c and 3126t targeting a 550-bp region of the ITS1, 5.8S rRNA gene, and ITS2. Sequences revealed 99% similarity to *C. cassiicola* in NCBI (Conner *et al.*, 2013).

Predisposing factors

The disease requires high humidity for infection and is favoured by substantial periods of high moisture (16–44 hours). In other words, the leaf wetness is likely a major environmental factor driving the disease for this fungal pathogen. Additional factors that increase target spot disease include higher planting rates, excessive N rates, narrow row spacing, vigorous growth, as well as hot weather (Kelly, 2017).

Disease cycle

Corynespora leaf blight of cotton may appear from infected seed. However, the initial inoculum, most commonly, comes from the air-borne conidia from other hosts or from the left-over stubble of cotton (Galbieri *et al.*, 2014). The disease multiplies in the host and overwinters as mycelia or conidiophores on the plant debris in field.

Control measures

The fungicides, Elatus (5-7.3 floz/a), Headline (6-12 floz/a), Priaxor (4-8 floz/a), Quadris (and generic azoxystrobin products, 6-9 floz/a), Stratego YLD (5 floz/a) and Twinline (7-8.5 floz/a), are used to control *Corynespora* leaf blight in cotton (Kelly, 2017).

5. Grey mildew

Occurrence & severity status

The grey mildew disease was reported for the first time on upland cotton (*Gossypium hirsutum* L.) in Auburn, Alabama, USA in the year 1890 and was named as 'Areolate mildew' of cotton (Atkinson, 1890). Subsequently, the disease was reported in Southern Nigeria (Farquharsoq, 1913), India (Butler, 1918), Combodia and Saigaon (Vincens, 1919), Uganda (Snowden, 1921), Australia (Birmingham and Hamilton, 1923), Ceylon (Gadd, 1926), Italy (Petri, 1931), South Africa (Khabiri, 1952), and Middle East and Brazil (Blank, 1953).In India, grey mildew is of great concern. In Maharashtra state of India, the disease is called as 'Dahiya Disease' due to its appearance as sprinkled curd on foliage (Gokhale and Moghe, 1965). It affects all the four commercially cultivated *Gossypium* species i.e. *Gossypium arboreum, G. herbaceum, G. hirsutum* and *G. barbadense* (Bell, 1981). The disease is very severe in the diploid Desi/ Asiatic *G. arboreum* and *G. herbaceum* cottons in India, particularly in low-lying moist localities at wet years (Wangikar and Shivankar1989). The disease initiates its occurrence in August and grows during the month of September and reaches at its peak level in the month of October (Venkatesh *et al*., 2015). Here, the seed cotton yield losses are ranging between 21 to 26 per cent (Kodmelwar, 1976). Further, under exclusive monoculture of 'Diploid' cotton, the losses may extend up to 90 percent (Sangitrao et. al., 1993). In intra-hirsutum hybrid H4 (tetraploid cotton), the yield losses of 62 to 68 per cent had been recorded in chemically unprotected crop, in the disease endemic area of Akola district of Maharashtra (Shivankar and Wangikar, 1992).

Symptoms

The disease first appears on the lower canopy of older leaves when bolls set. Irregular, angular or irregularly circular whitish translucent spots are formed by the veins of leaves. Later, the spots become dark, limited by the veins, 3 to 4 mm in size, with conidia in profusion giving a frosted appearance to the spots. The spots are hypophyllous, rarely amphigenous (Mulder and Holliday, 1976). Leaves become yellow, turn to brown, curled and defoliate prematurely. The development of symptoms on cotyledons, bracts, flower buds and bolls are also reported (Gokhale and Moghe, 1967; Rao and Jayasankar, 2015). Heavily infected cotyledons drop and the infection spots become blackish during the process of withering. Perithecia are formed on such cotyledons (Gokhale and Moghe, 1967).

Causal organism: *Ramularia gossypii* (Speg.) Cif. 1962

[Mycoflora Domingensis Integrata. Quadernodel Laboratorio Crittogamico del Istituto Botanico del l'Università di Pavia. 19: 1-539]

Synonyms

= *Cercosporella gossypii* Speg., (1886)

= *Septocylindrium gossypii* (Speg.) Subram., (1971)

= *Ramulariopsis gossypii* (Speg.) U. Braun, (1993)

= *Ramularia areolata* G.F. Atk. (1890)

= *Ramularia areola* G.F. Atk

= *Mycosphaerella areola* Ehrlich & F.A. Wolf 1932 (Teleomorph)

Morphological/ Morphophysiolgical characteristics

Fungal colonies in plant are amphigenous but mostly hypophyllous. Stroma is substomatal, size variable (mostly 30 μm diam.); primary mycelium internal. Conidiophores borne on the substomatal stroma, emerging through the stoma, forming a loosely arranged fascicle, short, 2-3 septate, rounded or flattened at the ends, branched at the base often bifurcate at the broad apex, bearing conspicuous conidia scars and measure (25-) 30-55 (-75) x 3,5-4 (-7) μm (Ehrlich and Wolf, 1932; Mulder and Holliday, 1976). Conidia are catenulate, hyaline, straight, cylindrical, usually pointed at both ends or occasionally rounded, (1-) 2 (-3)-septate, conidia sometimes branched, scars thickened, lying flat against the conidial wall, 14-37 x 2.5-5 μm. (Mulder and Holliday, 1976). Spermagonia are between 30 - 65 μm. Spermatia are numerous, rod-shaped, up to 7.6 μm in length (Mehta *et al.*, 2016). Ascoma (perithecia) are 40-60 μm in diameter, contained several hyaline, bitunicate asci with a distinct foot cell. Asci are 54.1 and 67.8 μm in length. Ascus contains 8 ascospores. Ascospore is hyaline, one septate, showing constriction at the septum, measured 6.6×18.2 μm (Mehta *et al.*, 2016)

The fungus grows on artificial media. In medium it sporulates abundantly. The conidia and spermogonia are produced in the culture (Gokhale and Moghe, 1967).The modified Kirchoffs medium (consisting of Potassium phosphate- 0.187g, Magnesium corbonate- 0.175g, Asparagine- 1.0g, Sucrose- 100g, Calcium sulphate- 0.180g, Ammonium molybdenate- 0.017g, Manganese sulphate- 0.024g, Ferric chloride- 0.014g, Biotin- 25 ppm, Agar- 20g, Distilled water- 1000 ml) + Cotton leaves decoction dextrose medium (50:50) is better for abundant growth and sporulation of pathogen, while, the media, viz., modified Richard's medium (Sucrose- 100g, Aspergine- 2g, Potassium dihydrogen phosphate- 5g, Magnesium sulphate- 2.5g, Ferric chloride- 0.02g, Biotin- 20 mg, Agar- 20g, Distilled water- 1000 ml) and V8 Juice medium(Vegetable juice- 200ml, Calcium carbonate- 3g, Agar- 20g, Distilled water- 1000ml) may yield moderate growth and sporulation of the fungus (Gaviyappanavar, 2012).

Molecular characteristics

Molecular variability of *Ramularia gossypii* (= *R. areola*) pathogen was studied by Ganiger *et al.* (2017). In their studies, it was found that phylogenetic analysis grouped *R. gossypii* isolates into two main clusters with 0.12 similarity coefficient. In cluster A, Dharwad, Arabhavi and Naragun disolates were present showing only 3-5.8 per cent divergence. In cluster B only one isolate Devihosur was present with nearly 12 per cent divergence with cluster A. The full length ITS rDNA region was amplified with ITS-1 and ITS-4 primers for the isolates of *R. gossypii*. DNA amplicon was observed at the region 560 bp,

Representative samples were sequenced and by using the NCBI BLAST programme confirm the isolates as *R. gossypii* (Ganiger *et al.*, 2017).

Predisposing factors

This pathogen survives mainly on plant debris and volunteer plants (Rao and Jayasankar, 2015). The disease mainly develops in the moist localities and also low-lying wet soils. The high humidity (90-91 % RH), temperature 23-27°C and susceptible cotton varieties favour in the spread of this disease (Mohan *et al.*, 2006).

Disease cycle

This pathogen survives mainly on plant debris and volunteer plants (Rao and Jayasankar, 2015). However, the ascospores have been considered as the primary source of inoculum (Mohan *et al.*, 2006). The diseased defoliated leaves with the development of sexual (perithecia) stage remain in plant debris. With the onset of monsoon, the ascospores shoot-out from perithecial bodies and cause initial infection on young seedlings. Secondary infection is caused by conidia which are airborne and play main role in disease spread in the field (Mohan *et al.*, 2006).

Management/Control measures

Use of host resistance

In India, the *Gossypium arboreum* genotypes 30805, G 135-49, 30814, 30826, 30838 are immune or highly resistant to grey mildew, while, *G. arboreum* cultivar AKA 8401, is susceptible (Mohan *et al.*, 2006). Resistant breeding with the use of the immune genotypes is under progress (Mohan *et al.*, 2006).

Cultural control

It is suggested to use deep ploughing, crop rotation with cereals, regional tolerant varieties to minimize the disease incidence. Further, removal and burning of crop residues are essential to minimize the inoculum density (Rao and Jayasankar, 2015).

Chemical control

First foliar spray with 3 gm wettable sulphur per one litre of water is needed to be done at initial stages of the disease. Alternatively, dusting of 8-10 kg of Sulphur powder per acre may be done for effective control of the disease. Carbendazim (0.05% a.i.) or Benomyl (0.05% a.i.) are also effective against the disease. In case of high disease intensity, the fungicide Hexaconazole (1l/ha) or Nativo-75 WG (300 g/ha) is suggested to spray against the grey mildew disease (Rao and Jayasankar, 2015).

6. *Mycosphaerella* (*Cercospora*) Leaf spot

Occurrence & severity status:

Mycosphaerella (*Cercospora*) Leaf spot, popularly known as *Cercospora* leaf spot, occurs all over the world (Chupp, 1954; Little, 1987a; Phengsintham *et al.*, 2013). The disease is less severe in cotton growing areas of India and considered as a minor disease of cotton. However, it is prevalent at the end of the growing season of Peru, causing severe leaf fall (Little, 1987a). Moreover, during favourable conditions this disease occurs heavily in cotton fields of Louisiana (Price *et al.*, 2013).

Symptoms

The initial symptoms are usually present in late season and occur as small, scattered, circular and red lesions on the leaves, which are more prominent on upper surface. As the disease progresses, the lesions enlarge and turn white to grey or light brown in the centre. Concentric zones are often present with a red or purple colour at the margins (Little, 1987a; Wade *et al.*, 2015). Spore masses form in the centre of the lesion that making it dark grey appearance on leaf (Kokane and Bhale, 2016). Later on, the spots become irregular surrounded by a chlorotic halo. Leaves turn yellow and defoliate prematurely.

Causal organism: *Mycosphaerella gossypina* (G.F. Atk.) Earle 1900

[Bulletin of the Alabama Agricultural Experimental Station 107: 309]

Synonyms

= *Cercospora gossypina* Cooke 1883

= *Sphaerella gossypina* G.F. Atk. 1891

Morphological/ Morphophysiolgical characteristics

Colonies are amphigenous. Stromata are rudimentary consisting of a few large brown cells. Conidiophores are in fascicles of 2-10or rarely with as many as 30, not branched, straight to curved, cylindrical, medium brown, paler and more narrow at the apex, slightly or abruptly geniculate, 1 -3-septate, 70-150 x 4-5 μm, very rarely up to 400 x 8.5 μm. Conidiogenous cells are terminal, on the main axis of the conidophores, with thickened but not prominent spore scars, straight to flexuous. Conidia are solitary, hyaline, acicular to obclavate, straight to curved, truncate base, subacute tip, septa indistinct, [differently 5-18-septate (Little, 1987a); 1-7- septate (Phengsintham *et al.*, 2013)], 60-130 x 2.5-4 μm. (Chupp, 1954).

Colonies on PDA after 3 weeks at 25°C are with pinkish grey mycelium, reaching 8–14 mm diam., hyphae, septate, constricted at the septa, distance between septa 13–40 μm, reddish or brownish, wall smooth. Conidia are not formed in culture (Phengsintham *et al.*, 2013).

Predisposing factors

Mycosphaerella leaf spot occurs while plants are under stress (Wade *et al.*, 2015). In most cases the disease appears when the cotton growing in less-fertile areas of the field. Cotton growing in areas deficient in potassium or drought-stressed is at risk to this disease. The disease favours wet, warm and humid weather. The disease becomes severe where the climatic conditions such as cloudy weather couples with intermittent rain followed by dry weather. Dense planting with high relative humidity favours the disease.

Disease cycle

The fungus survives as dormant mycelium in infected plant debris. Infection begins at the bottom of the plant, presumably, by wind-borne and rain-splash dispersed conidia, and by contaminated tools (Little, 1987a). Later the disease progresses towards leaves with new growth. When this fungus experiences favourable conditions, it may progress through 4 or 5 cycles in one season.

Management/Control measures

Cultural control

It is suggested to remove previous season crop debris, if disease was present. Otherwise, if no major pests or diseases were present in the previous season at the field, leaving of residues and ploughing to reduce inoculum levels and to improve soil condition in preventing moisture loss are suggested (Kokane and Bhale, 2016). Maintenance of plant vigour by having proper fertility and preventing drought stress through irrigation that helps delay primary infections and reduce the severity of disease outbreaks is necessary. In acid soil, addition of nitrogenous and phosphorus fertilizers and in sandy soils addition of potassium fertilizer are helpful to avoid nutrient deficiency stress, which can cause cotton to be vulnerable to disease. Crop rotation with maize or soybeans is suggested after every three years of cotton grown in one field.

Chemical control

Seed treatment: Seed dressing with Thiram before planting is necessary.

Foliar spray: The fungicides with active ingredients copper oxychloride, sulfur, maneb, mancozeb, chlorothalonil, propiconazole, and neem oil are effective to spray in field. Applications of appropriate foliar fungicides are required on susceptible cultivars of plant (Kokane and Bhale, 2016).

7. *Paramyrothecium* (*Myrothecium*) Leaf Spot

Occurrence & severity status

Paramyrothecium (*Myrothecium*) leaf spot or blight is a prevalent disease in most countries of Europe, the Middle East, Africa, South Asia, the Far East, New Zealand, China and USA (Srinivasan, 1994). It is caused by a polyphagous pathogen which is capable of infecting all the four cotton species (Dake 1980). This disease was first recorded in India by Munjal in 1960. Thereafter, its occurrence was noticed in all cotton growing areas of India (Suryanarayana, 1965; Shrivastava and Singh, 1973; Shrinivasan and Kannan, 1974; Raut *et al.* 1980). The disease is very common in India but it considered as a minor disease. However, in Madhya Pradesh the disease is now well established and has been observed to occur on a regular basis (Anonymous, 2004; Tomar, 2007). There, the disease has been considered as a major disease of cotton. During the years, 2002-03 and 2003-04, the incidence of the disease was 41.71% and 38.28 %, respectively (Tomar, 2005).This disease becomes severe under warm condition and causes losses up to the extent of 30% in India (Monga, 2011). In Brazil, particularly in the cotton fields in Maranhao State, the disease causes yield reduction of up to 60% (Meyer *et al.*, 2006).

Symptoms

The disease in cotton appears as small, circular, tan, reddish or brown coloured spots of 0.5-10 mm diameter that irregularly distributed near the margins of the leaves (Munjal, 1960; Meyer *et al.*, 2006). The spots become with brown centre and broad violet, purple or dark brown margin, surrounded by zones of translucent areas forming concentric rings of sporodochia. After a few days, the dark green sporodochia are surrounded by a rim of white hair like mycelia, particularly in the region where rings are formed. The spots may coalesce to from large necrotic patches and later the centre may drop off, leaving a shot hole appearance (Munjal, 1960; Tomar, 2005). Additionally, the spots may also appear on the petiole, bracts and bolls, causing damages to lint (Dake 1980; Raut *et al.* 1980;Tomar, 2005).

Causal organism: *Paramyrothecium roridum* (Tode) L. Lombard & Crous 2016

[Persoonia 36: 211]

Synonyms

= *Dacrydium roridum* (Tode) Link 1809

= *Myrothecium advena* Sacc. 1908

= *Myrothecium roridum* Tode 1790

Morphological/ Morphophysiolgical characteristics

The mycelium is hyaline becoming ochraceous due to accumulation of protoplasmic contents and aggregates to form sporodochia or stromatic mass in the intercellular spaces. Sporodochia are discoid or flat, circular or irregular in surface view, often confluent in larger masses, without setae, green at first becoming black white rimmed or fringed (Saccardo, 1886; Preston, 1943). From the sporodochia 3 or 4 hyphae are come out of the stomata or by rupture of epidermis and form a hymenium layer on which conidia are produced. Phialids are 12-16 x 2-4 µ in size. Conidia are abundantly produced, held together in a gelatinous mass, elliptic or oblong, one celled, light to olive green in colour, but in mass dark green to black, 2 to 3 guttulate, both ends rounded, 5-9 x 1.0- 2.5 µm in size (Munjal, 1960; Tomar, 2005; Meyer *et al.*, 2006; Tomar *et al.*, 2016).

Molecular characteristics

A highly sensitive real-time PCR (qPCR) assay to detect *P. roridum* from pure culture and infected samples of cotton plants has been reported by Chavhan *et al.* (2018). In their studies, a specific set of primer pair pMyro F/R was designed to target the 185 bp ITS region of rDNA of *P. roridum* species and validated using qPCR. The PCR primer pMyro-F/R specifically detected *P. roridum* from mycelial DNA and infected leaf tissue of cotton. The limit of detection, using SYBR Green dye, from purified DNA was as low as 0.10 pg. per IL of DNA, which revealed the high sensitivity of detection assay (Chavhan *et al.*, 2018).

Predisposing factors

The fungus grows and sporulates well at 25°C, pH 6 and in alternate light and dark photoperiod (Chauhan and Suryanarayana, 1970). The leaf spot disease caused by *P. roridum*, decreased subsequently with warm and dry weather condition (Raut *et al.*, 1980). The severity of the disease is associated with the built up of high humidity in the field (Shrivastava, 1980; Tomar *et al.*, 2009). Also, increased fertilization rates cause lush foliage that can lead to more outbreaks of the disease.

Disease cycle

Paramyrothecium roridum is a soil borne fungus which continues its life cycle on dead and decaying plant tissues in soil (Domsch *et al.*, 2007). The conidia of soil serve as a mechanism to spread the disease. Infection may occur by conidia that splashed from soil (Murakami *et al.*, 1998). In cotton, the red cotton bug, *Dysdercus cingudatus*, was shown to transmit the pathogen (Munjal, 1960). The conidia geminate to cause infection and produce conidia there in. By this way infection repeats

Management/Control measures

Use of host resistance

The cotton (*G. hirsutum*) hybrid LHH-144 (Ajit) has been found as resistant to the disease (Randhawa *et al.*, 2002). Further, Kirti (*G. hirsutum*) hybrid is moderately resistant (Singh *et al.*, 1993). Kumar (2014) has reported fifty seven germplasm lines as resistant to *Paramyrothecium* leaf spot disease which can be further used in breeding disease resistant programme.

Chemical control

Foliar spray of propiconazole at 0.1% concentration, while applied twice, is effective to control the leaf spot disease (Kumar, 2014). Other fungicides, like, Bavistin @ 0.2%, Topsin-M @ 0.1% and Brestanol @ 0.1%, are also found very effective to minimize the disease (Peshney *et al.*, 1999). However, the inhibitory effect of thiophanate methyl has been reported as superior over all the other fungicides tested (Patidar, 2016).

8. Root rot

Root rot of cotton is mainly of two types, one is Black root or black heart disease another is *Phymatotrichopsis* root rot or Texus root rot. The first one is mainly prevailed in Indian subcontinent and the second one is in North and South American countries.

i) Black root rot

Occurrence & severity status

Black root rot, also called as 'black heart' of cotton, is one of the most serious diseases of cotton particularly in the northern region of India where around 1.8 m. hectare area exists in the states of Punjab, Haryana, Rajasthan, western Uttar Pradesh and Jammu and Kashmir(Vasudeva, 1935). In Pakistan, this disease is very drastic in major cotton producing areas (Khan *et al.*, 2017). The disease affects both, the hirsutum and arboreum cotton species (Monga and Raj, 1994). It first appears in June and becomes vigorous during July. In August, the attack slows down and almost ceases by the end of September. It was also observed that incidence of root rot was maximum ie., 69.1% in 1993 when the moisture ranged between 3.8 to 13.4% and soil temperature varied from 29.7 to 36.9°C (Monga and Raj, ND). In Pakistan, the disease appears with 4.4 -5.0 % incidence and causes average yield loss at about 2.5% (Khan *et al.*, 2017).

Symptoms

Sudden and complete wilting of plants is the first visible symptom of the disease (Monga and Raj, ND). In that case, every leaf from top to bottom droops down. When the affected plants uprooted, which can be done easily, roots appear black and there are generally few lateral roots. The black layer can be removed by thumb and forefinger exposing a white centre. Subsequently, the complete root system becomes rotten. As the season gets warmer, the black layer is broken into shreds and normal growth resumes. Usually the disease spreads in field in patches. In subsequent seasons the patches may no longer be obvious as the pathogen spreads throughout the field. The causal fungus can infect the centre of the tap root causing a 'black heart' (Mass *et al.*, 2012).

Causal organism: *Macrophomina phaseolina* (Tassi) Goid. 1947

Synonyms

= *Rhizoctonia bataticola* (Taubenh.) E.J. Butler 1925

Morphological/ Morphophysiolgical characteristics

The causal fungus has been described earlier in preceding chapter. It produces sclerotia of black, irregular in shape and have an average dimension of 105 ± 2 μ. Highest number of sclerotia usually occur at a depth of 15-30 cm whereas the number decreases at 35-45 cm depth. The fungus also produces pycnidia (Monga and Raj, ND). The fungus is considered highly aerobic.

In medium, the fungus produces light grey to black coloured colony of radial to irregular shaped. Aerial mycelia are less, but some isolates may produce very high aerial mycelia. Sclerotial intensity varies from very less to high and is moderate in majority of the isolates. Shape of the sclerotia varies from oblong, ellipsoid, irregular and round. Sclerotial length/width ratio ranged from 1.1 to 1.8 (Sharma *et al.*, 2012).

Molecular characteristics

The *Macrophomina phaseolina* (*R. bataticola*) isolates from different agro climatic regions of India were analyzed with amplified fragment length polymorphism (AFLP) and found that five AFLP primer combinations provided a total 121 fragments. All fragments were found polymorphic with an average polymorphic information content value of 0.213. The dendrogram based on AFLP analysis showed that the maximum number of *M. phaseolina* isolates were very diverse and did not depend on geographical origin (Sharma *et al.*, 2012).

Predisposing factors

The fungus is seed borne (Gangopadhyay, *et al.*, 1970) and the seed may serve as the primary source of inoculum. The fungus is widely distributed in soil. Mostly, it causes severe disease symptoms under warm (28-30°C) condition with soil being the primary sources of inoculums (Dhingra and Sinclair, 1973b). Higher temperature and low moisture conditions favour infection.

Disease cycle

The fungus survives on decaying organic matter in soil or in seed. During favourable condition, it germinates into mycelium which in contact with cotton root infects the plant, or infected seeds may germinate along with the fungal mycelium. On infection the fungus kills the root cells, the hyphae continue to grow and colonize dead tissue, often forming sclerotia. The sclerotia produced in dead root tissues serve as inocula of the next season.

Management/Control measures

Use of host resistance

It is suggested to grow resistant varieties in disease prone areas. The hirsutum varieties-LH 900, LH 886 and F 505 show relatively slow progress of root rot (Monga and Raj, ND).

Cultural control

Proper time sowing and mixed cropping with sorghum may decrease disease severity.

Biocontrol

Gliocladium virens is effective against the pathogen (Monga and Raj, ND).

Chemical control

Fungicide spray is not much effective in root rot disease control. However, seed borne pathogen can be minimized by seed treatment with thirum. The fungicides, Captan at 100 ppm and mancozeb at 1000 ppm also effective to inhibit the growth of the causal fungus (Monga and Raj, ND).

ii) *Phymatotrichopsis* root rot

Occurrence & severity status

Phymatotrichopsis root rot, also known as 'Texas root rot', occurs in Mexico (northern), USA (south-western states including Arizona, Arkansas, California, Louisiana, Nevada, New Mexico, Oklahoma, Texas, Utah), Brazil and Venezuela. The disease affects *Gossypium herbaceum, G. hirsutum* and *G. barbadense* cottons. The disease in Texas and Arizona causes raw fibre loss of about 2.2- 3.5 % (Streets and Bloss, 1973). Each year, the loss of 2% cotton yield occurs in Texas (USA) due to root rot (Watkins, 1981). Hectares of cotton fields are found to be affected in Upland and Pima cotton in Arizona (Mulrean *et al.*, 1984). There, the yield losses are 10 and 13% for infested fields of Upland and Pima cottons, respectively (Mulrean *et al.*, 1984).

Symptoms

Symptoms are most likely to occur from June through September when soil temperatures reach 28 °C (82 °F). The first symptoms are slight yellowing or bronzing of the leaves. Leaves appear flaccid, develop visible wilt and usually within 3 days of these initial symptoms, become permanently wilted and die (Ezekiel and Taubenhaus, 1932; Streets and Bloss, 1973; Watkins, 1981). Leaves desiccate, but generally remain attached to the plant. Roots are usually extensively invaded by the pathogen by the time wilting occurred. Affected areas often appear as circular patterns of dead plants. The dead plants can be pulled from the soil with little effort. Root bark is decayed and brownish, and bronze coloured woolly strands of the fungus are frequently apparent on the root surface. If there is abundant water, brown to black wart-like sclerotia are also seen in the mycelial strands on the surface of roots. Discoloration of xylem elements in the roots and lower stems can often be found with infected cotton plants (Olsen *et al.*, 1983; Rush *et al.*, 1985).

Causal organism: *Phymatotrichopsis omnivora* (Duggar) Hennebert 1973 [Persoonia, 7 (2): 199]

Synonyms

= *Ozonium omnivorum* Shear 1907

= *Phymatotrichum omnivorum* Duggar 1916

Morphological/ Morphophysiolgical characteristics

The pathogen is an asexual, holoanamorphic (mitosporic) fungus. It produces root-like strands (rhizomorphs) that grow through the soil to contact with plant roots and proliferates around the hypocotyl, producing a cottony, mycelial growth. The mycelial strands are about 200 µm in diameter, bearing acicular hyphae, with distinctive cruciform branches emanating from the peripheral mycelium. It produces conidia on white to tan coloured spore mats, which is 10-20 cm in diameter and 0.6 cm thick, and composed of large celled branched fungal strands on the soil surface. Conidia are unicellular, hyaline, globose or ovate, 4.8-5.5 µm in diameter or 6-8 x 5-6 µm. The conidia appear sterile, and their role in the spread of the pathogen has not been documented. Sclerotia are irregular in shape, at first white, changing to buff, brown and black with age, small, 1-2 (-5) mm in diameter (Streets, 1937; Streetsand Bloss, 1973; Walter and Everett, 1982).

Molecular characteristics

Phylogenetic analyses of nuclear small- and large-subunit ribosomal DNA and subunit 2 of RNA polymerase II from multiple isolates indicate that it is neither a basidiomycete nor closely related to other species of *Botrytis* (*Sclerotiniaceae, Leotiomycetes*). The fungus (*P. omnivora*) is a member of the family *Rhizinaceae* and *Pezizales* allied to *Psilopezia* and *Rhizina* (Marek *et al.*, 2009). Further, ITS-rDNA sequences were amplified, using *P. omnivora*-specific primers (PoITSA 5'-CCTGCGGAAGGATCATTAAA-3' and PoITSB 5'-GGGGGTTTTCTTTGTTAGGG-3'), from sporemat herbarium specimens of the fungus. Based on the alignment of this sequence with ITS sequences from over one hundred other *P. omnivora* isolates, the herbarium specimen sequence was most similar to *P. omnivora* isolates from El Campo, TX (100 % identity, 302/302) and the ATCC 48084 isolate (99 % identity, 302/303), which belong to an ITS haplotype common in southern Oklahoma and throughout eastern and central Texas (Marek *et al.*, 2009).

The fungus, *P. omnivora*, is difficult to isolate. So, a multigene-based, field-deployable, rapid, and reliable identification method for the fungal plant pathogen has been developed with the use of real-time quantitative PCR (qPCR) and the Razor Ex Bio Detection system (Arif *et al.*, 2013). Specific PCR primers and probes were designed based on *P. omnivora* nucleotide sequences of the genes encoding rRNA internal transcribed spacers, beta-tubulin, and the second-largest subunit of RNA polymerase II (RPB2). The PCR products were cloned and sequenced to confirm their identity. All primer sets allowed early detection of *P. omnivora* in infected but asymptomatic plants. A modified rapid DNA purification method, which facilitates a quick (30-min) on-site assay capability for *P. omnivora* detection, was developed. Combined use of three target genes increased the assay accuracy and broadened the range of detection (Arif *et al.*, 2013).

Predisposing factors

The disease is favoured by calcareous clay loam soils with a pH range of 7.0 to 8.5 and the areas of high summer temperatures.

Disease cycle

The fungus, *P. omnivora*, is not seed borne. The primary inoculum is sclerotia, or strands surviving on roots of host plants for many years and often is found as deep in the soil as roots penetrate. The fungus generally invades new areas by continual slow growth through the soil from plant to plant. It may also be moved about on roots of infected plants moved to new areas. After heavy summer rain, the fungus may come to the soil surface and form large tawny mycelial mats on which conidia are borne. These spores germinate with difficulty or not at all and probably play no part in dissemination.

Management/Control measures

The disease is very difficult to control. The fungal behaviour in different soils and years are so erratic that several approaches are used to get satisfactory control.

Phytosanitary measures

It is recommended to prohibit the importation of soil from countries where *P. omnivora* occurs (OEPP/EPPO, 1990a).

Cultural control

Several measures, like, crop rotations of 3-4 years with monocotyledonous crops, organic amendments by the use of wheat, oats, and other cereal crop residues, deep ploughing up to 6 to 10 inches deep, using barrier crops with a resistant grass crop such as sorghum around an infected area in a field and the use of ammonia as nitrogenous fertilizer, are effective to minimize the disease up to a considerable limit (Texas Plant Disease Handbook).

Chemical Treatment

The fungus is highly sensitive to the systemic triazole fungicides (Whitson and Hine, 1986; Riggs and Lyda, 1988). The application of a systemic triazole fungicide deep in the soil near the root appears to offer the potential for disease reduction. Propiconazole and triadimenol are also effective fungicides while applied in subsurface drip irrigation to reduce plant mortality in cotton (Olsen and George, 1987). The fungus is also sensitive to the fungicide benomyl at the rate of 15.98g/L or 2 lbs/15gal water/are (Hine *et al.*, 1969; Lyda and Burnett, 1970); 15.98g/L or 2b../ 15gal water/area.

9. Rust disease

Three different rust diseases are common in cotton. They are as follows:

i) Cotton Rust (*Puccinia schedonnardii*)

Occurrence & severity status

The cotton rust (*Puccinia schedonnardii*) is a minor leaf disease of cotton, distributed and recorded in Canada (Manitoba, Alberta) and U.S.A. including border states from New York to Montana but the main distribution is toward the southwest into Mexico and South America (Parmelee, 1980). The disease affects leaves, bolls and involucral bracts (Mulder and Holliday, 1971a). The aecial stage occurs on cotton, *Gossypium* spp. and few other plant species, while, uredial and telial stages are on species of *Limnodea, Melica, Monanthochloe, Muhlenbergia, Schedonnardus* and *Triplasis* (Mulder and Holliday, 1971a).

Symptoms

This disease typically affects cotton leaves, bolls and involucral bracts (Mulder and Holliday, 1971a). It first appears as small yellow pustules on leaves then transform into larger, orange/red pustules which release aeciospores.

Causal organism: *Puccinia schedonnardi* Kellerm & Swingle, 1888

[Journal of Mycology 4 (9): 95]

Synonyms

= *Caeoma hibisciatum* Schw., 1832.

= *Aecidium gossypii* Ellis &Everh., 1896.

= *Puccinia muhlenbergiae* A. & H., 1902.

= *Puccinia hibisciata* Kellerm., 1903.

= *Dicaeoma hibisciatum* Arth., 1906.

Morphological/ Morphophysiolgical characteristics

The fungus is heteroecious, macrocyclic. Pycnia are amphigenous, grouped, reddish brown, surrounded by aecia. Aecia are amphigenous, but mostly hypophyllous, cupulate, in compact groups, pale yellow. Aeciospores are globoid to ellipsoid, 16-23 x 16-20 µm; wall finely verrucose. Uredinia are elliptical, amphigenous, pulvinate, pale yellow. Urediniospores are globoid to obovoid, sometimes ellipsoid, 17-26 x 15-22 µm; wall finely echinulate, pale yellow, pores 6-8 scattered. Telia are amphigenous, mostly hypophyllous, chocolate-brown, elliptical, scattered to gregarious, pulvinate. Teliospores are broadly ellipsoid to oblong, dark brown, slightly constricted at septum, apex round, base obtuse or slightly narrow, 21-37.5 x 17.5-25 µm; pedicel pale yellow to colourless or light cinnamon-brown, usually 1 to 3 times length of spore (Mulder and Holliday,1971a; Parmelee, 1980).

Predisposing factors

The rust is favoured by humid conditions and periods of prolonged wetness. The moisture on the leaf surfaces for longer period leads to disease occurrence and spore release and germination. The increased relative humidity underneath the canopy favours the disease.

Disease cycle

The pathogen overwinters as teliospores that are produced in telia on the alternate host. In the spring, the teliospores germinate to produce basidiospores. The basidospores are then windblown to the cotton host where they enter via stomata. On germination, the basidiospores produce a mycelium from which flask-shape pycnia as well as receptive hyphae are formed. The pycnia produce pycniospores that fertilize receptive hyphae of a different mating type. On mating, dikaryotic mycelia are formed, which grow through the cotton leaf to produce aecia. Aecia produce aeciospores that are released from the aecia and infect alternate hosts. In the alternate host, uredia are formed. The uredospores released by the uredium are able to reinfect grasses in a single season. The uredospores can become overwintering teliospores at late stage.

Management/Control measures

Phytosanitary measure

According to IPPC (2016) the cotton rust fungus, *P. schedonnardi*, is a quarantine pest.

Cultural control

It is suggested to utilize non-susceptible crops in a rotation to decrease infection rate in future cotton crops.

Chemical control

An application of Mancozeb as foliar spray before the disease infection can be used to prevent the disease, but little can be done after infection.

ii) Southwestern cotton rust (*Puccinia cacabata*)

Occurrence & severity status

The occurrence of 'Southwestern cotton rust' (*Puccinia cacabata*) was first reported from Baja California, Mexico, in 1893 and from Arizona in 1922. The disease occurs only in certain parts of Southern Arizona, New Mexico, West Texas, and Northern Mexico (Olsen and Silvertooth, 2001). The disease causes serious damages to cotton growth once every 3-4 years in the southwest cotton belt including New Mexico. Southwestern cotton rust is very erratic in occurrence and the degree of severity depends to a large extent on weather conditions and proximity of the telial hosts to cotton. The aecial state appears in USA and Mexico only, while, the telial stage occurs more widely including Argentina, Bahamas, Bolivia, Brazil and Dominican Republic (Mulder and Holliday, 1971b). This rust disease is very severe in the irrigated cotton of S.W. USA and N. Mexico where losses of 40% or more can occur under favourable climatic conditions. Heavy rust infection was occurred in USA and Mexico in the mid-2000s (Percy, 1993; Zhang *et al.*, 2017). In some locations in Arizona in 1930, yields were reduced by 75% (Percy, 1993). In 1959, losses in southeastern Arizona and the adjacent corner of New Mexico were varied from 50 to 75% (Percy, 1993). In Mexico, rust caused losses were 50% in some fields. In the major cotton production district at Delicias, Chihuahua, a rust epidemic in 1963 resulted in the loss of 100,000 bales (Percy, 1993).The southwestern cotton rust fungus lives on cotton and grama grass (*Bouteloua* spp.) (Olsen and Silvertooth, 2001).

Symptoms

The rust disease in cotton at first appears as small, somewhat inconspicuous, pale-green lesions on both leaf surfaces. The lesions subsequently develop into bright yellow spermogonial pustules, sometimes surrounded by a maroon zone. The spermogonial pustules may also appear on any of the aboveground plant parts. Later, cup-like aecia erupt through the lower leaf epidermis. Aecia are large and easily observed. They appear as orange-yellow, become brown with age, circular, and slightly raised lesions on the lower leaf surfaces, bracts, green bolls and stems of cotton plants. Carpel infection leads to dwarfing of the bolls and the attacks on the peduncle contribute to breakage and loss of bolls. Infection of cotton seedlings may cause death (Mulder and Holliday, 1971b; Percy, 1993; Nyvall, 1999; Olsen and Silvertooth, 2001).

Causal organism: *Puccinia cacabata* Arthur &Holw. 1925

[Proceedings of the American Philosophical Society 64: 179]

Synonyms

= *Aecidium gossypii* Ellis & Everh., 1896.

= *Uredo chloridis-berroi* Speg., 1925.

= *Puccinia stakmanii* Presley, 1943.

= *Uredo chloridis-polydactylidis* Viegas, 1945.

Morphological/Morphophysiolgical characteristics

The fungus is heteroecious, macrocyclic. Pycnia are amphigenous, numerous in raised circular groups; yellow to orange. Aecia are mainly hypophyllous, in groups surrounding pycnia, peridium orange, becoming yellow or colourless; peridial cells oblong, strongly echinulate. Aeciospores are globose to oblong, 18-25 x 13-19 µm, wall finely echinulate, yellow to hyaline. Uredia are mainly epiphyllous, rarely amphigenous, becoming confluent, early naked, cinnamon-brown. Urediospores are globose to broadly ellipsoid 23-31 x 19-24 µm; wall pale to cinnamon-brown, moderately echinulate; pores 3 (-4), distinct, equatorial. Telia are amphigenous, crowded, round elliptical or linear, sometimes confluent, naked early with ruptured epidermis. Teliospores are variable in shape and colour, rounded at both ends or slightly pointed at the apex, 27-39 x 19-26 µm; sometimes constricted at the septum; wall chestnut brown or slightly coloured (Mulder and Holliday, 1971b).

Predisposing factors

Rust is a very erratic disease because of the dependence on extended wet and high rainfall periods. Leaf wetness periods in excess of 16 hours coupled with high relative humidity and moderate temperature are the factors necessary for epidemics to occur.

Disease cycle

The disease spreads from overwintering teliospores on the wild grama grass (*Bouteloua* spp.) hosts to cotton by sporidia in the late spring and summer (Mulder and Holliday, 1971b). During summer rains, the spore stage on grama grass germinates to produce airborne sporidia which are carried up to eight miles and cause initial infections in cotton. There is no repeating spore stage on cotton. All new infections on cotton are dependent upon sporidia showers from grama grass. The spores produced on cotton can only infect grama grass (Olsen and Silvertooth, 2001).

Management/Control measures

Phytosanitary measure

According to IPPC (2016) the Southwestern cotton rust fungus is a quarantine pest.

Use of host Resistance

Currently, all the commercial Upland and Pima cottons are susceptible to rust infections under the natural field conditions (Zhang *et al.*, 2017). However, a few obsolete resistant Upland cotton lines of *G. hirsutum* were developed through the transfer of the resistance from diploid species *G. arboreum* and *G. anomalum* in the early 1970's (Blank, 1971). The expression of resistance in petioles and stems was independent of that in cotyledons, and was associated with the possession of the glandless and nectariless traits (Percy and Bird, 1985). Genetic analysis of resistance in cotyledons has shown that it is controlled by a single dominant locus, designated Pu (Percy and Bird, 1985). The resistance mechanism appears to be a hypersensitive reaction to the fungus (Percy, 1993).

Chemical control

Zineb and mancozeb are effective as a foliar spray before infection (Mulder and Holliday, 1971b; Olsen and Silvertooth, 2001). Application of any of the fungicides should be made prior to the first "spore showers" because these fungicides are only protective (Olsen and Silvertooth, 2001). Rust may be controlled by three to four aerial applications of mancozeb (0.2%) plus a sticker in 10 gals (38 l) water/acre, during July and August (Olsen and Silvertooth, 2001).

iii) Tropical rust (*Phakopsora desmium*)

Occurrence & severity status

Tropical rust is an emerging disease in cotton. It occurs in tropical and subtropical cotton-growing areas of the world. The disease causes significant yield losses of cotton cultivated in Brazil, India and Jamaica (Guerra *et al.*, 2013; Pindikur *et al.*, 2012). It is one of the devastating diseases of cotton in northern Karnataka (India) and has been reported to reduce the yield (Hillocks, 1991b) to the extent of 24% (Johnston, 1963). The disease appears in dry season during December-March and is prevalent in Karnataka, Andhra Pradesh and Gujarat states (Puri *et al.*, 1998). In Brazil, the tropical rust is an emerging disease in cotton that causes significant yield losses of cotton. The tropical cotton rust has also been found in the commercial crops in Venezuela (Malagutiand Lopez Pinto, 1974).

Symptoms

The tropical rust appears as ferruginous (rusty) pustules of uredia on the abaxial surface of leaves. The pustules at first appear on older leaves and spread to the new leaves rapidly. On reaching maturity, the uredia easily break the epidermis to release uredospores. The plants become prematurely defoliated as the disease develops (Kirkpatrick and Rothrock, 2001; Pindikur *et al.*, 2012). Symptoms may also appear on petioles, stem and boll (Pindikur *et al.*, 2012).

Causal organism: *Phakopsora desmium* (Berk. & Broome) Cummins 1945

[Bulletin of the Torrey Botanical Club 72: 206]

Synonyms

= *Aecidium desmium* Berk. & Broome 1873
= *Cerotelium desmium* (Berk. & Broome) Arthur 1925
= *Cerotelium gossypii* (Lagerh.) Arthur 1917
= *Kuehneola desmium* (Berk. & Broome) E.J. Butler 1918
= *Phakopsora gossypii* (Lagerh.) Hirats. f. 1955

Morphological/ Morphophysiolgical characteristics

Uredia are both epiphyllous (aecialuredia) and hypophyllous (uredialuredia), scattered in groups, slightly pulverulent, yellowish brown, up to 0.5 mm diam., in purplish spots 1-5 mm wide. Paraphyses are peripheral, clavate, usually incurved, wall smooth, colourless. Urediospores are ellipsoidal or obovoidal, 16-19 x 19-27 µm; wall light yellow to hyaline, echinulate; pores inconspicuous, 2, equatorial. Telia are rare, hypophyllous, scattered, inconspicuous, naked or somewhat pulverulent, light cinnamon brown. Teliospores are in crusts of laterally adherent spores, 5-8 spores in depth, angular to irregularly oblong, 24-32 x 10-14 µm, wall smooth, pale brown (Punithalingam, 1968a).

Predisposing factors

The disease favours dry season and cooler temperature of about 25°C.

Disease cycle

The tropical rust spreads by airborne urediospores (Punithalingam, 1968a). The uredospores germinate and produce germ tubes and appressoria that may directly penetrate the leaf cuticle. In about 20 days, the infection develops and forms closed uredia containing uredospores on the abaxial leaf surface. After 25 days, the uredia start to open and become fully open by 35 days and release uredospores (Araujo *et al.*, 2016).

Management/Control measures

Phytosanitary measure

According to IPPC (2016) the tropical rust fungus is a quarantine pest.

Use of host resistance

The commercial cotton hybrids available in India are mostly susceptible to this disease (Shetty *et al.*, 2011). However, the Bt cotton hybrids, viz., Mallika (BG-I), Chirutha, Tulasi-1179BG-I, JKCH-99 Bt, JK Indra, JK Durga, Maruti-96Bt, and UASR Bt-2, showed highly resistant during a field screening in 2009 (Shetty *et al.*, 2011). Further, Chattannavar and Govindappa (2009) reported 23 cotton genotypes as highly resistant to tropical rust in India. The supply of Si to cotton plants may also contribute to increase their resistance to rust due to the participation of the defense enzymes (Guerra *et al.*, 2013).

Chemical control

Tropical rust is mainly controlled with the use of fungicides, because resistant cotton cultivars are not yet available to growers (Kirkpatrick and Rothrock, 2001; Pindikur *et al.*, 2012). The fungicide, propiconazole (0.1%), while sprayed at 75, 90, 105 and 120 days after sowing is most effective to control the disease and avoid crop losses by 34.05% with the benefit cost ratio 1:82 in India (Bhattiprolu, 2015). Further, several other fungicides, such as, hexaconazole, difenconazole, chlorothalonil and carbendazim+mancozeb (SAAF) are effective in inhibiting uredospore germination of the fungus (Pindikur *et al.*, 2012).

10. *Stemphylium* Leaf Spot

Occurrence & severity status

Stemphylium Leaf Spot, also called '*Stemphylium* Leaf Blight', was first described by George Weber in 1930 in Florida, United States. The disease occurs in almost all the cotton producing areas of Africa, Asia, Oceania, Europe, North America, Central America and South America (Ellis and Gibson, 1975; Yu, 2015). It starts to appear around the fourth week of bloom with a heavy fruit/boll load which correspond to a heavy demand for potassium. The disease is normally associated with plant stress factors (drought, nutrient deficiencies, nematode and insect pressure) and yield loss can be severe from the stress complex. A severe epidemic of the disease was occurred in the state of Paraná, Brazil, during 1994 and 1995, causing up to 100% yield losses in some commercial fields of cultivar Paraná 3 (Mehta, 1998). Apart from cotton, the disease also affects other crops like, *Capsicum annuum* L. (Kim and Yu, 1996; Yu, 2001) and *Lycopersicum esculentum* Mill (Kim *et al.*, 1999).

Symptoms

The disease symptoms at first appear towards the margin on the upper leaves in plants of 50-55 days old as small, circular to irregular, dark brown to black spots, varying between 2 and 5 mm and becoming reddish brown with age. Later, the spots develop up to 10 (-20)mm in diameter, circular in shape and brown in colour with concentric zones and a whitish centre that cracks, giving the appearance of a frog's eye. At a later stage, the cracked area drops out, leaving a hole in the centre of the spot. In severely infected leaves, the spots coalesce, yellowing of the leaves occur, and in a few days the leaves desiccate and abscise (Mehta, 1998; Kelly, ND). Dark masses of conidiophores are formed on the edge of the spot on both leaf surfaces (Ellis and Gibson, 1975).

Causal organism: *Stemphylium solani* **G.F. Weber, 1930**
[Phytopathol. 20: 513]

Synonyms

= *Thyrospora solani* (G.F. Weber) Sawada, 1931.

Morphological/ Morphophysiolgical characteristics

Colonies are effuse, grey or greyish-brown, hairy or velvety. Mycelium is mostly immersed. Conidiophores arise singly or in groups, unbranched or occasionally branched, straight to flexuous, more or less cylindrical, erect, brown, with vascular swellings of 8-10 µm diam, 5 to 8 septa, bearing a single conidium at the apex, and measuring (80-95 to 117 (-350) × (4-) 5 to 6 (-8) µm. Conidia solitary; oblong, commonly smooth-walled, rarely minutely verruculose, pointed at the apex showing a small beak, rounded at the base, pale to golden brown, basal part dark brown to black showing a white pore where it attached to the conidiophore, with 3-5(-6) transverse and 1-2 (or several) longitudinal septa, mostly constricted at the median septum only, (32-) 35-55 (-58) × (17-)18-28 µm (Ellis and Gibson, 1975; Mehta, 1998; Yu, 2015).

The fungus shows considerable variation in culture. Colonies may grow slow on PDA, reaching 24 to 66 mm in diameter after 8 days of incubation (Mehta, 1998) or fast, reaching 45-50 mm in diam. after 7 days on PDA (Yu, 2015). The colonies are gray to dark gray with velvety centres of smooth and shiny mycelial mat, and produced a yellow pigment in the medium that turned deep red with age (Mehta, 1998). On artificial media, like, PDA, V8 juice agar, and maltose peptone agar, the fungus rarely sporulates. However, it sporulates well when 7- to 10-day-old colonies grown in culture plates with a thin layer of PDA (10 ml/plate), are scraped, washed with water, and exposed to direct sunlight for 2 days (4 h/day) (Mehta, 1998).

Molecular characteristics

A molecular analysis was performed to assess the genetic diversity among the *Stemphylium solani* isolates from cotton, and to verify their relationship with representative *S. solani* isolates from tomato. Random amplified polymorphic DNA (RAPD) markers and internal transcribed spacers of ribosomal DNA (rDNA) were used to compare 33 monosporic isolates of *S. solani* (28 from cotton and five from tomato). Amplifications of the ITS region using the primer pair ITS4/ITS5 resulted in a single PCR product of approximately 600 bp for all the isolates. Similarly, when amplified fragments were digested with eight restriction enzymes, identical banding patterns were observed for all the isolates. Hence, rDNA analysis revealed no inter-generic or intra-specific variation (Mehta, 2001).

Predisposing factors

The crop under irrigation with heavy fruit set and insufficient K in the fourth week of bloom are the most prone to *Stemphylium* infection, because of the high K demand at that time. Mehta (1998) observed that the disease epidemic was correlated with the excessive rains during the crop cycle. The monthly rainfall 108 to 348 mm and the monthly temperature 22 to 25°C favoured the disease epidemic during the 1995 to 1996 crop cycle (Mehta, 1998).

Disease cycle

The pathogen survives as conidia and mycelia in plant debris and these are believed to serve as the primary inoculum (Zheng et al., 2010). The fungus reproduces and spreads through the formation of conidia on conidiophores. The disease spreads quickly to additional hosts via either mycelium, when leaves of adjacent plants are touching or conidia, which can spread through rain or air. The disease can also be spread via infected seed (Zheng et al., 2010). Conidia can cause several stages of secondary infection throughout the growing season but infection is most severe following early fruiting (Cedeño and Carrero, 1997). The fungus overwinters as mycelia and conidia in plant debris (Zheng et al., 2010).

Management/Control measures

Cultural control

The disease can be managed by split potassium (K) applications, especially on sandy soils; applying half of the recommended dose at planting time and the other half around first square period. Petiole test may also be done to determine the requirement of foliar K application. If needed, foliar applications should be made during the first four weeks of bloom (Mulvaney, 2015).

Chemical control

The disease can control effectively by spraying twice with the fungicide fentin hydroxide-40% at 200 g a.i./ha-1 and fentin acetate-20% at 200 g a.i./ha-1 at a 15-day interval, starting the first application 57 days after sowing (Mehta and Oliveira, 1996).

11. Tobacco Streak Virus Disease

Occurrence & severity status

Tobacco Streak Virus (TSV) disease, also known as 'Cotton necrosis disease', occurs in North and South America, Europe, India, Japan, Australia, New Zealand and South America (Scott, 2001). The geographical distribution of TSV disease in cotton appears to be closely related to the distribution of the major alternative host, parthenium weed (Sharman, 2011). The disease is an emerging threat in India (Rageshwari et al., 2016a). A survey reveals that TSV incidence is highest in Telangana with mean PDI of 25.31, followed by Andhra Pradesh with mean PDI of 16.47 (Vinodkumar et al., 2017). The hybrid, RCH 659 BG

II cultivated in Dameera, Warangal district has been recorded with highest incidence (51.11) (Vinodkumar *et al.*, 2017). In India, the disease may cause up to 62.7% cotton yield loss (Rageshwari *et al.* 2017). It affects many other hosts, such as tomatoes, asparagus, beans, soybeans, grapes, strawberries and ornamental plants (Fulton, 1948; Fulton, 1985). It is generally more problematic in the tropics or warmer climates.

Symptoms

Tobacco streak virus (TSV), inciting cotton necrosis, exhibits different types of symptoms. Initially, the disease appears with chlorotic lesions, which may later turn into necrotic, purplish brown spots, surrounded by yellow halos. The spots may be chlorotic or necrotic, small or large, purplish or purplish brown or brownish black, irregular or ring-like, confined at the interlobular region /apex or distributed all over the lamina. Spots are varied in size and pattern. In cases of severe infection, drying of squares, stems and stunting of plants are also observed (Vinodkumar *et al.*, 2017). Symptoms are strongly influenced by temperature. For example, plants experiencing a temperature of 20° C will develop small necrotic spots while at temperatures above 30° C, necrotic spots will become large necrotic arcs (Gulati *et al.*, 2016).

Causal organism: *Tobacco streak virus* **(TSV)**

Virus

Tobacco streak virus (TSV), belonging to genus Ilarvirus (subgroup I) of family Bromoviridae family, is a multipartite, single-stranded, positive-sense, RNA virus. The virus is characterized by icosahedral particles measuring 27-35 nm in diameter. Its genome consists of three single stranded positive sense RNA species of 2.9, 2.7 and 2.2 kb designated as RNA-I, RNA-2 and RNA-3 respectively. The coat protein, coded by RNA-3, is required for infectivity (van Vloten-Doting, 1975; Brunt *et al*, 1996).

Transmission

Tobacco streak virus has been reported to be transmitted through mechanical means, infected seeds and through thrips species (Kaiser *et al.* 1982; Sharman 2009; Jagtap *et al.* 2012a). In cotton, the disease is found to be transmitted by *Thrips palmi* (cotton thrips) and *Thrips tabaci* (onion thrips) (Jagtapa *et al.*, 2012b).

Molecular detection

Systemic nature of TSV in cotton has been detected by serological and molecular approaches (Rageshwari *et al.*, 2016b). The serological detection of TSV is accomplished through DAC-ELISA. Molecular detection includes characterization of coat protein and movement protein through Reverse Transcriptase (RT)-PCR. Further, latent infection in apparently healthy cotton plants and symptomless leaves of infected plants is confirmed using DAC-ELISA, RT PCR and Immuno- capture RT PCR. This unveils systemic nature of TSV in cotton (Rageshwari *et al.*, 2016b). Vinodkumar *et al.* (2017) have reported that the RT-PCR products produce amplicons of ~900 bp on 1.2% agarose gel. The coat protein (CP) gene sequences of the six TSV isolates originating from different hosts and locations are amplified. The resulting amplicons are cloned and sequenced to assess molecular variability. The sequence and phylogenetic analyses reveal that the CP gene among TSV isolates of different hosts and locations is highly conserved (99-100%), suggesting a common origin (Bhat *et al.*, 2002).

Predisposing factors

Environmental factors greatly influenced the establishment of TSV in cotton. Minimum temperature (22.81°C), relative humidity (81.42%) and leaf wetness (23.94 h) favour TSV incidence (Vinodkumar *et al.*, 2017).

Disease cycle

The TSV virus requires a living plant to survive for a period of time. The virus may be transmitted by thrips vector, mechanical damage or pollen. The TSV consists of a ss-RNA genome encapsulated by spheroid particles made up of coat protein subunits. The coat protein protects the viral genome and plays a role in cell to cell movement (Gulati *et al.*, 2016). Once the virus has penetrated into the host cells, it uncoats and releases its viral genomic RNA into the cytoplasm and expresses proteins replicate. After replication, the dsRNA genome is synthesized from genomic ssRNA (Anonymous, ND7). Subgenomic RNA4 is translated producing capsid proteins and the new virus particles are assembled (Annonymous, ND7).

Management/Control measures

No control methods are known for TSV. General farm hygiene to minimise the presence of the major alternative host of TSV, parthenium weed, is advised and susceptible rotational crops such as, mung beans, are suggested to grow (Sharman, 2011).

12. Vascular (*Fusarium*) Wilt

Occurrence & severity status

Vascular wilt, also known as *Fusarium* wilt, is one of the major diseases of cotton found wherever this crop is grown. The disease was first identified in cotton fields in Alabama and described in southern United States in 1892

(Atkinson, 1892). It is widespread in the USA, the former Soviet Union (Menlikiev, 1962) and China. In Africa, Tanzania is the worst affected country (Hillocks, 1981). In India the disease has been reported from Maharashtra, Tamil Nadu, Karnataka and other states (Agropedia, ND). The disease is capable of causing significant economic loss in localized areas and for individual farmers in areas where the disease is endemic. It is highly virulent on cotton (*Gossypium hirsutum*) in America, Asia and Africa. In Angola, it causes serious yield and quality losses (Ragazzi, 1992). The disease may also affect other crops, like, brinjal, chilli, tobacco and bhendi (Agropedia, ND).

Symptoms

The disease may appear at any stage of crop development, since, plants at all stages of maturity are potentially susceptible. The fungus infects via roots and enters vascular tissues, from where it spreads upwards through the plant. Typical internal symptoms of continuous, dark, vascular discolouration are usually evident. When inoculum density is high or infection initiates from the seed, plants may be killed at the seedling stage. In seedlings, symptoms begin as vein darkening on the cotyledons followed by necrosis, wilt and shedding, resulting in bare stems. In case of older plants, the disease appears in patches. First symptoms are normally seen in the field between 1 and 2 months after planting, with a period of increased susceptibility around the onset of flowering. The infected older plants remain stunted in comparison with uninfected neighbours. Lower leaves often display symptoms first that spreads upwards in the plant. Interveinal leaf yellowing and localised wilting occurs, followed by leaf browning, abscission and death. Affected plants therefore appear stunted and defoliated (Brayford, 1992). As the infection becomes fully systemic, all the leaves are affected, become necrotic and the plant dies from moisture stress.

Causal organism: *Fusarium oxysporum* f. sp. *vasinfectum* (G.F. Atk.) W.C. Snyder & H.N. Hansen, 1940

[American Journal of Botany 27: 66]

Synonyms

= *Fusarium vasinfectum* G.F. Atk., 1892

Morphological/Morphophysiolgical characteristics

The fungus colonizes both inter and intra cellularly in the host tissue as mycelium. The mycelium plugs the xylem vessels partially or completely. Mycelium is white and septate and produces macro- and micro-conidia, and intercalary and terminal chlamydospores which are stable overwintering structures. The fungus produces violet pigmentation in PDA and PSA (Cano and Solsoloy, 1994).

Molecular characteristics

Molecular characterisation is necessary to identify *Fusarium oxysporum* f.sp. *vasinfectum* (FOV) infection accurately. Since, different wilt agents are frequently found together in the same cotton field and appear a complex community of *Fusarium* spp. Thus, sometimes, it is difficult to distinguish morphologically FOV, often coexist with it on various organs of cotton (Miller and Weindling, 1939; Fulton and Bollenbacher, 1959; Mertley and Snow, 1978; Roy and Bourland, 1982; Batson and Borazjani, 1984; Colyer, 1988). A polymerase chain reaction assay was developed for the detection of FOV (Moricca *et al.*, 1998). In that PCR assay, the primers Fov1 (5^1-CCCCTGTGAA CATACCTTACT- 30) and Fov 2 (5^1- CCAGTAACGAGGG TTTTACT-30) were selected based on small nucleotide differences in internal transcribed spacer sequences between 18S, 5·8S and 28S ribosomal DNAs. These primers unambiguously amplified a 400-bp DNA fragment of all the FOV isolates tested but did not amplify any other isolates of mycoflora associated with cotton (Moricca *et al.*, 1998).

Predisposing factors

Severity of the disease is closely related to the disease resistance of cotton and to soil temperature (Ma *et al.*, 1980). The disease is more severe in heavy soil with soil temperature 20-30 °C during the crop season, but mean temperatures about 23°C is most favoured. Higher temperatures above 28°C reduce symptom expression. The pathogen survives in association with non-hosts and other crops that allows *F. oxysporum* f. sp. *vasinfectum* to persist in soil for years or decades (Smith and Snyder, 1974). The pathogen also persists in seed, which can occur by contamination of cotton bolls in the field (Bennett *et al.*, 2008; Elliot, 1923). The disease is a stress-related pathogen, thus, the fungus can rapidly colonise the stressed plants.

Disease cycle

The fungus is soil borne and may be transmitted by seed. In soil, it can survive for several years in a dormant state (chlamydospores) in plant debris or in the soil. The fungus may also sustain on the outer surface of roots of many crops and weeds. It invades the plants through the roots, especially through root wounds caused by nematodes (*Meloidogyne* spp.), and subsequently infects the vascular system, resulting in wilt symptoms (Hillocks, 1992c). Once the disease has established, it can spread across fields, farms and regions. Spores are effectively spread over long distances in infected soil attached to boots, vehicles, farm equipment and in water. It can also be transferred in infected plant material, including seed.

Physiologic Races/Strains

Fusarium oxysporum f. sp. *vasinfectum* (FOV), the causal organism of vascular (*Fusarium*) wilt, is composed of six recognized races. Although, eight nominal races were initially identified, the molecular classification has eliminated redundancy between races 4 and 7 and races 3 and 5 (Chen et al., 1985; Ibrahim, 1966; Nirenberg et al., 1994). The race 1 was characterized in the United States (Armstrong and Armstrong, 1958) and has worldwide distribution. Race 1 is distinguished from race 2 by DNA sequence analysis (O'Donnell et al., 2009) as well as by pathogenicity on plants other than cotton, including okra (Armstrong and Armstrong, 1958; Elliot, 1923). These two genotypes are particularly devastating on cotton when the root knot nematode, *Meloidogyne incognita*, is present (Garber et al., 1979). Race 3, initially reported on Pima cultivar of cotton in Egypt (Fahmy, 1927), is distributed in Egypt, Sudan, Israel, Uzbekistan, China and the United States (Dishon and Nevo, 1970; Fahmy, 1927; Hillocks, 1992; Kim et al., 2005). Race 4 was first reported in India in 1960 (Armstrong and Armstrong, 1960) and was later detected in California in 2001 (Kim et al., 2005). This race 4 genotype causes economic losses independent of nematode pressure on both Pima and Upland cultivars (Ulloa et al., 2006). Race 5 was described in Sudan in 1966 (Ibrahim, 1966) but later it was found same as the previously described race 3 (Nirenberg et al., 1994). Race 6, which was first reported in Brazil in1978, is genetically similar to races 1 and 2 but is distinct in its pathogenicity on certain cotton cultivars as well as okra (Armstrong and Armstrong, 1978). Races 7 and 8 were initially described in China in 1985 (Chen et al., 1985) but race 7 was later found to be the same as race 4.

In addition, several isolates of FOV, causing *Fusarium* wilt of cotton in Australia, have distinct morphological and genetic differences from FOV races 1, 2, 3, 4, 6 and 8 (Davis et al., 1996; Kim et al., 2005; Wang et al., 2010). Further, four more genotypes, designated as, LA108, LA110, LA112 and LA127/140, from the south-eastern United States are highly virulent on some commercial cotton cultivars (Holmes et al., 2009). These genotypes have unique partial translation elongation factor (EF-1α) sequences (Holmes et al., 2009). LA108 and LA110 are particularly aggressive on upland varieties and the occurrence of these genotypes is not rare (Holmes et al., 2009).

Management/Control measures

Use of host resistance

Host resistance is the most effective approach for managing *Fusarium* wilt of cotton. Although cotton cultivars with resistance to all races and genotypes of FOV do not exist, resistance to specific genotypes of FOV is available. *Fusarium* wilt resistance in Pima lines, is controlled by two dominant genes with additive effects (Hillocks, 1992c; Smith and Dick, 1960), while, resistance in Upland cotton is controlled by one major dominant gene and other minor genes which are quantitatively inherited (Hillocks, 1992; Kappelman, 1971; Smith and Dick, 1960). Greenhouse and field disease assays indicate that Phytogen 800, a Pima cultivar, is the most resistant cultivar to FOV race 4 (Ulloa *et al.*, 2006). Some of the Sea Island (*G. barbadense*) varieties are almost immune to the disease and the more resistant Upland cottons may have been derived from crosses or introgression with *G. barbadense*. The *G. barbadense* variety, Seabrook Sea Island, is a useful source of resistance (Wilhelm, 1981). More emphasis is now placed on selection for resistance to root-knot nematode in material with some resistance to *Fusarium* wilt as the best way to improve the level of resistance to the disease complex (Kappelman, 1975; Hyer *et al.*, 1979). This has resulted in a series of lines derived from Auburn 56 known as Auburn RNR lines (Shepherd, 1982). Commercially successful, wilt-resistant varieties have also been produced In Tanzania, UK77 and UK91 are upland cottons derived from Albar 51 (Hillocks, 1984; Hillocks and Bridge, 1992). Some of the Giza and Barakat varieties in Egypt are wilt resistant (Abdel-Raheem *et al.*, 1974: Plantwise, ND). The Egyptian long-staple varieties Ashmouni and Menoufi have been used in wilt resistance breeding programmes in the former Soviet Union (Wilhelm, 1981). 'Pima' cultivars, derived from *G. barbadense* are grown in parts of the USA, Israel, Peru and elsewhere. Some 'Pima' cultivars are wilt-resistant, such as L-60 DSV-UNP from Peru (Rodriguez-Galvez and Maldonado, 1998).

Cultural Control

Crop rotation with barley and other crops is effective measure to reduce the soil inoculum (Goshaev, 1971). Further, it is suggested to clean the crop residues and apply proper weeding in cotton field. Since, the pathogen remains dormant in the form of chlamydospores and by localized infection of the roots of a number of non-host plants and weeds (Wood and Ebbels, 1972; Smith and Snyder, 1974; Smith and Snyder, 1975). Soil solarization is also effective in decreasing the population of soilborne pathogens and has been shown to decrease the incidence of *Fusarium* wilt in Israel (Katan *et al.*, 1983). Field plots heavily infested with FOV were covered with polyethylene (0.03 mm thick) sheets for 7 weeks at the end of summer (Katan *et al.*, 1983).This is especially true in fields infested with race 4, which can cause devastating yield losses in certain susceptible Acala, non-Acala Upland, and Pima varieties. It is also suggested to use *Fusarium*-free seed produced in disease-free fields at all times. Avoidance of gin trash moving that originated in infested cotton fields to noninfested fields is needed. Washing soil from equipment with pressurized water will help to limit the spread of *Fusarium* and should be considered in sites where race 4 has been confirmed.

Seed treatment: A hot water treatment at 60 °C for 20 min decreased seed infestation from 10.5% to 4.7% without decreasing seed vigor and germination. Alternatively, seed may be submerged in thiophanate (0.35 g a.i./L) at 75 and 80 °C for 20 min for complete elimination FOV from seed. However, in latter case, seed vigor and germination may be negatively affected.

Chemical control

Fusarium wilt may also be controlled by the combination of a new systemic fungicide, viz., Miejuncuzhangji, with carbendazim while apply as a foliar spray (Li and Guo, 1999; Plantwise, ND). Further, control of the root knot nematode is important to manage *Fusarium* wilt caused by most genotypes (races 1, 3, and 8) of the causal fungus. Application of nematicides in fields infested with the nematode is often necessary.

13. *Verticillium* wilt

Occurrence & severity status

Verticillium wilt is a widespread disease that occurs in most cotton-producing areas. The disease was first reported from Arlington (Virginia) in 1914 (Carpenter, 1914, 1918). Subsequently, it was discovered in Mississippi in 1929 (Miles and Persons, 1932). The disease is of major importance in the lower Mississippi Valley and the irrigated areas of the Southwest. Losses in yield from it of 10 to 15 percent over large areas are not uncommon (Presle, 1953). *Verticillium* wilt of cotton has also been reported from northern Alabama. There, it is estimated that the disease reduced yields by an average of 1.5% (Land *et al.*, 2016). The cotton wilt was reported to occur in Greece in 1932 by Sarejanni (1936) immediately following importation of cotton seed from the United States. Occurrence of the wilt was noticed in Brazil in 1933, where appreciable economic loss in the cotton crop was being experienced (Viegas and Krug, 1935). *Verticillium* wilt had become the most important disease of cotton in Arizona (Annonymous, 1949). A severe wilt disease outbreak with over 50 percent infection occurred in the province of Sul do Save (Mozambique) in 1949 (Cabral, 1951). In India the disease is more prevalent in the state Tamil Nadu. In China, the disease mostly occurs in the Jiangsu coastal area (Li *et al.*, 2017), causing severe loss of lint cotton (Chen *et al.*, 1985).

Symptoms

In *Verticillium* infection, defoliation is the main symptom, but sometimes, sectorial chlorosis and necrosis of leaf tissue are the only visible symptoms. Infection at seedling stage causes yellowish cotyledons that quickly dry out. Young plants with three to five true leaves suffer considerable stunting. The leaves appear darker green than those of a normal plant and become somewhat crinkled between the veins. The outstanding symptom is the chlorotic areas on the leaf margins and between the principal veins, which make it mottled look. In older plants, the symptoms usually occur in the lower leaves first. They spread to the middle and upper leaves of the plant later in the season. The chlorotic areas gradually become larger and paler (Presle, 1953). Vascular staining may be present, but stunting may be the only effect of the disease. The amount of stunting apparently depends on the stage of development of the plant when it becomes infected. In general, *Verticillium* wilt occurs before the squaring stage and reaches a peak at the boll-setting stage with yellow mottled or defoliating symptoms (Hu *et al.*, 2015). Severely affected plants shed all the leaves and most of the bolls (Presle, 1953).The wilting symptom may not be seen or may be absent in this disease. Only partial wilting of some shoots or leaves may be visible. However, after several weeks of vegetative growth of the infected plants, an irreversible wilting of the whole plant may occur and that followed by plant death. Microsclerotia are formed in senescing diseased tissues.

Causal organism: *Verticillium dahliae* **Kleb. 1913**

[Mycologisches Centralblatt 3: 66]

Synonyms

= *Verticillium albo-atrum* var. *dahliae* (Kleb.) R. Nelson, (1950)

= *Verticillium albo-atrum* var. *chlamydosporale* Wollenw. 1929

= *Verticillium albo-atrum* var. *medium* Wollenw. 1929

= *Verticillium tracheiphilum* Curzi, (1925)

= *Verticillium ovatum* G.H. Berk. & A.B. Jacks., (1926)

= *Verticillium dahliae* f. *chlamydosporale*, (Wollenw.) J. F. H. Beyma 1940

= *Verticillium dahliae* f. *medium*, (Wollenw.) J. F. H. Beyma 1940

Morphological/ Morphophysiolgical characteristics

Colonies of *V. dahliae* produce moderately fast-growing white mycelium which turns black from the centre when microsclerotia are formed. Aerial mycelium is often limited. Conidiophores are abuntant, more or less erect, hyaline, verticillately branched. Phialides are subtended in whorls, 3-4 phialides arising at each node, variable in size, mainly 16-35 x 1-2.5 µm, sometimes secondarily branches. Conidia are arising singly at the apices of the phialides, ellipsoidal to irregularly sub-cylindrical, hyaline, 0 (-1)-septate, 2.5-8 x 1.5-3 μm. Resting mycelia are dark brown, only formed in association with microsclerotia. Chlamydospores are absent. Microsclerotia arise from a single hypha by repeated budding, dark brown to black, torulose or botryoital, consisting of swollen almost globular cells, very variable in shape, elongate to irregularly spherical, very variable in size, 15-50 (-100) μm diam (Hawksworth and Talboys, 1970).

Cultures grow rapidly on potato-dextrose agar and malt agar at 23°C. The prostrate hyphae are hyaline and first produced. Mycelia are flocculose and white, rather more densely compacted on PDA than MA, hyaline, whitish to cream in reverse after 1 week, later becoming black with the formation of microsclerotia; hyaline sectors arising very frequenty in the generally white colonies. Microsclerotia arise centrally in cultures (Hawksworth and Talboys, 1970). Microsclerotia are abundant producing a colony with a black reverse (Land *et al.*, 2016).

Molecular characteristics

A molecular study was done in order to confirm *V. dahliae* on Cotton in Alabama (Land *et al.*, 2016). In their studies, DNA was isolated using the DNeasy Plant Mini Kit from Qiagen Inc. (Germantown, MD). The primers ITS1 and ITS4 were used to amplify the internal transcribed spacer (ITS) region (500 bp), which was sequenced at the Auburn University Genomics and Sequencing Lab. Sequence results were cross-referenced with NCBI GenBank. GenBank Megablast showed 99% shared identity between the sequences of *V. dahliae* (Accession No. KT318493) -the cotton isolate under study and *V. dahliae* VdLs.17 (Accession No.CP010980.1) (Land *et al.*, 2016).

Predisposing factors

Cool and wet weather favours the disease. Further, the disease is favoured by heavy application of organic matter to the soil (Presle, 1953).

Disease cycle

The disease is monocyclic. The pathogen can exist in the forms of mycelia, chlamydospores or microsclerotia in soil and crop debris and can persist in soil for long period or until a new cycle of infection begins (Chen *et al.*, 1985). Spread of disease may occur by dissemination of debris of earlier infected plants in soil. The pathogen can also be disseminated with cotton seed and as microsclerotia associated with the lint (Hawksworth and Talboys, 1970). Their germination is stimulated by root exudates. On germination, hyphae produced from germinating microsclerotia or conidia infect roots at or just behind the tips and progress into the cortex and towards the developing vascular tissues. Activity of certain tylenchid nematodes may predispose some hosts to infection by

Verticillium. Once the fungus spreads in the xylem vessels by mycelia, it produces conidia which are transported into the shoot with the transpiration stream. Symptoms are developed from the occlusion of vessels and toxin production. Microsclerotia are formed in senescing diseased tissues for repetition of the disease in the next year.

Physiologic Races/ strains

There are two races of *V. dahliae* that infect cotton, which are classified as defoliating or nondefoliating based on symptoms (Hu *et al.*, 2015). Symptoms of the infected cotton plants include defoliation by the former one while, the latter one causes stunting and wilting (Hanson, 2000).

Management/Control measures

Use of host resistance

The resistant varieties may be sown as and where they are suited. The varieties of Egyptian, Pima, sea island, and some South American cotton (*G. barbadense*) have a high degree of resistance or tolerance (Presle, 1953). Use tolerant varieties reduce losses, but they do not prevent inoculum from increasing. These may include most modern Acala and Pima varieties (UC IPM, 2017).

Cultural control

The concentration or density of inoculum in soil is a major factor in choosing management strategies for *Verticillium* wilt. Where the density is low, the disease may be prevented by a regular rotation with non susceptible crops such as corn, wheat or barley or with dicotyledonous crops such as sorghum, safflower or alfalfa (UC IPM, 2017). Further, it is suggested to delay first irrigation if disease pressure is high (more than 10 microsclerotia per gram of soil) and air temperatures are cool (UC IPM, 2017).

Biological Control

It was found that two strains (G-6 and G-4) of *Trichoderma virens* significantly reduced the disease-severity ratings in *V. dahliae*-inoculated plants of two cotton cultivars, Rowden and Deltapine 50. This result indicated that *T. virens* may induce a systemic resistance response in cotton (Hanson, 2000). It was also found that the strain G-4 produced the antibiotic gliovirin and strain G-6 produced the antibiotic gliotoxin (Howell *et al.*, 1993).

Chemical control

Spot application in soil of ceresan at 0.1% (6.5kg/hect) or Benomyl or Benlate at 0.05% and using acid delinted seeds after their treatment with vitavax (0.4%) may be helpful to manage the disease chemically.

B. Pests

1. Cotton Aphid/Melon aphid

Occurrence & severity

Cotton aphid is the most common aphid on cotton. It is now found in tropical and temperate regions throughout the world except extreme northern areas. It is common in North and South America, Central Asia, Africa, Australia, Brazil, East Indies, Mexico and Hawaii and in most of Europe. It is cosmopolitan in habitat. In India, this pest is distributed all over the country. The aphid infestations on cotton commence after the true leaves emerge and their field distribution is often clumped. Typically aphid infestations will develop along areas downwind from bunds and occasionally spotty within fields. High levels of aphid populations develop on late than early-planted cotton. Excessive nitrogenous fertilizer applications stimulate faster development of aphids (Vennila *et al.*, 2007a). In the United States, the aphid is regularly a pest in the southeast and southwest, but is occasionally damaging everywhere (Capinera, 2000). Earlier in the USA, *Aphis gossypii* caused more insect-related damage to cotton than any other pest in 1991. Of 13 million acres harvested, around 10 million acres were classified as infested with aphids, resulting in losses of over 360,000 bales (Head, 1992). Losses in Texas alone were around 333,000 bales, representing a yield loss of approximately 6%. Yield reductions of over 100 pounds of lint per acre are not uncommon (Price *et al.*, 1983). In California, USA, seed yield was reduced by 0.21 lbs seed per aphid-day for cotton planted early in the season (Godfrey and Wood, 1998). In India, yield losses in cotton due to sucking pests were between 20.90 and 26.30% (Kulkarni and Raodeo, 1986). Reductions in cotton yield due to sucking pests by 16.2 to 55.6% have been reported from Russia and Brazil (Moskovetz, 1941; Vendramin and Nakano, 1981). In cotton, in Zambia, *A. gossypii* damage caused up to 80% yield loss, while a survey of that country's farmers ranked it as the most serious cotton pest (Javaid *et al.*, 1987). In China, *A. gossypii* infestations are most serious in the seedling stage, particularly in the northern cotton zone. In a study, it was found that cotton seedlings were more sensitive to infestations, while damage on older plants was lower, in part due to compensatory growth effects when precipitation was sufficient (Zhang *et al.*, 1982). Direct feeding damage by *A. gossypii* on cotton is related to plant growth stage and level of aphid infestation. Aphid populations increase rapidly with favourable climatic conditions and plant nutritional quality. Levels of damage are influenced by the presence of natural enemies and biological control, pesticide efficacy, the presence of pesticide resistance in aphids, and compensatory growth in plants (Zhang *et al.*, 1982; Slosser *et al.*, 1989). The planting date was the most important variable, in a multiple regression analysis, affecting aphid density in dry land cotton in Texas, USA (Slosser *et al.*, 1989). Optimum temperature for population growth is around 20-25°C (Akey and Butler, 1989). Light intensity and day length significantly influence rate of population increase, while heavy rain can directly reduce populations by washing them off leaves (Ebert and Cartwright, 1997). The cotton aphid has a very wide host range. It can seriously affect watermelons, cucumbers, cantaloupes, squash and pumpkin. Other vegetable crops attacked include pepper, eggplant, okra and asparagus. It also affects citrus (Mendoza *et al.*, 2001).

Types of damage

The adults and nymphs of the cotton aphid feed on the underside of leaves or on the growing tips of shoots, sucking juices from the plant. The leaves may become chlorotic, crumpling, downward curling and die prematurely. The great deal of leaf curling and distortion in severe attacks hinders efficient photosynthesis. Honeydew is excreted by the aphid that causes sticky cotton due to deposits of honeydew on open bolls. This also allows sooty moulds to grow, resulting in a decrease in the quantity and quality of the produce. Younger plants suffer more attack than older plants. Aggregating populations are seen at the terminal buds and largest populations are found below leaves of lower third of plants where they are partially protected from sunlight and higher temperature (Vennila *et al.*, 2007a). The aphid transmits *Cotton anthocyanosis virus*, *Cotton curliness virus*, cotton blue disease, *Cotton leaf roll and purple wilt viruses* (Kennedy *et al.*, 1962; Brown 1992).

The initial symptom of *A. gossypii* attack is a yellowing of the leaves. Later, the leaves show downward crumpling. Leaves are shiny with honeydew or darkened by sooty mould growing on the honeydew. As populations continue to rise, aphids move to younger leaves, stems and flowers. Plants are covered with a black sooty mould which grows on the honeydew excreted by the aphid. Activity of ants on the aphid-infested plants is common (Vennila *et al.*, 2007a).

Scientific name & order: *Aphis gossypii* Glover, 1877 (Hemiptera: Aphididae)

Synonyms

= *Aphis bauhiniae* Theobald, 1918

= *Aphis citri* Ashmead, 1909

= *Aphis citrulli* Ashmead, 1882

= *Aphis cucumeris* Forbes, 1883

= *Aphis cucurbiti* Buckton, 1879

= *Aphis minuta* Wilson, 1911

= *Aphis monardae* Oestlund, 1887

= *Cerosypha gossypii* Glover, 1877

Biology

Egg: The egg is oval, yellow when first laid but soon turn glossy black.

Nymph: Nymph is small, yellow to green to nearly black and remained on the ventral surface of the leaf and on the terminal shoot. It often has a dark head, thorax and wing pads and the distal portion of the abdomen is usually dark green. The body appears dull because it is dusted with wax secretions (UC IPM). The nymph is mostly wingless (Vennila *et al.*, 2007a). The nymphal period lasts for about seven to nine days.

Adult: Adult is yellowish brown to black, 1.25 mm long with a pair of black cornicles on the 6th abdominal segment and yellowish green abdominal tip. Both apterate (0.9-1.8 mm) and alate (1.1-1.8 mm) occur together (Vennila *et al.*, 2007a). The apterate has an ovoid body in varying shades of green (or dark purplish green). The legs are yellow, as are the antennae which are three quarters of the length of the body. The apices of the femora, tibia and tarsi are black. The cylindrical black siphunculi are wide at the base and one fifth of the body length. The alate has a fusiform body. Its head and thorax are black, the abdomen yellowish-green with black lateral spots and the antennae are longer than those of the apterous female. The life span of a parthenogenic female is about twenty days.

Life cycle

Sexual reproduction of the cotton aphid is not important. Females continue to produce offspring without mating so long as the weather is favourable for feeding and growth. The aphid lives in colonies and the alate as well as apterous females multiply parthenogenetically and viviparously. Female gives birth to 8-22 nymphs per day. Nymphal period lasts for 7-9 days and the adults live for 12-20 days. In all, the cotton aphid has 12-14 generations per year (Vennila *et al.*, 2007a).

Molecular studies

Aphids are infected with a wide variety of endosymbionts which provides nutrients not obtained in sufficient quantities from plant phloem (Douglas, 1998; Oliver *et al.*, 2010). The study on the bacterial diversity of the cotton aphid *A. gossypii* by targeting the V4 region of the 16S rDNA using the Illumina MiSeq platform revealed that bacterial communities of *A. gossypii* are generally dominated by the primary symbiont *Buchnera*, together with the facultative symbionts *Arsenophonus* and *Hamiltonella* (Zhao *et al.*, 2016).

Management /Control measures

Host-Plant Resistance

Cultivar selection appears to influence aphid population growth. Pima cultivars appear to be more susceptible to aphid infestations and associated damage. Within the Acala cotton cultivars, hairy-leaf varieties, which comprise the majority of the market, are more susceptible to aphids than are smooth-leaf varieties (UC IPM). This indicates that glabrate cotton supported fewer aphids than more pubescent cotton (Dunnam and Clark, 1938; Weathersbee *et al.*, 1994). Resistance has also been documented in *Gossypium hirsutum* and *G. arboreum* (Chakravarthy and Sidhu, 1986; Reed *et al.*, 1999).

Cultural Control

Early plantation and limited use of water and fertilizers, especially nitrogenous fertilizers, are helpful to avoid aphid attacks. Since, it is established that higher cotton aphid numbers consistently develop on late-planted cotton (late April to early May) when compared to early-planted cotton (early April). Additionally, aphids prefer cotton plants that are well watered and highly fertilized (UC IPM). Cotton plants fertilized with high nitrogen are better hosts for *A. gossypii*, and this results in greater damage (Cisneros and Godfrey, 2001; Nevo and Moshe, 2001).

Biological Control

Natural enemies: During the pre-squaring period of the crop, natural control of aphids is generally strong. The parasitic wasp *Lysiphlebus testaceipes* and a group of aphid predators the lady beetles, *Hippodamia convergens* and *Coccinella novemnotata franciscana*, and the predatory larvae of syrphid flies are important natural enemies.(UC IPM). In Egypt, it was found that the release of *Chrysoperla carnea* in field at a ratio of 1:5 (predator: aphid) eliminated the aphid in 12 days. Further, a single release of *Coccinella undecimpunctata* at a ratio of 1:50 was able to get 99.7% of aphid control (Zaki *et al.*, 1999).

A novel approach to pest management has been done in China, where cotton plants were physically wounded, and some wounds were infected with a bacterium (*Pseudomonas gladioli* D-2251). Wounded plants had fewer aphids. Aphid infestation frequency on infected plants was further reduced if the bacterium was present (Li *et al.*, 1998).

Chemical control

Seed treatment with the insecticides of neonicotinoid group such as imidacloprid and thiamethoxam is effective to control aphids during early growth of crop. The seed treatment is suggested to do with imidacloprid 60 FS @ 10 ml/kg seed or with thiamethoxam 70 WS @ 5 g/kg seed.

Stem smearing: It is suggested to smear about 1 inch area of shoot tips with the solution of Imidacloprid 17.8 SL diluting twenty times, i.e. @1ml insecticide/ 20 ml water. In this method only 30-40 ml insecticide is required per acre and is very effective to reduce the aphid attacks.

2. Bollworms

i) American cotton bollworm (*Helicoverpa zea*)

Occurrence & severity

American Cotton bollworm, also called as 'Cotton bollworm' or 'Corn earworm', is a problem in growing cotton. The cotton bollworm is found in temperate and tropical regions of North America, with the exception of northern Canada and Alaska as it cannot overwinter in these areas (Mau and Kessing, 1992; Capinera, 2015b). It is a serious pest in the southern cotton growing areas in USA (Michaud, 2013). The cotton bollworm, *Helicoverpa zea*, is found in the eastern United States. At there, also, it also does not overwinter successfully (Capinera, 2015b). They live in Kansas, Ohio, Virginia, and southern New Jersey, but survival rate is mainly affected by the severity of the winter. American Cotton bollworm moths regularly migrate from southern regions to northern regions depending on winter conditions (Capinera, 2015b). They are also found in Hawaii, the Caribbean islands, and most of South America, including Peru, Argentina, and Brazil (Blanchard, 1942; Charles *et al.*, 1993). This has also been reported from China in 2002 (Lu and Liang, 2002). The American Cotton bollworm, which is more popularly known as 'Corn earworm', is the second-most important economic pest species in North America, next to the codling moth (Hardwick, 1965b). The estimated annual cost of the damage is more than US$100 million (Capinera,2001). The moth's high fecundity, ability to lay between 500 and 3,000 eggs, polyphagous larval feeding habits, high mobility during migration, and a facultative pupal diapause have led to the success of this pest (Fitt, 1989: Capinera, 2001).

Types of damage

The cotton bollworm larvae feed on leaf tissue and small squares, then move down the plant and damage the larger squares and bolls (Michaud, 2013).

Scientific name & order: *Helicoverpa zea* (Boddie, 1850)
[Lepidoptera: Noctuidae]

Synonyms

= *Heliothis zea* Boddie, 1850

= *Heliothis umbrosus* Grote, 1862

= *Heliothis ochracea* Cockerell, 1889

= *Helicoverpa stombleri* (Okumura & Bauer, 1969)

= *Heliothis stombleri* Okumura & Bauer, 1969

Biology

Egg: Egg is sub spherical, radially ribbed, size 0.50 -0.52 mm high and 0.55-0.59 mm in diameter, white to pale green to yellowish and grey before hatching, stuck singly to the plant substrate. Egg develops a dark red or brown ring around the top after 24 hours and completely darkens before hatching. The egg hatches in two or three days during warm weather or at 20-30°C (Neunzig, 1964; Michaud, 2013). The egg of this species is distinguishable from looper egg by deeper ridges and a more hemispherical shape (Flint, 1985).

Larva: Following hatching, larva feeds on the reproductive structure of the plant. Young larva is difficult to find until it is about three to four days old. At this stage, it is about ¼ inch (6.35 mm) long and brownish coloured with some scattered hairs. Larva develops through four to six instars (Mau and Kessing, 1992). The full-grown larva is about 1½ inches (about 40 mm) long with a light-coloured head capsule. However, it usually has orange head, black thorax plate, and a body colour that is primarily black (Zimmerman, 1958). Its body can also be brown, pink, green, and yellow with many thorny microspines (Zimmerman, 1958). The predominant body colour may range from pink or green to various shades of tan or dark brown. A series of dark stripes run lengthwise on the body (Michaud, 2013). Larva has 5 pairs of prolegs. Mature larva migrates to the soil, where it pupates.

Pupa: Pupa is light to dark brown in colour depending on maturity. It measures 5.5 mm wide and 17 to 22 mm long (Zimmerman, 1958; Hill, 1983a). Larva pupates 5 to 10 cm below the soil surface (Zimmerman, 1958). Duration of pupal period is 12 to 16 days.

Adult: The adult is a medium-sized, cream (Michaud, 2013) or brown coloured, 20-25 mm long and stout-bodied moth. It has forewings that are yellowish brown in colour and has a dark spot located in the centre of its body (Kogan *et al.*, 1978). The moth has a wingspan ranging from 32 to 45mm (Kogan *et al.*,

1978). Adult moth is nocturnal and hides in vegetation during the day (Kogan *et al.*, 1978). The adult moth collects nectar or other plant exudates from a large number of plants and lives for 12 to 16 days. Female can lay up to 2,500 eggs in its lifetime (Hill, 1983a).

Life cycle

The life cycle can be completed in 28-30 days at 25°C and in the tropics there may be up to 10-11 generations per year. All stages of the insect are to be found throughout the year if food is available, but development may be slowed or stopped by either drought or cold. In the northern USA there are only two generations per year and in Canada only one generation. In Hawaii, the entire life cycle occurs in 55-70 days with several generations.

Management/Control measures

Quarantine measure

Helicoverpa zea has recently been added to the EPPO A1 list of quarantine pests, and is also considered as a quarantine pest by APPPC. The addition to the EPPO list harmonizes it with EU Directive Annex I/A1 (EPPO/CABI, 1996).

Use of host resistance

In cotton, gossypol glands on the calyx crowns of flower buds confer considerable resistance to *H. zea* (Calhoun *et al.*, 1997). The use of Bt cotton is increasing. The need for this technology is questionable given the generally low levels of bollworm infestation. Bollworms must ingest the Bt gene (Bollgard™) before they will die and it may take up to five days from time of ingestion to death. However, there has been some recent evidence of resistance to the Bollgard™ gene evolving in *H. zea* populations (Michaud, 2013).

Cultural Control

Various cultural practices can be used to kill the different instars, including deep ploughing, manipulation of sowing dates, mechanical destruction, and trap crops are also used to kill different instars (Hardwick, 1965b).

Biological Control

The most frequently tried method of achieving biological control is by releasing artificially reared parasites or predators, especially, *Trichogramma* spp. However, releases in cotton have not been consistently effective against heliothine populations. Alternatively, *Microplitis croceipes* may be more effective because it is less affected by organophosphate pesticides and synthetic pyrethroids.

The use of entomophagous pathogens such as *Bacillus thuringiensis* and *Heliothis* NPV is common biocontrol agents to manage the pest (Lambert *et al.*, 1996: Capinera, 2015b). In addition, *B. thuringiensis* toxin has also been used commercially. It is reported that, a commercial formulation of the nuclear polyhedrosis virus *Baculovirus heliothis* gave control that was equal to chemical methods. However, the cost of virus applications was higher than chemical control methods (Martinez and Swezey, 1988).

Chemical Control

Bollworm management is based on scouting for eggs or small larvae. Chemical treatment is recommended when 10 eggs or five small worms per 100 plants are present during early bloom in late July and early August (Michaud, 2013). Chemical control of the larvae has been the most widely used and generally successful method of pest destruction on most crops, but it is not easy because of larvae feeding within plant structures. COPR (1983) includes a list of 29 insecticides effective for the control of the pest. The chemicals that recommended for control include sulprofos, profenofos, thiodicarb, chlorpyrifos, acephate, amitraz and pyrethroids. Several *Bacillus thuringiensis* sprays are also recommended for its control (Anon., 1997).

ii) Old World bollworm (*Helicoverpa armigera*)

Occurrence & severity

Old World bollworm (*Helicoverpa armigera*) is also known as 'Cotton bollworm', 'African cotton bollworm', 'Scarce bordered straw worm', 'Corn earworm', 'Tobacco budworm', 'Tomato grub', 'Tomato worm' and 'Gram pod borer'. It has been reported as an economically important pest or a key pest in Africa, Asia, Europe and the former USSR and Oceania (Oerke *et al.*, 1994). Previously, *H. armigera* was reported as partly responsible for a major portion of cotton crop losses (Ridgway *et al.*, 1984). However, later it was found to cause major damage in cotton yield all over the worlds. In Africa, *H. armigera* can reduce yields substantially. In the years between 1978 and 1983 at Cote d'Ivoire the cotton crop losses in the south of the country were primarily due to *H. armigera* and those were about 60% (Moyal, 1988). In Zimbabwe, potential crop losses due to *H. armigera* were 1175 kg/ha (Gledhill, 1976). In Tanzania, the economic loss of cotton was estimated over $US 20 million (Reed and Pawar, 1982). In India, *H. armigera* is the predominant bollworm on cotton, causing 14-56% damage (Kaushik *et al.* 1969, Manjunath *et al.* 1989, Jayaraj, 1990). At Andhra Pradesh in India, the problems in controlling *H. armigera* were first encountered in 1987. More than 30 insecticide treatments were applied, yet the average yield was decreased from 436 kg/ha in 1986/87 to 186 kg/ha in 1987/88. That reduction was about 61% (Armes *et al.*, 1992). In Thailand, *H. armigera* is the principal cotton pest since the mid-1960s. Losses due to *H. armigera* were at least 31% in 1975-79 (Mabbett *et al.*, 1980). In China, losses due to *H. armigera* larvae increased with plant age. Crop losses were substantial regardless of soil fertility (Sheng, 1988). The pest (*H. armigera*) has been reported on over 180 cultivated hosts and wild species in at least 45 plant families (Venette *et al.*, 2003). The larvae feed mainly on the flowers and fruit of high value crops, and thus high economic damage can be caused at low population densities (Cameron, 1989; CABI, 2007). The most important crop hosts of which *H. armigera* is a major pest are cotton, pigeonpea, chickpea, tomato, sorghum and cowpea. Other hosts include groundnut, okra, peas, field beans (*Lablab* spp.), soyabeans, lucerne, *Phaseolus* spp., other Leguminosae, tobacco, potatoes, maize, flax, a number of fruits (Prunus, Citrus), forest trees and a range of vegetable crops. A wide range of wild plant species support larval development. The important wild plant species in India include *Acanthospermum* spp., *Datura* spp., *Gomphrena celosioides* and, in Africa, *Amaranthus* spp., *Cleome* sp. and *Acalypha* sp. (Majunath *et al.*, 1989; Matthews, 1991).

Types of damage

The insect causes bore holes at the base of flower buds, the latter being hollowed out. Bracteoles are spread out and curled downwards. Leaves and shoots may also be consumed by larvae. Larger larvae bore into maturing green bolls, as such the young bolls fall after larval damage.

Scientific name & order: *Helicoverpa armigera* Hubner, 1808 [Lepidoptera: Noctuidae]

Synonyms

= *Chloridea armigera* Hubner

= *Chloridea obsoleta* Duncan & Westwood, 1841

= *Helicoverpa commoni* Hardwick, 1965

= *Heliothis armigera* Hubner, 1805

= *Heliothis rama* Bhattacherjee & Gupta, 1972

= *Noctua armigera* Hubner, 1805

Biology

Egg: Egg is yellowish-white when first laid, later changing to dark brown just before hatching, gum drop-shaped, and 0.4 to 0.6 mm in diameter. The apical area surrounding the micropyle is smooth, the rest of the surface sculptured in the form of approximately 24 longitudinal ribs, alternate ones being slightly shorter, with numerous finer transverse ridges between them. Eggs are laid on plants which are flowering, or are about to produce flowers (Dominguez Garcia-Tejero, 1957; Hardwick, 1965b; Cayrol, 1972; Delatte, 1973; King, 1994; Bhatt and Patel, 2001; CABI, 2007).

Larva: Larval colour darkens with successive moults for the six instars typically observed for *H. armigera*. Coloration can vary considerably due to diet content. Coloration ranges from bluish green to brownish red (Fowler and Lakin, 2001). The first and second larval instars are generally translucent and yellowish to reddish-brown in colour, without prominent markings. Head, prothoracic shield, supra-anal shield and prothoracic legs are very dark-brown to black, as are also the spiracles and tuberculate bases to the setae. Prolegs are present on the third to sixth, and tenth, abdominal segments. The larvae have a spotted appearance due to sclerotized setae, tubercle bases, and spiracles (King, 1994; Bhatt and Patel, 2001). A characteristic pattern develops in subsequent instars. The full grown larvae are brownish or pale green with brown lateral stripes and a distinct dorsal stripe. Larvae are long and ventrally flattened but convex dorsally. The head is brown and mottled. The prothoracic and supra-anal plates and legs are pale-brown, only claws and spiracles remaining black. The skin surface consists of close-set, minute tubercles. Crochets on the prolegs are arranged in an arc. The final body segment is elongated. Larval size in the final instar ranges from (3.0-) 3.5 to 4.0 (-4.2) cm in length (Dominguez Garcia-Tejero, 1957; Hardwick, 1965b; Cayrol, 1972; Delatte, 1973; King, 1994).

Pupa: Pupa is mahogany-brown or dark tan to brown, 14 to18 (-22) mm long, and 4.5 to 6.5 mm wide. Body is rounded both anteriorly and posteriorly, with two tapering parallel spines at posterior tip. Pupae typically are found in soil (Dominguez Garcia-Tejero, 1957; Hardwick, 1965b; Cayrol, 1972; Delatte, 1973; King, 1994; Sullivan and Molet, 2007).

Adult: Adult is a stout-bodied moth of typical noctuid appearance, with 3.5-4 cm wing span; broad across the thorax and then tapering, 14-18 (-19) mm long; colour variable, but male usually greenish-grey, yellowish-brown, light yellow or light brown and female orange-brown. Forewings have a line of seven to eight blackish spots on the margin, a broad, irregular, transverse brown band and a black or dark brown kidney-shaped marking near the centre. Hind wings are creamy white or pale-straw colour with a dark brown or dark gray band on outer margin; they have yellowish margins and strongly marked veins and a dark, comma-shaped marking in the middle. Antennae are covered with fine hairs (Dominguez Garcia-Tejero, 1957; Hardwick, 1965b; Cayrol, 1972; Delatte, 1973; King, 1994; Brambila, 2009).

Life cycle

The bollworm (*H. armigera*) overwinters in the soil as pupa. Moth emerges in May to June depending on latitude and lays eggs singly on a variety of host plants on or near floral structures. Plants in flower are preferred more than those that are not in flower (Firempong and Zalucki, 1990b). In certain cases, the moth also lays eggs on leaf surfaces. Female moth tends to choose pubescent (hairy) surface for oviposition rather than smooth leaf surface (King, 1994). Tall plants are also attracted more for oviposition than shorter plants (Firempong and Zalucki, 1990b). The number of larval instars varies from five to seven, with six being most common (Hardwick, 1965b). Larva drops off the host plant and pupates in the soil, then emerges as adult to start the next generation. The bollworm exhibits overlapping generations, it is difficult to determine the number of completed generations. Typically two to five generations are achieved in subtropical and temperate regions and up to 11 generations may occur in tropical areas (Tripathi and Singh, 1991; King, 1994; Fowler and Lakin, 2001). Temperature and availability of suitable host plants are the most important factors influencing the number of generations (King, 1994). In South Africa, the oviposition period is 10-23 days, with an average of 730 eggs per female. Eggs hatch in 3 days at 22.5°C, and in 9 days at 17.0°C. The larval period lasts 18 days at 22.5°C and 51 days at 17.5°C, development thresholds being 14 and 36°C (EPPO/CABI, ND; Pearson, 1958; King, 1994). In South and South-East Asia, development times are generally similar to those in South Africa (Jayaraj, 1982; Tripathi and Singh, 1991; King, 1994). The moths, under adverse conditions, can migrate long distances (King, 1994; Zhou *et al.* 2000; Casimero *et al.*, 2001; Shimizu and Fujisaki, 2002; CABI, 2007). Adults can disperse distances of 10 km during "non-migratory flights" and hundreds of kilometres (up to 250 km) when making "migratory flights," which occur when host quality or availability declines (Saito, 1999; Zhou *et al.*, 2000; Casimero *et al.*, 2001; Fowler and Lakin, 2001).

Management/Control measures

Quarantine measure

The pest (*H. armigera*) is listed as an A2 quarantine pest by EPPO (OEPP/EPPO, 1981). The pest is a serious outdoor pest in Mediterranean countries and there is ample probability to reach the limits of its natural distribution in the EPPO region. Quarantine status arises from the risk of introduction into glasshouse crops in northern Europe. EPPO recommends (OEPP/EPPO, 1990b) that imported propagation material should be derived from an area where *H. armigera* does not occur or from a place of production where *H. armigera* has not been detected during the previous 3 months.

Use of host resistance

The planting of varieties that are resistant or tolerant to *H. armigera* has received major attention, particularly for cotton. The varieties with glabrous leaves are tolerant to this pest. Since, fewer eggs are laid by this insect on plants having the glabrous leaf character in cotton. In recent years, genetic engineering techniques have enabled genes carrying the toxic element of *Bacillus thuringiensis* to be introduced into cotton. A novel tool is now available to control *H. armigera* in the form of cotton genetically engineered to express an insecticidal protein, Cry1Ac, derived from the bacterium *B. thuringiensis* (Berliner) (Bt). These transgenic varieties (Bt cotton) provide effective control of *H. armigera* and other bollworms such as *Earias vittella* (F.) and pink bollworm, *Pectinophora gossypiella* (Saunders). Bt cotton varieties have now been registered for commercial use in the United States, Australia, Mexico, Colombia, Argentina, China, India, and South Africa (Ravi *et al.*, 2005).

Cultural Control and Sanitary Methods

Cultural manipulations of the crop or cropping system and land management have been tried as tactics to manage *H. armigera* populations. Amongst them, the ploughing cotton stubble to reduce overwintering populations of pyrethroid-resistant *H. armigera* (Fitt and Forrester, 1987), and post-harvest cultivation to destroy pupae of bollworms have some beneficial effects in reducing the pest attack. It is also suggested to keep sufficient spacing between the plants so that the target larvae would be more accessible to insecticides or microbial formulations applied by conventional means.

Biological Control

Use of local biological control agents to suppress *H. armigera* populations to below an economic threshold without the use of insecticides would be a major advantage, both in ecological and economic terms, particularly if this was sustainable. To this end, substantial efforts have been made either to introduce exotic natural enemies or to augment existing populations of parasitoids and predators to achieve satisfactory levels of control. Because of the need to produce very large numbers of parasitoids or predators simultaneously and economically, emphasis has been placed on *Trichogramma* spp. which are most amenable to mass rearing. Although these and a number of other parasitic species has been field evaluated against *H. armigera*, results have not so far been encouraging, especially in agrosystems where insecticide applications against *H. armigera* or other pests are consistently necessary (Waterhouse and Norris, 1987).

Considerable efforts have been made to develop the most promising agents, *B. thuringiensis* and *Helicoverpa armigera nuclear polyhedrosis virus* (HaNPV) into commercially viable products. Both these agents, particularly HaNPV, have some impact on *H. armigera* populations, although seldom reaching the epizootic proportions necessary to achieve effective control. Field tests with artificially produced Bt and HaNPV have so far had only limited success, mainly because of rapid degradation by UV light, insufficient titres ingested by larvae, and lack of virulence. However work is continuing to overcome these constraints stimulated by increasing resistance to insecticides and awareness of the environmental threats they pose (King and Jackson, 1989).

Use of Pheromone traps

Pheromone traps using (Z)-11-hexadecenal and (Z)-9-hexadecenal in a 97:3 ratio have been used to monitor populations of *H. armigera* (Pawar *et al.*, 1988; Loganathan and Uthamasamy, 1998; Loganathan *et al.*, 1999; Visalakshmi *et al.*, 2000; Zhou *et al.*, 2000). The pheromone at the dose of 1.0 mg/septum attracts the most males (Loganathan and Uthamasamy, 1998). Rubber septa impregnated with this sex pheromone component (1 mg/septum) are effective in capturing males for 11 days in the laboratory (Loganathan *et al.*, 1999). Captures of *H. armigera* in the field are lower with 15-day-old lures than with fresh lures and replacing of lures should be done every 13 days (Loganathan *et al.*, 1999). Traps are suggested to place 1.5 to 1.8 m (5 to 6 ft) above the ground (Kant *et al.*, 1999; Zhou *et al.*, 2000; Aheer *et al.*, 2009a).

Chemical Control

Fifty-four percent of the total insecticides used on all crops in India are used on cotton, and most of these are directed against *H. armigera* (Mohan and Manjunath 2002). As a consequence, this pest has evolved resistance to many insecticides in India (Armes *et al.* 1996, Kranthi 1997). The considerable selection pressure which *H. armigera* has experienced, particularly to the synthetic pyrethroids which were used predominantly in the early 1980s, has resulted in the development of resistance to the major classes of insecticides in many of the areas where these have been used. Now, all rely on a strict temporal restriction in the use of pyrethroids and their alternation with other insecticide groups to minimize selection for resistance. In an investigation, the insectcides, like, indoxacarb (at 0.0075%), spinosad (0.009%), profenophos + cypermethrin (0.044%) and endosulfan (0.07%), are found as effective against the pest (Babariya *et al.*, 2010). In another investigation, Tracer (spinosad) 240SC @ 60 ml /acre has been found as the most effective in restricting the pest infestation followed by Steward (indoxacarb) 150SC @ 150 ml/ acre, and Lanante (methomyl) 40SP @ 300 g/ acre (Ahmed *et al.*, 2004).

3. Dusky cotton bug

Occurrence & severity

Dusky Cotton Bug is also called as 'Cotton seed bug'. It is a widespread species, occurring in the Middle East, Africa, Asia, Europe, Central and South America. Its occurrences are more common in Southern Europe (Bosnia and Herzegovina, Bulgaria, Croatia, France, Greece, Italy, Portugal, Spain, former Yugoslavia and Albania), Afrotropical realm, Neotropical ecozone and Oriental ecozone. Earlier, the dusky cotton bug was used to consider as a minor pest of cotton, but its importance has grown in the last few years, as populations are becoming larger. It has now attained the status of a major pest of cotton crops that affects lint as well as the seed quality of cotton in Pakistan (Abbas *et al.*, 2015). This cotton seed bug is found heavily infesting feral cotton in Matthew town, Great Inagua, Bahamas. It is considered as a major economic threat to cotton in the United States (Smith and Brambila, 2008). The cotton seed bug is also a serious pest of cotton in Egypt, and in Southeast Asia and Africa. Weight loss can reach 15%, seed germination may be severely reduced, and the oil quality of the seed can be affected. It has been documented as an important pest of cotton in the Mediterranean region and in coastal Africa (Adu-Mensah and Kumar, 1977). Sometimes this pest attack leads to 6.8, 32 and 6% reductions in cotton yield, seed weight and oil content, respectively (Sewify and Semeada, 1993). A number of generations throughout the year have been observed. The bug also lives on other plants, such as, *Abutilon, Abelmoschus* (okra), *Cola, Eriodendron, Gossypium, Malva, Sphaeralcea, Hibiscus, Pavonia, Sida, Dombeya, Sterculia* and *Triumfetta* (Samy, 1969; Dimetry,1971; Hussain, 2012). However, of these, *Gossypium* and *Abelmoschus* (okra) appear to be the most preferred hosts (Kirkpatrick, 1923; Adu-Mensah and Kumar, 1977; Hussain, 2012).

Types of damage

Nymphs and adults of the cotton seed bug suck the sap gregariously from immature seeds (Vennila *et al.*, 2007c). They feed mainly on the seeds of cotton and causes multiple injuries to cotton seed that includes reduction in seed weight used for oil extraction and viability of seeds (Peacock, 1913), injuries to the embryo radical and cotyledons (Kirkpatrick, 1923; Pearson, 1958) reducing the value of seed cotton and staining the lint when insect bodies are crushed in the ginning process (Henry, 1983). The staining of the lint is main damage, since it decreases the market value of the cotton (Hussain, 2012). It also causes a continuous decrease in cotton-seed weight, germination and oil quality (Henry, 1983). After feeding, greasy spots appear on the bolls that become disfigured following the injection of toxic saliva (Schaefer and Panizzi, 2000). It also feeds on leaves, and young stem and petiole tissues to obtain additional moisture

(Holtz, 2006). The bugs suck the sap in groups from the immature seed which consequently do not ripen and remain light weight. The bugs also cause nuisance to workers during picking (Vennila *et al.*, 2007c; Hussain, 2012).

Scientific name & order: *Oxycarenus hyalinipennis (Costa 1843) [Hemiptera: Lygaeidae]*
Synonyms:

= *Aphanus hyalinipennis* Costa

= *Aphanus tardus var. hyalipennis* Costa 1847

= *Cymus cincticornis* Walker 1870

= *Oxycarenus cruralis* Stal 1856

= *Oxycarenus leucopterus* Fieber 1852

[**N.B.** Another species, *Oxycarenus laetus* Kirby (Hemiptera: Lygaeidae), is also found as a serious threat to cotton due to early cultivation of Bt. cotton in Pakistan (Khan *et al*, 2014). This is a common pest of some economic crops such as cotton and okra.]

Biology

Egg: The eggs, laid on the lint of half opened bolls of cotton, are white and cigar shaped. They are laid in clusters of (2-) 3-18 (Hussain, 2012). A female of this bug lays about twenty eggs.

Nymph: Nymph is round when young (Hussain, 2012). Nymph is dusky, greyish brown bug, with pointed head, hyaline wings and pink to red abdomen (Smith and Brambila, 2008). There are 5 instars.

Adult: Adults are about 3.8 mm in males, of 4.3 mm in females (USDA), tapered anteriorly, rounded and truncate posteriorly. Males and females have similar coloration, but males are slightly smaller than females (Slater 1972). Body of these bugs is black with hemelytra (forewings) hyaline, hind wings whitish or, translucent white wings (Slater 1972) or, dirty white semi-transparent wings (Hussain, 2012). Thorax and head are black with minute silvery dots, with brownish-black antennae. The second antennal segment usually is partially pale yellow. Pronotum is blackish-brown. Corium is usually yellowish-whitish and hyaline. Femora are black, while tibiae are brown with a yellow-white band (USDA).

Life cycle: Each female lays about 20 eggs, singly or in small groups on open or damaged cotton bolls and the emerging nymphs, which tend to aggregate, feed on the seeds and mate thereon. The pest may also feed on leaves and young stems, and move between host plants. This species goes through five nymphal stages. A generation lasts about twenty days. At the end of the breeding season, the cotton seed bug seeks a hibernaculum. Overwintering adults are not completely inactive, but do not feed or mate until Malvaceae seeds are available again. Seeds of various host plants are available at different periods throughout the year, maintaining the cotton seed bug populations (Sweet, 2000). It completes 3-4 (in some cases 6-7) generations per year. The last generation hibernates on the branches or leaves of grass and weeds.

Molecular characters: In an investigation (Khan *et al.*, 2017) described the genetic barcode of dusky bug of cotton from Pakistan. The insect (cotton dusky bug) samples were collected from cotton fields in Faisalabad. COI gene was amplified from genomic DNA of bug and cloned into pTZ57R/T vector (Fermentas). The clone was sent to Macrogen (South Korea) for Sanger sequencing. The phylogenetic analysis and pairwise multiple sequence alignment showed that the insect samples of cotton dusky bug grouped with two species of Oxycarenus genus and highest sequence identity was 91.1% with *Oxycarenus hylinipennis* (Khan *et al.*, 2017).

Management/Control measures

Cultural control

It is suggested to inspect and remove the wild plants that grow around crops. The easiest control is to remove alternate host plants like okra etc. before and near the main crop

Use of light trap: UV-light traps are needed to place for monitoring in areas where preferred host plants, like cotton or okra, are grown.

Biological control

The entomopathogenic fungi, like, *Beauveria bassiana* and *Metarhizium anisopliae*, have shown some promise. In Africa astigmatic mites were found on the bugs, which became sluggish and soon died; spiders were also reported to attack this pest.

Spraying of plant extracts: The neem extract can control the dusky cotton bug (Nurulain *et al.* 1989) better than malathion after 72 hrs. Neem caused various developmental deformities and mortality.

Chemical control: As the pest usually infests the boll or the leaf litter beneath the plant, effective pesticide applications are problematic. During outbreaks, control may be obtained with chemicals, like some organophosphates that have contact and systemic properties, applied early in the morning while the insects are less active. Organophosphates have proved as most effective and gave 79.3% pest mortality followed by Pyrethroids (53.6%) and Neonicotinoids (53.4%) (Shah *et al.*, 2016). Spraying of chlorpyriphos @ 1000ml per 100 litres of water is also sudggested to manage the pest (Hussain, 2012).

It is to be noted that the dusky cotton bug, *O. hyalinipennis* has developed resistance to a number of conventional (bifenthrin, deltamethrin, lambda-cyhalothrin, profenofos, triazophos) and novel chemical (emamectin benzoate, spinosad, chlorfenapyr, imidacloprid, and nitenpyram) insecticides. Regular assessment of resistance to insecticides and integrated management plans like judicious use of insecticides and rotation of insecticides along with different modes of action are required to delay resistance development in *O. hyalinipennis* (Ullah *et al.*, 2016).

4. Weevils

i) Cotton grey weevil (*Myllocerus undecimpustulatus*)

Occurrence & severity

Cotton grey weevil (*Myllocerus undecimpustulatus*), also called as Ash weevil or Sri Lanka weevil or yellow-headed ravenous weevil, is a short snouted weevil (Marshall, 1916). This weevil pest is active from April to November and passes winter in the adult stage, hidden in debris (Atwal, 1976). The grey weevil is polyphagus and a minor pest of more than twenty economic valued crops including cotton, vegetables, palms, and ornamentals in India and Pakistan (Pruthi and Batra, 1960; O'Brien et al. 2006; Josephrajkumar et al., 2011). This weevil is native to southern India, Sri Lanka and Pakistan. The species has been also reported in Florida (Thomas, 2000). It is particularly abundant in northern Miami-Dade and southern Broward Counties. In Florida, the subspecies feeds on above 103 plant species including native ornamentals and fruit crops, tropical fruit trees, and palms (Thomas 2005).

Types of damage

Larvae feed on the roots of plants; however, the exact level of damage they cause is unknown. When adult weevils feed on leaves, they feed inward from the leaf margins (or edges), causing the typical leaf notching. Damage can be ranged from notching on the leaf margins in an irregular pattern to much more extensive feeding along the leaf veins. The adults prefer new plant growth. Intense feeding by numerous weevils may cause plant decline or stunting. Young seedlings may not survive when a large amount of feeding damage occurred (UF/IFAS, ND1).

Scientific name & order: *Myllocerus undecimpustulatus* Faust, 1891
(Coleoptera: Curculionidae)

Synonyms

= *Myllocerus fausti* Formánek, 1927

= *Myllocerus maculosus* Desbrochers des Loges, 1899

= *Myllocerus marmoratus* Faust, 1897

= *Myllocerus pistor* Faust, 1897

= *Myllocerus undatus* Marshall (?)

[**N.B.** *Myllocerus undecimpustulatus undatus* Marshall, the Sri Lankan weevil, is a plant pest with a wide range of hosts. This weevil spread from Sri Lanka to India and then Pakistan where many subspecies of *Myllocerus undecimpustulatus* Faust are considered pests of more than 20 crops. *Myllocerus undecimpustulatus*, a species native to southern India, and then again as *Myllocerus undatus* Marshall native to Sri Lanka, finally as *Myllocerus undecimpustulatus undatus* Marshall to show its status as a subspecies]

Biology

Egg: The female lays on an average 360 eggs over a period of 24 days. The eggs hatch in 3-5 days (Atwal, 1976). The eggs are laid directly on organic material at the soil surface or on soil rich in organic matter (George *et al.*, 2015). Eggs are less than 0.5 mm, ovoid and usually laid in clusters of 3-5. The eggs are white or cream-coloured at first, then gradually turn brown when they are close to hatching (UF/IFAS, ND1).

Grub: Grub (larva) is small and apodous. Its size ranges from 1.09 ± 0.05 mm as first instar grub to 4.0 ± 0.05 mm as fourth instar grub, beige-white with a reddish brown head. It burrows into the soil where it feeds on plant rootlets (Butani, 1979). The grub completes its development in one to two months (Atwal, 1976; UF/IFAS, ND1).

Pupa: Pupation occurs in the soil inside earthen cells and takes about one week (Atwal, 1976).

Adult: Adult weevils vary in length from approximately 6.0 to 8.5 mm; the female weevil is slightly larger than the male by 1.0 to 2.0 mm. The weevils are whitish in colour with unevenly distributed black markings and yellowish colouration on their rostrum and head with dark-mottled dark elytra. The prothoracic and mesothoracic femora (front and middle) bear two spines (bidentate), while the metathoracic femora (hind) have three spines (tridentate).

The shoulders of the front of the elytra are strongly angled and broader than the prothorax (O'Brien et al. 2006; UF/IFAS, ND1).

Life cycle: In southern India, the pest is active from April to November and passes winter in the adult stages under the debris (Atwal, 1976). Oviposition occurs in soil close to roots. A single female lays on an average of 360 eggs over a period of 24 days. Grubs feed on roots. Pupation occurs in soil inside the earthen cells and takes about one week. Life cycle is usually completed in 6-8 weeks (Atwal, 1976).

Management/Control measures

Cultural Control: Cultural practices may be of value. Atwal (1976) has reported that frequent hoeing and inter culture may disturb and kill the grubs of the cotton grey weevil. Further, the cotton grey weevil has a marked preference for arhar, *Cajanus cajan*, which can be sown as a trap crop (Atwal, 1976). The adult weevils can be removed from the trap plants by vigorous shaking a branch over an open, inverted umbrella. The collected weevils can then be dumped into a container of soapy water.

Chemical control

Apparently several insecticides are effective against adult weevils in the Genus Myllocerus (Budhraja *et al.*, 1984; Singh *et al.*, 1991; Sinha and Marwaha, 1995). However, the chemical treatment may be of little or no economic value. The insecticides provide only limited management of this pest. Chemical control of the adults is difficult because of their ability to fly, hide, or feign death and drop to the ground. Chemical control of the eggs, larvae, and pupae is more difficult due to their location on or in the soil. In spite of that, it is suggested to spray monocrotophos 36 sl@ 1 ml/litre in the month of August-September when the population is at its peak. This can control the pest population build up (Paunikar, 2015).

ii) Cotton Grey weevil (*Myllocerus subfasciatus*)

Occurrence & severity

Cotton Grey weevil (*Myllocerus subfasciatus*) is more popularly known as 'Brinjal Grey weevil' or 'Eggplant ash weevil'. It is an important pest of eggplant that also attacks other solanaceous crops and cotton in Asia and Southeast Asia. In cotton, this is a minor pest, but, an economically important weevil occurring in South India and adjacent areas (Subramanian, 1958; Mohandas *et al.*, 2004). Roots of the same host plants are eaten by the larvae. Damage occurs in summer and rainy season crops, the summer damage being most severe (Subramanian, 1958).

Types of damage

The adult feeds on leaf, causing marginal notching and defoliation. Grubs feed on roots causing wilting of plants. Symptoms include leaf margins notched by brown coloured weevils, wilting and death of plants in patches, especially during fruit bearing stage.

Scientific name & order: *Myllocerus subfasciatus* Guerin-Meneville (Coleoptera: Curculionidae)

Description: The adult is a brown weevil, lays about 500 eggs in soil. The eggs hatches in about a week and the grubs are small, white, apodous, fully fed in 2-2.5 months. The grub pupates in soil in earthen cocoon. The pupal period lasts for 10-12 days.

Management/Control measures

The control measures are same as described earlier in this chapter with other grey weevil. However, certain measures which are practiced in different crops and suggested for this weevil are given hereunder.

Cultural control

It is suggested to collect and destroy the adults, and to apply Neem cake @ 500 kg/ha at the time of last ploughing in two equal split doses at 30 days interval.

Biocontrol

The toxin, Bt toxins (25 ng/mL concentrations as soil drenching) from *Bacillus thuringiensis* ssp. *tenebrionis* is effective against *M. subfasciatus* (Mohandas *et al.*, 2004).

Chemical control

Application of carbofuran 3 G @15 kg/ha on 15 days after planting in endemic areas is effective to manage the pest. Futher, spraying of carbaryl 50 WP @3g + wettable sulphur 2g/litre mixture is also suggested to control the pest attack.

iii) Green weevil (*Myllocerus viridanus*)

Occurrence and severity status

Green weevil (*Myllocerus viridanus*) is a short snouted weevil, also known as 'Ash Weevil'. The weevil Pest occurs in areas of Eurasia, however, in Asia it is mostly found in South East Asia, like Sri Lanka and India (Tamil Nadu). The pest also reported from other parts of India, such as, Ranchi, Jharkhand (Chattopadhyay, 2015). In cotton this is a minor pest, although, it is one of the major Insect pests of drumstick (*Moringa olifera* Lamk.). Prevalence of the weevil is observed throughout the year. High incidence of weevil occurs from November to January. However, the peak period of its incidence is in the month of August.

Type of damage

Adults feed on the leaves from the edges in a serrated manner, causing notches of leaf margins. In severe attack they may cause defoliation, leaving only the midrib. Grubs feed on roots and cause wilting and drying of plants.

Scientific name & order: *Myllocerus viridanus* (Fabricius, 1775) [Coleoptera: Curculionidae]

Synonym

= *Myllocerus angustifrons* Faust, 1897

Description

Adult is small light green or greenish-white in colour, 5 mm in length. Elytra are with brown and white spots.

Management/Control Measures

The weevil can be minimized by collecting and destructing the adults. Spraying of Carbaryl 0.2% or Malathion 0.1% is effective to manage the pest. Further, it is suggested to incorporate in soil with dust of 0.1% Chloropyriphos or drenching of the soil with 0.1% Chlorpyriphos to control the pest effectively.

iv) Ash weevil (*Myllocerus discolor*)

Occurrences and severity status

Ash weevil (*Myllocerus discolour*), also known as 'Apple Weevil' or 'Litchi weevil', is a polyphagous pest with a wide range of host like maize, sugarcane, sunflower, citrus, mango, jute, cotton, brinjal, soyabean, litchi, mulberry (*Morus alba*), *Dalbergia sisoo* and ber. It is distributed in Maharashtra (Nagpur, Kolhapur), Madhya Pradesh (Indore), Assam, Andhra Pradesh, Jammu and Kashmir, Karnataka and Odisha in India (Paunikar, 2015). The Ash weevil is a severe pest of *ber* (*Ziziphus jujube*) which is an economically important crop in the state of Tripura, northeast India (Das and Das, 2016). However, in cotton this is a minor pest.

Type of damage: Type of damage is same as other *Myllocerus* weevils. Adult notches on the leaves and leaf margins and the grubs feed on roots resulting in wilting of plants.

Scientific name & order: *Myllocerus discolour* **Boheman, 1834**

Description: *Myllocerus discolor* is a short snouted weevil. Body colour is ferruginous brown, with patches of fawn-colored scaling and mottled black, dense covering of greyish to green, somewhat shining scales with black cuticular marks (Kiyanthy and Mikunthan, 2009), in some cases the weevil appears as metallic green in colour (Mazumder *et al.*, 2014). Head is produced into rectangular snout bearing 12 segmented, geniculate antennae laterally. Antenna is 12 segmented, the first being the longest. Eyes are black, prominent, but much smaller in size, present laterally at the base of rostrum (Mazumder *et al.*, 2014). Elytra are prominent covering the abdomen completely and covered by green scales completely. Legs are similar in size, covered by brownish hairs and scales, a pair of claws at the tip. Mesothorax is rectangular and hairy. Eggs are small, ovoid and cream coloured. Grubs are C shaped, creamy white with brown heads and legless.

5. Cotton Leaf Roller

Occurrence & severity

Cotton Leaf Roller (*Syllepte derogata*), also known as 'bhindi leaf roller', is a common pest of malvaceous plants (CABI, 2007). It is more important as a sporadic pest of cotton (Khoo *et al.*, 1991). Sometimes the pest causes serious damage of cotton in India (Sidhu and Dhawan, 1979). Odebiyi (1982) has described *S. derogata* as a serious pest of cotton in south-west Nigeria, which occasionally reaches outbreak level. 10-14% of an unprotected farm may be attacked, although attacks are usually restricted to less than 10% in the north. Apart from cotton, the pest causes damage to other fibre crops like, *Corchorus capsularis* (white jute*), C. olitorius* (tossa jute) (CABI, 2007) and *Hibiscus cannabinus* (mesta/kenaf) in Kagoshima in Japan and in Pakistan (ZamanandKarimullah, 1987; Hiramatsu *et al.*, 2001). The pest is widely distributed in the Comoros, the Democratic Republic of the Congo, Ghana, Reunion, Madagascar, the Seychelles, South Africa, the Gambia, Australia, Fiji, Papua New Guinea, Samoa, the Solomon Islands, the Andaman Islands, Bali, India, Sri Lanka, Malaysia, Myanmar, Singapore, Sri Lanka, Vietnam, China and Japan (De Prins and De Prins, 2018).

Type of damage

The caterpillars roll leaves and eat the leaf margins, causing the leaves to curl and droop (CABI, 2007). The rolled leaves are in the form of trumpets fastened by silken threads and each roll contains more than one active glistening green larva with a black head. The plant becomes stunted and defoliates. This defoliation results in the premature ripening of bolls and impairs bud formation.

Scientific name & order: *Syllepte derogata* (Fabricius, 1775) (Lepidoptera: Crambidae)

(nbair, 2013)

Synonyms

= *Sylepta derogata* (Fabricius, 1775.)

= *Haritalodes derogata* (Fabricius, 1775.)

= *Phalaena derogata* Fabricius, 1775

= *Botys multilinealis* Guenée, 1854

= *Botys otysalis* Walker, 1859

= *Zebronia salomealis* Walker, 1859

Biology

Egg: Eggs are laid both singly and in batches on both sides of the leaf, but more on the abaxial surface. Newly laid eggs are colourless but later change to dirty white towards the time of eclosion. These are measured as 0.76 ± 003 mm x 0.51 ± 001mm (Anioke, 1989).

Larva: The larva moults 4 or 5 times. The general characters of all the instars are similar. The first instar larva is light green and the body is slender and slimy. It has a small dark brown head and enlarged prothorax. Each thoracic segment bears a pair of legs. Four abdominal prolegs are located on the 3rd to the 6th segments. The last abdominal segment bears the 5th abdominal prolegs or claspers. The body is covered with numerous setae. The second instar larva is green. Two small brown spots are located on the dorsal aspect of the prothorax on the position of the D1 setae. In case of third instar larva the epicranial and frontal sutures become distinct and the head capsule is slightly bilobed. The posterior portion is brownish while the anterior portion is light brown. The two brown spots on the prothorax start to develop into thoracic shields. The fourth instar larva is deep green. The bilobed head capsule and thoracic shields are black. The prothoracic segment becomes slightly flattened on the ventral surface. The thoracic legs are black with white claws. The abdominal prolegs are dirty white. The stematae become visible by the sides of the head near the antennae. The fifth instar larva is also green but towards pupation turns light green. The black thoracic shields start to degenerate from the posterior portion. The mandibles are stout with 5 teeth, 3 of which are more prominent. The sixth instar larva is similar to the proceeding instar but for the prothoracic shields which continued to degenerate, thereby exposing the greater part of the prothoracic segment (Anioke, 1989). Full-grown larva reaches a length of about 15 mm.

Pre-pupa. The larva shortens and loses its characteristic green colour and becomes pinkish. It becomes inactive (Anioke, 1989).

Pupa: Pupation takes place in the rolled leaf. The pupa is with more or less convex head. The thoracic segments are not visible on the ventral aspect having been concealed by the appendages. A dorsomedial ridge runs from the mesothorax and extends to the last abdominal segment. The spiracles have thick round to oval and dark-brown peritremes. The abdomen is conical in shape. The 9th and 10th segments are not distinctly separated. Pairs of proleg scars are conspicuous on the ventral surface. On the last segment is attached a black cremaster which is conical in shape. It bears six curved hooks. The pupa is dark-brown towards the time of adult emergence (Anioke, 1989).

Adult: The adult is with wingspan of 28–40 mm. It is a small delicate moth of silvery white to gold or pale brown colour with complex dark brown sinuous lines on the wings (Anioke, 1989; Don and Stella, 2013). Both the wings and the body are covered with scales. The abdomen of the male is slender and tapers gently towards the tip. In the female the abdomen is stouter and blunter at the tip. Female moths are darker in colour than the males (Anioke, 1989).

Life cycle: The number of eggs produced by a mated female is about 443 during its life span. Unmated female lays fewer, 207, unviable eggs. The pre-oviposition period is 2.5 days, while, the oviposition and post-oviposition periods are 6.5 and 3.0 days, respectively. The life cycle completes in about 34 (33.9) days (Anioke, 1989).

Molecular Studies

The complete mitochondrial (mt) genome of Cotton Leaf Roller, *Syllepte derogata*, is determined as 15,253 bp in length (GenBank accession number: KC515397) containing 37 typical animal mitochondrial gene and an A + T-rich region. The gene order of *S. derogata* mtDNA was different from the insect ancestral gene order in the translocation of trnM, as shared by previously sequenced lepidopteran mtDNAs (Zhao *et al.*, 2016).

Management/Control measures

Host Plant Resistance

Use of resistant variety, if available, is probably the best measure for the insect control. In this regard, a few works have been done in both China (Zhang and Sun, 1988) and India (Yein, 1983).

Biocontrol

The biocontrol agent, *Bacillus thuringiensis*, as well as the biopesticide, azadirachtin, are found effective to minimize the pest attack.

Chemical control

The insect damage can be minimized by the application of insecticides. Several insecticides with the active ingredients deltamethrin, methoxyfenozide, spinosad, indoxacarb and teflubenzuron are found effective to control the pest. Yein and Barthakur (1985) found synthetic pyrethroids, fenvalerate, permethrin and deltamethrin (at 0.1 kg a.i./ha), as effective chemicals to control the pest in India. However, there are certain different views. Since, Dhawan *et al.* (1988) have found that the pyrethroids fenvalerate, permethrin, cypermethrin, deltamethrin and flucythrinate are less effective than monocrotophos, endosulfan and carbaryl in India.

6. Cotton mealybug

Occurrence & severity

Cotton mealy bug (Fig. 29A) is wide-spread in tropical and subtropical regions. It appears to be native to North America, where it was first discovered in 1898 by Tinsley (1898a) from New Mexico. Later, this was reported to have spread in different parts of USA (McKenzie, 1967). Also, it has spread to various other countries, damaging plants in a variety of habitats ranging from dry arid areas to tropical regions. It is an important plant pest worldwide (Williams and Granara de Willink, 1992; Hodgson *et al.*, 2008). The mealy bug was reported to be seriously infesting cotton plants in the Punjab and Sindh regions of Pakistan (Arif *et al.*, 2007). By 2006, it had spread to a large number of cotton-growing districts where the mealy bug not only had a dramatic impact on the plant, but also reduced the value of the cotton. The population density of this invasive pest was varied on cotton (*Gossypium* spp.) in surveyed regions in Pakistan (Dhawan *et al.*, 2009b). In India and Pakistan, earlier the mealybug was considered as a secondary pest of cotton that maintained at low population levels by chemical applications that required controlling the other major pest. However, with the emergence of transgenic (Bt) cotton and the reduced need for chemical applications, the

Fig. 29 (A). Pest attacks in cotton. Cotton mealybug *(Phenacoccus solenopsis)*

mealy bug has emerged as a major pest. It causes serious damage to cotton in Pakistan (Saeed *et al.*, 2007; Dhawan *et al.*, 2009a,b) and India (Jhala *et al.*, 2008; Bhosle *et al.*, 2009). It has also been reported to be spread to southern China around 2005 (Wang *et al.*, 2009; Wu and Zhang, 2009). In Brazil, particularly at the regions of the Southwest and Middle São Francisco, Bahia and also in the regions of the Agreste and Semi-arid of the Paraiba State, high infestations of cotton mealy bugs have occurred during the cotton season of 2007 and 2008 (Domingues da Silva, 2012). The pest also has the ability to spread rapidly to uninfested areas by natural carriers such as the wind, rain and water-ways, on farm equipment, and by clinging to clothing and animals. International trade plays a major role in the spread of this pest to new regions of the world. It was discovered to infest ornamentals in Nigeria (Akintola and Ande, 2009). Hodgson *et al.* (2008) inferred that the infestation in Nigeria may have originated from South America. The cotton mealy bug is a polyphagous insect feeding on more than 200 plant species (Goeden, 1971; Pinto and Frommer, 1980; Sharma, 2007; Arif *et al.*, 2009; Jagadish *et al.*, 2009b). These plant species are assigned to approximately 60 families. However, the insect has a preference for Asteraceae, Euphorbiaceae, Fabaceae, Malvaceae and Solanaceae. It has also a wide morphological diversity, biological adaptations and ecological adjustments that give it a high capacity to feed on different host plants (Hodgson *et al.*, 2008).

Types of damage

The mealybug damages the plant by extracting sap, which stresses the plant. Various parts, such as, leaves, stems, roots, of the plant are infested resulting in chlorosis and deformation. The extraction of sap results in the leaves of the plant turning yellow and becoming crinkled or malformed, which leads to loss of plant vigour, foliage and fruit-drop, and potential death of the plant, if not treated. Phloem feeding affects the growing regions of the plant often resulting in bunched and stunted growth (Dhawan *et al.*, 2009b; Jagadish *et al.*, 2009a). Outbreaks of the mealy bug on cotton cause significant damage like fewer, smaller bolls and yield reduction (Kumar *et al.*, 2014). In addition, the high numbers of developing mealy bugs produce large amounts of honeydew that fall onto the lower leaves producing a substrate for the development of sooty mould, which inhibits photosynthesis within the plant (Arif *et al.*, 2012). The honeydew attracts ants. The foraging ants enter into a mutualistic association with the mealy bugs by collecting the honeydew and keeping the area clean of the excess waste product, simultaneously, protecting the mealy bugs from potential natural enemies. The production of honeydew and its occurrence on the lint can also interfere with the processing of the cotton by making the ginning process more difficult and may harm the marketing of production (Arif *et al.*, 2012).

Scientific name & order: *Phenacoccus solenopsis* Tinsley, 1898 (Hemiptera: Pseudococcidae)

[Zootaxa 3802(1): 109-121]

Synonyms

= *Phenacoccus cevalliae* Cockerell, 1902

= *Phenacoccus gossypiphilous*

[**N.B.** In some cases, the mealy bug specimens of cotton in India have different morphological traits from that of *Phenacoccus solenopsis* of other places. Hence, the Indian specimens were referred as a new species, viz., *Phenacoccus gossypiphilous* (Abbas *et al.*, 2005, 2008, 2009). However, Hodgson *et al.* (2008) concluded from a comprehensive morphological study that there were no significant differences in specimens from the Indian subcontinent compared to those from the Neotropics; and thus, considered the name *P. gossypiphilous* as a synonym of *P. solenopsis*.]

Biology

Egg: Eggs are pale-yellow. A female lays about 150-600 eggs in a white waxy ovisac.

Nymph: First instar nymph (crawlers) is separated from the other stages by possessing six-segmented antennae, lack of circulus, and quinquelocular pores on the head, thorax and abdomen. Its body is oval in outline, 710–730 μm long, 359–380 μm wide; anal lobes short; a pair of oral collar tubular ducts present on head; with 2 pairs of indistinct ostioles; 18 pairs of indistinct cerarii present, most with only one spinosecerarian seta; claw with a well-developed denticle (Hodgson *et al.*, 2008). Second-instar nymph is oval in outline, size of female is little larger than male, female size 0.75-1.1 mm long, 0.36-0.65 mm wide and male size 0.72-0.79 mm long, 0.35-0.4 mm wide; anal lobes shallow; antennae 6 segmented; legs of normal length; circulus absent; with 18 pairs of distinct cerarii; oral collar tubular ducts present on venter; quinquelocular pores absent; claw with a well-developed denticle. The second instar nymphs are distinguished by having 18 pairs of (distinct in female and anteriorly not very distinct in male cerarii around the margin of the body, the lack of quinquelocular pores on the body and the claw with a distinct denticle (Hodgson *et al.*, 2008). Third instar nymph is oval in outline, 1.02-1.73 mm long, 0.82-1.00 mm wide; anal lobes shallow; antennae seven-segmented ; legs of normal length; circulus present; with 18 pairs of distinct cerarii; oral collar tubular ducts present on venter; multilocular and quinquelocular pores absent; claw with a well-developed denticle (Hodgson *et al.*, 2008).

Pre-Pupa: Pre-Pupal body length is about 1.35-1.38 mm, width across anterior abdomen 525-550 μm. Body is not clearly divided into tagmata. Antennae are apparently 9 segmented, quite short, about 350 μm long, extending posteriorly to about procoxae. Legs are rather well developed, although tibio-tarsal articulation sometimes indistinct. Body is with many setose setae, longest along abdominal margin. Anterior wing bud is quite small, barely reaching mesocoxae; multilocular disc pores fairly evenly distributed over dorsum and venter; oral collar tubular ducts present (Hodgson *et al.*, 2008).

Pupa: Pupal body is quite small, total-body length about 1.43-1.48 mm, width across anterior abdomen 475-500 μm. Body not clearly divided into tagmata; antennae 10 segmented, extending posteriorly past anterior spiracles. Legs are rather well developed, with tibio-tarsal articulation usually distinct. Body is with frequent setose setae, longest along abdominal margin; loculate pores abundant, each with mainly 10 loculi. Anterior wing bud is long, extending posteriorly to level with metacoxae; multilocular disc pores distributed mainly in segmental bands, particular on abdomen; oral collar tubular ducts absent (Hodgson *et al.*, 2008).

Adult Female: Female body is pale yellow to almost orange. Dorsum is with a series of dark dorsal markings as follows, A pair of "exclamation marks" are present on head, about six pairs of transverse markings across pro- and mesothorax, absent on the metathorax, and with pairs of dark transverse markings on each abdominal segment; also with a submarginal line of dash marks on thorax and abdomen. Venter is with dark circulus. Body is oval in outline, 3.0–4.2 mm long, 2.0–3.1 mm wide. Anal lobes are moderately developed. Antennae are 9 (rarely 8) segmented. Legs are well developed, all metacoxa without pores. Circulus is present, oval, occasionally slightly constricted laterally, rather variable in size. Cerarii are distinct, numbering 18 pairs. Oral collar tubular ducts are restricted to venter. Quinquelocular pores are absent. Multilocular disc pores are absent dorsally, present medially on venter of posterior 4 (occasionally 5) abdominal segments. Claws are with a small denticle each (Hodgson *et al.*, 2008).

Adult Male: Head is somewhat twisted but probably with a distinct postero-ventral bulge with ventral eyes. Body is quite small, total-body length about 1.41 mm. Antennae are about 2/3rds total-body length, 10 segmented. Body is with few setae, all hair-like setae, fleshy setae apparently absent apart from on antennae and legs; length of fleshy setae on antennae a little less than twice width of antennal segment. Loculate pores are mainly with 4, but occasionally 5 loculi. Abdomen is with glandular pouches and glandular pouch setae on segments VII and VIII. Penial sheath is with a distinct constriction towards apex. Wings are about 4/5ths of total-body length (Hodgson *et al.*, 2008).

Life cycle

Females of this bisexual species are capable of producing 150-600 pale-yellow eggs in a white, waxy ovisac (Lu *et al.*, 2008). The first instar nymphs (crawlers) disperse to settle primarily on the leaves as well as the stems, leaf petioles, and bracts of fruiting cotton (Ben-Dov, 1994). Upon hatching, females undergo three immature stages prior to reaching adulthood, whereas males undergo first, second, prepupa and pupa stages prior to adulthood. The period of development from crawler to adult stage is approximately 25-30 days, depending upon the weather and temperature. This species is capable of producing multiple generations annually. This mealybug has been reported to be capable of surviving temperatures ranging from 0-45°C, throughout the year (Sharma, 2007). The location on the plant appears to be influenced by humidity (Hodgson *et al.*, 2008).

Management/Control measures

SPS measures: Several countries have incorporated measures to inspect plant material from locations where the mealybug pest is known to occur. China has initiated a notice of inspection and quarantine for *P. solenopsis* (Ministry of Agriculture, 2009).

Cultural control and sanitary measures: It is important to give attention to the field borders for plants that can serve as an alternate host for the mealybug. Such plants should be removed to prevent the mealybugs from overwintering and infesting crops in the future. Trap plants may be planted that initially attract the mealybugs and can be targeted for control treatments to protect the primary crop. It is necessary to sanitize equipment and check clothing items to prevent the transfer of the pest into new locations. Small populations of *P. solenopsis* can be controlled by inspection of plants, removing infested twigs and handpicking the specimens from newly-infested plants. Soap applications are often effective against targeted, small populations of the mealybug. Use of sticky traps placed throughout the field is an effective means to survey for the presence and population density of the pest.

Biological control: Conservation of natural enemies of mealy bug and their effective utilization in its control may apply to manage the pest population in lower level. Several parasitoids and predators have been documented to attack *P. solenopsis*. Three parasitic wasps (*Chalcaspis arizonensis*, *Cheiloneurus* sp. and *Aprostocetus minutus*) are discovered as attacking the *P. solenopsis* mealybug on cotton in Texas, USA (Fuchs *et al.*, 1991). In India, an unidentified species of the solitary endoparasitoid, *Aenasius* sp., has also been reported to attack *P. solenopsis* (Sharma, 2007; Tanwar *et al.*, 2008). Hayat (2009) has described a new species of parasitoid, *Aenasius bambawalei*, to be associated with *P. solenopsis*, which has been documented as a very effective candidate for biological control. This parasitoid is found parasitizing cotton mealybugs (Muniappan, 2009) and reported to parasitize up to 72% of the *P. solenopsis* populations infesting cotton plants grown in some districts in India (Muniappan, 2009; Pala Ram *et al.*, 2009). Further, the pest is attacked by the encrypted edendoparasitoid, *Aenasius arizonensis* which may kill 60-70% or more of the mealybug population. The coccinellid predators *Brumoides suturalis* (Fabricius) and *Nephus regularis* also attack the pest. In India, the coccinellid *Hyperaspis maindroni*) is also identified to be associated with *P. solenopsis* (Patel *et al.*, 2009). Most predators feed on the eggs or crawlers within the mealybug's ovisac and reduce the number of mealybugs available to extract sap and weaken the plant. Other potential predators, such as the larvae of the lacewing, *Chrysoperla carnea*, are found to consume 30 mealybug eggs daily in developmental laboratory tests (Kaur *et al.*, 2008).

Chemical control

The use of insecticides is the most effective control against the mealybug when applications are timed to coincide with the crawler stage. The toxicity of a variety of insecticides (Dhawan *et al.*, 2008b; 2009a) was evaluated for efficacy against *P. solenopsis*. The insecticides, Spinosad and Spirotetramat, are able to control the pest and cause the least reduction in parasitization by natural enemies. In addition, the insecticides, carbamates and imidacloprid are also very effective. The new molecule, spirotetramate is found to be most toxic to one-day-old nymphs and adult female. Particularly among the conventional insecticides, profenophos and thiodicarb are the best insecticides for mealy bug control. Acephate is found to be least toxic against both nymph and adult of mealy bug (Dhawan *et al.*, 2008a).

IPM System

A management strategy to control *P. solenopsis* in India with the use of cultural, mechanical, biological and chemical control measures has recently been developed (Tanwar *et al.*, 2007). In that system it is recommended a survey for the mealybug prior to planting, targeting and chemically treating small populations, removal of alternate host plants and ant colonies, using recommended insecticides for optimal effectiveness on the plants and around their root system, providing an attractive habitat for native and exotic natural enemies, and using a variety of sanitation methods to prevent spread of the pest to new fields.

7. Midges

There are different species of gall midges known to damage cotton. These are *Dasineura gossypii* Felt, *Contarinia gossypii* Felt and *Porricondyla gossypii*

Coquillet. These midges cause different degrees of crop injury as and where they prevailed.

i) Flower bud maggot (*Dasineura* gossypii)

Occurrence & severity

Flower Bud Maggot (*Dasineura gossypii*) is also called as 'Cotton midge' or 'Cotton Flower Bud Fly'. This cotton midge is known to occur in Eurasia, mostly in South Asian countries. It is most prevalent in India. However, it was earlier considered as a minor pest of cotton (Mani, 1934).The flower bud maggot (*D. gossypii*) becomes a potential pest for first time during 2009 in Karnataka with severe incidence in farmer's field at Hesarur village. More than 90% fruiting body damage was recorded in largely cultivated Bt cotton cultivars (Chakraborty *et al.*, 2015). Later, in Kakol and Konanatambagi villages (Haveri District) also the midge was noticed in severe form (Udikeri *et al.*, 2011a). This flower bud maggot occurs seriously in southern part of India especially in Karnataka (Udikeri *et al.*, 2011a). The year 2013-14, once again, evidenced the very impact of devastation of this pest in cotton growing districts viz., Haveri, Davangeri, Chitradurga, Dharwad and part of Belagavi. MRC7351 Bt cotton genotype experienced severe infestation of flower bud maggot which resulted in 60-70 per cent reduction in the seed cotton yield (Lamani, 2014).The incidence of flower bud maggot is increasing day by day in Karnataka. Currently more than 90 per cent fruiting body damage has been recorded in largely cultivated Bt cotton cultivars viz., Kanaka and Neeraj (Chakraborty *et al.*, 2015).

Types of damage

The maggots (larvae) of this midge feed on anthers, staminal column and outer wall of style leading to decaying of inner content of the flower buds which fail to grow properly, leading to twisted petals with decaying and drying of the flower bud and drop. In most cases 5-20 maggots can be found to cause damage in single flower bud under favourable conditions (Udikeri *et al.*, 2011a; Chakraborty *et al.*, 2015).

Scientific name & order: *Dasineura gossypii* (Felt, 1916) (Diptera: Cecidomyiidae)

Biology

Egg: Freshly laid eggs are white or hyaline, elongate, and cylindrical to ellipsoidal, with one end round and the other end rather pointed. The eggs are minute, measured 0.29 mm in length and 0.10 mm in width. Eggs are inserted inside the tender squares by the female flower bud maggot with its long, sharp ovipositor. Egg incubation period is about 1-2 (1.5) days (Chakraborty *et al.*, 2015).

Larva: The newly hatched, first instar larvae are white, transparent, elongate, slightly oval. The length and width of first instar larva are 0.66 mm and 0.12 mm, respectively. The second instar larvae are milky white to slightly cream coloured; the length and width are 1.28 mm and 1.15 mm, respectively. Third instar larvae are orange in colour, spindle shaped; the length and width are 1.72 mm and 0.35 mm, respectively. Larval stage completes in three instars. The total larval period is 4 to 5 days (Chakraborty *et al.*, 2015).

Pupa: Pupae are orange in colour and covered by a white, silken cocoon. The pupal period last for 4.4 days. The length and width of pupae are about 1.64 mm and 0.30 mm, respectively. Pupation usually occurs inside the bract but sometimes in soil also (Chakraborty *et al.*, 2015).

Adult: Both adult male and female are minute, soft bodied, orange in colour, weak flier and short lived. Males and females can be easily distinguished by the presence of ovipositor. Females are larger than males in size, about 2.21 x 0.44 mm, with swollen abdomen, a long (sometimes as long as body) sharp ovipositor, and less hair-like sensory receptors on moniliform antennae. Males are smaller in size, about 1.82 x 0.42 mm, with more hair-like sensory receptors on moniliform antennae. The longevity of female is about 1.2 to 1.3 days, while, that of male is 1.1-1.2 days (Chakraborty *et al.*, 2015).

Life cycle: Female of *D. gossypii* prefers to lay eggs in square tips, where it lays 42-45. Eggs are inserted inside the tender squares by the female flower bud maggot with its long, sharp ovipositor. There are 3 instars in the larval stage and thereafter pupation takes place on bracts or in soil. The mean durations of different life stages of this midge, viz. eggs, first instar maggot, second instar maggot, third instar maggot, pupal period, adult stage (male) and adult stage (female) are of 1.5, 1.2, 1.7, 2, 4.4, 1.2 and 1.3 days, respectively. The lifecycle completes within 10-13 days. The number of generations per year is 8-10 (Chakraborty *et al.*, 2015; Nandihalli *et al.*, 2015).

Management/Control measures
Malathion 50 EC @ 2.0 ml/l is found as most effective followed by profenophos 50 EC @ 2.0 ml/l. But, the efficacy of these insecticides is not up to the expected level. However, when these insecticides are applied in combination with DDVP 78 EC @ 0.25 ml /l give effective control of the pest (Udikeri *et al.*, 2011a). Most midges including *D. gossypii* pupate in soil and targeting the pupae in soil and emerging adults can be a good control method. In a study, the soil application with three insecticides viz., Phorate 10 G, Carbofuran 3 G and DDVP 76% EC, against maggots of cotton flower bud maggot, it is found that DDVP 76% EC is the most effective with the lowest adult emergence (3.33 %) followed by Phorate (5.56 %) as against 55 % in control (Babar *et al.*, 2016).

ii) Cotton Gall Midge (*Contarinia gossypii*)

Occurrence & severity

Cotton Gall Midge (*Contarinia gossypii*), also known as 'Cotton Flower-bud Maggot', occurs mainly in West Indies and Florida, U.S.A. It was reported to cause serious damage on the midge's first appearance in Antigua at the end of 1907 (Ballou, 1919). In 1909, it was reported from Montserrat, in 1916 from Tortola in the British Virgin Islands (Anon., 1918) and in 1923 from St. Croix in the American Virgin Islands (Wilson, 1923). Later, Wolcott (1933) reported its probable presence in Porto Rico and Haiti, occurring in small numbers. It was reported from Florida, U.S.A., during 1932 (Rainwater, 1934). The insect was injurious to cotton in the British-West Indies (Felt, 1908). The midge attacks for about three months starting from December to February. This is the usual period of attacking, although the midge may appear as early as in October (Ballou, 1919; Barnes, 1949). Cotton (*Gossypium* spp.) of both wild and cultivated species is the food plants of this midge (Ballou, 1909).

Types of damage

There is a characteristic odour attached to badly infested buds and squares. The inner parts of attacked blooms darken and decay becoming watery, as do the insides of infested squares. The larvae feed in among the developing anthers and stamens causing the death of the bud. The bracts surrounding the base of the buds usually flare back instead of remaining close round the base. As many as 43 larvae have been found in a single bud. In attacked bolls, the lint becomes discoloured, the bolls become soft and watery and the seeds are eaten. Excessive dropping of buds of cotton is caused by numerous yellow "jumping" larvae. The base of the squares near the bracts has a tendency to crinkle and become detached from the remaining parts producing a somewhat corrugated appearance (Ballou, 1909; Barnes, 1949).

Scientific name & order: *Contarinia gossypii* Felt, 1908 (Diptera: Cecidomyiidae)

[Entomological News Philadelphia, 19: 210-211]

Biology

Egg: The midge lays its eggs in the young flower buds.

Larva: The larva is yellowish or pinkish, jumping habit.

Pupa: The mature larvae pupate in soil.

Adult

Female adult: Female adult is about 1.05 mm in length. Antanae are as long as the body, sparsely haired, pale yellowish, 14 segments. Palpi are quadriarticulate; the first segment irregularly fusiform; the second narrowly oval and half longer than the first; the third half longer than the second, more slender; all rather thickly clothed with coarse setae. Abdomen appears to be fuscous greenish yellow and posterior margins of the segments are more thickly clothed with coarse setae. Tarsi are slightly darker. Ovipositor is yellowish, as long as the body when extended, the terminal lobes very long, slender (Felt, 1908).

Male adult: Male adult is about 1 mm in length. Antenae are about twice the length of the body, thickly haired, light brown, with 14 segments. Palpi are quadriarticulate; the first segment apparantly short, stout, irregularly subquadrate. The second segment is little longer, broadly ovate. The third one is fully half longer than the second, more slender. The fourth one is as long as the third, more slender all rather thickly clothed with corse setae. Face is fuscous, yellowish; eyes large, black. Mesonotumis is dark brown, submedian lines yellowish. Abdomen is greenish yellow, posterior segments with coarse setae. Wings are hyaline, costa pale straw. Halters are yellowish transparent. Coxae are yellowish, femora and tibae pale yellowish straw, tarsi slightly darker. In genitalia, basal clasp segment is rather long, broad and tapering to rounded apex; terminal clasp segment long and slightly tapering to an obtusely rounded apex (Felt, 1908).

Life cycle

The midge lays its eggs in the young flower buds. The larvae feed in among the developing anthers and stamens causing the death of the bud. The larvae leave the buds by "jumping" and pupate in the soil. The egg and larval stages occupy 12-14 days. The pupal stage requires from 10 to 14 days and the adult midges live 2-3 days (Ballou, 1909).

Control measures

The cultural methods, like, keeping the surface of the soil well pulverised, dry and free from weeds, the removal of wild cotton and handpicking the infested flower buds are effective to minimize the midge attacks (Ballou, 1909). Further it is suggested that that the practice of early planting, which allows the formation of the bolls before the oviposition of the midges, would enable a crop free from the midge attack (Ballou, 1912). Soil may be fumigated with the use of a soil fumigant (Vaporite) to kill the pupae in soil (Ballou, 1909).

iii) Cotton Red Maggot.

Occurrence & severity

Cotton Red Maggot is referred as a West Indian species that was described by Coquillet (1905) from specimens reared from larvae living in cotton stems without producing any enlargement of the stem (Barnes, 1949).The cotton red maggot was first observed in 1903 in Barbados, where it has on occasion seriously damaged cotton and it is also present although not of a economic importance, in Monserrat (Callan, 1940). This midge is now only of minor importance, there being few records and little literature concerning it (Dash, 1917; Bovell, 1921). Bourne (1921) recorded that injury to stems had been noted in three or four instances during 1919- 20.

Types of damage

The presence of Cotton Red Maggots is indicated by a discoloured and shrunken area on stem. The larvae attack the stem of the cotton. The larvae occur in numbers beneath the bark and feed on cambium (Callan, 1940), soft tissue of the bark and developing wood. In cases of severe attack when the feeding extends all-round the stem, it causes the death of the entire portion beyond that spot (Barnes, 1949).

Scientific name & order: *Porricondyla gossypii* Coquillet, 1905 (Diptera: Cecidomyiidae)

Biology

Egg: Eggs are laid in wounds on the stem (Barnes, 1949).

Larva: The larvae were originally described as yellowish-white with the median portion orange-red. When full grown they are orange-red to reddish in colour and about 4 mm. in length. They are gregarious (Barnes, 1949).

Pupa: Pupate in the soil.

Adult: In general, adult is tiny fragile flies with long antenna, weak veined wing and without tibial spurs (Gagné 1981). In female adult, the head is with 14 flagellomeres. Circumfila is consisting of two rings interconnected by two longitudinal threads, fourth flagellomere with short neck, node almost twice as long as wide. Terminalia is with ovipositor beginning with segment VI, one third as long as abdomen, slightly protrusible. Tergite IX is presumably not enlarged. Basicercus is slightly longer than disticercus. In male adult, circumifila is lacking posterior extensions and the parameters being tusk shaped and discrete, whereas the female dicerurine-like in that 14 flagellomeres and the ninth tergite not enlarged (described as Neotropical Porricondylinae; Jaschhof, 2014).

Control measures

It is suggested to cut down and burn off the affected stems as soon as the pest detected and to destruct all cotton plants after the harvest of the cotton crop (Ballou, 1909). But, prevention of wound on the plants is also apparent as the eggs are laid in wounds on the stem (Callan, 1941).

8. Mirid bugs

Cotton is attacked by several mirid species, which cause significant losses in quality and quantity of cotton at different phonological stages. Mirid bugs have been found to be associated with early season damage to cotton crops in most of the production centres in New South Wales and Queensland in Australia. There, mostly the green mirid (*Creontiades dilutus*) and the *yellow mirid* (*Campylomma livida*) are found amongst the mirids (Adams *et al*., 1984). The mirid bugs have not been historically referred as key pests of cotton but nowadays these have assumed key status only after widespread cultivation of Bt cotton hybrid. Recently, both the occurrence of these bugs and the damage caused by these insects has progressively increased in cotton crops (Lu *et al*., 2010). Three species of mirid bugs are found to infest cotton in India. The three species viz., *Poppiocapsidea biseratense* (=*Creontiades biseratense*), *Campylomma livida* and *Hyalopeplus linefer* are infesting cotton since 2005. These bugs have been restricted to Central and South India. The dominant species is *P. biseratense*. It is most potential, widespread and also becoming a production constraint every year (Udikeri *et al*., 2008).The apple dimpling bug or yellow mirid, *C. livida* is a dominant species in Maharastra, however, it is also noticed in Karnataka. In Karnataka, *P. biseratense* is number one pest and that followed by Tamila Nadu. The mirid bug, *H. linefer*, is recorded in Karnataka and Maharastra but not as regular pest (Udikeri, 2008; Udikeri *et al*., 2011b). Almost all the cultivated species of Bt cotton hybrids are affected by mirid bugs (Saravanan *et al*., 2017).The incidence of mirid bugs begins during September and exists till December; however the peak incidence is during October/November months. The maximum incidence is noticed during second fortnight of November in Karnataka (Udikeri *et al*., 2008).

i) Brown mirid bug (*Poppiocapsidea biseratense*)

Occurrence & severity

The brown mirid bug has recently emerged as a predominant pest in Bt cotton in central and southern parts of India. In recent years, this bug has emerged as a key sucking pest in South India causing a severe damage to Bt cotton. The green colour morph of this mirid species is appearing in Karnataka since 2005 and posing a threat to the Bt cotton cultivation in several parts of Karnataka

(Patil *et al.*, 2006). The pest has also been reported from Tamil Nadu, Andhra Pradesh and Maharashtra (Surulivelu and DharaJyoti, 2007; Udikeri *et al.* 2010). Presently, this mirid bug is appearing in severe form throughout the Karnataka, with most aggressive status in Haveri district (Udikeri *et al.*, 2008; Rohini *et al.*, 2009). The brown mirid bug incidence starts from 50 days old cotton crop and remains at peak during November last week. Its population starts declining from second week of December, reaching by December end and by March it disappeared totally (Ravi, 2007). The occurrence of the pest is negatively correlated with maximum temperature and minimum temperature, though; it has positive correlation with relative humidity and rainfall during cropping period of Bt cotton (Saravanan *et al.*, 2017). The mirid bug causes 20 to 55 per cent cotton square shedding and 15 to 33 per cent boll shedding of Bt cotton in seven days period in Karnataka (Patil *et al.*, 2006).This mird bug has been considered as a great nuisance in cotton production in Karnataka (Rohini *et al.*, 2009; Udikeri *et al.*, 2010). There has been huge loss due to this pest in North Karnataka particularly in Haveri district as the pest causes heavy shedding of squares and tiny bolls (Shyadaguppi, 2011). This brown mirid bug has several alternate hosts, viz., pigeon pea, sunflower, safflower, sorghum, maize, bajra, dhaincha, castor, amaranthus, pig weed and Indian mallow (Ravi, 2007). However, the maximum incidence of brown mirid bugs population occurs on pigeon pea and maize (Ravi and Patil, 2008).

Types of damage

The adult female prefers to insert the eggs on petiole, followed by bracts, and flower petals. One to two days old bolls with dried petals intact provide a good habitat to the insect for feeding and sheltering. Both nymphs and adults are found to suck the sap by piercing their stylet into the plant tissues, squares and small tender bolls (Khan *et al.*, 2004). They release the pectinase enzyme and other chemicals that destroy the cells in the feeding zone (Prakash and Bheemanna, 2014). The characteristic symptoms of feeding on the flower bud shows oozing out of yellow fluid from the buds and staining of this yellow fluid on the inner surface of the bracts. Infested tender bolls have number of black patches on all sides of the outer surface of boll rind (Udikeri *et al.*, 2008). The affected portion rapidly turns to dull in colour, then becomes blackens and ultimately dies (Prakash and Bheemanna, 2014). Feeding on matured bolls leads to parrot beaking and improper opening (Udikeri *et al.*, 2008).

Scientific name & order: *Poppiocapsidea biseratense* **(Distant) (Hemiptera: Miridae)**

[nbair (2013) http://www.nbair.res.in/insectpests/Poppiocapsidea-biseratense.php]

Synonym

= *Creontiades biseratense* (Distant)

Biology

Egg: Eggs are transparent nacreous white, cigar shaped, laid singly or in groups on the petiole, bracts, flower petals, squares, tender bolls and leaf tissues. The length of egg is 0.93- 1.06 mm and breadth is 0.25- 0.28 mm. Two days before hatching, egg turns to pinkish colour, and develops a pair of shiny red spots at the anterior end representing compound eyes. Hook shaped egg cap is formed at the anterior end which aids in hatching. The incubation period of eggs is 5-7(-9) days (Prakash and Bheemanna, 2014; Ravi *et al*., 2015).

Nymph: The freshly hatched nymphs are transparent yellow in colour with the tip of antennae having reddish tinge and thorax with brownish or light reddish in colour and depending on feeding habit some turn to light green colour. Later, the nymphs are green or brown in colour and present inside the square, when disturbed, rapidly move. Nymphal development period is ranged from 11-12 days. The mirid bug (*P. biseratense*) has five nymphal instars (Prakash and Bheemanna, 2014).The duration of first instar varied from 2-3 (av. 2.50) days. The duration of second instar nymphal period is 2-3 (av. 2.57) days. The duration of third instar nymphal period is 2-3 (av. 2.70) days. The duration of fourth instar nymphal period is 3-4 (av. 3.67) days. The duration of fifth instar nymphal period is 3-5 (av. 3.83) days. The duration of total nymphal period is 12-18 (av. 15.27) days. The final instar nymph is measured about 4.70 mm in length and 2.27 mm in breadth (Ravi *et al*., 2015).

Adult: The newly formed adults are light green in colour. After 10 to 14 hours, the body colour changes to reddish brown. The fully grown adults are brownish colour with swift flying activity. Males are brownish in colour and smaller in size and it is 5.23 mm and 2.62 mm in length and breadth, respectively. The mean life span of male adult bug on cotton plants is 11.50 days. The abdomen of female bug is slightly broader and elongate than male. The female bugs live longer as compared to males with a mean adult longevity of 19.1 days on cotton plant. Female measured is 6.24 mm and 2.96 mm in length and breadth, respectively (Prakash and Bheemanna, 2014).

Life cycle

The newly paired male and female mate after two to three days of emergence and their mating last for 3-4 hours. The average pre-oviposition, oviposition and post ovipositional periods are 2.60, 10.30 and 6.20 days, respectively. A female lays about 75- 185 eggs. Total life cycle of male varies from 27- 39 (av. 34.92) days, while, that of female varies from 32 to 50 (av.42.72) days (Ravi *et al*.,

2015). The longevity of adult males ranges from 10 to 15 (av. 11.50) days, while, that of females ranges from 16.0 to 22.0 (av. 19.1) days (Prakash and Bheemanna, 2014).

Management/Control measure

Biocontrol

Use of natural enemies: Ravi and Patil, (2008) reported that the peak incidence of *Geocoris* sp. was observed during November (2.0 bugs/plant) and also noticed spiders feeding on nymphs and adults of brown mirid bug.

Chemical control

The isecticides, Fipronil 5 SC, Acephate 75 SP, Acetamiprid 20 SP and Imidacloprid 200SL are found effective to control mirid bugs (Udikeri *et al.*, 2008; Venkateshalu *et al.*, 2015). However, amongst them, fipronil and acephate are very effective. In the study, Fipronil 5 SC @ 1.0 ml/l has recorded significantly lowest mirid bugs population at 3 days after second spray. The next best treatments were acephate 75 WP @ 1 g/l and fipronil 5 SC+table salt @ 0.5 ml+5.0 g/l (Venkateshalu *et al.*, 2015).

ii) Yellow mirid (*Campylomma livida*)

Occurrence & severity:

Yellow mirid (*Campylomma livida*), popularly known as Dimple bug or Apple dimpling bug, is an indigenous species of Australian mirids (Lloyd *et al.*, 1970). However, there are certain disputes on its initial occurrence. According to Schuh (1984), *C. livida* was originally described from India and later recorded from Sri Lanka and the Philippines. Further, *C. livida* does not occur in Australia (Malipatil, 1992). The yellow mirid (*C. livida*) is considered as a pest of cotton and occurs regularly in cotton fields (Evenson and Basinski, 1973; Bishop, 1980; Adams and Pyke, 1982; Adams *et al.*, 1984; SIRATAC, 1982-1986). Apart from this, the yellow mirid both at its nymphal and adult stages is predacious on eggs of *Helicoverpa* spp. in cotton fields (Room, 1979a) and the adults prey on young *Heliocoverpa* caterpillars (Room, 1979b). There is also evidence that this mirid species is predacious on mites in orchards (Readshaw, 1975). The yellow mirid is also polyphytophagous. The plants on which the damage by this bug has been well documented are apple, cotton, sunflower, lucerne, rose and other ornamentals (Malipatil, 1992). In India, it mostly damages cotton in Maharastra. In an investigation in Queensland on the damage caused to cotton by *C. livida* and *Creontiades dilutus* (two mirid bugs occurred naturally), there were 54% fewer squares in 1976-77 and 35% fewer in 1977-78 on untreated plants in the field than on plants protected against the mirids with a mixture of DDT and toxaphene (Bishop, 1980).

Types of damage

The yellow mirid, both at its nymphal and adult stages, causes significant damage to squares and bolls of cotton. Feeding on the terminal growth, squares, flowers and bolls of cotton plants with the piercing/sucking mouthparts causes excessive shedding of flowers, small squares and immature bolls (Nagrare *et al.*, ND). The feeding results in small, dark, sunken lesions on the surface of the boll and in severe cases deformed bolls are formed due to lack of fertilisation of some ovules. This symptom is often referred to as "parrot beaking" of bolls.

Scientific name & order: *Campylomma livida* Reuter, 1885 (Hemiptera: Miridae)

Synonym

= *Campylomma morosa* Ballard 1921

= *Ragmus morosus* Ballard 1921

Biology

Egg: Eggs are cylindrical, slightly recurved and laterally compressed; shining white in colour, turning to yellow as it matures. These are laid singly, preferentially on the leaf petiole. Eggs hatch within 4 to 5 days (Kranthi *et al.*, 2009).

Nymph: There are five nymphal instars, each of about 2 to 3 days of duration. The wing pads start to develop at the 3rd instar (Kranthi *et al.*, 2009).

Adult: Adult is macropterous, small oval, with body length 2.59-2.60 mm and width 1.00-1.10 mm. long, with long legs and antennae (Kranthi *et al.*, 2009), pale yellow coloured bugs. Head is pale yellow, more than twice broader than long, weakly rounded in front, frons declivous anteriorly, posterior margin weakly produced behind on pronotal margin. Eyes are brown, occupying entire height of head in lateral view. Antennal fossula sponge is separated from eyes. Antennae are segmented, pale yellow, with short, stout, black 1st segment and basal black ring at 2nd segment, labium reaching metacoxae. Pronotum is greenish yellow, wider than long, lateral margins rounded, linear with brown spine-like setae on anterolateral margin, posterior margin weakly truncate. Metathoracic scent gland evaporatory area is pale yellow with prominent opening. Mesoscutum is exposed, scutellum triangular, flat. Hemelytra is pale yellow, ovoid, with short golden yellow setae, cuneus longer than broad, membrane hyaline, with two cells, inner cell small triangular and outer cell elongate. Male genital capsule is with projection on left side ventrally Legs, meso and meta femora are with short spine-like setae with their bases black, tibiae long, with black spines. Parempodiais is setiform, weakly convergent apically, pulvilli small. Male genitalia: Pygophore is broad, conical, with thumb-like lateral process towards left side. Phallotheca is weakly C-shaped, with broad base and narrow apex. Endosoma is S-shaped, elongate, terminated by two blades, varying in length, with small spicules (Yeshwanth, 2013). The adults can live for 3 to 5 weeks and a female can lay up to 80 eggs in the life time (Kranthi *et al.*, 2009).

iii) Green mirid (*Creontiades dilutus*)

Occurrence & severity

The green mirid is endemic to Australia, and found throughout the continent including Tasmania (Hill, 2017). This insect is found throughout the hot and arid interior of the continent and is particularly abundant in these regions during winter at southern hemisphere. In summer months the interior of the country is very hot and dry and there are very few plants available for green mirid to feed on. Thus, in the summer conditions it migrates to the eastern cropping regions. These regions are ideal for mirid growth and development and, thus, large populations are built up. The green mirid is highly polyphagous having been recorded from over 100 host plants, mostly in Fabaceae (Hereward and Walter, 2012). This bug causes considerable damage to many agricultural crops, including cotton, lucerne, and soybean (Malipatil and Cassis, 1997; Hereward, 2012). In cotton, currently, particularly after the introduction of Bt cotton, green mirid is the most serious insect pest in Australia. Since, prior to the introduction of Bt cotton, broad spectrum insecticides were used to control *Helicoverpa armigera* and that incidentally controlled green mirid bugs. The introduction of Bt cotton has greatly reduced pesticide application in cotton crop. As such, the population of mirid bug has increased and it becomes the primary target of pesticide application in cotton.

Types of damage

Adults and nymphs pierce cotton plant tissue and release a chemical that destroys cells in the feeding zone. The growing points are killed, resulting in increased branching. Squares, buds, flowers and small bolls are shed on the insect feeding that decreases yield potential. Boll feeding in cotton reduces lint yield and quality.

Scientific name & order: *Creontiades dilutus* **Stal (1859) (Hemiptera: Miridae)**

Biology

Egg: Eggs are laid singly within plant tissue, making them virtually impossible to see. The only visible evidence is an oval shaped operculum (egg cap) which used to 'breathe'. The operculum is light brown in colour. The respiratory horns are short and thick on egg caps of green mirids (Khan and Quade, 2008).

Nymph: Newly hatched nymphs are pale or light green in colour, changing to yellowish green with age, about 1.4 to 1.6 mm in length with long antennae and legs. Nymphs have pear-shaped body. Antennae are four segmented, with red or reddish brown tip. The second instars are 2.1 to 2.4 mm in length, while, 3rd instars are 3 to 3.2 mm long. The wing pads start to develop at the 3rd instar. The 4th instar nymph is about 4.4 to 4.6 mm long. In the 5th instar, the green mirid develops a light brown tinge on the hind legs. The size of this nymph is about 5.7 to 5.9 mm long (Khan and Quade, 2008).

Adult: Adults are elongated, about 7 to 9 mm in length, light yellow-green colour, often with red markings, with clear wings folded flat on the back, long legs and antennae. Antennae are nearly as long as the body. Distal half of the femur of hind legs is tinged with light brown. The ring behind the head (collar) is white. The scutellum is heart-shaped (Khan and Quade, 2008).

Life cycle

The Mirid lays eggs singly, preferably on the leaf petiole within a crop. The egg is inserted into the plant tissue with an oval egg cap projecting above the leaf or petiole surface. Egg hatches after 7-10 days depending on temperature; at 30-32° C (average temperature) egg hatches within 4-5 days and there are five nymphal instars, each of about 2-3 days duration. Under summer conditions, a generation (egg to adult) can be completed in about three weeks. At 30°C, green mirid takes 9 to 11 days to develop from 1st instar to adult. When the weather remains cloudy and temperatures are around 32°C for a few days, green mirid population explodes within a short time frame, faster than when temperatures are cooler or hotter. Adults can live for 3-4 (-5) weeks and a female can lay up to 80 eggs in her life time (Khan and Quade, 2008).

Molecular studies

In order to determine taxonomic identity of *C. dilutus* molecular characterization is used as a tool. Earlier, it was believed that *C. dilutus* had been recorded in the USA during 2006, but that is not established in molecular studies. Hereward (2012) has used sequence data (Cytochrome Oxidase 1 (CO1) and 28S ribosomal gene) and established that the insects concerned are highly unlikely to be *C. dilutus*. Subsequent taxonomic work has also confirmed that the USA species is indeed a separate species, *Creontiades signatus* (Hereward, 2012). Using *C. signatus* as an out-group, further phylogenetic analyses show that *C. dilutus* and *C. pacificus* are well differentiated according to the sequence of both genes. The CO1 sequences also indicate low levels of genetic diversity in *C. dilutus*, especially in comparison to *C. pacificus*. The low CO1 diversity indicates that more variable markers will be required for further analyses of gene flow in this species, and consequently 12 microsatellites are developed by enrichment (Hereward, 2012). The complete mitochondrial genome of the green mirid, *C.dilutus*, comprises 15,864 bp and has a GC content of 22.3%. The layout of the 13 mitochondrial protein-coding genes follows the ancestral insect arrangement, and 22 tRNA's are detected as well as the small and large rRNA's. Phylogenetic analysis of available mirid mitogenomes places *Creontiades* closer to *Adelphocoris* than the other four genera (Hereward, 2016).

Management/Control measures

Use of trap crop

In agricultural areas the highest densities are found in Lucerne and this observation lead to the suggestion that Lucerne can be grown next to cotton as a trap crop for green mirids (Mensah.and Khan, 1997).

Use of pheromone trap

Synthetic pheromone trap as prepared by mixing of hexyl hexanoate and (E)-2-hexenyl hexanoate in 5:1 ratio is the most effective mean in field trapping of green mirids that come to pheromone traps only between 18:00 and 06:00 h, due to their nocturnal habit (Lowor *et al.*, 2009).

Chemical control

In Australia, recent trials have shown that reduced rate of Indaxacarb or Fipronyl combined with salt rendered effective control of mirid bugs (Khan, 2003).

iv) Tarnished plant bug (TPB)

Occurrence & severity

The tarnished plant bug, *Lygus lineolaris*, is most commonly found in the eastern half of North America. This is a very serious pest of cotton in Alabama. Damage caused by plant bugs may occur any time during the fruiting season. The occurrence of the tarnished plant bug is now common all over the state, infestations are more consistent, and serious damage is as apt to occur in July as in June (Freeman, 2011). The first plant bug damage to cotton usually occurs in early to mid-June and is caused by migratory adults entering fields. Plant bugs may also damage large squares, especially as nymphs become common in late June and July. This insect can be found across North America, from northern Canada to southern Mexico. The insect causes significant yield losses in cotton (*Gossypium hirsutum*). It is the principal mirid pest of this crop in the eastern and southern USA (Schwartz and Foottit, 1992). Yield losses of up to 32% by *L. lineolaris* were reported in Mississippi, USA, with the highest insect densities recorded delaying crop maturity by two weeks (Baker *et al.*, 1993). Yield reductions of 42 and 55% were reported in plots with high infestations in Arkansas (Tugwell *et al.*, 1976). Losses of around 21% were attributed mainly to *L. lineolaris* in the control plots of insecticide trials in Mississippi (Scott *et al.*, 1986). The tarnished plant bug is a highly polyphagous species. Its foodplants include a wide range of crops throughout Canada, the USA and Mexico (Young, 1986).

Types of damage

The tarnished plant bug affects leaves and fruiting structures (squares, bolls and blooms) of cotton, causing considerable yield losses. It can damage cotton throughout most of the growing season; however, economic damage is most likely to occur during the period from first square through to early bloom. It causes damage primarily by feeding on small squares. After being fed upon, the small squares turn yellowish, dry up, turn dark brown, and finally, cause them to abort ('blasted squares') within a few days, leaving an abscission scar (Layton, 1995). In addition to feeding on pinhead squares in June, plant bugs sometimes feed in the terminal of the plant. High plant bug (*L. lineolaris*) populations can further reduce yields by delaying maturity and altering fruiting patterns, while the injection of salivary enzymes also disrupts the growth of plant tissue (Layton, 1995). When the toxins are injected into the meristematic tissue of the plant terminal, devastating physiological changes occur. Heavy infestations can sometimes kill terminal shoots, leading to a loss of apical dominance and the development of numerous secondary terminals ('crazy cotton'). This is generally of limited economic importance, but can delay fruiting and boll maturity (Scales and Furr, 1968; Hanny *et al.* 1977). Plant bugs may also damage large squares, especially as nymphs. Feeding on larger squares at the bloom stage can damage developing anthers. When less than 30% of the anthers are damaged there is little or no effect, but malformed bolls result from higher damage levels (Layton, 1995). On opening the bloom, darkened anthers and warty spots can be seen on the petals. This type of damage is referred to as a "dirty bloom". The bugs prefer soft and immature bolls. Boll injury appears as small, dark sunken spots on the outside of the boll, while, internally, the damage appears as brownish discoloration on and near the developing seeds. A warty growth may also be present where the bug penetrated the boll wall. Feeding on the bolls themselves may cause seed damage, discoloration of lint and decreased lint weight (Bacheler *et al.*, 1990; Layton, 1995). Severe boll feeding can cause the young bolls to shed, but, more often, localized lint and seed damage causes hardlocked bolls (Freeman, 2011).

Scientific name & order: *Lygus lineolaris* **Palisot de Beauvois, 1818 (Hemiptera: Miridae)**

Synonyms

= *Capsus flavonotatus* Provancher, 1872

= *Capsus lineolaris* Palisot de Beauvois, 1818

= *Capsus oblineatus* Say, 1832

= *Capsus strigulatus* Walker, 1873

= *Lygus pratensis* var. *rubidus* Knight, 1917

Biology

Egg: Tarnished plant bug eggs are laid in plant tissue and require 1 to 2 weeks to hatch (Freeman, 2011).

Nymph: The five nymphal stages, which are yellowish-green or green with 5 black dots on the back, are completed in 2 to 3 weeks (Freeman, 2011). The 1st and 2nd instars look like aphids, but are more active and move faster. The 3rd-5th instars larvae have five black dots and developing wing pads on their backs (liu et al., 2003).

Adult: The tarnished plant bug is a typical plant bug with piercing-sucking mouthparts. The adult is predominantly brown with accents of yellow, orange or red, with a light-coloured or pale yellow "V" shape on the scutellum on the back (dorsal) and measures in length 3.7-4.6 mm (male) and 3.8-5.0 mm (female) (Schwartz and Foottit, 1998; Liu et al., 2003; Freeman, 2011).

Life cycle

The tarnished plant bug overwinters as an adult in the north of America but active all year in the south (Young, 1986). It utilizes cotton plants as one of its main reproductive hosts. Female lays eggs in the first row of cotton plants and later occupies more plants in the field (Jones and Allen, 2013). The female usually lays eggs in May after the overwintering period. The eggs hatch in 10-21 days and nymphs begin to develop around June (Day, 2006). Nymphs develop through 5 instars in 3 weeks; summer life cycle from egg to adult takes approximately 4 (3- 5) weeks. Summer adults live 1-2 months. There are 3 generations per year in the north, and more in the south (Young, 1986; Freeman, 2011).

Management/Control measures

Use of host resistance

Use of nectariless cotton varieties may be beneficial than conventional varieties, since, nectariless cotton varieties have resistance to *L. lineolaris* (Shepherd *et al.*, 1986; Scott *et al.*, 1988), giving higher yield due to reduced damage by the plant bug (Hardee and Bryan, 1997; Platt and Stewart, 1999). However, the yields in a nectariless cotton variety are not better than in a Bt transgenic variety (Platt and Stewart, 1999).

Cultural control

Mowing and maintenance of weed plants can control for the population of the adult pest within the fields (Fleury *et al.*, 2010). Further, use of white sticky traps has been found as the most effective measure in collecting the tarnished plant bug (Legrand and Los, 2003).

Chemical Control

Chemical control of the plant bug is considered when pinhead square damage reaches 20 percent. Since, it is estimated that when damage exceeds 20 percent, plant bug populations are usually greater than 50 bugs per 100 row feet, and adults are easily observed (Freeman, 2011). The plant control is being done by a range of insecticides. Application of insecticides 3–5 times each year is practiced to control this insect. Imidacloprid , which is a insecticide of neonictinoid family, has been used to control population of *L. lineolaris*. This insecticide causes interference and blockage of the nicotinergic pathway in the central nervous system of insects (Zhu and Luttrell, 2014). In an experiment in the Mississippi Delta, early season multiple-insecticide applications (imidacloprid) provided a 13.8% yield increase over control, in situations where *L. lineolaris* was regarded as one of the major pests (Hopkins and Donaldson, 1996). The chemical control measure has been found as effective in many cases, but not in all due to insecticide resistance and/or inadequate coverage of plants (Snodgrass and Elzen, 1995).

9. Pink Bollworm

Occurrence & severity

Pink bollworm (*Pectinophora gossypiella*) is native to Asia, but has become an invasive species in most of the world's cotton-growing regions. This bollworm species is considered as one of the most destructive insect pests of cotton worldwide (Henneberry and Naranjo, 1998). The pink bollworm was firstly described by Saunders in 1842 as *Depressaria gossypiella* in India (Hennereberry, 2007). However, the origin of this bollworm remains unknown

(Ingram, 1994). Nevertheless, it is assumed that the origin of this pest is Indonesia-Malaysia (Common, 1958; Wilson, 1972). It is believed that pink bollworm was at first introduced into Egypt in 1906-1907 within the infected seeds imported from India. Between 1911 and 1913 it was spread from Egypt to other countries such as Mexico, Brazil and Philippine islands (USDA-APHIS 1977). It was recorded by nearly all the cotton growing countries of the world (CAB Institute of Entomology, 1990). It had reached the cotton belt in the southern United States by the 1920s. Texas was its first landfall in the U.S. in 1917. Pink bollworm was found in eastern Arizona by 1926, and the main cotton producing region of central Arizona 3 years later (Spears 1968, Noble 1969). It is a major pest in the cotton fields of the southern California deserts in the south-western United States (Henneberry and Naranjo, 1998). It is distributed throughout tropical America, Africa, Asia, Australasia, including subtropical regions, Pakistan, Egypt, USA (Arizona) and Mexico. Schwartz (1983) calculated that the potential loss, without control, was 61% due to pink bollworm in the USA and estimated losses of 9% were suggested where the pest was controlled. Frisbie *et al.* (1989) indicated that economically damaging thresholds were reached if boll infestation rose above 5-15%. Agarwal and Katiyar (1979) calculated 20.2% crop loss due to pink bollworm based on field trials in Delhi. In the Yangtze valley (China), Cai *et al.* (1985) reported that the pink bollworm reduced cotton yield by about10%. In the Wuhan region, it reduced fibre yield by 17-26% (Luo *et al.*, 1986). Yuan and Wu (1986) estimated specific losses due to this bollworm species (*P. gossypiella*) of 0.0467 g/larvae due to direct injury and 0.0544 g of cotton due to indirect damage in the Shanghai region. In Sudan, Darling (1951) estimated that 10.7% of the potential cotton yield could be lost following infestation by pink bollworm. The attack of pink bollworm in cotton is severe in India when the maximum temperature is greater than 33° C, morning relative humidity less than 70 % and evening relative humidity greater than 40 % during the standard weeks of 40, 41 and 43, and less than 12°C minimum temperature between the standard weeks 48 and 49 (Vennila *et al.*, 2007b). The pink bollworm (*P. gossypiella*) is an oligophagous pest. Apart from cotton, it also affects okra, mesta/kenaf/deccan hemp (*Hibiscus cannabinus*) and roselle (*H. sabdariffa*) in Egypt. Further, it prefers okra over cotton towards the end of the season when the cotton boll surface becomes hard (Khidr *et al.*, 1990).

Types of damage

Pink bollworms damage squares and bolls, the damage to bolls being the most serious. The female moth lays eggs in a cotton boll and when the larvae emerge from the eggs they inflict damage through feeding. The emerged larvae when attack the bud of less than 10 days old, shedding of bud occurs and the larvae die. But with older bud, larvae can complete development. The larvae burrow into bolls, through the lint, to feed on seeds. As the larvae burrow within a boll, lint is cut and stained, resulting in severe quality loss. Larvae in flower bud spin webbing that prevents proper flower opening leading to "rosetted-bloom". Ten to twenty days old bolls are attacked from under bracteoles. Larvae feed on the developing seeds. While in younger bolls entire content may be destroyed, in older bolls development could be completed on three to four seeds. Interloculi movement is also seen. Several larvae can infest a single boll (Vennila *et al.*, 2007b). Under dry conditions, yield and quality losses are directly related to the percentage of bolls infested and the numbers of larvae per boll. With high humidity, it only takes one or two larvae to destroy an entire boll because damaged bolls are vulnerable to infection by boll rot fungi (UC-IPM, 2015).

Scientific name & order: ***Pectinophora gossypiella*** **(Saunders, 1844) (Lepidoptera: Gelechiidae)**

Synonyms

= *Depressaria gossypiella* Saunders, 1844

= *Gelechia gossypiella*

= *Platyedra gossypiella*

= *Ephestia gossypiella*

= *Gelechiella gossypiella*

= *Gelechia umbripennis* Walsingham, 1885

Biology

Egg: Eggs are pearly iridescent white, elongated-oval, very small, 0.4-0.6 mm in length and 0.2-0.3 mm wide, flattened at the base and sculptured with longitudinal lines. They are usually laid under the calyx of green bolls singly, or in groups of (4-) 5-10 (Vennila *et al.*, 2007b; UC-IPM, 2015).

Larva: The new born larva is a dull white or pale coloured tiny caterpillar with dark brown heads, and measuring about 1-2 mm long when they first hatch. Larvae usually pass through four instars. The first and second instars are difficult to see against the white lint of the bolls. Pink colouration develops from third instar onwards and mature larvae are 12-15 mm long with a prominent pinkish coloration, having wide conspicuous transverse pink bands along its dorsum. The caterpillar is with eight pairs of legs (Vennila *et al.*, 2007b; UC-IPM, 2015).

Pupa: Pupae are light brown when fresh, gradually become dark brown or reddish-brown as the pupation proceeds. Pupa measures up to (7-) 8-10 mm in length (Vennila *et al.*, 2007b; UC-IPM, 2015).

Adult: Adults are small, greyish brown, inconspicuous moths with blackish bands on the forewings and the hind wings are silvery grey. When their wings are folded, they have an elongated slender appearance. The wing tips are conspicuously fringed (UC-IPM, 2015). The adults are measuring about 12-20 mm across the wings (USDA, 1948). The head is reddish brown in colour with pale, iridescent scales. Antennae are brown and the basal segment bears a pecten of five or six long, stiff, hair-like scales. The labial palpi are long and curved upwards: the second segment bears a slightly furrowed hairy brush on the underside that becomes smooth distally and the terminal segment is shorter than the second. The proboscis is scaled. Forewings are elongated-oval, pointed at the tips and bearing a wide fringe. The ground colour of the forewings is brown and they have fine dark scales that form vague patches in the region of the medial cells and at the wing base. The hind wings are broader than the fore wings and silvery grey with a darker, iridescent hind margin. Legs are brownish black. The abdomen is ochreous toward the upper side, dark brown laterally and covered with ochreous-brown scales on the underside. In the genitalia, the male uncus is broad at the base, tapering to a point and the aedeagus has a hooked tip. The female ovipositor is weakly sclerotized (USDA, 1948).

Life cycle: Eggs are laid in any of the sheltered places near the cotton bolls at the time of flowering. Once the bolls are 15 days old, these become favoured sites for oviposition. Young larvae emerge after 3-5 (-6) days, entering the cotton bolls shortly after emergence where they feed internally within the pod, making a small hole to the exterior to allow air to penetrate. Larvae usually pass through four instars. Larval cycle lasts for 9-14 days in hotter regions. The mature larvae are either 'short-cycle' and will go on to pupate or 'long cycle' to enter a state of diapause. In short cycle, after the larvae emerge from the top of the boll. Pupation takes place in the ground, about 50 mm below the surface (Saunders, 1843) and adults emerge after about 9 (or 8-13) days. The life cycle is completed in 25-31 days (Green and Lyon, 1989). There may be four to six generations per year. In long cycle, the larvae enter diapause, spin a tough thick walled, closely woven, spherical cell referred as " hibernaculum" with no exit hole. Always, the longterm larvae occur during end of crop season, where there are mature bolls present and larvae often form their hibernaculae inside seeds. Hibernacula may occupy single seeds or double seeds. The pink bollworm hibernates as full fed larvae during cold weather. Diapause larvae often spin up in the lint of an open boll and if still active in ginnery, will spin up on bales of lint, bags of seed or in cracks and crevices. Moths emerging from the hibernating larvae are long lived with females and males alive for 56 and 20 days, respectively (Vennila *et al.*, 2007b).

Molecular characters

The population genetic structure, distribution, and genetic diversity of *P. gossypiella* in cotton growing zones of India using partial mitochondrial DNA cytochrome oxidase-I (*COI*) gene was studied (Sridhar *et al.*, 2017). The overall haplotype (*Hd*), number of nucleotide differences (*K*), and nucleotide diversity (π) were 0.3028, 0.327, and 0.00047, respectively which suggest that entire population exhibited low level of genetic diversity. Zone-wise clustering of population revealed that central zone recorded low level of *Hd* (0.2730) as compared to north (0.3619) and south (0.3028) zones. The most common haplotype (H1) reported in all 19 locations could be proposed as ancestral/ original haplotype. This haplotype with one mutational step formed star-like phylogeny connected with 11 other haplotypes. The phylogenetic relationship studies revealed that most haplotypes of populations are closely related to each other. Haplotype 5 was exclusively present in Dharwad (South zone) shared with populations of Hanumangarh and Bathinda (North zone). From this study, it is also suggested that there is a low genetic diversity among the pink boll worm populations of India and the most common haplotype (H1) is the ancestral haplotype of all other haplotypes (Sridhar *et al.*, 2017).

Resistance

The deployment of Bt cotton which was introduced in 1996 for selective control of caterpillars led to dramatic regional reductions in abundance of pink bollworm, and associated crop damage and insecticide use (Naranjo and Ellsworth, 2010). However, later in parts of India, the pink bollworm is now resistant to first generation transgenic Bt cotton (Bollgard cotton) that expresses a single Bt gene (Cry1Ac) (Bagla, 2010). It is assumed by Monsanto scientists that the Bollgard II product has no additional toxin to combat pink bollworm, since, Bt toxin is active for 90 days only, while pink bollworm is a late season pest (Jebaraj, 2010). However, all transgenic varieties of cotton give a better yield than conventional strains in China (Lin, 2000) and Mexico (Godoy-Avila *et al.*, 2000). A problem related to this is the resistance developed by the larvae. A strategy for overcoming this as suggested is the use of refuges in which nontransgenic plants are planted nearby (Tabashnik *et al.*, 1999), or else a strategy of interplanting one nontransgenic row in five is used (Simmons *et al.*, 1998).

Management/Control measures

Quarantine measure

Pink bollworm is considered as a quarantine pest in the USA and Russia.

Use of resistance

Much recent research has centred on the use of genetically manipulated cotton which enables pest larvae to become infected with the *Bacillus thuringiensis* toxin (Perlak *et al.*, 2001).

Cultural control

Several cultural methods are suggested by different workers during different time. It is suggested to destroy cotton sticks after harvest or remaining bolls, burn the affected bolls under heaps of cotton sticks the debris of cotton ginning factories, to turn the heaps over to expose bolls to sunlight after the final picking, let livestock graze on unwanted bolls. This will help reduce the pest attack next season. It is suggested to expose the seed of the next crop to sunlight (sun drying) for 8 hours one week before sowing irrigation should be avoided after 30th September. Deep ploughing of the field after harvest by deep furrow turning plough is effective to kill the pupae and overwintering larvae. Application of nitrogenous fertiliser after 15th of August should be avoided. Late planting of crops has been used as a cultural control method where the end of diapause is triggered by day length. Larvae that emerge before the crop is ready then have no food supply (Frisbie *et al.*, 1989). Use of crop rotation, giving a 3-year break from cotton production has also been used successfully in areas not highly susceptible to constant re-infestation from other areas (USDA, 1948). In the mechanical method, the stalks and the dried infected bolls are collected manually and are disposed by burning to destroy the larvae within the bolls. This method killed about 90% of the larvae attached to the stalks (Noble, 1969). However, the disadvantages of this method are that collecting the infected bolls is a laborious work.

Biological Control

The parasitoid, *Bracon kirkpatrick*, is only found as effective to control pink bollworm populations effectively (Sankaran, 1974; Greathead, 1989). More recently nematodes have been used as control agents in the USA (Henneberry *et al.*, 1996; Gouge *et al.*, 1999). Further, it is suggested to use *Trichogramma chilonis* cards to destroy eggs before they hatch. Use 10-15 cards per acre.

Pheromonal Control

Synthetic pheromones have been employed extensively in the detection and control of *P. gossypiella*. Trapping with the synthetic pheromone gossyplure has been widely used and is reported to have resulted in a 60-80% reduction of the pest population in China (Gao *et al.*, 1992). Pheromone trapping has also been used in India for an attempted eradication programme (Simwat *et al.*, 1988). The pheromone has been found to enhance the efficacy of insecticides in India (Dhawan and Simwat, 1993). Busoli (1993) reported that the use of the sex pheromones gossyplure and virelure were more economically viable than the use of conventional insecticides. Early-season use of pheromone coupled with insecticides applied at low thresholds is generally most profitable, especially at low pink bollworm population densities (Frisbie et al., 1989). Moreover, early trials using gossyplure to saturate the cotton environment with pheromone in an attempt to disrupt the location of females by males proved inconclusive (Frisbie et al., 1989). In Egypt, Boguslawski and Basedow (2001) found that mating disruption gave more effective control than insecticides.

Chemical control

Although, insecticidal control is hindered by the internal feeding of larvae and development of resistance against the insecticides, there several insecticides are found as effective against the pest. The effective insecticides as reported by different workers are as, asymethrin (Dhawan et al., 1992), chlorpyriphos (Mahar et al.,1987; Dhawan et al., 1989; Green and Lyon, 1989), synthetic pyrethroids (Dhawan et al., 1990a; Butter et al., 1990; Gao et al., 1992), fenpropathrin (Mahar et al., 1987), teflubenzuron (Green and Lyon, 1989), carbaryl (Gao et al., 1992), cyhalothrin and fluvalinate (Thangaraju et al., 1993) and fenvalerate (Tadas et al., 1994). Application of insecticides may be done as singly or in combination of both systemic and protectant ones. In addition, the insecticides Chlorantraniliprole @ 50 ml/acre, Spinetoram @ 100-120 ml/acre, Lambda-cyhalothrin @330 ml/100l of water and Spinosad @ 80ml/100l of water are also effective to control the pest. These insecticides may be reapplied after 7-10 days to get satisfactory protection,

Integrated Pest Management

Combinations of biological and chemical controls have also proved successful. Tuhan et al. (1987) found that application of *Trichogramma brasiliense* in combination with chemical insecticides gave good control of pink bollworm in India, and *Bacillus thuringiensis* has been found to be effective in combination with chemical insecticides in Egypt (Hussein et al., 1990). Cultural controls, intensive monitoring with pheromone baited traps for adult males, boll sampling and pheromone applications for mating disruption may also be able to augment the efficacy of pink bollworm control and to reduce insecticide use (Walters et al., 1998). Moreover, the control methods such as biological control, cultural, behavioral, genetic engineering and host plant resistance can be integrated together in harmony for integrated pest management system (IPM) to control *P. gossypiella* (Henneberry, 2007).

10. Red cotton bugs

Red Cotton Bugs (*Dysdercus* spp.), also called as 'cotton stainers' or 'red cotton stainers', are thought to be the most serious pests of cotton (van Doesburg, 1968). A number of species of the genus *Dysdercus* are found as serious pests of cotton all over the world as and where cotton is cultivated and they prevailed. These cotton stainers damage cotton plants by sucking the sap and destroy the cotton bolls by staining them with excrement. In piercing the boll they introduce microorganisms, like, cotton staining fungus, *Nematospora gossypii*, which cause the bolls to rot, or the lint to become discoloured; this greatly reduces yields in cotton-growing countries (Maxwell-Lefroy, 1908; Freeman, 1947; Van Doesburg, 1968; Fuseini and Kumar, 1975; Iwata, 1975; Ahmad and Kahn, 1980; Ahmad and Schaefer, 1987; Yasuda, 1992).

i) Red Cotton Bug (*Dysdercus koenigii*)

Occurrence & severity

The species, *Dysdercus koenigii*, is a serious pest of cotton crops, although cotton is not its preferred choice of hosts. It is commonly found in India (Panizzi and Grazia, 2015), Pakistan and south-eastern Asia (Waqar *et al.*, 2013). In Pakistan, in year 2011, there was severe lint staining problem and decrease in market price followed by conversation on causes of cotton staining. It is thought that major cause of cotton staining is red cotton bug, *D. koenigii* (Waqar *et al.*, 2013) and considered as a destructive pest of cotton due to severe lint staining problem (Ahmad and Mohammad 1983; Ashfaq *et al.*, 2011).

Types of damage

The adults and older nymphs feed on the emerging bolls and the cotton seeds as they mature, transmitting cotton staining fungi as they do so (Waqar *et al.*, 2013). Affected bolls open badly with their lint stained with the excreta or body juices.

Scientific name & order: *Dysdercus koenigii* **(Fabricius, 1775) (Hemiptera: Pyrrhocoridae)**

Biology

Egg: Eggs are oval and creamy white which turns to yellowish orange before hatching or creamy-yellow in colour (Kamble, 1971).

Nymph: Nymph has five growth stages (instars). The first instar nymphs emerge as light orange which turns to blood red within one day. Body length is about 1.58 mm and width 0.94 mm. The antennal length is shorter than the body and wing pads are absent. The second instar nymphs are similar in appearance to that of first instar but with body length 3.02 mm and width 25 mm. The third instar is different from first and second instar because of emergence of wing pad on thorax. Initially moulted nymphs are orange red in colour which changed to reddish colour within one day. Three pairs of unclear dorsal spots are developed on the abdomen. The fourth instar is crimson and cylindrical with larger, darker wing pads and the fifth instar is similar to fourth one, but with prominent dark wing pads, black antennae and legs. The average body size of fifth instar nymph is 12.22 mm in length and 4.98 mm in width (Waqar *et al.*, 2013).

Adult: The adult insect is also crimson red in colour, with a pair of black spots on the forewings. Hind wings are membranous and broader than fore wings. The membranous hind wings are concealed under the forewings when the insect is at rest. Males are about 14 mm (0.6 in) in length and females are a little larger (15.30 mm in length). Fore wings in female are measured 11.22 mm, while in male they are measured 10.04 mm (Waqar *et al.*, 2013).

Life cycle: The female has pre-oviposition period is about 7days (Waqar *et al.*, 2013). It usually lays three batches of seventy or eighty eggs in damp soil, under plant litter or in crevices (Shuklaand Upadhyay, 2006). The eggs take about six days to hatch. The nymphs pass through five instar stages and become winged adults after fifty to sixty days. The adults survive for a few weeks (male lives for about 27-56 days and female lives for 32- 61 days at $28\pm2°C$; according to Waqar *et al.*, 2013) and there are several generations throughout the year (Kamble, 1971). The adults and later stage instars feed on immature cotton bolls and on the developing and ripening seed. By their presence in the bolls, they admit fungi such as *Eremothecium gossypii* which indelibly stains the lint (Kamble, 1971).

Control measures

The insecticides, like, Pyrethrum powder, Alpha cypermethrin-based products, Lambda-cyhalothrin-based products and Cypermethrin-based products may be effective to control the red cotton bug.

ii) Red Cotton Bug (*Dysdercus cingulatus*)

Occurrence & severity

The red cotton bug (*Dysdercus cingulatus*) is a species of true bug, commonly known as 'red cotton stainer'. The species is thought to be the most serious pest of cotton in the Southeast Asian countries (van Doesburg, 1968; Kohno and Bui Thi, 2004). The species is native to two of the major cotton producing countries, like, China and India. It is distributed in all the cotton growing regions of India (David and Ananthakrishnan, 2004; Sahayaraj and Ilayaraja, 2008).It also occurs in Sri Lanka, north-eastern India, Bangladesh, Thailand, the Philippines, Sumatra, Borneo, Papua New Guinea and northern Australia (Schaefer and Panizzi, 2000). Apart from cotton, the red cotton bug feeds on a number of other crop plants including okra, musk mallow, hibiscus, white jute, citrus and maize. It also attacks trees including silk cotton tree (*Bombax ceiba*), kapok (*Ceiba pentandra*), teak (*Tectona grandis*) and the portia tree (*Thespesia populnea*)

Types of damage

Both adults and nymphs suck the sap from leaves, green bolls and seeds of partially opened bolls (Vennila et al., 2007c). However, *D. cingulatus* is principally a seed feeder (Van Doesburg, 1968; Ahmad and Schaefer, 1987). They pierce the outer covering of the seeds and suck the plant saps within the seed. Constant feeding of the seeds may weaken them, causing it to dry up eventually killing the cotton bolls. Affected bolls open badly with their lint stained with the excreta or body juices, this behaviour lends the red cotton bug infamous title "cotton stainer" (Schaefer and Panizzi, 2000). This also results in major economic losses (Grazia et al., 2015). Quality of the lint is affected and the attacked seeds become unfit for either sowing or oil extraction. Boll rot is caused by the secondary infection due to bacteria wherein rotting of the entire contents of the boll occur following the initial discolouration of the lint to yellow or brown (Vennila et al., 2007c). They may also inadvertently introduce the fungus *Atospora gossypii* which results in internal boll disease (Frazer, 1944).

Scientific name & order: *Dysdercus cingulatus* Fabricius, 1775 (Hemiptera: Pyrrhocoridae)

Synonyms

Dysdercus megalopygus Breddin, 1909

Cimex cingulata Fabricius, 1775

Lygaeus cingulatus (Fabricius) Fabricius, 1794

Biology

Egg: Eggs are laid singly or in small, loose clusters in moist soil or in crevices in the ground. They are spherical, shiny, smooth surfaced and pale yellow in colour (Vennila et al., 2007c; Verma et al., 2013).

Nymph: Nymphs after hatching are wingless pale orange which later turns red coloured with black median dorsal spots on the inter-tergal membrane of 3/4, 4/5 and 5/6 abdominal segments. There is a pair of white dorsal spot on each of the third, fourth and fifth tergal plates on the abdomen. Nymphs moult 5 times with wings developing from the third instar; the fifth and final moult gives rise to fully mature adult with developed wings (Vennila *et al.*, 2007c; Verma *et al.*, 2013).

Adult: Adults (Fig. 29B) are 12-13 (-18) mm in length and have deep red legs, an elongated sucking mouth part and antennae. Forewings are half membranous and half sclerotized (Grazia *et al.*, 2015). The sclerotized region of the forewings is red in colour with black spots and the membranous portion of the forewings is black. The eyes are black in colour. Legs are dark red and the femora have varying amounts of black. There is also a black spot in each forewing. Transverse white bands can be seen in the collar, just behind the head and on the underside along the margin demarcating each thoracic and abdominal segment. The transverse bands along the posterior margins of each thoracic and abdominal sterna, the collar behind the head and the spots at the base of the head are white in colour (Schaefer and Panizzi, 2000; Vennila *et al.*, 2007c).

Fig. 29 (B). Pest attacks in cotton. Red cotton bug (*Dysdercus cingulatus*) adult bug.

Life cycle

The red cotton bug undergoes partial (hemi-) metamorphosis in life cycle. The eggs are laid under the soil in cracks and are covered with loose earth or with small dry leaves. Each female lays about 100-130 eggs. Egg period lasts 7-8 days (Vennila *et al.*, 2007c). Nymphs after hatching are wingless with their abdomen red with central row of black spots and row of white spots on either side. Nymphs pass through five moults with wings developing from the third instar and attaining full form after the fifth. The first-instar nymphs remained underground until the first moult (Encarnacion, 1970). Nymphal period lasts 35 to 39 days with an average of 37.4 days (Verma *et al.*, 2013). The development is completed in 50-90 days. Males are smaller than the females and the swollen abdomen can differentiate females from males (Vennila *et al.*, 2007c).

Management/Control measures

Cultural control

Some cultural methods may also reduce the damage caused the red cotton bug. For example, it is suggested to remove debris and ratoon cotton to prevent red cotton bug attack. It is better to avoid planting crops near other possible host plants. Cotton crop needs to be planted annually, if crops are maintained all year round, it increases the chance for cotton stainer populations to persist. Further, the removal and destruction of all standing cotton by a fixed date, as soon as the cotton has been picked and ceased to bear any profitable yield is an effective measure to control *D. cingulatus* in cotton (Pomeroy and Golding, 1923). The elimination of trees, such as *Bombax*, and other wild malvaceous plants is also strongly recommended.

Biocontrol

Most literature suggests an assassin bug *Phonoctonus* spp. as a common predator of *D. cingulatus* (Schaefer and Ahmad, 1987). Sahayaraj and Borgio (2010) observed a green muscardine fungus, viz., *Metarhizium anisopliae*, as an effective biocontrol agent for the pest. Application of the fungus can cause 75% to 92.30% mortality of the *D. cingulatus*. Further, the crude metabolites of the fungus is also capable of causing 45% mortality against *D. cingulatus* (Sahayaraj and Tomson, 2010). The entomopathogenic fungus, *Isaria fumosorosea*, isolated from Southern Tamil Nadu (India) has also been suggested to use as biopesticide for the control of the red cotton bug and other insect pests (Moorthi *et al.*, 2012). Furthermore, fresh custard apple leaf extract can be sprayed on the crop plants as an insect repellent/insecticide

Chemical Control

The insecticides, like, Pyrethrum powder, Alpha cypermethrin-based products, Lambda-cyhalothrin-based products and Cypermethrin-based products are mostly used to control the red cotton bug.

iii) Red Cotton Bug (*Dysdercus suturellus*)

Occurrence & severity

Red Cotton Bug (*Dysdercus suturellus*), is also known as 'cotton stainer'. The bug is mostly known as Cotton strainer or red bug in USA. The cotton stainer derives its name from its habit of staining cotton an indelible brownish yellow. The cotton stainer of the United States is known only from Florida, Georgia, and portions of South Carolina and Alabama. Except in Florida, it occurs in small numbers in other states (Hunter, 1912). The cotton stainer is most common in southern Florida and in Cuba. Outside of the United States this insect is known from and the upper West Indian Islands, namely, Bahamas, Cuba, Puerto Rico and Jamaica (Hunter, 1912). In view of its scarcity in north Florida, it must be presumed rare in the other three states mentioned. Earlier this red cotton bug was the most destructive cotton pest in Florida (Morrill, 1910). Currently, the cotton stainer is a minor pest of cotton. This species lost its importance on cotton during the mid-twentieth century, primarily because of the elimination of cotton waste that provided breeding and overwintering sites, and also due to effective chemical controls.

Types of damage
Nymphs and adults feed on seeds in developing cotton bolls (Sprenkel 2000). The feeding activities of the cotton stainers on cotton produce a stain on the lint which reduces its value. The staining of cotton lint was thought that the stain had come from excrement of the bugs. However, recent studies indicate that the stain primarily is a result of the bug puncturing the seeds in the developing bolls causing a juice to exude that leaves an indelible stain. Feeding by puncturing flower buds or young cotton bolls usually causes a reduction in size, or the fruiting body may abort and drop on the ground (Mead and Fasulo, 2014).

Scientific name & order: ***Dysdercus suturellus*** **Herrich-Schaeffer, 1842 (Hemiptera: Pyrrhocoridae)**

Biology
Egg: The eggs are oval, pale yellow in colour and when magnified show a finely reticulated surface. These are laid singly or in small, loose clusters in sand, earth, debris, upon the food plants or decaying vegetable matter. Each female deposits about 100 eggs. The egg hatches in about a week (Hunter, 1912; Mead and Fasulo, 2014).

Nymph: Nymphs resemble the adult in form and coloration, although the general colour of the body is somewhat more reddish (Hunter, 1912). There are five nymphal stages or instars. The first usually is spent underground. The nymphs are generally red. The fourth and fifth instars have dark wing pads, and the dividing lines between abdominal segments become very distinct as maturity is approached (Mead and Fasulo, 2014).

Adult: The adult is narrow, around 10-15 mm long, long legged, has a red head with black antennae of which the first segment is longer than the second. The thorax has a white pronotal collar. The forepart of the thorax is red, varying from light to dark. The hinder portion of the thorax and of the wing-covers varies from dark brown to black, the latter being crossed with narrow lines of light yellow. The underside of the body is bright red, with the segments outlined by narrow light-yellow bands. The sternites of the abdomen are red, bordered with white posteriorly. The corium, the leathery base of the wings, is dark brown margined with cream, giving the insect a large cross-shaped pattern. All tibiae and tarsi; the femora or thighs are red. The beak is red, except the last joint which is black (Hunter, 1912; van Doesburg, 1968; Mead and Fasulo, 2014).

Life cycle: The red cotton bug (*D. suturellus*) is a true bug and does not undergo metamorphosis. The eggs are laid singly or in small groups in sand, leaf or plant debris, and hatch in about a week. The nymphs pass through five instars. The first instar lives underground after which the nymph climbs the host plant. The duration of each of the first four stages typically averages four to five days during midsummer, but the fifth stage commonly takes about twice as long. All five stages require from 21 to 35 days to complete development (Sprenkel, 2000). The life cycle can vary from about a month to three and a half months, depending primarily upon temperature differences. There are several generations each year (Mead and Fasulo, 2014).

Management/Control measures

In order to prevent the pest, it is suggested that no cotton or cotton seed or other host plant debris that can serve as breeding material is to be left on the ground. Further, it is suggested for small infestations that colonies of cotton stainers on plants can be shaken into a bucket of soapy water to kill them. If possible, 'Tangle foot' around tree trunks may be used to keep the young bugs from crawling up to fruits and blossoms. Small heaps of seeds, fruits, or bits of sugarcane can be used as baits to attract cotton stainers. Then the insects can be killed with a spray of soapy water (Mead and Fasulo, 2014). Insecticides, like, Pyrethrum powder, Alpha cypermethrin-based products, Lambda-cyhalothrin-based products and Cypermethrin-based products, can also be used.

11. Spider Mites

Spider mites are tiny eight-legged arthropods in the family Tetranychidae and are more closely related to ticks and spiders than insects. Several species of spider mites occur on cotton. The most important are *Tetranychus cinnabarinus* (carmine spider mite) and *T. urticae* (two spotted spider mite).

i) Carmine spider mite

Occurrence & severity

Carmine spider mite, also known as 'red spider mite' or 'cotton red spider', is one of the most damaging polyphagous pest mite in agriculture and forestry of warmer regions around the world. This spider mite utilizes stylet to suck plant sap, causing mechanical damage to the host tissue (Sun and Meng, 2001; Zhang *et al.*, 2003).

Types of damage

The mite feeds mainly on the underside of the leaves, causing chlorosis, leaf curling and premature leaf drop. Cotton in the early stage as well as at the late stage is attacked by this mite. The mites both at nymphal and adult stages are generally found on the under surface of leaves forming fine webs. Mites puncture the leaf tissue and the oozing plant sap is sucked. Removal of plant sap with chlorophyll and other plant pigments results in the characterizing blocking with reddish bronze discoloration of leaves. Leaves become hard and crisp. Severe infestation leads to pre mature defoliation of leaves.

Scientific name & order: *Tetranychus cinnabarinus* Boisduval, 1867 (Acari: Tetranychidae)

Synonyms

= *Acarus cinnabarinus*

= *Eutetranychus cinnabarinus*

= *Eutetranychus cucurbitacearum*

= *Eutetranychus dianthica*

= *Tetranychus cucurbitacearum* (Sayed)

[N.B. There is some dispute in considering Carmine spider mite (CSM), *Tetranychus cinnabarinus*, as a separate mite species. The CSM and two-spotted spider mite (TSM), *Tetranychus urticae*, are widely distributed polyphagous pest mites and both are polymorphic in morphology, and are very similar in external morphologies. Therefore, many researchers considered them as two forms, red and green, of a single species, *Tetranychus urticae* (Robinson, 1961; Dupont, 1979; Ehara, 1989; Gotoh and Tokioka, 1996; Bolland *et al.*, 1998; Ros and Breeuwer, 2007; Renata *et al.*, 2011; Auger *et al.*, 2013). In 1956, Boudreaux had first separated CSM from TSM as an independent species based on experimental results of breeding and morphological characteristics. The separation of CSM and TSM as two entirely different species with complete reproductive isolation was further confirmed by Kuang and Chen in 1990, performing a comprehensive comparative study of the two species, focusing on the aspects of hybridization, changes in body colour, body size, external morphological features, ultrastructure, physiology, biochemistry and ecology. However, many taxonomists still questioning the two species just were red and green forms of *T. urticae* (Smith *et al.*, 1969; Jordaan, 1977; Brandenburg and Kennedy, 1981; Goka *et al.*, 1996; Zhang and Jacobson, 2000; Sugasawa *et al.*, 2002; Li *et al.*, 2009). Auger *et al.* (2013) suggested considering *T. cinnabarinus* as a synonym of the polymorphic species *T. urticae* to which it constitutes the red form. The phylogenetic analysis showed a significant genetic diversity of *T. urticae* that was organized in two well defined phylogenetic lineages. Further studies on DNA sequence analysis, corroborated the presence of two major cytochrome oxidase I (COI) lineages of *T. urticae* (Hinomoto *et al.*, 2001; Xie *et al.*, 2006) and that the colour, either green or red, was an inconsistent criterion to separate *T. urticae* samples (Hinomoto *et al.*, 2001; Xie *et al.*, 2008). While, on the contrary, in a few studies using sequences of a fragment of the COI from red and green mites collected in China, *T. cinnabarinus* was regarded as a valid species (Li *et al.*, 2010). Therefore, Xu *et al.* (2014) stated that the published genomic information of *T. urticae* (TSM) (Grbic *et al.*, 2011) cannot be fully utilized when investigating CSM (Xu *et al.*, 2014). Although, in an earlier study, while re-examined the sequences of the most studied fragment (450 nucleotides length) of the COI deposited in Genbank and labelled as *T. cinnabarinus*, together with sequences of *T. urticae* and closely related species (de Mendonça *et al.*, 2011), it was found that all the sequences matched 100% with at least one *T. urticae* Genbank entry, often to several entries at 99%-100%, and in some cases they matched with related species, *T. truncatus* or *T. kanzawai* (Auger *et al.*, 2013). Thus, Auger *et al.* (2013) stated that phylogenetic analyses of DNA sequences of the red and green mites of *T. urticae* do not support two distinct monophyletic groups according to the colour.]

Biology

Egg: Eggs are small, shiny spherical, white when laid and turn pink or orange as the embryo develops (Evans and Browning 1955; Hussey *et al.*, 1969) or straw coloured with reddish tinge (Davis, 1961) or brownish (Hussey and Parr, 1958) or pale to rosy-amber or slightly smoky (Dillon, 1958), always with a trace of red (Boudreaux, 1956) and colour more intense when laid by unmated females (Boudreaux, 1956). Eggs hatch in 3 days. They are approximately 0.1 mm diameter and laid singly on the underside of the leaf surface or attached to the silken webs spun by the adults (Mau and Kessing, 2007).

Larva: Larvae are slightly larger than the egg, pinkish, and have three pairs of legs. This stage lasts a short time, perhaps a day (Mau and Kessing, 2007).

Nymph: Nymph is light brown in colour with two eye spots and four pairs of legs. There are two nymphal stages, the protonymph and deutonymph. The nymphal stage differs from the larval stage by being slightly larger, reddish or greenish, and having 4 pairs of legs. This nymphal stage lasts about 4 days (Mau and Kessing, 2007).

Adult: Adult females are more or less elliptical, around 0.5 mm long, reddish, may be yellowish-red (Dosse, 1952), orange red (Dillon, 1958), brownish-red/reddish-brown (Hussey and Parr, 1958; Dosse, 1966; Kuang and Cheng, 1990), dark brownish red (Van de Bund and Helle, 1960), dark red (Boudreaux, 1963; Gotoh and Tokioka, 1996), carmine red (Boudreaux and Dosse, 1963a) or deep red (Veerman, 1970). Young female body colour is amber (Dillon, 1958) or pale-amber in colour with orange tint (Van de Bund and Helle, 1960) with two large pigmented dark spots on body. Body is separated into two distinct parts, gnathosoma and idiosoma. The adult female may live for up to 24 days and lay 200 eggs (Mau and Kessing, 2007). The males are slightly smaller and wedge shaped. They have a black spot on either side of their relatively colourless bodies (Mau and Kessing, 2007).

Life cycle

The complete life cycle from egg, through larva and nymph (protonymph and deutonymph), to mature adult lasts about 9-19 days. Per year, 20 and more generations may be completed. The spider mite life cycle starts with a small, round egg. There are three active immature stages, each separated by a resting stage (quiescence period) before the final moult to adult. The life cycle of spider mites is temperature driven and proceeds more rapidly at warmer temperatures (Steinkraus *et al.*, ND). Mites of both sexes often mated several times with different individuals during the first few days following emergence (David, 1975).

Control

Use of plant extract

The acetone and pentane extracts of neem (*Azadirachta indica*) seed kernels while sprayed directly on adult female mites caused repellency and reduction of fecundity and also mortality of adults (Mansour and Ascher, 1983). Hence, neem seed kernel extracts may be effective to manage the spider mite problem in field.

Use of Chemicals

Currently, in field crops, the control and prevention of spider mites mainly depends on spraying chemical insecticides and acaricides. Application of dicofol or wettable sulphur is mostly suggested to use for managing the spider mites. However, the development of pesticide resistance is a problem for this mite (He *et al.*, 2009; Jia *et al.*, 2011).

ii) Two spotted spider mite

Occurrence & severity

The two-spotted spider mite, *Tetranychus urticae*, is greenish and a major pest of cotton (Jeppson *et al.*, 1975; Migeon and Dorkeld, 2013). It was originally native only to Eurasia, but now it has acquired a cosmopolitan distribution (Raworth *et al.*, 2002). It is considered as a temperate zone species but also found in the subtropical regions. The two spotted spider mite prefers hot, dry weather of the summer and fall months, but may occur anytime during the year. Earlier, the spider mites were considered a late season pest in U.S.A. However, nowadays, it has become a significant early season pest of cotton, *Gossypium hirsutum* L., in the midsouthern U.S.A. (Gore *et al.*, 2013). As regards the economic impact, up to a 45% yield loss was observed in Mississippi for the infestations initiated at the three leaf stage of cotton development when those infestations persisted for four weeks (Smith, 2010). Similarly, a 35% yield reduction was observed from infestations that were initiated during mid-season (Furr and Pfrimmer, 1968). In Australia, yield reductions ranged from 13 to 48% for early-season infestations and 7 to 34% for midseason infestations (Wilson, 1993). Gore *et al.* (2013) reported yield reductions ranged from 20 (first flower) to 30% (three leaf) for infestations that were initiated early-season. This spider mite is extremely polyphagous and apart from cotton, it can feed on hundreds of plants, including most vegetables and food crops, such as, peppers, tomatoes, potatoes, beans, maize and strawberries, as well as, ornamental plants such as roses (Thomas and Denmark, 2009).

Types of damage

The two-spotted spider mite has needle-like piercing-sucking mouthparts. It feeds by penetrating the plant tissue with its mouthparts and is found primarily on the underside of the leaf. The mite causes injuries to the epidermis and resulting in yellow, brown blotch accompanied by dryness and leaf fall. Severe mite-feeding results in the reduction of the quality and quantity of the crop. The spider mites spin fine strands of webbing on the host plant (UF/IFAS, ND2).

Scientific name & order: *Tetranychus urticae* **Koch 1836 (Acari: Tetranychidae)**

[Species 2000 & ITIS Catalogue of Life: April 2013]

Synonyms

= *Tetranychus cinnabarinus* (Boisduval, 1867)

= *Tetranychus piger* Donnadieu, 1875

= *Tetranychus reetalius* Basu 1963

= *Tetranychus rosarum* (Boisduval, 1867)

= *Tetranychus russeolus* Koch, 1838

= *Tetranychus telarius* (Linnaeus, 1758)

Biology

Egg: Egg is pearl-like, spherical, less than 0.1 mm in diameter, smooth, clear white translucent and shining when laid. It becomes pale creamy to yellowish as developed and yellowish-orange before hatching, but never becomes reddish (Gasser, 1951; Boudreaux, 1956; Dillon, 1958; Van de Bund and Helle, 1960; Hatzinikolis, 1970; Veerman, 1974; Raworth et al., 2002). Eggs are usually found on the underside of leaves. About 82-83 eggs are laid by a female during autumn at temperature of 25°C but it lays 57-58 eggs during winter with prevailing temperature 13-14°C. The eggs are deposited per day are only 1-2 in winter but it is higher (11-12) during summer at temperature 30-32°C (Naher et al., 2008). Larvae hatch from egg in about 5 days under optimum conditions of 25-30 °C and 45- 55% relative humidity (Tehri, 2014).

Larva: Larva is round, about the same size as the egg and has 3 pairs of legs. It is colourless with red eyespots when hatched, but after feeding, it turns yellowish, pale green or pinkish with black spots on the dorsum (back) (Beers and Hoyt, 1993).The larval stage lasts 1 to 3 days depending on temperature.

Nymph: The nymphs are pale green with darker markings and have four pairs of legs. There are two nymphal instars, protonymph and deutonymph. Before each molt there is a short quiescent stage. Each nymphal stage lasts 1 to 3 days depending on temperature. The protonymph is larger and more oval. The two dorsal spots are more pronounced, and the green colour is slightly deeper. The deutonymph is slightly larger than the preceding stage, and males can be distinguished from females at this stage by the smaller size and more pointed abdomen (Beers and Hoyt, 1993).

Adult: The two spotted spider mite displays sexual dimorphism. The adult female is oval in shape, about 0.4-0.6 mm long, pale green or greenish-yellow with two darker patches (feeding spots) on the body and12 pairs of long dorsal setae. Overwintering females are orange to orange-red in colour. The feeding dark spots in active female are often trifid (Pritchard and Baker, 1955; Boudreaux, 1956) or divided into 2 or 3 smaller ones (Van de Bund and Helle, 1960); a second pair may also be present (Dillon, 1958; Boudreaux and Dosse, 1963b).The male is elliptical and smaller than the female (0.3 mm long). It has narrower and more pointed body than the female. The axis of knob of aedeagus is parallel or forming a small angle with axis of shaft (UF/IFAS, ND2).

Life cycle

The two spotted spider mite disperses individually by walking from one plant to another (Bell et al., 2005), or aerially by wind (Kennedy and Smitley, 1985).

It reproduces by two major kinds of parthenogenetic reproduction: (1) thelytoky, in which fertilized eggs yield diploid female offspring and (2) arrhenotoky, in which unfertilized eggs yield haploid males (Tehri, 2014). Its life cycle is composed of the egg, the larva, two nymphal stages (protonymph and deutonymph) and the adult. The length of time from egg to adult varies greatly depending on temperature. The spider mite completes its development from egg to adult within 7–8 days (Helle and Sebelis 1985) while at 21-23 °C, it takes about 10-14 days (Meyer, 1981). The adult female lives two to four weeks and is capable of laying 57-83 or more eggs during her life. There are many overlapping generations per year and all the life stages are present throughout the year, depending on the environmental conditions (Helle and Sabelis 1985).

Molecular characters

The whole genome sequence of *T. urticae* has been reported (Grbic *et al.*, 2011). The completely sequenced and annotated spider mite genome, representing the first complete chelicerate genome at 90 megabases, is the smallest sequenced arthropod genome. The spider mite genome shows unique changes in the hormonal environment and organization of the Hox complex and also reveals evolutionary innovation of silk production. It is also found strong signatures of polyphagy and detoxification in gene families associated with feeding on different hosts and in new gene families acquired by lateral gene transfer (Grbic *et al.*, 2011).

Control

Kim *et al.* (1999) have noticed that the compound Tebufenpyrad is highly effective against different stages of *T. urticae*. El-Ela (2014) has suggested from Egypt that Challenger36 SC (@ 40cc/100l of water; active ingredient-cypermethrin), Ortus 5 SC (@ 50cc/100l of water; active ingredient - fenpyroximate) and Vertimec 1.8 EC (@ 40 cc/100 l water; active ingredient-abamectin) can be used successfully in controlling the spider mite (*T. urticae*) in the presence of the predators (insects, mites and spiders) because of their lower and negligible side effect on these natural enemies.

12. Spotted bollworm

i) Spiny bollworm (*Earias insulana*)

Occurrence & severity

Spiny bollworm (*Earias insulana*), also known as 'Egyptian stem borer', 'Egyptian bollworm', 'spiny bollworm' or 'cotton spotted bollworm', is found in most of the Africa, southern Europe, India, Iran, Japan, Taiwan, the Philippines and Australia. The pest occurs more frequently around the Mediterranean and in Africa. It is one of the most important pests of cotton. In Iran, the pest is recorded to be spread in cotton farming areas (Esmaili *et al.*, 1995: Mirmoayedi and Maniee, 2009). Faseli (1977) reported that in the south Khorrasan region of Iran the spiny bollworm caused about 80% of the damage to cotton. In the semi-arid region of Turkey, it is found that a 1% increase in infestation ratio would reduce 2.5-6% of cotton yield (Unlu and Bilgic, 2004). Stam and Elmosa (1990) report that spiny bollworm was the most damaging pest in the Syrian cotton agro-ecosystem from 1980 to 1983. It is also regarded as an important pest in most of the cotton-growing areas of India, being more common in the Punjab and less common in further south (Dhawan *et al.*, 1990b). However, it is reported that even in areas with ideal conditions, cotton may not be the first choice of host for spiny bollworm (Pearson, 1958). Moreover, in certain cases (as in Indian Punjab during 1985-1986), the damage by spiny bollworm may not contribute significant loss of yield in cotton (Dhawan *et al.*, 1990b). The adults of spiny bollworm show seasonal polymorphism, depending on the temperature. In some parts of the distribution area, two distinct forms can be recognized - the bright green summer moth and the brownish-yellow autumn moth. The cold winter in northern Iran also effectively keeps populations of spiny bollworm in check (Heidari *et al.*, 1981). The spiny bollworm is oligophagous on Malvaceae, including cotton and okra (*Hibiscus esculentus* Linnaeus) but sometimes feed on rice, sugarcane and occasionally corn.

Type of damage

Initially, the caterpillars (larvae) tunnel into the cotton buds by destroying vessels and growing points. The larvae tunnel downwards for 3.5-5 cm until the woody part is reached and then left to infest another young shoot. Only soft growing tissue is attacked. They destroy all the pith cells as well as some parts of the xylem and adjoining parenchyma and phloem, preventing the upward movement of sap. The main shoots are more heavily infested than the side shoots (Nasr and Azab, 1969a).The larvae cause adjacent blooms, young leaves and eventually whole shoots to turn blackish-brown and die off. Extensive tunnelling results in wilting of the top leaves and the collapse of the apex of the main stem. The whole apex turns blackish-brown and dies. If only the apical bud is attacked, the damage may not be noticed until the main stem divides (twinning) when the axilliary buds take over growth (Kashyap and Verma, 1987; Reed, 1994).This can result in bunched growth of young plants and death of the growing point in the mature plant. Examination of the affected plant parts reveals a number of small holes either on, or near to, leaf or flower buds. On cotton, the damaged flower buds sometimes spread their bracteoles prematurely (flared squares). After the square and boll formation, the infestation is noticed on these parts also. The larvae feed on the bolls, which become brown and fall off. Drying and drooping of terminal shoots during pre-flowering stage, formation of side shoots, shedding of squares and young bolls, flaring up of squares during square formation, plugged entrance hole on boll with excreta, drop off affected bolls in early stage and bad boll opening are seen as symptoms due to infestation of the pest. Secondary invasion by fungi and bacteria may conceal the spiny bollworm infestation. The bollworm can transmit *Xanthomonas citri* subsp. *malvacearum,* causing angular leaf spot disease of cotton (Borker *et al.*, 1980). It may be responsible for the establishment of the black fungus infection (caused by *Rhizopus nigricans*) as present in cotton crop (Nasr and Azab, 1969b).

Scientific name & order: *Earias insulana* Boisduval, 1833 (Lepidoptera: Nolidae)

Synonyms

= *Tortrix insulana* Boisduval, 1833

= *Earias siliquana* Herrich-Schaffer, 1851

= *Earias smaragdinana* Zeller, 1852

= *Earias frondosana* Walker, 1863

= *Acontia xanthophila* Walker, [1863]

= *Earias simillima* Walker, 1866

= *Earias anthophilana* Snellen, 1879

= *Earias tristrigosa* Butler, 1881

Biology

Egg: Eggs are sky blue coloured, singly laid, crown shaped and sculptured. A female lays up to 200 eggs.

Larva: Larva is initially grey, later grey-blue or brown with yellow spots and dorsum showing a white median longitudinal streak. The head is dark and shiny and the final length is (13-)15 to18 mm. The last two thoracic segments and all the abdominal segments have two pairs of fleshy tubercles (finger shaped processes) one dorsal and the other lateral.

Pupa: Pupa is brown and boat shaped that covered in a boat shaped tough silken cocoon attached to plant parts or on the fallen leaves or fruiting bodies. Typically, the pupal stage takes 9–15 days, but may extend to up to two months if development is delayed by low temperatures.

Adult: Adult is small buff coloured. The adult moth has an overall wingspan of about 20-22 mm. It is covered with a soft, dense coating of scales. Forewings are uniformly silvery green or yellow-green or sometimes brown and with a diagonal green stripe. Presumable these colour morphs are seasonal, representing local adaptations to ambient heat, humidity and foliage density. The hind wings are dull white with a brown sub terminal line.

Life cycle

Each female deposits up to 200 eggs at about 25-28°C on the young shoot, leaf bracts, shoot tips, buds and squares of host plant. The larva hatches in 3 days. The larva of the first generation bores into terminal cotton bud, that of later generations into flower buds, flowers and newly-set bolls. Larval development can take 8-25 days (or 3 weeks) during summer, but requires twice as long at 19°C. The mature larva spins felt-like white cocoon, which it attaches to plant parts or to plant debris on the ground and pupates therein. The pupal stage normally lasts about 9-15 days, but it may extend to up to two months if development is delayed by low temperatures. There is no true diapause. In some areas, the insect moves between crops with different growing seasons (e.g. okra and cotton), so there is no interruption to their food supply, and the population can build up over a long period. Normally, during favourable time, it completes its life-cycle by 20-25 days.

Management/Control measures

Use of host-Plant Resistance

Considerable resistance to *Earias* has been recorded in several wild species of *Gossypium* (Anson *et al.*, 1948). *Gossypium hirsutum* has been reported to be more susceptible than both *G. barbadense* (Badway, 1974) and *G. aboreum* (Butani, 1974).

Cultural Control

It is suggested to collect and destruct the infested bolls. Further, collection and destruction of plant debris and trash before sowing are beneficial to manage the pest. Burning or ploughing cotton fields to a depth of 30 cm or more, in order to destroy all residual material and any pests remaining thereon after harvest, is effective to minimize the pest population. After harvest, cotton plants should be uprooted and destroyed in order to eliminate the food source for the spiny bollworm and thereby to interrupt population build-up.

Biological control

Several egg parasitoides reduce pest populations, but they are often killed by the extensive use of chemical pesticides. Coccinellidae and hemipteran predators are more important predators in cotton. Although parasitoids can control Earias, large numbers of those are necessary and that high levels have to be artificially maintained. Stam and Elmosa (1990) have found that parasitoids are relatively unimportant in controlling lepidopterous pests in cotton in Syria.

Use of *Bacillus thuringiensis* serovar *kurstaki* 5% WP @ 750- 1000 g/ha is effective against the spiny bollworms (TNAU Agritech Portal, ND2). Insecticidal toxins from *B. thuringiensis* (Bt) can now be deployed either as sprays or as transgenic plants. However, Bt. sprays are not generally competitive with chemical insecticides and seem unlikely to displace them (Roush, 1994). Further, neem formulations can be applied against the pest.

Use of pheromone trap

The sex pheromone, (E,E)-10,12-Hexadecadienal which has been identified as a component of the sex pheromone of the female spiny bollworm moth, *Earias insulana* from the abdominal tip extracts of female moths may be used as a trap (Hall *et al*, 1980). Further, traps baited with virgin females proved to be very efficient in detecting fluctuations in adult population density and in indicating adult peaks (Kehat and Bar, 1975). Cork *et al.* (1988) identified six pheromone components detected by the male moth. A series of trials in Pakistan, using a slow release 'twist-tie' formulation, containing the major components of *E. insulana* pheromone, has successfully disrupted mating and controlled the pest (Critchley *et al.*, 1987; Chamberlain *et al.*, 1992, 1993; Hall *et al.*, 1994). Nakache *et al.* (1992) have observed similar results in a trial in Israel and concluded that the use of pheromones is a viable control method.

Chemical control

Several insecticides are used to control the insect pest. Earlier, diazinon was used for the control of *E. insulana*. Now, spinosad is mostly suggested instead of diazinon, in any integrated control of spiny bollworm. The field sprayed with 200 ppm spinosad had the minimum number of blind damaged bolls (Mirmoayedi *et al.*, 2010). Chlorfluazuron and teflubenzuron, and possibly other benzoylphenylureas, have potential for controlling spiny bollworm larvae under field conditions (Horowitz *et al.*, 1992). Spraying any one of the following insecticides, such as, Carbaryl 5% DP 20 kg/ha, Phosalone 35% EC 1714ml/ha, Chlorantraniliprole 18.5% SC 150 ml/ha, Profenofos 50% EC 1500-2000 ml/ha, Flubendiamide 39.35% SC 100-125 ml/ha, Triazophos 40% EC 1500-2000 ml/ha or Indoxacarb 14.5% SC 500 ml/ha, may be effective to manage the pest chemically.

ii) Spotted bollworm (*Earias vittella*)

Occurrence & severity

Spotted bollworm (*Earias vittella*), also called as spiny bollworm, is a widely distributed species in tropics and subtropics of the old world and Australasia. The spotted bollworm mainly occurs in the regions with comparatively greater rainfall. As a relatively hardy species among insects, the spotted bollworm tolerates a wide range of environmental conditions and so is prevalent in many regions of the world. It is active throughout the year on different host plants under field conditions (Abdul-Nasr *et al.*, 1973; Arif and Attique, 1990). In India it is more common in Punjab and Rajasthan. It is also a common pest in southern region of India. In Pakistan, on cotton crop, their initial attack is noticed in June and July. The attack on the bolls is generally higher than buds. Maximum infestation is recorded during August and September (Qureshi and Ahmed, 1991). Cotton is the main host of this insect but other malvaceous plants, namely, Okra, Hibiscus, Anthaea, Abutilon, Malvastrum etc. are also common hosts of this bollworm. In Pakistan, among bollworms, *Earias* spp. are most abundant on cotton in Sindh as compared with other bollworm species (Leghari and Kalro, 2002) and cause 1.79 to 2.38% (Abro *et al.*, 2003) or differently, 3.8 to 12.6% damage (Chang *et al.*, 2002). Hiremath and Thontadarya (1984) reported that under controlled conditions, one larva *E. vittella* damaged nearly 30% of the bolls on a plant, while, two damaged around 45%, four damaged nearly 60% and six damaged about 75%. Bourgeois (1970) reported that the farmers in Cambodia were starting to give up growing cotton, partly because of infestation by *E. vittella*.

Types of damage

The attacks are similar for all *Earias* species and are often described in general terms rather than for individual species. The newly hatched larvae wandering for few hours bore into tender shoot, flower bud or bolls. The young larvae enter the terminal bud of the vegetative shoot and channel downward from the growing point, or directly penetrate the internode. Only soft growing tissue is attacked by the larvae. The extensive attacks of young larvae lead growing shoots to drooping and withering of the top shoot. Then, the whole apex turns blackish-brown and is generally killed. The result is bunched growth in young plants and killing of the growing point in the mature plant. If only the apical bud is attacked, damage may not be noticed until the main stem divides ('twinning') as the axilliary buds take over (Pearson, 1958; Kashyap and Verma, 1987). In later stages buds, flowers and bolls are also damaged. Flower buds open up prematurely causing "flared squares". The bolls are attacked only when unripe. The larvae usually bore deeply, filling the tunnel opening with excrement. The tunnel often enters bolls from below, entering at a slight angle to the peduncle (Pearson, 1958). Small bolls up to 1 week old turn brown rot and drop, whereas bigger bolls of 2-4 weeks may not drop but open prematurely and be so badly damaged as to be unharvestable. The bolls of up to 6 weeks of age are vulnerable (Butani, 1976; Sidhu and Sandhu, 1977). In damaged bolls pulp is eaten up and lint is stained. The larvae do not confine their feeding to just one plant or boll. As a result, partial damage is seen at many of their favourite feeding sites. The tendency for secondary invasion by fungi and bacteria may also conceal the infestation. Borker *et al.* (1980) reported that *Earias* spp. could transmit *Xanthomonas citri* subsp. *malvacearum*, causing bacterial blight or angular leaf spot of cotton.

Scientific name & order: *Earias vittella* Fabricius, 1794 (Lepidoptera: Nolidae)

[Species 2000 & ITIS Catalogue of Life]

Synonyms

= *Aphusia speiplena* Walker, 1858

= *Earias fabia* (Stoll, 1781)

= *Earias huegeli* Rogenhofer

= *Micra partita* Walker, 1865

= *Noctua fabia* Stoll, 1781

Biology

Egg: Eggs are spherical, crown shaped, sculptured, less than 0.5 mm diameter and light bluish-green or deep sky blue in colour. These are laid singly on flower buds, bolls, peduncles and bracteoles of cotton plant; the favoured region being young shoots (Vennila *et al*, 2007d). Each female deposited on an average of 199 eggs during her life time. The incubation period is about 3-5 days (Rehman and Ali, 1981; Sundraraj and David, 1987; Singh and Bichoo, 1989; Syed *et al.*, 2011, Sewak, 2016).

Larva: Larvae are brownish with longitudinal white stripes on dorsal side having orange dots on prothorax without finger shaped process on the cream coloured or pale yellow body. There are 5 (differently 6) larval instars. The larval duration for 1st, 2nd, 3rd, 4th and 5th instars is about 1.6, 2.0, 2.5, 2.5 and 3.0 days, respectively. The average of total larval period is about 11.5 days (Sewak, 2016). However, the shortest larval period of 9.6 days was recorded on cotton during August, while, the duration of larval period in October was 14-15 days (Syed *et al.*, 2011).

Pupa: Pupation occurred in a dirty white boat-shaped cocoon. The pupal period varies between 6 and 14 (differently, 9.6-14.4) days, depending upon the seasons and host plant (Nayar*et al*., 1976: Rehman and Ali, 1981; Atwal, 1984; Sundraraj and David, 1987; Singh and Bichoo, 1989; Al-Mehmmady, 2000; Syed *et al*., 2011; Sewak, 2016).

Adult: Adult moths are small and off-white or buff coloured. Forewings are pea green with a wedge shaped white band running from base to out margin, brown stripe along the middle of each forewing or yellowish forewings with elongated green streak in the middle. Hindwing is pale whitish. The wingspan is about 1.5 cms. The moth has a green thorax. The longevity of male and female adult is about 4.2 and 9.5 days, respectively. The adult female after emergence took an average period of about 1.8 days to lay the eggs (Sewak, 2016). The females lay 150 to 200 eggs.

Life cycle

Duration of life cycle ranges between 30.4 to 44.6 days on cotton (Syed *et al*., 2011). However, Sundararaj and David (1987) have observed that the total development period of *E. vittella* on cotton is 27.9 days. The life stages, the egg, larval, pupal and adult are completed in about 3-5, 10-15, 6-14 and 4-10 days, respectively. Each mated female produces about 150-200 eggs in its egg laying period of 8-10 days. Pre-copulation, pre-oviposition, oviposition and post oviposition periods of *E. vittella* adults vary when feeding on different host plants as larvae. The temperature variation affects all the above biological parameters e.g. the larval period (Syed *et al*., 2011).

Management/Control measures

Use of host resistance

Considerable resistance to *Earias* has been recorded in several wild species of *Gossypium* (Anson *et al*., 1948). *Gossypium hirsutum* has been reported to be more susceptible than both *G. barbadense* (Badway, 1974) and *G. aboreum* (Butani, 1974). The cultivars which possess high levels of tannin and gossypol (Sharma and Agarwal, 1984; Mohan *et al*., 1994), frego-bract and okra-leaf characters (Thombre, 1980) and red pigmentation (Duhoon and Singh, 1980) have been found to be more tolerant than many commercial cultivars. Hirsute varieties (Agarwal and Katiyar, 1974) and glandless varieties (Brader, 1969) have been found to be more susceptible. The tall plants having bigger top leaves and bolls in clusters are attacked more by spotted bollworms (Singh *et al*., 1974), while, the dwarf varieties with early flowering habit can escape the damage (Wankhede and Sadaphal, 1977).

Cultural Control

Crop sanitation is an effective cultural method for this insect. As such, after harvesting, cotton plant stumps are to be removed from the field (Sawhney and Nadkarny, 1942). Removal of possible alternative host plants, such as, malvaceous weeds, is also recommended (Christidies and Harrison, 1955). The timing of sowing may also help to reduce infestation. In India, it is found that early sowings are beneficial (Ilango and Uthamasamy, 1989). In China, Meng *et al.* (1973) observed more *E. vittella* infesting cotton boll when two sowings of cotton are made per year instead of one. Further, reduced damage by *Earias* spp. is observed when the crop is closely spaced (Bishara 1969, Abdel-Fattah *et al.*, 1976) and after deep ploughing (Faseli, 1977).

Biological Control

The use of natural enemies in any integrated pest management system is suggested for cotton. Since, a large number of parasitoids have been recorded preying on *E. vittella*, and they apparently discriminate little between the different *Earias* species (Reed, 1994). Patel and Bilapate (1984) found 10-20% parasitization of *E. vittella* by *Rogas aligharensis* (*Aleiodes aligharensis*) and 15-20% by a Bracon species in cotton field in India. Sekhon and Varma (1983) recorded parasitism rates of more than 25% in eggs and 37% in pupae of *Earias* species, including *E. vittella* in the Punjab. Khan and Verma (1946) reported a very high population of *Bracon greeni* from July to September in India. However, in Pakistan, *Trichogramma australicum* and *B. greeni* while applied in cotton field showed limited success against *Earias* species (Habib and Mohyuddin, 1981). Krishnamoorthy and Mani (1985) investigated the use of the egg parasitoid *T. achaeae* and the predator Chrysopascelestes (*Brinckochrysa scelestes*), but found that the predator ate parasitized and normal eggs indiscriminately, precluding their use together.

Use of pheromone trap

The components of the female sex pheromones of *E. vittella* have been identified (Cork *et al.*, 1988). They also identified six pheromone components detected by the male moth. Subsequently, synthetic formulations have shown promise in Pakistan, disrupting mating and achieving season-long control (Critchley *et al.*, 1987; Qureshi and Ahmed, 1989; Chamberlain *et al.*, 1992, 1993; Hall *et al.*, 1994).

Use of plant products

A marked decline in egg hatchability was observed in *E. vittella* when the eggs were exposed to the vapours of palmarosa (*Cymbopogon martini*) or citronella (*C. winterianus*) oils. The main chemical ingredients are geraniol and citronellol, with geraniol present in both oils (Singh *et al.*, 1989). These observations provide evidence favouring the use of bio-active principles of the essential oils as potential agents for checking the population build-up of *E. vittella* in cotton in the tropics (Marimuthu *et al.*, 1997). Further, neem oil reduces the egg hatchability. When the concentration of the neem derivatives increased, the hatchability of egg was reduced (Thara *et al.*, 2009). Gajmer *et al.* (2002) observed that the hatchability of egg was lower at higher concentration than at lower concentration of methanolic extract of neem on *E. vittella*. Chandrasekaran *et al.* (2003) also reported decreased hatchability, while increasing the concentration of neem products on the eggs. Several other plant extracts/oils, like, clove oil and garlic bulb vapour, also affect the hatchability of eggs of this insect (Pathak and Krishna, 1993; Gurusubramanian and Krishna, 1996; Krishna, 1996).

Chemical control

Chemical control involves timing of spray of insecticides with egg-laying or before larvae bore into the bolls. The insecticides, like, parathion 0.025%, endrin 0.02%, carbaryl 0.1%, endosulphan 0.05%, sevimol and quinalphos 0.02%, have been found to be effective. These insecticides can be sprayed at the rate of about 2.0-2.5 litre/hectare of crop.

13. Stink bugs

Stink bugs have been considered as pests of cotton since the early 1900s (Morrill 1910). These bugs (Hemiptera: Pentatomidae) are true bugs with piercing-sucking mouthparts and incomplete metamorphosis. Stink bug infestations can cause substantial economic losses through reduced yield, loss of fibre quality, and increased control costs. The immature stages (nymphs) and adults of several species may damage cotton. However, primarily, three stink bug species – the green stink bug (*Chinavia hilaris*), the brown stink bug (*Euschistus servus*) and the brown marmorated stink bug (*Halyomorpha halys*) mostly attack cotton in several places (Munyaneza and McPherson 1994; Kamminga *et al.*, 2012, 2014). Stink bugs feed on developing seeds in young bolls, causing to become hardened, discoloured and unharvestable (Wene and Sheets, 1964; Barbour *et al.*, 1988). They cause considerable crop losses in cotton in various countries. In USA alone, losses during the year 2005 due to stink bugs were estimated to be over $63 million with control costs exceeding $11.5 million (Williams, 2006).

i) Green stink bug

Occurrence & severity

The green stink bug is one of the most damaging native stink bug species in the United States. It is a pest of economic importance in a variety of commodities, including cotton (*Gossypium hirsutum* L.) in Tennessee of United States (Kamminga *et al.*, 2012). The green stink bug is polyphagous and feeds on a variety of plants, but prefers woody plant tissue (McPherson 1982). It is found in orchards, gardens, woodlands and crop fields throughout North America on a wide variety of plants from May until the arrival of frost.

Types of damage

Stink bugs damage cotton by puncturing squares, piercing the bolls and feeding on the developing seeds. Although stink bugs favour medium-sized bolls, they can feed on any size boll. Their feeding on young bolls (less than 10 days old) usually causes the bolls to shed. However, principal damage is caused to older bolls. On older bolls lint may be stained and matted, and seeds shrunken by stink bug feeding. Excessive stink bug feeding causes reduced yield, stained lint, poor colour grades, and reduced fibre quality. In addition to direct damage, stink bug feeding can transmit plant pathogens, like, bacteria and fungi, which cause boll rot (Annonymous, ND8).

Scientific name & order: *Chinavia hilaris* Say, 1832 (Hemiptera: Pentatomidae)

Synonyms

= *Pentatoma hilaris*

= *Nezara hilaris*

= *Acrosternum hilare*

= *Acrosternum hilaris*

= *Chinavia hilare*

Biology

Egg: Eggs are barrel-shaped, colour changes from light green to yellow and then to light pink before hatching in about 1 week (Miner 1966), measuring about 1.4 by 1.2 mm and usually laid in clusters on the under surfaces of leaves or around suitable food (Gomez and Mizell, 2013)

Nymph: Nymphs are wingless, oval-shaped, black and red in colour, predominantly black, become green on maturity and lack the triangular plate. They closely resemble the adults throughout their five nymphal stages (Underhill 1934, Miner 1966). The first instars do not feed and remain clustered together around the egg mass. Second instars are less gregarious and begin to feed. Third instars behave in a similar manner to second instars, but are slightly larger in size. Feeding by the fourth and fifth instars causes economic damage to the plants (Barbour *et al.* 1988). At early stages, they have a distinctive pattern of whitish spots on the abdominal segments, and short non-functional wing pads which at final instar make them to look somewhat adults (Gomez and Mizell, 2013). Developmental time of the nymphs can occur most rapidly at 27°C (Simmons and Yeargan 1988). Adulthood is reached in 36 days (Underhill 1934, Miner 1966). Nymphs have large stink glands on the underside of the thorax extending more than half-way to the edge of the metapleuron. They discharge large amounts of this foul-smelling liquid when disturbed.

Adult: The green stink bug is 1.3–2.0 cm long and 0.8 cm wide. It is shield-shaped with an elongate, oval form bug. It is typically bright green, with narrow yellow, orange, or reddish border around the margin of the abdomen, head, and thorax (Underhill 1934). It can be differentiated from the southern green stink bug species *Nezara viridula* by its black outermost three antennal segments and a pointed abdominal spine, instead of red antennal bands with a rounded abdominal spine (Miner 1966, McPherson 1982, McPherson and McPherson 2000). The adults have large stink glands on the underside of the thorax extending more than half-way to the edge of the metapleuron. They discharge large amounts of this foul-smelling liquid when disturbed.

Life cycle: Green stink bugs overwinter as adults in leaf litters or in deciduous woodlands (Underhill 1934). They become active during the first warm days of spring when temperatures surpass 18°C (McPherson 1982). Adults become most active when temperatures exceed 24°C, and are more prone to flight when temperatures exceed 27°C (Underhill 1934). After emerging from diapause, the adult females require a preovipositional period to develop as reproductively mature (Kamminga *et al.* 2009a, Nielsen and Hamilton 2009b). The ovipositional period of overwintered adults in eastern North America begins in the middle of June and ends the first week of September, with peak egg laying occurring in the third week of July (Underhill 1934, Javahery 1990). The green stink bug has a single generation in cooler northern areas (Javahery 1990), but in southern warm conditions it has two generations (Sailer 1953, Kamminga *et al.* 2009a). Newly emerged females begin copulating after 22 days, and lay their first egg masses about 3 weeks later (Miner 1966). Females can lay a new egg cluster every 8 –10 days. These eggs are deposited vertically in clusters of 1-72 (Underhill 1934) with an average of 32 eggs per egg mass (Miner 1966). The normal developmental period from egg to adult requires about 35-36 days, but varies with temperature (Gomez and Mizell, 2013).

Management/ Control measures
Cultural control: Trap cropping and the planting of resistant varieties may be effective to decrease crop injury by stink bugs (Kamminga *et al.*, 2012).

Blacklight Trapping: Blacklight trap catch may be useful for improving the timing of scouting and management methods for stink bugs (Kamminga et al. 2009a, Nielsen and Hamilton 2009b).

Pheromone Trapping: The green stink bug uses the pheromone methyl (E,Z,Z)-2,4,6-decatrienoate in its communication system and this may be used to attract the bug away from crop fields. Therefore, pheromone traps are used to monitor green stink bug populations effectively in different crops (Leskey and Hogmire 2005; Mizell and Tedders 1995). A yellow base pyramid trap with a pheromone lure is commonly used to monitor green stink bug populations (Leskey and Hogmire 2005, Hogmire and Leskey 2006). The aggregation pheromone blend for the green stink bug is a 95:5 cis:trans blend of (4S)-cis-Z-bisabolene epoxide and (4S)- trans-Z-bisabolene epoxide (Aldrich et al. 1989, 1993; McBrien et al. 2001).

Biological control: Stink bugs are largely protected from predators by their foul smell and bad taste. However, the green stink bugs are parasitized by the tachinid fly *Trichopoda pennipes* (Diptera) and by some species of Hymenoptera (Gomez and Mizell, 2013). They lay eggs on adults, hatching maggots, then burrow into stink bugs and feed from within.

Chemical Control: Stink bugs have become a major challenge to integrated pest management systems because control options are basically limited to the application of broad-spectrum insecticides such as organophosphates, carbamates, and pyrethroids (Kuhar et al. 2006, Herbert 2012). Dicrotophos, an organophosphate, has been cited as having a high toxicity to various stink bug species (Tillman and Mullinix 2004, Snodgrass et al. 2005), and it is suggested that it provides consistent control of the green stink bug (Willrich et al. 2003, Snodgrass et al. 2005, Kamminga et al. 2009c). However, neonicotinoids are generally effective for control of this stink bug and may be less disruptive to its natural enemies. (Lanham, 2012)

ii) Brown stink bug

Occurrence & severity

Brown stink bug occurs throughout the eastern North America with two subspecies, viz. *Euschistus servus servus* (Say) and *E. servus euschistoides* (Voltenhoven). The first one occurs throughout the south-eastern U.S.A. from Florida through Louisiana to California, while, the second subspecies occurs in Canada, especially in British Columbia and southern part, and in the northern part of the U.S.A (Munyanezaand McPherson 1994; Borges et al., 2001; Maw, 2011; Gomez et al., 2018). The brown stink bug is highly polyphagous insect. Apart from cotton, it damages several other crops, like, tomato, peach, pecan, maize, lucerne, soybean, sorghum, okra, millet, wheat, beans, peas, tobacco, mullein (Hall and Teetes, 1981, Panizzi *et al.*, 2000; Borges *et al.*, 2001; Gomez *et al.*, 2018; Mizell, ND). It also occurs on a variety of other hosts, such as shrubs, vines and many weeds of leguminous family (Gomez *et al.*, 2018).

Types of damage

Adult bugs feed by inserting their needle like mouthparts into the stems, leaves and seed pods. During feeding, they inject toxic substances into the plant parts that inhibit plant development in the area of the punctures (Gomez *et al.*, 2018). They mostly feed on developing seeds in young bolls, which become hardened, discoloured and unharvestable (Wene and Sheets, 1964; Barbour *et al.*, 1988). However, the damage caused by insect feeding itself is less (Barlow, ND). Their feeding may insert cotton boll rotting pathogens, like, *Pantoea agglomerans* strain Sc 1-R, which play a significant role in cotton yield and quality loss (Medrano *et al.*, 2016; Barlow, ND). Further, the insects not exposed to pathogenic microbes and deposited only nonpathogenic microbes cause insignificant damage to the boll tissue (Medrano *et al.*, 2016).

Scientific name & order: *Euschistus servus* (Say) (Hemiptera: Pentatomidae)

[**Subspecies:** *Euschistus servus servus* (Say, 1832) and *Euschistus servus euschistoides* (Vollenhoven, 1868)]

Biology

Egg: Eggs are barrel or kettle shaped, usually laid in groups of 14 or deposited in medium-sized clusters of below 50 eggs in loosely aligned rows (Munyaneza and McPherson 1994; Greene *et al.*, 2006). Egg surface appears rough to spiny with 26-39 rod shaped micropyles (Greene *et al.*, 2006). The egg colour is yellowish-translucent, but it starts turning toward a light pink before hatching (Gomez *et al.*, 2018).

Nymph: Nymphs are greenish-brown or yellowish tan, with dark brown plates down middle of abdomen becoming more obvious in later instars and giving the appearance of two to three large dark spots (Greene *et al.*, 2006). They resemble adults but are smaller and oval. The nymphs develop through five instars that require about 29 days for development (Gomez *et al.*, 2018).

Adult: Adults are brown or light brown or greyish yellow with dark brownish-grey punctures that becoming denser at the edges of the pronotum. Body length is 11.0–15.0 mm and width 7-8.5 mm. It is shield shaped or oval with the underside being slightly concave, the abdomen narrow and with piercing-sucking mouthparts. The ventral surface usually has a pinkish tinge. Cheeks are large, passing the clypeus in length and more pointed. Humeral angles ("shoulders") of pronotum are round or variable in shape but not with sharp pointed spines. Membranous parts of wings are not extending beyond tip of abdomen (McPherson 1982; Blatchley 1926; Greene *et al.*, 2006; Gomez *et al.*, 2018). The two subspecies are similar in size and colour. However, they differ in certain aspects as followed. The tip of the head of *E. servus euschistoides* appears notched due to the presence of longer juga than tylus, whereas the tip of the head of *E. servus servus* does not appear notched, due to the presence of equal or nearly equal juga and tylus. The last two antennal segments (fourth and fifth) are usually dark brown in *E. servus euschistoides* and yellowish-brown or reddish-brown in *E. servus servus*. Further, the edge of the abdomen is more or less completely covered by the front wing in *E. servus euschistoides*, whereas the edge of the abdomen is more exposed in *E. servus servus* (Paiero *et al.* 2013; Koch *et al.*, 2017).

Life cycle

The brown stink bug is a bivoltine species that overwinters as an adult in the fall (Rolston & Kendrick 1961, McPherson 1982). The adult bugs overwinter in protected areas such as ditch banks, fence rows, under boards and dead weeds, ground cover, stones, and under the bark of trees. They become active during the first warm days of spring when temperatures rise above 21°C (Gomez *et al.*, 2018). At that time, the bugs mate tolay eggs. Each female lays about 18 egg masses (clusters) over a period of above 100 days and approximately four to five weeks are required in developing from hatching to adult emergence. The first generation normally develops on wild (noncrop) hosts, while the second generation typically develops on cultivated crops (Ehler 2000; Gomez *et al.*, 2018).

Management/Control measures

Control measures are same as described with Green Stink bug. In addition, use of trap crops and accelerometers are suggested to monitor the appearance of brown stink bugs in field.

Use of trap crops

The brown stink bugs may be monitored by growing trap crops, like, buckwheat, sorghum, millet and sunflower on the exterior of cotton field as suggested in peach garden (Gomez *et al.*, 2018).

Use of accelerometers

Females of *E. servus* emit two distinct songs while the males emit four distinct songs. The detection of these two female and four male songs may be done by placing accelerometers. The characterization of these songs will allow for the development of a monitoring system in the field using strategically placed accelerometers to detect stink bug vibrational communication (Lampson *et al.*, 2010).

iii) Brown marmorated stink bug (BMSB)

Occurrence & severity

Brown marmorated stink bug is native to China, Japan, Korea and Taiwan (Hoebeke and Carter 2003, Lee *et al.* 2013). It is believed that the pest has been introduced into the United States from a single introduction originating from Beijing, China (Xu *et al.* 2014). Now, the pest is present in at least 41 states of USA (Leskey *et al.* 2014) and other countries in Canada (Fogain and Graff, 2011), Switzerland (Wermelinger *et al.*, 2008), Liechtenstein (Arnold, 2009), Germany (Heckmann, 2012), Italy (Pansa *et al.*, 2013), France (Callot and Brua, 2013), Hungary (Vétek *et al.*, 2014), Australia, New Zealand, Uruguay, Brazil, Argentina, Angola, Congo and Zambia (Zhu *et al.*, 2012). The brown marmorated stink bug is rapidly spreading and is one of the most economically damaging insect pests of cotton in a reduction of lint yield and quality at harvest (Kamminga *et al.*, 2014). It is a highly polyphagous plant feeder with a wide range of host plants.

Types of damage

Like other true bugs, brown marmorated stink bug feeds by sucking plant juices. Adults generally feed on fruit, whereas the nymphs feed on leaves, stems and fruit. Its feeding can cause wart-like growths on inner carpel walls, stained lint, and shriveled seeds resulting in economic loss to growers (Bundy *et al.*, 2000; Emfinger *et al.*, 2004; Wene and Sheets, 1964). Feeding injury increases as boll size increased (Kamminga *et al.*, 2014).

Scientific name & order: ***Halyomorpha halys*** **(Stal) (Hemiptera Pentatomidae)**

Synonymy

= *Dalpada brevis* Walker, 1867

= *Dalpada remota* Walker, 1867

= *Pentatoma halys* Stal, 1855

= *Poecilometis mistus* Uhler, 1860

Biology

Egg: Eggs are small, spherical, about 1mm diameter and pale green or light blue in colour, laid in triangular clusters of 20-30 eggs on the underside of leaves. As the embryo develops it may become visible through the egg, with the eyes appearing as two red spots. The top of the eggs have a circular operculum and a black-framed triangle at the top used by the nymphs to burst out of the egg. After the nymphs emerge, the eggs are opaque and white in colour. Egg incubation period is 4-5 days (Nielsen and Hamilton, 2009a; Penca and Hodges, 2018).

Nymph: First instar nymphs are approximately 2.4 mm in length with black heads, red eyes, and reddish orange abdomens with black markings. They aggregate around the egg mass and feed from eggshells, possibly acquiring endosymbionts (Taylor *et al.* 2014). Second instar nymphs emerge 3–5 days after first instars and appear dark, with rough spiny projections along the lateral edge of the thorax, lack of wings and dispersed from egg masses. They feed on host plants. Third, fourth, and fifth instar nymphs are dark brown coloured and larger in size with visible wing pads on the thorax. The third instars moult 12–13 days after eggs hatch, whereas, fourth and fifth instars emerge 19–20 and 26–27 days after egg hatch, respectively (Nielsen *et al.* 2008; Nielsen and Hamilton, 2009a; Penca and Hodges, 2018)

Adult: Adults are larger than those of most native stink bug species, the size ranging from 12 to 17mm long and 7 to 10 mm wide, without humeral (shoulder) spines and forming shield shape. They are of various shades of brown on both the top and undersides, with grey, off-white, black, copper, and bluish markings. Markings are unique to this species that include alternating light bands on the antennae and alternating dark and light bands on the thin outer edge of the abdomen. Males and females can be distinguished by their genitalia, found at the ventral tip of the abdomen. Females have a rounded ventral surface at the tip of the abdomen. Males have two prongs on either side of the abdominal tip. Legs are brown with faint white mottling or banding. The stink glands are located on the underside of the thorax, between the first and second pair of legs, and on the dorsal surface of the abdomen (Jacobs and Bemhard, 2008; Nielsen and Hamilton, 2009a; Penca and Hodges, 2018).

Life cycle: The brown marmorated stink bug is multivoltine with four to six generations in southern China (Hoffmann 1931), while, in northern China and mid-Atlantic United States, it has one to two generations per year (Yu and Zhang, 2007; Nielsen *et al.* 2008). In temperate region, the non reproductive adults overwinter in artificial and natural shelters, gradually emerge from these sites beginning in April and remain active until October. However, after emergence, they take some time to come out from diapause and to be reproductive mature before mating (Yanagi and Hagihara 1980; Nielsen *et al.* 2008). Increased temperatures and day length in the spring signal an end to the diapause period. Reproductively mature bugs can disperse into crops in search of food. Adult females are polyandrous and lay eggs in triangular clusters of 20-30 eggs (frequently, about 28) on the underside of leaves between June and September (Kawada and Kitamura 1983, Nielsen *et al.* 2008). Developmental period from egg to adult requires 32-35 days (at 30°C) with minimum and maximum developmental thresholds of 14 and 35°C, respectively (Nielsen *et al.* 2008).

Management/Control measures

Cultural control

Black light trap: Adults of brown marmorated stink bug are most attracted to white, blue, and black (UV) stimuli (Leskey *et al.* 2014). Black light traps are suggested to use in understanding the adult populations in field for monitoring (Nielsen *et al.* 2013).

Pheromone trap: The pheromone trap, with 2,4,6, E,E,Z methyl decatrienoate pheromone, has been suggested to use in baited traps for capturing adults of brown marmorated stink bugs season-long (Weber *et al.* 2014).

Biological control

Several natural enemies of *H. halys*, like arthropod predators, dipteran parasitoids and hymenopteran egg parasitoids, have been identified in Asia (Qiu 2007, Qiu *et al.* 2007, Leskey *et al.* 2012a, Leskey *et al.* 2013). So, the protection of natural enemies may be beneficial in managing the pest attack.

Chemical control

Several compounds including carbaryl, permethrin, insecticidal soap, petroleum oil, and acetamiprid are effective against different stages of brown marmorated stink bug. However, nymphs are more susceptible to insecticides than adults (Bergmann and Raupp, 2014). Amongst them, the pyrethroids (bifenthrin, permethrin, fenpropathrin, and beta-cyfluthrin), the neonicotinoids (dinotefuran, clothianidin, and thiamethoxam,), the carbamates (methomyl and oxamyl), the organophosphate (acephate), and the organochlorine (endosulfan) have been found most effective (Kuhar *et al.* 2012a,b,c,d; Krawczyk *et al.* 2012; Leskey *et al.* 2014). Unfortunately, these insecticides can harm on the natural enemy populations (Leskey *et al.* 2012b).

14. Thrips

i) Cotton seedling thrips/Onion thrips

Occurrence & severity

Cotton seedling thrips (*Thrips tabaci*) is most popularly known as 'onion thrips'. It is also called as 'potato thrips' or 'tobacco thrips'. This thripsis thought to have originated in the Mediterranean region but is now found throughout the world, starting from sea level up to 2000 meters above sea level, on all continents except Antarctica (Mau and Kessing., 1991). It occurs with its population peaks during July-August months in cotton fields of Central India (Vennila *et al.*, 2007e). In Guntur (South India), also, similar trend is found, where it occurs during *kharif* seasons with higher thrips population (Rajasekhar *et al.*, 2014). Hot and dry weather, i.e. water stress condition, favour this thrips populations (Lewis 1973), while, heavy rains wash out the thrips from plants (Harris *et al.* 1935, North and Shelton 1986). In cotton, it occurs alongside of jassids without any competition and their association is mutually exclusive. Moreover, there exists a direct relation between weediness of the field and the population of *T. tabaci* (Vennila *et al.*, 2007e). The thrips attacks on cotton seedlings and causes proliferation and branching. Early damage on seedling significantly reduces yields of fibre and seed (Klein *et al.*, 1986). The cotton seedling/onion thrips (*T. tabaci*) is highly polyphagous in nature. It has high reproductive rate, short generation time, high survival of cryptic (nonfeeding prepupa and pupa) instars, ability to reproduce without mating (parthenogenesis), ability to transmit plant pathogens and ability to develop resistance to insecticides (Morse and Hoddle 2006, Diaz-Montano *et al.* 2011).

Types of damage

Seedlings infested with thrips grow slow and the leaves become wrinkled, curl upwards and distorted with white shiny patches. It's feeding drains underlying cells, leaving air-filled spaces which impart a silvery sheen and cause distortion during growth and reduction in photosynthetic capacity; Rusty appearances in patches develop on undersurface of leaves (Vennila *et al.*, 2007e). The thrips attacks on cotton seedlings and causes proliferation and branching (Klein *et al.*, 1986). During the fruiting phase there is premature dropping of squares, and the crop maturity is delayed combined with yield reductions (Vennila *et al.*, 2007e). *Thrips tabaci* transmits all known isolates of *tomato spotted wilt virus* (TSWV), also affecting cotton worldwide (Sakimura, 1962; Groves *et al.*, 1998). Although, further studies show that some populations can transmit some TSWV isolates, but the efficiency of transmission is relatively low and there is considerable variation between populations (Wijkamp *et al.*, 1995; Chatzivassiliou *et al.*, 1999).

Scientific name & order: *Thrips tabaci* Lindeman, 1889 [Thysanoptera: Thripidae]

Synonyms

= *Heliothrips tabaci*

= *Limothrips allii* Gillette, 1983

= *Thrips allii* Sirrine & Lowe, 1894

= *Thrips bremnerii* Moulton, 1907

= *Thrips dianthi* Moulton, 1936

= *Thrips hololeucus* Bagnall, 1914

Biology

Egg: Eggs are minute; length and width of eggs are 0.23 mm and 0.08 mm, respectively. They are kidney-shaped, white or yellow, with orange tinge on maturation and eventually, with reddish eye spots (Patel *et al.* 2013). They are laid singly and inserted them into leaf tissue (Rudea and Shelton, 1995). Incubation period is 4-5 days (Fekrat *et al.* 2009), while, that is 5–10 days under cooler field conditions (Alston and Drost, 2008).

Nymph: Nymphs are very small, with sizes (0.5-) 0.7 to 0.9 (-1.2) mm, white to pale yellow or yellow in colour, resemble adults but wingless. The abdomen is divided into eight distinct segments and has a large posterior segment that is conical in shape. Their eyes have darker colouration and they have short antennae (Rudea and Shelton, 1995; Patel *et al.* 2013). The first instar is small, 0.35–0.38 mm in length, semitransparent and dull white, changing later to yellowish white. The second instar is larger and yellow (Patel *et al.* 2013). Duration of the first instar varies from 2 to 3 days, and the second instar can range from 3 to 4 days (Pourian *et al.* 2009).

Pupa: The pupae are about 1.0–1.2 mm in length and yellowish white that changing to yellow before adult emergence. They appear as an intermediate form between the nymph and the adult. They have short antennae, and the wing buds are visible but short and not functional (Rudea and Shelton, 1995; Patel *et al.* 2013).The pupal period varies from 3-10 days (Patel *et al.* 2013).

Adult: Adults are elongated, slender, measuring female about 1.0–1.3 mm in length (Orloff *et al.* 2008) and male 0.7 mm in length, with body colour pale yellow to dark brown (Rudea and Shelton, 1995). The forewings and hindwings are fringed and pale in colour. Mouthparts are piercing-sucking, antennae are 7-segmented, and eyes are gray (Patel *et al.* 2013). Adult longevity varies from (16-) 28 to 30 (-42) days (Changela 1993; Patel *et al.* 2013). Females exhibit a

1 (one) week preoviposition period and can lay eggs up to 3 weeks (Alston and Drost 2008). Adults often fly and have attraction to white and yellow colours (Rueda and Shelton 1995).

Life cycle

The adult thrips overwinters in the soil (Larentzaki *et al.* 2007) or on weeds during the off season and migrate to cotton as soon as the seedlings emerge above ground. Males are rare and the reproduction is parthenogenetic (Vennila *et al.*, 2007e). However, it can reproduce both asexually (parthenogenesis) and sexually. The most common reproductive mode is thelytoky, a parthenogenesis in which females are produced from unfertilized eggs. Other type of reproduction that occurs in this thrips is called arrhenotoky, a parthenogenesis in which males are produced from unfertilized eggs and the females from fertilized eggs. The thrips that reproduce via thelytoky differ genetically from those that reproduce via arrhenotoky (Toda and Murai, 2007; Kobayashi and Hasegawa, 2012). The adult stage overwinters in the soil in onion fields (Larentzaki *et al.* 2007) and in small grain and hay fields (Shelton and North, 1987). Eggs hatch in 4-5 days time, nymphal and pupal period lasts for 5-7 and 3-10 days, respectively (Patel *et al.* 2013). The preimaginal stage is spent in soil without feeding (Vennila *et al.*, 2007e). Adults survive for (16-) 28 to 30 (-42) days (Changela, 1993; Patel *et al.* 2013). Females exhibit one week preoviposition period and can lay eggs up to 3 weeks (Alston and Drost 2008). Life cycle of *T. tabaci* from egg to adult lasts for 13-19 days and it has about 15 overlapping generations per year. The thrips inhabit on leaves of cotton up to mid season and colonise on bolls during the late season (Vennila *et al.*, 2007e).

Management/Control measures

Use of host resistance

The Bt cotton hybrid, MRC-7351, is categorized as resistant in Dharwad (India), while, the three hybrids, viz., Chiranjivi, MRC-7918 and Bunny Bt, are found as moderately resistant (Vivek and Nandihalli, 2015).

Cultural control

It is suggested to maintain optimum plant stand during very early crop growth stage with proper weeding. The weed free condition in cotton fields from the beginning of crop growth reduces the spread of development of thrips (Vennila *et al.*, 2007e).

Spraying plant extracts

Spraying of 5% neem seed kernel extract or crude neem oil spray @ 1% suppresses thrips population during pre-squaring crop stage (Vennila *et al.*, 2007e).

Chemical control

The chemical insecticides, such as,Tolfenpyrad, 150 and 125 a.i./ha, Imidacloprid 200 SL @ 0.5 ml/lit, or Thiamethoxam 25 WG @ 1- 1.5 gm/lit of water can be sprayed if there is high degree of infestation (Vennila *et al.*, 2007; Kalyan *et al.*, 2014). However, amongst them, Tolfenpyrad, 150 and 125 a.i./ha is the most effective (Kalyan *et al.*, 2014). The efficacy of the insecticides may be further augmented by applying detergent / soap powder @ 1 gm/litre of spray (Vennila *et al.*, 2007e).

ii) Common blossom thrips (*Frankliniella schultzei*)

Occurrence & severity

Common blossom thrips, also known as 'cotton thrips', is considered to be originated from South America or Africa. It has worldwide distribution in tropical and subtropical areas, affecting wide host ranges (Palmer 1990; Vierbergen and Mantel 1991; Cluever and Smith, 2016). However, the major hosts of *Frankliniella* schultzei are cotton, groundnut, beans and pigeon pea (Kakkar *et al.*, 2010). In Argentina and Paraguay, it is the main pest at the beginning of the cotton season and destroys a large number of emerging plants. The blossom thrips occurs in two different colour morphs, a dark and a pale form (Sakimura 1969). These two forms are distributed across the globe. In Egypt, India, Kenya, Puerto Rico, Sudan, Uganda and New Guinea, mixed colonies of both colour morphs are found (Mound, 1968).

Types of damage

The thrips (*F. schultzei*) can cause direct and indirect damage to host plant. As regards the direct damage, both adults and nymphs of the thrips feed on young cotton plants and cause similar deformation symptoms to that of *Thrips tabaci* (Bournier, 1994). The thrips also feeds on flowers and pollen, leading to flower abortion. It may cause slight wounding in young bolls, but this damage does not seem to affect the maturation and quality of the seed cotton. The indirect damage by *F. schultzei* is due to the virus transmission. It is a vector of several tospoviruses. Amongst them, *tomato spotted wilt virus* (TSWV) is the most important, being transmitted in a persistent manner (Wijkamp, 1995; Thompson and van Zijl, 1996).

Scientific name & order: *Frankliniella schultzei* Trybom, 1910 (Thysanoptera: Thripidae)

Synonyms

= *Physopus schultzei* Trybom (1910)

= *Euthrips gossypii* Shiraki (1912)

= *Frankliniella dampfi* Priesner (1923)

= *Frankliniella africana* Bagnall (1926)

= *Frankliniella lycopersici* Andrewartha (1937)

= *Frankliniella ipomoeae* Moulton (1948)

= *Frankliniella schultzei nigra* Moulton (1948)

Biology

Egg: Eggs are very small, embedded in the foliage, and unlikely to be seen. (Cluever and Smith, 2016). They hatch after four days.

Nymph: Nymphs are light in colour, wingless, and are usually similar to that of other thrips (Cluever and Smith, 2016).

Pupa: Pupae appear as an intermediate form between the nymph and the adult. They have short antennae, with visible wingbuds, but not functional.

Adult: Adults exist in two colour morphs- the dark morph and the light-yellow morph. The two colour morphs are sympatric in distribution and interbreed freely, and the intermediate form is common. The adult females are 1.1-1.5 mm long, whereas adult males are 1.0-1.6 mm in length (Kakkar *et al.*, 2010). The adult has 4 major setae on the anterior margin of the pronotum. It has ctenidia (oblique rows of fine hairs) on tergite VIII that are anterior to the spiracles. The pedicel (base of antennal segment III) is smooth. Ocellar III setae (large pair nearest the ocelli) arise between the posterior ocelli. Metanotal campaniform sensilla (sensory structures that appear like two little circles) are absent. The comb on tergite VIII (row of microtrichia) is not well developed (Cluever and Smith, 2016).

Life cycle

The adult female thrips lays her eggs in the tissue of host plant. The eggs hatch after four days (differently in 5-8 days; Gahukar, 2004). There are two nymphal (larval) stages, followed by a non-feeding prepupal and a pupal stage before the adult emerges (Ananthakrishnan, 1984). The 1st and 2nd nymphal instars, prepupa and pupa take an average of 2.5, 2.5, 1.2, and 2.1 days respectively

(Pinent and Carvalho, 1998). The adult female and male longevity is approximately 13 days (Kakkar *et al.*, 2010). Life cycle usually takes about 2-5 weeks (Hill 1983c). However, at around 25 °C the life cycle takes about 12.6 or 13 days (Pinent and Carvalho, 1998).

Management/Control measure

Quarantine treatment

Gamma radiation is suggested to use in quarantine treatment. Since, irradiation at 250 Gy caused non-eclosion of eggs and pupae, failure of larval development and sterility of adults of *F. schultzei* (Yalemar *et al.*, 2001).

Cultural Control

In cotton, early sowing can reduce attacks at emergence when damage is greatest. Irrigation by flooding fields can destroy a large proportion of pupal populations in the soil (Bournier, 1994). Removal of weeds stops withering of weeds and migration of thrip pests. Marigold (*Tagetes patula*) may also be used as a trap crop for monitoring (Peres *et al.*, 2009).

Use of sticky trap

The thrips (*F. schultzei*) can be sampled using coloured sticky traps. The male thrips are most attracted to yellow sticky traps while female thrips are more attracted to pink sticky traps (Yaku *et al.* 2007).

Biological control

Two predatory mites, *Amblyseius cucumeris* and *A. swirskii* have potential to control the common blossom thrips (Kakkar *et al.* 2016).

Chemical Control

The insecticides, deltamethrin, dimethoate, imidacloprid, lambda-cyhalothrin, permethrin and pyraclofos are recorded to be effective against populations of the thirps (*F. schultzei*) (Ramello *et al.*, 1975; Nasseh and Link, 1990; Prudent, 1990; Peter and Sundararajan, 1991; Branco, 1996). Foliar application of diamide insecticide is more effective than soil application of the same. However, diamide insecticide with neonicotinoid (Imidacloprid) provides better management of the thrips (Seal *et al.*, 2014).

15. White fly

Occurrence & severity

Cotton whitefly (*Bemisia tabaci*), also called 'tobacco whitefly', 'sweetpotato whitefly', 'silverleaf whitefly', occurs worldwide in tropical, subtropical, and less predominately in temperate habitats (Greenberg *et al.*, 2000). This is a minor pest in many parts and occurs occasionally. In India, it is mostly distributed Bihar, Delhi, Haryana, Punjab, Rajasthan and Maharashtra. Cotton whiteflies remain active throughout the year moving from one plant to another. Several whitefly species infest cotton (Watve, 1971; Gamarra *et al.*, 2016). However of these, cotton whitefly (*B.tabaci* Biotype Bor MEAM1) is the most serious economic pest of cotton. An upsurge of this insect species has been noticed in the cotton system in India resulting in a heavy loss of seed cotton and reduced spinnability of cotton fibres (Sundaramurthy, 1992). They are very active from June to September but are most active during the hot and humid months of August and September. However, cold temperatures kill both the adults and the nymphs of the species (Greenberg *et al.*, 2000). It is a polyphagous pest which feed on several crops like tobacco, cassava, cabbage, cauliflower, melon, mustard, sweetpotato, brinjal, cotton, etc. and is important as a vector of leaf curl virus.

Types of damage

Whiteflies feed on the underside of the cotton leaves. They feed on sap of the leaves and release a fluid on which a sooty mould fungus grows. This affects photosynthesis and lowers the strength of the plant. Their feeding also spreads *Cotton Leaf Curl Virus*, causing disease in cotton. The virus infected plants are often indicative of *B. tabaci* colonization. Infested plants are shorter and turn yellowish brown with curled shiny leaves. They also cause shedding of buds and bolls, or poor boll opening.

Scientific name & order: ***Bemisia tabaci*** **Gennadius, 1889 (Hemiptera: Aleyrodidae).**

Synonyms

= *Bemisia argentifolii* Bellows & Perring

Biology

Egg: Eggs are laid in groups on the under surface of the leaves. These are very small, oval in shape, somewhat tapered towards the distal end, 0.21-0.25 mm in length and 0.096 -0.10 mm in width. The initial colour is pearly white or whitish-yellow that changes to dark brown near hatching. At 25°C, the eggs hatch in six to seven days (McAuslane, 2009).

Nymph: Nymphs develop through four instar stages. The first instar, commonly called a crawler, is the only mobile nymphal stage. The first instar nymph is oval in shape and measures approximately 0.27 mm in length and 0.14 mm in width. The dorsal surface of the crawler is convex while the ventral surface, appressed to the leaf surface, is flat. The crawler has three pairs of well-developed four-segmented legs, three-segmented antennae, and small eyes. It is whitish-green in colour and has two yellow spots, the mycetomes, visible in the abdomen through the integument (skin) (McAuslane, 2009). The mobile nymphs walk to find a suitable area on the leaf with adequate nutrients and moult into an immobile stage usually two to three days after eclosion from the egg. The second, third and fourth nymphal instars are immobile with atrophied legs and antennae, and small eyes. The nymphs secrete a waxy material at the margins of their body, which helps them to adhere with the leaf surface. These are flattened and oval in shape, greenish-yellow in colour, and range from 0.365 mm (second instar) to 0.662 mm (fourth instar) in length. The mycetomes are yellow. The second and third nymphal instars each last about two to three days (McAuslane, 2009). They feed by stabbing into the plant with their mouth-parts and sucking up plant juices (Johnson *et al.*, 2005). After the fourth instar, the nymph transforms into a pupal stage (also called red-eyed nymphal stage). There is no moult between the fourth nymphal instar and the pupal stage but there are morphological differences. The fourth and red-eyed nymphal stages combined lasts for five to six days (McAuslane, 2009).

Pupa: The pupae (red-eyed nymphal stage) bear 7 pairs of setae on their dorsum. The eyes are deep red coloured, with thick and yellow body. This is not a true pupal stage.

Adult

Adult whiteflies are soft, whitish-yellow while emerged, light yellow on maturation, and with white wings that is attributed by the secretion of wax across its wings and body (Brown *et al.*, 1995). The body of the female is measured 0.96 mm from the tip of the vertex (head) to the tip of the abdomen, while the body of male is somewhat smaller at 0.82 mm (McAuslane, 2009). Female adult lays 50 to 400 eggs randomly, either singly or in scattered groups, usually on the under-surface of leaves. The whitefly adult folds its wings tent-like over its body while feeding or resting (UF/IFAS, 2017).

Life cycle

These whiteflies mate shortly after emergence, reproduction being by arrhenotoky. Eggs are inserted into the underside of leaves. The emerging crawlers move away until setlling, losing their legs at the first moult and excreting copious amounts of honeydew throughout their feeding stages. It passes through the egg, four nymphal stages and the pupal stage. When feeding on cotton at 30°C, a life cycle is completed in 17 days. The adults can live for 5 to 60 days depending on the temperature and on the host plants on which they develop (Avila, 1986)

Physiological races/strains/ biotypes

The pest status of *B. tabaci* insects is complicated and it is generally accepted that, rather than one complex species, the pest is a complex of 11 genetic groups, which again composed of at least 34 morphologically indistinguishable species (Dinsdale *et al.*, 2010; De Barro *et al.*, 2011; Boykin and De Barro, 2014). Amongst them, the Middle East-Asia Minor 1 (MEAM1) cryptic species, formerly referred to as 'B biotype' and Mediterranean (MED) cryptic species, formerly referred to as 'Q biotype' are the two most widely distributed pests. These two species are the greatest threat to crops worldwide (Bethke *et al.*, 2009). A few biotypes, including the cosmopolitan MEAM1 cryptic species, the Pakistan K biotype and MED cryptic species, have become major pests, often within large mono-cropping areas where they are regularly exposed to insecticides. Exposure to insecticide treatments may have promoted other characteristics of these 'pest' biotypes, such as increased fecundity and host adaptability. A million ha of cotton, which comprises 60% of Pakistan's foreign exchange, is being affected by K biotype transmitted cotton leaf curl disease (CLCuV) (Mansoor *et al.*, 1993; Briddon and Markham, 2000).

Management/Control measures

The control of white fly is very difficult and an integrated effort is usually recommended.

Use of host resistance

Cotton resistance is associated with morphological features of the plant leaves. A heavy population of *B. tabaci* usually occurs on *G. hirsuium*. Among the different cultivars of cotton, highly pubescent types are more infested by a large population of *B. tabaci* than the glabrous types (Reddy *et al.*, 1986). The cultivars of cotton released in India for commercial exploitation are hairy and thus, become highly susceptible to *B. tabaci* (Sundaramurthy and Basu, 1989).

Cultural control

It is suggested to use best time of sowing 15 April to 15 May with proper plant to plant spacing to avoid maximum whitefly population. Other methods include the removal crop residues, weeds and alternate host plants, avoiding excessive use of nitrogenous fertilizer and reduction of the interval of irrigation. The installation of barriers by screens with very fine mesh or sticky yellow polyethylene sheets can reduce infestations.

Trap crops: Another important control is the use of trap crops. Squashes can act as trap crops for the cotton whitefly due to the flies' attraction is more to the squash crop (Schuster, 2004).

Use of botanicals

Spraying of 5% neem seed kernel extract or neem oil at 5 ml/l of water also effective against white fly.

Biocontrol

Several natural enemies have been found to prey whiteflies. There are also four species of predators that are commercially available for control of *B. tabaci*. These are, *Delphastus pusillus*, *Macrolophus caliginosus*, *Chrysoperla carnea*, and *C. rufilabris* (Hoddle, 1999). *Delphastus pusillus* is a species of small, shiny, black beetle which sucks out the contents of both adult and nymph of the cotton whitefly by piercing its exoskeleton, while, *C. rufilabris* is only able to feed on the immature stages or the larval stages of *B. tabaci* (Hoddle, 1999).

Chemical Control

Different insecticides, such as, bifenthrin, buprofezin, imidacloprid, fenpropathrin, amitraz, fenoxycarb, deltamethrin, azidirachtin and pymetrozine, are used to control cotton/silverleaf white fly all over the world (Dennehy *et al.*, 2010). However, the cotton white fly develops resistance to all groups of pesticides that have been used for its control. A rotation of insecticides that offer no cross-resistance is suggested to use in controlling *B. tabaci* infestations (Dennehy *et al.*, 2010; Cuthbertson *et al.*, 2012). Moreover, the insecticides, malathion, quinalphos and chlorpyrifos generally exhibit no or a very low level of resistance in *B. tabaci* and are effective to control the white fly (Ahmad *et al.*, 2010).

References

Abawi, G.S. and Pastor Corrales, M.A. (1990a). Root rots of beans in Latin America and Africa: Diagnosis, research methodologies and management strategies. Centro Internacional de Agricultura Tropical (CIAT), Cali, Colombia, P-114.

Abawi, G.S. and Pastor-Corrales, M.A. (1990b). Seed transmission and effect of fungicide seed treatments against *Macrophomina phaseolina* in dry edible beans. Turrialba, 40: 304–339.

Abbas, G., Arif, M.J. and Saeed, S.(2005). Systematic status of a new species of genus *Phenacoccus cockerell* (Pseudococcidae) a serious pest of cotton *Gossypium hirsutum* L. in Pakistan. Pakistan Entomologist, 27(1): 83-84.

Abbas, G., Arif, M.J., Saeed, S. and Karar, H. (2008). Increasing menace of a new mealybug, *Phenacoccus gossypiphilous*, to the economic crops of southern Asia. In: Proceedings of the XI International Symposium on Scale Insect Studies, Oeiras, Portugal, 24-27 September 2007 (eds, by Branco, M., Franco, J. C., Hodgson, C. J.). Lisbon, Portugal: ISA Press, P.322.

Abbas, G., Arif, M.J., Saeed, S. and Karar, H. (2009). A new invasive species of genus *Phenacoccus cockerell* attacking cotton in Pakistan. International Journal of Agriculture and Biology, 11(1): 54-58.

Abbas, M., Hafeez, F., Farooq, M. and Ali, A. (2015).Dusky Cotton Bug *Oxycarenus* spp. (Hemiptera: Lygaeidae): Hibernating Sites and Management by using Plant Extracts under Laboratory Conditions. Polish Journal of Entomology, 84(3): 127-136.

Abe, T. and Kono, M. (1957). Studies on the white root rot of tea bush IV. On the toxicities of cultural filtrate of the fungus. Scientific Reports of the Saikyo University of Agriculture, 8: 74–80.

Abdel-Fattah, M.I., Hosny, M.M. and El-Saadany, G. (1976). The spacing and density of cotton plants as factors affecting populations of the bollworms, *Earias insulana* Boisd. and *Pectinophora gossypiella* (Saund.). Bulletin of the Entomological Society of Egypt, 60: 85-94.

Abeygunawardena, D.V.W. and Wood, R.K.S. (1957). Factors affecting the germination of sclerotia and mycelial growth of *Sclerotium rolfsii* Sacc. Transactions of the British Mycological Society, 40(2): 221-231.

Abdel-Raheem et al. (1974). In: vascular cotton wilt (*Fusarium oxysporum* f.sp. *vasinfectum*) – Plantwise. https://www.plantwise.org/KnowledgeBank/Datasheet.aspx?dsid=24715

Abdul-Nasr, Megahed, S.M.M. and Mabtouk, A.A.M. (1973). A study on the host plants of the spiny bollworm *Earias insulana* (Boisd). Bull. Soc. Ent. Egypt, 56: 151-161.

Aboshosha, S., Attaalla, S., El-Korany, A. and El-Argawy, E. (2007). Characterization of *Macrophomina phaseolina* isolates affecting sunflower growth in El-Behera governorate, Egypt. International Journal of Agriculture and Biology, 9: 807-815.

Abro, G.H., Syed, T.S. and Dayo, Z.A. (2003). Varietal resistance of cotton against *Earias* spp. Pak. J. biol. Sci., 6: 1837-1839.

Abrol, D.P. and Kotwal, D.R. (1996). Insect pollinators of linseed (*Linum usitatissimum*Linn.) and their effect on yield components. J. Anim. Morphol. Physiol., 43: 157–161.

Accordi, S.M. (1989). The survival of *Ceratocystis fimbriata* f.sp. *platani* in the soil. Informatore Fitopathologica, 39: 57-62.

Acosta, P.P. (1986). Estimation of losses caused by *Melampsora lini* (Pers.) Lev. using sister lines of linseed. Boletin Genetico, Instituto de Fitotecnia, Castelar, Argentina, 14: 35-40.

Adam, T. (1986). Contribution à la connaissance des maladies du niébé (*Vigna unguiculata* (L.) Walp.) au Niger avec mention spéciale au *Macrophomina phaseolina* (Tassi) Goïd. Université de Renne I. Thèse de doctorat. P.117.

Adams, G.D., Foley, D.H. and Pyke, B.A. (1984). Notes on the pest status of and sampling methods for sap-sucking bugs in cotton. Aust. Cotton Growers Res. Conf., Toowoomba, pp. 60-166.

Adams, G.D. and Pyke, B.A. (1982). Sap-sucking bugs-are they pests? Aust. Cottongrower, 3(4):49-50.

Adamson W.C., Martin J.A., Minton N.A. (1975). Reaction of kenaf and Roselle on land infested with root knot nematodes. Plant Dis. Rep., 59: 130-132.

Adegbite, A.A., Agbaje, G.O. and Abidoye, J. (2008).Assessment of yield loss of roselle *(Hibiscus sabdariffa* L.) due to root-knot nematode, *Meloidogyne incognita* under field conditions. Journal of Plant Protection Research, 48(3): 267-273.

Adegbite, A.A., Agbaje, G.O., Akande, M.O., Amusa, N.A., Adetumbi, J. A. and Adeyeye, O.O. (2005). Expression of resistance to *Meloidogyne incognita* in kenaf cultivars (*Hibiscus cannibinus*) under field conditions. World Journal of Agricultural Sciences, 1(1): 14-17.

Adeniji, M.O. (1970). Reaction of kenaf and Roselle varieties to the root-knot nematode in Nigeria. Plant Dis. Rep., 54: 547–549.

Adeoti, A.A. (1991). The biology and control of Coniella leaf spot of kenaf, *Hibiscus cannabinus* L., induced by *Cconiella musaiaensis* B. Sutton var *hibisci* B. Sutton. Phd thesis, Ahmadu Bello University, Zaria, Nigeria

Adhikari, B.N., Hamilton, J.P., Zerillo, M.M., Tisserat, N., Levesque, C.A. and Buell, C.R. (2013). Comparative genomics reveals insight into virulence strategies of plant pathogenic oomycetes. PLoS One, 8, e75072.

Adu-Mensah, K. and Kumar, R. (1977). Ecology of *Oxycarenus* species (Heteroptera: Lygaeidae) in southern Ghana. Biological Journal of the Linnean Society, 9(4): 349-377.

Aegerter, B., Gordon, T. and Davis, R. (2000). Occurrence and pathogenicity of fungi associated with melon root rot and vine decline in California. Plant Dis., 84: 224–230.

Agarwal, R.A. and Gupta, G.P. (1983).Insect pests of fibre crops.pp- 147-164. in P.D. Srivastava et al. (eds). Agricultural Entomology, vol-II, All India Scientific Writers' Society, New Delhi, India.

Agarwal, R.A. and Katiyar, K.N. (1974). Ovipositional preference and damage by spotted bollworm (*Earias fabia* Stoll) in cotton. Cotton Development, 4: 28-30.

Agarwal, R.A. and Katiyar, K.N. (1979). An estimate of losses of seed kapas and seed due to bollworms on cotton in India. Indian Journal of Entomology, 41(2): 143-148.

Agrios, G.N. 1988. Plant Pathology, 3rd. ed. Academic Press, Inc.: New York. 803pp.

Agrios, G.N. (2005). Plant Pathology. 5th Edition. Academic Press, New York

Agropedia, (ND). Fusarium wilt of Cotton. Agropedia, ICAR, NAIP. Website: http://agropedia.iitk.ac.in/content/fusarium-wilt-cotton

Aheer, G. M., Ali, A. and Akram, M. (2009a). Effect of weather factors on populations of *Helicoverpa armigera* moths at cotton-based agro-ecological sites. Entomological Research, 29: 36-42.

Aheer, G.M., Shah, Z. and Saeed, M. (2009b). Seasonal history and biology of cotton mealybug, *Phenacoccus solenopsis* Tinsley. J. Agric. Res., 47(4): 423-431.

Ahemad, N. and Sultana, K. (1984). Fungitoxic effect of garlic on treatment of jute seeds. Bangladesh J. Bot., 13(2): 130-136.

Ahmad, I. and Kahn, N.H. (1980). Effects of starvation on the longevity and fecundity of red cotton bug, *Dysdercus cingulatus* (Hemiptera: Pyrrhocoridae) in successive selected generations. Appl. Entomol. Zool., 15: 182-183.

Ahmad, I. and Mohammad, F.A.(1983). Biology and immature systematics of red cotton stainer *Dysdercus koenigii* (Fabr.) (Hemiptera: Pyrrohocoridae) with a note on their phylogenetic value. Bull. Zool., 1:1-9.

Ahmad, I. and Schaefer, C.W. (1987). Food plant and feeding biology of the Pyrrhocoroidea (Hemiptera). Phytophaga, 1: 75-92.

Ahmad, J., Singh, B.R., Al-Khedhairy, A.A., Khan, J.A. and Musarrat, J. (2011). Characterization of Sunnhemp begomovirus and its geographical origin based on in silico structural and functional analysis of recombinant coat protein. African Journal of Biotechnology, 10(14): 2600-2610.

Ahmad, M., Arif, M.I. and Naveed, M. (2010). Dynamics of resistance to organophosphate and carbamate insecticides in the cotton whitefly *Bemisia tabaci* (Hemiptera: Aleyrodidae) from Pakistan. J Pest Sci., 83: 409-420.

Ahmad, T., and Mani, M.S. (1939). Two new chalcidoid parasites of the linseed midge, *Dasyneura Lini* Barnes. I. Biology and morphology of *Systasis dasyneurae* Mani. II. Description of the parasites. Indian J. Agric. Sci., 9: 531-539.

Ahmed, F.A., Alam, N. and Khair, A. (2013). Incidence and biology of *Corynespora cassiicola* (berk.& curt.) Wei. disease of okra in Bangladesh. Bangladesh J. Bot., 42(2): 265-272.

Ahmed, M.U. (1977). A review of plant parasitic nematodes in Bangladesh. A paper presented in a seminar in the Imperial College of Science and Technology, Univ. London, U.K., 3

Ahmed, S., Zia, K. and Shah, N-U-R.(2004). Validation of chemical control of gram pod borer, *Helicoverpa armigera* (Hub.) with new insecticides. International Journal of Agriculture & Biology, 6(6): 978-980.

Ainsworth, G. S., F. K. Sparrow and A. S. Sussman. (1973). The fungi. Vol. IV A, Academic press, New York. pp 567.

Akey, D.H. and Butler, G.D. Jr. (1989). Developmental rates and fecundity of apterous *Aphis gossypii* on seedlings of *Gossypium hirsutum*. Southwestern Entomologist, 14(3): 295-299.

Akintola, A.J. and Ande, A.T. (2008). First record of *Phenacoccus solenopsis* Tinsley on Hibiscus rosa-sinensis in Nigeria. Agric. J., 3(1): 1-3.

Akintola, A.J. and Ande, A.T. (2009). Pest status and ecology of five mealy bugs (Family: Pseudococcidae) in the Southern Guinea Savanna of Nigeria. Journal of Entomological Research, 33(1): 9-13.

Alan, R.S., Kobayashi, T. and Oniki, M. (1994). Angular Leaf Spot of Ramie, *Boehmeria nivea*, Caused by *Pseudocercospora boehmeriae* in Indonesia. Jpn. J. Trop. Agr., 38(1): 59-64.

Alasoadura, S. O. (1970). Culture studies on *Botryodiplodia theobromae* Pat. Mycopathologia et mycologia applicata, 42(1-2): 153–160.

Alconero, R. and Stone, E. (1969). *Phytophthora nicotianae* in roselle and kenaf in Puerto Rico. Plant Disease Reporter, 53(9): 702-705.

Aldrich, J.R., Lusby, W.R., Marron, B.E., Nicolaou, K.C., Hoffmann, M.P. and Wilson, L.T. (1989). Pheromone blends of green stink bugs and possible parasitoid selection. Naturwissenschaften, 76: 173-175.

Aldrich, J.R., Numata, H., Borges, M., Bin, F., Waite, G.K. and Lusby, W.R. (1993). Artifacts and pheromone blends from *Nezara* spp. and other stink bugs (Heteroptera: Pentatomidae). Zeit Naturforschung, 48C: 73-79.

Alegbejo, M.D. (2000) The Potential of Roselle as an Industrial Crop in Nigeria. NOMA, IAR Samaru, Zaria, 14: 1-3.

Alekseev, V.I. and Bukejs, A. (2011). Contributions to the knowledge of beetles (Insecta: Coleoptera) in the Kaliningrad region. 2. Baltic J. Coleopterol., 11(2): 209-231.

Alexandrovitch, O.R., Lopatin, I.K., Pisanenko, A.D., Tsinkevitch, V.A. and Snitko, S.M. (1996). A catalogue of Coleoptera (Insecta) of Belarus. Minsk, pp.1-103.

Alfenas, R.F., Bonaldo, S.M., Fernandes, R.A.S. and Colares, M.R.N. (2018). First report of *Choanephora cucurbitarum* on *Crotalaria spectabillis*: A highly aggressive pathogen causing a flower and stem blight in Brazil. Plant Disease, 102(7): 1456.

Aliyu, L. (2000). Roselle (*Hibiscus sabdariffa* L. var. *sabdariffa*) production as affected by pruning and sowing date. Journal of Applied Agricultural Technology, 6(1): 16-20.

Allen, D.J. (1971). Control of Zebra Disease of Agave Hybrids by Breeding for Resistance to *Phytophthora* spp. International Journal of Pest Management, 17: 42-46.

Allen, M.W. (1955). A review of the nematode genus Tylenchorhynchus. University of California Publications in Zoology, 61: 129-166.

Allen, S.J. (2007). Different Forms of Seed and Boll Damage in Australia. https://wcrc.confex.com/wcrc/2007/techprogram/P1980.HTM

Allen, S.J. and West, K.L.D. (1986). Phytophthora boll rot of cotton. Australas Plant Pathol., 15: 34.

Allen, S.J. and West, K-L.D. (1991). Predominance of race 18 of *Xanthomonas campestris* pv. *malvacearum* on cotton in Australia. Plant Dis., 75: 43-44.

Allison, C.C. (1934). Powdery mildew of flax in Minnesota. Phytopathology, 24: 305–307.

Al-Mehmmady, R.M. (2000). Biological studies on the okra moth, *Earias vittella* (Fab.) (Lepidoptera: Noctuidae) in Jeddah, Saudi Arabia. Res. Cent. Coll. Agric. King Saud Univ., Res. Bull., 96: 5-18.

Almeida, A.M.R., Abdelnoor, R.V., Calvo, E.S., Tessnman, D. and Yorinori, J.T. (2001). Genotypic diversity among Brazilian isolates of *Sclerotium rolfsii*. J. Phytopathol., 149: 493-502.

Almeida, A.M.R., Binneck, E., Piuga, F.F., Marin, S.R.R., do Valle, P.R.Z.R. and Silveira, C.A. (2008). Characterization of powdery mildew strains from soybean, bean, sunflower and weeds in Brazil using rDNA ITS sequences. Tropical Plant Pathology, 33(1): 20-26.

Al-Subhi, A., Hogenhout, S.A., Al-Yahyai, R.A. and Al-Sadi, A.M. (2017). Classification of a new phytoplasmas group 16SrII-W associated with Crotalaria witches broom disease in Oman based on multigene sequence analysis. BMC Microbiology, 17: 221.

Alston, D.G. and Drost, D. (2008). Onion thrips (*Thrips tabaci*). ENT-117-08PR. Utah Pests Fact Sheet, Utah State University Extension. Utah State University extension and Utah Plant Pest Diagnostic Laboratory, Logan, UT.

Altschul, S.F., Madden, T.L., Schaffer, A.A., Zhang, J., Zhang, Z., Miller, W. and Lopman, D.J. (1997). Gapped BLAST and PSI-BLAST: a new generation of protein database search programs. Nucleic Acids Res. 25 (17): 3389-3402.

Alvarez, L.A., Perez-Sierra, A., Armengol, J. and García-Jimenez, J. (2007). Characterization of *Phytophthora nicotianae* isolates causing collar and root rot of lavender and rosemary in Spain. Journal of Plant Pathology, 89 (2): 261-264.

Aly, A.A., Abdel-Sattar, M.A., Omar, M.R. and Abd-Elsalam, K.A. (2007). Differential antagonism of *Trichoderma* sp. against *Macrophomina phaseolina*. Journal of Plant Protection Research, 47: 91-102.

Aly, A.A., Hussein, E.M., Asran, A.A., El-Hawary, O.G.H., Mosa, A.A. and Mostafa, M.H. (2013). Non-differential interaction between isolates of *Rhizoctonia solani* and flax cultivars, Icel. Agric. Sci., 26: 37-44.

Aly, A.A., Mohamed, H.I., Mansour, M.T.M. and Omar, M.R. (2013). Suppression of Powdery Mildew on Flax by Foliar Application of Essential Oils. Journal of Phytopathology, 161(6): 376-381.

Alyokhin, A., Drummond, F., Sewell, G. and Storch, R. (2011). Differential Effects of Weather and Natural Enemies on Coexisting Aphid Populations. Environmental Entomology, 40(3): 570-580.

Amano, K. (1986). Host range and geographical distribution of the powdery mildew fungi. Jpn. Sci. Soc. Press, Tokyo, P.741.

Amate, J., Barranco, P. and Cabello, T. (1998). Identification of larvae of the principal noctuid pest species in Spain (Lepidoptera: Noctuidae). Boletiacute n de Sanidad Vegetal, Plagas, 24(1): 101-106.

Ambrosov, A. and Neofitova, V. (1978). The control of diseases of flax in concentrated and specialized areas. Sb. Nauch. Rabot BezNIIZR 'Zashchita Rastenii', 2: 3-9.

Ambuja, H., Aswathanarayana, D.S., Govindappa, M.R., Naik, M.K. and Patil, M.G. (2017). Survey and biological characterization of leaf curl disease of Mesta (*Hibiscus sabdariffa* L.). Journal of Pharmacognosy and Phytochemistry, 6(6): 1949-1954.

Amin, A.A.H. and Abdin, M.I. (1997). Preleminary analysis of field population of black cutworm, *Agrotis ipsilon* (Hufn) and some measurements for its field life table in Egypt. Proceedings Beltwide cotton Conferences, New Orleans, LA, USA, January 6-10, 1997, 2: 1116-1118.

Ammar, E., Awadallah, K. and Rashad, A. (1979). Ecological studies on *Ferrisia virgata* CKLL on Acalypha shrubs in Dokki, Giza (Homoptera, Pseudococcidae). Deutsche Entomologische Zeitschrift, 26: 207-213.

Amselem, J., Cuomo, C.A., van Kan, J. A. L., Viaud, M., Benito, E.P., Couloux, A., Coutinho, P.M., de Vries, R.P. and Dyer, P.S. (2011). Genomic Analysis of the Necrotrophic Fungal Pathogens *Sclerotinia sclerotiorum and Botrytis cinerea*. .PLOS Genetics, 7 (8): e1002230. doi:10.1371/journal.pgen.1002230. ISSN 1553-7404

Amusa, N.A. (2004). Foliar Blight of Roselle and its Effect on Yield in Tropical Forest Region of Southwestern Nigeria. Mycopathologia, 157(3): 333-338.

Amusa, N.A., Adegbite, A.A. and Oladapo, M.O. (2005). Vascular wilt of roselle (*Hibiscus sabdariffa* L. var. *sabdariffa*) in the humid forest region of south-western Nigeria. Plant Pathol J., 4: 122-125.

Amusa, N.A., Kogbe, J.O.S. and Ajibabe, S.R. (2001). Stem and foliar blight in Roselle in the tropical forest of south western Nigeria. J.Hortic.Sci.Biotechnol., 76: 681-684.

Anand, G.P.S. (1968). Particle shape of Southern Sunnhemp mosaic virus. Curr.Sci., 37(24): 706.

Andersen, T.F. (1996). A comperative taxonomic study of *Rhizoctonia* sensu *lato* employing morphological, ultrastructural and molecular methods. Mycol. Res., 100: 1117-1128.

Anderson, N.A. (1973). The *Rhizoctonia* complex in relation to seedling blight of flax. Proc of the Flax Institute of the United States. Fargo, ND, Flax Institute of the United States, 43: 4-5.

Anderson, N.A. (1977). Evaluation of the *Rhizoctonia* complex in relation to seedling blight on flax. Plant Disease Reporter, 61: 140-142.

Anderson, N.A. and Stretton, H.M. (1978). Genetic control of pathogenicity in *Rhizoctonia solani*. Phytopathology, 68: 1314-1317.

Anderson, P.A., Lawrence, G.J., Morrish, B.C., Ayliffe, M.A., Finnegan, E.J. and Ellis, J.G. (1997). Inactivation of the flax rust resistance gene M associated with loss of a repeated unit within the leucine-rich repeat coding region. Plant Cell, 9: 641-651.

Anderson, R.V. and Potter, J. (1991).Stunt nematodes: *Tylenchorhynchus, Merlinius*, and related genera.In William R. Nickle.Manual of Agricultural Nematology. New York: CRC Press. pp.359-360.

Anderson, W.H. (1948). Larvae of some genera of Calendrinae (= Rhynchophorinae) and Stromboscerinae. Annals of the Entomological Society of America, 41: 413-437.

Andkeeva, N.V. (1930). The Hemp Flea-beetle and its economic Importance. Bull. Serv. Chernoz. oblastn. sel. –khoz. oputin. Stantz. (Bull North Black Soil Zone Reg. Agric. Expt. Stn.), 1: 20-23.

Andresch, W. and Schwarz, M. (2003). Clothianidin seed treatment (Poncho Reg.)-the new technology for control of corn rootworms and secondary pests in US-corn production. Pflanzenschutz Nachrichten Bayer, 56(1): 147-172.

Andruszewska, A. and Korbas, M. (1989). Badania nad chorob¹ pasma lnu i dzialaniem fungicydów zastosowanych do jej zwalczania. [in Polish]. Phytopathol. Pol., 10: 37-46.

Angel, O-A.S., Javier, H-M., Sergio, S-I. J., Victoria, A-E., Lauro, S-R. and Antonino, A-J. (2015). Distribución y Frecuencia de Organismos Asociados a la Enfermedad "PataPrieta" de la Jamaica (*Hibiscus sabdariffa* L.) en Guerrero, México.Revista Mexicana de Fitopatología. 33(2): 173-194.

Angelova, R. (1968). Characteristics of the bionomics of the hemp flea beetle, *Psylliodes attenuatus* Koch. Rastenievudni Nauki, 5(8): 105-114.

Anioke, S. C. (1989). The biology of *Sylepta derogata* Fabricius (Pyralidae), a lepidopterous pest of okra in Eastern Nigeria. Tropical Pest Management, 35(1): 78-82.

Annecke, D.P. and Moran, V.C. (1982). Insects and mites of cultivated plants in South Africa. Durban, South Africa: Butterworths.

Anon. (1918). Reports on the agricultural department, Tortola, 1915- 16 and 1916- 17, Barbados, 33: 14·

Anon. (1955). Annu. Rep., Sisal Res. Stn. Mlingano. Tanganyika Sisal Growers Assoc.

Anon. (1960). Index of Plant Diseases in the United States.AgricultureHandbook No. 165. United States Department of Agriculture.

Anon. (1961). Relative susceptibility of olitorius, capsularis, mesta and roselle. Ann. Rep., J.A.R.l. 1960-1961, Barrackpore p. 57.

Anon., (1997). Insect Control Guide. Ohio, USA: Meister Publishing Co., P.442.

Anonymous (1949). Arizona Agricultural Experiment Station.60th Annual Report for year ending June 30, 1949. P.58.

Anonymous (1955). New Plant diseases. Agr. Gaz. New South Wales, 66(6): 312.

Anonymous (1969). Annual report of Jute Agril Res. Instt, Barrackpore, pp70.

Anonymous (1970). Annual report of Jute Agril Res. Instt, Barrackpore, pp72.

Anonymous, (1979). List of Plant Diseases in Taiwan. Plant Protection Society, Republic of China.

Anonymous (1980). Research Report, 1978-80, of the Biological and Chemical Research Institute, New South Wales Department of Agriculture. Research Report, 1978-80, of the Biological and Chemical Research Institute, New South Wales Department of Agriculture. NSW Dep. Agric. Rydalmere Australia, P.80.

Anonymous (1996). Growing Flax: Production, Management and Diagnostic Guide. Flax Council of Canada. P.56.

Anonymous, (1998). Control of flax pests and diseases.Chapter 5. In Wang, J.Z., and Zhou, M.C. (eds), Flax Cultivation and Processing, China Agricultural Press, Beijing, pp.64–90.

Anonymous, (2004). All India coordinated cotton improvement project annual report, Khandwa. P 96.

Anonymous, (2006). Army Cutworm - Alberta Agriculture and Forestry - Government of Alberta Website: https://www1.agric.gov.ab.ca/$Department/deptdocs.nsf/all/prm2475

Anonymous, (2012a). Alternaria leaf spot in cotton - Department of Agriculture and Fisheries. Website: https://www.daf.qld.gov.au/business-priorities/plants/.../cotton/.../alternaria-leaf-spot

Anonymous, (2012b) Alternaria leaf spot of cotton | agropedia, website: https://www.agropedia.iitk.ac.in/content/alternaria-leaf-spot-cotton

Anonymous, (2012-13). Report on new insect pests of Ramie. In: Annual Report 2012-13, Central Research Institute for Jute and Allied Fibres (Indian Council of Agricultural Research), Barrackpore, Kolkata - 700120, West Bengal.
Anonymous (2013). Crop Calendar for Jute and Allied Fibres, ICAR-Central Research Institute for Jute and allied fibres, Barrackpore, P-12. website: www.crijaf.org.in/pdf/cropcalendar/JafCropCalendar_2013.pdf
Anonymous, (2013-14). Annual Report, Central Research Institute for Jute and Allied Fibres (Indian Council of Agricultural Research) Barrackpore, Kolkata, West Bengal
Anonymous. (2015a). Annual Report (2014-15), ICAR-Central Research Institute for Jute and Allied Fibres, Barrackpore, Kolkata, P.124.
Anonymous (2015b). Missouri Pest Monitoring Network (September 29, 2015). European Corn Borer I.D. Integrated Pest Management, University of Missouri, Division of Plant Sciences.
Anonymous (2016). Broad mite. Insect and Related Pests of Flowers and Foliage Plants.
Anonymous (2017a). Pest Alert: Cannabis or bhang aphid. IPPM, Oregon Department of Agriculture Fact Sheets and Pest Alerts.Oregon Department of Agriculture 635 Capitol St. NE Salem, OR 97301-2532. Website: https://www.oregon.gov/ODA/shared/Documents/Publications/IPPM/CannabisAphidAlert.pdf
Anonymous (2017b). European Corn Borer and *Bacillus thuringiensis*. Plant & Soil Sciences e Library
Annonymous, (2018). Annual report 2017-2018 of Central Research Institute for Jute and Allied Fibres, Barrackpore, West Bengal (India).
Anonymous (ND1). *Somena scintillans* Walker, 1856; List Spec. Lepid. Insects Colln Br. Mus. 7: 1734; TL: North India. Website: http://ftp.funet.fi/index/Tree-of-life/insecta/lepidoptera.
Anonymous (ND2). Chapter-7. Pdf- Sodhganga. Website: sodhganga.inflibnet.ac.in-bitstream
Anonymous (ND3). India: Crop information by indiaagronet-IndiaAgroNet.Com, https://www.indiaagronet.com
Anonymous, (ND4). *Cheilomenes sexmculata* (Fabricus)- Featured insects. Website: www.nbair.res.in/Featured_insects/Cheilomenes-sexmaculata.php
Anonymous (ND5). *Phorodon humuli* (Schrank):Hop aphid, Damson-hop aphid. Website: https://www.7.inra.fr/hyppz/RAVAGEUR/6phohum.htm
Annonymous, (ND6). Xanthomonas - List of Prokaryotic names with Standing in Nomenclature, website: https://www.bacterio.net/xanthomonas.html
Anonymous, (ND7). Ilarvirus. Viral Zone, Swiss Institute of Bioinformatics. Website: https://viralzone.expasy.org/136?outline=all_by_species
Annonymous, (ND8). Stink Bugs - Cotton Insect Management Guide - Texas A&M University. Website: https://cottonbugs.tamu.edu/
Ansari, M.,Eslaminejad, T., Sarhadynejad, Z. and Eslaminejad, T. (2013). An Overview of the Roselle Plant with Particular Reference to Its Cultivation, Diseases and Usages. European Journal of Medicinal Plants, 3(1): 135-145.
Anselmi, N. and Giorcelli, A. (1990). Factors influencing the incidence of *Rosellinia necatrix* Prill in poplars. European Journal of Forest Pathology, 20: 175-183.
Anson, R.R., Knight, R.L., Evelyn, S.H. and Rose, M.F. (1948). Anglo-Egyptian Sudan Progress Report of the Cotton Breeding Stations of the Empire Cotton Growing Corporation, 1946-47, pp.49-78.
Anwar, S. (2017). Association of Kenaf leaf curl betasatellite with *Mesta yellow vein mosaic virus* in naturally infected *Malvastrum coromandelianum*. Tropical Plant Pathology, 42 (1): 46–50.
Apeyuan, K. D., Nwankiti, A. O., Oluma, H. O. A. and Ekefan,E. J. (2016a). Comparative Effect of Foliar Application of Cow Dung, Wood Ash and Benlate on the Disease Initiation and Development of Roselle (*Hibiscus sabdariffa* L.) Leaf Spot Disease Caused by *Coniella*

musaiensis Var. *hibisci* .in Makurdi, Central Nigeria. Journal of Geoscience and Environment Protection, 4: 26-32.

Apeyuan, K.D., Nwankiti, A.O., Oluma, H.O.A. and Ekefan, E.J. (2016b). Effects of Insects on Disease Initiation and Development of Roselle (*Hibiscus sabdariffa* L.) leaf spot disease caused by *Coniella musaiensis* var. *hibisci* in Makurdi, Central Nigeria. Open Journal of Medicine & Healthcare (OJMH), 1(3): 1-10.

APPS, (2009). *Colletotrichum coccodes* (Wallr.) Hughes; 09 Pathogen of the month, Pathogen of the month- June 2009.

Araki, T. (1967). Soil conditions and the violet and white root rot diseases of fruit trees. Bulletin of the Natinal Institute of Agricultural Science, Nishihara, Series C (Plant Pathology and Entomology), 21: 101-110.

Araujo, L., Guerra, A.M.N.M., Berger, P.G. and Rodrigues, F.A. (2016). Infection process of *Phakopsora gossypii* in cotton leaves. Sci. agric. (Piracicaba, Braz.), 73(4): 301-395.

Archer, T.L., Musick, G.L. and Murray, R.L. (1980). Influence of temperature and moisture on black-cutworm (Lepidoptera: Noctuidae) development and reproduction. Canadian Entomologist, 112: 665-673.

Arif, M., Fletcher, J., Marek, S.M., Melcher, U. and Ochoa-Coronaa, F.M. (2013). Development of a Rapid, Sensitive, and Field-Deployable Razor Ex Bio Detection System and Quantitative PCR Assay for Detection of *Phymatotrichopsis omnivora* Using Multiple Gene Targets. Appl. Environ. Microbiol., 79(7): 2312-2320.

Arif, M., Lin, W., Lin, L. and Islam, W. (2017). *Cotton leaf curl Multan virus* infecting *Hibiscus sabdariffa* in China. Canadian Journal of Plant Pathology, 40(1): 128-131.

Arif, M.I. and Attique, M.R. (1990). Alternative hosts in carryover of *Earas insulana* (Boisd) and *Earias vitella* (F) in Punjab, Pakistan. The Pakistan Cotton, 34: 91-96.

Arif, M.I., Rafiq, M. and Ghaffar, A. (2009). Host plants of cotton mealybug (*Phenacoccus solenopsis*): a new menace to cotton agroecosystem of Punjab, Pakistan. International Journal of Agriculture and Biology, 11(2): 163-167.

Arif, M.I., Rafiq, M., Wazir, S., Mehmood, N.and Ghaffar, A. (2012).Studies on cotton mealybug, *Phenacoccus solenopsis* (Pseudococcidae: Homoptera), and its natural enemies in Punjab, Pakistan. International Journal of Agriculture and Biology, 14: 557-562.

Arif, M.J., Abbas, G. and Saeed, S. (2007). Cotton in danger. DAWN, March., Pakistan: Pakistan Herald Publication. Website: http://www.dawn.com/weekly/science/archive/070324/science3.htm

Armes, N.J.,Jadhav, D.R. andDeSouza, K.R. (1996). A survey of insecticide resistance in *Helicoverpa armigera* in Indian subcontinent. Bull. Entomol. Res., 86: 499-514.

Armes, N.J., Jadhav, D.R. and King, A.B.S. (1992). Pyrethroid resistance in the pod borer, *Helicoverpa armigera*, in southern India. Proceedings, Brighton Crop Protection Conference, Pests and Diseases, 1992 Brighton, November 23-26, 1992 Farnham, UK; British Crop Protection Council, pp.239-244.

Armstrong, G.M. and Armstrong, J.K. (1960).American, Egyptian, and Indian cotton-wilt Fusaria: their pathogenicity and relationship to other wilt Fusaria. Tech. Bull. S.C. Agric. Exp. Stn., 1219.

Armstrong, G.M. and Armstrong, J.K. (1968). Formae speciales and races of *Fusarium oxysporum* causing a tracheomycosis in the syndrome of disease. Phytopathology, 58: 1242-1246.

Armstrong, G.M. and Armstrong, J.K. (1978). A new race (race 6) of the cotton-wilt *Fusarium* from Brazil. Plant Dis. Rep., 62 (5): 421-423.

Armsrong, J.K. and Armstrong, G.M. (1950). The Fusarium wilt of Crotalaria. Phytopath. ,40: 785.

Armstrom, J.K. and Armstrong, G.M. (1951). Physiological races of Crotalaria wilt (*Fusarium udum* f. sp. *crotalariae*). Phytopath., 41: 714.

Armstrong, J.K. and Armstrong, G.M. (1958).A race of the cotton wilt Fusarium causing wilt of the yelredo soybean and flue-cured tobacco. Plant Dis. Rep., 42: 147-151.
Arnold, K. (2009). *Halyomorpha halys* (Stal, 1855), a stink bug species newly detected among the European fauna (Insecta: Heteroptera, Pentatomidae, Pentatominae, Cappaeini). Mitteilungen des Thuringer Entomologenverbandes, 16: 10.
Arnold, M.H. and Arnold, K.M. (1960). Bacterial blight. Progress Report Experimental Stations of the Empire Cotton Growing Group Corporation (Tanganyika Territory, Lake Province), pp.8-13.
Arora, G.S. and Chaudhury, M. (1982). On the lepidopterous fauna of Arunachal Pradesh & adjoining areas of Assam in north-east India: family arctiidae , Technical Monograph No. 6, Zoological Survey of India, Calcutta, P.68.
Arora, R., Jindal, V. Rathore, P., Kumar, R. Singh, V. and Baja, L. (2010). Integrated pest management of cotton in Punjab, India. In E.B. Radeliffe, W.D., Hutchisow and R.E. Conclado (eds). Rodcliffe's IPM World Text Book. Website: http://ipmworld.umn.edu/arora, University of Minnesota, St Paul, MN.
Arora, R., Singh, B. and Dhawan, A.K. (2012).Theory and practice of integrated pest management, Scientific Publishers , Jodhpur (India).
Arowosoge, O.G.E. (2008). Investment Potential of *Hibiscus sabdariffa*. Agricultural Journal, 3: 476-481.
Arve, S.S. (2009). Population dynamics, varietal screening, biology and chemical control of mealy bug, *Phenacoccus solenopsis* Tinsley on Hibiscus rosasinensis L. M.Sc. (Agri.) Thesis submitted to NAU, Navsari.
Ashby, S, F. (1927). *Macrophomina phaseoli* (Maubl.) Comb. Nov. The pycnidial stage of *Rhizoctonia bataticola* Taub. Bull. Trans. Brit. MycoL Soc., 12: 141-147.
Ashfaq, S., Khan, I.A., Saeed, M., Saljoki, R., Kamran, S., Manzor, F., Shoail, K., Habib, K. and Sadozai, A. (2011). Population dynamics of insect pests of cotton and their natural enemies. Sarhad J. Agric., 27: 251-253.
Ashour, A.Z.A. and Afify, A.H. (2000). Biocontrol of Flax Seedling Blight with Mixtures of *Pseudomonas* Spp. Pakistan Journal of Biological Sciences, 3(3): 368-371.
Ashraf-Uz-Zaman, M (2004). Management of jute diseases in the field. M.Sc. Thesis, Department of Plant Pathology, Bangladesh Agricultural University, Mymensingh, Bangladesh.
Ashwathi, S., Ushamalini, C., Parthasarathy, S. and Nakkeeran, S. (2017). Morphological and molecular characterization of *Fusarium* spp. associated with Vascular Wilt of Coriander in India. Journal of Pharmacognosy and Phytochemistry, 6(5): 1055-1059.
Aster Yellows, (ND). Aster Yellows - Publications Saskatchewan, publications.gov.sk.ca/documents/20/84064-aster-yellows.pdf
Aswathanarayana, D.S., Govindappa, M.R., Ambuja, H., Premchand, U. and Vinaykumar, H.D. (2016). Natural occurrence of mesta leaf curl virus disease in Southern India. Abstact presented In: National symposium on Recent trends in Plant Pathological research and education: UAS Raichur 5th-6th Jan, 2016, 5-6.
Atiq, M., Shabeer, A. and Ahmed, I. (2001). Pathogenic and cultural variation in *Macrophomina phaseolina*, the cause of charcoal rot in sunflower. Sarhad. J. Agric., 2: 253-255.
Atkinson, G.F. (1890). A new Ramularia on cotton. Botanical Gazette, 15: 166-168.
Atkinson, G.F. (1892). Some diseases of cotton. III. Frenching. Ala. Agric. Exp. Stn., Auburn Univ. Bull., 41: 19-29.
Atwal, A.S. (1976). Agricultural Pests of India and Southeast Asia. Kalyani Publishers, Delhi, India.
Atwal, A.S. (1991). Pest of fibre crops. In: Agricultural pests in India and South East Asia. Kalyani Publishers, Ludhiana and New Delhi, pp 292-293.

Auger, P., Migeon, A., Ueckermann, E.A., Tiedt, L. and Navajas, M. (2013). Evidence for synonymy between *Tetranychus urticae* and *Tetranychus cinnabarinus* (Pcari, Prostigmata, Tetranychidae): review and new data. Acarologia 53(4): 383-415.

Aurivillius, C. (1891). *Alcidodes affaber*.Nouv. Arch. Mus. Hist. Nat. Paris. 3 (3): 218.

Ausemus, E.R. (1943). Breeding for disease resistance in wheat, oats, barley, and flax. Bot. Rev., 9: 207-260.

Averre, C.W. (1966). Diseases of Kenaf in Dade County, Florida and their relationship to winter vegetable production. Mimeographed Report, SUB 67-1, University of Florida, USA

Avila, A.L. (1986). Taxonomy and biology of *Bemisia tabaci*. In: Cock, M.J. W. (ed), Bemisia tabaci-literature survey on the cotton whitefly with annotated bibliography. CAB International, Wallingford; FAD, Rome, pp.3-12.

Avozani, A., Reis, E.M. and Tonin, R.B. (2014). Sensitivity loss by *Corynespora cassiicola*, isolated from soybean, to the fungicide carbendazim. Summa Phytopathologica, 40(2): 273-276.

Awadallah, K., Ammar, E., Tawfik, M. and Rashad, A. (1979). Life history of white mealy bug *Ferrisia virgata* (CKLL) (Homoptera: Pseudococcidae). Deutsche Entomologische Zeitschrift, 26: 101-110.

Awaneesh, (2009). Linseed Gall-Midge, agropedia, website: agropedia.iitk.ac.in/content/linseed-gall-midge-0

Awasthy, R.C. and Venkatakrishna, N.S. (1977). Benefit evaluation of Tocklai recommendations III control of red spider. Two Bud, 24(2): 37-38.

Ayala, A. and Ramirez, C.T. (1964). Host-range, distribution, and bibliography of the reniform nematode, *Rotylenchulus reniformis*, with special reference to Puerto Rico. Journal of Agriculture of University of Puerto Rico, 48: 140-160.

Aycock, R. A. (1966). Stem rot and other diseases caused by *Sclerotium rolfsii*. North Carolina Agricultural Experiment Station, Raleigh, Tech. Bull. 174: pp.202.

Ayre, G.L. (1990). The response of flax to different population densities of the red backed cutworm,

Euxoa ochrogaster (Gn.) (Lepidoptera: Noctuidae). Can. Entomol., 122: 21-28.

Ayyar, T.V.R. (1922). Weevil fauna of South India with special reference to species of economic importance. Bull. Agri.Rres. Inst. Pusa, 125: 15.

Ayyar, T.V.R. (1963). Hand book of economic entomology for South India. Govt. of Madras. pp. 235-237.

Ayyar, T.V.R. and Margabandhu, V. (1934). Hymenopterous parasites of economic importance in South India. Madras agric. J., 22: 431-446.

Azaizeh, M. and Bashan, Y. (1984). Chemical control of *Xanthomonas campestris* pv. *vesicatoria* in inoculated pepper fields in Israel. Annals of Applied Biology, 120(5): 60-61.

Azam, M. (2007). Diversity, distribution and abundance of weevils (Coleoptera: Curculionidae) of districts Poonch and Rajouri (Jammu). Ph.D. Thesis, University of Jammu, Jammu, India.

Baayen, R., O'Donnell, K., Bonants, P., Cigelnik, E., Kroon, L., Roebroeck, E. and Waalwijk, C. (2000). Gene genealogies and AFLP analyses in the *Fusarium oxysporum* complex identify monophyletic and nonmonophyletic formaespeciales causing wilt and rot disease. Phytopathology, 90(8): 891-900.

Babar, K.S., Pawar, K.S., Chennegowda, S.C., Aware, P.K., Ravindran, S.S. and Parimi, S. (2016). Efficacy of insecticides on adults and soil inhabiting maggots of cotton flower bud midge (*Dasineura gossypii* Fletcher). J. Farm Sci., 29(4): 461-465.

Babariya, P.M., Kabaria, B.B., Patel, V.N. and Joshi, M.D.. (2010). Chemical control of gram pod borer, *Helicoverpa armigera* Hubner infesting pigeonpea. Legume Research, 33(3): 224-226.

Babu, B.K., Srivastava, A.K., Saxena, A.K. and Arora, D.K. (2007). Identification and detection of *Macrophomina phaseolina* by using species specific oligonucleotide primers and probe. Mycol., 99: 733-739.

Babu, B.K., Saikia, R. and Arora, D.K. (2010b). Molecular characterization and diagnosis of *Macrophomina phaseolina*: a charcoal rot fungus, In: Molecular Identification of Fungi. Eds., Gherbawy Y., Voigt K., Springer, New York, pp.179-193.

Bacheler, J.S., Bradley, J.R. Jr. andEckel, C.S.(1990). Plant bugs in North Carolina: dilemma or delusion? Proceedings - Beltwide Cotton Production Research Conferences Memphis, USA; National Cotton Council, pp.203-206.

Backman, P.A. and Brenneman, T.B. (1984). Compendium of Peanut Diseases. Amer. Phytopath. Soc., St. Paul, Minnesota.

Badawy, A. (1974). The susceptibility of certain American Upland and Sakel cotton varieties to bollworms infestation (Lepidoptera: Arctiidae and Gelechiidae). Bulletin de la Societe Entomologiqued' Egypte, 58: 261-266.

Badayeva, P.K. (1930). Flax diseases in Siberia. In: Morbi Plantarum, Vol. 19, Leningrad, pp. 192-199.

Bagga, H.S. (1970). Pathogenicity studies of organisms involved in the cotton boll-rot complex. Phytopathology, 60: 158-160.

Bagla, P. (2010). Hardy Cotton-Munching Pests Are Latest Blow to GM Crops. Science, 327 (5972): 1439-1439.

Bagnall, R.S. (1913). Brief descriptions of new Thysanoptera I. Annals and Magazine of Natural History, (8)12: 290-299.

Baker, C.J., Harrington, T.C., Krauss, U. and Alfenas, A.C. (2003). Genetic variability and host specialisation in the Latin American clade of Ceratocystis fimbriata. Phytopathology, 93: 1274-1284.

Baker, D.N., Reddy, V.R.,McKinion, J.M. andWhisler, F.D.(1993).An analysis of the impact of lygus on cotton. Computers and Electronics in Agriculture, 8(2): 147-161.

Bailey, K.L., Gossen, B.D., Gugel, R.K., and Morrall, R.A.A. (2003). Diseases of field crops in Canada. 3rd ed. The Canadian Phytopathological Society, Saskatoon, Saskatchewan.

Baiswar, P., Chandra, S., Kumar, R., Ngachan, S.V. and Munda, G.C. (2010). First report of powdery mildew caused by *Podosphaera* sp. on *Hibiscus sabdariffa* in India. Australasian Plant Disease Notes, 5: 123-125.

Baker JR. (1997). Cyclamen mite and broad mite. Ornamental and Turf Insect Information Notes. (2 May 2016).

Bako, L. and Nitri, I. (1977). Pokusy s ochranou proti obalovaci konopnemu (*Grapholitha sinana* Feld). Len a Konopi (Sumperk, Czech Rep.), 15: 13-31.

Balachowski, A.S. and Mesnil, L. (1936).Les Insectes Nuisibles aux Plantes Cultivees. Imprime Busson, Paris, 2: 1429-1430.

Ballou, H.A. (1909). The Flower-Bud Maggot of Cotton (*Contarinia gossypii* Felt), W. Ind. Bull., 10: 1-28.

Ballou, H.A. (1912). Insect Pests of the Lesser Antilles, Pamphl. Ser. Dep. Agric. W. Ind., 71: 210.

Ballou, H.A. (1919. Report by the Entomologist on a visit to the Northern Islands (St. Kitts-Nevis, Antigua, Montserrat), Typescript at the Commonwealth Institute of Entomology.

Ballou, H.A. (1920). A weevil attacking Agave. Agric. News. Barbados, 19(462): 10.

Baloch, G.M., Mushtaque, M. and Ghani, M.A. (1974). Natural Enemies of *Papaver* spp. and *Cannabis sativa*. Annual report, Commonwealth Institute of Biological Control, Pakistan station, pp. 56-57.

Baly, J.S. (1878). Description of New Species and Genera of Eumolpidae. Zoological Journal of the Linnean Society, 14(75): 246-265.

Bancroft, K. (1910). A handbook of the fungus diseases of West Indian plants.Geo. Pulman and Sons, Ltd.

Bancroft, C.K. (1914). A disease affecting sisal hemp plant (*Colletotrichum agaves* Cav.). Journal of the Board of Agriculture, British Guiana, 7(4): 181-182.

Bandopadhyay, A.K. (1981). Sunnhemp- an allied fibre of great promice. Nepal Jute, 38: 19-27.

Bandopadhyay, A.K. and Mitra,G. C. (1994). Incidence of disease caused by *Choanephora cucurbitarum* (Berk & Lav.) Thaxter in sunnhemp and other Crotalaria species. J. Ind. Bot. Soc., 73. (in Sarkar et al., 2015).

Bandopadhyay, A.K. and Som, D. (1984). Tip rot and leaf blight – A new fungal disease of jute from West Bengal. Current Science, 53(4): 219-220.

Bandyopadhyay, S., Gotyal, B.S., Satpathy, S., Selvaraj, K., Tripathi, A.N. and Ali, N. (2014). Synergistic effect of Azadirachtin and *Bacillus thuringiensis* against Bihar hairy caterpillar, *Spilosoma obliqua* Walker, Biopesticides International, 10(1): 71-76.

Banks, N. (1904). Class III, Arachnida, Order 1, Acarina, four new species of injurious mites. Journal of the New York Entomological Society 12: 53-56

Banttari, E.E. and Frederiksen, R.A. (1959).Transmission of oat blue dwarf virus to flax. Phytopathology, 49: 539.

Banttari, E.E. and Moore, M.B. (1962). Virus cause of blue dwarf of Oats and its transmission to Barley and Flax. Phytopathology, 52(9): 897-902.

Banttari, E.E. and Zeyen, R.J. (1969). Chromatographic purification of the oat blue dwarf virus. Phytopathology, 59: 183-186.

Banttari, E.E. and Zeyen, R.J. (1970). Transmission of Oat Blue Dwarf Virus by the Aster Leafhopper Following Natural Acquisition or Inoculation. Phytopathology, 60: 399-402.

Banttari, E.E. and Zeyen, R.J. (1971). Histology and ultrastructure of flax crinkle. Phytopathology, 61(10): 1249-1252.

Banttari, E.E. and Zeyen, R.J. (1972a). Evidence for multiplication of the oat blue dwarf virus in the aster leafhopper. Phytopathology, 62: 745.

Barbour, K.S., Bradley, J.R. Jr. and Bachelor, J.S. (1988). Phytophagous stink bugs in North Carolina cotton: an evaluation of damage potential, pp. 280 –282. In Proceedings, 1988 Beltwide Cotton Conferences, 3-8 January 1988, New Orleans, LA. National Cotton Council, Memphis, Tennessee (USA).

Bardin, M., Nicot, P.C., Normand, P. and Lemaire, J.M. (1997).Virulence variation and DNA polymorphism in *Sphaerotheca fuliginea*, causal agent of powdery mildew of cucurbits. European Journal of Plant Pathology, 103: 545–554.

Barlow, C. A. (1962). The Influence of Temperature on the Growth of Experimental Populations of *Myzus persicae* (Sulzer) and *Macrosiphium euphorbiae* (Thomas) (Aphididae). Can. J. Zool., 40: 146-156.

Barloy, J. and Pelhate, J. (1962). Premieres observations phytopathologique srelatives aux cultures de chanvreen Anjou. Annales des Epiphyties , 13: 117-149.

Barnes, H.F. (1949). Gall midges of economic importance. Vol. VI: In: Eds: Long, H.C., Agricultural and Horticultural Series, Crosby Lockwood & Son Ltd, London

Bashan, Y. (1986). Phenols in cotton seedlings resistant and susceptible to *Alternaria macrospora*. J. Phytopathol., 116: 1-10.

Bashi, E., Rotem, J., Pinnschmidt, H. and Kranz, J. (1983). Influence of controlled environment and age on development of *Alternaria macrospora* and on shedding of leaves in cotton. Phytopathology, 73: 1145-1147.

Batalova, T.S.; Kumacheva, E.M. (1983). Pasmo on fibre flax. Zashchita Rastenii, 3: 27.

Bateman, G.L., Ferguson, A.W. and Shield, I. (1997). Factors affecting winter survival of the florally determine white lupin cv. Lucane. Annals of Applied Biology, 130(2): 349-359.

Batson, W.E. and Borazjani, A. (1984). Influence of four species of *Fusarium* on emergence and early development of cotton. In: Brown JM, ed. Proceedings of the Beltwide Cotton Production Research Conference, National Cotton Council, Memphis, Tennessee, USA, pp.20.

Beaudoin, X. (1989a). Disease and pest control. In : Flax : Breeding and Utilization (Ed. G. Marshall), ECSC, EEC, EAEC, Brussels & Luxembourg, pp. 81- 88.
Beaudoin, X. (1989b). Le traitement des semences. C. r. ann. act. 1989, Inst. Tech. Agric. Lin, Paris, pp. 337-350.
Bell, A.A. (1981). Areolate mildew pp.32-35. In: Compendium of Cotton Diseases (Ed. G.M. Watkins), American Phytopathological Society, St. Paul, Minnesota, U.S.A., P.87.
Belot, J.L. and Zambiasi, T.C. (2007). Manual de identificacao das doencas, deficienciasminerais e injurias no cultivo do algodao. BoletimTecnico, No. 36, COODETEC-CIRAD-CA, Cascavel, P.95.
Belov, D.A. and Prudnikov, V.A. (2011). Derozal fungicide application for oil flax crops protection against the diseases. Agriculture and Plant Protection Scientific Practical Journal, 2(75): 35-37.
Ben-Dov, Y. (1994). A systematic catalogue of the mealybugs of the world (Insecta: Homoptera: Coccoidea: Pseudococcidae and Putoidae) with data on geographical distribution, host plants, biology and economic importance. Andover, UK, Intercept Limited, P.686.
Bennett, R.S., Hutmacher, R.B. and Davis, R.M. (2008).Seed transmission of *Fusarium oxysporum* f. sp. *vasinfectum* race 4 in California. J. Cotton Sci., 12: 160-164.
Bercio, H. and Folwaczny, B. (1979). The check-list of the beetles of Prussia. Verlag Parzeller & Co, Fulda: P.369. (in German)
Berezhkov, R.P. (1936). *Loxostege sticticalis* L. in the forest zone of western Siberia. (In German). Trav. Inst. Sci. Biol. Tomsk, **2**: 98-131.
Bergmann E.J. and Raupp, M.J. (2014). Efficacies of common ready to use insecticides against *Halyomorpha halys* (Hemiptera: Pentatomidae). Florida Entomologist, 97(2): 791-801.
Bethke, J.A., Byrne, F.J., Hodges, G.S., McKenzie, C.L. and Shatters, R.G. (2009). First record of the Q biotype of the sweetpotato whitefly, *Bemisia tabaci*, in Guatemala. Phytoparasitica, 371: 61-64.
Bhattiprolu, S.L. (2015). Estimation of crop losses due to rust (*Phakopsora gossypii* (Arth.) Hirat.f.) disease in Bt cotton hybrid. Journal of Cotton Research and Development, 29(2): 301-304.
Blackman, R.L. and Eastop, V.F. (1984). Aphids on the World's Crops. Wiley Interscience. Chichester, UK. P.466.
Blank, L.M. (1953). The leaf spot of cotton plants, year Book of Agriculture, United States Department of Agriculture, P.315.
Blatchley, W.S. (1926). Heteroptera or True Bugs of North America with Special Reference to the Faunas of Indiana and Florida. Nature Publishing Company, Indianapolis, USA, P.1116.
Barlow, V.M. (ND). Brown stink bug, *Euschistus servus* as a vector of cotton boll pathogens within cotton fields in Southern California. Website: https://ucanr.edu/sites/2017bugsymposium/files/260983.pdf
Barnes, H.F. (1949). Gall Midges of Economic Importance. Lockwood and Sons, Ltd, London, pp.142-145.
Bartlett, B.R. (1978). Pseudococcidae. In: Introduced Parasites and Predators of Arthropod Pests and Weeds: a World Review (Ed. Clausen CP), pp. 137–170. Agriculture Handbook no. 480, USDA, Washington (US).
Basu, C.R., Bhaumik, A.R. and Sengupta, T. (1981). Chrysomelidae (Coleoptera) of Tripura (India). Rec. Zoo. Surv. India, 78: 41-61.
Bateman, D.F. and Beer, S.V. (1965). Simultaneous production and synergistic action of oxalic acid and polygalacturonase during pathogensis by *Sclerotium rolfsii*. Physiopathology, 55: 204-211.
Baturo-Ciesniewska, A., Groves, C.L., Albrecht, K.A., Grau, C.R., Willis, D.K. and Smith, D.L. (2017). Molecular Identification of *Sclerotinia trifoliorum* and *Sclerotinia sclerotiorum* Isolates from the United States and Poland. Plant Disease, 101(1): 192-199.

Baurn, T.J., Lewis, S.A. and Dean, R.A. (1994). Isolation, characterization and application of DNA probes specific to *Meloidogyne arenarea*. Phytopathology, 84(5): 489-494.
Bauers and Paul, (1991). In: Pasmo disease of flax (*Mycospharerlla linicola*)-Plantwise. https://www.plantwise.org-datasheet
Beale, R.E. (1991). Studies of resistance in linseed cultivars to *Oidium lini* and *Botrytis cinerea*. Aspects Applied Biol., 28: 85-90.
Beas-Fernández, R., De Santiago-De Santiago, A., Hernandez-Delgado, S. and Mayek-Perez, N. (2006). Characterization of Mexican and non-Mexican isolates of *Macrophomina phaseolina* based on morphological characteristics, pathogenicity on bean seeds and endoglucanase genes. Journal of Plant Pathology, 53-60.
Beck, S.D. and Hanec, W. (1960). Diapause in the European corn borer, *Pyrausta nubilalis* (Hübn.). Journal of Insect Physiology. 4(4): 304-318.
Bedlan, (1984). In: Pasmo disease of flax (*Mycospharerlla linicola*)-Plantwise. https://www.plantwise.org-datasheet.
Beers, E.H. and Hoyt, S.C. (1993).Two spotted spider mite. Orchard Pest Management Online. Tree Fruit Research & Extension Center, Washington State University, Website: https://www.jenny.tfrec.wsu.edu/opm/displaySpecies.php?pn=260
Beeson, C.F.C. (1919). The food plants of Indian forest insects. Indian Forester., 45: 312-323.
Begum, H.A., Sultana, K., Alauddin, S. and Khardker, S.(1991).Control of root knot nematode disease of jute through soil amendment (in Bangladesh). Website: agris.fao.org-agris-search-search; http://www.barc.gov.bd/home.php
Bell, J.R., Bohan, D.A., Shaw, E.M. and Weyman, G.S. (2005). Ballooning dispersal using silk: world fauna, phylogenics, genetics and models. Bull. Entomol. Res., 95: 69-114.
Bendicho-Lopez, (1998). New distributional and food plant records for twenty cuban moths. J. Lep. Soc., 52(2): 214-216.
Bengtsson, M., Karpati, Z., Szöcs, G., Reuveny, H., Yang, Z. and Witzgall, P. (2006). Flight Tunnel Responses of Z Strain European Corn Borer Females to Corn and Hemp Plants. Environmental Entomology, 35(5): 1238-1243.
Berry, R.E. (1998). Insects and Mites of economic importance in the Noerthwest. 2nd Ed. P.221.
Bes, A. (1967). Konopljinsavijac (*Grapholitha delineana* Walk.) nova stetocinakonoplje. Zastita Bilja, 18: 399-400.
Bes, A. (1978). Prilog poznavanju izgledza ostecenja i stetnosti konopljinog savijaca-*Grapholitha delineana* Walk. Radovi Poljoprivrednog Fakuleta Univerzita u Sarajevu, 26(29): 169-189.
Besoain, C.A.X. (2013). Control biologico de Phytophthora entomates y pimientos desarrollados bajo invernadero. Pp: 65-68. In: Montealegre ARJ y Perez RLM (Eds). Control biologico de enfermedades de las plantasen Chile. Santiago, Chile, Facultad de Ciencias Agronomicas, Universidad de Chile, P.147.
Beute, K. and Rodriguez-Kabana, R. (1981). Effects of soil moisture, temperature, and field environment on survival of *Sclerotium rolfsii* in Alabama and North Carolina. Phytopathology, 71: 1293-1296.
Bhadauria, N.K.S., Jakhmola, S.S. and Bhadauria, N.S. (2001). Biology and feeding potential of *Menochilus sexmaculatus* on different aphids. Indian Journal of Entomology, 63(1): 66-70.
Bhagat, R.M. and Sharma, P. (2000). Status of *Agrotis ipsilon* Hufn. in Kangra valley of Himachal Pradesh. Insect Environment, 5(4): 166-167.
Bhat, A.I., Jain, R.K., Chaudhary, V., Reddy, M.K., Ramiah, M., Chattannavar, S.N. and Varmas, A. (2002). Sequence Conservation in the Coat Protein Gene of Tobacco streak virus Isolates
Causing Necrosis Disease in Cotton, Mung bean, Sunflower and Sunn-hemp in India. Indian Journal of Biotechnology, 1: 350·356.

Bhatt, N. J. and Patel, P.K. (2001).Biology of chickpea pod borer, *Helicoverpa armigera*. Indian Journal of Entomology, 63(3): 255-259.

Bhaskaran, R and Shanmugam, N. (1973). Laboratory evaluation of some fungicides against *Alternaria macrospora* Zimm. Madras Agricultural Journal, 60(7): 646-647.

Bhaskara Reddy, B.V., Sivaprasad, Y., Naresh Kumar, C.V.M., Sujitha, A., Raja Reddy, K. and Sai Gopal, D.V.R. (2012). First Report of *Tobacco streak virus* Infecting Kenaf (*Hibiscus cannabinus*) in India. Indian J Virol., 23(1): 80-82.

Bhatt, N.A. (2016). *Heliotropium ovalifolium* Forsk., a weed, as a host of *Utetheisa pulchella* L. (Lepidoptera: Erebidae). Biotic Environment, 21(4): 61-62.

Bhattacharya, B. (2013). Advances in jute agronomy and processing and marketing. PHI Learning Pvt. Ltd., New Delhi

Bhattacharyya, B., Pujari, D., Handique, G. and Dutta, S.K. (2013). Monograph on *Lepidiota mansueta* Burmeister (Coleoptera: Scarabaeidae). Published by Directorate of Research (Agri.), Assam Agricultural University, Jorhat-785013, Assam, pp1-33.

Bhattacharyya, B., Sharma, A.K., Gawande S.P. and Sreedevi, K. (2014). Occurrence of *Lepidiota* sp. (Coleoptera: Scarabaeidae) in western Assam. Current Biotica, 8(2): 195-196.

Bhattacharya, D., Dhar, T.K., Siddiqui, K.A.I. and Ali, E. (1994). Inhibition of seed germination by *Macrophomina phaseolina* is related to phaseolinone production. J. Appl. Bacteriol., 77: 129-133.

Bhattiprolu, S.L. (2010). Efficacy of *Pseudomonas fluorescens* against bacterial blight and leaf spot diseases of cotton. Indian Journal of Agricultural Sciences, 80(3): 235-237.

Bhattiprolu, S.L. (2015). Estimation of crop losses due to rust (*Phakopsora gossypii* (Arth.) Hirat.f.) disease in Bt cotton hybrid. Journal of Cotton Research and Development, 29(2): 301-304.

Bhattiprolu, S.L and PrasadaRao, M.P. (2009a). Estimation of crop losses due to Alternaria leaf spot in cotton. Journal of Indian Society for Cotton Improvement, 14 (2): 151-154.

Bhattiprolu, S.L and PrasadaRao, M.P. (2009b). Management of Alternaria leaf spot on cotton by biological approach. Journal of Cotton Research and Development, 23(1): 135-137.

Bhosle, B.B., Sharma, O.P. and More, D.G. (2009).Management of mealybugs (*Phenacoccus solenopsis*) in rainfed cotton (*Gossypium hirsutum*). Indian Journal of Agricultural Sciences, 79(3): 199-202.

Bhuiyan, S.R. (1981). Spiral borer *Agrilus acutus* Thumb, a potential danger for Kenaf and Mesta (*Hibiscus cannabinus* and *Hibiscus sabdariffa*). Jute and Jute Fabrics-Bangladesh (in AGRIS, 1982)

Bhuyan, K.K., Saikia, G.K., Deka, M.K., Phukan B. and Barua, S.C. (2017).Evaluation of indigenous biopesticides against Red Spider Mite, *Oligonychus coffeae* (Nietner) in tea. Journal of Entomology and Zoology Studies, 5(2): 731-735.

Bianco, L.F., Martins, E.C., Toloy, R.S., Colletti, D.A.B., Teixeira, D.C. and Wulff, N.A. (2014a). First Report of Phytoplasmas Group 16SrI and 16SrXV in *Crotalaria juncea* in Brazil. Plant Disease, 98(7): 990.

Bianco, L.F., Martins, E.C., Colettii, D.A.B., Toloy, R.S. and Wulff, N.A. (2014b). Detection of 16SrIX phytoplasma (HLB phytoplasma) in Sunn Hemp (*Crotolaria juncea*) in Sao Paulo State, Brazil. Journal of Citrus Pathology, 1(1); 9.12P. htpps:/scholarship.org/uc/item/04f8q2t4.

Bigger, M. (1966). The biology and control of termites damaging field crops in Tanganyika. Bulletin of Entomological Research, 56 (3): 417-444.

Bila, J., Mortensen, C.N., Andersen, M., Vicente, J.G. and Wulff, E.G. (2013). *Xanthomonas campestris* pv. *campestris* race 1 is the causal agent of black rot of Brassicas in Southern Mozambique. African Journal of Biotechnology, 12(6): 602-610.

Bingham, C.T. (1905). The Fauna of British India, Including Ceylon and Burma Butterflies. 1 (1st ed.). London: Taylor and Francis, Ltd.
Birchfield, W. and Martin, W.J. (1968). Evaluation of nematicides for controlling nematodes of sweetpotatoes. Plant Disease Reporter, 52: 127-131.
Bird, L.S. (1959). Loss measurements caused by bacterial blight of cotton. Phytopathology, 41: 315.
Birmingham, W.A. and Hamilton, I.G. (1923). Disease of the cotton plant. Agril. Gazzette, 3: 805-810.
Bisby F.A., Roskov Y.R., Orrell T.M., Nicolson D., Paglinawan L.E., Bailly N., Kirk P.M., Bourgoin T., Baillargeon G., Ouvrard, D. (red.) (2011). Species 2000 & ITIS Catalogue of Life: 2011 Annual Checklist. Species 2000.
Bishara, I. (1969). Effect of agricultural factors on cotton yield and bollworm attack. Technical Bulletin, Plant Protection Department, UAR Ministry of Agriculture, No. 2: 28.
Bishop, A. L. (1980). The potential of *Campylomma livida* Reuter, and *Megacoelum modestum* Distant, (Hemiptera: Miridae) to damage cotton in Queensland. Australian Journal of Experimental Agriculture and Animal Husbandry, 20(103): 229-233.
Bisiach, M. (1988). White rot. In: Compendium of Grape diseases (ed Pearson, R.C. & Goheen, A.C.) pp22-23. APS Press, St. Paul, Minnesota.
Biswas, C., Dey, P., Bera, A., Satpathy, S. and Mahapatra, B.S. (2013). First Report of Bacterial Leaf Spot Caused by *Xanthomonas campestris* pv. *olitorii* on Jute Grown for Seed in India. Plant Disease, 97(8): 1109
Biswas, C., Dey, P., Mandal, K., Mitra, J., Satpathy, S. and Karmakar, P.G.(2014). First Report of a 16Sr I-B Phytoplasma Associated with Phyllody and Stem Fasciation of Flax (*Linum usitatissimum*) in India, Plant Disease, 98(9): 1267.
Biswas, C., Dey, P., Meena, P.N., Satpathy, S. and Sarkar, S.K. (2018). First report of a Subgroup 16SrLX-E ('Candidatus Phytoplasma phoenicium'- Related) phytoplasma associated with phyllody and stem fasciation of Sunn hemp (*Crotalaria juncea* L.) in India. Plant Disease, 102(7): 1445.
Biswas, C., Dey, P. and Satpathy, S. (2014). First report of bacterial leaf blight of jute (*Corchorus capsularis* L.) caused by *Xanthomonas campestris* pv. *capsularii* in India. Archives of Phytopathology and Plant Protection, 47(13): 1600-1602.
Biswas, C., Dey, P., Selvarajan, R., Bera, A., Mitra, S. and Satpathy, S.(2016). First report of Corchorus golden mosaic virus (CoGMV) infecting ramie (Boehmeria nivea) in India. Plant Disease, 100 (2): 541.
Biswas, A.C., Kabir, M.Q. and Ahmed, Q.A. (1968). Differential response of wild species of jute to stem rot (*Macrophomina phaseoli*). Agr Pakistan, 11: 165-167.
Blackman, R.L. and Eastop, V.P. (1985).Aphids on the World's Crops. John Wiley & Sons, NY. P-470.
Blake, J. H., Mueller, J. D. and Lewis, S. A. (1994). Diseases of Kenaf in South Carolina. Plant Dis., 78: 102.
Blanchard, R. A. (1942). Hibernation of the corn earworm in central and northeastern parts of the United States. USDA Tech. Bull., 83: 13.
Blank, L.M. (1971). Southwest cotton rust. Proc. Beltwide Cotton Prod. Res. Conf., pp.76-77.
Blumer, S. (1967). Echte Mehltupilze (Erysiphaceae) Ein Bestimmungsbuch fur die in Europa vorkommenden Arten. Veb. Gustav Fischer Verlag Jena.
Blunck, H. (1920). Die niederen Tierischenfeindeunserer Gespinstpflanzen. III. Landw. Zeig., 40: 259-260.
Boehm, E.W.A. and Bushnell, W.R. (1992). An ultrastructural pachytenekaryotype for *Melampsora lini*. Phytopathology, 82: 1212-1218.

Boccas, B. and Pellegrin, F. (1976). Evaluation de la resistance region de q_uelq_ues varietes de roselle au *Phytophthora parasitica* Dast. Cot. Fib. Trop., 31(2): 231-234.

Bocsa, I. and Karus, M. (1997). Der Hanfanbau: Botanik, Sorten, Anbau und Ernte. Muller Verlag, Heildelberg. 173 pp. [1998 English translation: The Cultivation of Hemp: Botany, Varieties, Cultivation and Harvesting. Hemptech, Sebastopol, CAI

Boedijn, K. B. (1962). The genus Cercospora in Indonesia. Nova Hedwigia, 3(4): 411-437

Boerema, G.H. and Howeler, L.H. (1967). *Phoma exigua* Desm. and its varieties. Persoonia, 5(1): 15-28.

Boguslawski, C.V. and Basedow, T. (2001). Studies in cotton fields in Egypt on the effects of pheromone mating disruption on *Pectinophora gossypiella* (Saund.) (Lep.,Gelechiidae), on the occurrence of other arthropods, and on yields. Journal of Applied Entomology, 125(6): 327-331.

Boldirev, V.F. (1923). Instructions for the Control of *Phytometra gamma* L., and its Larva, the Flax Worm.(In Russian), pp.29. Abs. Rev. Appl. Entomol. 1924 Series A, **12**: 23.

Bolland, H.R., Gutierrez, J. and Flechtmann, C.H.W. (1998). World catalogue of the spider mite family (Acari: Tetranychidae), with references to taxonomy, synonymy, host plants and distribution. Brill Academic Publishers, Leiden, Netherlands.

Bolley, H. L. (1901). Flax Wilt and Flax Sick Soil. No. Dak.Exp . Sta. Bull., No. 50: 60.

Bolley, H.L. (1931). Flax production in Argentina. N.D. Agric.Exp. Stn. Bull. 253.

Bolley, H.L. and Manns, T.F. (1932). Fungi of flaxseed and flax sick soil, North Dakota Experiment Station Bulletin, 259: 1-57.

Bolton, N.D., Thomma, B.P.H.J. and Nelson, B.D. (2006). *Sclerotinia sclerotiorum* (Lib.) de Bary: biology and molecular traits of a cosmopolitan pathogen. Molecular Plant Pathology, 7(1): 1-16.

Bonnemaison, L. and Bournier, A. (1964). Les thrips du lin [Flax Thrips]: *Thrips angusticeps* Uzel et *Thrips linarius* Uzel (Thysanopteres). Annales des Epiphyties (Paris), 15(2): 97-169.

Boquel, S., Giodanengo, P. and Ameline, A. (2011). Probing Behavior of Apterous and Alate Morphs of two Potato-Colonizing Aphids. Journal of Insect Science, 11(164): 1-10.

Borges, M., Zhang, A., Camp, M. and Aldrich, J. (2001). Adult diapause morph of the brown stinkbug, *Euschistus servus* (Heteroptera). Neotropical Entomology, 30(1): 179-182.

Borlaug, N.E. (1945). Variation and Variability of *Fusarium lini*. Minn. Agr. Exp. Sta. Tech. Bull., 168: P.40.

Borkar, S.G., Verma, J.P. and Singh, R.P. (1980). Transmission of *Xanthomonas malvacearum* (Smith) Dowson, the incitant of bacterial blight of cotton through spotted bollworms. Indian Journal of Entomology, 42(3): 390-397.

Bos, L. and Parlevliet, J.E. (1995). Concepts and terminology on plant/pest relationships: Towards consensus in plant pathology and crop protection. Annual Review of Phytopathology, 33: 69-102.

Booth, C. (1971). The Genus Fusarium. Commonwealth Mycological Institute, Kew, U.K., P.237.

Booth, C. (1978). *Fusarium udum*.CMI Descriptions of Pathogenic Fungi and Bacteria, P.575.

Booth, C. and Waterston, J.M. (1964). *Fusarium oxysporum*. CMI Descriptions of Pathogenic Fungi and Bacteria, 28: 1-2.

Borker, S. G. and Yamlembam, R. A. (2017). Bacterial Diseases of Crop plants. CRC Press, Taylor and francis Group, USA.

Boswell, K.F. and Gibbs, A.J. (eds). (1983). Viruses of legumes 1983. Description and keys from VIDE. Austral. Natl. Univ., Canbera.

Boudreaux, H.B. (1956). Revision of the two spotted spider mite (Acarina, Tetranychidae) complex, *Tetranychus telarius* (Linnaeus). Ann Entomol Soc Am., 49: 43-49.

Boudreaux, H.B. (1963). Biological aspect of some phytophagous mites. Annu. Rev. Entomol., 8: 137-154.
Boudreaux, H.B. and Dosse, G. (1963a). Concerning the names of some common spider mites. Advances in Acarology. Ithaca, Usa: Comstock Publishing Associates, pp.350-364.
Boudreaux, H.B. and Dosse, G. (1963b). The usefulness of the taxonomic characters in females of the genus *Tetranychus* Dufour (Acari :Tetranychidae). Acarologia, 5: 13-33.
Bouhot, D. (1968). Le *Macrophomina phaseoli* sur les plantes cultivées au Sénégal. Agr. Tropic., 23: 1172–1181.
Boulanger, J., Follin, J.C. and Bourely, J. (1984). Les hibiscus textiles en A friquetropicale, 1ere partie: conditions particulieres de production du kenaf et de la roselle. Cot. Fib. Trop, 5th Ed.
Bourgeois, A. (1970). Experimentation with cotton in Cambodia. II. Phytosanitary tests. Cotonet Fibres Tropicales, 25(2): 205-212.
Bourne, B.A. (1921). Report of the Assistant Director of Agriculture on the Entomological and Mycological Work carried out during the Season under Review.Rep. Dep. Agric. Barbados, 1919-1920, pp.10-31.
Bournier, J.P. (1994). Thysanoptera, pp, 381-391, In G. A. Matthews and J. P. Tunstall [eds.], Insect Pests of Cotton. CAB International, Wallingford UK
Bovell, J.R. (1921). Insect Pests and Fungoid Diseases, etc., Rep. Dep. Agric. Barbados 1918-1919, pp.22-27.
Bowdoin, H.C., Settlage, S.B., Orzco, B.M., Nagar, S. and Robertson, D. (1999). Geminiviruses: models for plant DNA replication, transcription and cell cycle regulation. Crit. Rev. Biochem. Mol. Biol., 18: 71-106.
Bowers, G.R. and Russin, J.S. (1999). Soybean disease management. In: Soybean production in the mid-south. L. G. Heatherly and H. F. Hodges. CRC Press
Boyce, H.R. and McKeen, C.D. (1967). Some observations on vectors and transmission of tobacco etch virus. Proc. Entomol. Soc. Ontario, 97: 68-71.
Boykin, L.M., and De Barro, P.J. (2014). A practical guide to identifying members of the Bemisia tabaci species complex: and other morphologically identical species. Front Ecol. Evolu., 2: 1-5.
Boyle, L.W. (1934). Histological Characters of Flax Roots in Relation to Resistance to Wilt and Root Rot. USDA Tech. Bull., 458: 18.
Brader, L. (1969). La faune des cottoniers sans glandes dans la partiemeridionale du Tchad. II Les chenilles de la capsule. Cotonet Fibres Tropicales, 29: 333-336.
Brakke, M.K. and Van Pelt, N. (1970). Properties of infectious ribonucleic acid from wheat streak mosaic virus. Virology, 42: 699-706.
Brambila, J. (2009). *Helicoverpa armigera* - Old World Bollworm, Field Screening Aid and Diagnostic Aid. Website: http://caps.ceris.purdue.edu/screening/helicoverpa_armigera
Branco, M.C. (1996). Chemical control of thrips (*Frankliniella schultzei*) in tomato. Horticultura. Brasileira, 14: 62.
Brandenburg, R.L. and Kennedy, G.G. (1981). Differences in dorsal integumentary lobe densities between *Tetranychus urticae* Koch and *Tetranychus cinnabarinus* (Boisduval) (Acarina: Tetranychidae) from Northeastern North Carolina. Int J Acaro., 7: 231-234.
Braun, U. (1987). A monograph of the Erysiphales (powdery mildews). Beihefte zur Nova Hedwigia, 89: P.700.
Braun, U. and Cook, R.T.A. (2012). Taxonomic manual of the Erysiphales (Powdery Mildews). CBS Biodiversity Series, 11: 707.
Braun, U. and Takamatsu, S. (2000). Phylogeny of *Erysiphe, Microsphaera, Uncinula* (Erysipheae) and *Cystotheca, Podosphaera, Sphaerptheca* (Cystotheceae) inferred from rDNA ITS sequences - some taxonomic consequences. Schlechtendalia, 4: 1-33.

Braun, U., Takamatsu, S., Heluta, V., Limkaisang, S., Divarangkoon, R., Cook, R., and Boyle, H. (2006). Phylogeny and taxonomy of powdery mildew fungi of *Erysiphe* sect.*Uncinula* on *Carpinus* species. Mycol. Prog., 5: 139-153.

Brayford, D. (1992). *Fusarium oxysporum* f. sp. *vasinfectum*. IMI Descriptions of Fungi and Bacteria. 1120: 1-2.

Brayford, D. (1996). *Fusarium oxysporum* f. sp. *lini*. IMI Descriptions of Fungi and Bacteria No. 1267. Mycopathologia, 133: 49-51.

Brentzel, W.E. (1926). The pasmo disease of flax. J. Agric. Res., 32(1): 25.

Briant, A.K. and Martyn, F.B. (1929). Diseases of cover crop. Tropic. Agr., 6: 258-260.

Briddon, R.W. and Markham, P.G. (2000). Cotton leaf curl virus disease. Virus Res., 71: 151-159.

Brink, M. (2011). Boehmeria nivea (L.) Gaudich. [Internet] Record from PROTA4U.Brink, M. & Achigan-Dako, E.G. (Editors).PROTA (Plant Resources of Tropical Africa / Ressources végétales de l'Afriquetropicale), Wageningen, Netherlands.

Brink, M. and Achigan-Dako, F.G. (eds) (2012). Plant Resources of Tropical Africa 16 Fibres. PROTA Foundation/ CTA Wageningen, Neitherlands, P-602.

Brink, M. and Escobin, R.P. (eds). (2003). Plant Resources of South-East Asia, No 17, Backhuys Publishers, Leiden .P.456.

Broadfoot, W.C., and Stakman, E.C. (1926). Physiologic specialization of *Fusarium lini*, Bolley. Phytopathology, 16: 84.

Brodeur, J. and Mcneil, J.N. (1994). Seasonal Ecology of *Aphidius nigripes* (Hymenoptera: Aphidiidae), a Parasitoid of *Macrosiphum euphorbiae* (Homoptera: Aphididae). Environmental Entomology. 23: 292-298.

Brown, H.B. and Ware, J.D.(1958). Cotton (3rd Ed.). McGrawHill Co., New York. P.566.

Brown, J.K. (1992). Virus diseases. In: Cotton diseases. Hillocks RJ, ed. Wallingford, UK: CAB International, 275-329.

Brown, J.K., Frohlich, D.R. and Rosell, R.C. (1995). The sweetpotato/silverleaf whiteflies: biotypes of *Bemisia tabaci* (Genn.), or a species complex? Annu. Rev. Entomol., 40: 511-534.

Brown, R.D. and Jones, V.P. (1983). The Broad Mite on Lemons in Southern California. California Agriculture, 37(7/8): 21-22.

Brown, S.J. (1976). Plant Pathology, pp.151-174. In Arnold, M.H. (eds), Cotton Research corporation: Agricultural research for development-The Namulonge Contribution. Cambridge University Press, London

Brudea, V. and Gheorghe, M. (1989). Results of experiments on control of the flax thrips (*Thrips linarius* Uzel) in Suceava district. (In Romanian). Cercetari Agronomice in Moldova, 22: 67-70.

Brunt, A. A., Crabtree, K., Dallwitz, M. J., Gibbs, A. J. and Watson, L. (1996). Viruses of plants. Descriptions and lists from the VIDE database. CAB International. Wellingford, U K., P.1484.

Bruton, B.D. and Fish, W.W. (2012). *Myrothecium roridum* leaf spot and stem canker on watermelon in the southern Great Plains: Possible factors for its outbreak. Online. Plant Health Progress doi:10.1094/PHP-2012-0130-01-BR.

Budhraja, K., Singh, O.P., Misra, U.S., Dhamdhere, S.V., and Deole, J.Y. (1984). Seasonal incidence, host plants, and efficacy of some insecticides against *Mycllocerus maculosus* infesting hybrid sorghum. Indian Journal of Agricultural Sciences, 54(5): 418-421.

BugGuide, (ND). Species Macrosteles quadrilineatus-Aster Leafhppper-BugGuide.Net. Department of Entomology, Iowa State University. Website: https://bugguide.net-node-view

Bull, C.T., Manceau, C., Lydon, J., Kong, H., Vinatzer, B.A. and Saux M. F.L. (2010b) *Pseudomonas cannabina* pv. *cannabina* pv. nov., and *Pseudomonas cannabina* pv. *alisalensis* (Cintas Koike and

Bull, 2000) comb. nov., are members of the emended species *Pseudomonas cannabina* (ex Sutic & Dowson 1959) Gardan, Shafik, Belouin, Brosch, Grimont & Grimont 1999. Systematic and Applied Microbiology 33: 105-115.

Bundy, C.S., McPherson, R.M. and Herzog, G.A. (2000). An examination of the external and internal signs of cotton boll damage by stink bugs (Heteroptera: Pentatomidae). J. Entomol. Sci., 35:402-410.

Burke, M., Scholl, E.H., Bird, D.McK., Schaff, J.E., Coleman, S., Crowell, R., Diener, S., Gordon, O., Graham, S., Wang, X., Windham, E., Wright, G.M. and Opperman, C.H. (2015). The plant parasite *Platylenchus coffeae* carries a minimal nematode genome. Nematology, 17: 621-637.

Burkejs, A. (2009). To the knowledge of flea beetles (Coleoptera: Chrysomelidae: Alticinae) of the Lativian fauna. Acta Zoologica Lituanica, 19(3): 109-119.

Busoli, A.C. (1993). Control of *Pectinophora gossypiella* (Saunders) and *Heliothis* spp. in cotton crop, using pheromones and by the mating disruption method. Anais da Sociedade Entomologica do Brasil, 22(1): 139-148.

Butani, D.K. (1974). Insect pests of cotton. XVII. - Effects of cotton varieties, cultural practices and fertiliser on infestation by bollworms. Coton Fibres Trop., 29(2): 237-240.

Butani, D.K. (1976). Spotted bollworms of cotton, *Earias* spp. (Noctuidae: Lepidoptera). Cotton Development, 6: 17-22.

Butani, D.K. (1979). Insects and Fruits. Periodical Expert Book Agency, India.

Butler, E.E. and Bracker, C. (1970). Morphology and cytology of *Rhizoctonia solani*. In: J.R. Parmeter Jr. (Ed.), *Rhizoctonia solani*: Biology and Pathology, University of California Press, Berkeley, pp. 32-51.

Butler, E.J. (1905). Pilzkrankheiten in Indien im Jahre 1903. Zeitsch fur Pflanzenkr, 15: 44-48.

Butler, E.J. (1918). Fungi and Disease in Plants.Thacker, Spink and Co., Calcutta, P.547.

Butter, N.S., Kular, J.S. and Singh, T.H. (1990). Effectiveness of new synthetic pyrethroids against cotton bollworms. Journal of Research, Punjab Agricultural University, 27(4): 620-622.

Buttner, P., Koch, F., Voigt, K., Quidde, T., Risch, S., Blaich, R., Brückner, B. and Tudzynski, P. (1994). Variations in ploidy among isolates of *Botrytis cinerea*: implications for genetic and molecular analyses. Current Genetics, 25(5): 445-450.

CAB International Institute of Entomology.(1990). *Pectinophora gossypiella* (Saunders). International Institute of Entomology Distribution Maps of Pests Series A (Agricultural), Map No. 13, 3rd rev.

CABI/EPPO. (2006). *Scyphophorus acupunctatus*. Distribution maps of plant pests, No. 66. CABI Head Office, Wallinford, UK.

CABI. (2007). CABI Crop protection compendium. Commonweath Agricultural Bureau, International. Website: http://www.cabicompendium.org/.CABI, (2018). *Meloidogyne arenaria* (peanut root-knot nematode)-CABI. Website: https://www.cabi.org-isc-datasheet

CABI, (ND). *Agrotis ipsilon* (black cutworm). Invasive Species Compendium. Website: https://www.cabi.org/isc/datasheet/3801

Cabral, A.O. (1951). 'Wilt' na Provincia do Sul do Save. Agron. lusit., 13: 13-18.

Caffrey, D.J. and Worthley, L.H. (1927). A progress report on the investigations of the European corn borer. USDA Bulletin, Washington, D.C., 1476: 154.

Cai, S.H., Xiong, Y.Q., Ke, D.X. and He, B.J. (1985). Studies on the dynamics of pink bollworm population and the damage on cotton. Insect Knowledge (Kunchong Zhishi), 22(2): 64-69.

Calderon-Urrea, A., Vanholme, B., Vangestel, S., Kane, S.M., Bahaji, A., Pha, K., Garcia, M., Snider, A. and Gheysen, G. (2016). Early development of root-knot nematode *Meloidogyne incognita*. BMC Development Biology, 16:10; DOI 10.1186/s12861-016-0109-x

Calhoun, D.S., Jones, J.E., Dickson, J.I., Caldwell, W.D., Burris, E., Leonard, B.R., Moore, S.H. and Aguillard, W. (1997). Registration of 'H1244' cotton. Crop Science, 37(3): 1014-1015.
Callan, E. McC. (1940). Some economic aspects of the gall midges (Diptera, Cecidomyidae) with special reference to the West Indies. Tropical Agriculture, 17(4): 63-66.
Callan, E. McC. (1941).The Gall Midges (Diptera, Cecidomyidae) of Economic Importance in the West Indies. TfOP. Agriculture, Trin., 18: 117-127.
Callaway, J.C. and Laakkonen, T.T. (1996).Cultivation of Cannabis oil seed varieties in Finland. J International Hemp Association, 3: 32-34.
Callot, H. and Brua, C. (2013). *Halyomorpha halys* (Stal, 1855), the marmorated stink bug, new species for the fauna of France (Heteroptera: Pentatomidae). LO Entomologiste, 69: 69-71.
Cameron, P.J. (1989). *Helicoverpa armigera* (Hubner), a tomato fruit worm (Lepidoptera: Noctuidae). Tech. Commun. Commw. Inst. Biol. Control, 10: 87-91.
Campo, S., Gilbert, K.B., and Carrington, J.C. (2016). Small RNA-Based Antiviral Defense in the Phytopathogenic Fungus *Colletotrichum higginsianum*. PLoS Pathog. 2016 Jun; 12(6): e1005640.Published online 2016 Jun 2. doi: 10.1371/journal.ppat.1005640
Camprag, D., Jovanic, M. andSekulic, R. (1996). Stetocinekonoplje i integralne mere suzbijanja. Zbornik Radova, 26/27: 55-68.
Cano, L.C. and Solsoloy, A.D. (1994).Occurrence, morphology and pathogenecity of *Fusarium oxysporum* f. sp. *vasinfectum* causing wilt cotton. USM CA. Journal, 5(2): 65-73.
Capinera, J.L. (2001). Handbook of vegetable pests. Academic Press, New York, P. 800.
Capinera, J.L. (2006). Common Name: Black Cutworm. Entomology and Nematology, University of Florida. Website: https://www.entnemdept.ifas.ufl.edu/creatures/veg/black_cutworm.htm.
Capinera, J.L. (2015a). Black cutworm, *Agrotis ipsilon* (Hufnagel) (Insecta: Lepidoptera: Noctuidae). UF/IFAS Extension, University of Florida. Website https://www.researchgate.net/publication/237761357
Capinera, J. L. (2015b). Corn Earworm, *Helicoverpa* (=*Heliothis*) *zea* (Boddie) (Lepidoptera: Noctuidae). EDIS, U.S. Department of Agriculture, UF/IFAS Extension Service, University of Florida, IFAS, Florida A & M University Cooperative Extension Program, and Boards of County Commissioners Cooperating.
Capinera, J.L. (2017a). Green peach aphid-featured creatures, Entomology & Nematology, UF/IFAS, University of Florida. Website: http://entnemdept.ufl.edu/creatures/veg/aphid/green_peach_aphid.htm
Capinera, J.L. (2017b). European corn borer *Ostrinia nubilalis* (Hubner). Featured Creatures, UF/IFAS, Publication Number: EENY-156, University of Florida.
Capinera, J.L. (2000). Melon aphid or cotton aphid- *Aphis gossypii* Glover. Publication Number: EENY-173, University of Florida; website: entnemdept.ufl.edu/creatures/veg/aphid/melon_aphid.htm
Capoor, S.P. (1950). A mosaic disease of sunnhemp in Bombay. Curr. Sci., 19: 22.
Capoor, S.P. (1962). Southern sunnhemp mosaic virus: A strain of tobacco mosaic virus. Phytopath., 52: 393-397.
Capoor, S.P. and Varma, P.M. (1948). Enation mosaic of *Dolichos lablab* Linn., a new virus disease. Curr. Sci., 17(2): 57
Capoor, S.P., Varma, P.M. and Uppal, B.N. (1947). A mosaic disease of *Vigna catjang* Walp. Curr. Sci., 17: 57-58.
Carpenter, G.H. (1920). Injurious insects and other animals observed in Ireland during the years 1916, 1917, and 1918. Econ. Proc. R. Dublin Soc., 2(15): 259–272.
Carpenter, C.W. (1914). The Verticillium wilt problem. Phytopath.,4: 393.

Carpenter, C.W. (1918). Wilt diseases of ' okra ' and the Verticillium wilt problem. Jour. Agr. Res. (U. S.), 12: 529-546.

Carrillo, R., Cornejo, C., Neira, M., Balocchi, O. Mundaca, N. and Cisternas, E. (2001). Larvae of noctuids associated to permanent pastures in Valdivia, Chile, during winter time. Agro. Sur., 29(1): 27-31.

Casimero, V., Nakasuji, F. and Fujisaki, K. (2001). The influence of larval and adult food quality on the calling rate and pre-calling period of females of the cotton bollworm, *Helicoverpa armigera* Hubner (Lepidoptera :Noctuidae). Appl. Entomol. Zool., 36(1): 33-40.

Cate, P. (1984). The pests of flax (In German). Der Pflanzenarzt, 37: 27-28.

Cavara, F. (1889). Materiaux de mycologieLombarde. Revue Mycologique, 11: 173-193.

Cayrol, R.A. (1972). Famille des Noctuidae.Sous-famille des Melicleptriinae, *Helicoverpa armigera* Hb. In: Balachowsky AS, ed. Entomologie appliquee a l'agriculture, Vol. 2, Paris, France: Masson etCie, pp.1431-1444.

Ceapoiu, N. (1958). Cinep Studiu monographic.Editura Academiei Republicii Populare Romine, Bucharest. P.652.

Cedeno, L. and Carrero, C. (1997). First Report of Tomato Gray Leaf Spot Caused by *Stemphylium solani* in the Andes Region of Venezuela, Plant Disease, 81(11): 1332-1332.

Cedeno, L., Mohali, C.S., Palacious-Pru, E. and Quintero, K. 1995. Passion fruit die-back caused by *Lasiodiplodia theobromae* in Venezuela. Fitopatologia Venezolana, 8(1): 7-10.

Cerny, K. (2011). A review of the subfamily Arctinae (Lepidoptera: Arctidae) from Phillipines. Entomofauna, 32(3): 29-92.

Chairin, T., Pornsuriya, C.,Thaochan, N. and Sunpapao, A. (2017). *Corynespora cassiicola* causes leaf spot disease on lettuce (*Lactuca sativa*) cultivated in hydroponic systems in Thailand. Australasian Plant Dis. Notes, 12: 16.

Chakraborty, P., Prabhu, S.T., Balikai R.A. and Udikeri, S.S. (2015). Biology of cotton flower bud maggot, *Dasineura gossypii* Fletcher- an emerging pest on Bt cotton in Karnataka, India. J. Exp. Zool. India, 18(1): 143-146.

Chakravarthy, A.K. and Sidhu, A.S.(1986). Resistance to insect pest damage in four cotton varieties in Ludhiana. Insect Science and its Application, 7(5): 647-652.

Chamberlain, D.J., Ahmad, Z., Attique, M.R. and Chaudhry, M.A. (1993). The influence of slow release PVC resin pheromone formulations on the mating behaviour and control of the cotton bollworm complex (Lepidoptera: Gelechiidae and Noctuidae) in Pakistan. Bulletin of Entomological Research, 83(3): 335-343.

Chamberlain, D.J., Critchley, B.R., Campion, D.G., Attique, M.R., Rafique, M. and Arif, M.I. (1992). Use of a multi-component pheromone formulation for control of cotton bollworms (Lepidoptera: Gelechiidae and Noctuidae) in Pakistan. Bulletin of Entomological Research, 82(4): 449-458.

Chandrasekaran, M., Balasubramanian, G. and Kuttalam, S. (2003). Ovicidal action and ovipositional deterrence of certain neem products against bhendi fruit borer (*Earias vittella* Fabricius). Madras Agric. J., 90(4-6) : 376-379.

Chan YH and Sackston W E (1973). Non-specificity of the necrosis inducing toxin of *Sclerotium bataticola*. Canadian Journal of Botany, 51: 690-692.

Chandra, G. (2017). Online guidance for the students of Zoology Termites (Insecta: Isoptera), IASZoology.com, www.iaszoology.com/termites

Chandra, S., Singh, B.P., Nigam, S.K. and Srivastava, K.M. 1975. Effect of some naturally occurring plant products on southern sunnhemp mosaic virus (SSMV). Current Science, 44: 511-512.

Chang, M.S., Chang, M.A., Lakho, A.R. and Tunio, G.H. (2002). Screening of newly developed cotton strains at Mirpurkhas against bollworm complex. Sindh Baloch. J. Plant Sci., 4: 135-139.

Changela, N.B. (1993). Bionomics, population dynamics and chemical control of thrips (*Thrips tabaci* Lindeman) on garlic, pp. 82-83. MS thesis, Sardarkrushinagar, India.
ChannaBasavanna, G.P. (1966). A contribution to the knowledge of Indian eriophyid mites (Eriophyoidea: Trombidiformes: Acarina). University of Agricultural Sciences, Hebbal, Bangalore, India, P.154.
Charchar, M.J.D., Anjos, J.R.N. and Ossipi, E. (1999). Ocorrencia de nova doenca do algodoeiroirrigado, no Brasil, causadapor *Sclerotinia sclerotiorum*. Pesquisa Agropecuaria Brasileira, Brasilia, 34(6): 1101-1106.
Charles, M., Robert, W. P. and Matthews, M. (1993). Biosystematics of the Heliothinae (Lepidoptera: Noctuidae). Annual Review of Entomology, 38(1): 207-225.
Chase, A.R. (1987). Compendium of ornamental foliage plant diseases. Amer. Phytopathol. Soc. Press, St. Paul, Minn.
Chattannavar, S.N. and Govindappa, N. H. (2009). Field study of cotton genotypes for screening against bacterial blight and rust diseases. Karnataka J. Agric. Sci., 22(1): 226-228.
Chattannavar, S.N., Hosagoudar, G.N., Ashtaputre, S.A and Ammajamma, R. (2009).Evaluation of cotton genotypes for grey mildew and Alternaria blight diseases. Journal of Cotton Research and Development, 23(1): 159-162.
Chattannavar, S.N., Kulkarni, S and Khadi, B.M. (2006). Chemical control of Alternaria blight of cotton. Journal of Cotton Research and Development, 20(1): 125-126.
Chattannavar, S.N., Sharamila, A.S., Patil, S.B and Khadi, B.M. (2004). Reaction of Bt cotton genotypes to foliar diseases. International Symposium on 'Strategies on Sustainable Cotton Production' – A Global Vision 3, Crop Production, 23-25 November 2004, University of Agricultural Sciences, Dharwad, India (Abstr.): pp.353-354.
Chatterjee, A. and Ghosh, S. K. (2007a). A new monopartite begomovirus isolated from *Hibiscus cannabinus* L. in India. Archives of Virology, 152(11): 2113–2118.
Chatterjee, A. and Ghosh, S.K. (2007b). Association of a satellite DNA beta molecule with mesta yellow vein mosaic disease. Virus Genes, 35: 835–844.
Chatterjee, A. and Ghosh, S.K. (2008). Alterations in biochemical components in mesta plants infected with Yellow vein mosaic disease. Braz.,J.,Plant Physiol., 20(4): 267-275.
Chatterjee, A., Roy, A. and Ghosh, S.K. (2008). Acquisition, transmission and host range of a begomovirus associated with yellow vein mosaic disease of mesta (*Hibiscus cannabinus* and *H. sabdariffa*). Aus Plant Pathol., 37: 511-519.
Chatterjee. A., Roy, A., Padmalatha, K.V., Malathi, V.G., Ghosh, S.K. (2005a). Yellow vein mosaic disease of kenaf (Hibiscus cannabinus) and Roselle (H. sabdariffa): a new disease in India caused by a Begomovirus. Indian J. Virol. 16: 55-56.
Chatterjee, A., Roy, A., Padmalatha, K.V., Malathi, V.G. and Ghosh, SK (2005b) Occurrence of a Begomovirus with yellow vein mosaic disease of mesta (*Hibiscus cannabinus* and *Hibiscus sabdariffa*). Australas Plant Pathol., 34: 609-610.
Chatterjee, A., Sinha, S.K., Roy, A., Sengupta, D.N. and Ghosh, S.K. (2007). Development of diagnostics for DNA A and DNA â of a Begomovirus associated with mesta yellow vein mosaic disease and detection of geminiviruses in mesta (*Hibiscus cannabinus* L. and *H. sabdariffa* L.) and some other plant species. Journal of Phytopathology, 155: 683-689.
Chatterjee, P.B. (1965). The utilization of *Bacillus thuringiensis* Berliner in the control of *Anomis sabulifera* Guenee (Lepidoptera: Noctuidae) on jute plant (*Corchorus olitorius* Linnaeus). J. Invert. Path., 7(4): 512-513.
Chatterji, S.M., Das, L.K. and Singh, B. (1988). Effect of different degrees of infestation by jute semilooper, *Anomis sabulifera* Guen. on fibre yield of jute crop. Sci. Cult., 52: 136-137.
Chatterjee, S.M. et al. (1978). Effect of environmental factors on the incidence of major pests of jute. J. Ent. Res., 2(2): 163-168.

Chattopadhyay, S. (2015). Observation of foliage infestation of teak seedlings by the weevil insect pests, *Myllocerus viridanus* Fabricius (Coleoptera: Curculionidae) from Ranchi, Jharkhand. Journal of Experimental Zoology, 18(1): 499-500.

Chaturvedi, Y. and Khera, S. (1979). Studies on taxonomy, biology and ecology of nematodes associated with jute crop. Technical Monograph, Zoological Survey of India, 2: P.105.

Chatzivassiliou, E.K., Nagata, T., Katis, N.I. and Peters, D. (1999). Transmission of tomato spotted wilt tospovirus by *Thrips tabaci* populations originating from leek. Plant Pathology, 48(6): 700-706.

Chaudhury, J., Singh, D. P. and Hazra, S.K. (ND). Sunnhemp (*Crotalaria juncea*, L). Central Research Institute for Jute & Allied Fibres (ICAR), Barrackpore, West Bengal, India. Website: https://www.doc-development-durable.org/

Chauhan, L.S., and Srivastava, K.N. (1975). Estimation of loss of yield caused by blight disease of linseed. Indian J. Farm. Sci., 3: 107-109.

Chauhan, M.S. and Suryanarayana, D. (1970).Effect of temperature, pH and light on growth and sporulation of *Myrothecium roridum* the causal organism of leaf spot disease of cotton in Haryana state. Indian Phytopath., 23: 660-663.

Chavhan, R.L., Hinge, V.R., Kadam, U.S., Kalbande, B.B. and Chakrabarty, P.K. (2018). Real-time PCR assay for rapid, efficient and accurate detection of *Paramyrothecium roridum* a leaf spot pathogen of *Gossypium* species. J. Plant Biochem. Biotechnol., 27(2): 199-207.

Chen, Q.Y., Ji, X.Q. and Sun, W.J. (1985). Identification of races of cotton Fusarium wilt in China. Scientia Agric. Sinica, 6: 1-6.

Chesters. C.G.C. and Hornby, D. (1965). Studies on *Colletotrichum coccodes*: II. Alternative host tests and tomato fruit inoculations using a typical tomato root isolate. Trans. Br. mycol. Soc. 48(4): 583-594.

Chinery, M. (1993). Insects of Britain and Northern Europe. Harper Collins Publishers Ltd, London.

Chitwood, B.G. (1949). Root-knot nematodes-Part-1. A revision of the genus *Meloidogyne* Goeldi, 1887. Proceedings of the Helminthological Society, Washington, 16: 90-104.

Choudhury, K.C.B. and Pal, A.K. (1981). Infection of sunnhemp seed with *Fusarium* spp. Seed Sci. and Technol., 93(3): 729-732.

Choudhury, K.C.B. and Pal, A.K. (1982). Infection of sunnhemp seed by *Macrophomina phaseolina*. Seed Sci. and Technol., 10(1): 151-153.

Chaudhury, N.C. (1933). Jute and its substitute. 3rd ed. W. Newsman & Co. Ltd, Calcutta, p. 24-27.

Chaurasia, S., Chaurasia, A. K., Chaurasia, S. and Chaurasia, S. (2014). Pathological Studies of Sclerotium rolfsii causing Foot-rot disease of Brinjal (*Solanum melongena* Linn.). International Journal of Pharmacy & Life Sciences, 5(1): 3257-3264.

Chen, H.F., Lu, P.W. and Zhang, J.W. (1990). Insect pests of bastfiber crops (In Chinese). Encyclopedia of Chinese Agriculture, China Agricultural Press, Beijing, P.247.

Chen, L-F., Brannigan, K., Clark, R. and Gilbertson, R.L. (2010). Characterization of Curtoviruses Associated with curly top diseases of tomato in California and Monitoring for these viruses in beet leafhoppers. Plant Disease, 94(1): 99-108.

Cherian, M.C. (1932). Pests of ganja. Madras Agricultural Journal, 20: 259-265.

Chet, I., Henis, Y. and Mitchell, R. (1967). Chemical composition of hyphal and sclerotial walls of *Sclerotium rolfsii* Sacc. Canadian Journal of Microbiology, 13(2): 137-141.

Chibber, H.M. (1914). A list of diseases of economic plants occurring in the Bombay Presidency. Bull. Dept. Agr. Bombay, 65: 17.

China Coordinating Team for Control of Root Rot Nematodes with Rugby (1990). Study on the control of root-rot nematodes in ramie using Rugby. China's Fiber Crops, 3: 21-25.

Cho, S. E., Choi, Y. J, Han, K. S., Park, M. J. and Shin, H. D. (2016). First Report of Powdery Mildew Caused by *Golovinomyces orontii* on *Lactuca sativa* in Korea. Plant Disease, 100(5): 1015.
Cho, W.D. and Shin, H.D. (eds), (2004). List of Plant Diseases in Korea, 4th edn. Korean Society of Plant Pathology
Choi, Y-J., Hong, S-B.and Shin, H-D. (2005). A re-consideration of *Pseudoperonospora cubensis* and *P. humuli* based on molecular and morphological data. Mycological Research, 109(7): 841-848.
Chowdhury, A.N. (1962). *Podagrica bowringi* Baly. As a major pest of *Hibiscus cannabinus* commonly known as 'mesta'. Indian J. Ent., 23(2): 152.
Chowdhury, H., Kar, C.S., Gotyal, B.S., Dhyani, S.K. and Tripathi, M.K. (2012). Feeding inhibitory effect of some plant extracts on jute hairy caterpillar (*Spilosoma obliqua*). Indian J. Agric. Sci., 82(1): 59-62.
Chowdhary, S. (1948). J. Indian Bot. Soc., 26: 227 (in Holliday, 1980).
Chowdhury, S. (1955). Notes on fungi of Assam-I. Lloydia, 18: 82-87.
Chowdhury, S. (1957). A Cercospora leaf spot of ramie in Assam .Transactions of the British Mycological Society, 40 (2): 260-262.
Christensen, J.J., Henderson, L. and Aragaki, M. (1953). Dissemination of *Septoria linicola*. Phytopathology. 43: 468.
Christidies, B.G. and Harrison, G.J. (1955).Cotton Growing Problems. McGraw-Hill, New York, USA.
Chupp, C. (1953). A monograph of the fungus genus Cercospora. 667 pp, by the author, Ithaca.
Chupp, C. (1954). A monograph of the fungus genus *Cercospora*. P.667.
Chupp, C. (1964). Some Colletotrichums on potato and tomato. Mycologia, 56: 393-397.
Chupp, C. and Sherf, A.F. (1960). Vegetable diseases and their control. The Ronald Press Company, New York, P.693.
Ciliberti, N., Fermaud, M., Roudet, J. and Rossi, V. (2015). Environmental Conditions Affect *Botrytis cinerea* Infection of Mature Grape Berries More Than the Strain or Transposon Genotype. Phytopathology, 105(8): 1090-1096.
Cilliers, C.J. and Bedford, E.C.G. (1978). Citrus mealy bugs. In: Bedford ECG, ed. Citrus Pests in the Republic of South Africa. Science Bulletin, Department of Agricultural Technical Services, Republic of South Africa, No. 391, 89-97.
Cisneros, J.J. and Godfrey, L.D. (2001). Midseason pest status of the cotton aphid (Homoptera: Aphididae) in California cotton: is nitrogen a key factor. Environmental Entomology, 30(3): 501-510.
Cizej, M.R. and Zolnir, M. (2003).Rastlin, Zrece [Hop flea beetle (*Psylliodes attenuata* Koch), one of the most common hop pests in Slovenia], Zbornikpredavanj in referatov 6 Slovenskega Posvetovanje o Varstvu Slovenije, 4-6 marec, 2003, pp.233-238.
Clara, F. M. and Castillo, B. S. (1950). Leaf spot of ramie, *Boehmeria nivea* (Linn.) Gaudich. Philippine Journal of Agriculture , 15: 9-21.
Clinton, P.K.S. and Peregrine, W.T.H. (1963). The zebra complex of sisal hybrid no. 11648. East Africa Agricultural and Forestry Journal, 29(2): 110-113.
Cluever, J.D. and Smith, H.A. (2016). Common blossom thrips: *Frankliniella schultzei* (Trybom). ENY-893. Morphology - UF's EDIS, University of Florida, Florida, USA.
Colbran, R.C. (1954). Problems in tree replacement. Ii. The effect of certain methods of management on the nematode fauna of an orchard soil. Journal of Australian Institute of Agricultural Science, 20: 234-237.
Colhoun, J. (1946). The relation between the contamination of flax seed with *Polyspora lini* Laff. and *Colletotrichum linicola* Pethybr. & Laff. and the incidence of disease in the crop. Annals of Applied Biology, 33(3): 260-263.

Colhoun, J. (1959). Testing for resistance to *Polyspora lini* Laff.in flax breeding. Transactions of the British Mycological Society, 42(3): 370-377.

Colhoun, J. (1960). Physiologic specialization in *Polyspora lini* Laff. Trans. Br. Mycol. Soc., 43: 150–154.

Colhoun, J. and Muskett, A.E. (1943). "Pasmo" disease of flax. Nature, 151: 223-224.

Colyer, P.D. (1988). Frequency and pathogenicity of *Fusariums* pp. associated with seedling diseases of cotton in Louisiana. Plant Disease Reporter, 72: 400-402.

Common, I.F.B. (1958). A revision of the pink bollworms of cotton [*Pectinophora* Busck (Lepidoptera: Gelechiidae)] and related genera in Australia. Aust. J. Zool., 6: 268-306.

Conlin, K.C. and McCarter, S.M. (1983). Effectiveness of selected chemicals in inhibiting *Pseudomonas syringae* pv. *tomato* in vitro and in controlling bacterial speck. Plant Dis., 67: 639-603.

Conner, K.N. (2004).Alabama Cooperative Extension System.Auburn University, Auburn, 1379.

Conner, K.N., Hagan, A.K.and Zhang, L. (2013). First Report of *Corynespora cassiicola*-Incited Target Spot on Cotton in Alabama. Plant Disease, 97(10): 1379.

Conners, I.L. (1967). An Annotated Index of Plant Diseases in Canada. Publ. 1251, Can. Dept. Agric., Res. Br., Ottawa, P.381.

Cook, A. (1981). Diseases of tropical and subtropical field, fiber, and oil crops. McMillan Publ., NY. P.450.

Cook, B.G., Pengelly, B.C., Brown, S.D., Donnelly, J.L., Eagles, D.A., Franco, M.A., Hanson, J., Mullen, B.F., Partridge, I.J., Peters, M. and Schultze-Kraft, R..(2005).Tropical forages: an interactive selection tool. *Crotalaria juncea*. CSIRO, DPI & F (Qld), CIAT, and ILRI, Brisbane, Australia

Cook, C.G. and Riggs, J.L. (1995). First report of powdery mildew on kenaf caused by *Leveillula taurica* in lower Rio Grande Valley of Texus. Plant Dis., 79: 968.

Cook, G.E., Boosalis, M.G., Dunkle, L.D. and Odvody, G.N. (1973). Survival of *Macrophomina phaseolina* in corn and sorghum stalk residues. Plant Dis. Rep., 57: 373-375.

Cooke, M. C. (1879). Some exotic fungi. Grev., 7: 94-96.

Coosemans, J. (1977). Interaction and population dynamics of *Pratylenchus penetrans* (Cobb) and *Verticillium* spp on flax. Parasitica , 33: 53-58.

COPR, (1983). Pest Control in Tropical Tomatoes. London, UK, COPR.

Coquillet, D.W.(1905. A New Cecidomyiid on Cotton, Canad. Ent., 37: 200.

Corato, U. de, (1996). Research on stem rot of kenaf caused by *Sclerotinia sclerotiorum* in Basilicata. Informatore Fitopatologico, 46(9): 30-32.

Corato, U. de, Carlucci, A., Frisullo, S. and Lops, F. (1999).The wilting of kenaf by Fusarium oxysporum [*Hibiscus cannabinus* L. - Basilicata] . AGRIS, FAO, Rome

Cork, A., Chamberlain, D.J., Beevor, P.S., Hall, D.R., Nesbitt, B.F., Campion, D.G. and Attique, M.R. (1988). Components of female sex pheromone of spotted bollworm, *Earias vittella* F. (Lepidoptera: Noctuidae): identification and field evaluation in Pakistan. Journal of Chemical Ecology, 14(3): 929-945.

Correa, S.E., Ortiz, G.C.F., Torres, de la C.M., Bautista, M.C.C., Rivera, C.M.C., Lagunes, E.L.C. and Hernández, S.J.H. (2011). Etiología de la manchaacuosa de la jamaica (*Hibiscus sabdariffa*) en Tabasco, México. Revista Mexicana de Fitopatología, 29: 165-167.

CosmeBojorques, R., Karunakaran, M., James, D., McCreightand, R .S and Garcia, E. (2011). *Podosphaera xanthii* but not *Golovinomyces cichoracearum* infects Cucurbits in a Greenhouse at Salinas, California. Cucurbit Genetics Cooperative Report, 33-34: 24-28.

Costa, M.L.N., Dhingra, O.D., DA Silva, J.L. (2005). Influence of internal seedborne *Fusarium semitectum* on cotton seedlings. Fitopatol Bras., 30: 183-186.

Cotton, R.T. (1924). A contribution toward the classification of the weevil larvae of the subfamily Calendrinae, occurring in North America. Proceedings of the United States National Museum No. 2542, 66(5): 1-11.

Cotty, P. J. (1987). Evaluation of cotton cultivar susceptibility to Alternaria leaf spot. Plant Disease, 71: 1082-1084.
Coutinho, W.M., Suassuna, N.D., Luz, C.M., Suinaga, F.A. and Silva, O.R.R.F. (2006). Bole rot of sisal caused by *Aspergillus niger* in Brazil. Fitopatol. Bras., 31(6): 605.
CPC (2004). Crop Protection Compendium.CABInternational, Wallingford, UK.
Cranham, J.E. (1982). Integrated control of damson-hop aphid, *Phorodon humuli*, on English hops: A review of recent work. Agriculture and Environment, 7(1): 63-71.
Creelman, D.W. (1965). Summary of the prevalence of plant diseases in Canada in 1964. A compilation. Canadian Plant Disease Survey, 45(2): 37-83.
CRIJAF (1977, 1986, 1987- 1989, 1990-1992). Annual Reports, ICAR-Central Research Institute for Jute and Allied Fibres, Barrackpore, West Bengal (India).
Critchley, B.R., Campion, D.G., Cavanagh, G.G., Chamberlain, D.J. and Attique, M.R. (1987). Control of three major bollworm pests of cotton in Pakistan by a single application of their combined sex pheromones. Tropical Pest Management, 33(4): 374.
Crous, P.W., Slippers, B., Wingfield, M.J., Rheeder, J., Marasas, W.F.O., et al. (2006). Phylogenetic lineages in the Botryosphaeriaceae. Studies in Mycology, 55: 235-253.
Cruickshank, R.H. (1983). Distinction between *Sclerotinia* species by their pecticzymograms. Transactions British Mycological Society , 80: 117-119.
Cui, X., Tao, X., Xie, Y., Fauquet, C.M., and Zhou, X. (2004). A DNA beta associated with Tomato yellow leaf curl China virus is required for symptom induction. J. Virol., 78: 13966-13974.
Cullen, D.W., Lees, A.K., Toth, I.K. and Duncan, J.M. (2002). Detection of *Colletotrichum coccodes* from soil and potato tubers by conventional and quantitative real-time PCR. Plant Pathology, 51: 281-292.
Curzi, M. (1932).*Corticium rolfsii* Curzi. Bollettinodella Stazione di PatologiaVegetale di Roma N.S., 11(4): 306-373.
Custodio, P.T. (1944). Henequen, sisal and similar species of Agave, their cultivation and industrial uses in the Republics of Mexico and El Salvador. Bol. Direcc. Agric. Peru, 16(48/51): 41-172.
Cuthbertson, A.G.S., Buxton, J.H., Blackburn, L.F., Mathers, J.J., Robinson, K.A., Powell, M.E., Fleming, D.A. and Bell, H.A. (2012). Eradicating *Bemisia tabaci* Q biotype on poinsettia plants in the UK. Crop Protection, 42: 42-48.
Czencz, K. (1985). Thrips pests of cultivated flax (Atermesztett lent karositotripszek). Novenyvedelem, 21 (7): 293-298. (Hungarian). Abs. Rev. Appl. Entomol. 1987 Series A, 75: 766-767.
Czyzewska, S. and Zarzycka, H. (1961).Ergebnisse der Bodeninfektions versuchean *Linum usitatissinum, Crarnbealys sinica, Cannabis sativa* und *Cucurbita pepo* var. *oleiferamiteinigen* Fusarium-Arten. Instytut Ochrony Roslin, Reguly, Polen. Report, 41: 15-36.
Dake, G.N. (1980). Effect of *Myrothecium roridum* on the germination of cotton seeds. Indian Phytopathol., 33: 591-593.
Damm, U., O'Connell, R.J., Groenewald, J.Z. and Crous, P.W. (2014).The *Colletotrichum destructivum* species complex – hemibiotrophic pathogens of forage and field crops. Studies in Mycology, 79: 49-84.
Damm, U., Woudenberg, J.H.C., Cannon, P.F. and Crous, P.W. (2009). *Colletotrichum* species with curved conidia from herbaceous hosts. Fungal Diversity. 39: 45-87.
Danilevski, A.C. and Kuznetsov, V.I. (1968). Leafroller family Tortricidae, fruit moth tribe Laspeyresia Fauna USSR, Lepidopterous Insects, 5(1): 261-263.
Darling, H.S. (1951).Pink bollworm, *Platyedra gossypiella* (Saund.), as a pest of cotton at Zeidab, northern Sudan. Bulletin of Entomological Research, 42: 157-167.

Das, C.R. and Raychaudhury, S.P. (1963). Further studies on host range and properties of sunnhemp mosaic virus. Indian Phytopath., 16(2): 214-222.

Das, C. R. and Sen Gupta, P. K. (1964). Occurrence of stem-gall disease of jute caused by *Physoderma corchori* Lingappa in West Bengal. Ind. Phytopath., 17: 180-181.

Das, G.M. (1947). *Ferrisia virgata* Ckll. (Coccidae), a pest on jute *Corchorus olitorius* L. in Bengal. Sci. & Cult., 12: 453-254.

Das, G.M. (1948). Insects and mite pests of jute. Sci. & Cult., 14 (5):186·190.

Das, G.M. (1959). Bionomics of the Tea Red Spider, Oligonychus coffeae (Nietner). Bulletin of Entomological Research, 50(2): 265-274.

Das, J. and Das, A. (2016). *Myllocerus discolor* - a pest of ber (*Ziziphus jujube*) in Tripura, India. Asian J. Biol. Life Sci., 5(2): 103-104.

Das, L.K. and Roychaudhuri, D.N. (1979). *Phyasalis minima* (Solanaceae)- a new host plant of yellow mite, *Polyphagotarsonemus latus* (Banks). Sci. & Cult., 45: 169-170.

Das, L.K. and Singh, B. (1976). The effect of *Bacillus thuringiensis* Berliner on the gut of the jute semilooper, Anomis sabulifera Guen. Sci. & Cult., 42: 567-569.

Das, L.K. and Singh, B. (1977). Economic control measures against the major pests of jute. PANS, 23(2): 159-161.

Das, L.K. and Singh, B. (1985). Number of sprays suitable against yellow mite, *Polyphagotarsonemus latus* (Banks) of jute. Sci. & Cult., 51: 376-377.

Das, L.K. and Singh, B. (1986). Effective control measures against pest complex of capsularies jute, *Corchorus capsularis*. Jute Dev. J, 6(1): 25-27.

Das, S. and Das, S.N. (1986). Host range of *Pratylenchus coffeae*. Indian Journal of Nematology, 16: 180-184.

Das, S., Ghosh, R., Paul, S., Roy, A. and Ghosh, S.K. (2008a). Complete nucleotide sequence of a monopartite begomovirus associated with yellow vein mosaic disease of mesta from North India. Arch Virol., 153: 1791-1796.

Das, S., Roy, A., Ghosh, R., Paul, S., Acharyya, S. and Ghosh, S.K. (2008b). Sequence variability and phylogenetic relationship of betasatellite isolates associated with yellow vein mosaic disease of mesta in India. Virus Genes, 37: 414-424.

Dasgupta, B., Mohanty, B. and Datta, P. (2012). Survival of *Phytophthora parasitica* causing foot and leaf rot of betelvine under different soil pH, moisture and temperature regimes. SAARC Journal of Agriculture, 10(1): 29-43.

Dasgupta, K. and Ghosh, I. (2008). Jute Insect Pest Identification: An expert system based approach. Proceeding of UGC Sponsored National Level seminar on AI & Its impact on modern IT world. DOI/UGC-NLSAI-GM-978-81-920386-4-3-08; agrocomp. atwebpages.com-PDF.

Dasgupta, N.N., De, M.L. and Raychoudhury, S.P. (1951). Structure of sunnhemp (*Crotalaria juncea Linn*) mosaic virus with electron microscope. Nature.168: 114.

Dash, J.S. (1917). Report of the Assistant Superintendent of Agriculture on the Entomological and Mycological Work carried out during the Season under Review. Rep. Dep. Agric. Barbados 1916-1917, pp.56- 60.

da Silva, M.T.B. (1987). Bioecology, damage and control of *Rachiplusia nu* (Guenée, 1852) in flax. (In Spanish). Revista do Centro de Ciencias Rurais, 17: 351–367.

Datnoff, L.E., Nemec,S. and Pernezny, K. (1995). Biological control of Fusarium crown and root rot of tomato using *Triciroderma harzianum* and *Glomus intraradices*. Biological Control, 5:427-431.

Datta, A.K. and Dutt, D.K. (1968). Studies on the anatomical changes in the stem of mesta (*Hibiscus cannabinus* Linn.)-a fibre plant-due to the tunnelling activities of the larva of *Agrilus acutus*Thunb. (Buprestidae: Coleoptera). Indian Agric., 12 (2): 148-161.

David, B.V. and Ananthakrishnan T. N. (2004). General and Applied Entomology. Tata McGraw - Hill publishing company Limited, New Delhi.
David, B.V. and Kumarswami, T. (1960). Drino (Prosturnia) inconspicua Mg (Tachinid, Diptera), a parasite of *Utethasia pulchella* Linn in South India. Madras agric. J., 47: 481.
David, J.C. (1988). *Alternaria gossypina*. CMI Descriptions of Pathogenic Fungi and Bacteria, no. 953. CAB International Mycological Institute, Kew, Survey, England.
David, J. C. (1991). *Alternaria linicola*. IMI Descriptions of Fungi and Bacteria, No 1075. Mycopathologia, 116: 53-54.
David, J.T. (1975). The energy relations of the mites *Tetranychus cinnabarinus* Boisduval and *Phytoseiulus persimilis* Athias-Henriot. PhD Thesis, University of Leicester.
Davide, R.G. (1988). Nematode Problems Affecting Agriculture in the Philippines. Journal of Nematology, 20(2): 214-218.
Davidyan, G.E. (2003). Pests: *Phylliodes attenuatus* (Koch)- Hop Flea Beetle. Interactive Agricultural Ecological Atlas of Russia and Neighbouring countries.Economic plants and their diseases, pests and weeds.AgroAtlas - Pests - *Psylliodes attenuatus* (Koch) - Hop Flea Beetle http://www.agroatlas.ru/en/content/pests/Psylliodes_attenuatus/index.html
Davis, D.W. (1961). Biology of *Tetranychus multisetis* the polychaetous form of *T. cinnabarinus* (Acarina: Tetranychidae). Ann. ent. Soc. Am., 54: 30-34.
Davis, M. R. and Raid, R. N., eds. (2002). Compendium of Umbelliferous Crop Diseases. St. Paul: The American Phytopathological Society, pp.58-59.
Davis, R.D., Moore, N.Y. and Kochman, J.K. (1996). Characterisation of a population of *Fusarium oxysporum* f. sp. *vasinfectum* causing wilt of cotton in Australia. Aust. J. Agric. Res., 47: 1143-1156.
Day, W.H. (2006). The Effect of Rainfall on the Abundance of Tarnished Plant Bug Nymphs [*Lygus lineolaris* (Palisot)] in Alfalfa Fields. Transactions of the American Entomological Society, 132(3/4): 445-450.
Dayal, R. (1997). Chytrids of India. M.D. Publications Pvt. Ltd., New delhi
De, B.K., Chattopadhya, S.B. and Arjunan, G. (1992). Effect of potash on stem rot diseases of jute caused by *Macrophomina phaseolina*. Journal of Mycopathological Research, 30: 51-55.
De, T.K. (1986). IMI 313683, IMI records for geographical unit West Bengal. Herb IMI, Kew, London.
De Barro, P.J., Liu, S.S., Boykin L.M. and Dinsdale, A.B. (2011). *Bemisia tabaci*: a statement of species status. Annu. Rev. Entomol., 56: 1-19.
De Conti, B., Bueno, V., Sampaio, M. and Lenteren, J. (2011). Development and survival of *Aulacorthum solani*, *Macrosiphum euphorbiae* and *Uroleucon ambrosiae* at six temperatures. Bulletin of Insectology, 64(1): 63-68.
deMendonça, R., Navia, D., Diniz, I., Auger, P., Navajas, M. (2011). A critical review on some closely related species of *Tetranychus* sensu stricto (Acari: Tetranychidae) in the public DNA sequences databases. Experimental and Applied Acarology, 55 (1): 1-23.
de Souza, J.T., Jesus, E.S. and de Jesus Santos, A.F. (2017). Putative pathogenicity genes of *Aspergillus niger* in sisal and their expression in vitro. Revista,Brasileira de CienciasAgrarias, 12(4): 441-445.
de Tempe, J. (1963a). Health testing of flaxseed. Proc. Internat. Seed Test Assoc., 28(1): 107-131.
de Tempte, J. (1963b). Inspection of seeds for adhering pathogenic elements. Proc. Internat. Seed Test Assoc., 28: 153-165.
de Wit, P. J. G. M. (1992). Molecular Characterization of Gene-For-Gene Systems in Plant-Fungus Interactions and the Application of Avirulence Genes in Control of Plant Pathogens. Annual Review of Phytopathology, 30 (1): 391-418.
Deacon, J. (2006). Fungal Biology. Blackwell Publishing, New York.

Debraj, Y. and Singh, R. (2010). Seasonal biology of Bihar hairy caterpillar, *Spilactia obliqua* (Lepidoptera: Arctidae) on castor, a primary food plant of eri silkworm. Uttar Pradesh Journal of Zoology, 30 (1): 17-20.

Debraj, Y., Singh, T.K. and Singh, J. (1997). Seasonal occurrence, life-history and feeding efficiency of certain predators of the cabbage aphid, *Brevicoryne brassicae* (Linn.) in north east India. Journal of Apphidology, 11(1): 143-146.

Decognet, V., Cerceau, V. and Jouan, B. (1994). Control of *Phoma exigua* var. *linicola* on flax by seed and foliar spray treatments with fungicides. Crop Protection, 13(2): 105-108.

Delattre, R. (1973). Pests and diseases in cotton growing. Phytosanitary handbook. Parasites et maladies en culture cotonniere. Manuel phytosanitaire. Paris, Institut de Recherches du Cotonet des Textiles Exotiques. France, P.146.

Dempsey, J.M. (1975). Hemp. pp.46-89 in Fiber Crops. University of Florida Press, Gainesville, FL

Denmark, H.A. (1980). Broad mite, *Polyphagotarsonemus latus* (Banks) (Acarina: Tarsonemidae) on Pittosporum. Fla. Dept. Agric. and Consumer Services Division of Plant Industry.Entomology Circular No. 213.

Dennehy, T.J., Degain, B.A., Harpold, V.S., Zaborac, M., Morin, S., Fabrick, J.A., Nichols, R.L., Brown, J.K., Byrne, F.J. and Li, X. (2010). Extraordinary resistance to insecticides reveals exotic Q biotype of *Bemisia tabaci* (Gennadius) in the New World. J. Econ. Entomol., 103: 2174-2186.

Dennis, R.W.G. (1960). British cup fungi and their allies: An introduction to the Ascomycetes. Ray Society, London. P-280.

Dennis, R.W.G. (1970). Fungus flora of Venezuela and adjacent countries. Kew Bull .Addit. Ser. III, 531 pp, Cramer, Lehre.

De Prins, J. and De Prins, W. (2018). *Haritalodes derogata* (Fabricius, 1775). Afromoths. http://www.afromoths.net/species/show/13382

Desai, B.B., Kotecha, P.M.,and Shalunkhe, D.K. (1997). Seeds Handbook: Biology, Production, Processing and Storage. Marcel Dekker Inc., New York.

Desai, S.A., Siddaramaiah, A.L., Hegde, R.K. and Kulkarni, S. (1984). Efficacy of some fungicides against *Fusarium lateritium* f. sp. *crotalariae* causing premature wilting of sunnhemp *in-vitro*. Curr. Res., 13: 10-12.

Deshmukhi, S.D., Singh, K.M. and Singh, R.N. (1992). Pest Complex and their succession on linseed *Linum usitatissimum* L. Indian.J.Ent., 54(2): 168-173.

Deshpande, V.P., Lingappa, S. and Kulkarni, K.A. (1994). Ovipositon behavious of cotton shoot weevil, *Alcidodes affaber* Aurivillus in cotton. Karnataka Journal of Agricultural Sciences, 7(4): 427-430.

Devaiah, M.G., Thippeswamy, C., Govind, R. and Thimmaih, C. (1981). Biology of cotton shoot weevil *Alcidodes affaber* Aurivillus (Coleoptera: Curculionidae) on *Hibiscus panduraeformis* Burm., Curr. Res., 10: 68-69.

Dewdney, M.M. (2015). Alternaria Brown spot 1. EDIS New Publications RSS. Web.22.

Dey, P. K. (1933). An Alternaria blight of the linseed plant. Indian Journal of Agricultural Science 3(5): 881-896.

Dhandapani, N. and Jayaraj, S. (1982). Effect of Chilli Seedling Root Dip in Insecticides for the Control of Sucking Pests. Pestology, 6(3): 5-10.

Dhar, T.K., Siddiqui, K.A.I. and Ali, E. (1982). Structure of phaseolinone, a novel phytotoxin from *Macrophomina phaseolina*. Tetrahedron Lett., 23(5): 5459–5462.

Dhawan, A.K. and Saini, S. (2009). Study on biology of mealy bug, *Phenacoccus solenopsis* Tinsley on cotton in Punjab. Symposium Abstracts. Proceedings of the National Symposium on IPM Strategies to Combat Emerging Pests in the Current Scenario of Climate Change held at CAU, Pasighet (Arunachal Pradesh) on January 28-30, 2009, pp-35-36.

Dhawan, A.K., Saini, S. and Singh, K. (2008a). Evaluation of novel and conventional insecticides for management of mealy bug, *Phenacoccus solenopsis* Tinsley in Punjab. Pesticide Research Journal, 20(2): 214-216.
Dhawan, A.K., Saini, S., Singh, K. and Bharathi, M. (2008b). Toxicity of some new insecticides against *Phenacoccus solenopsis* (Tinsley) [Hemiptera: Pseudococcidae] on cotton. Journal of Insect Science Ludhiana, 21(1): 103-105.
Dhawan, A.K., Sidhu, A.S. and Simwat, G.S. (1989). Management of bollworm through chlorpyriphos in cotton system. Journal of Research, Punjab Agricultural University, 26(4): 599-603.
Dhawan, A.K. and Simwat, G.S. (1993). Management of pink bollworm (*Pectinophora gossypiella*) through a sprayable formulation of gossyplure. Indian Journal of Agricultural Sciences, 63(3): 193-194.
Dhawan, A.K., Simwat, G.S. and Sidhu, A.S.(1988). Testing of synthetic pyrethroids for control of cotton leaf roller, *Sylepta derogate* F. Journal of Research, Punjab Agricultural University, 25(1): 70-72.
Dhawan, A.K., Simwat, G.S. and Sidhu, A.S. (1990a). Management of bollworms on cotton with synthetic pyrethroids. Journal of Insect Science, 3(2): 158-161.
Dhawan, A.K., Simwat, G.S. and Sidhu, A.S. (1990b). Square shedding due to bollworm in different varieties of *Gossypium arboreum*. J. Res. Punjab Agri. Univ., 27: 606-610.
Dhawan, A.K., Simwat, G.S. and Sidhu, A.S. (1992). Field evaluation of asymethrin (Chinmix) for bollworm control on cotton. Indian Journal of Plant Protection, 20(1): 24-26.
Dhawan, A.K., Singh, K. and Singh, R. (2009a). Evaluation of different chemicals for the management of mealy bug, *Phenacoccus solenopsis* Tinsley on Bt cotton. Journal of Cotton Research and Development, 23(2): 289-294.
Dhawan, A.K., Singh, K.,Aneja, A. andSaini, S. (2009b). Distribution of mealybug, *Phenacoccus solenopsis* Tinsley in cotton with relation to weather factors in South-Western districts of Punjab. Journal of Entomological Research, 33(1): 59-63.
Dhingra, O.D. and Sinclair, J.B. (1973a). Location of *Macrophomina phaseolina* on soybean plants related to cultural characteristics and virulence. Phytopathol., 63: 934-936.
Dhingra, O.D. and Sinclair, J.B. (1973b).Variation among the isolates of *Macrophomina phaseolina* (*Rhizoctonia bataticola*) from different regions. Phytopathol Z, 76: 200-204 .
Dhingra, O.D. and Sinclair, J.B. (1974). Isolation and partial purification of phytotixin produced by *Macrophomina phaseolina*. Phytopathologische Zeitschrift., 80(1): 35-40.
Dhingra, O.D. and Sinclair, J.B. (1977). An annotated bibliography of *Macrophomina phaseolina* 1905-1975. Universidade Federal de Vicosa, Minas Gerais, Brazil, P.277.
Dhingra, O.D. and Sinclair, J.B. (1978). Biology and pathology of *Macrophomina phaseolina*. Minas Gerais, Brazil, Universidade Federal de Vicosa., P.166.
Diaz-Montano, J., Fuchs, M., Nault, B.A., Fail, J. and Shelton, A.M. (2011). Onion thrips (Thysanoptera: Thripidae): A global pest of increasing concern in onion. J. Econ. Entomol., 104: 1-13.
Dickson, D.W. and Hewlett, T.E. (1988). Efficacy of fumigant and nonfumigant nematicides for control of *Meloidogyne arenaria* on peanut. Annals of Applied Nematology, 2: 95-101.
Diederichsen, A., Rozhmina, T. and Kudrjavceva, L. (2008). Variation patterns within 153 flax (*Linum usitatissimum* L.) genebank accessions based on evaluation for resistance to fusarium wilt, anthracnose and pasmo. Plant Genetic Resources: Characterization and Utilization, 6(1): 22-32.
Dillard, H.R. (1987). Vegetable crops, Tomato anthracnose. Cooperative Extension , Cornell University Fact sheet, New York State , pp.735-770.
Dillard, H.R. (1992). *Colletotrichum coccodes*: the pathogen and its hosts. In: Colletotrichum: biology, pathology and control, pp.225-236 (Eds J A Bailey & M J Jeger). CAB International, Wallingford.

Dillard, H.R. and Cobb, A.C. (1998). Survival of *Colletotrichum coccodes* in infected tomato and in soil. Plant Dis. 82(2): 235-238.
Dillon, L.S. (1958). Reproduction isolation among certain spider mites of the *Tetranychus telarius* complex, with preliminary notes. Ann. Entomol. Soc. Am., 51: 441-448.
Dimetry, N. Z. (1971). Studies on the host preference of the cotton seed bug *Oxycarenus hyalinipennis* (Costa) (Lygaeidae: Hemiptera). Z. Ang. Ent., 68: 63-67.
Dinsdale, A.B., Cook, L., Riginos, C., Buckley, Y.M. and De Barro, P.J. (2010). Refined Global Analysis of *Bemisia tabaci* (Hemiptera: Sternorrhyncha: Aleyrodoidea: Aleyrodidae) Mitochondrial Cytochrome Oxidase 1 to Identify Species Level Genetic Boundaries. Ann. Entomol. Soc. Am., 103: 196-208.
Dippenaar, M.C., du Toit, Cl. N. and Botha-Greeff, M.S. (1996). Response of hemp (*Canabis sativa* L.) varieties to conditions in Northwest Province, South Africa. J. International Hemp Association, 3(2): 63-66.
Dishon and Nevo, D. (1970).The appearance of Fusarium wilt in the pima cotton cultivar. Hassadeh, 56: 2281-2283.
Dixon, A.F.G. (1985). Aphid Ecology.Blackie/Chapman & Hall, NY, P.157.
Dixon, L.J., Schlub, R.L., Pernezny, K. and Datnoff, L.E. (2009). Host specialization and phylogenetic diversity of *Corynespora cassiicola*. Phytopathology, 99: 1015-1027.
Dizon, T.O., Reyes, T.T., San Pedro, J., Cabangbang, R.P. (1988). Etiology of stem rot of ramie in the Philippines. Philippine Phytopathology, 24(1&2): 29-35.
Dmitriev, A.A., Krasnov, G.S., Rozhmina, T.A., Novakovskiy, R.O., Snezhkina, A.V., Fedorova, M.S., Yurkevich1, O.Y., Muravenko, O.V., Bolsheva, N.L., Kudryavtseva, A.V. and Melnikova, N.V. (2017). Differential gene expression in response to *Fusarium oxysporum* infection in resistant and susceptible genotypes of flax (*Linum usitatissimum* L.). BMC Plant Biology, 17 (Suppl 2): 253.
Dodds, P.N., Lawrence, G.J. and Ellis, J.G. (2001a). Contrasting modes of evolution acting on the complex N locus for rust resistance in flax. Plant J., 27: 439-453.
Dodds, P.N., Lawrence, G.J., Pryor, T. and Ellis, J.G. (2001b). Six aminoacid changes confined to the leucine-rich repeat â-strand/â-turn motifdetermine the difference between the P and P2 rust resistance specificities in flax. Plant Cell, 13: 163–178.
Doeksen, J. (1938). Bad heads of flax caused by *Thrips lini*. (In Dutch). Ladureau.Tijdschrift Pl. Ziekten, 44: 1-44.
Doidge, E.M., Bottomley, A.M., van der Plank, J.E. and Pauer, G.D. (1953). A revised list of plant diseases in South Africa. So.African Dept. Agr. Sci. Bull., 345: 122.
Domingues da Silva, C.A. (2012). Occurrence of new species of mealybug on cotton fields in the States of Bahia and Paraíba, Brazil. Bragantia Campinas, 71(4): 467-470.
Dominguez Garcia-Tejero, F. (1957). Bollworm of tomato, *Heliothis armigera* Hb. (= *absoleta* F). In: Dossat SA, ed. Plagas y Enfermedades de las Plantas Cultivadas, pp.403-407. Madrid, Spain.
Domsch, K.H., Gams, W. and Anderson, T. (2007). Compendium of soil fungi. 2nd Ed. Eching Germany. IHW-Verlag, Eching, P-672.
Don, H-E .and Stella, C. (2013). *Haritalodes derogata* (Fabricius, 1775) Cotton Leaf Roller. Australian Caterpillars and their Butterflies and Moths. Website: http://www.lepidoptera.butterflyhouse.com.au/spil/derogata.html
Dosse, G. (1952). The greenhouse spider mite, *Tetranychus urticae* Koch forma dianthica and its control. Pflanzenschutz-Nachrichten Bayer, 5: 239-267.
Dosse, G. (1966). Beiträgezum diapause -Problem von *Tetranychus urticae* Koch und *Tetranychus cinnabarinus* Boisduval-Komlexim Libanon (Acarina, Tetranychidae). Pflanzenschutz berichte, 34: 129-138.

Douglas, A. (1998). Nutritional interactions in insect-microbial symbioses: aphids and their symbiotic bacteria *Buchnera*. Annu. Rev. Entomol., 43: 17-37.
DPR (2017). Legal pest management practices for cannabis growers in California. Version: September 22, 2017. Website: https://www.co.monterey.ca.us/home/showdocument?id=50196
Drenth, A. and Guest, D.I. (2004). Diversity and management of Phytophthora in Southeast Asia. ACIAR Monograph, No. 114: 238.
Du, Z,G., Tang, Y,, He, Z,F, and She, X. (2015). High genetic homogeneity points to a single introduction event responsible for invasion of Cotton leaf curl Multan virus and its associated betasatellite into China. Virol J., 12: 163.
Duan, Y-B., Ge, G-Y.and Zhou, M-G, 2014.Molecular and biochemical characterization of *Sclerotinia sclerotiorum* laboratory mutants resistant to dicarboximide and phenylpyrrole fungicides. Journal of Pest Science, 87(1): 221-230.
Duges, E. (1886). Metamorphoses of some Mexican Coleoptera. Annales de la Societe Entomologique de Belgique, 30: 27-45.
Duhoon, S.S. and Singh, M. (1980).Resistance to spotted bollworms, *Earias* spp. in cotton, *Gossypium arboreum* Linn. Indian Journal of Entomology, 42(1): 116-121
Duke, J.A. (1983). *Hibiscus cannabinus* L., Handbook of Energy Crops. In website: https://www.hort.purdue.edu/newcrop/duke_energy/Hibiscus_cannabinus.html
Duncan, L.W., Inserra, R.N., Thomas, W.K., Dunn, D., Mustika, I., Frisse, L.M., Mendes, M.L., Morris, K. and Kaplan, D.T. (1999). Molecular and morphological analysis of isolates of *Pratylenchus coffeae* and closely related species. Nematropica, 29(1): 61-80.
Dunnam, E.W. and Clark, J.C. (1938). The cotton aphid in relation to the pilosity of cotton leaves. Journal of Economic Entomology, 31: 663-666.
Dupont, L.M. (1979). On gene flow between *Tetranychus urticae* Koch, 1836 and *Tetranychus cinnabarinus* (Boisduval) Boudreaux, 1956 (Acari: Tetranychidae): synonymy between the two species. Entomol Exp Appl., 25: 297-303.
Durnovo, Z.P. (1933). Results of work on the maize moth and other pests of newly cultivated annual fibre plants, pp. 85-106 in Bolyezni i Vredit. nov.lubyan. Kultur, Diseases and Pests of newly cultivated Fibre Plants. Institut Novogo Lubianogo Syriia (Institute of New Bast Raw Materials), Moscow.
Durnovo, Z.P. (1935). Character of damage caused to ripening flax by *Aphthona euphorbiae* Schr. (In Russian). Plant Protection fasc., 2: 104-106.
duToit, L.J., D.A. Glawe, and G.Q. Pelter (2004). First report of powdery mildew of onion (*Allium cepa*) caused by *Leveillula taurica* in the Pacific Northwest (PDF). Plant Health Progress.
Dutt, D.K. (1969). Bionomics of *Agrilus acutus* (Thunb.) (Col. Buprestidae) on mesta (*Hibiscus cannabinus*) in India. Bull. Ent. Res., 58(3): 421-430.
Dutt, N. (1952a). On the forecasting of epidemic out-breaks of the major insect pests of jute. Proc. zool. Soc., Calcutta 5(1): 71-81.
Dutt, N. (1952b). *Nupserha bicolour* Thoms.sub sp. *postbrunnea* Breun - A new pest on jute (*Corchorus olitorius* Linn.). Nature, 170: 287-288.
Dutt, N. (1958). *Anomis sabulifera* Guen and *Apion corchori* Marsh – Incidences and control. Jute Bulletin, 21(5): 121-128.
Dutt, N. (1958b). On the forecasting of epidemic outbreaks of the major pests of jute. Proc. Zool. Soc. Bengal, 5(1): 71-81.
Dutt, N. and Bhattacharjee, S.P. (1960). Observations on the incidence of the spiral borer, *Agrilus acutus* Thunb. (Col., Buprestidae), in *Hibiscus cannabinus* and *H. sabdariffa*. Indian J. Agric. Sci., 30(1): 39-47.

Dutt, N. and Ganguli, R.N. (1956).Studies on *Pseudococcus filamentosus* Ckll var. Gr. with special reference to its damage to jute crop.Proc. 43rd Indian Sci. Congr., 111: 295-96.

Dutt, N. and Kundu, B.C. (1960). A note on the stunted disease of ramie, *Boehmeria nivea* Gaud. Proc. 47th Indian Sci. Congr., 111: Abstracts 558.

Dutt, N. and Mitra, S.D. (l954). Gall formation and damage in *Hibiscus cannabinus* L. Sci. & Cult. 20(1): 45.

Ebert, T. A., and Cartwright, B. (1997). Biology and ecology of *Aphis gossypii* Glover (Homoptera: Aphididae). Southwest Entomol., 22(1): 116-153.

Eckenrode, C.J. and Webb, D.R. (1989). Establishment of various European corn borer (Lepidoptera: Pyralidae) races on selected cultivars of snap beans. J. Econ. Entomol., 82: 1169-1173.

Edgerton, C.W. (1912). The role of the cotton boll. La. Agr. Exp. Sta. Bull., 137: 1-85.

Edirisinghe, W.H.M.V.P. (2016). Characterization of Flax Germplasm for Resistance to Fusarium Wilt Caused by *Fusarium oxysporum* f. sp. *lini*. M.Sc. Thesis, University of Saskatchewan, 51 Campus Drive, Saskatoon, Saskatchewan, S7N 5A8, P.120.

Eduardo, O-H., Javier, H-M., Víctor, C-M., Alejandro, C.M-A., Juan, C-T. and Humberto, V-H. (2014). Biocontrol of *Phytophthora parasitica* and *Fusarium oxysporum* by *Trichoderma* spp. in *Hibiscus sabdariffa* plants under field and greenhouse conditions. African Journal of Agricultural Research, 9(18): 1338-1345.

Edwards, D.I. and Wehunt, E.J. (1973). Hosts of *Pratylenchus coffeae* with additions from Central American banna-producing areas. Plant Disease Reporter, 57(1): 47-51.

Edwards, M. C. and Weiland, J. J. (2010). First infectious clone of the propagatively transmitted Oat blue dwarf virus. Archives of Virology, 155 (4): 463–470.

Edwards, M.C. and Weiland, J.J. (2014). Coat protein expression strategy of oat blue dwarf virus, Virology, 450-451: 290-296.

Edwards, M.C., Zhang, Z. and Weiland, J.J. (1997). Oat Blue Dwarf Marafivirus Resembles the Tymoviruses in Sequence, Genome Organization, and Expression Strategy, Virology, 232(1): 217-229.

Ehara, S. (1989). Recent advances in taxonomy of Japanese tetranychid mites. Shokubutu-boeki, 43: 358-361.

Ehler, L.E. (2000). Farmscape ecology of stink bugs in Northern California. Thomas Say Publications: Memoirs. Entomol. Soc. Amer. Lanhan, MD., P.59.

Ehrlich, E. and Wolf, F.A. (1932). Areolate Mildew of Cotton. Phytopathology, 22: 229-240.

Eilam, T., Bushnell, W.R., Anikster, Y. and McLaughlin, D.J. (1992). Nuclear DNA content of basidiospores of selected rust fungi as estimated from fluorescence of propidium iodide-stained nuclei. Phytopathology, 82: 705–712.

Eisenback, J.D., Hirschmann, H., Sasser, J.N. and Triantaphyllou, A.C. (1981). A guide to the four most common species of root-knot nematodes, (*Meloidogyne* species) with a pictorial key. A Coop. Publ. Depts. Plant Pathol. And Genetics and U.S. Agency International Dev., Raleigh, NC.

Eisenback, J.D. and Triantaphyllou, H.H. (1991). Root knot nematodes: *Meloidogyne* species and races, pp. 191-274. In: Nickle, W.R., ed. Manual of Agricultural Nematology. New York, USA: Marcel Dekker.

Elad, Y. and Evensen, K. (1995). Physiological Aspects of Resistance to *Botrytis cinerea*. http://admin.apsnet.org/publications/phytopathology/backissues/Documents/1995Articles/Phyto85n06_637.PDF

El-Ela, A.A.A. (2014). Efficacy of five acaricides against the two-spotted spider mite *Tetranychus urticae* Koch and their side effects on some natural enemies. The Journal of Basic & Applied Zoology, 67(1): 13-18.

Elena, K. and Paplomatas, E.J. (1998). *Phytophthora boehmeriae* boll rot: a new threat to cotton cultivation in the mediterranean region. Phytoparasitica, 26(1): 20-26.

Elliot, C. (1951). Mannual of bacterial plant pathogens, 2nd rev. Edn. Chronica Botanica, Waltham, Mass
Elliot, J.A. (1923). Cotton wilt: a seed-borne disease. J. Agric. Res. (Washington, D.C.), 23: 387-393.
Ellis, J.B. and Everhart, B.M. (1888). New species of fungi from various localities. Journal of Mycology, 4(1): 9-10.
Ellis, J.G., Lawrence, G.J., Luck, J.E. and Dodds, P.N. (1999). Identification of regions in alleles of the flax rust resistance gene L that determine differences in gene-for-gene specificity. Plant Cell, 11: 495–506.
Ellis, M.B. (1966). *Dematiaceous hypomycetes*. VII. Curvularia, Brachysporium, etc. Mycological Papers 106: 1-57.
Ellis, M.B. and Gibson, I.A.S. (1975). *Stemphylium solani*. CMI Descriptions of Pathogenic Fungi and Bacteria, P.472.
Ellis, M.B. and Holliday, P. (1970). *Alternaria macrospora*. CMI Descriptions of Pathogenic Fungi and Bacteria. P.246.
Ellis, M.B. and Holiday, P. (1971). *Corynespora cassiicola* (Berk.& Curt.) Wei. Commonwealth Mycological Institute Descriptions of Fungi and Bacteria, 31: 303.
El-Zik, K.M., Thaxton, P.M. and Creech, J.B. (2001). A pedigree analysis of cotton cultivars affected by bronze wilt. pp.105-107. In Proc. Beltwide Cotton Conf., Anaheim, CA. 9-13 Jan. 2001. Natl. Cotton Counc.Am., Memphis, TN.
Ekins, M.G. (2000). Genetic diversity in Sclerotinia species. PhD Thesis, The University of Queensland, Australia.
Ekins, M.G., Aitken, E.A.B. and Goulter, K.C. (2005). Identification of *Sclerotinia* species. Australasian Plant Pathology, 34: 549-555.
Emfinger, K.D., Leonard, B.R., Willrich, M.M., Siebert, J.D., Fife, J.H. and Russell, J.S. (2004). Defining boll and yield tolerance to late season insect pests in LA, pp.1744-1746. In Proc. Beltwide Cotton Conf., San Antonio, TX. 5-9 Jan. 2004. Natl. Cotton Counc.Am. Memphis, TN.
Encarnacion, D. T. (1970).Biology of the cotton stainer, *Dysdercus cingulatus* Fabricius (Pyrrhocoridae, Hemiptera). Philippine Entomologist, 1(5): 341-349.
Endo, T., El Guilli, M., Farih, A. and Tantaoui, A. (2012). Identification of powdery mildew fungus on Moroccan cucurbitaceous plants. Al Awamia, pp.125-126.
Engelbrecht, C.J.B. and Harrington, T.C. (2005). Intersterility, morphology and taxonomy of *Ceratcystis fibriata* on sweet potato, cacao and sycamore. Mycologia, 97(1): 57-69.
Engelhard, B. (2005). The Impact of Weather Conditions on the Behaviour of Powdery Mildew in Infecting Hop (Humulus). Acta Hort (ISHS), 668: 111-116.
Ephrath, J. E., Shteinberg, D., Drieshpoun, J., Dinoor, A. and Marani, A. (1989). *Alternaria alternata* in cotton (*Gossypium hirsutum*) cv. Acala: Effects on gas exchange, yield components and yield accumulation. Neth. J. Plant Pathol., 95: 157-166.
Eppler, A. (1986). Untersuchungen zur Wirtswahl von *Phorodon humuli* Schrk. I. Besiedelte Pflanzenarten. Anzeigerfur Schadlingskunde, Pflanzenschutz, Unzweltschutz, 59: 1-8.
EPPO/CABI (1996). Quarantine pests for Europe.2nd edition (ed. Smith, I.M., McNamara, D.G., Scott, P.R. and Holderness, M.) CAB International, Wallingford, U.K.
EPPO/CABI, (ND). Data Sheets on Quarantine Pests *Helicoverpa armigera*. Website: http://www.veksthusinfo.no/dokument/1372918475.pdf
Erental, A., Dickman, M.B. and Yarden, O. (2008). Sclerotial development in *Sclerotinia sclerotiorum*: awakening molecular analysis of a "Dormant" structure. Fungal Biology reviews, 22(1): 6-16.
Erlandson, M.A. (1990). Biological and biochemical comparison of *Mamestra configurata* and *Mamestra brassicae* nuclear polyhedrosis virus isolates pathogenic for the bertha

armyworm, *Mamestra configurata* (Lepidoptera: Noctuidae). J. Invert. Pathol., 56: 47-56.
Ermolaev, M.F. (1940). The biology of *Thrips linarius* Uzel and control measures against it. (In Russian). Bulletin of Plant Protection, 3: 23-34.
Erwin, D.C. and Ribeiro, O.K. (1996). Phytophthora Diseases Worldwide. APS Press, St. Paul. MN, USA
Esbenshade, P.R. and Triantaphyllou, A.C. (1990). Isozyme phenotypes for the identification of *Meloidogyne* species. Journal of Nematology, 22(1): 10-15.
Escalante-Estrada, Y.I., Osada-Kawasoe, S. and Escalante-Estrada, J.A. (2001).Variabilidad patogenica de *Phytophthora parasitica* D. Enjamaica (*Hibiscus sabdariffa* L.). Revista Mexicana de Fitopatología, 19: 84-89.
Eslaminejad, P.T., Ansaria, M. and Eslaminejad, T. (2012). Evaluation of the potential of *Trichoderma viride* in the control of fungal pathogens of Roselle (*Hibiscus sabdariffa* L.) *in vitro*. Microbial Pathogenesis, 52(4): 201-205.
Eslaminejad, T. and Zakaria, M. (2011). Morphological characteristics and pathogenicity of fungi associated with Roselle (*Hibiscus Sabdariffa*) diseases in Penang, Malaysia. Microbial Pathogenesis, 51(5): 325-337.
Esmaili, M., Mirkarimi, A. and Azemayeshfard, P. (1995).Agricultural Entomology. 3rd Edn., Tehran University publications, Tehran, pp.378-380.
Esterberg, L.K. (1932). Sugar beet web worm (*Loxostege sticticalis* L.) in the district of Nizhni-Novgorod in 1929–1930. Plant Protection, 8: 275–292.
Evans, K., Drost, D., and Frank, E. (2008). Onion powdery mildew. Utah State University Extension. Website: https://utahpests.usu.edu-onion-pm08.
Evangelisti, E., Govetto, B., Minet-Kebdani, N., Kuhn, M.L., Attard, A., Ponchet, M., Panabières, F. and Gourgues, M. (2013). The Phytophthora parasitica RXLR effector penetration-specific effector 1 favours *Arabidopsis thaliana* infection by interfering with auxin physiology. New Phytol., 199(2): 476-189.
Evans, G.O. and Browning, E. (1955). Some British mites of economic importance. British Museum Economic Series No. 17.
Evans, K., Drost, D. and Frank, E. (2008). Onion Powdery Mildew. Utah State University Extension, USA.
Evenson, J.P. and Basinski, J.J. (1973). Bibliography of cotton pests and diseases in Australia. Cotton Grower Rev., 50: 79-86.
Ezekiel, W.N. and Taubenhaus, J.J. (1932). Leaf temperatures of cotton plants with *Phymatotrichum* root rot. Science, 75: 391-392.
Ezzine, A., Chahed, H., Hannachi, M., Hardouin, J., Jouenne, T. and Marzouki, M.N. (2016). Biochemical and molecular characterization of a new glycoside hydrolase family 17 from *Sclerotinia sclerotiorum*. Journal of New Sciences - Agriculture and Biotechnology, 28 (8): 1610-1621.
Fahmy, E.T. (1927). The Fusarium wilt disease of cotton and its control. Phytopathology, 17: 749-767.
Fakir, G.A. (1977). *Corynespora cassiicola*, a new seed borne pathogen of Jute. Bangladesh. J. Jute Fib. Res., 2: 51-55.
Fan, Y. and Petitt, F.L. (1994). Biological control of broad mite, *Polyphagotarsonemus latus* (Banks), by *Neoseiulus barkeri* Hughes on pepper. Biological Control, 4: 390-395.
Fang, L. and Subramanyam, B. (2003). Activity of spinosad against adults of *Rhyzopertha dominica* (F) (Coleoptera: Bostrichidae) is not affected by wheat temperature and moisture. J. Kans. Entomol. Soc., 76: 529–532.
Farkas, H.K. (1960). Über die Eriophyiden (Acarina) Ungarns I. Acta Zoologica Academiae Scientarium Hungaricae, 6: 315-339.

Farkas, H.K. (1965). Family Eriophyidae, Gallmilben. p. 84. In: Die Tierwelt Mitteleuropas, Band III, Lief 3, P.155.
Farr, D.F., Aime, M.C.,Rossman, A.Y. and Palm, M.E. (2006). Species of *Colletotrichum* on Agavaceae. Mycological Research, 110: 1395-1408.
Farr, D.F., Bills, G.F., Chamuris, G.P. and Rossman, A.Y.. (1989). Fungi on plants and plant products in the United States. American Phytopathological Society Press, St. Paul, Minn., P.1252.
Farr, D.F. and Rossman, A.Y. (2009). Fungal Databases, Systematic Botany & Mycology Laboratory, USDA: ARS. Website: http://nt.ars-grin.gov/fungaldatabases.
Farr, D.F. and Rossman, A.Y. (2017). Fungal Databases, Systematic Mycology & Microbiology Laboratory, ARS, USDA [Internet] Washington, DC: United States Department of Agriculture; Website: http://nt.ars-grin.gov/fungaldatabases/
Farr, D.F., Rossman, A.Y., Palm, M.E. and McCray, E.B. (1980). Fungal Databases, Systematic Mycology and Microbiology Laboratory, ARS, USDA. Website: https://www.scirp.org/(S(351jmbntvnsjt1aadkposzje))/reference/ReferencesPapers.aspx?ReferenceID=1626320
Farrag, E.S., Ziedan, E.H., Ibrahiem, E., Elsharony, T. and Khalaphallah, R. (2015) Molecular detection, isolation and identification of antifungal metabolite produced by biocontrol isolate of *Pseudomonas aeruginosa*. Journal of Basic and Applied Research International, 9(2): 2395-3446.
Farquharsoq, C.O. (1913). Plant disease in southern Nigeria. Annual report of Agricultural Department south province, Nigeria, pp.41-45.
Faseli, M.D. (1977). Investigations on the biology, ecology and control of *Earias insulana* Boisd. (Noctuidae). Entomologie et Phytopathologie Appliquees (Iran), 43(6-7): 39-54.
Fauquet, C.M., Mayo, M.A., Maniloff, J., Desselberger, U. and Ball, L.A. (2005). Virus Taxonomy. Classification and Nomenclature of Viruses. 8th Report of the International Committee on the Taxonomy of Viruses. Elsevier, Academic Press, San Diego, California.P.1259.
Fegan, M. and Prior, P. (2005). Bacterial Wilt Disease and the *Ralstonia solanacearum* Species Complex, page 449. C. Allen et al., eds. The American Phytopathological Society. St. Paul, MN.
Fekrat, L., Shishehbor, P., Manzari, S. and Nejadian, E.S. (2009). Comparative development, reproduction and life table parameters of three population of *Thrips tabaci* (Thysanoptera: Thripidae) on onion and tobacco. J. Entomol. Soc. Iran, 29: 11-23.
Felt, E.P. (1908). *Contarinia gossypii* n.sp. Entomological News Philadelphia, 19: 210-211.
Ferraris, T. (1935). Parassiti Vegetalidella Canapa. Riv. Agric., Rome, p.715.
Ferguson, A.W., Fitt, B.D.L. and Williams, I.H. (1997). Insect injury to linseed in south-east England. Crop Protection, 16: 643 -652.
Ferguson, M.W., Lay, C.L. and Evenson, P.D. (1987). Effect of pasmo disease on flower production and yield components of flax. Phytopathology. 77: 805-808.
Ferrin, D.M. and Rohde, R.G. (1992). *In-vivo* expression of resistance to metalaxyl by a nursery isolate of *Phytophthora parasitica* from *Chantharanthus roseus*. Plant Disease, 76: 82-84.
Ferris, G.F. (1950). Atlas of the scale insects of North America, Volume V: The Pseudococcidae (Part-1). Stanford Unversity Press.
Fery, R.L.and Dukes, P.D. (2002). Southern Blight (*Sclerotium rolfsii*) of cowpea: yield loss estimates and source of resistance. Crop Protection, 21(55): 403-408.
Fincham, J.R.S. and Day, P.R. (1965). Fungal Genetics, Blackwell, Oxford, P.326.
Finlayson, C., Alyokhin, A. and Porter, E. (2009). Interactions of native and non-native lady beetle species (Coleoptera: Coccinellidae) with aphid-tending ants in laboratory arenas. Environmental Entomology, 38(3): 846-855.
Finlow, R.S. (1917). Historical notes on experiments with jute in Bengal. Agricultural Journal of India, 12: 3-29.

Firempong, S. and Zalucki, M. (1990b). Host plant selection by *Helicoverpa armigera* (Hubner) (Lepidoptera: Noctuidae); role of certain plant attributes. Australian Journal of Zoology, 37: 675-683.

Fitt, B.D.L., Cook, J.W., and Burhenne, S. (1991a). *Verticillium* on linseed (*Linum usitatissimum*) in the United Kingdom. Aspects Applied Biol., 28: 91-94.

Fitt, B. D. L. and Ferguson, A. W. (1990). Responses to pathogen and pest control in linseed. Proceedings of the Brighton Crop Protection Conference-Pests and Diseases, pp.733-738.

Fitt, B.D.L., Jouan, B., Sultana, C., Paul, V.H., and Bauers, F. (1991d). Occurrence and significance of fungal diseases on linseed and fiber flax in England, France and Germany. Aspects Applied Biol., 28: 59–64.

Fitt, B.D.L., and Vloutoglou, I. (1992). Alternaria diseases of linseed. In Chelkowski, J. and Visconti, A. (eds), Alternaria, Biology, Plant Diseases and Metabolites, Elsevier, Amsterdam, pp. 289-300.

Fitt, G.P. (1989). The ecology of Heliothis species in relation to agroecosystems. Annual Review of Entomology, 34:17-52.

Fitt, G.P. and Forrester, N.W. (1987). Overwintering populations of *Heliothis* in the Namoi Valley and the importance of cultivation of cotton stubble. Australian Cotton Grower, 8(4): 7-8.

Fitton, M. and Holliday, P. (1970). *Myrothecium roridum*. CMI Descriptions of Pathogenic Fungi and Bacteria. P.253.

Flachs, K. (1936). Krankheiten und Schadlinge unserer Gespinstpflanzen. Nachrichten uber Schadlingsbekampfung, 11: 6-28.

Flax Council of Canada (ND1). Chapter 7: Field Insect Pests/Flax council of Canada. https://flaxcouncil.ca/growing-flax/chapters/field-insect-pests/

Flax Council of Canada. (ND2). Chapter 8: Diseases | Flax Council of Canada, https://www.flaxcouncil.ca/growing-flax/chapters/diseases/

Fletcher, T.B. (1917). Report of the Proceedings of the Second Entomological Meeting, Held at Pusa on the 5th to 12th February 1917, Pblished by Superintendent Government Printing, India, Calcutta. Website: Krishikosh krishikosh.egranth.ac.in/bitstream/1/2030077/1/14773.pd

Fletcher, T.B. (1919). Annotated list of Indian crop pests.Rep. Proc. 3rd ento. Meeting Pusa, P.199.

Fletcher, T.B., (1920). Life-histories of Indian insects : *Microlepidoptera* In: Memoirs of the Department of Agriculture in India : Entomological series, 6: 1-9.Published for the Imperial Dept. of Agriculture in India by Thacker, Spink, Calcutta.

Fleury, D., Mauffette, Y., Methot, S. and Vincent, C. (2010). Activity of *Lygus lineolaris* (Heteroptera: Miridae) Adults Monitored around the Periphery and inside a Commercial Vineyard. European Journal of Entomology, 107(4): 527-534.

Flint, M.L. (1985). Corn Earworm, *Heliothis zea.* pp. 51-55. In: Integrated Pest Management for Cole Crops and Lettuce. University of California Publication, 3307. P.112.

Flor, H.H. (1931). Physiologic specialisation of *Melampsora lini* in *Linum usitatissimum*. J. Agric. Res., 51: 119-137.

Flor, H.H. (1944). Relation of rust damage in seed flax to seed size, oil content, and iodine value of oil. Phytopathology, 34: 348-349.

Flor, H.H. (1946). Genetics of pathogenicity in *Melampsora lini*. J. Agric. Res., 73: 335-337.

Flor, H.H. (1954). Identification of Races of Flax Rust by Lines with Single Rust-Conditioning Genes. U.S. Dep. Agric. Tech. Bull. 1087. P.25.

Flor, H.H. (1955). Host–parasite interaction in flax rust–its genetics and other implications. Phytopathology, 45: 680–685.

Flor, H.H. (1956). The complementary genic systems in flax and flax rust. Adv. Genet., 8: 29–54.
Flor, H.H. (1971). Current status of the gene-for-gene concept. Ann. Rev. Plant Pathol., 9: 275–296.
Flores-Moctezuma, H.E., Montes-Belmont, R., Jiménez-Pérez, A.A. and Nava-Juárez, R.R. (2006). Pathogenic diversity of Sclerotiumrolfsii isolates from Mexico, and potential control of southern blight through solarization and organic amendments. Crop Protection, 25: 195-201.
Fogain, R. and Graff, S. (2011). First records of the invasive pest, *Halyomorpha halys* (Hemiptera: Pentatomidae), in Ontario and Quebec. Journal of the Entomological Society of Ontario, 142: 45-48.
Fortuner, R. and Luc, M. (1987). A reappraisal of Tylenchina (Nemata). 6. The family Belonolamidae Whitehead, 1960. Revue de Nematologie, 10: 183-202.
Fowler, G. A. and Lakin, K.R. (2001). Risk Assessment: The old world bollworm, *Helicoverpa armigera* (Hubner), (Lepidoptera: Noctuidae). USDA-APHIS-PPQ-CPHST-PERAL.
Franzen, D. (2004). Fertilizing flax. North Dakota State University Extension Publication, SF-717.
Frank, M. (1988). Marijuana Groiwer's Insider's Guide. Red Eye Press, Los Angeles, CA. P.371.
Frank, M. and Rosenthal, E. (1978). Marijzlana Grozuer's Guide. And/Or Press, Berkeley, CA. P.330.
Franssen, C.J.H. (1955). The bionomics and control of *Thrips augusticeps*. (In Dutch). Tijdschrift Pl. Ziekten, 61: 97-102.
Franssen, C.J.H. and Huisman, P. (1958). The Bionomics and Control of *Thrips angusticeps* Uzel on Flax in Holland. (In Dutch). 103 pp. Abs. Rev. Appl. Entomol.1846 Series A, 48: 356.
Franssen, C.J.H. and Kerssen, M.C. (1962). Aerial control of thrips on flax in the Netherlands. (In Dutch). Agricultural Aviation, 4: 50–54. Abs. Rev. Appl. Entomol. 1964 Series A, 52: 6.
Franssen, C.J.H. and Mantel, W.P. (1961).The damage cause to flax by thrips and its prevention. (In Dutch). Tijdschrift Pl. Ziekten, 67: 39-51.
Franssen, C.J.H. and Mantel, W.P. (1963). Het voorspelen van schade door de kortvleugelige Vroege akkertrips in het voorjaar. Landbouwkundig tijdschr, 75: 121-152.
Frazer, H.L. (1944). Observations on the method of tranmission of internal boll disease of cotton by the cotton stainer bug. Ann Appl Biol., 31: 271-290.
Frederiksen, R.A. (1961). Studies on the transmission, effect, and control of 2 viruses on *Linum usitatissimum* L. PhD Thesis, University- of Minnesota, USA
Frederiksen, R.A. (1964). Simultaneous infection and transmission of two viruses in flax by *Macrosteles fasciforms*. Phytopathology, 54: 1028–1030.
Frederiksen, R.A. and Culbertson, J.O. (1962). Effect of pasmo on the yield of certain flax varieties. Crop Sci., **2**: 434–437.
Fredericksen, R.A. and Goth, R.W. (1959). Crinkle, a new virus disease of flax. Phytopathology, 49: 538.
Freeman, B.L, (2011).Tarnished Plant Bugs in Cotton. Alabama Cooperative Extension System ANR-0180, Alabama A&M University and Auburn University.www.aces.edu/pubs/docs/A/ANR-0180/ANR-0180.pdf
Freeman, P. (1947). A revision of the genus *Dysdercus* Boisduval (Hemiptera: Pyrrhocoridae), excluding the American species. Trans R Ent. Soc London., 98: 373-424.
Frisbie, R.E., El-Zik, K.M. and Wilson, L.T. (1989). The future of cotton IPM. Integrated pest management systems and cotton production New York, USA; John Wiley and Sons, Inc., pp.413-428.
Frison, E.A., Bos, L., Hamilton, R.I., Mathur, S.B. and Taylor, J.D. (1989). FAO/IBPGR Technical guidelines for the safe movement of legume germplasm. International Board for Plant Genetic Resources , FAO, Rome.

Frisullo, S., Corato, U.de, Lops, F. and Trombetta, N.M., (*1995*). Diseases of kenaf in Basilicata. Informatore Fitopatologico, 45(1), 37-41.

Fritzsche, R. (1958). Contributions to the biology, ecology and control of the flax flea beetles. (In German).Nachrichtenblattfür den Deutschen Pflanzenschutzdienst, 12: 121-133.

Fritzsche, R. and Hoffmann, G.H. (1959).Infestation of adults of *Aphthona euphorbiae* Schrk. and *Longitarsus parvulus* Payk. by *Entomophthora* sp. (In German). Beitr.Ent., 9: 517–523.

Fritzsche, R. and Lehmann, H. (1975). Effect of micro-climate on the feeding activity of flax flea beetles. (In German). Arch. Phytopath. Pflanzen., 11: 153–159.

Fu, X.Y., Liu, Sh.Y., Jiang, W.T. and Li, Y. (2015). *Erysiphe diffusa*. A newly recognized powdery mildew pathogen of Wisteria sinensis. Plant Disease, 99(9): 1272.

Fuchs, T.W., Stewart, J.W., Minzenmayer, R. and Rose, M. (1991). First record of *Phenacoccus solenopsis* Tinsley in cultivated cotton in the United States. Southwestern Entomologist, 16(3): 215-221.

Fukui, T. (1918). New diseases of useful plants in Japan (in Japanese). J. Plant Prot. (Byochugaizasshi), 5: 732-734.

Fulmer, A.M., Walls, J.T., Dutta, B., Parkunan, V., Brock, J. and Kemerait Jr., R.C. (2012). First Report of Target Spot Caused by *Corynespora cassiicola* on Cotton in Georgia. Plant Disease, .96(7): 1066.

Fulton, N.D. and Bollenbacher, K. (1959). Pathogenicity of fungi isolated from diseased cotton seedlings. Phytopathology, 49: 684–689.

Fulton, R.W. (1948). Hosts of the tobacco streak virus. Phytopathology, 38: 421-428.

Fulton, R.W. (1985). Tobacco streak virus. CMI/AAB descriptions of plant viruses no. 307. Association of Applied Biologists, Wellesbourne.

Furr, R.E., and Pfrimmer, T.R. (1968). Effects of early, mid-, and late-season infestations of two spotted spider mites on the yield of cotton. J. Econ. Entomol., 61: 1446-1447.

Fuseini, B.A. and Kumar, R. (1975). Ecology of cotton strainers (Heteroptera: Pyrrhocoridae) in southern Ghana. Biol. J. Linn. Soc., 7: 113-146.

Gadd, C.H. (1926). Report of the division of mycology. Administration representative, Director of Agriculture, Ceylon for 19: D13- D15.

Gagne, R.J. (1981). Cecidomyiidae: pp257-292. In: Mcalpine, J.F., B.V. Peterson, G.E. Siiewell, H.J..Teskey; J.R. Vockeroth and D.M. Wood (Eds). Manual of Neartic Diptera. Ottawa, Research Branch Agriculture Canada, 1: P.674.

Gahukar, R.T. (2004). Bionomics and management of major thrips species on agricultural crops in Africa. Outlook on Agriculture, 33(3): 191-199.

Gaikwad, B.B. and Pawar, V.M. (1983). Life history of jute semilooper, *Anomis sabulifera* Guen. Bulletin of Entomology, 24(2): 133-135

Galano, C.D., Gapsin, R.M. and Lim, J.L. (1996). Efficacy of *Paecilomyces lilacinus* isolates for the control of root-knot nematode [*Meloidogyne incognita* (Kofoid and White) Chitwood] in sweetpotato. Annals of Tropical Research, 18: 4-12.

Galbieri, R., Araújo, D.C.E.B., Kobayasti, L., Girotto, L., Matos, J.N., Marangoni, M.S., Almeida, W.P. and Mehta, Y.R. (2014). Corynespora Leaf Blight of Cotton in Brazil and Its Management. American Journal of Plant Sciences, 5: 3805-3811.

Galindo-González, L. and Deyholos, M.K. (2016). RNA-seq Transcriptome Response of Flax (*Linum usitatissimum* L.) to the Pathogenic Fungus Fusarium *oxysporum* f. sp. *lini*. Front. Plant Sci., 7: 1766.

Gajmer, T., Singh, R., Saini, R.K. and Kalidhar, S.B. (2002). Effect of methanolic extracts of neem (*Azadirachta indica* A. Juss) and bakain (*Melia azedarach* L.) seeds on oviposition and egg hatching of *Earias vittella* (Fab.) (Lep., Noctuidae). J. Appl. Entomol., 126 (5): 238-243.

Gallup, C.A., Sullivan, M.J. and Shew, H.D. (2006). Black Shank of Tobacco. The Plant Health Instructor. DOI: 10.1094/PHI-I-2006-0717-01

Gama, E.V.S., Silva, F., Santos, I., Malheiro, R., Soares, A.C.F., Pereira, J.A. and Armond, C. (2015). Homeopathic drugs to control red rot disease in sisal plants. Agron. Sustain. Dev., 35: 649-656.

Gamarra, H., Carhuapoma, P., Mujica, N., Kreuze, J. and Kroschel, J. (2016).Greenhouse whitefly, *Trialeurodes vaporariorum* (Westwood 1956). In: Kroschel, J.; Mujica, N.; Carhuapoma, P.; Sporleder, M. (eds.). Pest distribution and risk atlas for Africa. Potential global and regional distribution and abundance of agricultural and horticultural pests and associated biocontrol agents under current and future climates. Lima (Peru). International Potato Center (CIP), pp.154 168.

Ganesan, P. and Gnanamanickam, S.S. (1987). Biological control of *Sclerotium rolfsii* Sacc. in peanut by inoculation with *Pseudomonas fluorescens*. Soil Biology and Biochemistry, 19(1): 35-38.

Ganiger, M., Ashtaputre, S.A. and Kulkarni, V.R. (2017). Molecular variability of *Ramularia areola* isolates causing grey mildew of cotton. J. Farm Sci., 30(2): 216-219.

Ganopadhyay, S., Wyllie, T.D. and Luedders, V.D. (1970).Charcoal rot of soybean transmitted by seed. Plant Dis. Rept., 54: 1088-1090.

Gao, H., Qi, G., Yin, R., Zhang, H., Li, C. and Zhao, X. (2016). Bacillus cereus strain S2 shows high nematicidal activity against Meloidogyne incognita by producing sphingoshine. Scientific Reports, 6: 28756.

Gao, J., Luoping, Guo, C., Li, J., Liu, Q., Chen, H., Zhang, S., Zheng, J., Jiang, C., Dai, Z. and Yi, K. (2012). AFLP analysis and zebra disease resistance identification of 40 sisal genotypes in China. Mol Biol Rep, 39(5): 6379-6385.

Gao, J., Yang, F., Zhang, S., Li, J., Chen, H., Liu, Q., Zheng, J., Xi, J. and Yi, K. (2014). Expression of a hevein-like gene in transgenic Agave hybrid No. 11648 enhances tolerance against zebra stripe disease. Plant cell, tissue, and organ culture, 119(3): 579-585.

Gao, Z.R., Zhao, H.Y. and Jiang, Y.F. (1992). A study on the occurrence, damage and control of the pink bollworm in Henan Province. Plant Protection, 18(4): 29-30.

Garassini, L.A. (1935). El "pasmo" del Lino Phlyctaena? Linicola Speg. Ensayo a campo de resistencia varietal y estudiomorfolgico y fisiolgico del parasito. Rev. Fac. Agron. La Plata, 20: 170–261.

Garber, R.H., Jorgenson, E.C.C., Smith, S. and Hyer, A.H. (1979). Interaction of population levels of *Fusarium oxysporum* f. sp. *vasinfectum* and *Meloidogyne incognita* on cotton. J. Nematol., 11: 33-37.

Garcia, L.E. and Sanchez-Puerta, M.V. (2012). Characterization of a root-knot nematode population of *Meloidogyne arenaria* from Tupungato (Mendoza, Argentina). Journal of Nematology, 44: 291-301.

Gardener, J.C.M. (1934). Immature stages of Indian Coleoptera (14) (Curculionidae). Indian Forest records, 20(2): 1-48.

Garibaldi, A., Gilardi, G., Ortega, S.F. and Gullino, M.L. (2016). First report of a leaf spot caused by *Boeremia exigua* var. *linicola* on Autumn sage (*Salvia greggii*) in Italy and Worldwide. Plant Disease, 100(8): 1777.

Garland, J.A. (1998). Pest Risk Assessment of the pink mealybug *Maconellicoccus hirsutus* (Green), with particular reference to Canadian greenhouses. PRA 96–21. Canadian Food Inspection Agency, Ottawa (CA).

Garnsey, S.M., Lee, R.F. and Brlansky, R.H. (1981). Rate zonal gradient centrifugation of citrus tristeza virus. Phytopathology, 71: 875.

Gasser, R. (1951). Zur Kenntnis der gemeinen Spinnmilbe *Tetranychus urticae* Koch. Mitt. Schweiz. Entomol. Ges., 24: 217-262.

Gaumann, E. (1959). Die rostpilzemitte leuropas. Beitragezur Kryptogamenflora Schweiz, 12 (7): 1407.

Gaviyappanavar, R.S. (2012). Studies on grey mildew of cotton caused by *Ramularia areola* Atk. and its management. M.Sc thesis, University of Agricultural Sciences, Dharwad. http://krishikosh.egranth.ac.in/bitstream/1/86011/1/th10401.pdf

Gavloski, J., Carcamo, H. and Dosdall, L. (2011). Insects of Canola, Mustard, and Flax in Canadian Grasslands. In Arthropods of Canadian Grasslands (Volume 2): Inhabitants of a Changing Landscape. Edited by K. D. Floate. Biological Survey of Canada. pp.181-214.

Gavloski, J. and Derksen, H. (2015). Scouting for Aster Leafhoppers. Website: cropchatter.com/tag/aster-yellows/

Gawande, S.P., Gotyal, B.S., Tripathi A.N. and Sharma A.K. (2015). In: Annual Report (2014-15), ICAR- Central Research Institute for Jute and Allied Fibres, Barrackpore, West Bengal

Gawande, S.P. and Sharma, A.K. (2016). Conservation and utilization of ramie (*Boehmeria nivea* L.Gaud.) Germplasms for identification of resistant sources against Anthracnose leaf spot. Vegetos-An International Journal of Plant Research, 29: 137-141

Gawande, S.P., Sharma A.K. and Satpathy S. (2011).Integrated management of Cercospora leaf spot of ramie.JAF News. pp.16.

Gawande, S.P., Sharma A.K. and Satpathy S. (2012). Indian Red Admiral Caterpillar (V. Indica): A new insect pest of ramie. JAF News. Pp. 22.

Gawande, S.P., Sharma A.K. and Satpathy S. (2014). New record of Indian red admiral caterpillar (*Vanessa indica* Herbst.) as a pest of ramie (*Boehmeria nivea* L. Gaud) from Assam. Current Biotica, 8(1): 93-96.

Gawande, S. P., Sharma, A.K. and Satpathy, S. (2016). Occurrence of new pest and disease in ramie (*Boehmeria nivea* L.Gaud) Germplasm under North Eastern conditions. Biotech Today, 6(1): 21-28.

Gawande, S. P.; Sharma, A. K.; Selvaraj, K.; Gotyal, B. S.; Satpathy, S. (2015). Leaf folder, *Pleuroptya* sp. (Lepidoptera: Crambidae): a new insect pest of ramie, *Boehmeria nivea* L. Gaud. Current Biotica, 9(1): 86-87.

Gawande, S. P., Sharma A.K., Selvaraj, K. and Satpathy S (2013). Record of *Pachenephorous bretinghami* Baly: A new insect pest of Ramie in Assam. JAF News. Central Research Institute for Jute And Allied Fibres (Indian Council Of Agricultural Research) Barrackpore, Kolkata, West Bengal, pp. 15

George, J., Morse, W.C. and Lapointe, S.L. (2015). Morphology and Sexual Dimorphism of the Weevil *Myllocerus undecimpustulatus undatus* (Coleoptera: Curculionidae). Annals of the Entomological Society of America, 108(3): 325–332.

Georghiou, G.P. and Papadopoulos, C. (1957). A Second List of Cyprus Fungi. Government of Cyprus, Department of Agriculture.

Gerginov.L., (1974). *Grapholitha delineana* Walk, a new hemp pest in Bulgaria. Rastenie vud Nauk 11(1): 147-154.

Gerson, U, (2016). Plant pests of the Middle East. Jerusalem, Israel: The Hebrew University of Jerusalem. http://www.agri.huji.ac.il/mepests

Gevens, A. J., Maia, G. and Jordan, S.A. (2009). First Report of Powdery Mildew Caused by *Golovinomyces cichoracearum* on *Crotalaria juncea* ('Tropic Sun' Sunn hemp). Plant Disease, 93(4): 427.

Ghaffar, A. (1976). Inhibition of fungi as affected by oxalic acid production by *Sclerotium delphinii*. Pak. J. Bot., 8: 69-73.

Gheorghe, M. (1987). Aspects concerning the ecology and control of the flax thrips *Thrips Linarius* Uzel. (In Romanian). Analel eInstitutului de Cercetaripentru Cerealesi Plante Technice Fundulea, 54: 355–361.

Gheorge, M., Brudea, V., Bigiu, L. and Popescu, F. (1990). Elements of integrated control of diseases and pests of flax. (In Romanian).AnaleleInstitutului de Cercetaripentru Protectia Plantelor, Academia de Stiinte Agricolesi Silvice, 23: 203–207.

Gheorghe, M. and Doucet, I. (1987). Behaviour of some flax cultivars to seed treatments with carbamate insecticides (In Romanian). Probl. Protect. Plant, 15: 91–93.

Gholve, V.M., Jogdand, S.M and Jagtap, G.P. (2012). In vitro evaluation of fungicides, bioagents and aqueous leaf extracts against Alternaria leaf blight of cotton. Scientific Journal of Veterinary Advances, 1: 12-21.

Ghose, S.K. (1961). Studies of some coccids (Coccoidea: Hemiptera) of economic importance in West Bengal, India. Indian Agriculturist, 5: 57-78.

Ghose, S.K. (1970). Predators, parasites and attending ants of the mealybug, *Maconellicoccus hirsutus*. Plant Protection Bulletin, India , 22: 22-30.

Ghose, S.K. (1971a). Assessment of loss in yield of seeds of roselle (*Hibiscus sabdariffa* var. *altissima*) due to the mealy-bug, *Maconellicoccus hirsutus*. Indian Journal of Agricultural Sciences, 41: 360–362.

Ghose, S.K. (1971b). Morphology of various instars of both sexes of the mealy-bug, *Maconellicoccus hirsutus*. Indian Journal of Agricultural Sciences, 41: 602–611.

Ghose, S.K. and Ghosh, A.B. (1990). Morphology of different instars of some mealy bugs (Pseudococcidae, Homoptera). Environment and Ecology, 8(1A): 137-142.

Ghose, S.K. and Paul, P.K. (1972). Observations on the biology of the mealy bug, *Ferrisia virgata* (Cockerell) (Pseudococcidae: Hemiptera). Proceedings of the Zoological Society (Calcutta), 25: 39-48.

Ghosh, A.B. and Ghose, S.K. (1989). Description of all instars of the mealy bug *Nipaecoccus viridis* (Newstead) (Homoptera, Pseudococcidae). Environment and Ecology, 7: 564-570.

Ghosh, N.K. and Chakraborti, S. (1988). The genus Aceria Keifer (Acarina : Eriophyoidea) from India with descriptions of three new species and key to Indian species. Acarologia, 29(4): 377-387.

Ghosh, K. and Ghosh, T. (1971). Ramie cultivation in India. Jute Bulletin, April-May, 3(1&2): 15-18.

Ghosh, R., Paul, S., Roy, A., Mir, J.I., Ghosh, S.K., Srivastava, R.K., and Yadav, U.S. (2007). Occurrence of begomovirus associated with yellow vein mosaic disease of kenaf (*Hibiscus cannabinus*) in northern India. Online Plant Health Progress, doi:10.1094/PHP-2007-0508-01-RS.

Ghosh, R., Paul, S., Das, S., Palit, P., Acharyya, S., et al. (2008). Molecular evidence for existence of a New World begomovirus associated with yellow mosaic disease of *Corchorus capsularis* in India. Australian Plant Disease Notes, 3: 59-62.

Ghosh, R., Palit, P., Ghosh, S.K. and Roy, A. (2011). A New World virus alters biochemical profiling of jute plants (*Corchorus capsularis*) upon infection. International Journal of Science & Nature, 2: 883-885.

Ghosh, R., Palit, P., Paul, S., Ghosh, S.K. and Roy, A. (2012). Detection of Corchorus golden mosaic virus Associated with Yellow Mosaic Disease of Jute (*Corchorus capsularis*). Indian I Virol., 23: 70-74.

Ghosh, T. (1983). Handbook of Jute. FAO Plant production and Protection paper, 51, F.A.O., Rome, P.219.

Ghosh, T. and George, K. V. (1955). *Sclerotium rolfsii* on jute and its perfect stage. Indian J. Agric . Sci., 25: 171-173.

Ghosh, T., Mohan, K.V.J. and Prakash G. (1977). Sunnhemp Variety. Jute Agricultural Research (ICAR), Barrackpore, West Bengal, India.

Ghumary, S.R. and Bisen, M.S. (1967). Sunnhemp has satisfied M.P. farmers. Indian Farming. 16: 17-18.

Gilkeson, L.A. and Hill, S.B. (1987). Release rates for control of green peach aphid (Homoptera: Aphididae) by the predatory midge *Aphidoletes aphidimyza* (Diptera: Cecidomyiidae) under winter greenhouse conditions. Journal of Economic Entomology, 80: 147-150.

Gill, K.S. (1987b). Linseed. Indian Council of Agricultural Research, New Delhi, P.386.

Gitman, L.S. (1935). A list of fungi and bacteria on new bast crops in the USSR. Za Novoe Volokno, 6: 36-39.

Gitman, L.S. (1968b). Little-known diseases of hemp. Znshchita Rastenii Mosk., 13(3): 44-45.

Gitman, L.S. and Boytchenko, E. (1934). Cannabis. pp.45-53. In: A Manual of the Diseases of the New Bast-fibre Plants.Inst. New Bast Raw Materials, Moscow, P.124.

Glazewska S. (1971). Roslinyzywicielskiegrzyba Peronosporahumili (Miy. et Tak.) Skel. Pamietnik Pulawski Prace Inst. Upr. Nowoz. Gleb., 49: 191-204.

Gledhill, J.A. (1976). Crop losses in cotton caused by Heliothis and Diparopsis bollworms. Rhodesia-Agricultural-Journal, 73: 135-138.

Godfrey, L.D. and Wood, J.P. (1998). Mid-season cotton aphid infestations in California: effects on cotton yield. Proceedings Beltwide Cotton Conferences, San Diego, California, USA, 2: 1056-1058.

Godoy-Avila, S., Palomo Gil, A. and GarcfaHernßndez, J.L. (2000). Evaluation of transgenic cotton (*Gossypium hirsutum* L.) varieties resistant to pink bollworm (*Pectinophora gossypiella* S.) I. Yield. *ITEA Produccion Vegetal*, 96(3): 157-164.

Goeden, R.D. (1971). Insect ecology of silver leaf nightshade. Weed Science, 19: 45-51.

Goh, T.K. and Hyde, K.D. (1999). Generic separation in the *Helminthosporium*-complex, based on sequence analysis of the rDNA. Unpublished (GenBank accession: AF163983.1)

Goidanich, G. (1955). Malattie Crittogamichedella Canapa. Associazione Produttore Canapa, Bologna-Naples, pp.21.

Goka, K., Takafuji, A., Toda, S., Hamamura, T., Osakabe, M., et al. (1996). Genetic distinctness between two forms of *Tetranychus urticae* Koch (Acari: Tetranychidae) detected by electrophoresis. Exp Appl Acarol., 20: 683-693.

Gokhale, V.P. and Moghe, P.G. (1965). Preliminary investigations on dahiya disease of cotton caused by *Ramularia areola* Atk. in Vidarbha. Nagpur Agricultural College Magazine, 38: 27-31.

Gokhale, V.P. and Moghe, P.G. (1967). First record of perithecial stage of *Ramularia areola* Atk., on cotton in India. Indian Phytopath., 20: 174-175.

Golato, C. (1970).*Cercospora malayensis* Stevens et Solheim on *Hibiscus sabdariffa* L. (Malvaceae) in Ghana. Rivista di Agricoltura Subtropicale e Tropicale, 64(1-3): 15-19.

Goldansaz, S. and McNeil, J. (2006). Effect of wind speed on the pheromone-mediated behavior of sexual morphs of the potato aphid, *Macrosiphum euphorbiae* (Thomas) under laboratory and field conditions. Journal of Chemical Ecology, 32(8): 1719-1729.

Gomez, C. and Mizell, R.F. (III).(2013). Featured Creatures-green stink bug.UF/IFAS, University of Florida, EENY-431.

Gomez, C., Mizell, R.F. (III) and Hodges, A.C.(2018). Featured Creatures- brown stink bug. UF/IFAS, University of Florida, EENY-433.

Gomez-Leyva, J.F., Martinez Acosta, L.A., LopezMuraira, I.G., Silos Espino, H., Ramirez-Cervantes, F. and Andrade-Gonzalez, I. (2008). Multiple shoot regeneration of roselle (*Hibiscus sabdariffa* L.) from a shoot apex system. International Journal of Botany, 4(3): 326-330.

Gonzalez, H.H., Del Real, J.I.L., and Solis, J.F. (2007). Manejo de plagasdel agave tequilero. Colegio de Postgraduados. Montecillo, Mexico, P.123.

Gopalan, M. (1976a). Studies on salivary enzymes of *Ragmus importunitas* Distant (Hemiptera: Miridae). Curr. Sci., 45: 188-189.

Gopalan, M. (1976b). Effect of infestations of *Ragmus importunitas* Distant (Hemiptera: Miridae) on the growth and yield of sunnhemp. Indian J. Agric. Sci., 46 (12): 588-591.

Gopalan, M. and Basheer, M. (1966). Studies on the biology of *Ragmus importunitas* D. (Miridae: Hemiptera) on sunnhemp. Madras Agric. J., 53: 22-33.

Gopalan, M. and Subramaniam, T. R. (1978). Effect of infestation of *Ragmus importunitas* Distant (Hemiptera :Miridae) on respiration, transpiration, moisture content and oxidative enzymes activity in sunn-hemp plants (*Crotalaria juncea* L.), Current Science, 47 (4): 131-134.

Gore, J., Cook, D.R., Catchot, A.L., Musser, F.R., Stewart, S.D., Leonard, B.R., Lorenz, G., Studebaker, G., Akin, D.S., Tindall, K.V. and Jackson, R.E. (2013). Impact of Two spotted Spider Mite (Acari: Tetranychidae) Infestation Timing on Cotton Yields. The Journal of Cotton Science, 17: 34-39.

Gosh, T. and Mukherjee, N. (1958). Tip rot of Mesta (*Hibiscus cannabinus* Linn.). Current Science, 27: 67-69.

Goshaev, D. (1971). The role of preceding crops of cotton in the suppression of Fusarium wilt. Review of Plant Pathology, 51: 2513.

Goth, R., and Ostazeshi, S. (1965). Sporulation of *Macrophomina phaseoli* on propylene oxide-sterilized leaf tissue. Phytopathology, 55: 1156.

Gotoh ,T. and Tokioka, T. (1996). Genetic compatibility among diapausing red, nondiapausing red and diapausing green forms of the two-spotted spider mite, *Tetranychus urticae* Koch (Acari: Tetranychidae). Japanese Journal of Entomology, 64: 215-225.

Gotyal, B., Krishnan, S., Meena, P. and S. Satpathy (2015). Host Plant Resistance in Cultivated Jute and Its Wild Relatives Towards Jute Hairy Caterpillar *Spilosoma obliqua* (Lepidoptera: Arctiidae). Florida Entomologist, 98(2):721-727.

Gotyal, B.S., Satpathy, S., Selvaraj, K. and Ramesh, B.V. (2013). Comparative Biology of Bihar Hairy Caterpillar, *Spilosoma obliqua* on Wild and Cultivated Species of Jute. Indian Journal of Plant Protection, 41(3): 219- 221.

Gotyal , B. S., Satpathy, S., Selvaraj, K., Meena, P. N. and Naik, R. K. (2014). Mealybug (*Phenacoccus solenopsis* Tinsley) - A New Pest in Jute Crop: Management Strategies. Popular Kheti , 2(3): 123-125.

Gouge, D.H., Lee, L.L. and Henneberry, T.J. (1999). Parasitism of diapausing pink bollworm *Pectinophora gossypiella* (Lepidoptera: Gelechiidae) larvae by entomopathogenic nematodes (Nematoda: Steinernematidae, Heterorhabditidae). Crop Protection, 18(8): 531-537.

Government of Saskatchewan (ND). Sclerotinia | Disease | Government of Saskatchewan https://www.saskatchewan.ca/business/agriculture-natural-resources.../sclerotinia

Grady, K.A. and Lay, C.L. (1994b). Registration of "Prompt" Flax. Crop Science , 34: 308.

Grandori, R. (1946). An experiment on the control of flax flea beetles.(In Italian).Boll. Zool. Agric. Bachic, 13: 3-7.

Grant, C.A., McLaren, D., Irvine, R.B. and Duguid, S.D. (2016). Nitrogen source and placement effects on stand density, pasmo severity, seed yield, and quality of no-till flax. Canadian Journal of Plant Science, 96(1): 34-47.

Grant, L. (2008). Effect of Pasmo on Flax in Manitoba and Inference of the Sexual State of the Fungus by Molecular Polymorphism. M.Sc. Thesis, Faculty of Graduate Studies , The University of Manitoba, Canada, P. 207.

Grazia, J., Panizzi, A. and Schaefer, C. (2015). True Bugs (Heteroptera) of the Neotropics. Springer, Dordrecht, Holland, P.902.

Grbic, M., van Leeuwen, T., Clark, R.M., Rombauts, S., Rouze, P., Grbic, V., Osborne, E.J., Dermauw, W., et al. (2011). The genome of *Tetranychus urticae* reveals herbivorous pest adaptations. Nature. 479(7374): 487-492.

Greathead, D.J. (1989). The potential of biological control in integrated pest management of insect pests in cotton. In: Green MB, Lyon DJ de B. eds. Pest Management in Cotton. Chichester, UK: Ellis Harwood.

Green, M.B. and Lyon, D.J. de B.(Editors), (1989). Pest management in cotton. Chichester, West Sussex, UK; Ellis Horwood Limited, P.259.

Greenberg, S.M., Legaspi, B.C., Jones, W.A. and Enkegaard, A. (2000). Temperature-Dependent Life History of *Eretmocerus eremicus* (Hymenoptera: Aphelinidae) on Two Whitefly Hosts (Homoptera: Aleyrodidae). Environmental Entomology, 29(4): 851-860.

Greene, J.K., Bundy, C.S., Roberts, P.M. and Leonard, B.R. (2006). Identification and Management of Common Boll-Feeding Bugs in Cotton. Clemson University in cooperation with Louisiana State University, New Mexico State University and the University of Georgia, EB 158.

Griot, M. (1944). A caterpillar that eats out the capsules of flax.(In Spanish). Rev. Argent. Agron., 44-57. Abs. Rev. Appl. Entomol. 1944 Series A, 32: 340.

Grigaliunaite, B. (1997). *Lietuvos grybai*. Miltenieėiai (Erysiphales). [in Lithuanian]. Valstieėiø laikraðtis LTD. Vilnius, 3(1): 210.

Grote, A.R. (1882). New moths with partial catalogue of Noctuidae. Bulletin of the United States Geological and Geographical Survey of the Territories, 6: 566.

Groth, J.V., Comstock, V.E., and Anderson, N.A. (1970). Effect of seed color on tolerance of flax to seedling blight caused by *Rhizoctonia solani*. Phytopathology, 60: 379-380.

Groves, J.W.and Skolko, A.J. (1944). Notes on seed-borne fungi. II. Alternaria. Canadian Journal of Research, Section C, 22 (5): 217-234.

Groves, R.L., Kennedy, G.G., Walgenbach, J.F. and Moyer, J.W. (1998). Inoculation of Tomato Spotted Wilt Virus into Cotton. Plant Disease, 82(8): 959-959.

Growing Flax, (ND). Growing Flax-Production, Management & Digonostic. Flax Council of Canada, Winnipeg, Canada. https:///flaxcouncil.ca/growing-flax/

Grubben,G. J. H. and Denton, O.A.(eds) (2004). Plant Resources of tropical Africa 2 Vegetables. PROTA Foundation/ CTA Wageningen, Neitherlands, P-668.

Gruzdeviene, E and Dabkevièius, Z. (2003). The control of flax anthracnose [*Colletotrichum lini* (West.) Toch.] by fungicidal seed treatment. Journal of Plant Protection Research, 43(3): 206-212.

Guala, G. and Döring, M. (2018). Integrated Taxonomic Information System (ITIS). National Museum of Natural History, Smithsonian Institution. Checklist Dataset https://doi.org/10.15468/rjarmt accessed via GBIF.org on 2018-04-10.

Guan, Y.M., Wu, L.J., Wei, Y.J. and Zhang, Y.Y. (2016).First Report of *Colletotrichum dematium* Causing Anthracnose on *Lycopus lucidus* in Northeastern China. Plant Disease, 100(12): 2535.

Guenee, A. (1852). Ommatophoridae, in Boisduval & Guenee, Histoire naturelle des insectes, Volume 7, Part iii, pp.41-42, No. 1370, Butterfly House.

Guenee, A. (1854). In: Boisduval & Guenée, 1854. Histoire Naturelle des Insectes. Species Général des Lépidoptéres. Tome Huitiéme.Deltoideset Pyralites Hist. nat. Ins., Spec. gén. Lépid., 8: 1-448.

Guerber, J.C., Correll, J.C. and Johnston, P.R. (2003). Characterization of diversity in *Colletotrichum acutatum sensu* lato by sequence analysis of two gene introns, mtDNA and intron RFLPs, and mating compatibility. Mycologia, 95: 872-895.

Guerra, A.M.N.deM., Rodrigues, F.A., Berger, P.G., Barros, A.F., Rodrigues, Y.C. and Lima, T.C. (2013). Resistencia do algodoeiro a ferrugem tropical potencializadapelosilicio (Cotton resistance to tropical rust mediated by silicon). Bragantia, 72: 279-291.

Gulati, A., Alapati, K., Murthy, A., Savithri, H.S. and Murthy, M.R. (2016). Structural studies on tobacco streak virus coat protein: Insights into the pleomorphic nature of ilarviruses. J Struct Bio., 193(2): 95-105.

Gupta, G. and Bhattacharya, A.K. (2008). Assessing toxicity of post-emergence herbicides to the *Spilarctia obliqua* Walker (Lepidoptera: Arctiidae). Journal of Pest Science, 81: 9-15.

Gupta, R., Tara, J.S. and Pathania, P.S. (2013). First report of yellow tail tussock moth, *Somena scintillans* Walker (Lepidoptera: Lymantridae) on apple plantations in Jammu. Journal of Insect Science, 26(1): 130-133.

Gupta, R.L. (1955). Life history of *Laspeyresia tricentra* Meyr. Proc. 42 nd. Indian Sci. Congr., 111: 294-295.

Gupta, S.K. (1985). Handbook plant mites of Iindia. Zoological Survey of India, Calcutta, West Bengal.

Gurusubramanian, G. and Krishna, S.S. (1996). The effects of exposing eggs of cotton insects to volatiles of *Allium sativum* (Lillaceae). Bulletin of Entomological Research, 86: 29-31.

Guthrie, D., Whitam, K., Batson, B., Crawford, J. and Jividen, G. (1994). Boll rot. Cotton Physiology Today, Vol. 5, No.8. Website: Boll Rot - The National Cotton Council https://www.cotton.org/tech/physiology/cpt/pest/upload/CPT-Sep94-REPOP.pdf

Ha, C., Coombs, S., Revill, P., Harding, R., Vu, M., et al. (2008). Molecular characterization of begomoviruses and DNA satellites from Vietnam: additional evidence that New World geminiviruses were present in the Old World prior to continental separation. J Gen Virol., 89: 312-326

Ha, C., Coombs, S., Revill, P., Harding, R., Vu, M., et al. (2006). Corchorus yellow vein virus, a New World geminivirus from Old World. J Gen Virol., 87: 997-1003.

Hachler, M. and Brunetti, R. (2002). Flight prediction of the black cutworm, *Agrotis ipsilon* Hufn. (Lepidoptera, Noctuidae), a pest of seed corns in the Tessin, Revue Suisse d'Agriculture, 34(2): 45-53.

Habib, R., and Mohyuddin, A.I. (1981). Possibilities of biocontrol of some pests of cotton in Pakistan. Biologica, 27: 107-113.

Hague, N.G.M. and Gowen, S.R. (1987). Chemical control of nematodes. pp.131-178. In: R.H. Brown and B.R. Kerry, eds. Principles and practice of nematode control in crops. New York and Sydney: Academic Press.

Halbert, S.E., Irwin, M.E., and Goodman, R.M. (1980). Alate aphid (Homoptera: Aphididae) species and their relative importance as field vectors of soybean mosaic virus. Ann. Appl. Biol., 97: 1-9.

Halffter, G. (1957). Plagas que afectan a las distintas plagas de agave cultivadas en Mexico. Secretaria de Agricultura y Ganaderia, Direction de Defensa Agricola, SAG, Mexico City, Mexico.

Halikatti, G. and Patil, S.B. (2015). Biology of shoot weevil, *Alcidodes affaber* (Aurivillius) on Bt cotton under laboratory condition. Journal of Experimental Zoology, 18(2): 871-874.

Hall, D.G. and Teetes, G.L. (1981). Alternate host plants of sorghum panicle-feeding bugs in southeast central Texas. Southern Entomologist, 6(3): 220-228.

Hall, D.R., Beevor, P.S., Lester, R. and Nesbitt, B.F. (1980). (E,E)-10,12-Hexadecadienal: A component of the female sex pheromone of the spiny bollworm, *Earias insulana* (Boisd.) (Lepidoptera: Noctuidae). Experientia, 36(2): 152-154.

Hall, D.R., Chamberlain, D.J., Cork, A., Desouza, K., McVeigh, L.J., Ahmad, Z., Krishnaiah, K., Brown, N.J., Casagrande, E. and Jones, O.T. (1994).The use of pheromones for mating disruption of cotton bollworms and rice stemborer in developing countries. Proceedings - Brighton Crop Protection Conference, Pests and Diseases, 3: 1231-1238.

Hall, G. (1991). *Pseudoperonospora cannabina.* IMI descriptions of fungi and bacteria no 1067. Mycopathologia, 115(3): 233-234.

Hall, G. (1994). *Phytophthora nicotianae.* IMI descriptions of fungi and bacteria, 1200: 1-2.

Hall, W.J. (1921). The hibiscus mealy bug (*Phenacoccus.hirsutus*). Bulletin Ministry of Agriculture Egypt Technical and Scientific Service Entomological Section, 17: 1–28.

Halley, S. (2007). Pasmo Disease on Flax. Website: https://www.ag.ndsu.edu/.../flax/.../Scott%20Haley%20Pasmo%20Disease%20on%20Fla.

Halley, S., Bradley, C.A., Lukach, J.R., McMullen, M., Knodel, J.J., Endres, G. and Gregoire, T. (2004). Distribution and severity of pasmo on flax in North Dakota and evaluation of fungicides and cultivars for management. Plant Dis., 88: 1123-1126.

Hammouda, A.M. (1988). Fungal diseases of vegetable marrow and their control in the southern region of Oman (Dhofar). Tropical Pest Management, pp.156-158.

Hampson, G.F. (1892). The Fauna of British India including Ceylon and Burma: Moths. I. Taylor & Francis, London, (Moths Vol-II. Digital Library of India, P-558).

Hampson, G.F. (1894). The Fauna of British India, including Ceylon and Burma, Moths Voll. II. Taylor & Francis, London. P.609.

Han, K.S., Choi, S.K., Kim, H.H., Lee, S.C., Park, J.H., Cho, M.R. and Park, M.J. (2014). First report of *Myrothecium roridum* causing leaf and stem rot disease on *Peperomia quadrangularis* in Korea. Mycobiology, 42(2): 203-205.

Hanny, B.W., Cleveland, T.C. and Meredith, W.R. Jr. (1977).Effects of tarnished plant bug, (*Lygus lineolaris*), infestation on presquaring cotton (*Gossypium hirsutum*). Environmental Entomology, 6(3): 460-462.

Hanson, L.E. (2000). Reduction of Verticillium wilt symptoms in cotton following seed treatment with *Trichoderma virens*. The Journal of Cotton Science, 4: 224-231.

Hardee, D.D. and Bryan, W.W. (1997). Influence of Bacillus thuringiensis-transgenic and nectariless cotton on insect populations with emphasis on the tarnished plant bug (Heteroptera: Miridae). Journal of Economic Entomology, 90(2): 663-668.

Hardwick, D.F. (1965a). The ochrogaster group of the genus *Euxoa* (Lepidoptera: Noctuidae), with description of a new species. The Canadian Entomologist, 97(7): 673-678.

Hardwick, D.F. (1965b). The corn earworm complex. Memoirs of the Entomological Society of Canada, 40: 1-247.

Hardwick, N. V. and Mercer, P. C. (1989). Fungicide spray timing for disease control in linseed. Tests of Agrochemicals and Cultivars. Annals of Applied Biology, 114 (Supplement) (10): 48-49.

Harris, H.M., Drake, C.J. and Tate, H.D. (1935). Observation on the onion thrips (*Thrips tabaci* Lind.). Iowa State Coll. J. Sci., 10: 155-171.

Harris, W.V. (1934). The sisal weevil. Entomological Circular, Tanganyika Department of Agriculture, 3.

Harris, W.V. (1936). The sisal weevil. East Africa Agricultural and Forestry Journal, 2(2): 114-126.

Harrison, C.J. (1937). The occurrence and treatment of red spider on tea in north east India. Memory of Tocklai Experimental Station, Indian Tea Association, 2: 26.

Hart, E. (1926). Factors affecting the development of flax rust. Phytopathology, 16: 185-205.

Hartfield, C.M., Campbell, C.A.M., Hardie, J., Pickett, J.A. and Wadhams, L.J. (2001). Pheromone traps for the dissemination of an entomopathogen by the damson-hop aphid *Phorodon humuli*. Biocontrol Science and Technology, 11(3): 401-410.

Hartman, M. (1996). Diseases. In: Rowland J. (ed.) Growing Flax: Production, management, and diagnostic guide. The Flax Council of Canada, pp.39- 42.

Hartowicz, L.E., Knutson, H., Paulsen, A., et al. (1971). Possible biocontrol of wild hemp. North Central Weed Control Conference Proceedings, 26: 69.

Hasan, M.M., Meah, M.B., Ali, M.A., Okazaki, K. and Sano, Y. (2015). Characterization and Confirmation of Corchorus Golden Mosaic Virus Associated with Jute in Bangladesh. J Plant Pathol Microb., 6: 256.

Hasan, M.M. and Sano, Y. (2014). Genomic variability of Corchorus Golden Mosaic Virus originating from Bangladesh. Int. J. Phytopathol., 3 (2): 81-88.

Hassan, N., Elsharkawy, M.M., Shimizu, M. and Hyakumachi, M. (2014a). Control of Root Rot and Wilt Diseases of Roselle under Field Conditions. .Mycobiology, 42(4): 376-384.
Hassan, N., Shimizu, M., and Hyakumachi, M. (2014b). Occurrence of Root Rot and Vascular Wilt Diseases in Roselle (*Hibiscus sabdariffa* L.) in Upper Egypt. Mycobiology, 42(1): 66-72.
Hattingh, V., Cilliers, C.J. and Bedford, E.C.G. (1998). Citrus mealy bugs. In: Bedford ECG, Berg MA van den, Villiers EA de, eds. Citrus Pests in the Republic of South Africa. 2nd edition (revised). Agricultural Research Council, Republic of South Africa, 391: 112-120.
Hatzinikolis, E.N. (1970). Contribution a l'etude de l'espece *Tetranychus telarius* (Linnaeus, 1758) (complex) (Acarina: Tetranychidae). Ann. Inst. Phytopathol. Benaki, 9: 207-218.
Haque, S.M.A., Sultana, K., Islam, M. N. and Rahaman, M.L.(2008). Effect of different trap crops against root knot nematode disease of jute. J.Innov. Dev. Strategy, 2(3): 42-47.
Hawksworth, D.L. and Talboys, P.W. (1970). *Verticillium dahliae*. CMI Descriptions of Pathogenic Fungi and Bacteria, 256: 1-2.
Hayat, M. (2009). Description of a new species of Aenasius Walker (Hymenoptera: Encyrtidae), parasitoid of the mealybug, *Phenacoccus solenopsis* Tinsley (Homoptera: Pseudococcidae) in India. Biosystematica, 3: 21-26.
Haydock, P.P.J. and Pooley, R.J. (1997). Evaluation of insecticides for control of flax beetle in linseed. Tests of Agrochemicals and Cultivars, 18: 4–5.
Hayward, A.C. (1964). Characteristics of *Pseudomonas solanacearum*. J. Appl. Bacteriol., 27: 265.
He, L., Xue, C.C., Wang, J.J., Li, M., Lu, W.C., et al. (2009). Resistance selection and biochemical mechanism of resistance to two acaricides in *Tetranychus cinnabarinus* (Boisduval). Pestic Biochem Physiol., 93: 47-52.
Head, R.B. (1992). Cotton insect losses 1991. In: Proceedings Beltwide Cotton Conference. National Cotton Council of America, Memphis, TN, pp621-625.
Heckmann. R. (2012). First evidence of *Halyomorpha halys* (Stal, 1855) (Heteroptera: Pentatomidae) in Germany. Heteropteron H, 36: 17-18.
Heffes, T.A., Coates-Beckford, P.L. and Hutton, D.G. (1991).Effects of *Meloidogyne incognita* on growth and nutrient content of *Amaranthus viridis* and two cultivars of *Hibiscus sabdariffa*. Nematropica, 21: 7-18.
Heidari, M., Bayat-Assadi, H. and Ghelichabai, M. (1981). Effect of low temperatures on hibernating larvae and pupae of *Earias insulana* [Spiny bollworm, cotton]. Entomologie et phytopathologie appliquees, 43: 39-54.
Hein, G.L., Campbell, J.B., Danielson, S.D. and Kalisch, J.A. (1993). Management of the Army cutworm and Pale western cutworm. NedGuide, G93-1145A.., U.S. Department of Agriculture and University of Nebraska, Lincoln. Website; NebGuide.pdf
Helle, W. and Sabelis, M.W. (1985). Spider mites: Their biology, natural enemies and control. Volume 1 Part A. Elsevier, Amsterdam, P.406.
Hendrix, F.F. and Campbell, W.A. (1973). Pythium as plant pathogens. Annu. Rev. Phytopathol., 11: 77-98.
Henneberry, T.J. (2007). Integrated Systems for Control of the Pink Bollworm *Pectinophora gossypiella* in Cotton . In: Vreysen, M., Robinson, A. and Hendrichs, J. (eds) Area-Wide Control of Insect Pests: From Research to Field Implementation. Springer, Dordrecht, the Netherlands, pp.567-579.
Henneberry, T.J., ForlowJech, L. and Burke, R.A. (1996). Pink bollworm adult and larval susceptibility to steinernematid nematodes and nematode persistence in the soil laboratory and field tests in Arizona. Southwestern Entomologist, 21(4): 357-368.
Henneberry, T.J. and Naranjo, S.E. (1998). Integrated management approaches for pink bollworm in the southwestern United States. Integrated Pest Management Rev., 3: 31-52.

Henry, A.W. (1934). Observations on the variability of *Polyspora lini* Lafferty. Can. J. Res. 10, 409-413.
Henry, A.W. and Ellis, C. (1971). Seed infestation of flax in Alberta with the fungus causing browning or stem-break. Canadian Plant Disease Survey, 51: 76-79.
Henry, T.J. (1983). Pests not known to occur in the United States or of limited distribution, No. 38: Cotton Seed Bug. USDA– APHIS–PPQ, 6.
Heppner, J.B. (2003). Arthropods of Florida and Neighboring Land Areas: Lepidoptera of Florida. Florida Department of Agriculture, 17(1): 1-670.
Herbert, D.A. (2012). Insects: soybeans, pp. 4 – 61– 4 –76. In D. A. Herbert and S. Hagood (eds.), Field crops 2012. Virginia Cooperative Extension Publication No. 456- 016.
Hereward, J.P. (2012). Molecular ecology of the green mired *Creontiades dilutus* Stal (Hemiptera: Miridae) - movement and host plant interactions across agricultural and arid environments. PhD Thesis, The University of Queensland.
Hereward, J.P. (2016). Complete mitochondrial genome of the green mired *Creontiades dilutus* Stal (Hemiptera: Miridae). Mitochondrial DNA Part B Resources , 1(1): 321-322.
Hereward, J.P. and Walter, G.H. (2012). Molecular Interrogation of the Feeding Behaviour of Field Captured Individual Insects for Interpretation of Multiple Host Plant Use. PLOS ONE. 7 (9): e44435.
Hernandez, M.J. and Romero, C.S. (1990). Identificacion del agente causal de "pataprieta de la jamaica (*Hibiscus sabdariffa*, L.)" y pruebas de fungicidas para su control bajocondiciones de invernadero. Revista Chapingo, (67-68): 50-54.
Hewitt, H.G. (1998). Fungicides in Crop Protection. CAB International, Wallingford, UK, P.221.
Higa, S.Y. and Namba. R. (1971). Vectors of the Papaya Mosaic Virus in Hawaii. Proc. Hawaiian Entomol. Soc. 21(1): 93-96.
Higgins, B.B. (1917). A Colletotrichum leaf spot of turnips. Journal of Agricultural Research. 10:157-161.
Highland, H.A. (1956). The biology of *Ferrisiana virgata*, a pest of azaleas. Journal of Economic Entomology, 49: 276-277.
Hill, D.S. (1983a). *Heliothis zea* (Boddie). pp. 367. In: Agricultural Insect Pests of the Tropics and Their Control. Cambridge University Press.
Hill, D. S. (1983b). *Polyphagotarsonemus latus* (Banks).pp. 504. In: Agricultural Insect Pests of the Tropics and Their Control. Cambridge University Press, P-746.
Hill, D.S. (1983c). Tropical crops and their pest spectra. In: Agricultural insect pests of the Tropics and their control. Cambridge University Press, London, P.614.
Hill, D.S. (1994). Agricultural Entomology. Timber Press, 133 SW 2nd Ave., Suite 450, Portland, OR., P-635.
Hill, L. (2017). Migration of green mirid, *Creontiades dilutus* (Stal) and residence of potato bug, *Closterotomus norwegicus* (Gmelin) in Tasmania (Hemiptera: Miridae: Mirinae: Mirini). Crop Protection, 96: 211-220.
Hillocks, R.J. (1981). Cotton Disease Research in Tanzania. Tropical Pest Management, 27(1): 1-12.
Hillocks, R.J. (1984). Production of cotton varieties with resistance to Fusarium wilt with special reference to Tanzania. Tropical Pest Management, 30: 234-246.
Hillocks, R.J. (1991a). Alternaria leaf spot of cotton with special reference to Zimbabwe. Tropical Pest Management, 37(2): 124-128.
Hillocks, R.J. (1991b). Cotton Diseases, Commonwealth Agricultural Bordeaux Press, P.400.
Hillocks, R.J. (1992a). Bacterial blight. pp.39-85. In R. J. Hillocks (ed.) Cotton diseases. CAB Int. Wallingford Oxon, UK.
Hillocks, R.J. (1992b). Fungal disease of the boll. In: Hillocks, R.J. (Ed.) Cotton Diseases. Wallington: CAB International, Chap. 7, pp.239-261.

Hillocks, R.J. (1992c). Fusarium wilt. In: R.J. Hillocks (Ed.), Cotton Diseases, C.A.B. International, Wallingford, UK., pp.127-160.

Hillocks, R.J. and Bridge, J. (1992). The role of nematodes in Fusarium wilt of cotton. Afro-Asian Journal of Nematology, 21(1-2): 35-40.

Hillocks, R.J. and Chinodya, R. (1988). Current status of breeding cotton for resistance to bacteria blight in Zimbabwe. Tropical Pest Management, 34(3): 303-308.

Hillocks, R. J., and Chinodya, R. (1989).The relationship between Alternaria leaf spot and potassium deficiency causing premature defoliation of cotton. Plant Pathol., 38: 502-508.

Hine, R.B., Johnson, D.L. and Wenger, C.J. (1969). Persistency of 2 benzimidazole fungicides in soil and their fungistatic activity against *Phymatotrichum omnivorum*. Phytopathology, 59: 798-801.

Hinomoto, N., Osakabe, M., Gotoh, T., Takafuji, A. (2001). Phylogenetic analysis of green and red forms of the two-spotted spider mite, *Tetranychus urticae* Koch (Acari: Tetranychidae), in Japan, based on mitochondrial cytochrome oxidase subunit I sequences. Appl. Entomol. Zool., 36: 459-464.

Hiramatsu, A., Sakamaki, Y., Kusigemati, K., 2001. A list of pest-insects on the Kenaf in Kagoshima City with seasonal abundance of some major pest-insects. Bulletin of the Faculty of Agriculture, Kogoshima University 51: 1-7.

Hirata,K. (1942). On the shape of the germ tubes of Erysipheae. Bull. Chiba College. Hort., 5: 34-49.

Hirata,K. (1955). On the shape of the germ tubes of Erysipheae II. Bull. Fac Agr., Niigata Univ., 7: 24-36.

Hirata, K. (1966). Host Range and Geographical Distribution of the Powdery Mildews. Niigata University Press, Niigata, Japan, P.472.

Hiratsuka, N. (1897). Report of Investigations of the Flax Wilt Disease. Hokkai no Skekusan (Resources in Hokkaido), 84(25): 365-416.

Hiremath, I.G. and Thontadarya, T.S. (1984). Estimation of damage due to cotton spotted bollworm. Current Research, University of Agricultural Sciences, Bangalore, 13(7/9): 55-56.

Hjortstam, K., Larsson, K.-H., Ryvarden, L. and Eriksson, J. (1988). The Corticiaceae of North Europe. Fungiflora, Oslo, 8: 1450-1631.

Hllaing, N.Z., Tin, T.W. and NwayOo, W.N. (2017). The biology of *Micraspis discolor* and study of its predation efficiency on black aphid. Proceedings of 105th The IIER International Conference, Bangkok, Thailand, 5th-6th June 2017:pp 86-89.

Hoddle, M.S. (1999). The Biology and Management of the Silverleaf Whitefly, *Bemisia argentifolii* Bellows and Perring (Homoptera: Aleyrodidae) on Greenhouse Grown Ornamentals. Website: Website: https://www.biocontrol.ucr.edu

Hodgson, C.J., Abbas, G., Arif, M.J., Saeed, S. and Karar, H. (2008). *Phenacoccus solenopsis* Tinsley (Sternorrhyncha: Coccoidea: Pseudococcidae), an invasive mealybug damaging cotton in Pakistan and India, with a discussion on seasonal morphological variation. Zootaxa, 1913: 1-35.

Hodgson, W.A., Pond, D.D., and Munro, J. (1974). Diseases and Pests of Potatoes. Canada Department of Agriculture Publication 1492, P.70.

Hoebeke, E.R. and Carter, M.E. (2003). *Halyomorpha halys* (Stal) (Heteroptera: Pentatomidae): a polyphagous plant pest from Asia newly detected in North America. Proc. Entomol. Soc. Washington, 105: 225-237.

Hoerner, G.R.. (1940). The infection capabilities of hop downy mildew. J. Agric. Res., 61:331-334.

Hoes, J.A. (1975). Diseases of flax in western Canada. Oilseed and Pulse Crops in Western Canada: A Symposium.Calgary, Western Co-operative Fertilizers, pp.415–423. Hoes, J.A. and Dorrell, D.G. (1979). Detrimental and protective effects of rust in flax plants of varying age. Phytopathology, 69: 695-698.

Hoes, J.A. and Kenaschuk, E.O. (1980). Post seedling resistance to rust in flax. Can. J. Plant Pathol., 2: 125-130.
Hoes, J.A. and Kenaschuk, E.O. (1992). Host–pathogen specificity in post seedling reaction of *Linum usitatissimum* to *Melampsora lini*. Can. J. Bot., 70: 1168-1174.
Hoes, J.A. and Tyson, I.H. (1963). A naturally occurring North American race of *Melampsora Lini* attacking flax variety Ottawa 770B. Plant Dis. Rep., 47: 836.
Hoes, J.A. and Tyson, I.H. (1966). Races of flax rust in the Canadian prairies in 1963 and 1964. Plant Dis. Rep., 50: 62–63.
Hoes, J.A. and Zimmer, D.E. (1976). New North American races of flax rust, probably products of natural hybridization. Plant Dis. Rep., 60: 1010-1013.
Hoffman, G.M. (1958). Das Auftreteneiner Anthraknose des Hanfes in Mecklenburg und Brandenburg. Nachrichtenbl.Deutsch.Pflanzenschutz-dienst NF, 12: 96-99.
Hoffman, G.M. (1959). Untersuchungenuber die Anthraknose des Hanfes (*Cannabis sativa* L.) Phytopathologische Zeitschrift , 35: 31-57.
Hoffmann, W.E. (1931). A pentatomid pest of growing beans in South China. Peking Nat. Hist. Bull., 5: 25-26.
Hogmire, H.W. and Leskey, T.C. (2006). An improved trap for monitoring stink bugs (Heteroptera: Pentatomidae) in apple and peach orchards. Journal of Entomological Science, 41: 9-21.
Holliday, P. (1980). Fungous Diseases of Tropical Crops. Cambridge University Press, Cambridge, UK.
Holmes, E.A., Bennett, R.S., Spurgeon, D.W., Colyer, P.D. and Davis, R.M. (2009). New genotypes of *Fusarium oxysporum* f. sp. *vasinfectum* from the southeastern United States. Plant Dis., 93: 1298-1304.
Holmes, S.J.I. (1976). Pasmo disease of linseed in Scotland. Plant Path., 25: 61.
Holtz, T. (2006). Qualitative analysis of potential consequences associated with introduction of cotton seed bug (*Oxycarenus hyalinipennis*) into the United States. Animal and Plant Health Inspection Service, Plant Protection and Quarantine and Risk Analysis Laboratory, Washington, D.C.
Homma, Y. (1927). Powdery mildew on the plants in green house. Engei (Horticulture). 19: 11-14 (in Japanese).
Homma, Y. (1928). On the powdery mildew of flax. Bot. Mag. (Tokyo), 42: 331-334.
Homma, Y. (1937). Erysiphaceae of Japan. J. Fac. Agric. Hokkaido Imp. Univ., 38: 461.
Hong, C.F., Tsai, S.F., Yeh, H.C. and Fan, M.C. (2013). First Report of *Myrothecium roridum* Causing Myrothecium Leaf Spot on *Dieffenbachia picta* 'Camilla' in Taiwan. Plant Disease, 97(9): 1253.
Hoog, G.S. and de Hemanides-Nijhof, E.J. (1977). The black yeasts and allied hypomycetes. Studies in Mycology, pp.151-122.
Hoog, G. S. de, Guarro, J., Gené, J. and Figueras.M.J. (2000). Atlas of clinical fungi, 2nd ed. Centraal bureau voor Schimmel cultures, Utrecht, The Netherlands.: P.1126.
Hoper, H., Steinberg, C. and Alabouvette, C. (1995). Involvement of Clay Type and pH in the Mechanisms of Soil Suppressiveness to Fusarium Wilt of Flax. Soil BiolBiochem., 27(7): 955-967.
Hopkins, A. and Donaldson, F. (1996).Early-season insect control with Provado in the Mississippi Delta. Proceedings Beltwide Cotton Conferences, Nashville, TN, USA, January 9-12, 1996. Volume 2: 945-948.
Hopkinson, D. and Materu, M.E.A. (1970a).The control of the sisal weevil (*Scyphophorus interstitialis* Gyll. Curculionidae, Coleoptera) in sisal in Tanzania.III-Trials with insecticides in field sisal. East Africa Agricultural and Forestry Journal, 35(3): 273-277.
Hopkinson, D. and Materu, M.E.A. (1970b).The control of the sisal weevil (*Scyphophorus interstitialis* Gyll.Curculionidae, Coleoptera) in sisal in Tanzania.IV-Field trials with

insecticides in bulbil nurseries. East Africa Agricultural and Forestry Journal, 35(3): 278-285.
Hopkins, J.C.F. (ý1931). *Alternaria gossypina* (Thum.) comb.nov. causing a leaf spot and boll of cotton. www.sciencedirect.com/science/article/pii/S0007153631800282/pdf?md5..
Hora, T.S., Chenulu, V.V. and Munjal, R.L. (1962). Studies on assessment of losses due to Melampsora lini on linseed. Indian Oilseeds J., 6: 196-198.
Horak, A. (1991). Strategies for control of flax flea beetles (*Aphthona euphorbiae*, *Longitarsus parvulus*) in linseed in Czechoslovakia. Aspects Applied Biol., 28: 133–136.
Horn, J., Lauster, S., Krenz, B., Kraus, J., Frischmuth, T. and Jeske, H. (2011). Ambivalent effects of defective DNA in beet curly top virus-infected transgenic sugarbeet plants. Virus Research, 158(1-2): 169-178.
Horowitz, A.R., Klein, M., Yablonski, S. and Ishaaya, I. (1992). Evaluation of benzoylphenylure as for controlling the spiny bollworm, *Earias insulana* (Boisd.), in cotton. Crop Protection, 11(5): 465-469.
Horsfall, J.L. (1924). Life history studies of *Myzus persicae* Sulzer. Pennsylvania Agric. Agricultural Experiment Station Bulletin, 185: pp.16.
Hosagoudar, G.N., Chattannavar, S.N. and Kulkarni, S. (2008). Biochemical studies in Bt and non-Bt cotton genotypes against Alternaria blight disease (*Alternaria macrospora* Zimm.). Karnataka J. Agric. Sci., 21(1): 70-73.
Houston, B.R. and Knowles, P.F. (1949). Fifty years survival of flax Fusarium wilt in the absence of *Fusarium* culture. Plant Dis. Rep., 33: 38–39.
Howard, R.J., Garland, J.A. and Seaman, W.L., eds. (1994). Diseases and Pests of Vegetable Crops in Canada. Entomological Society of Canada, Ottawa, Ontario, P.554.
Howell, C.R., Stipanovic, R.D. and Lumsden, R.D. (1993). Antibiotic production by strains of *Gliocladium virens* and its relation to the biocontrol of cotton seedling diseases. Biocontrol Sci. and Tech., 3: 435-441.
Hsieh, W.H. and Goh, T.K. (1990). *Cercospora* and similar fungi from Taiwan. Maw Chang Book Company, Taipei, Taiwan , P.167.
Hu, C.C. and Wang, L.C. (1965). A study of the annual lifecycle of the tea red spider mite *Oligonychus coffeae* (Nietner). J Agr Assoc China, 50: 1–4.
Hu, X.P., Gurung, S., Short, D.P.G., Sandoya, G.V., Shang, W-J., Hayes, R.J., Davis, R.M. and Subbarao, K.V. (2015). Nondefoliating and defoliating strains from cotton correlate with races 1 and 2 of *Verticillium dahliae*. Plant Dis., 99(12): 1713-1720.
Huang, M., Koh, D.C., Weng, L.J., Chang, M.L., Yap, Y.K., Zhang, L. and Wong, S.M. (2000). Complete nucleotide sequence and genome organization of hibiscus chlorotic ringspot virus, a new member of the genus Carmovirus: evidence for the presence and expression of two novel open reading frames. J Virol. 74(7): 3149-3155.
Huang, Q., Zhu, Y.Y., Chen, H.R., Wang, Y.Y., Lie, Y.L., Lu, W.J. and Ruan, X.Y. (2003). First report of pomegranate wilt caused by *Ceretocystis fimbriata* in Yunnan, China. Plant Disease, 87:1150.
Huang, Y., Takanashi, T., Hoshizaki, S. and Ishikawa, Y. (2002). Female Sex Pheromone polymorphism in Adzuki Bean Borer, *Ostrnia scapulalis*, is similar to that in European Corn Borer, *O. nubilalis*. Journal of Chemical Ecology, 28(3): 533-539.
Hughes, R.D. (1963). Population Dynamics of the Cabbage Aphid, *Brevicoryne brassicae*. J. Anim. Ecol., 32: 393-424.
Huiting, H.F. and Ester, A., (2007). Effects of seed coatings with thiamethoxam on germination and flea beetle control in flax. Commun Agric Appl Biol Sci., 72(3): 595-601.
Hukkinen, Y. and Syrjanen, V. (1940). Contribution to knowledge of the Thysanoptera of Finland. (In German). Annales Ent. Fenniae, 6: 115-128.

Humza, M., Iqbal, B. and Ali, S. (2016). Management of cotton leaf curl disease and its vector through in vivo evaluation of organic nutritional amendments, organic oils and insecticides. J. Plant Pathol. Microbiol., 7(12): 387-391.

Hunter, W. D. (2012).The cotton stainer, (*Dysdercus suturellus* H.-Schf.) .In: Bureau of Entomology-Circular No. 149. [Howard, L.O., Entomologist and Chief of Bureau.] , U. S. Department Of Agriculture, Washington . Website: https://archive.org/stream/cottonstainerdys2149hunt/cottonstainerdys2149hunt_djvu.txt

Hussain, M.L. (1925). Annual report of the Entomologist to Government of Punjab, Layallpur for the year ending 30 June, 1924. Report of Department of Agriculture, Punjab 1923-1924 (2)1: 59-90.

Hussain, S.I. (2012). Dusky Cotton Bug.Plantwise. Website: www.plantwise.org/KnowledgeBank/FactsheetForFarmers.aspx?pan=20147801357

Hussein, E.M. , Mansour , M.T.M., Hassan , M.E.M., Elkady, E.A. and Kasem, K.K. (2011). Use of serology, SDS-PAGE, and RAPD analysis to evaluate resistance of flax to powdery mildew. Egypt. J. Agric. Res., 89 (1): 17-34.

Hussein, N.M., El-Hamaky, H.M.A., Refaei, A.F. and Hegazy, M.A. (1990). Joint action of certain insecticides, *Bacillus thuringiensis* and their mixtures on the pink bollworm infestation in cotton plantation in Egypt. Mededelingen van de FaculteitL and bouw wetenschappen, Rijksuniversiteit Gent, 55(2a): 307-312.

Hussey, N.W. and Parr, W.J. (1958). A genetic study of the colour forms found in populations of the greenhouse red spider mite, *Tetranychus urticae* Koch. Ann. Appl. Biol., 46: 216-220.

Hussey, N.W., Read, W.H. and Hesling, J.J. (1969).The pests of Protected Cultivation. Edward Arnold, London.

Hussian, T. (1984). Prevalence and distribution of *Xanthomonas campestris* pv. *malvacearum* races in Pakistan and their reaction to different cotton lines. Trop. Pest Manage., 30: 159-162.

Hutson, J.C. (1930). Half Yearly Rep. Entomology Division, Dept. Agric., Ceylon, Jan.-June, 1930, CAB International.

Hutt, H.J., Van Emden, H.F. and Baker, T. (1994). Stimulation of plant growth and aphid population by a formulation ingredient of cymbush (cypermethrin). Bulletin of Entomological Research, 84: 509-513.

Hyer, A.H., Jorgenson, E.C., Garber, R.H. and Smith, S. (1979). Resistance to Root-Knot Nematode in Control of Root-Knot Nematode-Fusarium Wilt Disease Complex in Cotton. Crop Science, 19(6): 898-901.

Iamamoto, M.M. (2007). Doenças do algodoeiro. Fundacao de Apoio a Pesquisa, Ensino e Extensao, Jaboticabal, Brasil, pp.62.

Ibrahim, F.M. (1966). A new race of the cotton-wilt Fusarium in the Sudan Gezira. Emp. Cotton Grow. Rev., 43: 296-299.

ICAR (2013). *Somena scintillans* (Walker). ICAR-National Bureau of Agricultural Insect Resources.

ICJC. (1946). Control measures for some major jute pests, stem-rot disease and suggestion for improvement of retting technique. Agric Res. Brochure No 1. pp.10. Indian Central Jute Committee, Calcutta: 1-6.

Ide, J.Y. (2006). Inter- and intra-shoot distributions of the ramie moth caterpillar, *Arcte coerulea* (Lepidoptera: Noctuidae), in ramie shrubs. Applied Entomology and Zoology, 41(1): 49-55.

Ikata, S. and Yoshida, M. (1940). A new Anthracnose of Jute-plant. [in Japanese]. Annals of the Phytopathological Society of Japan, 10(2-3): 141-149

Ilango, L. and Uthamasamy, S. (1989). Influence of sowing time on the incidence of bollworms and its influence on boll rot complex of cotton. Madras Agricultural Journal, 76(10): 571-573.

Ilyas, M.B. and Synclair, J.B. (1974). Effects of plant age upon development of necrosis and occurrence of intra xylem sclerotia in soybean infected with *Macrophomina phaseolina*. Phytopathology, 64: 156-157.
Ingram, W R. (1994). Pectinophora (Lepidoptera: Gelechiidae), pp.107-149. In: G. A. Matthews and J. P. Tunstall [eds.], Insect Pests of Cotton. CAB International, Wallingford, UK.
IPPC (2016). Quarantine Pests – IPPC. https://www.ippc.int/static/media/files/.../2016/02/11/Sudan_Quarantine_Pests.pdf
Irene, V. (1994). Epidemiology of *Alternaria linicola* on linseed (*Linum usitatissimum* L.), PhD thesis, University of Nottingham.U.K., P-378.
Ishaque, M. and Kabir, A.K.M.F. (1967). A preliminary study on the polyhedrosis virus of jute semilooper. Pakistan J. Sci., 19(5): 205-208.
Islam, M.R. (1992). Control of flax diseases through genetic resistance. Zeitschrift Pflanzenkrheit Pflanzenschutz, 99: 550-557.
Islam, M.S., Haque, M.S., Islam, M.M., Emdad, E.M., Halim, A., et al. (2012). Tools to kill: genome of one of the most destructive plant pathogenic fungi. *Macrophomina phaseolina*. BMC Genomics, 13: 493.
Islam, M.S., Uzzal, M.S.I., Mallick, K. and Monjil, M.S. (2013). Management of seed mycoflora of mesta (*Hibiscus sabdariffa*) by seed washing, garlic extract and know in. Progressive Agriculture, 24(1&2): 1-7.
Iqbal, U. (2010). Biology and management of charcoal rot of mung bean (*Vigna radiate* L.) Wilczek and mash bean (*Vigna mungo* L.) Hepper. PhD Thesis, Department of Plant Pathology, Faculty of Crop and Food Sciences, PirMehr Ali ShahArid Agriculture University Rawalpindi, Pakistan
Ivanica, V., Andrei, E. And Barnaveta, E. (1992). The behaviour of some sunflower hybrids at the attack of *Erysiphe cichoracearum* D.C. f. sp. Helianthi Jacz. Cercetari Agronomice in Moldova, 25(1): 201-204.
Iwata, K. (1975). Shizenkansatsusha no shuki (Memoirs on Nature by an Observer). Asahi Shimbun Co., Tokyo, P.584.
Jacobs, J.M., Pesce, C., Lefeuvre, P. and Koebnik, R. (2015). Comparative genomics of a cannabis pathogen reveals insight into the evolution of pathogenicity in *Xanthomonas*. Plant Sci., 6: 431.
Jacobs, S.B. and Bernhard, K.M. (2008). Brown marmorated stink bug *Halyomorpha halys*. NP-15. The Pennsylvania State University Entomology Notes. Website: http://www.ento.psu.edu/extension/factsheets/pdfs/brMarmoratedStinkBug.pdf.
Jacobson, L.A. (1962). Diapause in eggs of the pale western cutworm *Agrotis orthogonia* Morr. (Lepidoptera: Noctuidae). The Canadian Entomologist, 94(5): 515-522.
Jacobson, L.A. (1965). Mating and oviposition of the pale western cutworm, *Agrotis orthogonia* Morrison (Lepidoptera: Noctuidae), in laboratory. The Canadian Entomologist, 97(9): 994-1000.
Jacobson, L.A. (1970). Laboratory ecology of red-backed cutworm, *Euxoa ochrogaster* (Lepidoptera: Noctuidae). The Canadian Entomologist, 102(1): 85-89.
Jaczewski, A.L.A. (1927). Karmannyiopredelitel' gribov. II. Muchnisto-rosjanyegriby. P.491.
Jagadish, K.S., Shadhanaikural, A., Chandru, R. and Shadakshari, Y. (2009a). Biochemical and morphological changes due to mealybug *Phenacoccus solenopsis* Tinsley (Homoptera: Pseudococcidae) infestation on sunflower (*Helianthus annuus* L.). Insect Environment, 15(1): 28-30.
Jagadish, K.S., Shankaramurthy, M., Kalleshwaraswamy, C.M., Viraktamath, C.A. and Shadakshari, Y.G. (2009b). Ecology of the mealy bug, *Phenacoccus solenopsis* Tinsley (Hymenoptera: Pseudococcidae) infesting sunflower and its parasitization by *Aenasius* sp. (Hymenoptera: Encyrtidae). Insect Environment, 15(1): 27-28.

Jagtap, G.P., Jadhav, T. H. and Utpal, D. (2012a). Occurrence, distribution and survey of tobacco streak virus (TSV) of cotton. Scientific Journal of Crop Science, 1(1): 16-19.

Jagtapa, G.P., Jadhava, T.H. and Utpal, D. (2012b). Host range and transmission of Tobacco streak virus (TSV) causing cotton mosaic disease. Scientific Journal of Veterinary Advances, 1(1): 22-27.

Jakhmola, S.S. (1974). Chemical control of linseed bud-fly, *Dasyneura lini* Barnes (Diptera: Cecidomyiidae). Indian J. Agric. Sci., 43: 1078-1080.

Jakhmola, S.S. and Yadev, H.S. (1983). Susceptibility of linseed cultivars to bud fly, *Dasyneura Lini* Barnes. Indian J. Entomol., 45: 165-170.

Jakhmola, S.S., Kaushik, U.K. and Kaushal, P.K. (1973). Note on the effect of sowing and nitrogen levels on the infestation of linseed bud-fly, *Dasyneura lini* Barnes (Diptera: Cecidomyiidae). Indian J. Agric. Sci., 43: 621-623.

Jameson, J.D. (1950). Dept of Agriculture, Uganda, Record of Investigations No. 1, for the period 1st April, 1948 to 31st March, 1949. Uganda: Department of Agriculture.

Jameson, M. L., Mico, E.andGalante, E. (2007). Evolution and phylogeny of the scarab subtribe Anisopliina (Coleoptera: Scarabaeidae: Rutelinae: Anomalini). Systematic Entomology. 32 (3): 429-449.

Jain, N.K., Gupta, B.N., Sasmal, P.K. and Datta, A.N. (1965). Fertility trials with roselle (*Hibiscus sabdariffa* L. var. *altissima* Kert.). Trop. Agriculture (Trinidada), 42(1): 88-92.

Jana, T., Sharma, T.R., Prasad, R.D. and Arora, D.K. (2003). Molecular characterization of *Macrophomina phaseolina* and *Fusarium* species by a single primer RAPD technique. Microbiological Research, 158 (3): 249-257.

Jana, T., Singh, N.K., Koundal, K.R. and Sharma, T.R. (2005). Genetic differentiation of charcoal rot pathogen, *Macrophomina phaseolina*, into specific groups using URP-PCR. Can. J. Microbiol., 51: 159-164.

Jankauskiene , Z., Gruzdeviene, E. and Endriukaitis, A. (2005). Protection of Fibre Flax Crop Against Flea Beetles and Seedling Blight Using Compound Seed-Dressers, Journal of Natural Fibers , 1(4): 37-57.

Jaschhof, M. (2014). A revision of the types of Neotropical Porricondylinae (Diptera: Cecidomyiidae). Zootaxa, 3779(4): 463-469.

Javahery, M. (1990). Biology and ecological adaptation of the green stink bug (Hemiptera: Pentatomidae) in Que´bec and Ontario. Annals of the Entomological Society of America, 83: 201-206.

Javaid, I., Zulu, J.N., Matthews, G.A. and Norton, G.A. (1987). Cotton insect pest management on small scale farms in Zambia I. Farmers' perceptions. Insect Science and its Application, 8(4-6): 1001-1006.

Jayaraj, S. (1982). Biological and ecological studies of Heliothis. In: Reed W, Kumble V, ed. Proceedings of the International Workshop on Heliothis Management. ICRISAT Center, Patancheru, India, 15-20 November 1981 International Crops Research Institute for the Semi-Arid Tropics Patancheru, Andhra Pradesh India, pp.17-28.

Jayaraj, S. (1990). The Problem of *Helicoverpa armigera* in India and Its Integrated Pest Management. In: Proceedings of the National Workshop, Jayaraj, S., Uthamasamy, S., Gopalan, M., Rabindra, R. J. (Eds.), Centre for Plant Protection Studies, Tamil Nadu Agricultural University, Coimbatore, Tamil Nadu, India, February 1988, PP.1-16.

Jebaraj, P. (2010). Bt cotton ineffective against pest in parts of Gujarat, admits Monsanto. The Hindu.New Delhi, March 06, 2010.

Jendek, E. (2004). Revision of *Agrilus acutus* (Thunberg, 1787) and related species (Coleoptera: Buprestidae). Zootaxa, 507: 1-19.

Jensen, D.D. (1950). A Crotalaria mosaic and its transmission by aphids. Phytopath. 40: 512-515.

Jeppson, L.R., Keifer, H.H. and Baker, E.W. (1975). Mites injurious to economic plants. University of California Press, USA.P.614.
Jha, A.K., De, R.K. and Sarkar, S. (2014). Screening of sisal germplasm against zebra disease. In: Book of Abstracts International Conference on Natural Fibres (Theme: Jute and Allied Fibres), organized by the Indian Natural Fibre Society, Kolkata, 1-3 August 2014, pp.125.
Jhala, R.C., Bharpoda, T.M. and Patel, M.G. (2008). *Phenacoccus solenopsis* Tinsley (Hemiptera: Pseudococcidae), the mealy bug species recorded first time on cotton and its alternate host plants in Gujarat, India. Uttar Pradesh Journal of Zoology, 28(3): 403-406.
Jia, F.L., Chen, Y.J., Chen, J., Wang, D.D. and Dai, G.H. (2011). Biological activity of extracts from 8 species of plants against *Tetranychus cinnabarinus*. Chinese Agricultural Science Bulletin, 27: 286–291.
Johnston, A. (1963). In: Quarterly report for July-September 1963, Plant Protection Committee of the South–East Asia and Pacific Region. FAO Publication, Bangkok, Thailand, P.22.
Johnson, D.A., Tew, T.L., and Banttari, E.E. (1977). Factors affecting symptoms in barley infected with the oat blue dwarf virus. Plant Dis. Rep., 61: 280–283.
Johnson, F.A., Short, D.E. and Castner, J.L. (2005). Sweetpotato/Silverleaf Whitefly Life Stages and Damage. Entomology and Nematology Department special publication 90 (revised ed.). Gainesville, Florida: Florida Cooperative Extension Service, Institute of Food and Agricultural Sciences, University of Florida.
Johnston, P. R. and Jones, D. (1997). Relationships among *Colletotrichum* isolates from fruit-rots assessed using rDNA sequences. Mycologia, 89: 420-430.
Jones, G.D. and Allen, K.C. (2013).Pollen Analyses of Tarnished Plant Bugs. Palynology, 37(1): 170-176.
Jones, D.R. and Behncken, G.M. (1980). Hibiscus chlorotic ringspot, a widespread virus disease in the ornamental *Hibiscus rosa-sinensis*. Australian Journal of Plant Pathology, 9: 4.
Jones, G.H. (1928). An Alternaria disease of the cotton plant. Annals of Botany, 42: 935-947.
Jones, J.P. (1961). A leaf spot of cotton caused by *Corynespora cassiicola*. Phytopathology, 51: 305-308.
Jordaan LC (1977) Hybridization studies on the *Tetranychus cinnabarinus* complex in South Africa (Acari: Tetranychidae). J. Entomol. Soc. S. Afr., 40: 147-156.
Josephrajkumar, A., Rajan, P., Mohan, C. and Thomas, R. J. (2011) . First record of Asian grey weevil (*Myllocerus undatus*) on coconut from Kerala, India. Phytoparasitica, 39: 63-65.
Joshi, L.M. (1960). Preleminary studies on *Uromyces decoratus*, the causal organism of Sunnhemp rust. Indian Phytopath., 13: 90-95.
Joshi, M.D., Butani, P.G. Patel, V.N. and Jeyakumar, P. (2010). Cotton mealy bug, *Phenacoccus solenopsis* Tinsley- a review. Agric. Rev., 31(2): 113-119.
Jourdheuil, P. (1960). Observations sur les altises du lin; remarques sur la biologie et les méthodes de lutte. C.R. Acad. Agric. France, 46: 477-480.
Jourdheiul, P. (1963). Famille des Chrysomelidae, sous-famille des Halticinae. In Balachowsky, A.S. (ed.), Entomologie Appliquee a l'Agriculture, pp.762-854.
Jovaisiene, Z. and Taluntyte, L. (2000). *Septoria linicola* (Speg.) Garassini Lietuvoje. [in Lithuanian]. Botanica Lituanica, 6(1): 97-100.
Judelson, H.S. (2012). Dynamics and innovations within oomycete genomes: insights into biology, pathology, and evolution. Eukaryotic Cell, 11(11): 1304-1312.
Juliatti, F.C. and Polizel, A.C. (2003). Manejo integrado de doenças na cotonicultura brasileira EDUFU, Uberlândia, MG, Brazil.
Juneja, R.C., Nayyar, V.L. and Mukerji, K.G. (1976). Further additions to plant diseases of Delhi. Angewandte Botanik, 50(1-2): 43-47.
Junell, L. (1967). Erysiphaceae of Sweden. Symbola Botanica Upsaliensis, 19: 117.

Juniora, A.M., Peronti, A., Dias, A., Morais, A. and Pereira, P. (2013). First report of *Maconellicoccus hirsutus* (Green, 1908). Website: http://www.scielo.br/scielo.php?pid=S1519-69842013000200413&script=sci_arttext,

Kabir, F.A.K.M.F. (1975). Jute pests of Bangladesh. Bangladesh Jute Research Institute, Sher-e-Bangla Nagar, Dhaka 15, Bangladesh Govt. Printing Press, Dacca, P-59.

Kabir, A.K.M.F.and Khan, S.A., (1969). Biology and life-history of jute hairy caterpillar, *Diacrisia obliqua* Walker, in East Pakistan. Pakist. J. Zool, Lahore, 1(1): 45-48.

Kaiser, W.J., Wyatt, S.D. and Pesho, G.R. (1982). Natural hosts and vectors of tobacco streak virus in eastern Washington. Phytopathology, 72: 1508-1512.

Kakkar, G., Kumar, V., Seal, D.R., Liburd, O.E. and Stansly, P.A. (2016). Predation by *Neoseiulus cucumeris* and *Amblyseius swirskii* on *Thrips palmi* and *Frankliniella schultzei* on cucumber. Biological Control., 92: 85-91.

Kakkar, G., Seal, D.R. and Jha, V.K. (2010). Common blossom thrips, *Frankliniella schultzei* Trybom (Insecta: Thysanoptera: Thripidae). Tropical Research and Education Center, Entomology and Nematology Department, UF/IFAS Extension, Gainesville, FL 32611. EENY 477

Kalshoven, L.G.E. (1951). The Pests of Cultivated Plants in Indonesia. Part II. Van Hoeve, The Hague, Netherlands.

Kalshoven, L.G.E. (1981). Pests of crops in Indonesia (revised & translated by van der P.A. Laan). : Ichtiar Baru, Jakarta, Indonesia, P.701.

Kalyan, R.K., Saini, D.P., Babu, S.R. and Urmila (2014). Evaluation of different doses of tolfenpyrad against aphids and thrips in cotton. Journal of Cotton Research and Development, 28(2): 293-296.

Kappelman A.J. (Jr). (1971). Inheritance of resistance to Fusarium wilt in cotton. Crop Sci., 11: 672-674.

Kappelman, A.J. (Jr). (1975). Fusarium wilt resislance in cot ton (*Gossypium hirsutum*). Pl. Dis. Rep., 59: 803-805.

Kamaruzzaman, M., Bhuiyan, A.A. and Faruque, M.O. (2017). Isolation and Molecular Characterization of a Wild Type *B. cinerea* from Infected Bottle Gourd (*Lagenaria siceraria*) in China. Journal of Advances in Microbiology, 7(4): 1-10.

Kamble, S.T. (1971). Bionomics of *Dysdercus koenigii* Fabr. (Hemiptera: Pyrrhocoridae). Journal of the New York Entomological Society, 79(3): 154-157.

Kamminga, K.L., Herbert, D.A. Jr., Kuhar, T.P. and Brewster, C.C. (2009a). Predicting black light trap catch and flight activity of *Acrosternum hilare* (Hemiptera: Pentatomidae) adults. Environmental Entomology, 38: 1716-1723.

Kamminga, K.L., Herbert, D.A. Jr., Kuhar, T.P., Malone, S. and Koppel, A. (2009c). Efficacy of insecticides against *Acrosternum hilare* and *Euschistus servus* (Hemiptera: Pentatomidae) in Virginia and North Carolina. Journal of Entomological Science, 44: 1-10.

Kamminga, K.L., Herbert, D.A., Toews, M.D., Malone, S. and Kuhar, T. (2014). Arthropod Management & Applied Ecology: *Halyomorpha halys* (Hemiptera: Pentatomidae) Feeding Injury on Cotton Bolls. The Journal of Cotton Science, 18: 68-74.

Kamminga, K.L., Koppel, A.L., Herbert, D.A. (Jr.), and Kuhar, T.P. (2012). Biology and Management of the Green Stink Bug. J. Integ. Pest Mngmt. 3(3): 1-8.

Kamoun, S. (2006). A catalogue of the effector secretome of plant pathogenic oomycetes. Annu Rev Phytopathol., 44: 41-60.

Kangatharalingam, N. (1987). Variability in *Rhizoctonia solani* Kuhn Pathogenic on Flax. Ph.D. Thesis. South Dakota State University, Brokkings, P.99.

Kant, K., Kanaujia, K.R. and Kanaujia, S. (1999). Rhythmicity and orientation of *Helicoverpa armigera* (Hubner) to pheromone and influence of trap design and distance on moth trapping. Journal of Insect Science, 12: 6-8.

Kaplan, I. and Thaler, J. (2012). Phytohormone-mediated plant resistance and predation risk act independently on the population growth and wing formation of potato aphids, *Macrosiphum euphorbiae*. Arthropod-Plant Interactions, 6(2): 181-186.

Kareem, K.T., Oduwaye, O.F. , Olanipekun, S.O. , Adeniyan, N.A. and Oyedele, A.O. (2017). Yellow Vein Mosaic disease in kenaf (*Hibiscus cannabinus* l.) under different sowing dates in two agroecologies. Nig. J. Biotech., 33: 83-88.

Kasanis, B. and Varma, A. (1975). Sunn-hemp mosaic virus. CMI/AAB Descriptions of Plant Viruses No. 153. Commonwealth Agricultural Bureaux, Slough.

Kashyap, R.K. and Verma, A.N. (1987). Management of spotted bollworms (*Earias* spp.) in cotton - a review. International Journal of Tropical Agriculture, 5(1): 1-27.

Katan, J., Fishler, G. and Grinstein, A. (1983). Short-and long-term effects of soil solarization and crop sequence on Fusarium wilt and yield of cotton in Israel. Phytopathology, 73: 1215-1219.

Katke, M. (2008). Seasonal Incidence, Biology and Management of Grape Mealy Bug, *Maconellicoccus hirsutus* (Green) (Homoptera: Pseudococcidae). M.Sc (Ag.) Thesis, Department Of Agricultural Entomology College of Agriculture, Dharwad University of Agricultural Sciences, Dharwad (India).

Kator, L., Hosea, Z. Y. and Oche, O. D. (2015). *Sclerotium rolfsii*; Causative organism of southern blight, stem rot, white mold and sclerotia rot disease. Annals of Biological Research, 6(11): 78-89.

Katsuki, S. (1965). Cercosporae of Japan. Trans. Mycol. Soc. Japan, Spec. Issue 1, pp.100.

Kaur, R., Kaur, R. and Brar, K.S. (2008). Development and predation efficacy of *Chrysoperla carnea* (Stephens) on mealy bug, *Phenacoccus solenopsis* (Tinsley) under laboratory conditions. Journal of Insect Science (Ludhiana), 21(1): 93-95.

Kaur, S., Dhillon, G.S., Brar, S.K., Vallad, G.E., Chand, R. and Chauhan, V.B. (2012). Emerging phytopathogen *Macrophomina phaseolina*: biology, economic importance and current diagnostic trends. Crit Rev Microbiol., 38(2): 136-151.

Kaushik, V.K., Rathode ,V.S. and Sood, N.K. (1969). Incidence of bollworms and losses caused to cotton in Madhya Pradesh. Indian J. Entomol., 31: 175-177.

Kawada, H. and Kitamura, C. (1983). The reproductive behavior of the brown marmorated stink bug, *Halyomorpha mista* Uhler (Heteriotera: Pentatomidae): observation of mating behavior and multiple copulation. Appl. Entomol. Zool., 18: 234-242.

Kay, I.R. (1986). Tomato Russet Mite: A Serious Pest of Tomatoes. Queensland Agricultural Journal, 112(5): 231-232.

Kay, I.R. and Shepherd, R.K.,(1988). Chemical Control of the Tomato Russet Mite on Tomatoes in the Dry Tropics of Queensland. Queensland Journal of Agriculture and Animal Sciences, 45(1): 1-8.

Kaydan, M. and Gullan, P. (2012). A taxonomic revision of the mealy bug genus *Ferrisia* Fullaway (Hemiptera: Pseudococcidae), with descriptions of eight new species and a new genus. Zootaxa, 3543: 1-65.

Kebdani, N., Pieuchot, L., Deleury, E., Panabières, F., Le Berre, J.Y. and Gourgues, M. (2010). Cellular and molecular characterization of *Phytophthora parasitica* appressorium-mediated penetration. New Phytol., 185(1): 248–257.

Kehat, M. and Bar, D. (1975).The use of traps baited with live females as a tool for improving control programs of the spiny bollworm, *Earias insulana* Boisd., in cotton fields. Phytoparasitica, 3(2): 129-131.

Keinath, A.P. and DuBose, V.B. (2017). Management of southern blight on tomato with SDHI fungicides. Crop Protection, 101: 29-34.

Kelly, H.M. (2017). Target Spot in Cotton – How to identify it and management options. UTcrops News Blog. Website: news.utcrops.com/2017/07/target-spot-cotton-identify-management-options.

Kelly, H.M. (ND). UT Crops Disease Field Guide. Website: guide.utcrops.com/cotton/cotton-foliar-diseases/alternaria-leaf-spot

Kempanna, C., Yaraguntaiah, R.C. and Govindu, H.C. (1960). Occurrence of *Colletotrichum curvatum* Braint & Martyn on *Crotalaria juncea* Linn, in Mysore. Curr. Sci. 29 : 357-358.

Kenaschuk, E.O. and Rashid, K.Y. (1993). AC Linora flax. Can. J. Plant Sci., 73: 839–841.

Kenaschuk, E. O. and Rashid, K.Y. (1994). AC Mc Duff Flax, Canadian journal of Plant Science, 74(4): 815-816.

Kenaschuk, E.O., Rashid, K.Y. and Gubbels, G.H. (1996). AC Emerson flax. Can. J. Plant Sci., 76: 483-485.

Kendrick, B. (1985). The Fifth Kingdom.Mycologue Publications, Waterloo, Canada, P.364.

Kennedy, G.G. and Smitley, D.R. (1985). Dispersal. In: Spider mites, their biology, natural enemies and control. (Helle W. and Sabelis M.W. eds). Vol. 1A Elsevier, Amsterdam, The Netherlands. pp.233-242.

Kennedy, J.S., Booth, C.O. and Kershaw, W.J.S. (1959). Host finding by aphids in the field. Annals Applied Biology, 47: 424-444.

Kennedy, J.S., Day, M.F. and Eastop, V.F. (1962). A conspectus of aphids as vectors of plant viruses. Wallingford, UK: CAB International.

Kenneth Horst, R. (Re.). (2001). Westcott's Plant disease hand book. Revised copy, 6th Edition, Springler Science and Business Media, New York.

Kerr, A. (1953). Foot rot of flax caused by *Ascochyta linkola*. Trans. Br. mycol. Soc., 36: 61-73.

Kerr, H.B. (1960). The inheritance of resistance of *Linum usitatissimum* L. to the Australian *Melampsora lini* (pers.) Lev. race complex. Proc. Linn. Soc. N.S.W., 85: 273-321.

Kerzhner, I. M. and M. Josifov. (1999). Catalogue of the Heteroptera of the Palaearctic Region. The Netherlands Entomological Society. Ponsen and Looijen. Wageningen., 14: 1-577.

Keswani, C.L. and Mwenkalley, A.H. (1982). Korogwe leaf spot of sisal. FAO Plant Protection Bulletin, 30(3/4): 145-150.

Khabbaz, S.E., Ladhalakshmi, D. and Abdelmagid, A. (2017). Comparative studies between *Xanthomonas citri* subsp. *malvacearum* isolates, causal agent of the bacterial blight disease of cotton. World Journal of Agricultural Research, 5(2): 64-72.

Khabiri, E. (1952). Contribution a la mycoflorae de l' Iran, primiere. Review mycologiae, pp.17.

Khadhaira, A-H., Tewari, J.P., Howard,, R.J. and Paul, V.H. (2001). Detection of aster yellows phytoplasma in false flax based on PCR and RFLP. Microbiological Research, 156(2): 179-184.

Khairy, E.A. and Abd-el-Rehim, S.H.M.A. (1971). Occurrence of powdery mildews of roselle and mulberry in U.A.R. (Egypt). Phytopatholgia Mediterranea, 10(3): 269-271.

Khan, A. and Fakir, G.A. (1993). Association of seed borne pathogens with growing capsules and their entry into developing seeds in jute. Bangladesh J. Pl. Path., 9(1&2): 1-3.

Khan, A.H. (1959). Biology and pathogenicity of *Rosellinia necatrix* (Hart.) Berl. Biologia Lahore, 5: 199–245.

Khan, A.N., Shair, F., Malik, K., Hayat, Z., Khan, M.A., Hafeez, F.Y. and Hassan, M.N. (2017). Molecular Identification and Genetic Characterization of *Macrophomina phaseolina* Strains Causing Pathogenicity on Sunflower and Chickpea. Front Microbiol., 8: 1309.

Khan, A.Z., Amad, I., Shaheen, S., Hussain, K., Hafeez, F., Farooq, M. and Noor UlAyan, H. (2017). Genetic barcoding and phylogenetic analysis of dusky cotton bug (*Oxycarenus hyalinipennis*) using mitochondrial cytochrome c oxidase I gene.Cell MolBiol (Noisy-le-grand), 63(10): 59-63.

Khan, J.A., Siddiqui, M.K. and Singh, B.P. (2002). The natural occurrence of a begomovirus in sunnhemp (*Crotalaria juncea*) in India. New Disease Reports 4: 17.

Khan, M. (2003). Salt mixtures- an IPM option to manage mirids in cotton. The Australian Cottongrower, 24(3): 10-13.

Khan, M. Dave K., Mark H., Robert M., Hugh B. and Liwis W. (2004). Mirid bug management in Australian cotton. Australian Cotton Research and Development Corporation.

Khan, M. and Quade, A. (2008). Pictorial identification of mirids life cycle: Green Mirid. QDPI&F and Cotton Catchment Communities Cooperative Research Centre, Kingaroy, Qld, 4610.

Khan, M.A., Gogi, M.D. Bashir, M.H. Hussain, M. Zain-ul-Abdin and Rashid, M.A. (2014). Assessment of density-dependent feeding-damage by cotton dusky bug, *Oxycarenus laetus* Kirby (Hemiptera: Lygaeidae) in cotton. Turkish Journal of Agriculture, 38: 198-206.

Khan, M.A., Khan, S.A., Imaran-ul-haq and Khan, R.W. (2017). Root Rot Disease Complex of Cotton: A Menace to Crop in Southern Punjab and its Mitigation through Antagonistic Fungi. Pakistan-journal-of-zoology, DOI: http://dx.doi.org/10.17582/journal.pjz/2017.49.5.1817.1828

Khan, M.H. and Verma, P.M. (1946). Studies on *Earias* spp. (the spotted bollworms of cotton) in the Punjab. III. The biology of the common parasites of *Earias fabia* Stoll., *E. insulana* Boisd., and *E. cupreoviridis* Wlk. Indian Journal of Entomology, 7: 41-63.

Khattabi, N., Ezzahiri, B., Louali, L. and Oihabi, A. (2001). Effect of fungicides and *Trichoderma harzianum* on *Sclerotium rolfsii*. Phytopathol. Mediterr., 40: 143-148.

Khatua, D.C. and Maiti, S. (1977). A new leaf blight disease of *Hibiscus cannabinus* in West Bengal. Indian J. Mycol. Plant Patol., 7(1): 87.

Khazaeli, P., Zamanizadeh, H., Morid, B. and Bayat, H. (2010). Morphological and Molecular Identification of *Botrytis cinerea* Causal Agent of Gray Mold in Rose Greenhouses in Central Regions of Iran. International Journal of Agricultural Science and Research, 1(1): 19-24.

Khidr, A.A., Kostandy, S.N., Abbas, M.G., El-Kordy, M.W. and El-Gougary, O.A. (1990). Host plants, other than cotton, for the pink boll worm *Pectinophora gossypiella* and the spiny bollworm *Earias insulana*. Agricultural Research Review, 68(1): 135-139.

Khoo, K.C., Ooi, P.A.C. and Ho, C.T.(1991). Crop pests and their management in Malasia. Kuala Lumpur, Malaysia, Tropical press Sdn. Phd., P.242.

Kim, B.S., Cho, H.J., Hwang, H.S. and Cha, Y.S. (1999). Gray leaf spot of tomato caused by *Stemphylium solani*. Plant Pathol. J., 15: 345-347.

Kim, B.S and Yu, S.H. (1996). Leaf spot of pepper caused by *Stemphylium* spp. in Kyongbuk area. Plant Dis. Agric., 2: 40-41.

Kim, Y., Hutmacher, R.B. and Davis, R.M. (2005). Characterization of California isolates of *Fusarium oxysporum* f. sp. *vasinfectum*. Plant Dis., 89: 366-372.

Kim, Y., Lee, H., Lee, S., Kim, G. and Ahn, Y. (1999). Toxicity of tebufenpyrad to *Tetranychus urticae* (Acari: Tetranychidae) and *Amblyseius womersleyi* (Acari: Phytoseiidae) under laboratory and field conditions. J. Econ. Entomol., 92: 187-192.

Kimaro, D., Msanya, B.M. and Takamura, Y. (1994). Review of Sisal Production and Research in Tanzania. African Study Monographs, 15(4): 227-242.

King, A.B.S. (1994). Heliothis/Helicoverpa (Lepidoptera: Noctuidae) In: G. A. Matthews & J. P. Tunstall (eds), Insect Pests of Cotton. Wallingford, UK: CAB International, Wallingford, pp.39-106.

King, A. M. Q., Adams, M.J., Carstens, E.B. and Lefkowitz. E.J. (2012). Family Geminiviridae. In Virus taxonomy: Ninth Report of the International Committee on Taxonomy of viruses. Elsevier, New York. pp.351-373.

King, C. (2013). Pasmo on flax: Advances in resistance and management. https://www.topcropmanager.com/.../pasmo-on-flax-advances-in-resistance-and-mana.

King, E.G. and Jackson, R.D., eds, (1989). Proceedings of the Workshop on Biological Control of Heliothis: Increasing the Effectiveness of Natural Enemies November 1985, New Delhi. New Delhi, India: Far Eastern Regional Research Office, US Department of Agriculture

King, K.M. (1926). The Red backed Cutworm and its Control in the Prairie Provinces. Canada Department of Agriculture Pamphlet 69. pp13 . Abs. Rev. Appl. Entomol. 1926 Series A, 14, 430–431.

King, K.M. (1928). *Barathra configurata*, Wlk., an armyworm with important potentialities on the northern prairies. J. Econ. Entomol., 21: 279-293.

Kingsland, G.C. and Sitterly, W.R. (1986). Studies on fungicides for control of *Corynespora cassiicola* leaf spot of tomatoes in the Republic of Seychelles. Tropical Pest Management, 32(1): 31-34.

Kirchner, O. (1906). Hanf, *Cannabis sativa* L. pp.319-323. In: Die Krankheiten und Beschadigungen uhserer landwirtschaftlichen Kulturpflanzen. E. Ulmer, Stuttgart. P-637.

Kirkpatrick, T.W. (1923). The Egyptian cotton seed bug *Oxycarenus hyalinipennis* (Costa). Its Bionomics, damage and suggestions for remedial measures. Bulletin, Ministry of Agriculture, Egypt, Technical and Scientific Service. 35: pp.107.

Kiyanthy, S. and Mikunthan, G. (2009). Association of Insect Pests with Neem, *Azadiracta indica* with Special Reference to Biology of Ash Weevils, *Myllocerus* sp in Jaffna, Sri Lanka. American Eurasian Journal of Scientific Research., 4(4): 250-253.

Klein, M., Franck, A. and Rimon, D. (1986). Proliferation and branching of cotton seedlings: the suspected cause *Thrips tabaci*, the influence on yield, and tests to reduce damage. Phytoparasitica, 14(1): 25-37.

Kluth, S., Kruess, A. and Tscharntke, T. 2002. Insects as vectors of plant pathogens: mutualistic and antagonistic interactions. Oecologia 133: 193-199.

Knight, R.L. (1945). The theory and application of the backcross technique to cotton breeding. Journal of Genetics, 47: 76-86.

Knight, R.L. (1946). Breeding cotton resistant to blackarm disease. Empire Journal of Experimental Agriculture, 14: 153-174.

Knight, R.L. (1954). Cotton breeding in the Sudan.Part 1. Egyptian cotton. Empire Journal of Experimental Agriculture, 22: 68-80.

Knowles, P.F. and Houston, B.R. (1955). Inheritance of resistance to *Fusarium* wilt of flax in Dakota selection 48-94. Agron. J., 47: 131-135.

Knowles, P.F., Houston, B.R. and McOnie, J.B. (1956). Inheritance of resistance to *Fusarium* wilt of flax in Punjab 53. Agron. J., 48: 135-137.

Knox-Davies, P.S. (1966). Further studies on pycnidium production by *Macrophomina phaseoli*. South Afri J Agri Sci., 9: 595-600.

Kobayashi, K. and Hasegawa, E. (2012). Discrimination of reproductive forms of *Thrips tabaci* (Thysanoptera: Thripidae) by PCR with sequence specific primers. J. Econ. Entomol., 105: 555-559.

Koch, R.L., Pezzini, D.T., Michel, A.P. and Hunt, T.E. (2017). Identification, Biology, Impacts, and Management of Stink Bugs (Hemiptera: Heteroptera: Pentatomidae) of Soybean and Corn in the Midwestern United States. Journal of Integrated Pest management, 8(1): Website: https://doi.org/10.1093/jipm/pmx004

Kodmelwar, R.V. (1976). Effect of grey mildew disease caused by *Ramularia areola* Atk. on three varieties of cotton (*Gossypium arboreum* L.). J. Maharashtra Agricultural Universities, 1: 127-130.

Koehler, C.S., Barclay, L.W. and Kretchun, T.M. (1983). Pests in the Home Garden. California Agriculture, 37(9/10): 11-12.

Koerwitz, F.L. and Pruess, K.P. (1964). Migratory potential of the army cutworm. J. Kansas Entomol. Soc., 37: 234-239.

Kogan, J., Sell, D.K., Stinner, R.E., Bradley, Jr. J.R. and Kogan, M. (1978). The literature of arthropods associated with soybean. V. A bibliography of *Heliothis zea* (Boddie) and *H. virescens* (F.) (Lepidoptera: Noctuidae). International Soybean Program Series, 17, P.240.

Kohn, L.M., Petsche, D.M., Bailey, S.R., Novak, L.A. and Anderson, J.B. (1988). Restriction fragment length polymorphisms in nuclear and mitochondrial DNA of *Sclerotinia* species. Phytopathology, 78: 1047-1051.

Koike, S.T., Feng, C. and Correll, J.C. (2015). Powdery mildew, caused by *Leveillula taurica*, on spinach in California. Plant Disease, 99(4): 555.

Koike, S. and Saenz, G. (1996). Occurrence of powdery mildew, caused by *Erysiphe circhoraceaurm*, on endive and radicchio in California. Plant Dis., 80: 1080.

Kokub, D., Azam, F., Hassan, A., Ansar, M., Asad, M.J. and Khanum, A. (2007). Comparative growth, morphological and molecular characterization of indigenous strains isolated from different locations of Pakistan. Pak. J. Bot., 39: 1849-1866.

Kommedahl, T., Christensin, J.J. and Freederikson, R.A. (1970). A Half Century of Research in Minnesota on Flax Wilt Caused by *Fusarium oxysporum*. Minnesota Agr. Exp. Sta. Tech. Bull., P.34.

Kornejeva, E.M. and Loschakova, N.I. (1976). Zabolevanijel'na-dolgunca- pasmo. [in Russian]. Len ikonoplia, 6: 17-19.

Koszatarab, M. (1996). Scale insects of northeastern North America: Identification, biology, and distribution. Virginia Museum of Natural History.

Kowalska, A. and Niks, R.E. (1998). Quantitative resistance of flax to flax rust (*Melampsora lini*). Can. J. Plant Pathol., 20: 182-188.

Kirk, P.M. (1984). A monograph of the Chonephoraceae. Mycological Papers, 152: 1-61.

Kirk P.M. (2018). Species Fungorum (version Oct 2017). In: Roskov Y., Abucay L., Orrell T., Nicolson D., Bailly N., Kirk P.M., Bourgoin T., DeWalt R.E., Decock W., De Wever A., Nieukerken E. van, Zarucchi J., Penev L., eds. (2018). Species 2000 & ITIS Catalogue of Life.

Kirk, W.D. (1996): Thrips. Naturalists' Handbooks 25. Richmond Publishing Company, Slough, P.70.

Kirkpatrick, T.L. and Rothrock, C.S. (eds) (2001). Compendium of cotton diseases. 2nd ed. St. Paul (MN): APS Press.

Kohno, K. and Bui Thi, N. (2004). Effects of host plant species on the development of *Dysdercus cingulatus* (Heteroptera: Pyrrhocoridae). Applied Entomology and Zoology, 39: 183-187.

Kokane, C.D. and Bhale, N.L. (2016). A Review: To Detect and Identify Cotton leaf disease based on pattern recognition technique. International Journal of Engineering and Computer Science, 5(12): 19334-19338.

Kranthi, K.R., Kranthi, S., Rameash, K., Nagrare, V.S. and Barik, A. (2009). Advances in cotton IPM. Technical Bulletin, Central Institute for Cotton Research, Shankar Nagar, Nagpur (MS), India.

Kranthi, V. (1997). Insecticide resistance management strategies for central India. Central Institute of Cotton Research, Nagpur, India.Technical Bulletin.

Krawczyk, G., Enyeart, T.R. and Hull, L.A. (2012). Understanding biology and behavior of brown marmorated stink bug as a basis for development of management programs in fruit orchards. Pa. Fruit News, 92: 22-27.

Krishna, S.S. (1996). Plant volatiles and insects: Potentials in biotechnological research. In Biotechnological Perspectives in Chemical Ecology of Insects (T.N. Ananthakrishnan, ed.), pp.54-62.Oxford & IBH Publishing Co. Pvt. Ltd.; New Delhi, India.

Krishnamoorthy, A. and Mani, M. (1985). Feeding behaviour of Chrysopascelestes Banks on the parasitised eggs of some lepidopterous pests. Entomon, 10(1): 17-19.

Kroes, G.M.L.W., Loffler, H.J.M., Parlevliet, J.E., Keizer, L.C.P. and Lange, W. (1999). Interactions of *Fusarium oxysporum* f. sp. *lini*, the flax wilt pathogen, with flax and linseed. Plant Path., 48: 491-498.

Kroes, I., Loffler, H., Parlevliet, J., and Lange, W. (1997). Do races in *Fusarium oxysporum* f. sp *lini* exist? In: Aspects of resistance of flax and linseed (*Linum usitatissimum*) to *Fusarium oxysporum* f. sp. *lini*, pp.91-103.

Kroes, G.M.L.W., Rashid, K.Y., Hammond, J. and Lange, W. (1998c). Assessment of resistance To *Fusarium oxysporum* f.sp. *lini* in flax, race specific interactions and environmental factors. Proc. of the Flax Institute of the United States.Fargo, ND, Flax Institute of the United States, 57: 118-124.

Kroes, I., K., Rashid, K.Y. and Lange, W. (2002).Variation in *Fusarium* wilt (*Fusarium oxysporum lini*) in Europe and North America. Eur. J. Plant Pathol. In: Flax: The genus Linum. https://books.google.co.in-books.

Kroon, L. P.N. M., Brouwer, H., Cock, A. W. A. M. and Govers, F. (2012). The Genus *Phytophthora* Anno 2012. Phytopathology, 102: 348-364.

Kryachko, Z., Ignatenko, M., Markin, A. andZaets, V. (1965). Notes on the hemp tortrix. Zashchita Rastenii Vredit. Bolez., 5: 51-54.

Krylova, T.V. (1981). Rhizoctonisis of fiber flax. Mikologiya Fitopatologiya, 15: 511-513.

Krylova, T.V. and Voronova, V.G. (1981). Resistant varieties of fibre flax are the starting point in the control of diseases. Zashchita Rastenii, 1: 15.

Kuang, H.Y. and Chen, L.S. (1990).Studies on the differentiation of two sibling species *Tetranychus cinnabarinus* (Boisduval) and *T. urticae* Koch. Acta Entomologica Sinica, 33(1): 109-116.

Kudryavtseva, L.P. (1988). Establishing the breeding value of source material of flax for resistance to a combination of diseases. Sbornik Nauchnykh Trudov, Vsesoyuznyi Nauchno-Issledovatel'skii InstitutL'na, 25: 31-34.

Kudryavtseva, L.P. (1998). Intraspecific differentiation in flax anthracnosis pathogen by virulence. Mikologiya I Fitopatologiya, 32: 62-64.

Kuhar, T.P., Doughty, H., Kamminga, K., Wallingford, A., Philips, C. and Aigner, J. (2012a). Evaluation of insecticides for the control of brown marmorated stink bug in bell peppers in Virginia experiment 1, 2011. Arthropod Manag.Tests, 37E37.

Kuhar, T.P., Doughty, H., Kamminga, K., Wallingford, A., Philips, C. and Aigner, J. (2012b). Evaluation of insecticides for the control of brown marmorated stink bug in bell peppers in Virginia experiment 2, 2011. Arthropod Manag.Tests, 37E38.

Kuhar, T.P., Doughty, H., Kamminga, K., Wallingford, A., Philips, C. and Aigner, J. (2012c). Evaluation of insecticides for the control of brown marmorated stink bug in bell peppers in Virginia experiment 3, 2011. Arthropod Manag.Tests, 37E39.

Kuhar, T.P., Doughty, H., Kamminga, K., Wallingford, A., Philips, C. and Aigner, J. (2012d). Evaluation of insecticides for the control of brown marmorated stink bug in bell peppers in Virginia experiment 4, 2011. Arthropod Manag.Tests, 37E40.

Kuhar, T.P., Phillips, S.B., Straw, R.A., Waldenmaier, C.M. and Wilson, H.P..(2006). Commercial vegetable production recommendations. Virginia Cooperative Extension Publication, Blacksburg, VA.

Kuhn, J.G. (1858). Die krankheiten der Kulturegewachse, ihre ursachen und ihre Verhutung. Gustv Bosselmann, Berlin, P.312.

Kulkarni, S.N. and Raodeo, A.K. (1986).The incidence of sucking pest complex and its effect on plant growth and yield in cotton. Indian Journal of Plant Protection, 14(2): 75-81.

Kulthe, K.S. and Mali, V.R. (1979). Occurrence of tobacco mosaic virus in cowpea (*Vigna unguiculata*) in India. Tropical Grain Leg. Bull., 16: 8-13.

Kumar, A., Kumar, J., Khan, Z.A., Yadav, N., Sinha, V., Bhatnagar, D. and Khan, J.A. (2010). Study of betasatellite molecule from leaf curl disease of sunnhemp (*Crotolaria juncea*) in India. Virus Genes, 41: 432-440.

Kumar, D., Singh, B., Singh, S.V. and Tuhan, N.C. (1992). Screening of linseed strains against major insect pests. J. Insect Sci., 5: 190-192.

Kumar, R., Kranthi, K.R., Monga, D. and Jat, S.L. (2009). Natural parasitization of *Phenacoccus solenopsis* Tinsley (Hemiptera: Pseudococcidae) on cotton by *Aenasius bambawalei* Hayat (Hymenoptera: Encyrtidae). J. Biol. Control, 23(4): 457-460.

Kumar, R. and Mandal, R.K. (2010). Effect of different fungicides on *Phytophthora parasitica* var. *sabdariffae* and *Sclerotium rolfsii* infecting mesta. Pestology, 34: 23-27.

Kumar, R., Nagrare, V.S., Nitharwal, M., Swami, D. and Prasad, Y.G. (2014).Within-plant distribution of an invasive mealybug, *Phenacoccus solenopsis*, and associated losses in cotton. Phytoparasitica, 42 (3): 311-316.

Kumar, S., Singh, N.K., Tripathi, A.N. and Kumar, P (2016). Diseases of jute and Sunnhemp crops and their management, pp-345-367 . In: Eds; Chand, G and Kumar, S. Crop diseases and their management: Integrated approaches. Apple Academic Press, Canada, P-404.

Kumar, V. (2014). Studies on the myrothecium leaf spot disease of cotton caused by *Myrothecium roridum*Tode ex Fr. M.Sc. Thesis, CCSHAU, http://krishikosh.egranth.ac.in/handle/1/79952

Kundu, B.C. (1951). Origin of jute. Indian Journal of Genetics, 2: 95-99.

Kundu, B.C. (1964). Sunnhemp in India. Proc. Crop Sci. Soc., Florida, 24: 396-404.

Kundu, B.C., Basak, K.C. and Sircar, P.B. (1959). Jute in India-a monograph. Indian Central Jute Committee. Calcutta, pp.159-170.

Kunte, K. (2013). Butterflies of The Garo Hills. Dehradun: Samrakshan Trust, Titli Trust and Indian Foundation of Butterflies. p. 114.

Kunte, K. and Sengupta, A. (2018).Vanessa indica Herbst, 1794 – Indian Red Admiral. Kunte, K., S. Sondhi, and P. Roy (Chief Editors). Butterflies of India, v. 2.35. Indian Foundation for utterflies. http://www.ifoundbutterflies.org/sp/574/Vanessa-indica

Kuo, P.C., Hsieh, T.F., Lin, M.C., Huang, B.S., Huang, J.W. and Huang, H.C. (2015). Analysis of antifungal components in the galls of Melaphis chinensis and their effects on the control of anthracnose disease of Chinese cabbage caused by *Colletotrichum higginsianum*. Journal of Chemistry. http://dx.doi.org/10.1155/2015/850103.

Kurdiumov, N.V. (1917). Blue Flax Flea Beetle *Aphthona euphorbiae* Schrank.(In Russian). Proc. Poltava Agric. Expt. Sta., 30: 26.

Kushka, (2017). How to fight Fusarium in your cannabis plants. Website: https://www.dinafem.org ›› Fungi › How to fight Fusarium in your cannabis plants

Kwon, H.W., Kim, J.Y., Choi, M.A., Son, S.Y. and Kim, S.W. (2014). Characcterization of *Myrothecium roridum* isolated from imported Anthurium Plant Culture Medium. Mycobiology, 42(1): 82-85.

Kwon, J.-H., Lee, S.-T., Choi, Y.-J. and Lee, S.-D. (2015). Outbreak of Anthracnose on *Hibiscus cannabinus* Caused by *Colletotrichum* sp. in Korea. Plant Disease, 99 (11): 1643.

Kwon, J.-H., Kang,D.-W. and Lee, S.-Y (2016). First Report of Brown Leaf Spot Caused by *Alternaria alternata* on *Aronia melanocarpa* in Korea. Plant disease, 100(5): 1011.

Ladipo, J.L. (2008). Viruses associated with a Mosaic disease of *Crotolaria juncea* in Nigeria, Journal of Phytopathology, 121(1): 8-18.

Laemmlen, F.F. and Hall, D.H. (1973). Interdependence of a mite, *Siteroptes reniformis*, and a Fungus, *Nigrospora oryzae*, in the Nigrospora lint rot of cotton. Phytopathology, 63: 308-315.

Lafferty, H.A., Rhynehart, J.G. and Pethybridge, G.H. (1922). Investigations on flax diseases (Third report). J. Dept. Agric. Tech. Instr. Ireland, 22: 103-120.

Lafontaine, J.D. (1987). The Moths of America North of Mexico. Fascicle, 27(2): 112.

Lafontaine, J.D. (2004). Noctuoidea, Noctuinae, Agrotini. In: R.W. Hodges; D.R. Davis; D.C. Ferguson; E.G. Munroe & J.A. Powel (Eds). The moths of America North of Mexico. Lawrence, Allen Press, Fasc 25-1, P.385.

Laha, S.K. and Pradhan, S.K. (1987). Susceptibility of bast fibre crops to *Meloidogyne incognita*. Nemalol. Medit., 15: 163-164.

Lakhmanov, V.P. (1970). The injuriousness of the yellow spurge flea beetle. (In Russian). Zashchita Rastenii, 15: 10.

Lakshmanan, P., Jeyarajan, R. and Vidhyasekaran, P. (1990). A boll rot of cotton caused by *Corynespora cassiicola* in Tamil Nadu, India. Phytoparasitica, 18(2): 171-173.

Lamani, M.S. (2014). Status of flower bud maggot, *Dasineura gossypii* Fletcher (Cecidomyiidae: Diptera) on cotton and its management in northern Karnataka. M.Sc. Thesis, University of Agricultural Sciences, Dharwad.

Lamb, R.J. (1989). Aphids in flax. Canadian Agricultural Insect Pest Review, 67: 20.

Lamb, R.J. and Grenkow, L. (2008). Efficacy of herbivore-plant interaction: conversion of biomass from flax (Linaceae) to aphid, *Macrosiphum euphorbiae* (Hemiptera: Aphididae). The Canadian Entomologist, 140(5): 600-602.

Lambert B., Buysse, L., Decock, C, Jansens, S., Peins, C, Saey, BvSeurinck, J., van Audenhove, K., van Rie, J., van Vliet, A. and Peferoen, M. (1996). A *Bacillus thuringiensis* insecticidal crystal protein with a high activity against members of the family Noctuidae. Applied and Environmental Microbiology, 62: 80-86.

Lampson, B., Han, Y., Khalilian, A., Greene, J., Mankin, R.W. and Foreman, E. (2010). Characterization of Substrate-Borne Vibrational Signals of *Euschistus servus* (Heteroptera: Pentatomidae). American Journal of Agricultural and Biological Sciences, 5(1): 32-36.

Larentzaki, E., Shelton, A.M., Musser, F.R., Nault, B.A. and Plate, J. (2007). Overwintering locations and hosts for onion thrips (Thysanoptera: Thripidae) in the onion cropping ecosystem in New York. J. Econ. Entomol., 100: 1194-1200.

Laudon, G.F. and Waterston, J.M. (1964). *Phytophthora nicotianae* var *nicotinae*. CMI Descriptions of Pathogenic Fungi and Bacteria, 34: 1-2.

Land, C.J., Lawrence, K.S. and Newman, M. (2016). First Report of *Verticillium dahliae* on Cotton in Alabama. Plant Disease, 100(3): 655.

Landry, B. (2016): Taxonomic revision of the Spilomelinae (Lepidoptera, Pyralidae s. l.) of the Galápagos Islands, Ecuador. Revue suisse de Zoologie, 123(2): 315-399.

Lange, W.H. and Bronson, L. (1981). Insect pests of tomatoes. Ann. Rev. Entomol., 26: 345–371.

Langenbruch, G.A., Welling, M. and Hosang, B. (1985). Untersuchungen über den Maiszünslerim Ruhrgebiet. Nachrichtenbl. Deut. Pflanzenschutzd., 37: 150-156.

Lanham, M.D. (2012). Biology and Management of the Green Stink Bug. Website: https://www.entsoc.org/biology-and-management-green-stink-bug

Lars, W. and Ralf, N. (2003). Insecticide resistance in *Phorodon humuli*. Pest Management Science, 59(9): 991-998

Last, F.T. (1960). Effect of *Xanthomonas malvacearum* (E.F. Sm.) Dowson on cotton yields. Empire Cotton Growing Review, 37: 115-117.

Latunde-Dada, A.O. (1993). Biological control of southern blight disease of tomato caused by *Sclerotium rolfsii* with simplified mycelial formulations of *Trichoderma koningii*. Plant Pathology, 42: 522-529.

Laundon, G.F. and Waterston, J.M. (1965). Melampsora lini. CMI Descriptions of Pathogenic Fungi and Bacteria, No. 51. Surrey, England: Commonwealth Mycological Institute.

Laura, E.G. and Sanchez-Puerta, M.V. (2012). Characterization of a Root-knot Nematode population of *Meloidogyne arenarea* from Tupungato (Mendoza, Argentina). Journal of Nematology, 44(3): 291-301.

Lavoipierre, M.M.J. (1940). Hemitarsonemus latus (Banks) (Acarina), a Mite of Economic Importance New to South Africa. J. Entomol. Soc. Southern Africa. 3: 116-123.

Lawrence, G.J. (1988). *Melampsora lini*, rust of flax and linseed. Adv. Plant Pathol., 6: 313-331.

Lawrence, G.J., Dodds, P.N. and Ellis, J.G. (2007). Rust of flax and linseed caused by *Melampsora lini*. Molecular Plant Pathology, 8(4): 349-364.

Lawrence, G.J., Ellis, J.G. and Finnegan, E.J. (1994). Cloning a rust-resistance gene in flax. Advances in Molecular Genetics of Plant-Microbe Interactions, 3: 303-306.

Lawrence, G.J., Finnegan, E.J., Ayliffe, M.A. and Ellis, J.G. (1995). The L6 gene for flax rust resistance is related to the Arabidopsis bacterial resistance gene RPS2 and the tobacco viral resistance gene N. Plant Cell, 7: 1195-1206.
Lawrence, K. S., Lawrence, G. W. and van Santan, E. (2005). Effect of Controlled Cold Storage on Recovery of *Rotylenchulus reniformis* from Naturally Infested Soil.Science. gov (United States) identify *Rotylenchulus reniformis*: Topics by WorldWideScience.org. https://worldwidescience.org/topicpages/i/identify+rotylenchulus+reniformis.html
Layton, M.B. (1995). Tarnished plant bug: biology, thresholds, sampling, and status of resistance. Proceedings Beltwide Cotton Conference. Memphis, USA: National Cotton Council, pp.131-134.
Lebeda, A. and Mieslerova, B. (2010).Taxonomy, distribution and biology of lettuce powdery mildew (*Golovinomyces cichoracearum sensu stricto*). Plant Pathology, 60(3). Website: https://doi.org/10.1111/j.1365-3059.2010.02399x
Lee, D.H., Short, B.D., Joseph, S.V., Bergh, J.C. and Leskey, T.C. (2013). Review of the Biology, Ecology, and Management of *Halyomorpha halys* (Hemiptera: Pentatomidae) in China, Japan, and the Republic of Korea. Environ. Entomol., 42: 627-641.
Lee, H.B. (2012a). First report of powdery mildew caused by *Erysiphe arcuata* on lance leaf coreopsis (*Coreopsis lanceolata*) in Korea. Plant Disease, 96: 1827.
Lee, H.B. (2012b). Molecular Phylogenetic Status of Korean Strain of *Podosphaera xanthii*, a causal pathogen of powdery mildew on Japanese Thistle (*Cirsium japonicum*) in Korea. The Journal of Microbiology, 50(6): 1075-1080.
Lee, H.B., Kim, C.J., and Mun, H.Y. (2011). First report of *Erysiphe quercicola* causing powdery mildew on Ubame oak in Korea. Plant Disease, 95: 77.
Lee, I-M., Gundersen-Rindal, D., Davis, R.E., Bottner, K.D., Marcone, C. and Seemuller, E. (2004). *Candidatus Phytoplasma asteris*, a novel taxon associated with aster yellows and related diseases. International Journal of Systemic Bacteriology, 54: 1037-1048.
Lee, P. E., and Robinson. A. G. (1958). Studies on the six-spotted leafhopper, *Macrosteles fascifrons* (Stal.) and aster yellows in Manitoba. Can. J. Plant Sci., 38: 320-327.
Leefmans, S. (1923). Een ernstige, nog onbekende plaag van de Java-jute, de Spiraal boorder (*Agrilus acutus*Thunb.). Mededeeling van het Instituut voor Plantenziekten.
Lefroy, H.M. (1906). Indian Insect Pests, Calcutta. p. 151.
Lefroy, H.M. (1907). Insect pests of jute. Agric. J. India 2:100-15.
Lefroy, H.M. (1909). Indian insect life. A manual of the insects of the plains. Thacker Spink and Co. Calcutta, 2, P.463.
Leghari, M.A. and Kalro, A.M. (2002). Screening of insecticides against spotted bollworm, *Earias* spp. of cotton crop. Sindh Baloch. J. Plant Sci., 4: 71-73.
Legrand, A. and Los, L. (2003). Visual Responses of *Lygus lineolaris* and *Lygocoris* spp. (Hemiptera: Miridae) on Peaches. Florida Entomologist, 86(4): 424-428.
Le Guigo, P., Rolier, A. and Le Corff, J. (2012). Plant neighborhood influences colonization of Brassicaceae by specialist and generalist aphids. Oecologia, 169(3): 753-761.
Lei, Z.M. (1981). Preliminary observation on *Loxostege sticticalis*. (In Chinese). Shanxi Nongye Kexue, 10: 14–15.
Lekic, M., and Mihajlovic, L.(1971). *Grapholihtha sinana* Felder (Tortricidae, Lepidoptera) opasnastetocinakonopljenapodrucju Vojvodine. Savremena Poljoprivreda, 19 (3): 63-68.
Leonard, K.J. and Szabo, L.J. (2005). Stem rust of small grains and grasses caused by *Puccinia graminis*. Mol. Plant Pathol., 6: 99-111.
Lepengue, A.N., Atteke, C. Mouaragadja, I., Ake, S. and M'batchi, B. (2010). Caracteristiques Biologiques De *Phoma sabdariffae* Sacc. Et *Trichosphaeria* Sp., Deux Agents Pathogenes De La Roselle Au Gabon. Agronomie Africaine, 22(1): 1-9.
Lepengue, A.N., Mouaragadja, I., M'batchi, B. and Séverin, A.K.E. (2009). Physico chemical

characteristics of *Phoma sabdariffae* Sacc. culture filtrate, a Jamaican sorrel pathogenic agent. Sciences & Nature, 6(2): 95-105.
Lepengue, A.N., Yala, J.F., Lebamba, J., Mouaragadja, I., Kone, D. and M'batchi, B. (2013). *Phoma sabdariffae*'s impact on roselle (*Hibiscus sabdariffa* L. var. *sabdariffa*) fructification parameters in Gabon. International Journal of Innovation and Applied Studies, 4(1): 155-164.
Leskey, T.C., Hamilton, G.C., Biddinger, D.J., Buffington, M.L., Dieckhoff, C., Dively, G.P., Fraser, H., Gariepy, T., Hedstrom, C., Herbert, D.A., et al. (2014). Data sheet for *Halyomorpha halys* (Stal), (Hemiptera: Pentatomidae). In. Crop Protection Compendium. CAB International, Wallingford, United Kingdom.
Leskey, T.C., Hamilton, G.C., Nielsen, A.L., Polk, D.F., Rodriguez-Saona, C., Bergh, J.C., Herbert, D.A., Kuhar, T.P., Pfeiffer, D., Dively, G.P., et al. (2012a). Pest status of the brown marmorated stink bug, *Halyomorpha halys* in the USA. Outlooks Pest Manag., 23: 218-226.
Leskey, T.C., and Hogmire, H.W. (2005). Monitoring stink bugs (Hemiptera: Pentatomidae) in mid-Atlantic apple and peach orchards. Journal of Economic Entomology, 98: 143-153.
Leskey, T.C., Short, B.D., Butler, B.R. and Wright, S.E. (2012b). Impact of the invasive brown marmorated stink bug, *Halyomorpha halys* (Stal), in mid-Atlantic tree fruit orchards in the United States: case studies of commercial management. Psyche 2012 . 535062.
Leskey, T.C., Short, B.D. and Lee, D.H. (2013). Efficacy of insecticide residues on adult *Halyomorpha halys* (Stal) (Hemiptera: Pentatomidae) mortality and injury in apple and peach orchards. Pest Manag. Sci., 70: 1097-1104.
Leslie, J. (1993). Fungal Vegetative Compatibility. Annual review of phytopathology , 31(1): 127-150.
Levesque, C.A., Brouer, H., Cano, L., Hamilton, J.P., Holt, C., Huitema, E. Et al., (2010). Genome sequence of the necrotrophic plant pathogen *Pythium ultimum* reveals original pathogenecity mechanisms and effector repertoire. Genome Biol., 11, R73.
Lewartowski, R. and Piekarczyk, K. (1978). Characteristics of development, appearance, intensity and noxiousness of more important diseases and pests of industrial plants in Poland in 1976. (In Polish). Biuletyn Instiyutu Ochrony Roslin, 62: 151–221.
Lewis, T. (1973). Thrips. The biology, ecology and economic importance. Academic Press, New York, P.350.
Lewis, T. (1973). Thrips: Their biology, ecology and economic importance. Academic, London, United Kingdom.
Leyendecker, P.J. (1950). Plant disease survey for New Mexico. Plant Disease Reporter, 34: 39-44.
Li and Guo, (1999). In: vascular cotton wilt (*Fusarium oxysporum* f.sp. *vasinfectum*) - Plantwise https://www.plantwise.org/KnowledgeBank/Datasheet.aspx?dsid=24715
Li, G-K., Wang, S-L., Wang, J., Wang, J., Luo J-S. and Chu, Z-J. (2007). Preliminary Study on Flax Powdery Mildew of Xinjiang [J]; Xinjiang Agricultural Sciences. Website: https://www.en.cnki.com.cn/Article_en/CJFDTotal-XJNX200705012.htm
Li, G.Q., Xue, X.F., Zhang, K.J., Hong, X.Y. (2010). Identification and molecular phylogeny of agriculturally important spider mites (Acari: Tetranychidae) based on mitochondrial and nuclear ribosomal DNA sequences, with an emphasis on Tetranychus. Zootaxa, 2485: 1-15.
Li, J., Zhang, X.Y. and Qian, Y.J. (2010). Molecular characterization of Ramie mosaic virus isolates detected in Jiangsu and Zhejiang provinces, China. Acta Virol, 54(3): 225-228.
Li, P., Zhang, Q. and Cai, Q. (1998). Study on induced resistance of cotton to *Aphis gossypii* with *Pseudomonas gladioli*. Acta Gossypii Sinica, 10(4): 193-198.

Li, R.M. and Ma, H.G. (1993). Studies on the occurrence and control of ramie anthracnose. J. Plant Prot., 20(1): 83-89.
Li, T., Chen, X.L. and Hong, X.Y. (2009). Population genetic structure of *Tetranychus urticae* and its sibling species *Tetranychus cinnabaribus* (Acari: Tetranychidae) in China as inferred from microsatellite data. Ann Entomol Soc Am., 102: 674-683.
Li, X., Zhang, Y., Ding, C., Xu, W.and Wang, X. (2017).Temporal patterns of cotton Fusarium and Verticillium wilt in Jiangsu coastal areas of China. Scientific Reports, 7, Article number: 12581.
Li, Z.D. (1980). The theory and practice of mast crops. Shanghai Science and Technology Publishing House, pp.302-303.
Liang, X., Ding, S. and Wong, S. (2002a). Development of a kenaf (*Hibiscus cannabinus* L.) protoplast system for a replication study of *Hibiscus* chlorotic ringspot virus. Plant Cell Reports, 20(10): 982–986.
Liang, X., Liu, F., Zhou, Y., Wu, J., Ding, D., Huang, L. and Liu, D. (1994). A comprehensive study of ramie mosaic, Acta Agriculturae Universitis Jiangxiensis, Website: A comprehensive study of ramie- Acta....en.cnki.com.cn/Article_en/CJFDTotal...
Liang, X-Z, Lucy, A.P., Ding, S-W. and Wong, S-M. (2002b). The p23 Protein of *Hibiscus Chlorotic Ringspot Virus* is indispensable for host specific reaction. J. Virol., 76(23): 12312-12319.
Lin, Y. (2000). Adaptation of transgenic strains of insect-resistant cotton to different ecological environments. Chinese Journal of Applied Ecology, 11: 246-248.
Lingappa, B.T. (1955). Two new species of *Physoderma* from India. *Mycologia, 47: 109*-121.
Lister, R.M. and Thresh, J.M. (1955). A mosaic disease of leguminous plants caused by a strain of tobacco mosaic virus. Nature, 175: 1047-1048.
Little, S. (1987a). *Cercospora gossypina*. CMI Descriptions of Pathogenic Fungi and Bacteria, 914:1-2.
Little, S. (1987b). *Cercospora malayensis*. CMI Descriptions of Pathogenic Fungi and Bacteria. 916: 1-2.
Littlefield, L.J. and Bracker, C.E. (1971). Ultrastructure and development of urediospore ornamentation in *Melampsora lini*. Canadian Journal of Botany, 49(12): 2067-2073.
Littlefield, L.J. and Bracker, C.E. (1972). Ultrastructural specialization at the host-pathogen interface in rust-infected flax. Protoplasma, 74: 271-305.
Litzenberger, S.C. (1976). Guide for field crops in the tropics and the subtropics. Office of Agriculture, Technical Assistance Bureau, Agency of International development, Washington. Reprinted by: Peace Crop Program & Training Journal, 10: 321.
Liu, B., Gumpertz, M.L and Ristaino, J.B. (2007). Long-term effects of organic and synthetic soil fertility amendments on soil microbial communities and the development of southern blight. Soil Biology and Biochemistry, 39: 2302-2316.
Liu, H., Skinner, M., Parker, B.L. and Day, W. H. (2003). Recognizing Tarnished Plant Bug Damage. University of Vermont Entomology Research Laboratory. USDA-BIRL, Newark, DE.
Liu, L and Free, S.J. (2016). Characterization of the *Sclerotinia sclerotiorum* cell wall proteome. Molecular Plant Pathology, 17(6): 985-995.
Liu, S. and Yuan, L. (2011). Genetic diversity of *Fusarium oxysporum* f. sp. *lini* by inter simple sequence repeat (ISSR). ITME 2011 - Proceedings: 2011 IEEE International Symposium on IT in Medicine and Education. 2. 10.1109/ITiME.2011.6132089.
Liu, X.Y., Chen, S.L., Sun, Q.A., He, D.T. and Wu, Y.N. (1993). Evaluation of *Fusarium* wilt resistance of flax varieties. Scientia AgriculturaSinica, 26: 44-49.
Lock, G.W. (1958). The sisal weevil. Kenya Sisal Board Bulletin, 24.

Lock, G.W. (1969). Sisal: Thirty years' sisal research in Tanzania. Longmans, Green, and Co., London, UK, P.365.
Lock, G.W. (1985). On the scientific and practical aspects of sisal (*Agave sisalana*) cultivation, in Biología y Aprovechamiento Integral del Henequen y Otros Agaves (Cruz, C., Del Castillo, L., Robert, M. L., and Ordanza, R. N., eds.), Centro de Investigaciom Cientifica de Yucatan A. C., México, pp 99-119.
Lodos, N., Onder, F. and Simsek, Z. (1984). Study of overwintering insect fauna and research on flight activity and migration behavior of some other species at the spring emergence during the migration period of the sunn pest (*Eurygaster integriceps* Put.) to the plain of Diyarbakir (Karacadag). (iii). Species of Coleoptera: Chrysomelidae. (In Turkish). Bitki Koruma Bulteni, 24: 113-118.
Loganathan, M., Sasikumar, M. and Uthamasamy, S. (1999). Assessment of duration of pheromone dispersion for monitoring *Heliothis armigera* (Hub.) on cotton. Journal of Entomological Research, 23: 61- 64.
Loganathan, M. and Uthamasamy, S. (1998). Efficacy of a sex pheromone formulation for monitoring *Heliothis armigera* Hubner moths on cotton. Journal of Entomological Research, 22: 35-38.
Lloyd, N.C., Jones, E.L., Morris, D.S., Webster, W.J., Harris, W.B., Lower, H.F., Hudson, N.M. and Geier, P.W. (1970). Managing apple pests: a new perspective. J .Aust. Inst. Agric. Sci., 36: 251-258.
Loncaric, I., Oberlerchner, J.T., Heissenberger, B. and Moosbeckhofer, R. (2009). Phenotypic and genotypic diversity among strains of *Aureobasidium pullulans* in comparison with related species. Antonie Van Leeuwenhoek International Journal of General and Molecular Microbiology, 95(2): 165-178.
Long, D.L. and Timian, R.G. (1971). Acquisition Through Artificial Membranes and Transmission of Oat Blue Dwarf Virus by *Macrosteles fascifrons*. Phytopathology, 61: 1230-1232.
Loof, P.A.A. (1960). Tijdschr. Pl. Ziekt, 66: 29-90.
Lopez-Martínez, V., Alta-Tejacal, I., Andrade-Rodríguez, M.,Garica-Ramírez, M.D.J. and Rojas, J.C. (2011). Daily activity of *Scyphophorus acupunctatus* (Coleoptera: Curculionidae) monitored with pheromone-baited traps in a field of Mexican tuberose. Florida Entomologist, 94(4): 1091-1093.
Lopez-Perez, J.A., Escuer, M., Diez-Rojo, M.A., Robertson, L., Piedra Buena, A., Lopez-Cepero, J. And Bello, A. (2011). Host range of *Meloidogyne arenarea* (Neal, 1889) Chitwood, 1949 (Nematoda: Meloidogynidae) in Spain. Nematropica, 41(1): 130-140.
Louws, F.J., Wilson, M., Campbell, H.L., Cuppels, D.A., Jones, J.B., Shoemaker, P.B., Sahin, F. et al. (2001). Field control of bacterial spot and bacterial speck of tomato using a plant activator. Plant Dis., 85: 481-488.
Lowor, S.T., Del Socorro, A.P. and Gregg, P.C. (2009). Sex Pheromones of the Green Mirid, *Creontiades dilutus* (Stal) (Hemiptera: Miridae). International Journal of Agricultural Research, 4(4): 137-145.
Lu, Y., Wu, K., Jiang, Y., Xia, B., Li, P., Feng, H., Wyckhuys, K.A.G.and Guo, Y. (2010). Mirid Bug Outbreaks in Multiple Crops Correlated with Wide-Scale Adoption of Bt Cotton in China. Science, 328 (5982): 1151-1154.
Lu, Y.Y. and Liang, G.W. (2002). Spatial pattern of cotton bollworm (*Helicoverpa zea*) eggs with geostatistics. Journal of Huazhong Agricultural University, 21(1): 13-17.
Lu, Y.Y., Zeng, L., Wang, L., Xu, Y.J. and Chen, K.W. (2008). Precaution of solenopsis mealybug *Phenacoccus solenopsis* Tinsley. Journal of Environmental Entomology, 30: 386-387.
Lubbe, C.M., Denman, S., Cannon, P.F., Groenewald, J.Z., Lamprecht, S.C. and Crous, P.W. (2004). Characterization of *Colletotrichum* species associated with disease of Proteaceae. Mycologia, 96: 1268-1279.

Lubeck, M. (2004). Molecular characterization of *Rhizoctonia solani*. Applied Mycology and Biotechnology, 4: 205-224.
Lugger, O. (1890). A Treatise on Flax Culture. Minn. Agr. Exp. Sta. Bull., 13: P.38.
Luo, S.B., Yan, J.P., Chai, C.J., Liang, S.P., Zhang, Y.M., Zhang, Y. and Le, G.K. (1986). Control of pink bollworm, *Pectinophora gossypiella* with *Bacillus thuringiensis* in cotton fields. Chinese Journal of Biological Control, 2(4): 167-169.
Lupoli, R., Marion-Poll, F., Pham-Delegue, M. and Masson, C. (1990). Effetdemissions volatiles de feuilles de maissur les preferences de ponte chez *Ostrinia nubilalis* (Lepidoptera: Pyralidae). C. R. Acad. Sci. Paris, 311: 225-230.
Lyda, S.D. and Burnett, E. (1970). Influence of benzimidazole fungicides on *Phymatotrichum omnivorum* and *Phymatotrichum* root rot of cotton. Phytopathology, 60: 726-728.
Ma, A.N., Yang, W.Y. and Wang, W.X. (2003). A preliminary report on the efficacy of a plant pill insecticide for the trapping and killing of *Agrotis ipsilon* Hufnagel. Plant Protection (Beijing, China), 29(1): 34-36.
Ma, C., Liu, H.T. and Ji, X.Q. (1980). Observation of relationship between the growth and decline of cotton Fusarium wilt and the growth period of plant in North Henan cotton region. Plant Prot., 3: 18-20.
Ma, J., Hill, C. B., and Hartman, G. L. (2010). Production of *Macrophomina phaseolina* conidia by multiple soybean isolates in culture. Plant Dis., 94: 1088-1092.
Ma, L.-J., Geiser, D., Proctor, R., Rooney, A., O'Donnell, K., Trail, F., Gardiner, D., Manners, J. and Kazan, K. (2013). Fusarium pathogenomics. Annual review of microbiology, 67: 399-416.
Maas, S., Allen, S. and Weir, D. (eds). (2012). Cotton Symptoms Guide - The guide to symptoms of diseases and disorders in Australian cotton. The Australian Cotton Industry Development & Delivery Team, P.76.
Mabbett, T.H., Dareepat, P. and Nachapong, M. (1980). Behaviour studies on *Heliothis armigera* and their application to scouting techniques for cotton in Thailand. Tropical Pest Management, 26(3): 268-273.
Macedo, A. (1943). Agave diseases. Bol. Minist. Agrc., Rio de J., 32(7): 27-28.
MacGillivary, M.E. and Anderson, G.B. (1958). Development of four species of aphids (Homoptera) on potato. The Canadian Entomologist, 90(3): 148-155.
MacGillivray, M. E. and Anderson, G.B. (1964). The Effect of Photoperiod and Temperature on the Production of Gamic and Agamic Forms in *Macrosiphum euphorbiae* (Thomas). Can. J. Zool., 42: 491-510.
Mackauer, M. (1968). Insect parasites of the green peach aphid, *Myzus persicae* Sulz., and their control potential. Entomphaga, 13: 91-106.
MacNish, G.C. (1963). Diseases recorded on native plants, weeds, fields and fiber plants in Western Australia. J. Agric. W. Aust. Ser. 4: 401-408.
Madamba, C.P., Davide, R.G. and Palis, R.K. (1968). Plant parasitic nematodes on ramie and their control by soil fumigation. Philippine Phytopathology, 4: 9-10.
Madani, A.S., Marefat, A., Behboudi, K. and Ghasemi, A. (2010). Phenotypic and genetic characteristics of *Xanthomonas citri* subsp. *malvacearum*, causal agent of cotton blight, and identification of races in Iran. Australasian Plant Pathology, 39: 440-445.
Maddens, K. (1987). Efficacite de divers fongicides en traitements de semences de fin fibre. Meded. Fac. Landbwet. Rijksuniv. Gent, 52: 919-928.
Madden, R. and Darby, H. (2011). Managing Powdery Mildew of Hops in the Northeast. University of Vermont-Extension.University of Vermont. Website: http://www.uvm.edu/extension/cropsoil/wp-content/uploads/PowderyMildew.pdf
Maelzer, D.A. (1977). The Biology and Main Causes of Changes in Numbers of the Rose Aphid, *Macrosiphum rosae* (L.) on Cultivated Roses in South Australia. Austral. J. Zool., 25: 269-284.

Mahadevakumar, S. and Janardhana, G.R. (2016). Morphological and molecular characterization of *Sclerotium rolfsii* associated with leaf blight disease of *Psychotria nervosa* (wild coffee). Journal of Plant Pathology, 98(2): 351-354.

Mahadevakumar, S., Yadav, V., Tejaswini, G.S. and Janardhana, G.R. (2016). Morphological and molecular characterization of *Sclerotium rolfsii* associated with fruit rot of *Cucurbita maxima*. European Journal of Plant Pathology, 145(1): 215-219.

Mahaffee, W., Engelhard, B., Gent, D.H. and Grove, G.G. (2009). Powdery Mildew. In: W. Mahaffee, S. J. Pethybridge, & D. H. Gent (Eds.), Compendium of Hop Diseases and Pests (pp.25-31). St. Paul, Minnesota: The American Phytopathological Society.

Mahamed, P. (2016). Korogwe leaf spot causes havocon sisal fibre, The Guardian, February 4, 2016.

Mahar, A.N., Lohar, M.K. and Abro, G.H. (1987). Field evaluation on the efficacy of endosulfan, chlorpyrifos and fenpropathrin against cotton bollworms. Proceedings of Pakistan Congress of Zoology, 7: 113-118.

Mahmud, A.U., Hoque, A.K.M.A., Bhuiyan, M.R., Khan, M. A.I., Kabir, M.E. and Mahmud, A. (2014). Management of jute yellow mosaic virus disease through cultural practices. Archives of Phytopathology and Plant Protection, 47(19). https://www.tandfonline.com-doi-abs

Mai, W.F. and Mullin, P.G. (1996). Plant parasitic nematode. A Pictorial Key to Genera, 5th Ed. Cornell University Press, Ithaca, New York.

Majumdar, M. and Mandal, N. (1996). Evaluation of *Hibiscus sabdariffa* L. and *H. cannabinus* L. genotypes for field resistance against foot and stem rot caused by *Phytophthora parasitica* Dast. Journal of Mycopathological Research, 34(1): 41-45.

Majumdar, M. and Mandal, N. (2000). Screening of roselle (*Hibiscus sabdariffa* L.) genotypes for field resistance to foot and stem rot caused by *Phytophthora parasitica* Dast. Indian Agriculturist, 44(1/2): 75-78.

Mazumder, N., Dutta, S. K., Bora, P., Gogoi, S., Das, P. (2014). Record of litchi weevil, *Myllocerus discolor* (Coleoptera: Curculionidae) on litchi (*Litchi sinensis* Sonn) (Sapindaceae) from Assam. Insect Environment, 20(1): 29-31.

Malaguti, G. and Lopez Pinto, O. (1974).Cotton rust, *Phakopsora gossypii*, in commercial crops (Venezuela) (In Spanish). Agris. Website: http://agris.fao.org/agris-search/search.do?recordID=AG19750004142.

Malik, Y.P. (1998). Birds: An eco-friendly approach for insect-pests management in linseed. Insect Environment, 3: 104.

Malik, Y.P., Singh, B. and Pandey, N.D. (1991). Role of blooming period in linseed bud fly (*Dasyneura lini* Barnes) resistance. Indian J. Entomol., 53: 276-279.

Malik, Y.P., Singh, B., Pandey, N.D. and Singh, S.V. (1996a). Assessment of economic threshold level of bud fly, *Dasyneura lini* Barnes, in linseed. Indian J. Entomol., 58: 185-189.

Malik, Y.P., Singh, B., Pandey, N.D. and Singh, S.V. (1996b). Role of fertilizer and irrigation in management of the linseed bud fly. Indian J. Entomol., 58: 132-135.

Malik, Y.P., Singh, B., Pandey, N.D. and Singh, S.V. (1998). Infestation dynamics of linseed bud fly *Dasyneura lini* Barnes in relation to weather factors under irrigated conditions in central Uttar Pradesh. Ann. Plant Prot. Sci., 6: 80-83.

Malik, Y.P., Singh, S., Singh, B., Pandey, N.D. and Singh, S.V. (1996c). Determination of physico-chemical basis of resistance in linseed for bud fly. Indian J. Entomol., 57: 267-272.

Malipatil, M.B. (1992). Revision of Australian *Campylomma* Reuter (Hemiptera: Miridae: Phylinae). J. Aust. ent. Soc., 31: 357-368.

Malipatil, M.B. and Cassis, G. (1997). Taxonomic Review of *Creontiades* Distant in Australia (Hemiptera: Miridae: Mirinae). Australian Journal of Entomology, 36(1): 1–13.

Mall, N., Srivastava, P.N. and Singh, R. (2010). First record of host plants of aphids (Homoptera: Aphididae) from India. Journal of Aphidology, 24(1&2): 85-86.

Mallick, R.N. (1981b). Keep your sunnhemp crop free from the attack of hairy caterpillar. Jute Dev. J., 1(2): 36.
Mallick, R. N., Chatterjee, A., Das, P. C. and Chatterji, S.M. (1980). Antifeeding properties of jute leaf extracts against *Myllocerus discolour* Boh. Journal of Entomological Research, 4(2): 148-152.
Mani, M. (1989). A review of the pink mealybug- *Maconellicoccus hirsutus*. Insect Science and its Application, 10: 157–167.
Mani, M. S. (1934). Studies on Indian Itonididae (Cecidomyidae :Diptera). Rec. Indian Mus., 36: 371- 451.
Mani, M. and Satpathy, S. (2016). Jute and Allied Fibre Crops, In: Eds: Mani, M. and Shivaraju, C. Mealy bugs and their Management in Agricultural and Horticultural crops pp 283-286, Springer, New Delhi.
Manjunath, T.M., Bhatnagar, V.S., Pawar, C.S. and Sithanantham, S. (1989). Economic importance of *Heliothis* spp. in India and an assessment of their natural enemies and host plants, pp.197-228 In: Proceedings of the Workshop on Biological Control of Heliothis: increasing the effectiveness of natural enemies New Delhi, India.
Manoj, K.V., Seetha, R.P. and Satya, N.N.H. (2010). Management Economics of Foot and Stem Rot of Mesta Incited by *Phytophthora parasitica*. Indian Journal of Plant Protection, 38(2): 183-185.
Manolache, C., Sandru, I. and Romascu, E. (1966). Un noudaunator al culturilor de cinepa- moliacinepii (*Grapholitha delineana* Walk.- Lepidoptera- Tortricidae). Probleme Agricole, 6: 68-72.
Mansoor, S., Bedford, I.D., Pinner, M.S., Stanley, J. and Markham, P.G. (1993). A whitefly- transmitted geminivirus associated with cotton leaf curl disease in Pakistan. Pak. J. Bot., 25: 105-107.
Mansoor, S., Briddon, R.W., Bull, S.E., Bedford, I.D., Basir, A., Hussain, M., Saeed, M., Zafar, Y., Malik, K.A., Fauquet, C.M. and Markham, P.G. (2003). Cotton leaf curl disease is associated with multiple monopartite begomoviruses supported by single DNA beta. Arch. Virol., 148: 1969-1986.
Mansour, F.A. andAscher, K.R.S. (1983). Effects of neem (*Azadirachta indica*) seed kernel extracts from different solvents on the carmine spider mite, *Tetranychus cinnabarinus*. Phytoparasitica, 11(3-4): 177-185.
Mansour, M.T.M. (1998). Pathological studies on powdery mildew of flax in A.R.E. Ph.D. Thesis, Zagazig Univ., Moshtohor, P.148.
Mao, M.J., He, Z.F., Yu, H. and Li, H.P. (2008). Molecular characterization of Cotton leaf curl Multan virus and its satellite DNA that infects Hibiscus rosa-sinensis. Chin J Virol., 24: 64-68.
Marek, S.M., Hansen, K., Romanish, M. and Thorn, R.G. (2009). Molecular systematics of the cotton root rot pathogen, *Phymatotrichopsis omnivora*. Persoonia, 22: 63-74.
Marimuthu, S., Gurusubramanian, G. and Krishna, S.S. (1997). Effect of Exposure of Eggs to Vapours from Essential Oils on Egg Mortality, Development and Adult Emergence in *Earias vittella* (F.) (Lepidoptera: Noctuidae). Biological Agriculture & Horticulture, 14(4): 303-307.
Markevicius, V. and Treigiene, A. (2003). Lietuvosgrybai. Spuogagrybieciai (Sphaeropsidales). [in Lithuanian]. Valstieciølaikrastis LTD, Vilnius, 10(3): 66-67.
Markku, S. (2008). *Agrotis* genus. Agrotis-nic.funet.fi. Website: ftp.funet.fi-index-noctuidae- noctuinae
Marks, M. and Gevens, A. (2014). Hop Powdery Mildew (PDF). University of Wisconsin- Extension.University of Wisconsin. Website: http://www.plantpath.wisc.edu/wivegdis/ pdf/2015/Hop%20PM%20A4053-02.pdf

Marshal, G.A.K. (1916). Coleoptera, Rhynchophora: Curculionidae. In: Shipley A. E. (Ed.) The Fauna of British India, including Ceylon & Burma. Taylor & Francis, London, P.367.
Martelli, G.P. (1961). Septoriosi (pasmo) of flax in Italy. Phytopathol. Medit., 1: 66–70.
Martens, J.W., Seaman, W.L. and Atkinson, T.G. (1984). Diseases of field crops in Canada. Canadian Phytopathological Society, P.160.
Martin, A.L.D., Frederiksen, R.A. and Westdal, P.H. (1961). Aster yellows resistance in Flax, Canadian J. Plant Sci., 41(2): 316-319.
Martinez, R. and Swezey, S.L. (1988). Control of *Heliothis zea* (Boddie) larvae with a nuclear polyhedrosis virus (*Baculovirus heliothis*) in cotton, Leon, Nicaragua 1983. Revista Nicaraguense de Entomologia, 2: 13-18.
Martin-Felix, Y., Groenewald, J.Z., Cai, L., Chen, Q., Marincowitz, S., Barnes., I. et al. (2017). Genera of phytopathogenic fungi: GOPHY 1. Studies in Mycology, 86: 99-216.
Matsushima, T. (1975). Icones Microfungorum: a Matsushima lectorum. The University of Michigan, P.209.
Matthews, M. (1991). Classification of the Heliothinae. NRI Bulletin No. 44. Chatham, Kent: Natural Resources Institute.
Mau, R.F.L. and Kessing, J.L.M. (1991). *Thrips tabaci* Linderman. Knowledge Master. Department of Entomology, Honolulu, Hawaii.
Mau, R.F.L. and Kessing, J.L.M. (1992). *Helicoverpa zea* (Boddie). Crop Knowledge Master, Department of Entomology, Honolulu, Hawaii. http://www.extento.hawaii.edu/kbase/Crop/Type/helicove.htm#BIOLOGY
Mau, R.F.L and Kessing, J.L.M. (2007). *Tetranychus cinnabarinus* (Boisduval). Crop Knowledge Master. Website: https://www.extento.hawaii.edu/kbase/crop/Type/t_cinnab.htm
Maw, E. (2011).Checklist of the Hemiptera of British Columbia, 2011. Exported and modified from database underlying Maw H.E.L., Foottit, R.G., Hamilton, K.G.A., Scudder, G.G.E. 2000. Checklist of the Hemiptera of Canada and Alaska. NRC Press, Ottawa, P.220.
Maxwell-Lefroy, H. (1908). The red cotton bug (*Dysdercus cingulatus* Fabr.). Mem. Dept. Agric. India Entomol. Ser., 2: 47-58.
Mayek-Pérez, N., Garcia-Espinosa, R., López-Castañeda, C., Acosta-Gallegos, J. A. and Simpson, J. (2002). Water relations, histopathology, and growth of common bean (*Phaseolus vulgaris* L.) during pathogenesis of *Macrophomina phaseolina* under drought stress. Physiol. Plant Pathol., 60: 185-195.
Mayek-Perez, N., López-Castaneda, C., Gonzales-Chavira, M., Garcia-Espenosa, R., Acosta-Gallegos, J., Martinez de Vega, O. and Simpson, J. (2001). Variability of Mexican isolates of *Macrophomina phaseolina* based on pathogenesis and AFLP genotype. Physiol. Molec. Plant Pathol., 59: 257-264.
McAlpine, D. (1906). The rusts of Australia. Their structure, nature and classification. Department of Agriculture, R.S. Brain, Government Printer, Victoria.
McAuslane, H.J. (2009). Featured Creatures- sweetpotato whitefly B biotype or silverleaf whitefly.UF/IFAS, University of Florida, EENY-129.
McBrien, H.L., Millar, J.G., Gottlieb, L., Chen, X. and Rice, R.E. 2001. Maleproduced sex attractant pheromone of the green stink bug, *Acrosternum hilare* (Say). Journal of Chemical Ecology, 27: 1821-1839.
McCain, A. H. and Noviello, C. (1985). Biological control of *Cannabis sativa*. Proceedings of the VI International Symposium on Biological Control of Weeds, pp.635-642.
McClintock, N.C. and Tahir, I.M.E. (2004). *Hibiscus sabdariffa* L. In: Grubben GJH, Denton OA, editors. Vegetables/Legumes. Wageningen, Netherlands: PROTA. 2.
McGrath, M.T. (2004). Protectant Fungicides for Managing Powdery Mildew in Cucurbits: How Do They Stack Up? Webste: vegetablemdonline.ppath.cornell.edu/NewsArticles/Cuc_Cntct_Fcides.htm

McKay, G.J., Brown, A.E., Bjourson, A.J. and Mercer, P.C. (1999). Molecular characterisation of *Alternaria linicola* and its detection in linseed. European Journal of Plant Pathology, 105: 157-166.
Mckee, R and Enlow, C.R. (1931). Crotalaria, A new legume for the south. U.S. Dept. Agric. Cir. Bull. 137, pp. 30.
McKenzie, H.L. (1967). Mealybugs of California, with taxonomy, biology, and control of North American species (Homoptera: Coccoidea: Pseudococcidae). University of California Press, Berkeley and Los Angeles, and Cambridge University Press, London, P.526.
McLeod, R. (2005). Species *Agrotis ipsilon*. BugGuide, 8 Dec. 2005, bugguide.net/node/view/38914.
McMillan, R.T. (Jr.) (2010). Efficacy of Fungicides for the Control of *Myrothecium roridum* on Dieffenbachia picta 'Compacta'.Proc. Fla. State Hort. Soc., 123: 302–303.
McMillan, R.T. (Jr.). (2011). Efficacy of fungicides for control of *Colletotrichum gloeosporioides* on Dendrobiums. Proc. Fla. State Hort. Soc., 124: 314-316.
McPartland, J.M. (1983a). Fungal pathogens of *Cannabis sativa* in Illinois. Phytopathology 72: 797.
McPartland, J.M. (2002). Epidemiology of the Hemp Borer, *Grapholita delineana* Walker (Lepidoptera: Oleuthreutidae), a Pest of *Cannabis sativa* L. Journal of Industrial Hemp, 7(1): 25-42.
McPartland, J.M., Clarke, R.C. and Watson, D.P. (2000). Hemp Diseases and Pests Management and Biological Control - An Advanced Treatise. CABI Publishing, UK, P.251.
McPartland, J.M. and Cubeta, M.A. (1997). New species, combinations, host associations and location records of fungi associated with hemp (*Cannabis sativa*). Mycological Research, 101: 853-857.
McPartland, J.M., and Hillig, K.W. (2003).The hemp russet mite. Journal of Industrial Hemp, 8(2): 107-102.
McPartland, J.M. and Hillig, K.W. (2008). Cannabis Clinic Fusarium Wilt. Journal of Industrial Hemp, 13: 67-68.
McPherson, J.E. (1982). The Pentatomoidea (Hemiptera) of northeastern North America with emphasis on the fauna of Illinois. Southern Illinois University Press, Carbondale and Edwardsville, IL, P.240.
McPherson, J.E. and McPherson, R.M. (2000). Stink bugs of economic importance in North America & Mexico. CRC LLC, Boca Raton, FL.
McSorley, R. and Parrado, J.L. (1986). Relationship between height of kenaf and root galling by *Meloidogyne incognita*. Nematropica, 16(2): 205-211.
Mead, F.W. and Fasulo, T.R. (2014). Publication Number: EENY-33. (Originally published in 2004 and Revised in December 2014). University of Florida. cotton stainer – *Dysdercus suturellus* (Herrich-Schaeffer); website: entnemdept.ufl.edu/creatures/field/bugs/cotton_stainer.htm
Mederick, F.M. and Piening, L.J. (1982). *Sclerotinia sclerotiorum* on oil and fiber flax in Alberta. Canadian Plant Disease Survey, 62: 1.
Mediavilla, V., Spiess, E., Zurcher, B., Bassetti, P., Strasser, H.R., Konermann, M., Spahr, J., Christen, S., Mosimann, E., Aeby, P., Ott, A. and Meister, E. (1997). Erfahrungenausdem Hanfanbau 1996. Proceedings of 2nd Biorohstoff Hanf Symposium, pp.253-262.
Medina, J.C. (1959). Plantas Fibrosas da Flora Mundial.Sao Paulo: Industria Grafica Siqueira, P.913.
Medrano, E.G. and Bell, A.A. (2007). Role of *Pantoea agglomerans* in opportunistic bacterial seed and boll rot of cotton (*Gossypium hirsutum*) grown in the field. J Appl Microbiol, 102: 134-143.

Medrano, E.G., Bell, A.A. and Duke, S.E. (2016). Cotton (*Gossypium hirsutum* L.) boll rotting bacteria vectored by the brown stink bug, *Euschistus servus* (Say) (Hemiptera: Pentatomidae). Journal of Applied microbiology, 121(3): 756-766.

Medrano, E.G., Esquivel, J.F.and Bell, A.A. (2007). Transmission of cotton seed and boll rotting bacteria by the southern green stink bug (*Nezara viridula* L.). Journal of Applied Microbiology, 103: 436–444.

Meena, P.N., De, R.K., Roy, A., Gotyal, B.S., Mitra, S. and Satpathy, S. (2015). Evaluation of Elite Tossa Jute, *Corchorus olitorius* Germplasm against Stem Rot, *Macrophomina phaseolina* (Tassi) Goid in Jute. J. Appl. & Nat. Sci., 7 (2): 857 – 859.

Meena, P.L., Roy, A., Gotyal, B.S., Mitra, S. and Satpathy, S. (2014) Eco-friendly management of major diseases in jute (*Corchorus olitorius* L). Journal of Applied and Natural Science, 6 (2): 541-544.

Meeus, P. and Wittouck, D. (1999). Plant diseases of chicory and the use of fungicides for protection. Betteravier (Bruxelles), 33(349): 22.

Mehak N (2017). Stem-Rot of Jute: Symptoms and Control. www.biologydiscussion.com

Mehta, Y.R. (1998). Severe outbreak of Stemphylium leaf blight, a new disease of cotton in Brazil. Plant Dis., 82: 333-336.

Mehta, Y.R. (2001). Genetic diversity among isolates of *Stemphylium solani* from cotton. Fitopathologia Brasileira, 26: 703-709.

Mehta, Y.R., Bomfeti, C. and Bolognini, V. (2005). A semi-selective agar medium to identify the presence of *Xanthomonas axonopodis* pv.*malvacearum* in naturally infected cotton seed. Fitopatologia Brasileira, 30: 489-496.

Mehta, Y.R., Galbieri, R., Marangoni, M.S., Borsato, L.C., Rodrigues, H.P., Pereira, J. and Mehta, A. (2016). *Mycosphaerella areola*—The Teleomorph of *Ramularia areola* of Cotton in Brazil, and Its Epidemiological Significance. AJPS, 7(10): 1415-1422.

Mehta, Y.R., Motomura, K.F. and Almeida, W.P. (2005). Corynespora Leaf Spot of Cotton in Brazil. Fitopatologia Brasileira, 30: 131.

Mehta, Y.R., and Oliveira, M.A. (1996). Avaliacao da eficiencia de fungicidas no controle damanchapreta do algodoeiro. Fitopatol. Bras., 21(supl.): 371.

Melampsora lini-Science Direct; https://www.sciencedirect.com

Melike, C.K. and Ibrahim, O. (2010). Molecular characterization of *Rhizoctonia solani* AG4 using PCR-RFLP of the rDNA-ITS region. Turk. J. Biol., 34: 261-269.

Melo, E.M.P.F. and Ferraz, J.F.P. (1990). Influencia de várias fontes de carbono no crescimento de Rosellinia necatrix. Revista Ciencias Agrarias, 13: 151–156.

Mendoza, A. H.de., Belliure, B., Carbonell, E.A. and Real, V. (2001). Economic thresholds for *Aphis gossypii* (Hemiptera: Aphididae) on *Citrus clementina*. J Econ Entomol., 94(2): 439-444.

Meng, H.L., Chang, G.S., Du, S.L. and Ge, J.Y. (1973). On the activity rhythms and interspecific dominance of three species of spotted bollworms. Acta Entomologica Sinica, 16: 32-38.

Meng, Y., Zhang, Q., Ding, W. and Shan, W. (2014). *Phytophthora parasitica*: a model oomycete plant pathogen. Mycology, 5(2): 43-51

Menlikiev, M. Ya. (1962). Fusarium wilt of tine staple cotton and a study of *Fusarium oxysporum* f. sp. *vasinfectum* strs. as the causal agent of the disease in conditions of the Vaksh Valley. Izv. Akad. Nauk. Tadzh.SSR.Set. Biol. Sci., 4: 48-59.

Mercer, P.C. (1992). Diseases and pests of flax. In: The Biology and Processing of Flax (Eds. H. S. S. Sharma & C. F. Van Sumere). M Publications, Belfast, pp.111-155.

Mercer, P. C. and Hardwick, N. V. (1991). Control of seed-borne diseases of linseed. Aspects of Applied Biology , 28: 71-77.

Mercer, P. C., Hardwick, N. V., Fitt, B. D. L. and Sweet, J. B. (1991a). Status of linseed diseases in the UK Home Grown Cereals Authority Review OS 3, P.76.

Mercer, P. C., McGimpsey, H. C. and Ruddock, A. (1988). The control of seed-borne pathogens of linseed by seed treatments. Tests of Agrochemicals and Cultivars. Annals of Applied Biology, 112 (Supplement) (9): 30-31.

Mercer, P.C., McGimpsey, H.C., and Ruddock, A. (1989). Effect of seed treatment and sprays on the field performance of linseed. Tests of Agrochemicals and Cultivars, Annals of Applied Biology, 114 (Supplement) (10): 50-51.

Mercer, P. C., McGimpsey, H. C. and Ruddock, A. (1990). Effect of fungicide treatments on disease levels in linseed. Tests of Agrochemicals and Cultivars. Annals of Applied Biology, 116 (Supplement) (11): 44-45.

Mercer, P. C., McGimpsey, H. C and Ruddock, A. (1991b). Evaluation of chemical and biological agents against seed-borne diseases of linseed in N. Ireland. Tests of Agrochemicals and Cultivars. Annals of Applied Biology 118 (Supplement) (12): 44-45.

Mercer, P. C., Ritddock, A. and McGimpsey, H. C. (1992a). Evaluation of iprodione and Trichoderma viride against *Alternaria linicola*. Tests of Agrochemicals and Cultivars. Annals of Applied Biology, 120 (Supplement) (13): 20-21.

Mensah, R. K.and Khan, M. (1997). Use of *Medicago sativa* (L.) interplantings/trap crops in the management of the green mirid, *Creontiades dilutus* (Stal) in commercial cotton in Australia. International Journal of Pest Management, 43(3): 197-202.

Mersha, Z. (2017). Southern Blight - a disease becoming more prevalent in Missouri. Missouri Environment and Garden. Division of Plant Sciences, University of Missouri. Website: https://ipm.missouri.edu-MEG.

Mertley, J.C. and Snow, J.P. (1978). The boll rotting *Fusarium* species in Louisiana. In: Brown JM, ed. Proceedings of the Beltwide Cotton Production Research Conference, National Cotton Council, Memphis, Tennessee, USA, pp.29-30.

Meshram, M.K. and Raj, S. (1989). New sources of resistance to bacterial blight in exotic lines of upland cotton (*Gossypium hirsutum*). Indian Journal of Agricultural Sciences, 59(3): 169-170.

Meshram, M.K. and Raj, S. (1992). Effect of bacterial blight infection at different stages of crop growth on intensity and seed cotton yield under rainfed conditions. Indian Journal of Plant Protection, 20(1): 54-57.

Meyer, M.C., da Silva, J.C., Maia, G.L., Bueno, C.J. and Souza, N.L. (2006). Mancha de mirote cioem algodoeiro causada por *Myrothecium roridum* (Myrothecium leaf spot of cotton caused by *Myrothecium roridum*). Summa Phytopathol, 32(4): 390-393.

Meyer, M.K.P.S. (1981). Mite pests of crops in southern Africa. Science Bulletin 397, Dep. Agric. Fish South Africa.

Meyer, W.A., Sinclair, J.B. and Khare, M.N. (1974). Factors affecting charcoal rot of soybean seedlings. Phytopathology, 64: 845-849.

Meyrick, E. (1906). Descriptions of Indian Micro-Lepidoptera. II. in Journal of The Bombay Natural History Society, 17: 133-153.

Meyrick, E. (1907). Descriptions of Indian Micro-Lepidoptera V. Journal of the Bombay natural History Society, 18: 137–160.

Miah, M.A.M. (1974). Seed borne fungi of jute from Bangladesh and the methods for their detection. A report submitted to the Danish Government Institute of Seed Pathology for Developing Countries, Copenhagen, Denmark, P-27.

Michaud, J.P. (2013). Cotton Insects- Cotton Bollworm, *Helicoverpa zea*. Extension bulletin, Department of Entomology, Kansas State University, Manhattan KS 66506-4004; Website: https://entomology.k-state.edu › ... › Facts & Information on Cotton Pests

Migeon, A. and Dorkeld, F. (2013). Spider Mites Web: a comprehensive database for the Tetranychidae. Website: https://www.montpellier.inra.fr/CBGP/spmweb.

Mihail, J.D. (1989). *Macrophomina phaseolina:* Spatio-temporal dynamics of inoculum and of disease in a high susceptible crop. Phytopath., 79: 848–855.
Miles, D.E. and Persons, T.D. (1932). Verticillium wilt of cotton in Mississippi. Phytopath., 22: 767-773.
Miller, P.R., Weiss, F. and O'Brien, M.J. (1960). Index of plant Diseases in the United States. Agriculture Handbook No. 165. USDA, Washington, D.C., P.531.
Miller, P.R. and Weindling, R. (1939). A survey of cotton seedling diseases in 1939 and the fungi associated with them. Plant Disease Reporter, 23: 210–214.
Miller, R.D. (1999). Identification of the Pink Hibiscus Mealybug, *Maconellicoccus hirsutus* (Green) (Hemiptera: Sternorrhyncha: Pseudococcidae). Insecta Mundi., 13: 189-203.
Miller, W.E. (1982). *Grapholita delineana* (Walker), a Eurasian hemp moth, discovered in North America. Annals of the Entomological Society of America, 75(2): 184-186.
Millikan, C.R. (1944). J. Austral. Inst. Agric. Sci., 10: 129.
Millikan, C.R. (1945). Wilt disease of flax. J. Dept. Agr. Victoria, 43: 305-313, 354-361.
Millikan, C.R. (1948). Studies of strains of Fusarium lini. Proc. Roy. Soc. Victoria, 61: 1-24.
Millikan, C.R. (1951). Diseases of Flax and Linseed. Dept. of Agric., Victoria, Australia, P.140.
Milner, R.J. and Lutton, G.G. (1986). Dependence of *Verticillium lecanii* (Fungi: Hyphomycetes) on high humidities for infection and sporulation using *Myzus persicae* (Homoptera: Aphididae) as host. Environmental Entomology, 15: 380-382.
Miner, F.D. (1966). Biology and control of stink bugs on soybeans. Arkansas Agricultural Experiment Station Bulletin, 708: 1- 40.
Ministry of Agriculture, (2009).With regard to the hibiscus plant, *Phenacoccus* included in the notice of quarantine pests .Ministry of Agriculture, the State General Administration of Quality Supervision Bulletin 2009, 1147. http://www.chinagb.org/Article/standardCommodity/zhijiangonggao/200902/44969.html
Minton, E.B. and Garber, R.H. (1983). Controlling the seedling disease complex of cotton. Plant Dis., 67: 115-118.
Minton, N.A. and Adamson, W.C.(1979). Control of *Meloidogyne javanica* and *M. arenaria* on kenaf and roselle with genetic resistance and nematicides. J. Nematol., 11: 37-41.
Minton, N.A., Adamson, W.C. and White, C.A. (1970). Reaction of kenaf and roselle to three root knot nematode species. Phytopathology, 60: 1844-1845.
Mirmoayedi, A. and Maniee, M. (2009). Integrated pest management of cotton's spiny bollworm (*Earias insulana*) with spray of diazinon and relaese of green lacewings. J. Entomol., 6: 56-61.
Mirmoayedi, A., Maniee, M. and Yaghutipoor, A. (2010). Control of Cotton Spiny Bollworm, *Earias insulana* Boisduval, Using Three Bio-Insecticides, Bt, Spinosad and Neem-Azal. Journal of Entomology, 7: 89-94.
Mirza, M.S. and Ilyas, M.B. (1984). Sclerotinia stem rot of flax in Pakistan. Pak. J. Agric. Sci., 21(3-4): 246-247.
Mirzaee, M.R., Heydari, A., Zare, R., Naraghi, L., Sabzali F. and Hasheminasab, M. (2013). Fungi associated with boll and lint rot of cotton in Southern Khorasan province of Iran. Archives of Phytopathology and Plant Protection, 46(11): 1285-1294.
Mishra, P.R., Sontakke, B.K., Mukherjee, S.K. and Dash, P.C. (1996). Field evaluation of some linseed cultivars for resistance to bud-fly, *Dasyneura lini* Barnes. Indian J. Plant Protect., 24: 119-121.
Mishra, S.P. and Krishna, A. (2001). Assessment of yield losses due to bacterial blight in cotton, J. Mycol. Plant Pathol., 31: 232-233.
Misra, D.P. (1963). A new physiologic races of linseed rust in India. *Indian Phytopathol.,* **16:** 102-103.

Misra, D.P. and Lele, V.C. (1963). Physiologic races of linseed rust in India during 1960, 1961, and 1962. Prevalence and distribution. Indian Oilseeds J., 7: 336-337.

Mishra, C. (1995) Nematode problems of jute and allied fibre crops In: Eds, Swarup, G., Dasgupta, D.R. and J.S. Gill, J.S. Nematode Pest Management An Appraisal of Eco-friendly Approaches, Nematological Society of India,New Delhi, Chapter 28, pp.211-216.

Mishra, C. and Mandal, R. (1988). Occurrence of *Meloidogyne thamesi* and *Helicotylenchus mucronatus* on ramie, *Boehmeria nivea*. Indian Journal of Nematology, 18(1): 114.

Mishra, C.S. (1913). The red spider mite on jute (*Tetranychus bioculatus* Wodd Mason). Agric J India, 8: 309–316.

Mitchell, P.L. (2004). Heteroptera as vectors of plant pathogens. Neotropical Entomology, 33: 519-545.

Mitchell, S.J., Jellis, G.J., and Cox, T.W. (1986). *Sclerotinia sclerotiorum* on linseed: New or unusual records. Plant Path., 35: 403-405.

Mitra, M. (1934). Wilt disease of *Crotalaria juncea* Linn. (sunn-hemp). Indian J. Agr. Sci., 4: 701-714.

Mitra, M. (1937). An anthracnose disease of sunnhemp. Indian J Agric. Sci., 7: 443-449.

Mitra, S., Saha, S., Guha, B., Chakrabarti, K., Satya, P., Sharma, A. K., Gawande, S. P., Kumar, M. and Saha, M. (2013). Ramie: The Strongest Bast?bre of Nature, Technical Bulletin No. 8, Central Research Institute for Jute and Allied Fibres, ICAR, Barrackpore, Kolkata-120, P.38.

Mizell, R.F. (ND). Monitoring Stink Bugs with the Florida Stink Bug Trap. University of Florida. http://nfrec.ifas.ufl.edu/MizellRF/stink_bugs/stink_bugs.htm

Mizell, R.F. and Tedders, W.L. (1995). A new monitoring method for detection of the stink bug complex in pecan orchards. Proceedings of the Southeastern Pecan Growers Association, 88: 33- 40.

Mohan, K.S. and Manjunath, T.M. (2002). Bt cotton-India's first transgenic crop. J. Plant Biol., 29: 225-236.

Mohan, P., Mukewar, P.M., Singh, V.V., Singh, P., Khadi, B.M., Amudha, J. and Deshpande, V.G. (2006). Identification of sources of resistance to grey mildew disease (*Ramularia areola*) in diploid cotton (*Gossypium arboreum*). CICR Technical Bulletin No: 34, Central Institute for Cotton Research Nagpur

Mohan, S., Gopalan, M., SundaraBabu, P.C. and Sreenarayanan, V.V. (1994). Practical studies on the use of light trap and bait trap for management of *Rhyzopertha dominica* in rice warehouses. International Journal of Pest Management, 40(2): 148-152.

Mohandas, S., Saravanan, Y. and Manjunath, K. (2004). Biological control of *Myllocerus subasciatus* Guerin infesting brinjal (*Solanum melongena* l.) using *Bacillus thuringiensis* ssp. *tenebrionis*. Acta Hortic., 638: 503-508.

Monga, D. (2011). Changing scenario of cotton diseases in India-the challenge ahead. In: World cotton conference, 2011. India

Monga, D., Bhattiprolu, S.L and Prakash, A.H. (2013). Crop losses due to important cotton diseases. Central institute for cotton research, Nagpur, Technical Bulletin, 9.

Monga, D. and Raj, S. (1994). Progress of root rot in American (*Gossypium hirsutum*) and desi (*G. arboreum*) cotton varieties in northern region. National Symposium on current trends in the management of plant diseases. CCSHAU, Hisar, November 10-11, 1994.

Monga, D. and Raj, S. (ND). Root rot disease of cotton and its management. CICR Technical Bulletin No: 3, Central Institute for Cotton Research, Nagpur (India). Website: http://www.cicr.org.in/pdf/rootrot_disease.pdf

Mordue, J.E.M. (1967). *Colletotrichum coccodes*. CMI Descriptions of Pathogenic Fungi and Bacteria. 131.

Mordue, J.E.M. (1974). *Corticium rolfsii*. CMI Descriptions of pathogenic fungi and bacteria. 410: 1-2.
Mordue, J.E.M. and Holliday, P. (1976). *Sclerotinia sclerotiorum*. CMI Descriptions of Pathogenic Fungi and Bacteria. 513.
Moresco, E., Yuyama, M.M., Camargo, T.V. and Metha, Y.R. (2001). Manual de Identificacao e Manejo das Doenças do Algodoeiro. Facual, Cuiaba, MT, Brazil.
Morgan-Jones, G. (1967). *Ceratocystis fimbriata*. CMI Descriptions of Pathogenic Fungi and Bacteria. 141: 1-2.
Moricca, S., Ragazzi, A., Kasuga, T. and Mitchelson, K.R. (1998). Detection of *Fusarium oxysporum* f.sp. *vasinfectum* in cotton tissue by polymerase chain reaction. Plant Pathology, 47: 486-494.
Moore, D. (1988). Agents used for biological control of mealybugs (Pseudococcidae). Biocontrol News and Information, 9(4): 209-225.
Moore, W. D. (1946). New and interesting plant diseases. Transactions of the British Mycological Society, 29: 250-258.
Moorman, G.W. (ND). Pythium. PennState Extension. Website: http://extension.psu.edu/pests/plantdiseases/all-fact-sheets/pythium).
Moorthi, P.V., Balasubramanian, C., Avery, P.B., Banu, A.N. and Kubendran, T. (2012). Efficacy of fungal entomopathogens against red cotton stainer, *Dysdercus cingulatus* Fabricius (Hemiptera: Pyrrhocoridae). J Biopest, 5(2): 140-143.
Mori, Y., Sato, Y., and Takamatsu, S. (2000). Evolutionary analysis of the powdery mildew fungi using nucleotide sequences of the nuclear ribosomal DNA. Mycologia, 92: 74–93.
Morison, G.D. (1943). Notes on Thysanoptera found on flax. Ann. Appl. Biol., 30: 251-259.
Moriwaki, J., Tsukiboshi, T. and Sato, T. 2002. Grouping of *Colletotrichum* species in Japan based on rDNA sequences. J. Gen. Plant Pathol., 68: 307-320.
Morrill, A.W. (1910). Plant-bugs injurious to cotton bolls [Bulletin 86]. U.S. Dep. of Agriculture, USDA, Bureau of Entomology, Washington, D.C., P.110.
Morris, H. and Waterhouse, D.F. (2001). The distribution and importance of arthropod pests and weeds of Agriculture in Myanmar. P.73.
Morse, J.G. and Hoddle, M.S. (2006). Invasion biology of thrips. Ann. Rev. Entomol., 51: 67–89.
Morstatt, H. (1930). Leaf diseases of the Sisal Agave. Tropenpflanzer, 33(8): 307-312.
Morton, J. F. (1987). Roselle. Fruits of warm climates, Julia F. Morton, Miami: 281-286.
Moses, A.S. (1988). IMI 329478, IMI records for geographical unit Indian Subcontinent. Herb IMI, Kew, London.
Moskovetz, S.N. (1941). Virus disease of cotton and its control. In: Plant Viral Diseases, Moscow. Izvestiya Akademii Nauk Azerbaidzhanskoi SSR, Biologicheskikh Nauk, pp.173-190.
Mound, L.A. (1968). A review of R.S. Bagnall's Thysanoptera collections. Bulletin of the British Museum (Natural History) Entomology, Supplement, 11: 1-181.
Moyal, P. (1988). The borers of maize in the savannah area of Ivory Coast.Morphological, biological and ecological data. Control trials and plant-insect relations. Les foreurs du mais en zone des savanes en Cote-D'Ivoire.Donnéesmorphologiques, biologiques, écologiques.Essais de lutteet relation plante-insecte., pp.367.
Mpofu, S.I. and Rashid, K.Y. (2000a). Assessment of genetic variation among *Fusarium oxysporum* f. sp. *lini* isolates from western Canada based on vegetative compatibility grouping. Can. J. Plant Pathol., 22: 176.
Mpofu, S.I. and Rashid, K.Y. (2000b). Virulence of *Fusarium oxysporum* f. sp. *lini* on flax. Can. J. Plant Pathol., 22: 190.
Mpofu, S. and Rashid, K. (2001).Vegetative Compatibility Groups within *Fusarium oxysporum* f. sp. *lini* from *Linum usitatissimum* (flax) Wilt Nurseries in Western Canada. Canadian journal of Botany, 79: 836-843.

Mpunami, A. (1986). Purification and partial characterization of a virus associated with korogwe leaf spot disease of sisal. Master of Science Thesis in Botany and Plant pathology, Oregon State University, http://ir.library.oregonstate.edu/concern/graduate_thesis_or_dissertations/b5644t71g

Mtung'e, O.G., Luo, L., Liu, X., Mabagala, R.B., Diao, Y., Meng, Y. and Li, J. (2014). Prevalence and fungal isolates associated with Korogwe leaf spot disease (KLS) of sisal. Oral Technical Session: Diseases of Plants, 170-O, APS-CPS Joint Meeting, August 9-13, 2014, Minneapolis, Minnesota,The American Phytopathological Society

Mueller, J. D. and Lewis, S. A. (1993). Evaluation of Nematicides for Controlling *Meloidogyne incognita* and *Hoplolaimus columbus* on Kenaf (*Hibiscus cannabinus*). Nematropica, 23 (1): 91-97.

Mullen, J. (2001). Southern blight, Southern stem blight, White mold. The Plant Health Instructor. DOI: 10.1094/PHI-I-2001-0104-01.

Muhanna, Naglaa, A.S., Seham, S.M. R. and Gehad, M. Md. (2011). Mineral salts in controlling powdery mildew of squash. Egypt. J. Agric. Res., 89(3): 809-817.

Mukerji, K.G. (1968a). *Leveillula taurica*. CMI Descriptions of Pathogenic Fungi and Bacteria. 182.

Mukerji, K.G. (1968b). *Sphaerotheca macularis*. CMI Descriptions of Pathogenic Fungi and Bacteria. 188.

Mukherjee, N. and Basak, M. (1969).Eye-rot of *Hibiscus cannabinus* L. (Mesta). Indian J. Mycol. Res. 7(1): 37-38.

Mukherjee, N. and Basu, B. (1973). Fruit mould in jute. Sci & Cult., 39: 507-508.

Mukherjee, N. and Basu, B. (1974) Eye-rot of kenaf: Outbreaks and New Records, FAO Plant Protection bulletin, 22(4):.92. Website: Untitled-Southwest Florida Research and Education Center. https://swfrec.ifas.ufl.edu-database-pdf.

Mukherjee, N. and Ghosh, T. (1960). Corynespora sp. On jute. Sci & Cult., 26: 270-271.

Mukerjee, N. and Mishra, C.B.P. (1988). Diseases of jute and their management. Rev. Trop .Pl. Path., 5: 255–271.

Mulder, J.L. and Holliday, P. (1971a). *Puccinia schedonnardi*. Descriptions of Pathogenic Fungi and Bacteria. 293: 1-2.

Mulder, J.L. and Holliday, P. (1971b). *Puccinia cacabata*. CMI Descriptions of Pathogenic Fungi and Bacteria, 294: 1-2.

Mulder, J.L. and Holliday, P. (1976). *Ramularia gossypii*. CMI Descriptions of Pathogenic Fungi and Bacteria. 520: 1-2.

Mullen, J. (2001). Southern blight, Southern stem blight, white mold. The Plant Health Instructor. Doi: 10.1094/PHI-I-2001-0104-01. Website: https://www.apsnet.org/edcenter/intropp/lessons/fungi/Basidiomycetes/Pages/SouthernBlight.aspx

Muller, F.P. and Karl, E. (1976). Beitragzur Kenntnis der Bionomie und Morphologie der Hanfblattlaus, *Phordon cannabis* Passerini, 1860. Beitr. Ent., Berlin , 26: 455-463.

Mullick, R.N. (1986). Top shoot borer a serious enemy of sunnhemp. Indian Farmers digest, 19(3): 29-30.

Mulpuri, S., Soni, P.K. and Gonela, S.K. (2016). Morphological and molecular characterization of powdery mildew on sunflower (*Helianthus annuus* L.), alternate hosts and weeds commonly found in and around sunflower fields in India. Phytoparasitica, 44: 353–367.

Mulrean, E.N., Hine, R.B. and Mueller, J.P. (1984). Effect of Phymatotrichum root rot on yield and seed and lint quality in *Gossypium hirsutum* and *G. barbadense*. Plant Disease, 68(5): 381-383.

Mulvaney, M. (2015). Manage Potassium to Control Cotton Stemphylium Leafspot. Website: nwdistrict.ifas.ufl.edu/.../manage-potassium-to-control-cotton-stemphylium-leafspot/

Munch, S., Lingner, U., Floss, D.S., Ludwig, N., Sauer, N. and Deising, H.B. (2008). The hemibiotrophic lifestyle of *Colletotrichum*. Journal of Plant Physiology, 165: 41-51.

Mundkur, B.B. (1935). Influence of temperature and maturity on the incidence of sunnhemp and pigeon pea wilt at Pusa. Indian J. Agric. Sci., 5: 609-618.

Muniappan, R. (2009). A parasitoid to tackle the menace of the mealybug pest of cotton in India. IAPPS Newsletter, December (XII). http://www.plantprotection.org/news/NewsDec09.htm

Munjal, R.L. (1960). A commonly occurring leaf spot disease caused by *Myrotheoium roridum* Tode ex Fr. Indian Phytopathology, 13(2): 150-155.

Munson, G., Keaster, A.J. and Grundler, J.A. (1986). Corn cutworm control. Agricultural Guide 4150, University of Missouri-Columbia Extension Division, Columbia, MO.

Munyaneza, J. and McPherson, J.E. (1994). Comparative study of life histories, laboratory rearing, and immature stages of *Euschistus servus* and *Euschistus variolarius* (Hemiptera: Pentatomidae). Great Lakes Entomol, 26: 263-274.

Murakami, R. Shirata, A. and Inoue, H. (1998). A selective medium containing the toxins of *Myrothecium roridum* for isolation from soil. J. Seric. Sci. Jpn., 67(5): 381-387.

Muraleedharan, N. (2006). Sustainable cultivation of tea. Handbook of tea culture, Section 24, UPASI tea Research Foundation, Niran Dam, Valparai. Pp 1-12.

Mushtaque, M., Baloch, G.M. and Ghani, M.A. (1973). Natural enemies of *Papaver* spp and *Cannabis sativa*. Annual report, Commonwealth Institute of Biological Control, Pakistan station, pp.54-35.

Muskett, A.E. and Colhoun, J. (1945). Foot Rot (*Phoma* sp.) of Flax. Nature, 155: 367-368.

Muskett, A.E. and Colhoun, J. (1947). The Diseases of Flax Plant (*Linum usitatissimum* Linn.). W&G. Baird Ltd., Belfast., P.112.

Musser, F.R. and Shelton, A.M. (2005). The influence of post-exposure temperature on the toxicity of insecticides to *Ostrinia nubilalis* (Lepidoptera: Crambidae). Pest Manag Sci., 61: 508-510.

Mwenkalley, A.H. (1978). Some studies on korogwe leaf spot of sisal in Tanga region, Tanzania mainland. B.Sc. (Agric.) Special Project.Univ. of Dar-es-Salaam, Tanzania.

myFields, (ND). Pale Western Cutworm. myFields. Website: www.myfields.info/book/pale-western-cutworm

Nagrale, D.T. and Gawande, S.P. (2016). Effect of fungicides on growth and sporulation of *Alternaria alternata* (Fr.) Keissler, causing blight of Gerbera (*Gerbera jamesonii* H. Bolus Ex JD Hook). Progressive Agriculture, 16 (2): 152-155.

Nagrare, V.S., Deshmukh, A.J., Dharajothi, B., Amutha, M., Kumar, R., Kranthi, S. and Ktanthi, K.R. (ND). Sampling methodology for assessing field population of cotton infesting mirid *Campylomma livida* Reuter (Hemiptera: Miridae). World Cotton Research Conference on Technologies for Prosperity, Abstract No. Poster-106. Website: https://www.icac.org/meetings/wcrc/wcrc5/Pdf_File/150.pdf

Nagy, B. (1967). The hemp moth (*Grapholitha sinana* Feld, Lepid.:Tortricidae), a new pest of hemp in Hungary. Acta Phytopathologica Academiae Scientiarum Hungaricae, 2: 291-294.

Nagy, B. (1979). Different aspects of flight activity of the hemp moth, *Grapholitha delineana* Walk., related to intergrated control. Acta Phytopathologica Academiae Scientiarum Hungaricae, 14: 481-488.

Nagy, B. (1980). Interspecific sex-pheromone and sexual behaviour of the hemp moth, *Grapholitha delineana* Walk. Abstract, Conference on New Endeavours in Plant Protection, Budapest.

Naher, N., Islam, W., Khalequzzaman, M. and Haque, M.M. (2008). Study on the developmental stages of spider mite (*Tetranychus urticae* Koch) infesting country bean. J. bio-sci., 16: 109-114.

Naigch, B.B. and Vaishisth, K.S. (1963). A virus causing a typical mosaic disease of bean. Indian J. Microbiol., 3: 113-116.
Nair, M.R.G.K. (1986). Insects and mites of crops in India. ICAR, New Delhi, pp.122-126.
Naito, S. and Sugimoto, T. (1989). Choanephora rot in sugar beet's at flowering. Research Bulletin of Hokkaido Agricultural Experiment Station, 151:1-5.
Nakache, J., Dunkelblum, E., Kehat, M., Anshelevich, L. and Harel, M. (1992). Mating disruption of the spiny bollworm, *Earias insulana* (Boisduval) (Lepidoptera: Noctuidae) with the Shin Etsu twist-tie rope formulation. Bulletin of Entomological Research, 88: 369-373.
Nakahara, L M. (1984). New State record: *Thrips palmi* Karny. Hawaii Pest Report.Hawaii Dept. Agr., 4(1): 1-5.
Nakamura, H., Uetake, Y., Arakawa, M., Okabe, I. and Matsumoto, N. (2000). Observations on the teleomorph of white root rot fungus, *Rosellinia necatrix*, and a related fungus, *Rosellinia aquila*. Mycoscience, 41: 503–507.
Nandihalli, B.S., Patil, S.B., Poornima, M.H. and Megha, R. (2015), Biology and nature of damage of cotton flower bud maggot, *Dasineura gossypii* Fletcher. J. Exp. Zool. India, 18: 879-881.
Naranjo, S.E. and Ellsworth, P.C. (2010). Fourteen Years of Bt Cotton Advances IPM in Arizona. Southwestern Entomologist, 35(3): 437-444.
Narayanan, E.S. (1962). Insect pests of linseed and methods of their control. In: Richharia, R.H. (ed.), Linseed Monograph. Indian Central Oilseeds Committee, Hyderabad, India, pp.107-113.
Nariani, T.K., Kartha, K.K.and Prakash, N. (1970). Purification of the southern sunnhemp mosaic virus using butanol and differential centrifugation. Curr. Sci., 23: 539-541.
Narusaka, Y., Narusaka, M., Park, P., Kubo, Y., Hirayama, T., Seki, M. et al. (2004). RCH1, a locus in Arabidopsis that confers resistance to the hemibiotrophic fungal pathogen *Colletotrichum higginsianum*. Mol. Plant Microbe Interact., 17: 749-762.
Nasr, E. A. and Azab, A. K. (1969a). Behaviour and activity of the pink and the spiny cotton bollworms in Egypt (Lepidoptera). Bulletin de la Societe Entomologiqued' Egypte, 53: 235-243.
Nasr, E.A. and Azab.A.K. (1969b). *Rhizopus nigricans* Ehr. infection in cotton bolls in relation to bollworm infestation, age of bolls and the insecticides used on cotton plants (Lepidoptera). Bulletin of the Entomological Society of Egypt: Economic Series, 3: 97-102.
Nasseh, O.M. and Link, R (1990). Insecticidal control of pomegranate (*Punica granatum*) fruit dropping in the Yemen Arab Republic. Mitteilungen der Deutschen Gesellschaftfür Allgemeine und Angewandte Entomologie, In: agris.fao.org/agris-search/search. do? recordID=DE91U0600
National Research Council. (1979). Tropical legumes: resources for the future. National Academy of Science, Washington, D.C. pp. 272-278.
Nath, P. and Singh, A. K. (1996). Biology and insecticidal management of Bihar hairy caterpillar, *Spilosoma obliqua* infesting groundnut. Annals of Plant Protection Sciences, 4(1): 42-46.
Nattrass, R.M. (1943). The pasmo disease of flax in Kenya, *Sphaerella linorum* Wollenweber. East African Agric. J., 8: 223-226.
Nattrass, R.M. (1961). Host lists of Kenya fungi and bacteria. Mycological Papers, 81: 1-46.
Natwick, E. T. (2010). UC/IPM: UC management guidelines for bean aphid on sugarbeet [Online]. UC IPM pest management guidelines: Sugar beets, UC ANR Publication 3469, Insects and Mites. UC Statewide IPM Program, University of California, Davis. Website: http://www.ipm.ucdavis.edu/PMG/r735300311.html
Naqvi, S.F., Haq, M.I.U., Tahir, M.I., Khan, M.A. and Ali, Z. (2013). Morphological and biochemical characterization of *Xanthomonas campestris* (Pammel) Dowson pv sesame and it's management by bacterial antagonists. Pakistan Journal of Agricultural Sciences, 50(2): 229-235.

Nayar, K.K., Ananthakrishnan, T.N. and David, B.V. (1976). General and applied entomology. Tata McGraw Hill Publishing Company Limited, New Delhi. pp.268-269.
Nbair (2013). *Syllepte derogata*. ICAR-National Bureau of Agricultural Insect Resources. Last updated on: 30.05.17 Website: http://www.nbair.res.in/insectpests/Syllepte-derogata.php
NDSU (2017). Pasmo Crops- NDSU Agriculture and Extension, NDSU Crop Publications https://www.ag.ndsu.edu/crops/flax-articles/pasmo
Neal, D.C. and Gilbert, W.W. (1955). Boll diseases (rots). U. S. D. A. Farmers Bul., 1745: 32-34.
Neergaard, P. (1945). Danish species of Alternaria and Stempl lium. Oxford University Press, London, P.560.
Nejoa, O.J., Adeotia, A.Y.A., Popoolaa, A.R. and Kehindeb, I.A. (2007). Effect of foliar application of fungicides on incidence and severity of leaf spot disease of Roselle induced by *Coniella musaiaensis* B. Sutton var. *hibisci* Moor. Journal of Agricultural Research, 8(1). Website: https://www.ajol.info/index.php/mjar/article/view/116054
Nemri, A., Saunders, D.G., Anderson, C., Upadhyaya, N.M., Win, J., Lawrence, G.J., Jones, D.A., Kamoun, S., Ellis, J.G. and Dodds, P.N. (2014). The genome sequence and effector complement of the flax rust pathogen *Melampsora lini*. Front Plant Sci., 24 (5): 98.
Neofitova, V.K., Kukresz, L.M. and Portjankin, D.E. (1984). Infekcionnyj potencial patogenov l'na v poczve sevooborotov s razlicznym nasysczenijem l'nom. [in Russian]. pp.12-16. In: Zasczita Rastenij, vyp. 9. Bel, NIIZR, Minsk.
Netsu, O., Kijima, T. and Takikawa, Y. (2014).Bacterial leaf spot of hemp caused by *Xanthomonas campestris* pv. *cannabis* in Japan. Journal of General Plant Pathology, 80(2): 164-168.
Neunzig, H.H. (1964). The eggs and early-instar larvae of *Heliothis zea* and *Heliothis virescens* (Lepidoptera: Noctuidae). Annals of the Entomological Society of America, 57: 98-102.
Nevo, E. and Moshe, C. (2001). Effect of nitrogen fertilization on *Aphis gossypii* (Homoptera: Aphididae): variation in size, color, and reproduction. Journal of Economic Entomology, 94(1): 27-32.
Nielsen, A.L. and Hamilton, G.C. (2009a). Life history of the invasive species *Halyomorpha halys* (Hemiptera: Pentatomidae) in northeastern United States. Annals of the Entomological Society of America, 102(4): 608-616.
Nielsen, A.L. and Hamilton, G.C. (2009b). Seasonal occurrence and impact of *Halyomorpha halys* (Hemiptera: Pentatomida) in tree fruit. Journal of Economic Entomology, 102: 1133-1140.
Nielsen, A.L., Hamilton, G.C. and Matadha, D. (2008). Developmental rate estimation and life table analysis for *Halyomorpha halys* (Hemiptera: Pentatomidae). Environ. Entomol., 37: 348-355.
Nielsen, A.L., Holmstrom, K., Hamilton, G.C., Cambridge, J. and Ingerson-Mahar, J. (2013).Use of black light traps to monitor the abundance and spread of the brown marmorated stink bug. J. Econ. Entomol., 106: 1495-1502.
Nikam, N.D., Patel, B.H. and Korat, D.M. (2010). Biology of invasive mealy bug, *Phenacoccus solenopsis* Tinsley (Hemiptera: Pseudococcidae) on cotton. Karnataka J. Agric. Sci., 23(4): 649- 651.
Niu, S., Gil-Salas, F.M., Tewary, S.K., Samales, A.K., Johnson, J., Swaminathan, K, et al. (2014). Hibiscus Chlorotic Ringspot Virus Coat Protein is Essential for Cell to Cell and Long Distance Movement for Viral RNA Replication. PLoS One, 9(11): e113347. Doi:10.1371/journal.pone.0113347
Ng, L.C., Ismail, W.A. and Jusoh, M. (2017). *In vitro* Biocontrol Potential of Agro-waste Compost to Suppress *Fusarium oxysporum*, the Causal Pathogen of Vascular Wilt Disease of Roselle. Plant Pathology Journal, 16: 12-18.
Nirenberg, H.I., Ibrahim, G. and Michail, S.H. (1994).Race identity of three isolates of *Fusarium oxysporum* Schlect. f.sp. *vasinfectum* (Atk.) Snyd. & Hans.From Egypt and the Sudan. Z. Pflanzenkrankh. (Pflanzenpathol.) Pflanzenschutz, 101: 594-597.

Nishikado, Y., Kimura, K. and Miyawaki, Y. (1940). Two *Alternaria* species injurious to cotton fibre in balls. Ann. Phytopath. Soc. Japan, 10: 214-230.

Niu, X., Gao, H., Chen, M and Qi, J. (2016a). First Report of Anthracnose on White Jute (*Corchorus capsularis*) Caused by *Colletotrichum fructicola* and *C. siamense* in China. Plant Disease, 100(6): 1243.

Niu, X., Gao, H., Qi, J., Chen, M., Tao, A., Xu, J., Dai, Z. and Su, J. (2016b). *Colletotrichum* species associated with jute (*Corchorus capsularis* L.) anthracnose in southeastern China. Sci Rep., 6: 25179 (published online)

Newhook, F.J. (1942). Pasmo (*Sphaerella linorum*) on flax in New Zealand., N. Z. J. Sci. A., 24: 102-106.

Noble, L.W. (1969). Fifty years of research on the pink bollworm in the United States, USDA Agric. Handbook No. 357.

Noble, M. and Richardson, M.J. (1968). An Annotated List of Seedborne Diseases. Commonwealth Mycological Institute, Kew, U.K., P.191.

Nor Azizah, K., Madidhan, N.Z.A., Shahrizim, Z., Mohd, T.Y. and Nur Ain Izzati, M.Z. (2016). Morphological and molecular characterization of Curvularia and related species associated with leaf spot disease of rice in Peninsular Malayasia. Rend. Lincei Sci. Fis. Nat., 27(2): 205-214.

North, R.C. and Shelton, A.M. (1986). Ecology of Thysanoptera within cabbage fields. Environ. Entomol., 15: 520–526.

Noviello, C. and Snyder, W. C. (1962). Fusarium wilt of Hemp. Phytopathology, 52(12): 1315-1317.

Noviello, C., McCain, A.H., Aloj, B., Scalcione, M. andMarziano, F. (1990).Lottabiologicacontro *Cannabis sativa* mediante l'impiego di *Fusarium oxysporum* f, sp. *cannabis*. Annalidella Facolta di Scienze Agrariedella Universitadegli Studi di Napoli, Portici, 24: 33-44.

Nur-Ain-Izzati, M.Z., Mohd-Aizat, Z., Siti-Nordahliawate, M.S. and Mat-Rasid, I. (2014). Morphological and genetic variabilities of *Fusarium* species isolated from kenaf. Malaysian Applied Biology, 43 (1): 13-20.

Nurulain S.M., Tabassum R. and Naqvi S.N.H. (1989). Toxicity of neem fraction and malathion (57% E.C.) against dusky cotton bug (*Oxycarenus lugubris* Mostch.). Pakistan Journal of Entomology, 4(1-2): 13-14.

Nyvall ,R.F. (1989). Diseases of Flax. In: Field Crop Diseases Handbook. Springer, Boston, MA

Nyvall, R.F. (1999). Field Crop Diseases- Third edition.Wiley-Blackwell, USA, P.1021. Oakley, J.N., Corbett, S.J., Parker, W.E. and Young, J.E.B. (1996). Assessment of risk and control of flax flea beetles. Proc. Brighton Crop Protection Conference, Pests and Diseases 1996, pp.191–196.

O'Bannon, J.H. and Esser, R.P. (1975). Evaluation of citrus, hybrids, and relatives as hosts of *Pratylenchus coffeae*, with comments on other hosts. Nematologia Mediterranea, 3(2): 113-122.

O'Brien, C.W., Haseeb, M. and Thomas, M.C. (2006). *Myllocerus undecimpustulatus undatus* Marshall (Coleoptera: Curculionidae), a recently discovered pest weevil from the Indian Subcontinent. Fla. Dept. of Agriculture & Cons. Svs. Division of Plant Industry, Entomology Circular, No. 412: 1-4.

Odebiyi, J.A., (1982). Parasites of the cotton leaf roller, *Sylepta derogata* (F.) (Lepidoptera: Pyralidae), in south-western Nigeria. Bulletin of Entomollogical Research, 72: 329-333.

O'Donnell, K., Gueidan, C., Sink, S., Johnston, P.R., Crous, P.W., Glenn, A., Riley, R., Zitomer, N.C., et al. (2009).A two-locus DNA sequence database for typing plant and human pathogens within the *Fusarium oxysporum* species complex. Fungal Genet. Biol., 46: 936-948.

OEPP/EPPO. (1981). Data sheets on quarantine organisms No. 110, *Helicoverpa armigera*. Bulletin 11.

OEPP/EPPO (1990a). *Phymatotrichopsis omnivora*. Specific quarantine requirements. EPPO Technical Documents No. 108. Website: https://gd.eppo.int/taxon/PHMPOM

OEPO/EPPO, (1990b). Specific quarantine requirements. EPPO Technical Documents, No. 1008. Paris, France: European and Mediterranean Plant Protection Organization

Oerke, E.C., Dehne, H.W., Schonbeck, F. and Weber, A. (1994). Crop Production and Crop Protection: Estimated Losses in Major Food and Cash Crops. Elsevier, Amsterdam, P.808.

Okamoto, J., Limkaisang, S., Nojima, H., and Takamatsu, S. (2002). Powdery mildew of prairie gentian: Characteristics, molecular phylogeny and pathogenicity. J. Gen. Plant Pathol., 68: 200-207.

Oliveira, J.C., Albuquerque, G.M.R., Xavier, A.S., Mariano, R.L.R., Suassuna, N.D.and Souza, E.B. (2011). Characterization of Xanthomonas citri subsp. *malvacearum* causing cotton angular leaf spot in Brazil. Journal of Plant Pathology, 93 (3): 707-712.

Oliver, K. M., Degnan, P. H., Burke, G. R. & Moran, N. A. (2010).Facultative symbionts in aphids and the horizontal transfer of ecologically important traits. Annu. Rev. Entomol., 55: 247-266.

Olsen, M.W. and George, S. (1987). Applications of systemic fungicides through subsurface drip irrigation for control of Phymatotrichum root rot of cotton. Phytopathology, 77(12): 1748.

Olsen, M.W., Misaghi, IJ., Goldstein, D. and Hine, R.B. (1983). Water relations in cotton plants infected with *Phymatotrichum*. Phytopathology, 73(2): 213-216.

Olsen, M.W. and Silvertooth, J.C. (2001). Diseases and Production Problems of Cotton in Arizona. Cooperative Extension College of Agriculture and Life Sciences Report AZ1245. The University of Arizona Tucson, Arizona, P.20.

Omran, A.O., Frederikson, R.A. and Atkins, I.M. (1968). Heritability of seedling disease characteristics in flax. Crop Science, 8: 750-753.

Omura, H. and Honda, K..(2005). Priority of color over scent during flower visitation by adult Vanessa indica butterflies. Oecologia, 142: 588-596.

Ondrej, M. (1977). New ideas about fusariosis of flax. Rev. Plant Path., 57: 5521.

Ondrej M. (1983a). Phoma disease of potatoes and flax. Len a Konopi (Czech), Website: http://agris.fao.org/agris-search/search.do?recordID=CZ19830911825

Ondrej, M. (1983b). Possibilities of breeding flax for resistance to *Phoma exigua* var. *linicola* (Naoum. et Vass.) Maas. Len konopi., 19: 37-45 (in Czech).

Ondrej, M. (1985). Testovani genetických zdroju lnu na odolnost proti fuzariozam. Len a Konopi, 20: 71-76.

Ooi, K.H. and Salleh, B. (1999). Vegetative compatibility groups of *Fusarium oxysporum*, the causal organism of vascular wilt on roselle in Malaysia. Biotropia., 12: 31-41.

Ooi, K.H., Salleh, B, .Hafiza, M.H. and Zainal, A.A.A. (1999). Interaction of *Fusarium oxysporum* with *Meloidogyne incognita* on rosellle. Jurnal Perlindungan Tanaman Indonesia, 5(2): 83-90.

Orloff, S., Natwick, E.T. and Poole, G.J. (2008). Onion and garlic thrips: *Thrips tabaci* and *Frankliniella occidentalis*. How to Manage Pests. UC ANR Publication 3453. UC Pest Management Guidelines. University of California Agriculture and Natural Resources, CA.

Ortega-Acosta, S.A., Hernández-Morales, J., Ochoa-Martínez, D.L. and Ayala-Escobar, V. (2015a). First report of *Corynespora cassiicola* causing leaf and calyx spot on roselle in Mexico. Plant Disease, 99: 1041.

Ortega-Acosta, S.A., Hernandez-Morales, J., Sandoval-Islas, J.S., Ayala-Escobar, V., Soto-Rojas, L. and Alejo-Jaimes, A. (2015b). Distribution and frequency of organisms associated to disease "black leg" of roselle (*Hibiscus sabdariffa* L.) in Guerrero, Mexico. Revista Mexicana de Fitologia, 33(2): 173-194.

Ortega-Acosta, S.A., Velasco-Cruz, C., Hernández-Morales, J., Ochoa-Martínez, D.L. and Hernández-Ruiz, J. (2016). Diagrammatic logarithmic scales for assess the severity of spotted leaves and calyces of roselle. Revista Mexicana de Fitopatología, 34: 270-285.

Osorio-Hernandez, E., Hernandez-Morales, J., Conde-Martinez, V., Michel-Aceves, A.C., Cibrian-Tovar, J. and Vaquera-Huerta, H. (2014). Biocontrol of *Phytophthora parasitica* and *Fusarium oxysporum* by *Trichoderma* spp. in *Hibiscus sabdariffa* plants under field and greenhouse conditions. Afr. J. Agric. Res., 9: 1338-1345.

Oteifa, B.A. (1970). The reniform nematode problem of Egyptian cotton production. .Journal of Parasitology, 56: 255.

Ovsyannikova, E.I. and Grichanov, I.Ya. (2003). Pests *Grapholita delineana* Walker. - Eurasian Hemp Moth. Interactive Agricultural Ecological Atlas of Russia and Neighboring Countries. Economic Plants and their Diseases, Pests and Weeds. Website: http://www.agroatlas.ru/en/content/pests/Grapholita_delineana/

Oyen, L.P.A. (2011). *Agave sisalana* Perrine. [Internet] Record from PROTA4U. Brink, M. & Achigan-Dako, E.G. (Editors).PROTA (Plant Resources of Tropical Africa / Ressourcesvégétales de l'Afriquetropicale), Wageningen, Netherlands.<http://www.prota4u.org/search.asp>

Padaganur, G.M. (1979). The seed borne nature of *Alternaria macrospora* in cotton. Madras Agricultural Journal, 66: 325-326.

Padaganur, G.M and Siddaramaiah, A.L. (1979). Laboratory evaluation of fungicides against two cotton pathogens, *Alternaria macrospora* Zimm. and *Helminthosporium gossypii* Tucker. Pesticides, 13: 35-36.

Padwick, G.W. (1937). New plant diseases recorded in 1937. Int. Bull. Pl. Prot., 12: 122-123.

Paiero, S.M., Marshall, S.A., McPherson, J.E. and Ma, M.S. (2013). Stink bugs (Pentatomidae) and parent bugs (Acanthosomatidae) of Ontario and adjacent areas: A key to species and a review of the fauna. Canadian Journal of Arthropod Identification No. 1. Website: https://www.cjai.biologicalsurvey.ca/pmmm_24/pmmm_24.html

Paiva, F. deA., Asmus, G.L. and Araujo, A.E.de. (2001). Doenças. In: Embrapa Agropecuaria Oeste. Algodao: tecnologia de producao. Dourados: Embrapa CNPAO/Embrapa CNPA, pp.245-267.

Pakela, Y.P., Aveling, T.A.S. and Coutinho, T.A. (2002). Effect of plant age, temperature and dew period on the severity of anthracnose of cowpea caused by *Colletotrichum dematium*. African Plant Protection, 8(1&2): 65-68.

Pal, A.K. (1978). Studies on some root deseases of sunnhemp (*Crotalaria juncea* L.). PhD. Thesis. Banaras Hindu University, Varanasi, India.

Pal, A.K. and Basuchaudhury, K.C. (1980). Laboratory evaluation of fungicides to control some root diseases of sunnhemp. Indian J. Mycol. and Plant Pathol., 16: 212-215.

Pal, S., Srivastava, J.L. and Pandey, N.D. (1978). Effect of different dates of sowing on the incidence of *Dasyneura lini* Barnes (Diptera: Cecidomyiidae). Indian J. Entomol., 40: 433-434.

Pal, S.K. (1972). Efficacy of different insecticidal formulations on the control of white grub, *Aserica* sp. (Melolonthidae: Coleoptera) in desert area of Rajasthan. Indian Journal of Agricultural Research, 6(3): 215-220.

Palaniswamy, P., Underhill, E.W., Steck, W.F. and Chisholm, M.D. (1983). Responses of male redbacked cutworm, *Euxoa ochrogaster* (Lepidoptera: Noctuidae) to sex pheromone components in a flight tunnel. Environ. Entomol., 12: 748-752.

Pala Ram, Saini, R.K. and Vijaya, (2009). Preliminary studies on field parasitization and biology of solenopsis mealybug parasitoid, *Aenasius bambawalei* Hay at (Encyrtidae: Hymenoptera). Journal of Cotton Research and Development, 23(2): 313-315.

Palij, V.F. (1958). On the fauna and biocenology of Latvian flea beetles (Coleoptera, Chrysomelidae, Halticinae). Proceedings of the institute of biology of the Latvian SSR Academy of Sciences, 5: 69-89.

Palmateer, A.J., McLean, K.S., van Santen, E. and Morgan-Jones, G. (2003). Occurrence of Nigrospora Lint Rot Caused by *Nigrospora oryzae* on Cotton in Alabama. Plant disease, 87(7): 873.

Palmer, J.M. (1990). Identification of the common thrips of tropical Africa (Thysanoptera: Insecta). Tropical Pest Management, 36: 27-49.

Palti, J. (1988). The Leveillula Mildews. The Botanical Review, 54(4): 423-535.

Palumbo, J.C. and Kerns, D.L. (1994). Effects of imidacloprid as a soil treatment on colonization of green peach aphid and marketability of lettuce. Southwestern Entomologist, 19: 339-346.

Pande, Y.D. (1972). A Contribution to the Biology of *Utetheisa pulchella* Linn. (Lepid.,Arctiidae) on *Crotalaria burhea*. Journal of Applied Entomology, 70: 72-76.

Pandey, R.N., and Misra, D.P. (1992). Assessment of yield loss due to powdery mildew of linseed. Indian Bot. Reptr., 11: 62-64.

Pandit, N.C. (1985). Studies on the Bioecology of Major Pests of Jute. Annual Report, Central Research Institute for Jute and Allied Fibres, Barrackpore, Kolkata.

Pandit, N. C. (1995). White grub, *Rhinyptia meridionous* Arrow (Scarabaeidae: Coleoptera). A serious pest of ramie, *Boehmeria nivea* Guad. Environment and Ecology, 13(1): 245-246.

Pandit, N. C. (1998). Incidence and infestation pattern of *Nisotra orbiculata* (Mots.) on mesta, roselle, congo-jute, malachra and bhindi. Environment and Ecology, 16(3): 549-553.

Pandit, N.C. and Chakravorty, S. (1986). Relative toxicity of some insecticides to adults of mesta flea-beetle, *Nisotra orbiculta* (Mots.) (Coleoptera: Chrysomelidae). Indian J. Agric. Sci., 56: 136-138.

Pandit, N.C. and Chatterji, S.M. (1978). Bionomics of *Nisotra orbiculata* (Motsch) (Coleoptera: Chrysomelidae). Journal of entomological research, 2 (1): 49-54.

Pandit, N.C. and Pathak, S. (2000). Management of insect pests in mesta. Information Bulletin, 39pp. Central Research Institute for Jute and Allied Fibres, Barrackpore.

Pandit, N.C.and Pradhan, S.K. (1982). Incidence of *Eriophyses (Aceria) crotalariae* Chann. (Acari: Eryophyide) on sunnhemp (*Crotalaria juncea* L) from West Bengal. Sci. Cul., 48(11): 392-393.

Pandit, N.C. and Pradhan, S.K. (1991). Bionomics of the sunnhemp flea beetle *Longitarsus belgaumensis* Jac Coleoptera Chrysomelidae. Environment and Ecology, 94: 983-985.

Pandit, N.C. and Som, D. (1988). Culture of *Beauveria bassiana* and pathogenecity to insect pest of jute (*Corchorus capsularis* and *C. olitorius*) and mesta (*Hibiscus cannabinus* and *H. Sabdariffa*). Indian J. Agric. Sci., 58: 75-76.

Pandit, N.C., Som, D. and Chatterjee, S.M. (1979). *Beauveria bassiana* (Bals.) Vuile as a pathogen of *Nisotra orbiculata* Mots. infesting mesta, *Hibiscus cannabinus*. J. Entomol. Res., 3(1): 111-113.

Panizzi, A.R. and Grazia, J. (2015). True Bugs (Heteroptera) of the Neotropics. Springer, pp.522-523.

Panizzi, A.R., McPherson, J.E., James, D.G., Javahery, M. and McPherson, R.M. (2000). Stink bugs (Pentatomidae), pp. 421-474. In C. W.Schaefer and A. R. Panizzi (eds.), Heteroptera of economic importance.CRC Press, Boca Raton, FL.

Panja, B.N. (1999). Relative efficacy of different fungicide in vitro and in field for controlling twig blight disease of chilli caused by *Choanephora cucurbitarum* (Berk. And Rav.) Thaxter economically. Plant Protection Bulletin Faridabad, 51(1/2): 17-19.

Pansa, M.G., Asteggiano, L., Costamagna, C., Vittone, G. and Tavella, L. (2013). First discovery of *Halyomorpha halys* in peach orchards in Piedmont. (Primo ritrovamento di *Halyomorpha halys* nei pescheti piemontesi). Informatore Agrario, 69(37): 60-61.

Parish, D.L. and Statler, G.D. (1988). Slow rusting in flax. Proc.of the Flax Institute of the United States. Fargo, ND, Flax Institute of the United States, 52: 63-66.

Park, S-H., Choi, I-Y., Lee, W-H., Lee, K-J., Galea, V. and Shin, H-D. (2017). Identification and Characterization of *Cercospora malayensis* Causing Leaf Spot on Kenaf. Mycobiology, 45(2): 114-118.

Parker, J.R., Strand, A.L., and Seamans, H.L. (1921). Pale western cutworm (*Porosagrotis orthogonia* Morrison). J. Agric. Res., 22: 289-321.

Parker, W.B. (1913a). The hop aphids in the Pacific region. USDA Entomology Bulletin, 111: 9-39.

Parker, W.B. (1913b). The red spider on hops in the Sacramento valley of California. USDA Entomology Bulletin, 117: 1-41.

Parmelee, J.A. (1975). *Sphaerotheca macularis*. Fungi Canadenses, 63: 1-2.

Parmelee, J.A. (1980). *Puccinia schedonnardi*. Fungi Canadenses, 172: 1-2.

Parmeter, J.R. (Ed.) (1970). *Rhizoctonia solani*: Biology and Pathology, University of California Press, Berkeley, pp.149-160.

Parrado, J.L. (1958). Diseases of kenaf in Cuba. pp-113-123. In: Proceedings of the world conference on kenaf. International Cooperation Administration, Washington. 288p.

Partridge, D. (2017). *Macrophomina phaseolina*. Website: https://projects.ncsu.edu/cals/course/pp728/Macrophomina/macrophominia_phaseolinia.HTM .

Parris, G.K. (1959). A revised host index of Mississippi plant diseases. Mississippi State University, Botany Department. Miscellaneous Publication, 1: 1-146.

Parvin, N., Bilkiss, M., Nahar, J., Siddiqua, M.K. and Meah, M.B. (2016). RAPD analysis of *Sclerotium rolfsii* isolates causing collar rot of eggplant and tomato. Int. J. Agril. Res. Innov. & Tech., 6 (1): 47-57.

Passlow, T., Hooper, G.H.S. and Rossiter, P.D. (1960). Insecticidal control of Heliothis in linseed. Queensland J. Agric. Sci., 17: 117-120.

Patel, B.N., Patel, D.J. and Patel, R.C. (1983). Choanephora twig blight of *Crotalaria spectabilis*. Indian Phytopath., 36: 764-766.

Patel, D.J. (1982). Reniform nematode, *Rotylenchulus reniformis* Linford and Oliveira, 1940 in tobacco. Ph.D. Thesis, AAU, Anand (Gujrat, India).

Patel, H.P., Patel, A.D. and Bhatt, N.A. (2009). Record of coccinellids predating on mealy bug, *Phenacoccus solenopsis* Tinsley (Homoptera: Pseudococcidae) in Gujarat. Insect Environment, 14(4): 179.

Patel, J.S. and Ghosh, R.L.M. (1940). A review of agricultural investigations on jute in India. Indian Central Jute Committee. Agric. Res. Bull. Calcutta No.1: 49-50.

Patel, M.P. and Bilapate, G.G. (1984). Key mortality factors of spotted bollworm on cotton. Journal of Maharashtra Agricultural Universities, 9(2): 155-157.

Patel, N.V., Pathak, D.M., Joshi, N.S. and Siddhapara, M.R. (2013). Biology of onion thrips, *Thrips tabaci* (Lind.) (Thysanoptera: Thripidae) on onion *Allium cepa* (Linnaeus). J. Chem. Biol. Phys. Sci., 3: 370-377.

Patel, R. K. and Thakur, B. S. (2005). Insect pest complex and seasonal incidence in linseed with particular reference to bud fly (*Dasineura lini* Barnes). Journal of Plant Protection and Environment , 2(2): 102-107.

Patel, T. C. (1971). Studies on tuber rot of yam caused by *Botryodiplodia theobromae* Pat. M. Sc. (Agri.) thesis, Univ. of Udaipur, Udaipur.

Pathak, P.H. and Krishna, S.S. (1993). Effect of certain plant volatiles on the eggs of *Earias vittella* (F.) (Lepidoptera: Noctuidae). Annals of Forestry, 1: 165-167.

Patidar, P. (2016). Studies on the Myrothecium leaf blight of Cotton incited by *Myrothecium roridum* Tode ex. Fries. M.Sc Thesis, Rajmata Vijayaraje Scindia Krishi Vishwa Vidyalaya, Gwalior (M.P.).

Patil, B. V., Bheemanna, M., Patil, S. B., Udikeri, S. S. and Hosmani, A. (2006). Record of mirid bug, *Creontiades biseratense* (Distant) on cotton from Karnataka, India. Insect Environ., 11: 176-177.

Patil, S.D. (1962). *Physoderma corchori* Lingappa on stem, petioles and leaves of *Corchorus acutangulus* Lam. Indian Science Congress Association Part.3, Session. Forty-ninth. https://archive.org/.../2015.98788.Indian-Science-Congress-Association-Part-3-Session

Patschke, K., Gottwald, R. and Muller, R. (1997). Erste Ergebnissephytopathologischer Beobachtungenim Hanfanbauim Land Brandenburg. Nachrichtenblatt des Deutschen Pflanxenschutzdienstes, 49: 286-290.

Patzak, J. (2005). PCR detection of *Pseupernospora humuli* and *Podosphera macularis* in *Humulus lupulus*. Plant Protect. Sci., 41: 141-149.

Paul, N.C., Hwang, E.J., Nam, S.S., Lee, H.U., Lee, J.S., Yu, G.D., Kang, Y.G., Lee, K.B., Go, S. and Yang, J.W. (2017). Phylogenetic Placement and Morphological Characterization of *Sclerotium rolfsii* (Teleomorph: *Athelia rolfsii*) Associated with Blight Disease of *Ipomoea batatas* in Korea. Mycobiology, 45(3): 129-138.

Paul, S., Ghosh, R., Chaudhuri, S., Ghosh, S.K. and Roy, A. (2009). Biological and molecular variability of the begomoviruses associated with leaf curl disease of Kenaf in India. Journal of Plant Pathology, 91(3): 637-647.

Paul, S., Ghosh, R., Roy, A., Mir, J.I. and Ghosh, S.K. (2006). Occurrence of a DNA â containing begomovirus associated with leaf curl disease of kenaf (*Hibiscus cannabinus* L.) in India. Ausralian Plant Dis. Notes, 1 (1): 29-30.

Paul, V.H., Sultana, C., Jouan, B. and Fitt, B.D.L. (1991). Strategies for control of diseases on linseed and fibre flax in Germany, France and England. Production & Protection of Linseed, 1: 65-69.

Paunikar, S. (2015). *Myllocerus* spp. serious pest of tree seedlings in forest nurseries of North-Western and Central India. Biolife, 3(1): 353-354.

Pawar, V.P., Chaudhari, K.G., Kumar, P. and Nannavre, S. (2009). Morphological characterization of casual organism [*Sphaerotheca fuliginea* (Schlecht) Pollaci] on *Luffa cylindrica* powdery mildew. Shoudh Shamiksha Aur Mulyankan (Hindi), 2(7): 162.

Pawar, S.R., Desai, H.R., Bhanderi, G.R. and Patel, C.J. (2017). Biology of the Mealybug, *Phenacoccus solenopsis* Tinsley Infesting Bt Cotton. Int.J.Curr.Microbiol.App.Sci., 6(8): 1287-1297.

Pawar, C., Sithanantham, S., Bhatnagar, V., Srivastava, C. and Reed, W. (1988). The development of sex pheromone trapping of *Heliothis armigera* at ICRISAT, India. Tropical Pest Management, 34: 39-43.

Peacock, A.D. (1913). Entomological pests and problems of southern Nigeria. Bulletin of Entomological Research, 4: 191-220.

Pearson, C. A. S., Leslie, J.F. and Schwenk, F.W. (1986). Variable chlorate resistance in *Macrophomina phaseolina* from corn, soybean, and soil. Phytopathol., 76: 646–649.

Pearson, E.O. (1958). The Insect Pests of Cotton in Tropical Africa. Empre Cott.Gr. Crop and Commer. Insect. Ent London, pp.18-20, London, UK: CAB International.

Peetz, A.B., Mahaffee, W.F. and Gent, D.H. (2009). Effect of Temperature on Sporulation and Infectivity of *Podosphaera macularis* on *Humulus lupulus*. Plant Disease, 93(3): 281-286.

Peglion, V. (1917). Observations on hemp mildew (*Peronoplasmopara cannabina*) in Italy. Rend. Cl. Acad. Lincei, 114 [Ser. 5, 26(11)]: 618-620.

Pelayo-Sanchez, G., Yanez-Morales, M.J. and Solano-Vidal, R. (2017). First Report of Leaf Spot on *Coffea arabica* Caused by *Paramyrothecium roridum* in Mexico. Plant Disease, 101(6): 1044.
Pelozuelo, L., Malosse, C., Genestier , G., Guenego, H. and Frerot, B. (2004). Host-plant specialization in pheromone strains of the European corn borer *Ostrinia nubilalis* in France. J. Chem. Ecol., 30: 335-352.
Pena, J.E. (2013). Potential Invasive Pests of Agricultural Crops. United Kingdom: Oxfordshire.
Pena, J.E. and Campbell, C.W. (2005). Broad mite.EDIS. New Delhi, India: Malhotra Publishing House, 25: 944-949.
Pena, J.E., Osborne, L.S. and Duncan, R.E. (1996). Potential of fungi as biocontrol agents of *Polyphagotarsoneumus latus*. Entomophaga, 41: 27-36.
Penca, C. and Hodges, A. (2018). Featured creatures- brown marmorated stink bug .EENY-346.UF/IFAS, University of Florida. Website: https://www.entnemdept.ufl.edu/creatures/veg/bean/brown_marmorated_stink_bug.htm
Percy, R.G. (1993). Southwestern cotton rust. In: Watkins GM, ed. Compendium of cotton diseases. Minnesota, USA: APS Press, pp.37-39.
Percy, R.G. and Bird, L.S. (1985). Rust resistance expression in cotyledons, petioles, and stems of *Gossypium hirsutum* L. J. Hered., 76: 202-204.
Peregrine, W.T.H. (1963). Zebra disease of hybrid sisal. Kenya Sisal Bd Bull., 45: 38-39.
Peregrine, W.T.H. (1969a). A Note on Isolation and Inoculation Techniques used for Studies on Zebra Disease in Agave Hybrid No. 11648 in Tanzania. PANS, 15(4): 558-561.
Peregrine, W.T.H. (1969b). Investigations on chemical control of zebra disease in Agave hybrid no. 11648. Annals of Applied Biology, 63(1): 45-51.
Peres, F.S.C., Fernandes, O.A., Silveira, L.C.P. and Silva, C.S.B. 2009. Marigold as attractive plant for thrips in protected organic melon cultivation. Bragantia, 68(4): 953-960.
Perez-Jimenez, R.M. (2006). A review of the biology and pathogenicity of *Rosellinia necatrix*– the cause of white root rot disease of fruit trees and other plants. Journal of Phytopathology, 154: 257-266.
Perlak, F.J., Oppenhuizen, M., Gustafson, K., Voth, R., Sivasupramaniam, S., Heering, D., Carey, B., Ihrig, R.A. and Roberts, J.K. (2001). Development and commercial use of Bollgard cotton in the USA - early promises versus today's reality. Plant Journal, 27(6): 489-501.
Perpustakaan, S.B. (1991). Studies of Resistance of Six Kenaf Varieties to Phonia Leaf Rot. Yogyakarta, Universitas Gadjah Mada, Malaysia.
Perrin, R.M. and Ezeuh, M.I. (1978). The biology and control of grain legume olerthreutids (Tortricides). In S.R. Singh, H.F. van Emden and T.A. Taylor (eds) Pest of Grain Legumes: Ecology and Control, pp.201-207. Academic Press, London
Perryman, S. and Fitt, B.D.L. (1999). Effects of diseases on the yield of winter linseed. Aspects of App. Biol., 56: 211-218.
Persad, C. and Fortune, M. (1989). A new disease of sorrel (*Hibiscus sabdariffa* var. *sabdariffa*) caused by *Coniella musaiaensis* var. *hibisci* from Trinidad and Tobago. Plant Pathology, 38 (4): 615-617.
Peshney, N.L., Ninawe, B.N., Thakur, K.D. and Ingle, S.T. (1999). Chemical control of myrothecium leaf spot of cotton. 23: 7-10. Website: https://www.researchgate.net/publication/298662404_Chemical_control_of_myrothecium_leaf_spot_of_cotton
Petanovic, R., Magud, B. and Smiljanic, D. (2007). The hemp russet mite *Aculops cannabicola* (Farkas, 1960) (Acari: Eriophyoidea) found on *Cannabis sativa* L. in serbia: Supplement to the description. Arch. Biol. Sci., Belgrade, 59(1): 81-85.
Petch, T. (1917). Ann. Royal Botanic Gardens (Perademya), 6: 239.
Peter, C. and Sundrararajan, R., 1991, Chemical control of major insect pests of groundnut in Tamil Nadu. J. Insect. Sci., 4(1): 64-66.

Petersen, G.R., Russo, G.M. and Van Etten, J.L. (1982). Identiûcation of major proteins in sclerotia of *Sclerotinia minor* and *Sclerotinia trifoliorum*. Experimental Mycology, 6: 268-273.
Pethybridge, G. H., Lafferty, C. R., and Rhynehart, J. G. (1921). J. Dept. Agric. Irel., 21: 167.
Petri, L. (1931). Rassegna del casi fitopatologici Osservati nel 1930/ Review of phytopathological records in 1930, Bollettino della R. Stazione di Patologia Vegetate, 11(1): 1-50.
Petrovic-Obradovic, O. (2010). 14.35-*Macrosiphum euphorbiae* (Thomas, 1878) potato aphid (Hemiptera, Aphididae). BioRisk, 4: 930-931.
Phelan, P. L., Norris, K. H. and Mason, J. F. (1996). Soil-Management History and Host Preference by *Ostrinia nubilalis*: Evidence for Plant Mineral Balance Mediating Insect–Plant Interactions. Environmental Entomology, 25(6): 1329-1336.
Phengsintham, P., Braun, U., McKenzie, E.H.C., Chukeatirote, E., Cai, L. and Hyde, K.D. (2013). Monograph of Cercosporoid fungi from Thailand. Plant Pathology & Quarantine, 3(2): 67-138.
Philip, H.G. (1977). Insect Pests of Alberta. Alberta Agriculture Agdex , 612(1): 77.
Philip, N. (2007). Symptomology, Agronomy, and Economic Considerations in Aster Yellows Management. Website: www.umanitoba.ca/faculties/afs/MAC_proceedings/.../2007/ Philip_Northover.pdf
Photita, W., Taylor, P.W.J., Ford, R., Hyde, K.D. and Lumyong, S. (2005). Morphological and molecular characterization of Colletotrichum species from herbaceous plants in Thailand. Fungal Diversity, 18: 117-133.
Piao, C.S., Zhou, Y.S., Liu, L.C. and Zheng, Y.C. (1999). Determination of the toxicity of several acaricides to susceptible populations of Tetranychusurticae. Pl. Prot., 25: 20-22.
Pietkiewicz TA. 1958. Mikrofloranasionkonopi. Przegladliteratury. Poczn. Naukrol., Ser. A 77(4): 577-590.
Pinent, S.M.J. and Carvalho, G.S. (1998). Biologia de *Frankliniella schultzei* (Trybom) (Thysanoptera: Thripidae) emtomateiro (Biology of *Frankliniella schultzei* (Trybom) (Thysanoptera: Thripidae) in tomatoes). An. Soc. Entomol. Bras., 27(4): 519-524.
Pindikur, S.S., Rajanna, C.M., da Silva, J.A.T., Doijode, S. and Sunkad, G. (2012). *In vitro* and *in vivo* evaluation of fungal toxicants for the control of cotton rust caused by *Phakopsora gossypii* (Arth.) Hirat. The Asian and Australasian Journal of Plant Science and Biotechnology, 6: 7-13.
Pinkerton, A. and Bock, K.R. (1969). Parallel streak of sisal in Kenya. Experimental Agriculture, 5(1): 9-16.
Pinto, J.D. and Frommer, S.I. (1980). A survey of the arthropods on jojoba (*Simmondsia chinensis*). Environmental Entomology, 9(1): 137-143.
Piriyaprin, S., Manoch, L., Sunantapongsuk, V. And Somrang, A. (2007). Biological control of Pythium aphanidermatum and Sclerotium rolfsii, root rot of sunnhemp and mungbean by Gliocladium virens. Conference paper, In AGRIS since 2009.
Plaats-Niterink, A.J. van der. (1981). Monograph of the genus *Pythium*. Studies in Mycology, 21: 1-244.
Plantwise, (2015). Pest management decision guide: Green and yellow list; Ramie moth on ramie, www.plantwise.org/FullTextPDF/2015/20157802282.pdf
Plantwise, (ND). Vascular cotton wilt (*Fusarium oxysporum* f.sp. *vasinfectum*) – Plantwise. https://www.plantwise.org/KnowledgeBank/Datasheet.aspx?dsid=24715
Platt, W.D. and Stewart, S.D. (1999).Potential of nectariless cotton in today's cotton production system. 1999 Proceedings Beltwide Cotton Conferences, Orlando, Florida, USA, 3-7 January, 1999. 2: 971-974.
Ploetz, R.C., Palmateer, A.J., Geiser, D.M. and Juba, J.H. (2007). First Report of Fusarium Wilt Caused by *Fusarium oxysporum* on Roselle in the United States. Plant Disease, 91(5): 639.

Plonka, F. and Anselme, C. (1956). Les varietes de lin et leursprincipales maladies cryptogamiques, pp.165-168. Institut National de la Research Agronomique, Paris, France.

Pluke, R., Permaul, D. and Leibee, G. (1999). Integrated Pest Management and the Use of Botanicals in Guyana. Guyana: Richard Willium Hay Pluke, Bib. Orton IICA/CATIE, P-142.

Pokle, P.P. and Shukla, A. (2015). Chemical control of two spotted spider mite, *Tetranychus urticae* Koch (Acari: Tetranychidae) on tomato under polyhouse conditions. Pest Management in Horticultural Ecosystems, 21(2): 145-153.

Polat, Ý., Baysal, Ö.,Mercati, F., Gümrükcü, E., Sülü, G., Kitapcý, A., Araniti, F. and Carimi, F. (2018). Characterization of *Botrytis cinerea* isolates collected on pepper in Southern Turkey by using molecular markers, fungicide resistance genes and virulence assay. Infect Genet Evol., 60: 151-159.

Pollard, G.V. (1995). Pink or hibiscus mealybug in the Caribbean. CARAPHIN News, 12: 1–2.

Poltronieri, T.P.deS., Poltronieri, L.S., Verzignassi, J.R., Benchimol, R.L. and Carvalho, E.deA. (2012). Vinagreira: novo hospedeiro de *Corynespora cassiicola* no Pará. Summa Phytopathol. Botucatu, 38(2): 167.

Pomeroy, A.W.J. and Golding, F.D. (1923). Observations on the life histories of the cotton stainer bugs of the genus *Dysdercus* and on their economic importance in the Southern provinces of Nigeria. 2nd. Ann. Bull. Nigeria Agric. Dept., pp.23-58.

Pope, S.J. and Sweet, J.B. (1991). Sclerotinia stem rot resistance in linseed cultivars. Aspects Applied Biol., 28: 79-84.

Popescu, F., Doucet, I. and Marinescu, I. (1994). Geria - the first Romanian linseed variety resistant to wilt (*Fusarium oxysporum* f.sp. *lini* (Bolley) Snyder & Hansen). Romanian Agricultural Research, 2: 77-81.

Popov, K.I. (1941). The effect of ecological and agrotechnical conditions on the behaviour and injuriousness to flax of the flea beetles *Aphthona euphorbiae* Schrank and *Longitarsus parvulus* Payk.(In Russian). Trud. Obshch. Estestvoisp. Kazan, Gosud. Univ., 55: 157-203.

Popov, K.I. and Firsova, A.V. (1936). The influence of environmental conditions on the biology and injuriousness of *Aphthona euphorbiae* Schrank and *Longitarsus parvulus* Payk. (In Russian).
Plant Protection, 11: 94-102.

Pospisil, B. (1976). Breeding for resistance to flax anthracnose and the observation of its heredity. Len a Konopi, 14: 63-71.

Pott, J.N. (1976). A yucca borer, *Scyphophorus acupunctatus*, in Florida. Proceedings of the Florida State Horticultural Society, 88: 414-416.

Pourian, H.R., Mirab-balou, M., Alizadeh, M. and Orosz, S. (2009). Study on biology of onion thrips, *Thrips tabaci* Lindeman (Thysanoptera: Thripidae) on cucumber (var. Sultan) in laboratory conditions. J. Plant Prot. Res., 49: 390-394.

Powell, D.M. (1980). Control of the green peach aphid on potatoes with soil systemic insecticides: preplant broadcast and planting time furrow applications, 1973-77. Journal of Economic Entomology, 73: 839-843.

Powell, J.A. and Opler, P.A. (2009). Moths of Western North America. University of California Press., P.310.

Pradhan, S.K. and Saha, M.N. (1997). Effect of yellow mite (*Polyphagotarsonemus latus* Bank) infestation on the major nutrient contents of tossa jute (*Corchorus olitorius* L.) varieties. J Entomol Res., 21: 123-127.

Prakash and Bheemanna, M. (2014). Biology of mirid bug, *Poppiocapsidea* (=*Creontiades*) *biseratense* (Hemiptera: Miridae) on Bt cotton. International Journal of Plant Protection, 7(1): 45-49

Prakash, G. and Ghosh, T. (1964). Stem galls in cultivated Jute. Jute Bull., 26 (10): 237-239.

Prakash, S. (1989). Comparative efficacy of insecticides against top shoot borer *Laspeyresia Tricentra* Meyr - a pest of sunnhemp crop in Uttar Pradesh. Environment and Ecology, 7(1): 10-15.

Prakash, S. (1990). Estimation of loss in sunnhemp (*Crotalaria juncea*) fibre yield due to attack of top shoot borer, *Cydia* (*Laspeyresia*) *tricentra* Meyr. Jute dev. J., 10(1): 20-24.

Prasad, S.D., Basha, S.T. and Reddy, N.P.G.E. (2010).Molecular variability among the isolates of *Sclerotium rolfsii* causing stem rot of groundnut by RAPD, ITS-PCR and RFLP. Eur Asian Journal of Biosciences, 4: 80-87.

Prasad, S.N. (1967). Biology of the linseed blossom midge *Dasineura lini* Barnes. Cecidologia Indica, **2:** 31-41.

Prasada, R. (1948). Studies in Linseed rust *Melampsora lini* (Pers) Lev. in India. Indian Phytopath., 1: 1-18.

Prasadarao , J.A.V., Hanumantharao , A., Subbarao, D. and Ramaprasad, G. (1982). Granular systemic insecticides for control of green peach aphid (*Myzus persicae*) and their effect on yield and quality of tobacco grown in riverside lankas. Tropical Pest Management, 28(4): 381-384.

Prater, C.A., Redmond, C.T., Barney, W., Bonning, B.C. and Potter, D.A. (2006). Microbial Control of Black Cutworm (Lepidoptera: Noctuidae) in Turfgrass Using *Agrotis ipsilon* Multiple Nucleopolyhedrovirus. J. Econ. Entomol., 99(4): 1129-1137.

Presle, J.T. (1953). Verticillium wilt of cotton. In: Eds, HV Jordan, Yearbook of Agriculture, pp.301-303.

Preston, N.C. (1943). Observation on the genus *Myrothecium* Tode (I). The three classic species. Transactions of the British Mycological Society, 26: 158-168.

Price, J.F. (1979). Control of mealybugs on caladiums. Proceedings of the Florida State Horticultural Society, 92: 358-360.

Price, J.R., Slosser, J.E. and Puterka, G.J. (1983). Cotton aphid control. Chillicothe, TX, 1981. Insecticide and Acaricide Tests, 8: 197-198.

Price, T., Haggard, B., Lofton, J., Padgett, B. and Hollier, C. (2013).Cotton Leaf Spots. Louisiana Crops, Website: louisianacrops.com/2013/08/02/cotton-leaf-spots/

Prillieux, E.E.and Delacroix, M.G. (1891). *Endoconidium temulentum* nov. gen., nov. sp., Prillieux et Delacroix, champignon donnant au seigle des proprietes veneneuses. Bulletin de la Societe Mycologique de France, 7: 116-117.

Pring, D.R., Zeyen, R.L. and Banttari, E.E. (1973). Isolation and characterization of oat blue dwarf virus ribonucleic acid. Phytopathology, 63: 393-396.

Pritchard, A.E. and Baker, E.W. (1955). A revision of the spider mite family Tetranychidae. San Francisco: Pacific Coast Entomological Society. P.472.

PROSEA (2016). (PROSEA) *Boehmeria nivea* – Plant Use English uses. plantnet-project.org/en/ Boehmeria nivea (PROSEA)

Prudent, P. (1990). Efficacite de divers insecticides appliques en traitement de semence, contre les thrips du cotonnier au Paraguay. In : Methodes et responsabilites des entomologistesd' aujourd' hui. Résumes. Anon.. Gembloux : Faculte des Sciences Agronomiques, P.9.

Pruthi, H.S. (1936). Report of the Imperial Entomologist. pp.141-152. Scientific Report 1934-35. Agricultural Research Institute, Pusa, India.

Pruthi, H.S. (1937). Report of the Imperial Entomologist. Scientific Report 1936-37. Agricultural Research Institute, New Delhi, India.

Pruthi, H.S. and Batra, H.N. (1960). Important Fruit Pests of North-west India. Issued by The Indian Council of Agricultural Research. Government of India Press, Nasik Road, Delhi, India.

Pruthi, H.S. and Bhatia, H.L. (1937). A new cecidomyid pest of linseed in India. Indian J. Agric. Sci., 7: 797-808.
Pscheidt, J.W. and Ocamb, C.M. (Eds.). (2018). Pacific Northwest Plant Disease Management Handbook., Oregon State University. URL: https://pnwhandbooks.org/node/2870
Puinean, A.M., Foster, S.P., Oliphant, L., Denholm, I., Field, L.M., Millar, N.S., Williamson, M.S. and Bass, C. (2010). Amplification of a Cytochrome P450 Gene Is Associated with Resistance to Neonicotinoid Insecticides in the Aphid *Myzus persicae*. PLOS Genetics, Website: https://doi.org/10.1371/journal.pgen.1000999
Punithalingam, E. (1968a). *Phakopsora gossypii*. CMI Descriptions of Pathogenic Fungi and Bacteria. 172: 1-2.
Punithalingam, E. (1968b). *Uromyces decoratus*. CMI Descriptions of Pathogenic Fungi and Bacteria. 179: 1-2.
Punithalingam, E. (1976). *Botryodiplodia theobromae*. IMI Descriptions of Fungi and Bacteria, CABI Bioscience, Surrey, UK.
Punithalingam, E. (1980). Fibre plants (a) Jute (*Corchorus* spp.). In: Pnithalingam E (ed) Plant diseases attributed to *Botryodiplodia theobromae* Pat., pp 29–31.
Punithalingam, E. (1982). Conidiation and appendage formation in *Macrophomina phaseolina* (Tassi) Goid. Nova Hedwigia, 36: 249–290.
Punja, Z.K. (1985). The biology, ecology, and control of *Sclerotium rolfsii*. Annual Review of Phytopathology, 23: 97-127.
Punja, Z.K. and Damiani, A. (1996). Comparative growth, morphology and physiology of three *Sclerotium* species. Mycologia, 88: 694-706.
Punja, Z.K., and Grogan, R.G. (1981). Eruptive germination of sclerotia of *Sclerotium rolfsii*. Phytopathology, 71: 1092-1099.
Punja, Z.K. and Sun, L.J. (2001). Genetic diversity among mycelial compatibility groups of *Sclerotium rolfsii* (telemorph *Athelia rolfsii*) and *S. delphenii*. Mycol Res., 105(5): 537-546.
Purdy, L.H. (1979). *Sclerotiana sclerotiorum*: History and Symptomatology, Host Range, Geographic Distribution, and impact. Symposium on Sclerotiana. Phytopathology, 69 (8): 875-880.
Puri, S.N., Sharma, O.P., Murthy, K.S. and Raj, S. (1998). Handbook on Diagnosis and Integrated Management of Cotton Pests, National Centre for Integrated Pest Management, Pusa Campus, New Delhi, India, pp.52-53.
Purkayastha , R.P. and Sen-Gupta , M. (1973). Studies on conidial germination and appressoria formation in *Colletotrichum gloeosporioides* Penz.causing anthracnose of jute (*Corchorus olitorius* L.) / Unter such ungenüber Konidien keimung und Appressorien bild ung bel *Colletotrichum gloeosporioides* Penz., dem Erregereiner Anthraknose an Jute (*Corchorus olitorius* L.). Zeitschriftfür Pflanzenkrankheiten und Pflanzenschutz / Journal of Plant Diseases and Protection, 80(11/12): 718-724.
Purkayastha R.P. and Sen-Gupta M. (1975). Studies on *Colletotrichum gloeosporioides* inciting anthracnose of jute. Indian Phytopathol., 26: 650–653.
Purkayastha, S., Kaur, B., Dilbahi, N. and Chaudhury, A. (2006). Characterization of *Macrophomina phaseolina*, the charcoal rot pathogen of cluster bean, using conventional techniques and PCR-based molecular markers. Plant Pathol., 55: 106-116.
Purseglove, J.W. (1974). Tropical crops: Dicotylendons. Longman Group Limited, London.
Puszkar, L. and Jastrzebski, A. (1999).The biological protection of hops against damson-hop aphid and two spotted spider mite. Progress in Plant Protection, 39(2): 439-443.
Putnam, L.G. (1975). Insect pests of Brassica seed crops and of flax. In Harapiak, J.T. (ed.), Oilseed and Pulse Crops in Western Canada, Western Co-operative Fertilizers, Ltd., Calgary, Canada, pp.455-474.

Qiu, L.F. (2007). Studies on biology of the brown-marmorated stink bug, *Halyomorpha halys* (Stal) (Hemiptera: Pentatomidae), an important pest for pome trees in China and its biological control. Ph.D. dissertation. Chinese Academy of Forestry, Beijing, China.

Qiu, L.F., Yang, Z. and Tao, W. (2007). Biology and population dynamics of *Trissolcus halyomorphae*. Scientia Silvae Sinicae, 43: 62-65.

Qureshi, Z.A. and Ahmed, N. (1989). Efficacy of combined sex pheromones for the control of three major bollworms of cotton. Journal of Applied Entomology, 108(4): 386-389.

Qureshi, Z.A. and Ahmed, N. (1991). Sex pheromons as strategy to control pink bollworm of cotton in Sindh. The Pak. Cotton, 35: 129-144.

Raboudi, F., Makni, H. and Makni, M. (2011). Genetic Diversity of Potato Aphid, *Macrosiphum euphorbiae*, Populations in Tunisia Detected by RAPD. African Entomology, 19(1): 133-140.

Radadia, G.G., Pandya, H V., Patel, M.B. and Purohit, M.S. (2008). "Kapasana Mealy bugs (Chikto) ni Sankalit Niyantran Vyavastha", an information bulletin published in Gujarati by Main Cotton research Station, Navsari Agricultural University, Presented on 7th August, 2008 at Navsari.

Radewald, J.D., Maire, R.G., Shibhuya, F. and Sher, S.A. (1972). Control of stunt nematode (*Tylenchorhynchus claytoni*) on azalea (*Rhododendron indicum*). Plant Disease Reporter, 56(6): 540-543.

Radewald, J.D. and Takeshita, G. (1964). Desiccation studies on five species of plant-parasitic nematodes of Hawaii. Phytopathology, 54: 903-904.

Ragazzi, A. (1992). Different strains of *Fusarium oxysporum* f.sp. *vasinfectum*from cotton in Angola: biological aspects and pathogenicity. Journal of Plant Diseases and Protection, 99: 499–504.

Ragazzi, G. (1954). Nemici vegetali ed animali della canapa. Humus, 10(5):27-29.

Raghavendra, V.B., Siddalingaiah, L., Sugunachar, N.K., Nayak C. and Ramachandrappa, N.S. (2013). Induction of systemic resistance by biocontrol agents against bacterial blight of cotton caused by *Xanthomonas campestris* pv. *malvacearum*. International Journal of Plant Pathology, 2(1): 59-69.

Rageshwari, S., Renukadevi, P., Malathi, V.G., &Nakkeeran, S. (2016a). Occurrence, biological and serological assay of TSV infecting cotton in Tamil Nadu. Journal of Mycology and Plant Pathology, 46(2): 159-168.

Rageshwari, S., Renukadevi, P., Malathi, V.G., Amalabalu, P. and Nakkeeran, S. (2017). DAC-ELISA and RT-PCR based confirmation of systemic and latent infection by tobacco streak virus in cotton and parthenium. Journal of Plant Pathology, 99(2): 1-7.

Rageshwari, S., Vinodkumar, S., Renukadevi, P., Malathi, V.G. and Sevugapperumal, N. (2016b). Systemic infection of Tobacco streak virus in cotton. (Conference paper): International conference of Indian Virological Society on "Global perspectives in virus disease management", held in December 2016, at IIHR. Website: https://www.researchgate.net/publication/312590181_Systemic_infection_of_Tobacco_streak_virus_in_cotton

Rahman, A., Islam, K.S., Jahan, M.and Islam, N. (2016). Efficacy of three botanicals and a microbial derivatives acaricide (Abamectin) on the control of jute yellow mite, *Polyphagotarsonemus latus* (Bank). J. Bangladesh Agril. Univ., 14(1): 1–6.

Rahman, S. and Khan, M.R. (2006).Incidence of pests and avoidable yield loss in jute, *Corchorus olitorius* L., Annual Plant Protection Science, 2006, 14(2), 304–305.

Rahman, S. and Khan, M.R. (2012a). Incidence of pests in jute (*Corchorus olitorius* L.) ecosystem and pest–weather relationships in West Bengal, India. Archives of Phytopathology and Plant Protection, 45(5): 591-607.

Rahman, S. and Khan, M.R. (2012b). Field screening of popular jute (*Corchorus* spp.) varieties against the major pests in West Bengal, India. South Asian Journal of Experimental Biology, 2(5): 227-233.

Rai, A.N. (1980). IMI 246377, IMI records for geographical unit Indian Subcontinent. Herb IMI, Kew, London.
Rainwater, C.F. (1934). Insects and a Mite of Potential Economic Importance found on Wild Cotton in Florida. J. ceon. Ent., 27: 756-761.
Raj, S.K., Khan, M.S., Sneh, S.K. and Roy, R.K. (2007). Yellow vein netting of Bimili jute (*Hibiscus cannabinus* L.) in India caused by a strain of Tomato leaf curl New Delhi virus containing DNA. Australasian Plant Disease Notes, 2: 45–47.
Raj, S.K., Singh, R., Pandey, S.K. and Singh, B.P. (2003). Association of geminivirus with a leaf curl disease of sunnhemp (*Crotalaria juncea* L.) in India. European J. of Plant Pathology, 109(5): 467-470.
Rajasekhar, Y., Krishnayya, P.V. and Prasad, N.V.V.S.D. (2014). Influence of plant canopy and weather parameters on incidence of thrips (*Thrips tabaci* Lindemann) in Bt and non-Bt cottons. Journal of Research PJTSAU, 42(3): 13-18.
Raju, A.K., Rao, P.R.M., Apparao, R.V., Readdy, A.S. and Rao, K.K.P. (1988). Note on estimation of losses in yield of mesta due to mealy bug, *Maconellicoccus hirsutus*. Jute Development Journal, 8: 34–35.
Raju, B.C. (1985). Occurrence of chlorotic ringspot virus in commercial *Hibiscus rosa-senensis* cultivars. Acta Horticulture ae 164: VI International Symposium on Virus Diseases of Ornamental plants. Doi: 10.17660/ActaHortic.1985.164.30
Ram, S. (1968a). Effect of early sowing of sunnhemp on the incidence of its insects and pests. Indian J. Agriculture Sci., 38(1): 236-237.
Ram, S. (1968b). Record of parasites of sunnhemp top shoot borer, *Laspeyresia tricentra* Meyr. (Torticidae/ Lerpidoptera) in Uttar Pradesh. Indian J. Ent., 30(4): 254.
Ram, S. and Prakash, G. (1968). Trials with insecticides for the control of shoot borer, *Laspeyresia tricentra* Meyr. (Torticidae/ Lerpidoptera) of sunnhemp. Indian J. Agriculture Sci., 38(2): 310-313.
Ram, S. and Tripathi, R.L. (1969). Top shoot borer problem – pest of sunnhemp in U.P. Indian Farmers Digest, 2(11): 11,15.
Ramamurthy, V.V. and Ghai, S. (1988). A study on the genus *Myllocerus* (Coleoptera: Curculionidae). Orient Insects, 22: 377–500.
Ramello, J.C., Garcia, A. and Weht, S. (1975). Chemical control of the vector of tomato spotted wilt virus (TSWV) in the province of Tucuman (Argentine Republic). Revista Agronomica del Noroeste Argentino, 12(1/2): 119-139.
Ramesh-Chandra, (2004). Status of medicinal plants with respect to infestation of insect pests in and around Chitrakoot, District Satna (M.P.). Flora and Fauna Jhansi, 10(2): 88-92.
Ramezani, M.,Thomas Shier, W., Abbas, H.K., Tonos, J.L., Baird, R.E. and Sciumbato, G.L. (2007). Soybean Charcoal Rot Disease Fungus *Macrophomina phaseolina* in Mississippi Produces the Phytotoxin (H)-Botryodiplodin but No Detectable Phaseolinone. J. Nat. Prod., 70(1): 128-129.
Rana, M.K. (eds), (2017). Vegetable crop science, CRC Press.
Randhawa, L. S., Singh, T. H. and Indu, (2002). World's first leaf curl virus resistant cotton hybrid LHH 144. Journal-of-Research-Punjab-Agricultural University, 39 (3): 470.
Rangaswami, G. and Mahadevan, A. (1998). Diseases of crop plants in India. PHI Learning Pvt. Ltd., New Delhi, P-548.
Rangaswami, G. and Mahadevan, A. (2006). Diseases of crop plants in India (4rth Ed.). PHI India Ltd., New Delhi
Rangarani, A., Rajan, C.P.D., Harathi, P.N., Bhaskar, B. and Sandhya, Y. (2017). Evaluation of fungicides and heribicides on *Sclerotium rolfsii*, incitant of stem rot diseases in groundnut (*Arachis hypogaea* L.). Int. J. Pure App. Biosci., 5(3): 92-97.

Ranney, C.D., Hurshe, J.S. and Newton, O.H. (1971). Effect of bottom defoliation on microclimate and the reduction of boll rot of cotton. Agronomy Journal, Madison, Wisconsin, 63(2): 259-263.
Rao, G.N. (1974). Control of tea mites in South India-1. Planters' Chron., 69(5): 91-94.
Rao, V.G. (1962). The Genus *Phyllosticta* in Bombay-Maharashtra. Current Science- Zobodat. Website: www.zobodat.at/pdf/Sydowia_16_0275-0283.pd
Rao, A.V.B. and Jayasankar (2015). Grey mildew disease management in cotton. The Hindu, website: www.thehindu.com/sci-tech/science/grey-mildew-disease...cotton/article6830916.ece
Rao, V.G. (1973). Diseases of Fibre Crops in India. Website: https://www.researchgate.net/file.
Rashid, K.Y. (1991). Evaluation of components of partial resistance to rust in flax. Can. J. Plant Pathol., 13: 212-217.
Rashid, K.Y. (1997). Slow-rusting in flax cultivars. Can. J. Plant Pathol., 19: 19-24.
Rashid, K.Y. (1998). Powdery mildew on flax: A new disease in western Canada. Can. J. Plant Pathol., 20: 216.
Rashid, K.Y. (2000a). Pasmo disease in flax – Impact and potential control. Proc. Manitoba Agronomists Conference. Winnipeg, MB. pp.154-156.
Rashid, K.Y. (2000b). Sclerotinia on flax in western Canada – warning for a potential disease problem. Can. J. Plant Pathol., 22: 175-176.
Rashid, K.Y.(2003a). Diseases of flax. pp.147-154. In: Diseases of Field Crops in Canada, 3rd ed. K.L. Bailey, G.D. Gossen, R. K. Gugel, and R.A.A. Morrall, eds. University Extension Press, Saskatoon, SK, Canada.
Rashid, K.Y. (2003b). Principal diseases of flax, pp.92-123. In: Eds, Muir, A.D. and Westcott, N.D. Flax-The genus Linum, Taylor & Francis Ltd, New York.
Rashid, K. (2015). Pasmo Disease Management in Flax | Flax Council of Canada https://www.flaxcouncil.ca/tips_article/pasmo/
Rashid, K.Y. (2017). Resistance in flax cultivars and genotypes to powdery mildew, Agriculture and Agri-Food Canada, Morden Research and Development Centre, Morden, Manitoba. Website: https://www.umanitoba.ca/.../Rashid_MAC_PM_2017_Poster_Oct_24_2017_Portrait_LONG.pdf.
Rashid, K.Y., Desjardins, M. L. and Duguid, S. (2010). Diseases of flax in Manitoba and Saskatchewan in 2009. Can. Plant Dis. Surv., 90: 136-137.
Rashid, K.Y., Desjardins, M. L. and Duguid, S. (2011). Diseases of flax in Manitoba and Saskatchewan in 2010. Can. Plant Dis. Surv., 91: 128-129.
Rashid, K.Y., Desjardins, M. L and Duguid, S. (2013). Diseases of flax in Manitoba and Saskatchewan in 2012. Can. Plant Dis. Surv., 93: 161-162.
Rashid, K.Y., Desjardins, M. L., Duguid, S. and Northover, P.R. (2012). Diseases of flax in Manitoba and Saskatchewan in 2011. Can. Plant Dis. Surv., 92: 134-135 .
Rashid, K. and Duguid, S. (2005). Inheritance of resistance to powdery mildew in flax. Canadian Journal of Plant Pathology, 27(3): 404-409.
Rashid, K.Y. and Kenaschuk, E.O. (1992). Genetics of resistance to rust in the flax cultivars Vimy and Andro. Can. J. Plant Pathol., 14: 207-210.
Rashid, K.Y. and Kenaschuk, E.O. (1993). Effect of trifluralin on fusarium wilt in flax. Can. J. Plant Sci., 73: 893-901.
Rashid, K.Y. and Kenaschuk, E.O. (1994). Genetics of resistance to flax rust in six Canadian flax cultivars. Can. J. Plant Pathol., 16: 266-272.
Rashid, K.Y. and Kenaschuk, E.O. (1996b). Seed treatment for fusarium wilt control in flax.
Proc. of the Flax Institute of the United States. Fargo, ND, Flax Institute of the United States, pp.158–161.

Rashid, K.Y. and Kenaschuk, E.O. (1999). Three new rust resistance genes in flax introductions. Can. J. Plant Pathol., 21: 64-69.

Rashid, K. Y., Kenaschuk, E. O., Duguid, S. and Platford, R. G. (2000). Diseases of flax in Manitoba and eastern Saskatchewan in 1999, In Canadian Plant Disease Survey - Disease Highlights, The Canadian Phytopathological Society, Canada, 80: 92-94.

Rashid, K.Y., Kenaschuk, E.O., and Menzies, J.G. (1998a). Powdery mildew on flax, first encounter in western Canada. Proc. of the Flax Institute of the United States. Fargo, ND, Flax Institute of the United States, 57: 125-128.

Rashid, K.Y., Kenaschuk, E.O. and Platford, R.G. (1998a). Diseases of flax in Manitoba in 1997 and first report of powdery mildew on flax in western Canada. Can. Plant Dis. Surv., 78: 99-100.

Rashid, K.Y., Kenaschuk, E.O., and Platford, R.G. (1998b). Diseases of flax in Manitoba in 1997, and first report of powdery mildew on flax in western Canada. Can. Plant Dis. Surv., 78: 99-100.

Rashid, K.Y., Kenaschuk, E.O., and Platford, R.G. (2000). Diseases of flax in Manitoba in 1999. Can. Plant Dis. Surv., 80: 92-93.

Rashid, K.Y. and Kroes, G.M.I.W. (1999). Pathogenic variability in *Fusarium oxysporum* f. sp. *lini* on flax. Phytopathology, 89: S65.

Rasu, T., Sevugapperumal, N., Thiruvengadam, R. and Ramasamy, S. (2013). Morphological and genomic variability among *Sclerotium rolfsii* populations. The Bioscan, 8(4): 1425-1430.

Rataj, I.K. (1974). The influence of *Thrips linarius* Uzel on self-sterility.(In Russian). Len i Konoplya, 12: 91-105.

Rataj, K. (1957). Skodlivi cinitele pradnych rostlin. Prameny literatury, 2: 1-123.

Rataj, I.K. (1974). The influence of *Thrips linarius* Uzel on self-sterility. (In Russian). Len i Konoplya, 12: 91-105.

Ratnadass, A., Soler, A., Chabanne, A., Tullus, R.G., Teacher, P., Bellec, F.Le, Marnote, P., Streito, J.C. and Matocq, A. (2018). First record of *Moissonia importunitas* as a pest of rattle box (*Crotalaria* spp.) in Reunion Island (Hemiptera, Miridae). Bulletin de la Societe Entomologique de France, 123(1): 59-64.

Raut, J.G., Holey, N.R. and Moghe, P.G. (1980).Occurrence of Myrothecium leaf spot on cotton in Vidarbha. Indian Phytopathol., 33: 510–511.

Ravi, K.C., Mohan, K.S., Manjunath, T.M., Head, G., Patil, B.V., Greba, D.P.A., Premalatha, K., Peter, J. and Rao, N.G.V. (2005). Relative Abundance of *Helicoverpa armigera* (Lepidoptera:

Noctuidae) on Different Host Crops in India and the Role of These Crops as Natural Refuge for *Bacillus thuringiensis* Cotton. Environmental Entomology, 34(1): 59–69.

Ravi, P. R. (2007). Bioecology, loss estimation and management of Mirid bug *Creontiades biseratense* (Distant) (Hemiptera :Miridae) on Bt cotton. M. Sc. (Agri.) Thesis, Univ. Agric. Sci., Dharwad, Karnataka (India).

Ravi, P. R. and Patil, B. V. (2008), Biology of mirid bug, *Creontiades biseratense* (Distant) (hemiptera: miridae) on Bt cotton. Karnataka J. Agric. Sci., 21(2): 234-236.

Ravi, P.R., Patil, B.V., Narayanaswamy, K.S., Sowmya, E., Lepakshi, N.M. and Sajjan, P.S. (2015). Biology of mirid bug, *Creontiades biseratense* (Hemiptera: Miridae). IJBS, 2(2): 157-161.

Ravichandra, N.G. (2013). Plant Nematology, I.K. International Publishing House Private Ltd, New Delhi.

Rawat, R.R. and Kaushik, U.K. (1983). Crop losses due to insect pest in north-east M.P. Paper presented at All India Workshop on Crop losses. A.P. Agric. Uni. Hyderabad. PP.1-25

Rawlinson, C.J. and Dover, P.A. (1986). Pests and diseases of some new and potential alternative arable crops for the United Kingdom. Proc. 1986 Br. Crop Prot. Conf. Pests Dis., British Crop Protection Council, Farnham, UK , pp.721-732.

Raworth, D.A., Gillespie, D.R., Roy, M. and Thistlewood, H.M.A. (2002). *Tetranychus urticae* Koch, twospotted spider mite (Acari: Tetranychidae). In Peter G. Mason; John Theodore Huber. Biological Control Programmes in Canada, 1981-2000. CAB International, pp.259-265.

Raychaudhuri, S.P. (1947). A note on mosaic virus of sunn-hemp (*Crotalaria juncea* Linn.) and its crystallization. Curr. Sci., 16 (1): 26-28.

Raychaudhuri, S.P. and Pathanian, P.S. (1950). A mosaic disease of *Crotolaria mucronata* Desv. (C. striata DC). Curr. Sci., 19(7): 213

Readshaw, J.L.(1975). The ecology of tetranychid mites in Australian orchards. J. appl. Ecol., 12: 473-495.

Reddy, A.S., Reddy, O.C., Rosaiah, B. and BaskaraRao, T. (1986).Studies on the resurgence of spider mites and whiteflies of cotton. In: Proceedings of national seminar on resurgence of sucking pests, Tamil Nadu Agricultural University, Coimbaiore, India, pp.174-179.

Reddy, D.B. (1956). Sunnhemp and its insect fauna.Proc. 10th Int Congr. Ent. 3: 439-440.

Reddy, K.M.S., Revannavar, R. and Samad, A.S.N. (2001). Biology and feeding potential of aphid predators *Cheilomenes sexmaculata* (Coccinellidae: Coleoptera) and *Dideopsis aegrota* (Fab.) (Diptera: Syrphidae) on rose aphid, *Macrosiphum rosae* Linn. (Homoptera: Aphididae). Journal of Aphidology, 15(1&2): 83-85.

Reed, B.,Gannaway, J.,Rummel, D.R. and Thorvilson, H.G.(1999).Screening for resistance in cotton genotypes to *Aphis gossypii* Glover, the cotton aphid. 1999 Proceedings Beltwide Cotton Conferences, Orlando, Florida, USA, 2: 1002-1007.

Reed, W. (1994). *Earias* spp. (Lepidoptera: Noctuidae), pp.151–176. In: G. A. Matthews and J. P. Tunstall (ed.), Insect pests of cotton. CAB International, Ascot, United Kingdom.

Reed, W. and Pawar, C.S. (1982). Heliothis: a global problem. In: Reed W, Kumble V, ed. Proceedings of the International Workshop on Heliothis Management. ICRISAT Center, Patancheru, India, 15-20 November 1981 International Crops Research Institute for the Semi-Arid Tropics Patancheru, Andhra Pradesh India, pp.9-14.

Regupathy, A., Palanisamy, S., Chandramohan, N. and Gunathilagaraj, K. (1997). A guide on croppests. Soorya Desktop publishers, Coimbatore. Tamil Nadu, P.290.

Rehman, M.H. and Ali, H. (1981). Biology of spotted bollworm of cotton *Earias vittella* (F.). Pakistan J. Zool., 13: 105-110.

Ren, Y.Z., Liu, Y.Q., Ding, S.L., Li, G.Y. and Zhang, H. (2008). First Report of Boll Rot of Cotton Caused by *Pantoea agglomerans* in China. Plant Disease, 92(9): 1364.

Renata, S., de Mendonca, Navia, D., Diniz, I.R., Auger, P., et al. (2011). A critical review on some closely related species of *Tetranychus sensustricto* (Acari: Tetranychidae) in the public DNA sequences databases. Exp Appl Acarol., 55: 1- 23

Reyes-Franco, M.C., Hernandez-Delgado, S., Beas-Fernandez, R., Medina,-Fernandez, M., Simpson, J. and Mayek-Perez, N. (2006). Pathogenic variability within *Macrophomina phaseolina* from Mexio and other countries. J. Phytopathol., 154: 447-453.

Reza, P., Ali, G. and Shirin, F. (2013). Biological and molecular detection of *Hibiscus chlorotic ringspot virus* infecting Hibiscus rosa-sinensis in Iran. Phytopathologia Mediterranea, 52(3): 528-531.

Reza, M.M., Mehrdad, M. and Nakhei A. (2007).First report of powdery mildew on roselle caused by *Leveillula taurica* in Iran. Iranian Journal of Plant Pathology, 43 (4): 158.

Rhynehart, J.G. (1922). On the life-history and bionomics of the flax flea beetle (*Longitarsus parvulus* Payk.) with descriptions of the hitherto unknown larval and pupal stages. Scientific Proc. Royal Dublin Society, 16: 497-541.

Riaz, A., Khan, S.H., Iqbal, S.M. and Shoaib.M. (2007). Pathogenic variability among *Macrophomina phaseolina* (Tassi) Goid, isolates and identification of sources of resistance in mash against charcoal rot. Pak. J. Phytopathol., 19(1): 44-46.

Ridge G. and Shew, B. (2014). *Sclerotium rolfsii* (Southern blight of vegetables and melons). Website: https://wiki.bugwood.org/Sclerotium_rolfsii_(Southern_blight_of_vegetables_and_melons).

Ridgway, R/L., Bell, A.A., Veech, J.A. and Chandler, J.M. (1984). Cotton practices in the USA and world. In: Kohel RJ, Lewis CF, eds. Cotton Agronomy Monograph, 24. Madison, USA: American Society of Agronomy.

Riggs, J.L. and Lyda, S.D. (1988). Laboratory and field tests with triazole fungicides to control Phymatotrichum root rot of cotton. Proceedings, Beltwide Cotton Production Research Conferences, USA Memphis, USA; National Cotton Council of America, 45-48.

Roberts, J.K. and Pryor, A. (1995). Isolation of a flax (*Linum usitatissimum*) gene induced during susceptible infection by flax rust (*Melampsora lini*). Plant J., 8: 1-8.

Roberts, P. (1999). Rhizoctonia-forming fungi: a taxonomic guide. Royal Botanic Gardens, Kew, P.239.

Roberts, P.D. (2003). Southern blight. pp 20-21. In: Compendium of pepper diseases, Pernezny, K.L.,

Roberts, P.D., Murphy, J.F. and Goldberg, N.P. (eds). APS Press. St. Paul, MN.

Roberts, P.D., French-Monar, R.D. and McCarter, S.M. (2014). Southern Blight. pp. 43-44. In: Compendium of Tomato Diseases, 2nd edition, Jones, J. B., Zitter, T. A., Momol, M. T., and Miller, S. A. (eds.). APS Press. St. Paul, MN.

Robertson, A. (2015). *Cercospora* Leaf Spot on Roselle Hibiscus. Website: https://www.lsuagcenter.com/portals/our_offices/parishes/st%20helena/news/cercospora-leaf-spot-on-roselle-hibiscus

Robinson, A.F., Heald, C.M., Flanagan, S.L., Thames, W.H. and Amador, J. (1987). Geographical distribution of *Rotylenchulus reniformis*, *Meloidogyne incognita*, and *Tylenchulus semipenetrans* in the Lower Rio Grande Valley as related to soil type and land use. Annals of Applied Nematology, 1: 20-25.

Robinson, D.M. (1961). A species of *Tetranychus* Dufour (Acarina) from Uganda. Nature, 189: 857–858.

Robinson, G.S. (1975). Macrolepidoptera of Fiji and Rotuma, a taxonomic and biogeographic study. Faringdon: Classey, P.362.

Rodenhiser, H.A. (1930). Physiologic specialization and mutation in *Phlyctaena linicola* Speg. Phytopathology, 20(12): 931.

Rodriguez, G., Kibler, A., Campbell, P., and Punja, Z.K. (2015). Fungal diseases of *Cannabis sativa* in British Columbia, Canada. American Phytopathological Society Annual Meeting, Poster, P.529.

Rodríguez-Gálvez, E. and Maldonado, E. (1998). Reacción de TresCultivaresPeruanos de Algodonero Pima (*Gossypium barbadense* L.) a *Fusarium oxysporum* f. sp. *vasinfectum*. Fitopatología, 33(2): 127-132.

Rohini, R. S., Mallapur, C. P. and Udikeri, S. S. (2009). Incidence of mirid bug, *Creontiades biseratense* (Distant) on Bt cotton in Karnataka. Karnataka J. Agric. Sci., 22(3): 680-681.

Rojas, M.R., Gilberston, R.L., Russell, D.R. and Maxwell, D.P. (1993). Use of degenerate primers in the polimerase chain reaction to detect whitefly-transmitted geminiviruses. Plant Disease, 77: 340-347.

Rolston, L.H. and Kendrick, R.L. (1961). Biology of the brown stink bug, *Euschistus servus* Say. J. Kansas Entomol. Soc., 34: 151-157.

Room, P. M. (1979a). Insects and spiders of Australian cotton fields. N.S.W. Department of Agriculture: Sydney.

Room, P. M. (1979b). Parasites and predators of *Heliothis* spp. (Lepidoptera: Noctuidae) in cotton in the Namoi Valley, New South Wales. J. Aust. ent. Soc., 18: 223-228.

Ros, V.I.D. and Breeuwer, J.A.J. (2007). Spider mite (Acari: Tetranychidae) mitochondrial COI phylogeny reviewed: host plant relationships, phylogeography, reproductive parasites and barcoding. Exp Appl Acarol., 42: 239-262.

Rossem, G. van, Bund, C.F. van de, Burger, H.C. and GoffauL, J.W.de. (1981). Unusual infestations of insects in 1980. Entomologische Berichten, 41(1): 84-87.

Rost, H. (1938). Untersuchungen uber einige Krankheiten des Leins in Deutschland, Angewandte Botanik, 20(6): 412-430.

Roughley, N., Smith, L. and Allen, S. (2015). Integrated Disease Management for: Boll rot, seed rot & tight lock. Fact Sheet, Cottoninfo. https://www.cottoninfo.com.au/sites/default/files/documents/BollRot%20SeedRot.pdf

Roush, R.T. (1994). Managing pests and their resistance to *Bacillus thuringiensis*: Can transgenic crops be better than sprays? Biocontrol Sci. Technol., 4: 501-516.

Roy, A., Acharyya, S., Das, S., Ghosh, R., Paul, S., Srivastava, R.K. and Ghosh, S.K. (2009). Distribution, epidemiology and molecular variability of the begomovirus complexes associated with yellow vein mosaic disease of mesta in India. Virus Research, 141: 237-246.

Roy, K.W. and Bourland, F.M. (1982). Epidemiological and mycofloral relationships in cotton seedling disease in Mississippi. Phytopathology, 72: 868-872.

Roy, S., Gurusubramanian, G. And Nachimuthu, S.K. (2011a). Anti-mite activity of Polygonum hydroppier L. (Polygonaceae) extracts against tea red spider mite, Oligonychus coffeae Nietner (Tetranychidae: Acari). Int J Acarol., 37(6): 561-566.

Roy, S., Mandal, R.K., Saha, A.R. and Toppo, R.S. (2011). Symptomatology, variability and management of zebra disease of sisal in western Orissa. Journal of Mycopathological Research, 49(1): 53-58.

Roy, S., Mukhopadhyay, A. and Gurusubramanian, G. (2010b). Baseline susceptibility of *Oligonychus coffeae* to acaricides in North Bengal tea plantations, India. Int J Acarol., 36(5): 357-362.

Roy, S., Mukhopadhyay, A. and Gurusubramanian, G. (2011b). Anti-mite activities of *Clerodendrum viscosum* Ventenat (Verbanaceae) extracts on tea red spider mite, *Oligonychus coffeae* Nietner (Acari: Tetranychidae). Arch Phytopathol Plant Protect., 44(16): 1550-1559.

Roy, S., Muraleedharan, N. and Mukhopadhyay, A. (2014). The red spider mite, *Oligonychus coffeae* (Acari: Tetranychidae): its status, biology, ecology and management in tea plantations. Exp Appl Acarol. Doi 10.10007/s10493-014-9800-4

Royalty, R. N. and T. M. Perring. 1987. Comparative Toxicity of Acaricides to *Aculops lycopersici* and *Homeoprone matusanconai* (Acari: Eriophyidae, Tydeidae). J. Econ. Entomol. 80(2): 348-351.

Rozhmina , T.A. (2015). Dosti zheniya nauki i tekhniki APK, 12: 47-49 (in Russ.).

Rozhmina, T.A. and Loshakova, N.I. (2016). New sources of effective resistance genes to fusarium wilt in flax (*Linum usitatissimum* L.) depending on temperature. Agricultural Biology (Sel'skokhozyaistvennaya Biologiya), 51(3): 310-317.

Ruano, O., Almeida, W.P. de, Follin, J.C. and Girardot, B. (1988). Identification of sources of resistance to races 18 and 20 of *Xanthomonas campestris* pv. *malvacearum* in cotton genotypes. Fitopatologia Brasileira, 13(4): 328-333.

Rubies-Autonell, C. and Turina, M. (1994). Alfalfa mosaic virus (AMV) isolated from kenaf (*Hibiscus cannabinus*). Phytopathologia Mediterranea, 33(3): 234-239.

Rueda, A. and Shelton, A.M. (1995). Global crop pests-Onion thrips. Cornell International Institute for Food, Agriculture and Development, Cornell University.

Ruiz-Montiel, C., González-Hernández, H., Leyva, J. L., Llanderal, C. C., Cruz-López, L., and Rojas, J. C. (2003). Evidence for a male-produced aggregation pheromone in *Scyphophorus acupunctatus* Gyllenhal (Coleoptera: Curculionidae). J. Econ. Entomol., 96: 1126-1131.

Ruiz-Montiel, C., Rojas, J.C., Cruz-López, L. and González-Hernández, H. (2009). Factors affecting pheromone release by *Scyphophorus acupunctatus* (Coleoptera: Curculionidae). Environ Entomol., 38(5): 1423-1428.

Rush, C.M., Gerik, T.J. and Kenerley, C.M. (1985). Atypical disease symptoms associated with Phymatotrichum root rot of cotton. Plant Disease, 69(6): 534-537.

Ruszkowski, J.W. (1928). *Phytometra gamma* L., a serious field and vegetable pest. (In Polish). Poradnik Gospod., 39: 730-732.

Ryczkowski, A. (2011). Care & Disease of Agave Plants. Hunker. Website: https://homeguides.sfgate.com/care-disease-agave-plants-41542.html

Sa, J.O. (2009). Patogênese de Aspergillus niger e biocontrole da podridãovermelha do sisal por Trichoderma spp. MSc Dissertation, Universidade Federal do Recôncavo da Bahia. Cruz das Almas, BA.

Sabet, K. A. (1957). Studies in the bacterial diseases of Sudan crops. I. Bacterial leaf spot of jute (*Corchorus olitorius* L.). Annals of Applied Biology, 45(3): 516-520.

Scales, A.L. and Furr, R.E. (1968). Relationship between the tarnished plant bug and deformed cotton plants. Journal of Economic Entomology, 61: 114-118.

Saccaebo, P.A. (1884). Sylloge Fungoruxn. Vol. 8. P.860. Aim Arbor, Michigan: Edwards Brotliers, Inc

Saccardo, P.A. (1886). Sylloge Hyphomycetum. Sylloge Fungorum, 4: 1-807.

Sackston, W.E., and Gordon, W.L. (1945). Twenty-fourth Ann. Rep. of 1944 Survey. Canadian Plant Disease Survey, 29.

Saccardo, P. A. (1882). Sylloge Fungorum. Johnson Reprint Corporation. New York, N.Y. Vols. 1-25.

Saeed, S., Ahmad, M., Ahmad, M. and Kwon, Y.J. (2007). Insecticidal control of the mealy bug, *Phenacoccus gossypiphilous* (Hemiptera: Pseudococcidae), a new pest of cotton in Pakistan. Entomological Research, 37: 76-80.

Sagarra L.A. and Peterkin D.D. (1999). Invasion of the Carribean by the hibiscus mealybug, *Maconellicoccus hirsutus* Green [Homoptera :Pseudococcidae]. Pyroprotection. 80: 103-113.

Saharan, G., Mehta, N. and Sangwan, M. (2005). Fungal Diseases of Linseed. In: G. S. Saharan, N. Mehta, & M. S. Sangwan, Diseases of Oilseed Crops. pp.176-201. Indus Publishing.

Saharan, G.S. and Saharan, M.S. (1992). Studies on Physiological Specialization and Inheritance of Resistance in Linseed to Powdery Mildew Disease. 2nd Ann. Rept. Dept. Pl. Pathol. CCS HAU, Hisar, India, P.47.

Saharan, G.S. and Saharan, M.S. (1994a). Conidial size, germination and appressorial formation of *Oidium lini* Skoric, cause of powdery mildew of linseed. Indian J. Mycol. Pl. Pathol., 24: 176-178.

Saharan, G.S. and Saharan, M.S. (1994b). Studies on powdery mildew of linseed caused by *Leveillula taurica* (Lev.) Arnaud. Indian J. Mycol. Pl. Pathol., 24: 107-110.

Saharan, G.S. and Singh, B.M. (1978). New physiological races of *Melampsora lini* in India. Indian Phytopathol., 31: 450-454.

Sahayaraj, K and Borgio, J. F. (2010). Virulence of entomopathogenic fungus *Metarhizium anisopliae* (Metsch.) Sorokin on seven insect pests. Indian Journal of Agricultural Research, 44(3): 195-200.

Sahayaraj, K. and Ilayaraja, R. (2008). Ecology of *Dysdercus cingulatus* (Fab.). Egyptian Journal of Biology, 10: 122-125.

Sahayaraj, K. and Tomson, M. (2010). Impact of two pathogenic fungal crude metabolites on mortality, biology and enzymes of *Dysdercus cingulatus* (Fab.) (Hemiptera: Pyrrhocoridae). Journal of Biopesticides, 3(1 Special Issue): 163-167.

Sahu, K.R. (1999). Insect pest succession on linseed and management of bud fly, *Dasynema lini* Barnes and thrips, *Caliothrips indicus* (Bagnall), M.Sc. (Ag.) Thesis, Indira Gandhi Krishi Vishwavidyalaya, Raipur (India).

Sailer, R.I. (1953). A note on the bionomics of the green stink bug (Hemiptera: Pentatomidae). Journal of the Kansas Entomological Society, 26: 70-71.

Saito, O. (1999). Flight activity changes of the cotton bollworm, *Helicoverpa armigera* (Hubner) (Lepidoptera: Noctuidae), by aging and copulation as measured by flight actograph. Applied Entomology and Zoology, 35: 53-61.

Sajili, M.H., Din, B.N.M., Badaluddin, N.A., Suhaili, Z., Azizand, Z.F.A. and Jugah, K. (2017). Identification of *Coniella musaiaensis* as pathogen causing stem rot disease of *Hibiscus cannabinus* L. in Terengganu, Malaysia. World Applied Sciences Journal, 35 (8): 1342-1347.

Sakimura, K. (1962). The present status of thrips-borne viruses. pp.33-40. In: Maramorosch, K, ed. Biological Transmission of Disease Agents. Academic Press, New York, USA.

Sakimura, K. (1969). A comment on the color forms of *Frankliniella schultzei* (Thysanoptera: Thripidae) in relation to transmission of the tomato-spotted wilt virus. Pacafic Insects, 11: 761-762.

Salaheddin, K., Valluvaparidasan, V., Ladhalakshmi, D. and Velazhahan, R. (2010). Management of bacterial blight of cotton using a mixture of *Pseudomonas fluorescens and Bacillus subtilis.* Plant Prot Sci., 46: 41-50.

Salama, H.S., Salem, S.A., Zaki, F.N. and Abdel-Razek, A. (1999). The use of *Bacillus thuringiensis* to control *Agrotis ypsilon* and *Spodoptera exigua* on potato cultivation in Egypt. Archives of Phytopathology and Plant Protection, 32(5): 429-435.

Sharma, S.S. (2007). *Aenasius* sp. nov. effective parasitoid of mealy bug (*Phenacoccus solenopsis*) on okra. Haryana Journal of Horticultural Sciences, 36(3/4): 412.

Samrah, M., Rahaman, A., Phukan, A.K. and Gurusubramanian, G. (2006). Ovicidal, acaricidal and antifeedant activity of crude extracts of *Polygonum hydropipper* L. (Polygonaceae) against red spider mite and bunch caterpillar and its effect on *Stethorus gilviforns* Muslant. Uttarpradesh J Zool., 3(2): 127-135.

Samuels, G.J. (1996). Trichoderma: A review of biology and systematics of the genus. Mycological Research, 100(8): 923-935.

Samy, O. (1969). A revision of African species of *Oxycarenus* (Hemi. Lygaeidae). Trans. R. Ent. Soc. Lond., 121: 79-165.

Sanderson, F.R. (1963). An ecological study of pasmo disease (*Mycosphaerella linorum*) on linseed in Canterbury and Otago. New Zealand Journal of Agricultural Research, 6(5): 432-439.

Sanderson, F.R. (1965) Description and epidemiology of *Guignardia fulvida* sp. nov., The ascogenous state of *Aureobasidium pullulans* var. *lini* (Lafferty) Cooke, New Zealand Journal of Agricultural Research, 8(1): 131-141.

Sandru, I.D. (1972). Rezultate privind biologia si combatereamolieicinepei (*Grapholitha delineana* Walker-Lepidoptera-Tortricidae). Probleme Agricole, 24(4): 40-49.

Sangeetha, G., Anandan, A. and Usha Rani, S. (2012). Morphological and molecular characterisation of *Lasiodiplodia theobromae* from various banana cultivars causing crown rot disease in fruits. Archives of Phytopathology and Plant Protection, 45(4): 475-486.

Sangitrao, C.S., Moghe, P.G., Dahule, K.K., Shivankar, S.K. and Wangikar, P.D. (1993). Compendium of Grey Mildew of Cotton. Punjabrao Krishi Vidyapeeth, Akola, India, P.66.

Sankaran, T. (1974). Natural enemies introduced in recent years for biological control of agricultural pests in India. Indian Journal of Agricultural Sciences, 44(7): 425-433.

Santhakumari, P., Kavitha, K. and Nisha, M.S. (2002). Occurrence of collar rot in coral Hibiscus: a new record. Journal of Mycology and Plant Pathology, 32(2): 258-258.
Santos, A.F.dos, Luz, E.D.M.N. and Souza, J.T. (2005). *Phytophthora nicotianae*: agente etiologico da gomose da acacia-negra no Brasil. Fitopatologia Brasileira, 30: 81-84.
Santos, P.O., da Silva, A.C.M., Corrêa, E.B., Magalhães, V.C. and de Souza, J.T. (2014). Additional species of *Aspergillus* causing bole rot disease in *Agave sisalana*. Tropical Plant Pathology, 39(4): 331-334.
Sarejanni, J.A. (1936). La verticilliose du Coton en Grèce. Ann. Inst. Phytopath., Benaki, 2: 79-85.
Sardar, K.K. (1995). Kenaf mosaic. A new report. International Journal of Tropical Plant Diseases, 13(2): 245-247.
Sarkar, S.K. (2003). Investigations on seed mycoflora and seedling diseases of sunnhemp (*Crotalaria juncea* L) and their management. Ph.D. theses submitted to GBPUA&T, Pantnagar, India.
Sarkar, S.K. (2005a). Influence of crop age and weather parameters on vascular wilt of sunnhemp (*Crotalaria juncea* L.). New Agriculturist, 16 (1/2): 41-44.
Sarkar, S.K. (2005b). Diseases of sunnhemp- a review. Environment and Ecol., 23(4): 806-816.
Sarkar, S.K. (2010). Sunnhemp phyllody- an emerging threat in sunnhemp seed crop. In: Book of Extended Summaries / Abstract, National symposium on Emerging Trends in Pest management strategies under changing climatic scenario' 20-21 December, Orissa University of Agriculture and Technology, Orissa, India, pp. 6.
Sarkar, S.K., Hazra, S.K., Sen, H.S., Karmakar, P.G. and Tripathi, M.K. (2015). Sunnhemp in India. ICAR-Central Research Institute for Jute and Allied Fibres (ICAR), Barrackpore, West Bengal. pp.140.
Sarkar, S.K. and Tripathi, M.K. (2003). Summer crop of sunnhemp escape major pests and diseases. ICAR News, 9(4): 16.
Sarkar, S.K., Hazra, S.K., Sen, H.S., Karmakar, P.G. and Tripathi, M.K. (2015). Sunnhemp in India. ICAR-Central Research Institute for Jute and Allied Fibres (ICAR), Barrackpore, West Bengal. pp.140.
Sarkar, S.K. and Jana, A. (2017). Occurrence of leaf blight of Roselle (*Hibiscus sabdariffa* L.) in West Bengal. J. Mycopathol., 55 (3): 317-318.
Sarkar, S.K., Pradhan, S.K. and Tripathi, S.N (2000). Influence of boron, zinc and iron on the incidence of sunnhemp wilt (*Fusarium udum* f.sp. *crotalariae*). Journal of Mycology and Plant Pathology, 30(1): 116-118.
Sarma B.K. (1981). Diseases of Ramie (*Boehmeria nivea*). International Journal of Pest Management, 27(3): 370-374.
Sarma, B.K. (2009). Diseases of Ramie (*Boehmeria nivea*). Tropical Pest Management, 27(3): 370-374.
Sarr, M.P., Ndiaye, M'b, Groenewald, J.Z. and Pcrous, P.W. (2014). Genetic diversity in *Macrophomina phaseolina*, the causal agent of charcoal rot. Phytopathologia Mediterranea, 53(2): 250-268.
Saravanan, P., Karthikeyan, A., Padmanaban, B. and Nagasathiya, A. (2017). Weather factors responsible for the population dynamics of brown mirid bug, *Creontiades biseratense* distant (Hemiptera: Miridae) in Bt cotton. International Journal of Zoology and Applied Biosciences, 2(6): 370-376.
Sasaki A, Miyanishi M, Ozaki K, Onoue M, Yoshida K. (2005). Molecular characterization of a partitivirus from the plant pathogenic ascomycete *Rosellinia necatrix*. Arch Virol., 150(6): 1069-1083.
Sasser, J.N. and Carter, C.C. (1985). An advanced treatise on Meloidogyne. Vol. 1: Biology and control. North Carolina State University Graphics, P-422.

Sastry, K.S.M. and Vasudeva, R.S. (1963). Effect of host plant nutrition on the movement of sunnhemp mosaic virus. Indian Phytopath., 16: 143-150.

Sato, T., Iwamoto, Y., Tomioka, K. and Takaesu, K. (2002). Black band of Jew's marrow (*Corchorus olitorius*), trunk rot of satsuma mandarin (*Citrus unshiu*) and screw-pine (*Pandanus boninensis*) fruit-rot caused by *Lasiodiplodia theobromae* (abstract in Japanese). Jpn J Phytopathol., 68: 186.

Sato, T., Iwamoto, Y., Tomioka, K., Taba, S., Ooshiro, A. and Takaesu, K. (2008). Black band of Jew's marrow caused by *Lasiodiplodia theobromae*. J Gen Plant Pathol., 74: 91-93.

Satpathy S, Gotyal BS, Ramasubramanian T, Bhattacharyya SK, Laha SK (2009) Mealybug infestation in jute and mesta crop- a case study. In: National symposium on climate change, plant protection, and food security interface, West Bengal, Kalyani, 17–19 December 2009, P.78.

Satpathy, S., Gotyal, B.S., Ramasubramanian, T. and Selvaraj, K. (2013). Mealybug, *Phenacoccus solenopsis* Tinsley infestation on jute (*Corchorus olitorius*) and Mesta (*Hibiscus sabdariffa*). Insect Environment, 19(3): 187-188.

Sattar, M., Hamed, M. and Nadeem, S. (2007). Predatory potential of *Chrysoperla carnea* (Stephens) (Neuroptera: Chrysopidae) against cotton mealy bug. Pak. Entomol., 29(2): 103-106.

Saude, C., Melouk, H.A. and Chenault, K.D. (2004). Genetic Variability and mycelial compatibility groups of *Sclerotium rolfsii*. Phytopathology, 94: 592.

Saunders, K., Bedford, I.D., Briddon, R.W., Markham, P.G., Wong, S.M., and Stanley, J. (2000). A unique virus complex causes ageratum yellow vein disease. Proc. Natal. Acad. Sci. U S A, 97: 6890-6895.

Saunders, W.W. (1843). Description of a species of moth destructive to cotton crops in India. Transactions of the Entomological Society of London, 3: 284.

Sawada, K. (1916). A new stem rot of the jute plant caused by *Macrophomina corchori* sp. nov. Bull Agric. Exp .St. Formosa. 107. (vide; Mvcologia, 12: 82.).

Sawada, K. (1919). Descriptive catalogue of the Formosan fungi I: Report of the Department of Agriculture, Government Research Institute of Formosa, 19: 1-695.

Sawhney, K. and Nadkarny, NT. (1942). Results of an experiment to control cotton bollworms in Hyderabad State, 1937-40. Department of Agriculture Hyderabad Division, Bulletin No. 2: 40.

Saxena, G. and Mukerji, K. G. (2007).Management of Nematode and Insect-Borne Plant Diseases. New York: The Haworth Press.

Saxena, R.C. (1971). Bionomics of *C. indicus* (Bagnall) infesting Onion (Thripidae :Thysanoptera). Indian J.Ent., 33(3) : 342-345.

Schaefer, C.W. and Ahmad, I. (1987). Parasites and Predators of Pyrrhocoroidea Hemiptera, and Possible Control of Cotton Stainers By *Phonoctonus* Spp Hemiptera, Reduviidae. Entomophaga, 32(3): 269-275.

Schaefer, C.W. and Panizzi, A.R. (2000). Heteroptera of Economic Importance. CRC Press Inc., Boca Raton, Florida, pp.276-280.

Scheibelreiter, G.K. (1976). Investigations on fauna of *Papaver* spp. and *Cannabis sativa*. Annual report, Commonwealth Institute of Biological Control, European station, pp.34-35.

Scheifele, G. (1998). Final Report: Determining the feasibility and potential of field production of low THC industrial hemp (*Cannabis sativa* L.) for fibre and seed grain in northern Ontario. Kemptville College/University of Guelph, Thunder Bay, Ontario, Canada.<www. gov.on.ca:80/ OMAFRA.

Schieber, E., Sosa, O.N. and Escobar, P. (1961). Root knot nematode on kenaf in Guatemala. Plant Dis. Reptr., 45:119.

Schmidt, G.H., Ahmed, A.A.I. and Breuer, M. (1997). Effect of Melia azedarach extract on larval development and reproduction parameters of *Spodoptera littoralis* (Boisd.) and *Agrotis ipsilon* (Hufn.) (Lep. Noctuidae). Anzeiger fuumlaut r Schaumlaut dingskunde, Pflanzenschutz, Umweltschutz, 70(1): 4-12.

Schmidt, H.E. and Karl, E. (1970).Ein Beitragzur Analyse der Virosen des Hanfes under berucksichtigung der HanfplattlausalsVirus vektor. Zentralblatt Bakteriologle, Parasitenkunden, lnfecktionskrankheiten, Hygiene. Abt. 2, 125: 16-22.

Schmutterer, H. (1969). Pests of crops of Northeast and Central Africa.Gustav Fischer Verlag, Stuttgart & Portland.

Schnathorst, W.C. (1966). Eradication of *Xanthomonas malvacearum* from California through sanitation. Plant Disease Reporter, 50: 168-171.

Schneider, B., Cousin, M.T., Klinkong, S. and Seemuller, E. (1995). Taxonomic relatedness and phylogenetic positions of phytoplasmas associated with diseases of faba bean, sunnhemp, sesame, soybean and eggplant. Journal of Plant Diseases and Protection, 102(3): 225-232.

Schuh, R.T. (1984). Revision of the Phylinae (Hemiptera, Miridae) of the Indo-Pacific. Bulletin of the American Museum of Natural History, 177: 1-476.

Schumann, G.L. and D'Arcy, C.J. (2010). Essential Plant Pathology. 2nd Edition, The American Phytopathological Society Press, St. Paul.

Schuster, D.J. (2004). Squash as a trap crop to protect tomato from whitefly-vectored tomato yellow leaf curl. International Journal of Pest Management. 50 (4): 281.

Schwartz, M.D. and Foottit, R.G. (1992).Lygus bugs on the prairies. Biology, systematics, and distribution. Technical Bulletin - Agriculture Canada, No. 4E, 1-44.

Schwartz, M.D. and Foottit, R.G. (1998).Revision of the Nearctic species of the genus Lygus Hahn with a review of the Palaearctic species (Heteroptera: Miridae). Associated Publishers. Memoirs on Entomology, International Vol. 10, Gainesville, FL. P.428.

Schwartz, P.H. (1983). Losses in yield of cotton due to insects. Agriculture Handbook, USDA, 589: 329-358.

Schwencke, E.H. (1934). A new injury to sisal in East Africa. Tropenpflanzer, 37(8): 322-325.

Scott, J.G. (1995). Effects of temperature on insecticide toxicity, in Reviews in pesticide toxicology, Vol 3, ed by Roe, R.M. and Kuhr, R.J., Toxicology Communications, Raleigh, NC, USA, pp.111-135.

Scott, S.W. (2001). Tobacco streak virus. Descriptions of Plant Viruses. Clemson University, Department of Plant Pathology and Physiology.

Scott, W.P., Smith, J.W. and Snodgrass, G.L. (1986). Impact of early season use of selected insecticides on cotton arthropod populations and yield. Journal of Economic Entomology, 79(3): 797-804.

Scott, W.P., Snodgrass, G.L. and Smith, J.W. (1988). Tarnished plant bug (Hemiptera: Miridae) and predaceous arthropod populations in commercially produced selected nectariless cultivars of cotton. Journal of Entomological Science, 23(3): 280-286.

Seal, D.R., Kumar, V. and Kakkar, G. (2014). Common blossom thrips, *Frankliniella schultzei* (Thysanoptera: Thripidae) management and groundnut ring spot virus prevention on tomato and pepper in southern Florida. Florida Entomologist, 97(2): 374-383.

Secchi, V.A. (2001). Situation of soil pests in Rio Grande do Sul: a view of rural extension. Documentos Embrapa Soja, 172: 17-41.

Seifert, K.A., Gams, W., Corus, P.W. and Samuels, G.J. (2000). Molecules, morphology and classification: Towards monophyletic genera in Ascomycetes. Studies in Mycology, 45: 1-4

Seixas, C.D.S., Barreto, R.W. and Killgore, E. (2007). Fungal pathogens of *Miconia calvescens* (Melastomataceae) from Brazil, with reference to classical biological control. Mycologia, 99(1): 99-111.

Sekhon, B S. and Varma, G. C. (1983). Parasitoids of *Pectinophora gossypiella* [Lep.:Gelechiidae] and Earias spp. [Lep.: Noctuidae] in the Punjab. Entomophaga., 28: 45-53.

Sekhon, S.S., Sajjan, S.S. and Kanta, U. (1979). A note on new host plants of green peach aphid, *Myzus persicae* from Punjab and Himachal Pradesh. Indian J. Plant Protection, 7: 106.

Sellers, W.F. (1951). The limitations of biological control of the sisal weevil. East Africa Agricultural and Forestry Journal, 16(4): 175-177.

Selvaraj, K., Gotyal, B.S., Gawande, S.P., Satpathy, S. and Sarkar, S.K. (2016). Arthopod biodiversity on jute and allied fibre crops, pp.195-222. In Eds, Chakravarthy, A.K. and Sridhara, S. Economic and Ecological Significance of Arthopods in Diversified Ecosystems, Springer, Singapore.

Selvi, A.K., Delen, N., Gencer, R. and Teksür, P.K. (2016). Sensivity against some fungicides and molecular characterization of *Botrytis cinerea* isolates on grapes in Aegean region.

ISHS Acta Horticulturae 1144: III International Symposium on Postharvest Pathology: Using Science to Increase Food Availability. DOI: 10.17660/ActaHortic.2016.1144.41

Senapati, S. K.and Ghose, S. K. (1991). Biology and morphometrical studies of larvae of jute semilooper, *Anomis sabulifera* (Guenther), (Noctuidae: Lepidoptera). Annals of Entomology, 9(1): 35-39

Seney, H. (ed). (1984). Integrated pest management for cotton in Western region of the United States. The regents of the university California, California.

Sengonca, C., Griesbach, M. and Lochte, C. (1995). Suitable predator-prey ratios for the use of *Chrysoperlacarnea* (Stephens) eggs against aphids on sugar beet under laboratory and field conditions. *Zeitschriftfür Pflanzenkrankheiten und Pflanzenschutz*, 102: 113-120.

Severin, V. (1978). Einneues pathogens Bakteriuman Hanf—*Xanthomonas campestris* pathovar. *cannabis* (in German with English summary). Arch Phytopathol Pflanzenshutz, 14: 7-15.

Sewak, N. (2016). Biology of spotted bollworm, *Earias vittella* (Fab.) on okra. Journal of Global Research Computer Science & Technology, 4(2): 26-30.

Sewify G.H. and Semeada A.M. (1993). Effect of population density of the cotton seed bug, *O. hyalinipennis* (Costa) on yield and oil content of cotton seeds. Bulletin of Faculty of Agriculture, University of Cairo, 44: 445-452.

Shah, D. R. and Shukla, A. (2014). Chemical control of two spotted spider mite, *Tetranychus urticae* (Koch) (Acari: Tetranychidae) infesting gerbera. Pest Management in Horticultural Ecosystems, 20(2): 155-161.

Shah, S.I.A., Jan, M.T., Rafiq, M., Malik, T.H., Khan, I.R. and Hussain, Z. (2016). Efficacy of different groups of insecticides against dusky cotton bug, *Oxycarenus laetus* Kirby in field conditions of Pakistan. Journal of Agricultural and Research, 2(3): 1-17.

Shaima, P.D. (1985). Potential of phylloplane microorganisms for biological control of rust diseases. Front. Appl. Microbiol., 1: 201-231.

Shamsi, S. and Naher, N. (2014). Boll rot of cotton (*Gossypium hirsutum* L.) caused by *Rhizopus oryzae* Went & Prins. Geerl- a new record in Bangladesh. Bangladesh J. Agril. Res., 39(3): 547-551.

Shands, W.A., Hall, I.M. and Simpson, G.W. (1962). Entomophthoraceous fungi attacking the potato aphid in northeastern Maine in 1960. J. Econ. Entomol., 55: 174–179.

Shanker, C., Mohan, M., Sampathkumar, M., Lydia, Ch. and Katti, G. (2013). Functional significance of *Micraspis discolor* (F.) (Coccinellidae: Coleoptera) in rice ecosystem. Journal of Applied Entomology, 137(8): 601-609.

Shanmugapriya, V., Muralidharan, C.M. and Karthick, K. (2017). Biology and Bionomics of Zig Zag Beetle *Cheilomenes sexaculatus* Fabricus (Coleoptera: Coccinellidae). Int.J.Curr. Microbiol. App.Sci., 6(3): 541-548.

Sharma, A. and Pati, P.K. (2012). First record of the carmine spider mite, *Tetranychus urticae*, infesting *Withania somnifera* in India. Journal of Insect Science, 12(50): 50.

Sharma, A.K., Gawande, S.P., Karmakar, P.G. and Satpathy, S. (2014). Genetic Resource Management of Ramie (*Boehmeria* sp.): A Bast Fibre Crop of North Eastern India. Vegetos, 27(2): 279-286.

Sharma, H.C. and Agarwal, R.A. (1984). Factors imparting resistance to stem damage by *Earias vittella* F. (Lepidoptera: Noctuidae) in some cotton phenotypes. Protection Ecology, 6(1): 35-42.

Sharma, K.K., Sharma, J. and Jadhav, V.T. (2010). Etiology of Pomegranate wilt and its management. Fruit, vegetable and cereal science and biotechnology, 4 (special issue 2): 96-101.

Sharma, L.C. and Mathur, R.I. (1971). Variability in first single spore isolates of *Fusarium oxysporum* f.sp. *lini* in Rajasthan. Indian Phytopathol., 24: 698-704.

Sharma, M., Ghosh, R., Krishnan, R.R., Nagamangala, U.N., Chamarthi, S., Varshney, R. and Pande, S. (2012). Molecular and morphological diversity in *Rhizoctonia bataticola* isolates causing dry root rot of chickpea (*Cicer arietinum* L.) in India. African Journal of Biotechnology, 11(37): 8948-8959.

Sharma, M. and Kulshrestha, S. (2015). *Colletotrichum gloeosporioides*: An anthracnose causing pathogen of fruits and vegetables. Biosci. Biotechnol. Res. Asia, 12(2): 1233-1246.

Sharma, O.P. (1990). "Witch's Broom" disease of sunnhemp (*Crotalaria juncea* L.) by MLO. Indian J. Mycol. Pl. Pathol. 20(3): 264-65.

Sharma, R.K.(2007). Molecular characterization of *Sclerotium rolfsii* Sacc. PhD Thesis, Department of Plant Pathology College of Agriculture, Junagadh Agricultural University, Junagadh

Sharma, S., Tara, J. S., Kour, R. & Ramamurthy, V. V. (2012). Bionomics of *Alcidodes affaber* Aurivillius (Coleoptera: Curculionidae: Alcidodinae), a serious pest of Bhendi, *Abelmoschus esculentus* (L.) Moench. Munis Entomology & Zoology, 7(1): 259-266.

Sharma, S.R. and Varma, A. (1975). Three sap transmissible viruses from cowpea in India. Indian Phytopath., 23: 192-198.

Sharman, M. (2009). Distribution in Australia and seed transmission of tobacco streak virus in Parthenium hysterophorus. Plant Disease, 93(7): 708-712.

Sharman, M. (2011).Tobacco streak virus in cotton-scoping study. Queensland DEEDI, Series/ Report: DAQ0002, no.:03DAQ005

Shaw, D.E. (1984). Microorganisms in Papua New Guinea. Dept. Primary Ind., Res Bull., 33: 1344.

Shaw, F.L.F. (1912). The morphology and parasitism of *Rhizoctonia*. Mem. Dep. Agric. India (Bot. Ser.), 4: 115-153.

Shaw, F.J.F. (1921). Studies in the diseases of the Jute plant (I) *Diplodia corchori* Syd. Mem. Dep. Agric. India (Bot. Ser.), 11: 37-58.

Shaw, F.L.F. (1924). Studies on the diseases of jute plant (2) *Macrophoma corchori* Saw. Mem. Dep. Agric. India (Bot. Ser.), 13: 193-199.

Sheikh, A.H. and Ghaffar, A. (1979). Relation of sclerotial inoculum density and soil moisture to infection of field crops by *Macrophomina phaseolina*. Pak. J. Bot., 11: 185-189.

Shelton, A.M. and North, R.C. (1987). Injury and control of onion thrips (Thysanoptera: Thripidae) on edible podded peas. J. Econ. Entomol., 80: 1325-1330.

Sheng, C.F. (1988). Economic threshold of the third generation of *Heliothis armigera* in north China. Acta Entomologica Sinica, 31(1): 37-41.

Sheppard, J.W. (2005). Malt agar method for the detection of *Alternaria linicola* on *Linum usitatissimum*, International Seed Testing Association (ISTA), Bassersdorf, Switzerland, 7-17.

Shepherd, K.W. (1963). Studies of the genetics of host-pathogen interactions, with flax and its rust. PhD thesis, The University of Adelaide.

Shepherd, R.L. (1982). Registration of three germplasm lines of cotton (Reg. Nos.GP 164 to GP 166). Crop Science, 22: 692-693.

Shepherd, R.L., Jenkins, J.N., Parrott, W.L. and McCarty, J.C. (1986). Registration of eight nectariless-frego bract cotton germplasm lines. Crop Science, 26(6): 1260.

Sher, S.A. and Allen, M.W. (1953). Revision of the genus *Pratylenchus* (Nematoda: Tylenchidae). University of California Publications in Zoology, 57(6): 441-470.

Sherbakoff, C. (1949). Breeding for resistance to Fusarium and Verticillium wilts. The Botanical review, 15(6): 395-399.

Shetty, P.S., Sunkad, G., Naik, M.K. and Amaresh, Y.S. (2011). Sources of field resistance to rust of cotton caused by *Phakopsora gossypii*. Karnataka J. Agric. Sci., 24(3): 406-407.

Shimizu, K. and Fujisaki, K. (2002). Sexual differences in diapause induction of the cotton bollworm, *Helicoverpa armigera* (Hb.) (Lepidoptera: Noctuidae). Applied Entomology and Zoology, 37: 527-533.

Shivanathan, P., Shikata, E., Lizhi Yuan. and Maramorosch, K. (1983). Mycoplasmal disease of *Crotalaria juncea* (sunnhemp) in Sri Lanka. International Journal of Tropical Plant Diseases, 1: 83-85.

Shivankar, S.K. and Wangikar, P.D. (1992). Estimation of crop losses due to grey mildew disease of cotton caused by *Ramularia areola*. Indian Phytopath., 45: 74-76.

Short, G.E., Wyllie, T.D. and Ammon, V.D. (1978). Quantitave enumeration of *Macrophomina phaseolina* in soybean tissues. Phytopathology, 68: 736-741.

Short, G.E., Wyllie, T.D. and Bristow, P.R. (1980). Survival of *Macrophomina phaseolina* in soil and residue of soybean. Phytopathol., 70: 13-17.

Showers, W. B. (1997). Migratory Ecology of the Black Cutworm. Annu. Rev. Entomol., 42: 393-425.

Showers, W.B., Keaster, A.J., Raulston, J.R., Hendrix III W.H., Derrick, M.F., Mccorcle, M.D., Robinson, J.F., Way, M.O., Wallendrof, M.J. and Goodenough, J.L. (1993). Mechanism of southward migration of noctuid moth [*Agrotis ipsilon* (Hufnagel)]: a complete migrant. Ecology, 74: 2303-2314.

Shrivastava, N., Katiya, O.P., Shrivastava, S. and Das, S.B. (1994). Population dynamics of linseed Budfly *Dasyneura lini* Barnes (Diptera: Cecidomyiidae) on linseed (*Linum usitatissimum*). J. Oilseeds Res., 11: 160-164.

Shukla, A.K. (1992). Assessment of yield losses by various levels of rust infection linseed. Plant Dis. Res., 7: 157-160.

Shukla, G.S. and Upadhyay, V.B. (2006). Economic Zoology. Rastogi Publications, Meerut (India), P.496.

Shutova, N.N. and Strygina, S.P. (1969). The hemp moth. Zashchita Rastenif, 14(11): 49-50.

Shyadaguppi, M.V. (2011). Functional dimensions of mirid bug, *Creontiades biseratense* (Distant) (Miridae: Hemiptera) population dynamics in Bt cotton. M.Sc. Thesis. Department of Agricultural Entomology College of Agriculture, University of Agricultural Sciences, Dharwad.

Siddig, M.A. (1973). Barakat, a new long staple variety in the Sudan. Cotton Growing Review, 50: 307-315.

Siddiqui, K.A., Gupta, A.K., Paul, A.K. and Banerjee, A.K. (1979). Purification and properties of heat-resistant exotoxin produced by *Macrophomina phaseolina* (Tassi) Goid in culture. Experientia, 35: 1222–1223.

Siddiqi, M.R. (1972). *Pratylenchus coffeae* CIH. Descriptions of Plant-Parasitic Nematodes Set 1, 6. Wallingford, U.K.: CAB International, pp.3.

Sidhu, A.S.and Dhawan, A.K., (1979). Incidence of cotton leaf-roller (*Sylepta derogata* F.) on different varieties of cotton and its chemical control. Entomon, 4: 45-50.

Sidhu, A.S. and Sandhu, S.S. (1977). Damage due to the spotted bollworm (*Earias vittella* Fabr.) in relation to the age of bolls of hirsutum variety J 34. Journal of Research, Punjab Agricultural University, 14(2): 184-187.

Silantyev, A. (1897). Results of Investigations of the Hemp (Hop, Flax) and Beet Flea. Ministry Agriculture and Government Estates, St. Petersburg. pp.7.
Silfverberg, H. (2004). Enumeratio nova Coleopterorum Fennoscandiae, Daniaeet Baltiae. Sahlbergia, 9: 1-111.
Sill, W.H.Jr., King, C. L. and Hansing, E.D. (1954). Oats blue dwarf and red leaf in Kansas. Plant Dis. Reptr. 38: 695.
Silva, G.S. (2001). *Phytophthora nicotianae*, um novo patógeno da vinagreira (*Hibiscus sabdariffa*) no Brasil. Fitopatologia Brasileira, 26 (suplemento): 320.
Silva, G.S., Rêgo, A.S. and Leite, R.R. (2014). Doenças da vinagreira no Estado do Maranhão. Summa Phytopathologica, 40(4): 378-380.
Silva, M., Barreto, R.W., Pereira, O.L., Freitas, N.M., Groenewald, J.Z. and Crous, P.W. (2016). Exploring fungal mega-diversity: *Pseudocercospora* from Brazil. Persoonia, 37: 142-172.
Silva, N.M., Carvalho, L.H., Cia, E., Fuzatto, M.G., Chiavegato, E.J. and Alleoni, L.R.F. (1995). Seja o doutor do seualgodoeiro. Arquivo do Agrônomo, Piracicaba, SP: Potafos, 8: 17.
Silva, O.R.R.F.; Coutinho, W.M.; Cartaxo, W.V.; Sofiatti, V.; Silva Filho, J.L.; Carvalho, O.S.; Costa, L.B. (2008). Cultivo do sisal no nordeste brasileiro, Circular Tecnica 123. Campina Grande: Embrapa Algodão, 2008. 24p. http://ainfo.cnptia.embrapa.br/digital/bitstream/ CNPA-2009-09/22318/1/ CIRTEC123.pdf.
Silva, W.P.K., Deverall, B.J. and Lyon, B.R. (1998). Molecular, physiological and pathological characterization of *Corynespora* leaf spot from rubber plantations in Sri Lanka. Plant Pathol., 47: 267-277.
Silver, S., Quan, S. and Deom, C.M. (1996). Completion of the nucleotide sequence of sunn-hemp masaic virus: a tabamovirus pathogenic to legumes. Virus Genes, 13(1): 83-85.
Simmons, A.L., Dennehy, T.J., Tabashnik, B.E., Antilla, L., Bartlett, A., Gouge, D. and Staten, R. (1998). Evaluation of B.T. cotton deployment strategies and efficacy against pink bollworm in Arizona. Proceedings Beltwide Cotton Conferences, San Diego, California, USA, 2: 1025-1030.
Simmons, A.M. and Yeargan, K.V. (1988). Development and survivorship of the green stink bug, *Acrosternum hilare* (Hemiptera: Pentatomidae) on soybean. Environmental Entomology, 17: 527-532.
Simmons, E.G. (2007). Alternaria: an identification manual. CBS Biodiversity Series, 6. CBS Fungal Biodiversity Centre, Utrecht, The Netherland. P.775.
Simwat, G.S., Dhawan, A.K. and Sidhu, A.S. (1988). Effect of constant temperature and relative humidity on the termination of larval diapause in pink bollworm. Journal of Insect Science, 1(2): 133-135.
Singh, B. and Das, L.K. (1979). Semilooper (*Anomis sabulifera* Guen.) escalating on jute pods. Sci & Cult., 45(3): 121-123.
Singh, B. and Ghosh, T. (1981). Problems of root knot nematode in jute. Jute Dev. J., 1(1): 16-19.
Soltani, T., Nejad, R.F., Ahmadi, A.R. and Fayazi, F. (2013). Chemical control of Root-Knot Nematode (*Meloidogyne javanica*) on olive in the greenhouse conditions. J Plant Pathol Microb., 4: 183.
Singh, B.N. (1977). Development of *Utetheisa pulchella* Linn. on *Crotolaria juncea* and *Heliotropium indicum*. Indian Journal of Entomology, 38 (4): 398-400.
Singh, D., Siddiqui, M.S. and Sharma, S. (1989). Reproductive retardant and fumigant properties in essential oils against rice weevil (Coleoptera: Curculionidae) in stored wheat. Journal of Economic Entomology, 82: 727-733.
Singh, D.P. (1987). White cane rot- a new disease of ramie. Current Science, 56 (7): 312-313.
Singh, D.P. (1998). Ramie (*Boehmeria nivea*). Central Research Institute for Jute & Allied Fibres. Barrackpore, 24-Parganas (N), West Bengal, India.

Singh, D.P. (2010). Mesta (*Hibiscus Cannabinus* & *Hibiscus sabdariffa*). Technical bulletin, Central Research Institute for Jute & Allied Fibres, Barrackpore, West Bengal, India. Website: assamagribusiness.nic.in/mesta.pdf

Singh, H., Gupta, T. R, Singh, B. and Chhina, J. S. (2008). Chemical control of linseed gall midge (*Dasyneura lini* Barnes) in linseed (*Linum usitatissimum* L.). Agricultural Science Digest, 28(3): 210-212.

Singh, J., Singh, T.H., Kaley, H.S. and Singh, K. (1974). Influence of cotton plant morphology on bollworms incidence. Cotton Development, 4(2): 15-19.

Singh, K. (2007). Rhythmicity in life events of an aphidophagous ladybird beetle, *Cheilomenes sexmaculata*. Journal of Applied Entomology. 131(2): 85-89.

Singh, M. (2014). Status of Raw Jute (Jute and Mesta) in India, Directorate of Jute Development, Kolkata. P.46.

Singh, M.P. and Ghosh, S.N. (1970). Studies on *Maconellicoccus hirsutus* causing 'bunchy top' in mesta. Indian Journal of Science and Industry, 4: 99-105.

Singh, R.K., Dubey, S.R. and Srivastava, R.K. (2013). Status of kenaf (*Hibiscus cannabinus* L) diseases in the districts of north eastern plain zone of Uttar Pradesh. GJ B.A.H.S., 2(1): 72-73.

Singh, S., Rana, R.S., Sharma, K.C., Sharma, A. and Kumar, A. (2017). Chemical control of two spotted spider mite *Tetranychus urticae* (Acari: Tetranychidae) on rose under polyhouse conditions. Journal of Entomology and Zoology Studies, 5(6): 104-107

Singh, S.R., Jackai, L.E.N., Dos Santos, J.H.R. and Adalla, C.B. (1990). Insect pests of Cowpea. In: Singh S.R., editor. Insect pests of tropical food legumes. Wiley, Chichester, pp.43-89.

Singh, S.P. and Pandey, N.D. (1980). Chemical control of linseed gall midge, *Dasyneura lini* Barnes, attacking linseed. Indian J. Entomol., 42: 786-787.

Singh, U.C., Dhamdhere, S.V., Misra, U.S. and Bhadauria, N.S. (1991).Seasonal incidence and toxicity of some newer insecticides against *Myllocerus undecimpustulatus*. Indian Journal of Agricultural Research, 25(3): 119-124.

Singh, V.V., Narayanan, S.S. and Basu, A.K. (1993). 'Kjrti' hybrid cotton set to win over farmers in Maharashtra. Indian Farming, 43(1): 15-16.

Singh, Y. and Bichoo, A.S.L. (1989). Some biological and bionomical observation on *Earias fabia* (Stoll). Bull. Ent., New Delhi, 30: 84-91.

Singh, Y.P., Meghwal, H.P. and Singh, S.P. (2008). Biology and feeding potential of *Cheilomenes sexmaculata* Fabricius (Coleoptera: Coccinellidae) on mustard aphid. Annals of Arid Zone, 47(2): 185-190.

Singha, D., Singha, B. and Dutta, B. (2010). In vitro pathogenicity of *Bacillus thuringiensis* against the tea termites of Barak valley (Assam), India. Journal of Biological Contol., 24. 279-281.

Sinha, R.B.P. and Marwaha, K.K. (1995). Persistent and residual toxicity of certain insecticides against the adult of grey weevil, *Myllocerus undecimpustulatus* var. *maculosus*. Journal of Applied Biology, 5(1-2): 53-57.

SIRATAC, (1982-1986). SIRATAC Manual User Guides 1982/83-1986/87. SIRATAC Limited, Wee Waa, New South Wales.

Sivanesan, A. (1987). Graminicolous species of Bipolaris, Curvularia, Drechslera, Exserohilum and their teleomorphs. Mycological Papers, 158: 113-114

Sivanesan, A. and Holiday, P. (1972). *Rosellinia necatrix*. CMI Descriptions of pathogenic fungi and Bacteria, 352: 1-2.

Sivanesan, A. and Holliday, P. (1981). *Mycosphaerella linicola*, CMI Description of Pathogenic Fungi and Bacteria. Commonwealth Mycological Institute Kew, Surrey, England. 18: pp.2.

Skoracka, A., Smith, L., Oldfield, G., Cristofaro, M. & Amrine, J.W.Jr. (2010). Host-plant specificity and specialization in eriophyoid mites and their importance for the use of eriophyoid mites as biocontrol agents of weeds. Experimental and Applied Acarology, 51: 93-113.

Slater, J.A. (1972). The Oxycareninae of South Africa. Occ. Papers Univ. Conn. Biol. Sci. Ser., 2(7): 59-103.

Slosser, J.E., Pinchak, W.E. and Rummel, D.R. (1989). A review of known and potential factors affecting the population dynamics of the cotton aphid. Southwestern Entomologist, 14(3): 302-313.

Small, E., Marcus, D., Butler, G. and McElroy, A.R. (2007). Apparent Increase in Biomass and Seed Productivity in Hemp (*Cannabis sativa*) resulting from branch proliferation caused by the European Corn Borer (*Ostrinia nubilalis*). Journal of Industrial Hemp, 12(1): 15-26.

Smith, A.L. and Dick, J.B. (1960). Inheritance of resistance to Fusarium wilt in upland and sea Island cottons as complicated by nematodes under field conditions. Phytopathology, 50: 44-48.

Smith, F.F., Boswell, A.L. and Webb, R.E. (1969). Segregation between strains of carmine and green two-spotted spider mites. Proc Int Congr Acarol. 2nd, Sutton Bonington 1967, pp.155-159.

Smith, T.R. and Brambila, J. (2008). A major pest of cotton, *Oxycarenus hyalinipennis* (Heteroptera: Oxycarenidae) in the Bahamas. Florida Entomologist, 91(3): 479-482.

Smith et al. (1986). Reference cited in "Anthracnose of cotton (*Colletotrichum gossypii*) – Plantwise". https://www.plantwise.org/Knowledge Bank/Datasheet.aspx?dsid=25358

Smith, D., Chilvers, M., Dorrance, A., Hughes, T., Mueller, D., Niblack, T. and Wise, K. (2014). Charcoal Rot Management in the North Central Region. UW-Extension office & Cooperative Extension Publishing (website)

Smith, G.E. and Haney, A. (1973). *Grapholitha tristrigana* (Lepidoptera: Torttricidae) on naturalized hemp (*Cannabis sativa* L.) in east-central Illinois. Transactions Illinois State Academy Science, 66: 38-41.

Smith, G.S. and Wyllie, T.D. (1999). Charcoal rot. pp. 29-31 in: Compendium of Soybean Diseases. G. L. Hartman, J. B. Sinclair, and J. C. Rupe, eds. American Phytopathological Society, St. Paul, MN

Smith, I.M., Dunez, J., Phillips, D.H., Lelliott, R.A. and Archer, S.A. (1988). European Handbook of Plant Diseases. Blackwell Scientific Publications, Oxford, P.583.

Smith, J.F. 2010. Early-season management of two spotted spider mite on cotton and impacts of infestation timing on cotton yield loss. Ph.D. Dissertation, Mississippi State Univeristy, Starkville, MS.

Smith, L.J., Lehane, J., Kirkby, K.A., Lonergan, P.A., Cooper, B.R. and Allen, S.J. (2011-2012). Cotton Pathology, Website: https://www.csd.net.au/system/diseases/files/000/.../ Cotton_Pathology_2011_to_2012.pdf

Smith, S.N. and Snyder, W.C. (1974). Persistence of *Fusarium oxysporum* f. sp. *vasinfectum* in fields in the absence of cotton. Phytopathology, 65: 190-196.

Smith, S.N. and Snyder, W.C. (1975). Persistence of *Fusarium oxysporum* f. sp. *vasinfectum* in fields in the absence of cotton. Phytopathology, 65(2): 190-196.

Snodgrass, G.L., Adamczyk, J.J. Jr. and Gore, J. (2005). Toxicity of insecticides in a glass- vial bioassay to adult brown, green, and southern green stink bugs (Heteroptera: Pentatomidae). Journal of Economic Entomology, 98: 177-181.

Snodgrass, G.L. and Elzen, G.W. (1995). Insecticide resistance in a tarnished plant bug population in cotton in the Mississippi Delta. Southwestern Entomologist, 20(3): 317-232.

Snowden, J.D. (1921). *Report of the Govt. Botanist* for the period 1st April to 31st December, 1920. Am. Rept. Dept. Agric. Uganda, pp. 43-46.

Soares, D.J., Barreto, R.W. and Braun, U. (2009). Brazilian mycobiota of the aquatic weed *Sagittaria montevidensis.* Mycologia, 101(3): 401-416.
Sokolov, A.M. and Bezrukova, V.F. (1939). The injuriousness of the flax flea. (In Russian). Plant Protection, 18: 150-154.
Solomon, J.J. and Sulochana, C.B. 1973. Sunnhemp phyllody - A virus disease. Phytopathologische Zeitschrift. 78 (1): 62-68.
Song, Z-Q., Cheng, J-E., Cheng, F-X., Zhang, D-Y. and Liu, Y. (2017). Development and Evaluation of Loop-Mediated Isothermal Amplification Assay for Rapid Detection of *Tylenchulus semipenetrans* Using DNA Extracted from Soil. Plant Pathol. J., 33(2): 184-192.
Sood, N.K. and Pathak, S.C. (1990). Thickness of sepals- a factor for resistance in linseed against *Dasyneura lini* Barnes. Indian J. Entomol., 52: 28-30.
Sorauer, P. (1958). Handbunch der Pflanzenkrankheiten (Band 5). 26 Volumes. Paul Parey, Berlin.
Spaar, D., Kleinhempel, H. and Fritzsche, R. (1990). Ol-Und Fasrerpflanzen. Springer-Verlag, Berlin., P.248.
Sparrow, F.K. (1960). Aquatic phycomycetes. 2nd ed. Ann. Arbor, Michigan: University of Michigan Press.
Spears, J.F. (1968). The westward movement of the pink bollworm. Bull. Entomol. Soc. Am., 14: 118-119.
Spielmeyer, W., Lagudah, E., Mendham, N. and Green, A. (1998b). Inheritance of resistance to flax wilt (*Fusarium oxysporum* f.sp. *lini* Schlecht)in a doubled haploid population of *Linum usitatissimum* L. Euphytica, 101: 287-291.
Sprenkel, R.K., (2000). Cotton plant and pest monitoring manual for Florida, Florida. http://ifas.ufl.edu/NFREC
Springer, Y.P. (2006). Epidemiology, resistance structure, and the effects of soil calcium on a serpentine plant–pathogen interaction. Doctoral dissertation, University of California Santa Cruz.
Sreenivasaprasad, S., Mills, P.R., Meehan, B.M. and Brown, A.E. (1996). Phylogeny and systematics of 18 *Colletotrichum* species based on ribosomal DNA spacer sequences. Genome, 39: 499-512.
Sridhar, J., Naik, V.C.B., Ghodke, A., Kranthi, S., Kranthi, K.R., Singh, B.P., Choudhary, J.S. and Krishna, M.S.R. (2017). Population genetic structure of cotton pink bollworm, *Pectinophora gossypiella* (Saunders) (Lepidoptera: Gelechiidae) using mitochondrial cytochrome oxidase I (COI) gene sequences from India. Mitochondrial DNA Part A DNA Mapping, Sequencing, and Analysis , 28(6): 941-948.
Srinivasan, K.V. (1994). Cotton diseases. Indian Society for Cotton Improvement. Central Institute for Cotton Research, Mumbai, pp.157-311. Published by Indian Society for Cotton Improvement. Pearl Lino Service, Thane, P.312.
Shrinivasan, K.V. and Kannan, A. (1974). Myrothecium and Alternaria leaf spots of cotton in South India.Curr. Sci., 43: 489-490.
Srivastava, A.S. (1964). Pests of sunnhemp. Entomological Research during the last ten years in Uttar Pradesh. Research Memoir, 3: 29-32.
Srivastava, M.P. (1980). Occurrence of Myrothecium leaf spot of mung bean in Haryana (*Vigna mungo* L.; India). Indian Phytopathology, 33(1): 137.
Srivastava, M.P. and Singh, A. (1973). Myrothecium disease of cotton. Journal of Research - Haryana Agricultural University, Haryana, 3: 221-223.
Srivastava, N. and Verma, H.N. (1995). *Chenopodium murale* extract, a potent virus inhibitor. Indian Phytopath. 48(2): 177-179.
Srivastava, R.S., Singh, R. K., Kumar, N. and Singh S. (2010). Management of Macrophomina disease complex in jute (*Corchorus olitorius*) by *Trichoderma viride*. Journal of Biological Control, 24(1): 77–79.

Stam, P.A. and Elmosa, H. (1990). The role of predators and parasites in controlling populations of *Earias insulana*, *Heliothis armigera* and *Bemisia tabaci* on cotton in the Syrian Arab Republic. Bio Control, 35(3): 315-327.

Stary, P., Gerding, M., Norambuena, H. and Remaudiere, G. (1993). Environmental-research on aphid parasitoid biocontrol agents in Chile (Hym, Aphidiidae, Hom, Aphidoidea). Journal of Applied Entomology, 115(3): 292-306.

Statler, G.D. (1983). Pycnial morphology and mating in *Melampsora lini*. Canadian Journal of Plant Pathology, 5(2): 97-100.

Statler, G.D., Hammond, I.J. and Zimmer, D.E. (1981). Hybridization of *Melampsora lini* to identify rust resistance in flax. Crop Sci., 21: 219-221.

Steimel, J., Engelbrecht, C.J.B. and Harrington, T.C. (2004). Development and characterization of microsatellite markers for the fungus *Ceratocystis fimbriata*. Mol. Ecol. Notes, 4: 215-218.

Steinkraus, D., Zawislak, J., Lorenz, G., Layton, B. and Leonard, R. (ND). Spider Mites on Cotton in the Midsouth, University of Arkansas, Fayetteville, Ark. Website: Spider Mites on Cotton in the Midsouth - Cotton Incorporated. Website: https://www.cottoninc.com/fiber/.../Spider-Mites/SpiderMitesCottonMidsouth.pdf

Stephens, P. (2003). Control of white root rot (Rosellinia necatrix) in apple orchard. Horticultural Australia Ltd., Sydney, P-32.

Stevens, M. (1975). How to Grow Marijuana Indoors Under Lights, 3rd Ed. Sun Magic, Seattle. P.73.

Stoetzel, M. (1994). Aphids (Homoptera: Aphididae) of potential importance on citrus in the United States with illustrated keys to species. Proceedings of the Entomological Society of Washington, 96(1): 74-90.

Story, R.N. and Keaster, A.J. (1982). Development and evaluation of a larval sampling black cutworm (Lepidoptera: Noctuidae). Journal of Economic Entomology, 75(4): 604-610.

Straub, R.W., Weires, R.W. and Eckenrode, C.J. (1986). Damage to apple cultivars by races of European corn borer (Lepidoptera: Pyralidae). J. Econ. Entomol., 79: 359-363.

Strausbaugh, C.A., Wintermantel, W.M., Gillen, A.M. and Eujay, I.A. (2008). Curly top survey in the Western United States. Phytopathology, 98(11): 1212-1217.

Streets, R.B. (1937). Phymatotrichum (cotton or Texas) root rot in Arizona. Technical Bulletin, Arizona University College of Agriculture, No.71.

Streets, R.B. and Bloss, H.E. (1973). Phymatotrichum Root Rot. APS Monograph No. 8. St Paul, MN, USA: APS Press, P.38.

Strickland, E.H. and Criddle, N. (1920).The Beet Webworm (*Loxostege sticticalis* L.). Canada Dept. Agric., Crop Protection Leaflet, 12: pp2.

Struble, D.L. (1981a). A four-component pheromone blend for optimum attraction of redbacked cutworm males, *Euxoa ochrogaster* (Guence). J. Chem.Ecol., 7: 615-625.

Struble, D.L. and Jacobson, L.A. (1970). A sex pheromone in the redbacked cutworm. Journal of Economic Entomology, 63: 841-844.

Su, G., Suh, S.O., Schneider, R.W. and Russin, J.S. (2001). Host specialisation in the charcoal rot fungus, *Macrophomina phaseolina*. Phytopathology, 91: 120-126.

Suassuna, N.D., Chitarra, L.G., Asmus, G.L. and Inomoto, M.M. (2006). Manejo de doencas do algodoeiro. Circular Tecnica 97.Embrapa Algodao, Campina Grande, PB, Brazil.

Suassuna, N.D. and Coutinho, W.M. (2007). Manejo das principaisdoenças do algodoeiro no Cerrado Brasileiro. In: Freire, E. C. (Ed.) Algodao no Cerrado do Brasil. Brasilia, ABRAPA, Vol. 1, P.980.

Subramanian, T.R. (1959). Biology of *Alcidodes affaber* Aurivillus. Indian Journal of Agri. Sci., 29(4): 81-89.

Sugasawa, J., Kitashima, Y. and Gotoh, T. (2002). Hybrid affinities between the green and the red forms of the two-spotted spider mite *Tetranychus urticae* (Acari: Tetranychidae) under laboratory and semi-natural conditions. Appl Entomol Zool., 37: 127-139.

Sugimoto, K. (2002). Fluzinam (Frowncide)-a novel and effective method of application against white and violet root rot. Agrichemicals, 80: 14-16.

Sujatha, Soni, P. K. and Jawaharlal, J. (2015).Identification of *Podosphaera xanthii* causing powdery mildew on sesame (*Sesamum indicum* L.) M J. Oilseeds Res., 32(2): 183-185.

Subramanian, C.V. (1971). Hyphomycetes: an account of Indian species, except Cercosporae. Indian Council of Agricultural Research, New Delh, P-930.

Subramanian, T.R. (1958). Biology of *Myllocerus subfasciatus* Guerin. Jour Madras Univ., 28b(2/3): 69-76.

Sullivan, M. and Molet, T. (2007). CPHST Pest Datasheet for *Helicoverpa armigera*. USDA-APHIS-PPQ-CPHST. Revised April 2014.

Sultana, C. (1984). Lin etautresoléagineux. Cultivar.Juin 1984. Spécial Oleaoproteagineux., 173: 97–101.

Summers, T.E. and Seale, C.C. (1958). Root knot nematodes, a serious problem of Kenaf in Florida. Plant Disease Reporter, 42: 792-795.

Sun, P. and Yang, X.B. (2000). Light, Temperature, and Moisture Effects on Apothecium Production of *Sclerotinia sclerotiorum*. Plant Disease, 84(12): 1287-1293.

Sun, Q.T. and Meng, Z.J. (2001). Biological characteristics of *Tetranychus cinnabarinus*, a vegetable pest. Journal of Jilin Agricultural University, 23: 24-25.

Sundaramurthy, V.T. (1992). Upsurgence of Whitefly *Bemisia tabaci* Gen. in the Cotton Ecosystem in India. Outlook on Agriculture, 21(2): 109-115.

Sundaramurthy, V.T. and Basu, A.K. (1989). Behaviour of management system under stress and outbreak situations in the polycrop cotton agrosystem. National Seminar on Futuristic Approaches in Cotton Improvement, Hissar, India.

Sundararaj, R. and David, B.V. (1987). Influence of biochemical parameters of different hosts on the biology of *Earias vittella* (Stoll) (Lepidoptera: Noctuidae). Proc. Ind. Acad. Sci., (Animal Science), 19: 329-332.

Sundravadana, S., Alice, D., Kuttalam, S. and Samiyappan, R. (2006). Control of Mango Anthracnose by Azoxystrobin. Tunisian Journal of Plant Protection, 1: 109-111.

Surulivelu, T. and DharaJothi, B. (2007).Mirid bug, *Creontiodes biseratense* (Distant) damage on cotton in Coimbatore, http://www.cicr.gov.in.

Suryanaryana, D. (1965). A note on Myrothedum leaf spot disease of cotton. PAU. New Letter, 11: 8.

Sutic, D. and Dowson, W.J. (1959). An investigation of a serious disease of hemp (*Cannabis sativa* L.) in Jugoslavia. J Phytopathol., 34: 307-314.

Sutton, B.C. (1980a). *Lasiodiplodia* Ell.and Ev. apud Clendenin. In: Sutton, B.C. (ed), The coelomycetes. Fungi imperfecti with pycnidia, acervuli and stromata. Commonwealth Mycological Institute, Kew, P.191.

Sutton, B.C. (1980b). The Coelomycetes: Fungi Imperfecti with Pycnidia, Acervuli and Stromata. Commonwealth Mycological Institute, London, P.696.

Sutton, B.C. and Waterston, J.M. (1966). *Coniella diplodiella*. CMI Descriptions of pathogenic fungi and bacteria. 82: 1-2.

Sutton, B.C.S. and Shaw, M. (1986). Protein synthesis in flax following inoculation with flax rust. Can. J. Bot., 64: 13-18.

Swaine, G. (1971). Agricultural zoology in Fiji. Foreign Commonwealth Office, Overseas Development Administration, Overseas Research Publication 18.

Swart, W. J. (2001). First Report of Powdery Mildew of Kenaf Caused by *Leveillula taurica* in South Africa. Plant Disease, 85(8): 923.

Swathi, M. and Gaur, N. (2017). Effect of Border Crops and Insecticides on Management of Whitefly, *Bemisia tabaci* (Gennadius) Transmitted Yellow Mosaic Virus in Soybean. International Journal of Current Microbiology and Applied Sciences, 6(5): 613-617.

Sweet, M.H. II.(2000). Seed and chinch bugs (Lygaeidae). [in:] C.W. Schaefer and A.R. Panizzi (Eds.). Heteroptera of Economic Importance.CRC Press, Boca Raton, pp.143-264.

Syed, T.S., Abro, G.H., Khanum, A. and Sattar, M. (2011). Effect of Host Plants on the Biology of *Earias vittella* (Fab) (Noctuidae: Lepidoptera) Under Laboratory Conditions. Pakistan J. Zool., 43(1): 127-132.

Sztejnberg, A., Azaizia, H. and Lisker, N. (1989). Effects of tannins and phelonic extracts from plant roots on the production of cellulose and polygalacturonasa by *Dematophora necatrix*. Phytoparasitica, 17: 49-53.

Sztejnberg, A. and Madar, Z. (1980). Host range of *Dematophora necatrix*, the cause of white root disease in fruit trees. Plant Disease, 64: 662-664.

Tabashnik, B.E., Patin, A.L., Dennehy, T.J., Liu, Y.B., Miller, E. and Staten, R.T. (1999). Dispersal of pink bollworm (Lepidoptera: Gelechiidae) males in transgenic cotton that produces a *Bacillus thuringiensis* toxin. Journal of Economic Entomology, 92(4): 772-780.

Tadas, P.L., Kene, H.K. and Deshmukh, S.D. (1994). Efficacy of some newer insecticides against cotton bollworms. PKV Research Journal, 18(1): 138-139.

Tai, F.L. (1979). Sylloge Fungorum Sinicorum (in China). Science Press, Academia Sinica, Peking, P.1527.

Takamatsu, S., Niinomi, S., Harada, M., and Havrylenko, M. (2010). Molecular phylogenetic analyses reveal a close evolutionary relationship between *Podosphaera* (Erysiphales: Erysiphaceae) and its rosaceous hosts. Persoonia, 24: 38–48.

Takikawa, Y., Kijima, T., Yamashita, S., Doi, Y., Tsuyumu, S. andGoto, M. (1984). Bacterial leaf spot disease of hemp caused by *Xanthomonas campestris* pv. *cannabis* (abstract in Japanese). Ann Phytopathol Soc Japan, 50: 141.

Talhinhas, P., Sreenivasaprasad, S., Neves-Martins, J. and Oliveira, H. (2002). Genetic and morphological characterization of *Colletotrichum acutatum* causing anthracnose of lupins. Phytopathology, 92: 986-996.

Talukder, M. J. 1974. Plant diseases in Bangladesh, Bangladesh j. Agric. Res., 1: 61-68.

Tamaki, G., Annis, B. and Weiss, M. (1981). Response of natural enemies to the green peach aphid in different plant cultures. Environmental Entomology, 10: 375-378.

Tancik, J. (2017). Natural parasitism of the second generation European corn borer eggs *Ostrinia nubilalis* (Hubner) (Lepidoptera, Pyralidae) by *Trichogramma* spp. in sweet corn fields in Vojvodina, Serbia – short communication. Plant Protect. Sci., 53: 50-54.

Tanda, S. and Hirose, T. (2003). Renaming of Erysiphe Fungi on Maniokeibish and Flax First report of Odium sp. on Kenaf in Japan. J. Agri. Sci. Tokyo Univ. of Agric., 48(2): 50-58.

Tandingan, I.C. and Asuncion, E.V. (1990). Biological control of root rot nematodes in ramie using *Paecilomyces lilacinus* (Thom) Samson. USM Research and Development Journal, 1(1): 1824.

Tang, Y.F., He, Z.F., Du, Z.G., She, X.M. and Lan, G.B. (2015). Detection and identification of the pathogen causing kenaf (*Hibiscus cannabinus*) leaf curl disease in Hainan Province of China. Acta Phytopathol Sin., 45: 561-568.

Tanwar, R., Bhamare, V., Ramamurthy, V., Hayat, M., Jeukumar, P., Singh, A. and Bambawale, O. (2008). Record of new parasitoid on mealybug, *Phenacoccus solenopsis*. Indian Journal of Entomology, 70(4): 404-405.

Tanwar, R.K., Jeyakumar, P. and Monga, D. (2007). Mealybugs and their management. National Centre for Integrated Pest Management, New Delhi, Technical Bulletin, 19: 1-16.

Tao, Y., Zeng, F., Ho, H. Wei, J., Wu, Y., Yang, L. and He, Y. (2011). *Pythium vexans* Causing Stem Rot of Dendrobium in Yunnan Province, China. J Phytopathol., 159: 255-259.

Tariq, V.N., Gutteridge, C.S. and Jeffries, P. (1985). Comparative studies of cultural and biochemical characteristics used for distinguishing species within Sclerotinia. Transactions British Mycological Society, 84: 381-397.

Taylor, C.R. and Rodriguez-Kabana, R. (1999). Optimal rotation of peanuts and cotton to manage soil borne organisms. Agricultural Systems, 61(1): 57-68.

Taylor, A. L. and Sasser, J.N. (1978). Biology, identification and control of root-knot nematodes (*Meloidogyne* species). Raleigh: North Carolina State University Graphics.

Taylor, A.L., Sasser, J.N. and Nelson, L.A. (1982). Relationship of climate and soil characteristics to geographical distribution of *Meloidogyne* species in agricultural soils. A Coop Public. Depart. Plant Pathol., North Carolina State University and U.S. Agency Int. Devel. Raleigh.

Taylor, C.M., Coffey, P.L., DeLay, B.D. and Dively, G.P. (2014). The importance of gut symbionts in the development of the brown marmorated stink bug, *Halyomorpha halys* (Stal). PloS ONE, 9e90312.

Tehri, K. (2014). A review on reproductive strategies in two spotted spider mite, *Tetranychus urticae* Koch 1836 (Acari: Tetranychidae). Journal of Entomology and Zoology Studies, 2(5): 35-39

Teixeira de Sousa, A.J. and Whalley, A.J.S. (1991).Induction of mature stromata in *Rosellinia necatrix* and its taxonomic implications. Sydowia, 43: 281–290.

Teodoro, N.G. (1937). An enumeration of Philippine fungi. Department of Agriculture and Commerce, Manila. Technical Bulletin, 4: 1-585.

Teotia, T.P.S. and Pathak, M.D. (1956). Bionomics of sunnhemp shoot borer. *Enarmonia pseudonectis* Meyr (Eucosmidae/ Lepidoptera) in Uttar Prasdesh. Indian J. Ent., 18(3): 233-242.

Teran-Vargas, A.P., Azuara-Domínguez, A., Montecillo, C.d.P. and Azuara-Domínguez, A. (2012). Biological Effectivity of Insecticides to Control the Agave Weevil, *Scyphophorus acupunctatus* Gyllenhal (Coleoptera: Curculionidae), in Mexico. Southwestern Entomologist, 37(1): 47-53.

Tesfaendriasl, M. T., Swart, W. J. and Botha, W. (2004). The characterization of *Pythium* group G occurring on kenaf in South Africa. S. Afr. I. Plant Soil 2004, 21(1): 25-30.

Texas Plant Disease Handbook, (ND). Cotton Root Rot. Texas Plant Disease Handbook, https://plantdiseasehandbook.tamu.edu/problems-treatments/.../cotton-root-rot/

Thangaraju, D., Uthamasamy, S., Rangarajan, A.V. and Jayaraj, S. (1993). Effect of new insecticides against cotton bollworms. Madras Agricultural Journal, 80(8): 453-456.

Thara, S., Kingsley, S. and Revathi, N. (2009). Impact of Neem Derivatives on Egg Hatchability of Okra Fruit Borer, *Earias vittella* Fab. (Lepidoptera :Noctuidae). International Journal of Plant Protection, 2(1): 95-97.

Thaxton, P.M., Brooks, T.D. and El-Zik, K.M. (2001). Race identification and severity of bacterial blight from natural infestations across the Cotton Belt. pp.137-138. In Proc. Beltwide Cotton Conf., Anaheim, CA. 9-13 Jan. 2001. Natl. Cotton Counc.Am., Memphis, TN.

The Western Committee on Plant Diseases. (2012). Diseases of Oilseed Crops. In Guidelines For The Control of Plant Diseases In Western Canada (p. 6). Website: http://www.westernforum.org/Documents/WCPD/WCPD_documents/Current%20 Guideline%20 Files/Ch%204%20Diseases%20of%20Oilseeds.pdf

Thimmaiah, G., Gubbaiah, Kulkarni, K.A. and Bhat, K.M. (1975). Occurrence of shoot weevil *Alcidodes affaber* Aurivillius on cotton and bhendi. Curr. Res., 4: 105.

Thippeswamy, C., Govindan, R., Devaiah, M. C. and Thimmaiah, G. (1992). Life cycle of *Alcidodes affaber* Aurivillius (Coleoptera: Curculionidae) on Bhendi *Abelmoschus esculentus* (L.) Moench. Mysore J. of Agri. Sci., 26: 276-279.

Thippeswamy, C., Thimmaiah, G., Devaiah, M.C. and Govindan, R. (1980). Seasonal incidence of the cotton shoot weevil, *Alcidodes affaber* Aurivillus on lady's finger, *Abelmoschus esculentus* (Linnaeus) Moench. Madras Agric. J., 14(4): 505.

Thombre, M.V. (1980). Association of some morphological characters to confer resistance to bollworm in *Gossypium hirsutum* Cotton. Andhra Agriculture Journal, 27: 19-20.

Thomas, H.E., Wilhelm, S. and MaClean, N.A. (1953). Two root rots of fruit trees. U.S. Department of Agriculture. Year book of Agriculture, 702-704.

Thomas, M. C. (2000) Pest Alert: *Myllocerus* sp. (near) *undecimputulatus* Faust, a weevil new to the Western Hemisphere. http://doacs.state.fl.us/~pi/enpp/ento/weevil-pestalert.htm

Thomas, M.C. (2005). *Myllocerus undatus* Marshall, a weevil new to the Western Hemisphere. Florida Department of Agriculture and Consumer Services Pest Alert, DACS-P-01635. (http://freshfromflorida.s3.amazonaws.com/myllocerus-undatus.pdf).

Thomas, R.F. and Denmark, H.A. (2009). Twospotted Spider Mite, *Tetranychus urticae* Koch (Arachnida: Acari: Tetranychidae). Featured Creatures. of Florida / Institute of Food and Agricultural Sciences, UF/IFAS, EENY150. Website: https://www.edis.ifas.ufl.edu/pdffiles/IN/IN30700.pdf

Thompson, G.D., Busacca, J.D., Jantz, O.K., Kirst, H.A., Larson, L.L. and Sparks, T.C. (1995). Spinosyns: an overview of new natural insect management systems. ProcBeltwide Cotton Conf., 2: 1039-1043.

Thompson, G.J. and van Zijl, J.J.B. (1996). Control of tomato spotted wilt virus in tomatoes in South Africa. Acta Hortic., 431: 379-384.

Thompson and Zimmer, (1943). (Original not seen). Reference quoted in: Edirisinghe, W.H.M.V.P. (2016). Characterization of Flax germplasm ror resistance to Fusarium Wilt Caused by *Fusarium oxysporum* f. sp. *lini*. M.Sc Thesis, The College of graduate studies and research, Department of Plant Science, University of Saskatchewan, Saskatoon Website: https://ecommons.usask.ca-handle-ED

Thunberg, C.P. (1787). Museum Naturalium Academiae Upsalensis.4. Dissertation, Uppsala, pp.43-58.

Tillman, P.G. and Mullinix, B.G. Jr. (2004). Comparison of susceptibility of pest *Euschistus servus* and predator *Podisus maculiventris* (Heteroptera: Pentatomidae) to selected insecticides. Journal of Economic Entomology, 97: 800- 806.

Timian, R.G. and Alm, K. (1973). Selective Inbreeding of *Macrosteles fascifrons* for Increased Efficiency in Virus Transmission. Phytopathology, 63: 109-112.

Timm, R. W. and Ameen.M. (1960). Nematodes associated with commercial crops in East Pakistan. Agric. Pak., 11 (3): 355-366.

Timmer, L.M., Peever, T.L., Solel, Z. and Akimitsu, K. (2003). Diseases of citrus- Novel Pathosystems. Phytopathology Mediterranea, 42: 99-112.

Timmis, J. N. and Whisson, D. L. (1987). Molecular genetics and plant diseases. Journal of the Agricultural Society, University College of Wales, 68:.48-65.

Tinsley, J.D. (1898a).An ants'-nest coccid from New Mexico. The Canadian Entomologist, 30: 47-48.

Tiourebaev, K.S. (1999). Virulence and dissemination enhancement of a mycoherbicide. PhD. Thesis, Plant Pathology, Montana State University, Bozeman, Montana

Tiourebaev, K.S., Pilgeram, A.L., Anderson, T.W., Baizhanov, M.K. and Sands, D.C. (1998). *Fusarium oxysporum* f.sp. *cannabina* (SiC) as promising candidate for biocontrol of *Cannabis* sp. Phytopathology, 88 (9 supplement): S89.

TNAU, (2016). Pest of cotton, Leaf roller: *Sylepta derogate*. Crop Protection, TNAU Agritech Portal. Website: agritech.tnau.ac.in

TNAU Agritech Portal. (ND1). Bihar hairy caterpillar - TNAU Agritech Portal: Crop Protection. agritech.tnau.ac.in/crop_protection/.../crop_prot_crop_insect_pulsh_soyabean_1.html

TNAU Agritech Portal (ND2). Crop Protection, Cash Crop, Pest of Cotton. Website: https://www.agritech.tnau.ac.in/crop_protection/crop_prot_crop_insectpest%20_cotton.html

Tochinai, Y. and Takee, G. (1950). Studies on the physiologic specialization in *Fusarium lini* Bolley. J. Fac. Agr. Hokkaido Univ., 47: 193-266.

Toda, S. and Murai, T. (2007). Phylogenetic analysis based on mitochondrial COI gene sequences in *Thrips tabaci* (Thysanoptera: Thripidae) in relation to reproductive forms and geographic distribution. Appl. Entomol. Zool., 42: 309-316.

Tomar, D.S. (2005). Studies on the Myrothecium leaf blight of cotton. Ph.D. Thesis submitted to Dr. B.R. Ambedkar University, Agra. P.116.

Tomar, D.S. (2007). Survey and cyclic transmission of Myrothecium leaf blight of cotton in East Nimar of Madhya Pradesh. Indian Journal of Tropical Biodiversity, 15(2): 170-175.

Tomar, D.S., Shastry, P.P., Srivastava, A.K., Nayak, M.K. and Pathak, R.K. (2009). Effect of epidemiological factors on Myrothecium leaf blight of cotton. Journal of Oilseeds Research, 26: 440-441.

Tomar, D.S., Shastry, P.P. and Nayak, M.K. (2016). Morphological and histopathological studies of *Myrothecium roridum* on cotton. Journal of Cotton Research and Development, 30(1): 90-96.

Tourvieille de Labrouhe, D. (1982). Pénétration de *Rosellinia necatrix* (Hart.) Berl. dans les racines du pommieren conditions de contamination artificielle. Agronomie, 2: 553-560.

Transhel, V., Gutner, L. and Khokhryakov, M. (1933). A list of fungi found on new cultivated textile plants. Moscow Inst. Nov. Lubian. Syria, Trudy, 4: 127-140.

Triantaphyllou, A.C. (1979). Cytogenetics of root-knot nematodes. In: Lamberti, F., Taylor, C.E. eds. Root-knot nematodes (*Meloidogyne* species) systematic, biology and control. New York, USA, Academic Press, 85-114.

Triantaphyllou, A.C. and Hirschmann, H.(1960). Post-infection development of *Meloidogyne incognita* Chitwood, 1949 (Nematoda: Heteroderidae). Annales de l' Institut Phytopathologique, Benaki, 3: 3-11.

Tripathi, A.N., De, R.K., Meena, P.N. and Sharma, H.K. (2015a). Emergence of Leaf Blight in Mesta (*Hibiscus* spp.) caused by *Phoma exigua* (Desm.). Jaf News (ICAR-CRIJAF), 13 (1): 12.

Tripathi, A.N., De, R.K., Sharma, H.K. and Karmakar, P.G. (2015b). Emerging threat of *Sclerotinia sclerotiorum* causing white/cottony stem rot of mesta in India. New Disease Reports, 32: 19.

Tripathi, D.M., Bisht, R.S. and mishra, P.N. (2003). Bio efficacy of some synthetic insecticides and bio pesticides against black cutworm *Agrotis ispilon* infesting potato (*Solanum tuberosum*) in Garwal Himalaya. Indian Journal of Entomology, 65(4): 468-473.

Tripathi, H.S. (1973). Studies on some diseases of sunnhemp in Varanasi.M.Sc. Thesis Submitted to Banaras Hindu Unversity, India.

Tripathi, H.S., Singh, A.K. and Basu-Chaudhary, K.C.(1975). Two new fungal diseases of sunnhemp from Varanasi. Indian phytopathology, 28 (2): 263-265.

Tripathi, H.S., Vishwakarma, S.N. and Chaudhury, K.C.B. (1978). Evaluation of fungicides for control of leaf blight and tip rot and damping off and wilt of sunnhemp. Sci. and Cult., 44(1) : 30-33.

Tripathi, R.L. and Ram, S. (1969). Susceptibility of Tossa jute to spiral borer of mesta, *Agrilus acutus*Thunb (Buprestidae, Coleoptera). Indian J. Ent., 31(1): 80-81.

Tripathi, R.L. and Ram, S. (1971). A Review of Entomological Researches on Jute, Mesta, Sunnhemp and Allied Fibres. I.C.A.R. Technical Bulletin (Agric.), Indian Council of Agricultural Research, New Delhi, 36: 1-39.

Tripathi, R.L. and Ram, S. (1972). Loss in yield due to damage of jute semilooper, *Anomis sabulifera* (Guen) (Noctuidae: Lepidoptera). Indian J. Agric. Sci., 42(4): 334-336.

Tripathi, S. and Singh.T. (1991). Population dynamics of *Helicoverpa armigera* (Hubner) (Lepidoptera: Noctuidae). Insect Science Applications, 12: 367-374.

Trotus, E., Mincea, C. and Alexandrescu, S. (1994). Studies on the control of the flax flea beetle (*Aphthona euphorbiae* Schrank) by seed treatment. Cercetari Agronomice in Moldova, 27: 187-189.

Trouve, C., Ledee, S., Brun, J. and Ferran, A. (1996). Biological control of the hop aphid. A review of three years of tests in northern France. Phytoma, 486: 41-44.

Tsay, J.G. and Kuo, C.H. (1991). The occurrence of corynespora blight of cucumber in Taiwan. Plant Prot. Bull., 33: 227-229.

Tsuda, M. and Ueyama, A. (1985). Two new *Pseudocochliobolus* and a new species of *Curvularia*. Transactions of the Mycological Society of Japan, 26: 321-330.

Tu, C.C. and Cheng, Y.H. (1971). Interaction of *Meloidogyne javanica* and *Macrophomina phaseoli* in Kenaf Root Rot. Journal of Nematology, 31: 39-42.

Tu, C.C. and Kimbrough, J.W. (1978). Systematics and phylogeny of fungi in the *Rhizoctonia* complex. Botanical Gazette, 139: 454-466.

Tugwell, P., Young, S.C., Dumas, B.A. and Phillips, J.R. (1976). Plant bugs in cotton: importance of infestation time, types of cotton injury, and significance of wild hosts near cotton. Arkansas Agricultural Experimental Station Report, No. 227: pp.24.

Tuhan, N.C., Pawar, A.D. and Arora, R.S. (1987).Use of *Trichogramma brasiliensis* Ashmead against cotton bollworms in Sriganganagar, Rajasthan, India. Journal of Advanced Zoology, 8(2): 131-134.

Turian, G. (1954). Recent progress in the control of the *Coniella diplodiella* disease of vines. Rev Romande Agric Viticult Et Arboricult, 10(2): 12-14.

Turlier, M.-F., Eparvier, A. and Alabouvette, C. (1994). Early dynamic interactions between *Fusarium oxysporum* f.sp.*lini* and the roots of *Linum usitatissimum* as revealed by transgenic GUS-marked hyphae. Can. J. Bot., 72(11): 1605-1612.

Turner, J. (1987). Linseed Law: A Handbook for Growers and Advisers. Alderman Printing and Bookbinding, BASF UK Ltd. Hadleigh, Ipswich, P.356.

Tvelkov, S.G. (1970). *Sphacelotheca lini* a new species on fiber flax in the Novogorod region. Mikoli. Fitopatol., 4: 484.

Twinn, C.R. (1944). A summary of the more important insect pests in Canada in 1943, 74th Report Entomol. Soc. Ontario, pp.54-59.

Twinn, C.R. (1945). A summary of insect conditions of importance or special interest in Canada in 1944, 75th Report Entomol. Soc. Ontario, pp.45-49.

Tyler, B.M., Tripathy, S., Zhang, X., Dehal, P., Jiang, R.H., Aerts, A., Arredondo, F.D., Baxter, L., Bensasson, D. and Beynon, J.L. (2006). *Phytophthora* genome sequences uncover evolutionary origins and mechanisms of pathogenesis. Science, 313 (5791): 1261-1266.

UC-IPM, (2015). UC Management Guidelines for Pink Bollworm on Cotton - UC IPM. Website: ipm.ucanr.edu/PMG/r114301511.html

UC IPM, (2017). UC Pest Management Guidelines, Cotton: Verticillium Wilt, Pathogen: *Verticillium dahliae*. In: UC IPM Pest Management Guidelines: How to Manage Pests, UC ANR Publication 3444, Agriculture and Natural Resources, University of California. Website: ipm.ucanr.edu/PMG/r114100211.html

UC IPM, (ND). UC Management Guidelines for Cotton Aphid on Cotton. Website: ipm.ucanr.edu/PMG/r114300111.html

Udikeri, S.S. (2008). Mirid Menance-A potential emerging sucking pest problem in cotton. The International Cotton Advisory Committee Recorder, 26(4): 15.

Udikeri, S.S., Kranthi, S., Kranthi, K.R., Vandal, N., Hallad, A., Patil, S.B. and Khadi, B.M. (2011a). Species diversity, pestiferous nature, bionomics and management of mirid bugs and flower bud maggots: The new key pests of Bt Cotton. Paper presented in World Cotton Research Conference-5, 7-11 November, Mumbai, India, pp.203-209.

Udikeri, S.S., Kranthi, K.R., Patil, S.B. and Khadi, B.N. (2011b). Emerging pest of Bt cotton and dynamics of insects pests in different events of Bt cotton. Proceedings of 5th Asian Cotton Research and Development Network meeting. Lahore Pakisthan 23-25th Feb 2011. www.ICAC.org

Udikeri, S.S., Kranthi, K.R., Patil, S.B., Modagi, S.A and Vandal, N.B. (2010). Bionomics of mirid bug, *Creontiades biseratense* (Distant) and oviposition pattern in Bt cotton. Karnataka J. Agric. Sci., 23(1): 153-156.

Udikeri, S.S., Patil, S,B., Shaila, H.M., Guruprasad, G.S., Patil, S.S. and Kranthi, K.R. (2008). Miridmenance-a potential emerging sucking pest problem in cotton. Paper presented at the Fourth Meeting of Asian Cotton Research and Development Network, Anyang, China. http://www.icac.org/tis/regional_networks/documents/asian/papers/udikeri.pdf

UF/IFAS, (2017). Sweetpotato whitefly B biotype or silverleaf whitefly: *Bemisia tabaci* (Gennadius) or *Bemisia argentifolii* Bellows & Perring. Featured Creatures, University of Florida; Website: https://www.entnemdept.ufl.edu.

UF/IFAS (ND1). Featured Creatures.Sri Lankan weevil – *Myllocerus undecimpustulatus undatus* Marshall. entnemdept.ufl.edu/creatures/orn/sri_lankan_weevil.htm

UF/IFAS (ND2). Featured Creatures, twospotted spider mite, University of Florida. Website: twospotted spider mite – *Tetranychus urticae* Koch, Website: https://www.entnemdept.ufl.edu/creatures/orn/twospotted_mite.htm

Ullah, S., Shad, S.A. and Abbas, N. (2016). Resistance of Dusky Cotton Bug, *Oxycarenus hyalinipennis* Costa (Lygaidae: Hemiptera), to Conventional and Novel Chemistry Insecticides. J Econ. Entomol., 109(1): 345-351.

Ulloa, M., Hutmacher, R.B., Davis, R.M., Wright, S.D., Percy, R. and Marsh, B. (2006). Breeding and genetics: breeding for Fusarium wilt race 4 resistance in cotton under field and greenhouse conditions. J. Cotton Sci., 10: 114-127.

Underhill, G.W. (1934). The green stinkbug. Virginia Agriculture Experiment Station Bulletin, 294: 1-26.

Unlu, L. and Bilgic, A. (2004). The effects of the infestation ratio of spiny bollworm (*Earias insulana*) and pink bollworm (*Pectinophora gossypiella*) on cotton yield grown in semi arid region of Turkey. J. Applied Entomology, 128(9-10): 652-657.

Uppal, B.N. and Kulkarni, N.T. 1937. Studies in Fusarium wilt of sunnhemp. I. The physiology and biology of *Fusarium vasinfectum* Atk. Indian J. Agric. Sci., 7(3): 413- 442.

USDA, (1948). Pink Bollworm. Picture Sheet No. 21. Bureau of Entomology and Plant Quarantine.Agricultural Research Administration, U.S. Department of Agriculture.

USDA-APHIS, (1977). The sterile screwworm fly production plant. Mission, Texas. USDA, APHIS, U. S. Govt. Printing Off. 772–925/272/7. pp.6.

USDA - New Pest Response - Cotton Seed Bug. United States Department of Agriculture Animal and Plant Health Inspection Service

Uvarov, B.P. and Glazunov, V.A. (1916). A Review of Pests.(In Russian). Report on the work of the Entomological Bureau of Stavropol for 1914. Dept. Agric. of the Ministry of Agric., Petrograd, pp.13-54.

Vago, C. and Cayrol, R. (1955). A polyhedral virus of the gamma noctuid *Plusia gamma* L. (Lepidoptera).(In French). Ann. Épiphyt., 6: 421-432.

Valenzuela, I., Carver, M., Malipatil, M. and Ridland, P. (2009). Occurrence of *Macrosiphum hellebori* Theobald & Walton (Hemiptera: Aphididae) in Australia. Australian Journal of Entomology, 48: 125-129.

Van de Bund, C.F. and Helle, W. (1960). Investigations on the *Tetranychus urticae* complex in N. W. Europe (Acari: Tetranychidae). Entomol. Exp. Appl., 3: 142-156.

van der Aa, H.A, Boerema, G.H. and de Gruyte, J. (2000). Contributions towards a monograph of Phoma (Coelomycetes) VI -1.Section Phyllostictoides: Characteristics and nomenclature of its type species *Phoma exigua*. Persoonia, 17(3): 435-456.

Van Der Poorten, G. M. and Van Der Poorten, N. E. (2016). The Butterfly Fauna of Sri Lanka, pp.265-266.
Van der Werf, H.M.G. (1994). Crop physiology of fibre hemp (*Cannabis sativa* L.) Doctoral thesis, Wageningen Agricultural University, Wageningen, the Netherlands, P.152.
Van Doesburg, P.H.(Jr). (1968). A revision of the new world species of *Dysdercus* Guerin Meneville (Heteroptera, Pyrrhocoridae). Zool. Verh., 97: 1-215.
van Emden, H. and Harrington, R. (2007). Aphids as Crop Pests. Trowbridge, United Kingdom: CABI.
Vannacci, G. and Harman, G. E. (1987). Biocontrol of seed-borne *Alternaria raphani* and *Alternaria brassicicola*. Canadian Journal of Microbiology, 33: 850-856.
Vanterpool, T.C. (1947). *Selenophoma linicola* Sp. Nov. on flax in Saskatchewan. Mycologia, 39: 341-348.
Vanterpool, T.C. (1949). Flax diseases in Saskatchewan in 1948. 28th Ann. Rep. Can. Plant Dis. Surv., 28: 22-24.
vanVloten-Doting, L. (1975). Coat protein is required for infectivity of Tobacco streak virus: Biological equivalence of the coat proteins of Tobacco streak and Alfalfa mosaic viruses. Virology, 65: 215-225.
VaradaRajan. B.S. (1943). A mildew on iute. Sci. & Cult., 9: 351-352.
VaradaRajan.B.S. and Patel, J.S. (1943). Stem-rot disease of jute. Indian J. Agric. Sci., 13: 148-156.
Varma, A. (1986). Sunn-Hemp Mossaic Virus. pp249-266. In: Van Regenmortel, M.H.V., Fraenkel-Conrat, H. (eds) The Plant Viruses, Boston, MA.
Vasic, T., Bulajic, A., Krnjaja, V., Jevremovic, D., Zivkovic, S., Andjelkovic, B.(2014). First report of anthracnose on alfalfa caused by *Colletotrichum linicola* in Serbia, Plant Disease, 98 (9): 1276.
Vasic, T., Jevremovic, D., Andjelkovic, S., Markovic, J., Zornic, V., Babic, S. and Leposavic, A. (2016). Morphological and molecular identification of a new alfalfa parasite - *Colletotrichum linicola* in Serbia, VII International Scientific Agriculture Symposium, "Agrosym 2016", 6-9 October 2016, Jahorina, Bosnia and Herzegovina. Proceedings (eds, Kovacevic, D.), pp.1480-1485.
Vassilaina-Alexopoulou, P. and Mourikis, P.A. (1976). A new insect pest of the hemp in Greece *Cydia delineana*. Annales de Institut Phytopathologique Benaki, 11(3): 253-254
Vasudeva, R.S. (1935). Studies on the root-rot disease of cotton in the Punjab. Indian J. Agric. Sci., 5: 469-512.
Vaurie, P. (1971). Review of Scyphophorus (Curculionidae: Rhynchophorinae). Coleop. Bull., 25: 1-8.
Vawdrey., L.L., Grice, K.R.E. and Westerhuis, D. (2008). Field and laboratory evaluations of fungicides for the control of brown spot (*Corynespora cassiicola*) and black spot (*Asperisporium caricae*) of papaya in far north Queensland, Australia. Australasian Plant Pathology, 37(6): 552-558.
Vawdrey, L. L. and Peterson, R. A. (1990). Diseases of kenaf (*Hibiscus cannabinus*) in the Burdekin River Irrigation Area. Australasian Plant Pathology, 19 (2): 34.
Vawdrey, L.L. and Stirling, G.R. (1992). Reaction of kenaf and Roselle grown in the Burk in River Irrigation Area to root-knot nematodes. Aust. Plant Pathol., 21: 8–12.
Veech, J.A. (1992). Reproduction of Four Races of *Meloidogyne incognita* on *Hibiscus cannabinus*. Journal of Nematology, 24(4S): 717-721.
Veerman, A. (1970). The pigments of *Tetranychus cinnabarinus* Boisd. (Acari: Tetranychidae)-Comp. Biochem. Physiol., 36: 749-763.
Veerman, A. (1974). Carotenoid metabolism in *Tetranychus urticae* Koch (Acari:Tetranychidae). Comp. Biochem. Physiol. B: Comp. Biochem., 47: 101-116.

Vegheli, K., Balogh, I. and Polyak, D. (1995). Root rot fungi on grapevine in Hungery. Horticultural Science, 27: 48-53.

Velazquez-Fernandez, P., Zamora-Macorra, E.J., OchoaMartinez, D.L. and Hernandez-Morales, J. (2016).Virus asociados al amarillamiento de *Hibiscus sabdariffa* en Guerrero, Mexico. Revista Mexicana de Fitopatologia, 34: 200-207.

Vendramin, J.D. and Nakano, O. (1981). Assessment of damage by *Aphis gossypii* Glover, 1877 (Homoptera: Aphididae) to the cotton cultivar 'IAC-17'. Anais da Sociedade Entomologica do Brasil, 10(1): 89-96.

Venette, R.C., Davis, E.E., Zaspel, J., Heisler, H. and Larson, M.(2003). Mini Risk Assessment Old World bollworm, *Helicoverpa armigera* Hubner (Lepidoptera: Noctuidae). Cooperative Agricultural Pest Survey, Animal and Plant Health Inspection Service, US Department of Agriculture. http://www.aphis.usda.gov/plant_health/plant_pest_info/pest_detection/downloads/pra/harmigerapra.pdf.

Venkatesh, H., Chattannavar, S.N., Rajput, R.B. and Hiremath, J.R. (2015). Weather in relation to Grey mildew disease in cotton at Dharwad, Karnataka. Environment and Ecology, 33(4A): 1667-1671.

Venkatesh, I. (2014). Studies on Alternaria leaf spot of cotton. M.Sc. Thesis, Acharya N. G. Ranga Agricultural University, Hyderabad

Venkateshalu, Murthy, M. S. and Sushila, N. (2015). Management of mirid bug, *Poppiocapsidea* (=*Creontioides*) *biseratense* (Distant) on Bt cotton. Environment and Ecology, 33(3): 1139-1142.

Vennila, S., Biradar, V.K., Sabesh, M. and Bambawale, O.M. (2007a). Know Your Cotton Insect Pest APHIDS. Crop Protection Folder Series: 1 of 11. Central Institute for Cotton Research, Nagpur (India).

Vennila S., Biradar V.K., Sabesh, M. and Bambawale, O.M. (2007b). Know Your Cotton Insect Pest Pink Bollworm. Crop Protection Folder Series: 7 of 11, Central Institute for Cotton Research Nagpur.

Vennila, S., Biradar, V.K., Sabesh, M. and Bambawale, O.M. (2007c). Know Your Cotton Insects Stainers (Red and Dusky Cotton Bugs). Crop Protection Folder Series: 9, Central Institute for Cotton Research Post Bag No. 2, Shankar Nagar P. O. Nagpur 440 010, Maharashtra.

Vennila, S., Biradar, V.K., Sabesh.M. and Bambawale, O.M. (2007d). Know your cotton insect pest spotted and spiny bollworms. Crop Protection Folder Series: 5 of 11, Central Institute for Cotton Research, Nagpur.

Vennila, S., Deshmukh, A. J., Pinjarkar, D., Agarwal, M., Ramamurthy, V. V., Joshi, S., Kranthi, K. R. and Bambawale, O. M. (2010). Biology of the mealybug, *Phenacoccus solenopsis* on cotton in the laboratory. J. Insect Sci., 10: 1-9.

Vennila, S., Biradar, V.K., Sabesh, M., Bambawale, O.M. (2007e). Know Your Cotton Insect Pest Thrips. Crop Protection Folder Series: 3 of 11. Central Institute for Cotton Research, Nagpur

Vera, C.L., Duguid, S.D., Fox, S.L., Rashid, K.Y., Dribnenki, J.C.P. and Clarke, F.R. (2012). Comparative effect of lodging on seed yield of flax and wheat. Can. J. Plant Sci., 92: 39-43.

Vera, C.L., Irvine, R.B., Duguid, S.D., Rashid, K.Y., Clarke, F.R. and Slaski, J.J. (2014). Pasmo disease and lodging in flax as affected by pyraclostrobin fungicide, N fertility and year. Can. J. Plant Sci., 94: 119-126.

Verbeek, W.A. (1976). Annual report for the period 1 July, 1974 to 30 June, 1975. Secretary for Agricultural Technical Services, South Africa.

Verkley, G.J.M., Starink-Willemse, M., van Iperen, A. and Abeln, E.C.A. (2004). Phylogenetic Analyses of Septoria Species Based on the ITS and LSU-D2 Regions of Nuclear Ribosomal DNA. Mycologia, 96(3): 558-571.

Verma, H.N. and Awasthi, L.P. (1976). Sunnhemp rosette: A new virus disease of sunnhemp. Curr. Sci. 45(17): 642-643.

Verma, H.N. and Awasthi, L.P. (1978). Further studies on a rosette virus of *Crotolaria juncea*. Phytopathol. Z., 92(1): 83-87.

Verma, H.N. and Khan, M.M.A.A. (1984). Management of plant virus diseases by *Pseuderanthemum bicolor* leaf extract. Zeitschrift fur Pflanzenkrankheiten und Pflanzenschutz. 91: 266-272.

Verma, H.N. and Varsha (1995). Induction of systemic resistance by leaf extract of *Clerodendrum aculeatum* in sunnhemp against sunnhemp rosette virus. Indian Phytopath. 48 (2): 218-221.

Verma, J.P. (1986). Bacterial blight of cotton. CRC Press, Florida, USA.

Verma, J.P. and Singh, R.P. (1971). Epidemiology and control of bacterial blight of cotton. Proceedings of the Indian National Science Academy, B, 37: 326-331,

Verma, J.P. and Singh, R.P. (1974). Recent studies on the bacterial diseases of fiber and oil seed crops in India. In: Raychauhri S.P., Verma J.P. (eds). Current Trends in Plant Pathology, pp. 134-145. Luknow University, Luknow, India.

Verma, J.P. and Singh, R.P. (1975). Studies on the distribution of races of *Xanthomonas malvacearum* in India. Indian Phytopathol., 28: 459-463.

Verma, K.S. and Verma, A.K. (2002). Cutworm species associated with different crops in Himachal Pradesh. Insect Environment, 8(1): 23.

Verma, S., Haseeb, M. and Manzoor, U. (2013). Biology of red cotton bug, *Dysdercus cingulatus*. Insect Environment, 19(3): 140-141.

Vest, G., and Comstock, V.E. (1968). Resistance of flax to seedling blight caused by *Rhizoctonia solani*. Phytopathology, 58: 1161–1163.

Vetek, G., Papp, V., Haltrich, A. and Redei, D. (2014). First record of the brown marmorated stink bug, *Halyomorpha halys* (Hemiptera: Heteroptera: Pentatomidae), in Hungary, with description of the genitalia of both sexes. Zootaxa, 3780(1): 194-200.

Viegas, A.P. and Krug, H. P. (1935). A murcha do algodoeiro. Rev. Agric. S. Paulo, 10: 49-51.

Vierbergen, G. and Mantel, W.P. (1991). Contribution to the knowledge of *Frankliniella schultzei* (Thysanoptera: Thripidae). Entomologische Berichten (Amsterdam), 51: 7-12.

Viji, C.P. and Bhagat, R.M. (2001a). Bioefficacy of some plant products, synthetic insecticides and enteropathogenic fungi against black cutworm, *Agrotis ipsilon* larvae on maize. Indian Journal of Entomology, 63(1): 26-32.

Villiers, E.A. de and Stander, G.N. (1978). Control of the striped mealybug *Ferrisia virgata* on guavas. Citrus and Subtropical Fruit Journal, 541: 16-17.

Vincens, F. (1919). Rapport sommaire sur les travaux effectues au laboratoire de phyto-pathologie de l. Institut Scientifique de l' Indo-chine du l.er Janvier 1919 an l.er Juiller 1921. P.500.

Vinodkumar, S., Nakkeeran, S., Malathi, V.G., Karthikeyan, G. and Renukadevi, P. (2017). Tobacco streak virus: an emerging threat to cotton cultivation in India. Phytoparasitica, 45(5): 729-743.

Visalakshmi, V., ArjunaRao, P. and Krishnayya, P. (2000). Utility of sex pheromone for monitoring *Heliothis armigera* (Hub.) infesting sunflower. Journal of Entomological Research, 24: 255-258.

Vivek, U. and Nandihalli, B.S. (2015). Screening of Bt cotton genotypes against cotton thrips, *Thrips tabaci* (Lindeman) under field condition. Journal of Experimental Zoology, 18(1): 181-184.

Voicu, M., Popov, C., Ioan, I. and Luca, E. (1997). Halticines species from oil flax (*Linum usitatissimum* L.) crops in the Moldavian plain. (In Romanian).Cercetari Agronomice in Moldova, 30: 295–300. Abs. CAB Abstracts CD-ROM 1998–1999.

Volkl, W. and Stechmann, D.H. (1998). Parasitism of the black bean aphid (*Aphis fabae*) by *Lysiphlebus fabarum* (Hym.,Aphidiidae): the influence of host plant and habitat. Journal of Applied Entomology, 122: 201-206.

vonArx, J.A. (1957). Die Arten der Gattung Colletotrichum. Phytopathologische Zeitschrift., 29 (4): 413-468.

Wade, B., Kemerait, R., Brock, J., Hagan, A., Lawrence, K., Price, P., et al. (2015). Diagnosis and Management of Foliar Diseases of Cotton in the United States. UTcrops.com; Website: www.utcrops.com/cotton/Data%20Resources/Foliar%20Disease%20Bulletin.pdf

Wadud, M.A. and Ahmed, Q.A. (1962). Studies on fungous organisms associated with wilted jute plants. Mycopath Mycol Appl., 18: 107-114.

Walczak, F. (2002). Beware of noctuids and other soil pests. Ochrona Rosacute lin, 46(8): 8-10.

Wallace, M.M. and Diekmahns, E.C. (1952). Bole rot of Sisal. East African Agricultural Journal, 18(1): 24-29.

Walker, F. (1859). List of the Specimens of Lepidopterous Insects in the Collection of the British Museum. Supplement List Spec. Lepid. Insects Colln Br. Mus., 16: 1-253, 17: 255-508, 18: 509-798, 19: 799-1036.

Walker, F. (1863). Tortricites & Tineites. List of the specimens of lepidopterous insects in the collection of the British Museum, 28: 389.

Walker, G.P. (1982). The Dispersion and Abundance of the Potato Aphid [*Macrosiphum euphorbiae* (Thomas)] on Tomato (*Lycopersicon esculentum* Mill.). Ph.D. Dissertation, Ohio State University, Wooster.

Walker, G.P., Madden, L.V. and Simonet, D.E. (1984a). Spatial dispersion and sequential sampling of the potato aphid, *Macrosiphum euphorbiae* (Homoptera: Aphididae), on processing tomatoes in Ohio. Can. Entomol., 116: 1069–1075.

Walker, G. P., Nault, L.R. and D.E. Simonet, D.E. (1984b). Natural Mortality Factors Acting on Potato Aphid (*Macrosiphum euphorbiae*) Populations in Processing-Tomato Fields in Ohio. Environ. Entomol., 13(3): 724-732.

Walker, J.C. (1952). Sclerotinia Disease. In: Diseases of vegetable crops, McGraw and Hill Book Co., New York, P.529.

Wallace, T.P. and El-Zik, K.M. (1990). Quantitative analysis of resistance in cotton to three new isolates of the bacterial blight pathogen. Theoretical and Applied Genetics, 79(4): 443-448.

Walter J.W. and Everett J. (1982). Controlling cotton root rot on ornamental plants. Aggie Horticulture, Texas Agri Life Extension Service, Texas A&M System. Website: https://aggie-horticulture.tamu.edu/archives/parsons/.../cottonrootrot/cotton.html

Walters, K.F.A. and Lane, A. (1991). Incidence and severity of insects damaging linseed in England and Wales 1988-1989. Aspects Applied Biol., 28: 121-128.

Walters, M., Staten, R.T. and Roberson, R.C. (1998). Pink bollworm integrated management technology under field trial conditions in the Imperial Valley, CA. Proceedings Beltwide Cotton Conferences, San Diego, California, USA, 2: 1282-1285.

Wang, B., Brubaker, C.L., Summerell, B.A., Thrall, P.H. and Burdon, J.J. (2010). Local origin of two vegetative compatibility groups of *Fusarium oxysporum* f. sp. *vasinfectum* in Australia. Evol. Appl., 3: 505-524.

Wang, C. L. and Dai, Y.L. (2018). First report of sunnhemp Fusarium wilt caused by *Fusarium udum* f. sp. *crotalariae* in Taiwan. Plant Disease, 102(5): 1031.

Wang, H-S., Wang, M. and Fan, X-L. (2011). Notes on the tribe Nygmini (Lepidoptera: Erebidae) from Nanling National Nature Reserve, with description of a new species. Zootaxa: 2887: 57-68.

Wang, K.H. and McSorley, R. (2009). Management of nematodes and soil fertility with sunn hemp cover crop. Publication ENY-717. Univ. of Florida, IFAS Extension. http://edis.ifas.ufl.edu/ng043.

Wang, K.-H., Sipes, B.S. and Schmitt, D.P. (2002). Crotalaria as a cover crop for nematode management: A review. Nematropica, 32: 35-57.

Wang, L.H. (1971). Embryology and life cycle of *Tylenchorhynchus claytoni* Steiner, 1937 (Nematoda: Tylenchoidea). Journal of Nematology, 3(2): 101-107.
Wang, S-L., Li, G-K., Wang, J., Wang, J., Han, X-J., Chu, Z-J. (2006). Efficacy of Several Fungicides on Flax Powdery Mildew [J]; Xinjiang Agricultural Sciences. Website: https://www.en.cnki.com.cn/Article_en/CJFDTOTAL-XJNX200604016.htm
Wang, X.X., Chen, J., Wang, B., Liu, L.J. Huang, X., Ye, S.T. and Peng, D.X. (2011). First Report of Anthracnose on *Boehmeria nivea* Caused by *Colletotrichum higginsianum* in China. Plant Dis., 95 (10): 1318.
Wang, X.X, Chen, J., Wang, B., Liu L.J., Jiang, H., Tang, D. and Peng, D.X (2012). Characterization by suppression subtractive hybridization of transcripts that are differentially expressed in leaves of anthracnose-resistant ramie cultivar. Plant Molecular Biology Reporter, 30 (3): 547-545.
Wang, X.X., Wang, B., Liu, J. L., Chen, J., Cui, X. P., Jiang, H. and Peng, D. X. (2010). First report of anthracnose caused by *Colletotrichumgloeosporioides* on ramie in China. Plant Disease, 94 (12): 1508.
Wang, Y., Ma, J.Q., Yuan, Q.H., Li, X.L. and Miao, L.H. (2018). First report of *Colletotrichum linicola* causing anthracnose on Alfalfa in China, Plant Disease, 102(5): 1039.
Wang, Y., Meng, Y.L., Zhang, M., Tong, X.M., Wang, Q.H., Sun, Y.Y., Quan, J.L., Govers, F. and Shan, W.X. (2011). Infection of *Arabidopsis thaliana* by *Phytophthora parasitica* and identification of variation in host specificity. Mol Plant Pathol., 12(2): 187-201.
Wang, Y.P., Wu, S.A. and Zhang, R.Z. (2009). Pest risk analysis of a new invasive pest, *Phenacoccus solenopsis*, to China. Chinese Bulletin of Entomology, 46(1): 101-106.
Wangikar, P.D. and Shivankar, S.K. (1989). Investigation of *R. areola* causing Dahiya or Grey mildew of cotton in relation to physiological races, crop losses and control. Umpublised Ph.D. (Agri.) Thesis Dpt of Plant Pathology, Post Graduate Institute, P.K.V., Akola. P.191.
Wankhede, N.P. and Sadaphal, M.N. (1977). Response of the germplasms of upland cotton to population pressures and fertility levels and their reaction to bollworm damage. Indian Journal of Agricultural Sciences, 47(1): 8-14.
Waqar, J., Shafqat, S. and Nadir, N.M. (2013). Biology and bionomics of *Dysdercus koenigii* F. (Hemiptera: Pyrrhocoridae) under laboratory conditions. Pakistan Journal of Agricultural Science, 50(3): 373-378.
Ward, M. and Kalleshwaraswamy, C.M. (2017). Biology of Bihar hairy caterpillar, *Spilarctia obliqua* (Walker) (Erebidae: Lepidoptera) on field bean. Agric Update, 12(5): 1256-1260.
Ward-Gauthier, N., Beale, J., Amsden, B., and Dixon, E. 2015. Greenhouse hemp in Kentucky exhibits many common diseases. American Phytopathological Society Annual Meeting, Poster, P.502.
Waring, G.L. and Smith, R.L. (1986). Natural history and ecology of *Scyphophorus acupunctatus* (Coleoptera: Curculionidae) and its associated microbes in cultivated and native agaves. Annals of the Entomological Society of America, 79(2): 334-340.
Watanabe, T. (1947). Sen-i sakumotsubyogaku (fiber crop disease) (in Japanese). Asakura Publishing, Tokyo, pp.112-113.
Watanabe, T. (1948). Diseases of industrial crops (in Japanese). 256 pp, Yokendo, Tokyo.
Waterhouse, D.F. and Norris, K.R. (1987). Biological Control: Pacific Prospects. Inkata Press, Melbourne, P. 454.
Waterhouse, G.M. (1963). Key to the species of *Phytophthora* de Bary. Mycological Papers, 92: 1-22.
Waterworth, H.E., Lawson, R.H. and Monroe, R.L. (1976). Purification and Properties of *Hibiscus Chlorotic Ringspot Virus*. Phytopathology, 66: 570-575.

Watkins, G.M. (ed). (1981). Compendium of cotton diseases. 1st ed. The American Phytopathological Society, St. Paul, Minnesota, P.87.

Watson, M., El-Beheiry, M.M. and Guirguis, M.W. (1985). Laboratory and field studies on the effect of sequential application of pesticides on susceptibility and on some biological aspects of mite, *T. Urticae* (Boisd.). Acarologia, 26: 17-23.

Watve, C.M. (1971). Biology and control of the banded-wing whitefly, Trialeurodes abutiloneus (Haldeman), on cotton in Louisiana. LSU Historical Dissertations and Theses. 2096. https://digitalcommons.lsu.edu/gradschool_disstheses/2096

Way, M. J., Smith, P.M. and C. Potter, C. (1954). Studies on the bean aphid (*Aphis fabae* Scop.) and its control on field beans. Annals of Applied Biology, 41(1): 117-131.

Weathersbee, A.A. III, Hardee, D.D. and Meredith, W.R. Jr.(1994). Effects of cotton genotype on seasonal abundance of cotton aphid (Homoptera: Aphididae). Journal of Agricultural Entomology, 11(1): 29-37.

Weber, D.W., Leskey, T.C., Cabrera Walsh, G.J. and Khrimian, A. (2014). Synergy of aggregation pheromone with methyl decatrienoate in attraction of brown marmorated stink bug. J. Econ. Entomol., 107: 1061-1081.

Weber, F.G. (1930). Gray leaf spot of tomato caused by *Stemphylium solani* sp. nov. Phytopathology, 20: 513-518.

Wei, C.T. (1950). Notes on *Corynespora*. Mycol Papers, 34: 1-10.

Wei, W., Xiong, Y., Zhu, W., Wang, N., Yang, G. And Peng, F. (2016). *Colletotrichum higginsianum* mitogen-activated protein kinase ChMK1: Role in growth, cell wall integrity, colony melanisation, and pathogenicity. Front. Microbiol., 7: 1212.

Wei, Y-X., Zhang, H., Pu, J-J.andLiu, X-M. (2014). First report of target spot of cotton caused by *Corynespora cassiicola* in China. Plant Disease, 98(7): 1006.

Wellman, L.F. (1977). Dictionary of tropical American crops and their diseases. The Scarecorw Press, Inc. Metuchen, New Jersey. P.666.

Wene, G.P. and Sheets, L.W. (1964). Notes on and control of stink bugs affecting cotton in Arizona. J. Econ. Entomol., 57: 60-62.

Wermelinger, B., Wyniger, D. and Forster, B. (2008). First records of an invasive bug in Europe: *Halyomorpha halys* Stal (Heteroptera: Pentatomidae), a new pest on woody ornamentals and fruit trees. Mitteilungen der Schweizerischen Entomologischen Gesellschaft, 81(1/2): 1-8.

Westdal, P.H. (1968). Host range studies of oat blue dwarf virus. Can. J. Bot., 46(11): 1431-1435.

Westdal, P.H., Barrett, C.F. and Richardson, H.P. (1961). The six-spotted-leafhopper, *Macrosteles fascifrons* (Stal.) and aster yellows in Manitoba. Can. J. Plant Sci., 4l: 320-331. .

White, T. J., et al. (1990). In: PCR Protocols: A Guide to Methods and Applications. M. A. Innis, et al., eds. Academic Press, San Diego, P.315.

Whitehead, A.G. (1968). Taxonomy of Meloidogyne (Nematoda: Heteroderidae) with descriptions of four new species. Transaction of the Zoological Society of London, 31: 263-401.

Whiteside, J.O. (1955). Stem break (*Colletotrichum curvatum*) of sunnhemp in southern Rhodesia. Rhodesian Agr. J., 52: 417-425.

Whitson, R.S. and Hine, R.B. (1986). Activity of propiconazole and other sterol-inhibiting fungicides against *Phymatotrichum omnivorum*. Plant Disease, 70(2): 130-133.

Wickens, G.M. and Logan, C. (1957). Plant pathology. Progress Report Experimental Stations of the Empire Cotton Growing Group Corporation (Uganda), pp.32-37.

Wienk, J. F. (1967). A Note on the Transmission of *Aspergillus Niger* by Adult Sisal Weevils. Pest Articles & News Summaries. Section B. Plant Disease Control , 13(4): 392-395.

Wienk, J.F. (1968a). *Phytophthora nicotianae*: a cause of zebra disease in Agave Hydrid No 11648 and other Agaves. East African Agricultural and Forestry Journal, 33: 261-268.

Wienk, J.F. (1968b). Observations on the spread and control of zebra disease in Agave Hybrid No 11648. PANS (B), 14(2): 142-146.

Wiersema, H.T. (1955). Flax scorch. Euphytica, Neth. J. Plant Breeding, 4: 197–205.
Wightman J.A. (1972). Comparison of the Distributions of the Pink and the Green Biotypes of the Potato Aphid, *Macrosiphum euphorbiae* (Thos.), on Potato Plants. Plant Pathology, 21: 69-72.
Wijesekara, H.T.R. (2005). Taxonomic studies on the-genus *Colletotrichum* Corda and its Molecular Characterization. PhD Thesis, Division of Plant Pathology, Indian Agricultural Research Institute, New Delhi (India).
Wijkamp, I. (1995). Virus-vector relationships in the transmission of tospoviruses. PhD thesis, Wageningen University, The Netherlands.
Wijkamp, I., Almarza, N., Goldbach, R. and Peters, D. (1995). Distinct levels of specificity in thrips transmission of tospoviruses. Phytopathology, 85(10): 1069-1074.
Wilhelm, S. (1981). Sources and genetics of host resistance in field and fruit crops. In: Mace ME, Bell AA, Beckman CH (eds), Fungal Wilt Diseases of Plants. New York, Academic Press, pp.300-369.
Wilken, P.M., Steenkamp, E.T., Wingfield, M.J., de Beer, Z.W. and Wingfield, B.D. (2013). IMA Genome-F 1: *Ceratocystis fimbriata*: Draft nuclear genome sequence for the plant pathogen, *Ceratocystis fimbriata*. IMA Fungus, 4(2): 357–358.
Wilkerson, J.L., Webb, S.E., and Capinera, J.L. (2005). Vegetable Pests II: Acari - Hemiptera - Orthoptera -Thysanoptera. UF/IFAS CD-ROM. SW 181.
Williams, D.J. (1996). A brief account of the hibiscus mealybug *Maconellicoccus.hirsutus*, a pest of agriculture and horticulture, with descriptions of two related species from southern Asia. Bulletin of Entomological Research, 86: 617–628.
Williams, D.J. (2004). Mealy bugs of southern Asia. Natural History Museum Jointly with Southdene.
Williams, D.J. and Granara de Willink, M.C. (1992). Mealybugs of Central and South America. Wallingford, UK: CAB International, P.635.
Williams, D.J. and Watson, G.W. (1988). The scale insects of the Tropical South Pacific Region, Pt 2: The mealybugs (Pseudococcidae): CAB International Institute of Entomology, London, pp.260.
Williams, M.R. (2006). Cotton insect losses-2005. Proceedings of the Beltwide Cotton Conference, Jan. 3-6, National Cotton Council of America, San Antonio, Taxes, pp.1151-1204.
Williamson, B., Tudzynski, B., Tudzynski, P. and Van Kan, J. a. L. (2007). *Botrytis cinerea*: the cause of grey mould disease. Molecular Plant Pathology. 8 (5): 561-580.
Willrich, M.M., Leonard, B.R. and Cook, D.R. (2003). Laboratory and field evaluations of insecticide toxicity to stink bugs (Heteroptera: Pentatomidae). Journal of Cotton Science, 7: 156-163.
Willsie et al. (1942). In: W.Cranshaw (slides pre.), Hemp Insect Pest Management Challenges-Colorado State University, http://webdoc.agsci.colostate.edu/bspm/InsectInformation/Talks2017/6Hemp.pdf
Wilson, A.G.L. (1972). Distribution of pink bollworm, *Pectinophora gossypiella* (Saunders), in Australia and its status as a pest in the Ord irrigation area. J. Aust. Inst. Agric. Sci., 38: 95-99.
Wilson, C.E. (1923). Insect Pests of Cotton in St. Croix and Means of combating them. Bull. Virgin Is. (U.S.) Agric. Exp. Sea., 3: 20.
Wilson, F.D. and Menzel, M.Y. (1964). Kenaf (*Hibiscus cannabinus*), roselle (*Hibiscus sabdariffa*). Economic Botany, 18(1): 80–91
Wilson, F.D. and Summers, T.E. (1966). Reaction of kenaf, roselle and related species of Hibiscus to root-knot nematodes. Phytopathology, 56: 687-690.
Wilson, I. (1946). Observations on wilt disease of flax. Transactions of the British Mycological Society, 29(4): 221-231.

Wilson, L.J. (1993). Spider mites (Acari: Tetranychidae) affect yield and fiber quality of cotton. J. Econ. Entomol. 86: 566–585.
Win, N.K.K, Jung, H.Y and Ohga, S. (2011). Characterization of Sunn Hemp Witches Broom of Phytoplasma in Myanmar. Journal Faculty of Agriculture Kyushu University, 56(2) 217-221.
Wise, I.L. and Lamb, R.J. (1995). Spatial distribution and sequential sampling methods for the potato aphid, *Macrosiphum euphorbiae* (Thomas) (Homoptera: Aphididae), in oilseed flax. Can. Entomol., 127: 967-976.
Wise, I.L., Lamb, R.J. and Kenaschuk, E.O. (1995). Effects of the potato aphid *Macrosiphum euphorbiae* (Thomas) (Homoptera: Aphididae) on oilseed flax, and stage-specific thresholds for control. Can. Entomol., 127: 213-224.
Wise, I.L. and Soroka, J.J. (2003). Principal insect pests of flax. pp.142–145. In: Flax the genus *Linum*"(A.D. Muir, N.D. Westcott, eds.). London, New York, P.299.
Wolcott, G.N. (1933). An Economic Entomology of the West Indies. Ent. Soc. Puerto Rico, San Juan, P. 688.
Wong, A.L. and Willetts, H.J. (1975). Electrophoretic studies of Australasian, North America and European isolates of *Sclerotinia sclerotiorum* and related species. Journal of General Microbiology, 90: 355-359.
Wong, C.C., Daham, M.D. M., Aziz, A.M. A. and Abdullah, O. (2008). Kenaf germplasm introductions and assessment of their adaptability in Malaysia. J. Trop. Agric. and Fd. Sc., 36(1): 1-19.
Wood, C.M. and Ebbels, D.L. (1972). Host range and survival of *Fusarium oxysporum* f. sp. *vasinfectum* in North-western Tanzania. Cotton Grower Rev., 49: 79-82.
Woodhouse, E.J.(1913). Crop Pests Handbook for Bihar and Orissa (including West Bengal), Calcutta.
Woodruff, R.E. and Pierce, W.H. (1973). *Scyphophorus acupunctatus*, a weevil pest of Yucca and Agave in Florida (Coleoptera: Curculionidae). Entomology Circular, Florida Department of Agriculture and Consumer Services, 135: 1-2.
Wu, S., Xiong, J. and Yu, Y. (2015). Taxonomic resolutions based on 18S rRNA genes: a case study of subclass copepoda. PLoS One, 10(6):e0131498.doi:10.1371/journal.pone.0131498.
Wu, S.A. and Zhang, R.Z. (2009). A new invasive pest, *Phenacoccus solenopsis* threatening seriously to cotton production. Chinese Bulletin of Entomology. 46(1): 159-162.
Wu, Y.F., Cheng, A.S., Lin C.H. and Chen, C.Y. (2013). First Report of Bacterial Wilt Caused by *Ralstonia solanacearum* on Roselle in Taiwan. Plant Disease, 97(10): 1375.
Wulff, E.G., Mguni, C.M., Mortensen, C.N., Keswani, C.L. and Hockenhull, J. (2002). Biological control of black rot (*Xanthomonas campestris* pv. *campestris*) of Brassicas with antagonist strain of *Bacillus subtilis* in Zimbabwe. Europian Journal of Plant Pathology, 108(4): 317-325.
Wyllie, T.D. (1988). Soybean Diseases of the North Central Region. In: Wyllie, T.D. and Scott, D.H., editors. Charcoal rot of soybean-current status: APS, St. Paul. pp.106-113.
Wyllie, T. D. (1998). Charcoal rot. pp. 30-32. In: Sinclair J.B. & Backman, P.A. (eds), Compendium of Soybean diseases, 3rd ed. American Phytopathol. Soc., St. Paul, MN.
Xie, L., Hong, X.Y. and Xue, X.F. (2006). Population genetic structure of the two spotted spider mite (Acari: Tetranychidae) from China. Ann. Entomol. Soc. Am., 99: 959-965.
Xie, L., Xie, R.R., Zhang, K.J., Hong, X.Y. (2008). Genetic relationship between the carmine spider mite *Tetranychus cinnabarinus* (Boisduval) and the two-spotted mite *T. urticae* Koch in China based on the mtDNA COI and rDNA ITS2 sequences. Zootaxa, 1726: 18-32.
Xu, J., Fonesca, D.M., Hamilton, G.C., Hoelmer, K.A. and Nielsen, A.L. (2014). Tracing the origin of US brown marmorated stink bugs, *Halyomorpha halys*. Biol. Invasions., 16: 153-166.

Xu, Z. (2008). Overwinter survival of *Sclerotium rolfsii* and *S. rolfsii* var. *delphinii*, screening hosts for resistance to *S. rolfsii* var. *delphinii* and phylogenetic relationships among *Sclerotium* species. Graduate Theses and Dissertations, Iowa State University. http://lib.dr.iastate.edu/etd/10366

Xu, Z., Harrington, T.C., Gleason, M.L. and Batzer, J.C. (2010). Phylogenetic placement of plant pathogenic *Sclerotium* species among teleomorph genera. Mycologia, 102(2): 337-346.

Xu, Z., Zhu, W., Liu, Y., Liu, X., Chen Q, Peng, M., Wang, X., Shen, G. and Lin He, L. (2014). Analysis of Insecticide Resistance-Related Genes of the Carmine Spider Mite *Tetranychus cinnabarinus*. Based on a De Novo Assembled Transcriptome. PLoS ONE, 9(5): e94779.

Yaegashi, H., Nakamura, H., Sawahata, T., Sasaki, A., Iwamani, Y., Ito, T. and Kanematsu, S. (2013). Appearance of mycovirus-like double-stranded RNAs in the white root rot fungus, *Rosellinia necatrix*, in an apple orchard. FEMS Microbiol Ecol., 83: 49-62.

Yaku, A., Walter, G.H. and Najar-Rodriguez, A.J. (2007). Thrips see red - flower colour and the host relationships of a polyphagousanthophilic thrips. Ecological Entomology, 32: 527-535.

Yalemar, J.A., Hara, A.H., Saul, S.H., Jang, E.B. and Moy, J.H. (2001). Effects of gamma irradiation on the life stages of yellow flower thrips, *Frankliniella schultzei* (Trybom) (Thysanoptera: Thripidae). Ann. Appl. Biol., 138: 263-268.

Yamamoto, W. (1934). Cercospora-Artenaus Taiwan (Formosa) II. J. Soc. Trop. Agr., 6(3): 599-608.

Yanagi, T. and Hagihara, Y. (1980). Ecology of the brown marmorated stink bug. Plant Prot., 34: 315-326.

Yang, J., Ii, F., Chen, J., Shi, Q. and Zhang, Y. (2000). Virulence test of *Bacillus thuringiensis* on black cutworm. Journal of Fujian Agricultural University, 29(1); 65-68.

Yang, X., Guan, F-Z., Li, Z-G., Zhao, Y., Wang, X., Lu, Y., Song, Y., Zhang, L-G., Xiao, H. and Chen, H. (2011). RAPD markers linked to the powdery mildew resistance gene in flax line 9801-1. Acta Phytopathologica Sinica. Website: https://www.en.cnki.com.cn/Article_en/CJFDTOTAL-ZWBL201102016.htm

Yang, X., Li, Z-G., Guan, F-Z., Liu L-Y., Wu, G-W., Wang, X., Song, X-Y. and Zhao, D-S. (2007). A Study on the Occurrence Regularity of Flax Powdery Mildew. Plant Fiber Sciences in China. Website: https://www.en.cnki.com.cn/Article_en/CJFDTOTAL-ZGMZ200702009.htm

Yang, X., Zhao, Y., Guan, F-Z., Li, Z-G., Liu, L-Y., Wu, G-W., Wang, X., Song, X-Y.Liu, Z-J., Lu, Y., Li, T. and Kang, Q-H. (2008). Genetic analysis of resistance to powdery mildew in flax line 9801-1. Acta Phytopathologica Sinica. Website: https://www.en.cnki.com.cn/Article_en/CJFDTOTAL-ZWBL200806017.htm

Yaroslavtzev, G.M. (1931). A brief report on the pests of field cultures in 1930 according to the data of the State Service of dynamics and distribution of the injurious insects. (In Russian). Plant Protection, 8: 375-413.

Yasmin, S., Latif, M.A. and Akhter, N. (2013). Effect of Neem (*Azadirachta indica*) and other Plant Extracts on Yellow Mite of Jute. International Journal of Bio-Resource & Stress Management, 4(3): 412-417.

Yasuda, K. (1992). Cotton bug, pp.22-23. In: T. Hidaka (ed.), Insect Pests of Vegetables in the Tropics. Association for International Cooperation of Agriculture and Forestry, Tokyo, Japan.

Yein, B.R.(1983). Relative susceptibility of some cotton cultivars to insect pests. Journal of Research, Assam Agric. Univ., 4: 141-147.

Yein, B.R. and Barthakur, M.P.(1985). Comparative efficiency of different insecticides against *Earias vittella* and *Sylepta derogate* (Fab.) on cotton. Journal of Research, Assam Agricultural University, 6(1): 65-67.

Yeshwanth, H.M. (2013). Taxonomy of mirid bugs (Hemiptera: Miridae) of South India. PhD Thesis, Department of Agricultural Entomology, University of Agricultural Sciences, GKVK, Bangalore,

Yepsen, Jr., R.B. (1976). Organic plant protection. Rodale Press, Emmaus, Pa.

You, J.M., Wang, Q.H., Wang, G.J., Lin, X.M., Guo, J., Ali, L.Q. and Guo, X.L. (2016). First report of *Phoma exigua* causing leaf spot on Japanese Ginseng (*Panax japonicas*) in China. Plant Disease, 100(2): 534.

Young, O.P. (1986). Host plants of the tarnished plant bug, *Lygus lineolaris* (Heteroptera: Miridae). Annals of the Entomological Society of America, 79(4): 747-762.

Yu, G. and Zhang, J. (2007). The brown-marmorated stink bug, *Halyomorpha halys* (Heteroptera: Pentatomidae) in P.R. China. International Workshop on Biological Control of Invasive Species of Forests, 1: 70-74.

Yu, H. and Sutton, J.C. (1997). Effectiveness of bumble bees and honey bees for delivering inoculums of *Gliocladium roseum* to raspberry flowers to control *Botrytis cinerea*. Biol. Control, 10: 113-122.

Yu, S.H. (2001). Korean species of Alternaria and Stemphylium. Nat. Inst. Agric. Sci. Technol., Suwon, Korea.P.212.

Yu, S.H. (2015). Fungal Flora of Korea Volume 1, Number 2 Ascomycota: Dothideomycetes: Pleosporales: PleosporaceaeAlternaria and Allied Genera. National Institute of Biological Resources, Republic of Korea.P.149.

Yu, Y., Liu, H., Zhu, A., Zhang, G., Zeng, L. and Xue, S. (2012). A review of root lesion nematode: identification and plant resistance. Adv. Microbiol., 2: 411-416.

Yu, Y., Zeng, L., Yan, Z., Liu, T., Sun, K., Zhu, T. and Zhu, A. (2015). Identification of Ramie Genes in Response to *Pratylenchus coffeae* Infection Challenge by Digital Gene Expression Analysis. Int J Mol Sci., 16(9): 21989-22007.

Yu, Y., Zeng, L., Huang, L., Yan, Z., Sun, K., Zhu, T. and Zhu, A. (2016). First Report of Black Leaf Spot Caused by *Alternaria alternata* on Ramie in China. Journal of Phytopathology, 164 (5): 358-361.

Yu, Y-T., Chen, J.,Gao, C-S., Zeng, L-B., Li, Z-M.,Thu, T-T., Sun, K., Cheng, Y. and Sun, X-P. (2016). First report of brown root rot caused by *Pythium vexans* on ramie in Hunan, China. Canadian Journal of Plant Pathology, 38 (3): 405-410.

Yuan, L., Mi, N., Liu, S., Zhang, H. and Li, Z. (2013). Genetic diversity and structure of the *Fusarium oxysporum* f.sp. *lini* populations on linseed (*Linum usitatissimum*) in China. Phytoparasitica, 41(4) Website: https://www.researchgate.net/publication/257790765_ Genetic_ diversity_and_structure_of_the_Fusarium_oxysporum_fsp_lini_ populations_ on_linseed_Linum_usitatissimum_in_China

Yuan, Q.C. and Wu, W. (1986). The loss of yield and price of lint cotton due to boll damage by pink bollworm. Acta Phytophylactica Sinica, 13(2): 91-96.

Yurlova, N.A., de Hoog, S. and van den Ende, B.G. (1999). Taxonomy of *Aureobasidium* and allied genera , Studies in Mycology, 43: 63-69.

Zabrin, R. (1981). The fungus that destroys pot. War on Drugs Action Reporter, June, pp.61-62.

Zahi, S.E-Z ., Aref, S.A.E-S. and Korish, S.K.M. (2016). The cotton mealybug, *Phenacoccus solenopsis* Tinsley (Hemiptera: Pseudococcidae) as a new menace to cotton in Egypt and its chemical control. Journal of Plant Protection Research, 56 (2): 111-115.

Zaki, F.N., El-Shaarawy, M.F. and Farag, N.A. (1999).Release of two predators and two parasitoids to control aphids and whiteflies. Anz. Schadlingskde., Pflanzenschutz, Umweltschutz, 72 (1): 19-20.

Zalar,P., Gostincar,C., de Hoog,G.S., Ursic,V., Sudhadham,M. and Gunde-Cimerman,N. (2008). Redefinition of *Aureobasidium pullulans* and its varieties, Stud. Mycol., 61: 21-38.

Zaman, M. and Karimullah, (1987).Lepidoptera of jute cultivars in Peshawar. Pakistan Journal of Agricultural Research, 8: 290-297.

Zancan, W.L.A., Chitarra, L.G. and Chitarra, G.S. (2011). Fungosassociados a podridao de macas do algodoeironaregiao de Primavera do Leste, MT, Brasil: ocorrencia, controlequimico e influêncianaqualidade da Fibra. Bioscience Journal, Uberlandia, 27(4): 518-525.

Zancan, W.L.A., Chitarra, L.G. and Chitarra, G.S. (2013). Cotton in Brazil: Importance and Chemical Control of Bolls Rot. In: Fungicides- Showcases of Integrated Plant Disease Management from Around the World, Chapter-7, pp.135-152.

Zarzycka, H. (1976). Physiological races of *Colletotrichum lini* (Westend). Tochinai in Poland. Rev. Plant Path., 56: 3581.

Zawirska, I. (1960). Development of a population of *Thrips lini* Lad. on flax during the vegetative period. (In Polish). Biuletyn Instytutu Ochrony Roslin, 10: 51-67.

Zawirska, I. (1963). A contribution to the bionomics of *Thrips linarius* Uzel. (In Polish). Biuletyn Instytutu Ochrony Roslin, 19: 1-10.

Zelenay, A. (1960). Fungi of the genus Fusarium occuring on seeds and seedlings of hemp and their pathogenicity. Pracenauk. Inst. Ochr. Rosl., Pozan, 2(2): 248-249.

Zeng, L., Shen, A., Chen, J., Yan, Z., Liu, T., Xue, Z. and Yu, Y. (2016). Transcriptome Analysis of Ramie (*Boehmeria nivea* L. Gaud.) in Response to Ramie Moth (*Cocytodes coerulea* Guenée) Infestation. Biomed Res Int. 2016, 3702789. Published online 2016 Feb 29. doi: 10.1155/2016/3702789, PMCID: PMC4789370

Zeng L. B., Xue Z. D., Yu Y. T., et al. (2013). Changes of occurrence and control of *Cocytodes caerulea* Guenee. Hunan Agricultural Sciences, 10: 23-24.

Zeyen R.J. and Banttari E.E. (1972). Histology and ultrastructure of oat blue dwarf virus infected oats. Can. J. Bot.. 50: 2511-2519.

Zhao, J., Sun, Y., Xiao, L., Tan, Y. and Bai, L. (2016). Complete mitochondrial genome of Cotton Leaf Roller *Haritalodes derogata* (Lepidoptera: Crambidae). Mitochondrial DNA A DNA Mapp Seq Anal., 27(4): 2833-2834.

Zhalnina, LS. (1969). Porazhennost' konoplifuzariozompridlitel' nom primeneniiudobrenii. Khimiyasel'. Khoz., 7(11): 33-34.

Zhang, A., Amalin, D., Shirali, S., Serrano, M., Franqui, R., Oliver, J., Klun, J., Aldrich, J. and Meyerdirk, E. (2004). Sex pheromone of the pink hibiscus mealybug, *Maconellicoccus hirsutus*, contains an unusual cyclobutanoid monoterpene. Website: http://www.pnas.org/content/101/26/9601.

Zhang, D. (2014). Ramie diseases and its control (slides). In website Fibra. www.fibrafp7.net-Portals-4-Deyong.

Zhang, J., Sanogo, S., Percy, R.G., Wedegaertner, T. and Jones, D. (2017). Evaluation of Cotton for Resistance to Southwestern Cotton Rust (*Puccinia cacabata*). Beltwide Cotton Conferences, Dallas, TX, January 4-6, 2017. Website: https://www.researchgate.net/publication/317500382_Evaluation_of_Cotton_for_ Resistance_to_Southwestern_ Cotton_Rust_Puccinia_cacabata.

Zhang, J.F. and Sun, J.Z., (1988). Evaluation of resistance in cotton cultivars to *Sylepta derogata* Fabr. China Cottons, 6: 41-42.

Zhang, J.P., Wang, J.J., Zhao, Z.M., He, L. and Dou, W. (2003). In vitro inhibiting of esterases in *Tetranychus cinnabarinus* (Boisduval) by three insecticides. Acta Arachnologica Sinica, 12: 95-99.

Zhang, X. and Wong, S-M. (2009). *Hibiscus chlorotic ringspot virus* upregulates plant sulfite oxidase transcripts and increases sulfate levels in kenaf (*Hibiscus cannabinus* L.). Journal of General Virology, 90: 3042-3050.

Zhang, Y., Zhao, Z. and Cao, C. (1982). Studies on the damage of cotton caused by *Aphis gossypii* Glov. and the threshold for control. Acta Phytophylacica Sinica, 9: 229-236.

Zhang, Z., Ji, Z., Zhang, H., Huang, A. and Li, J. (1986). Identification of the pathogens of Crotalaria fusarium wilt and their pathogenecity. Acta Agriculture Boreali-Sinica, 1(1): 72-77.

Zhang, Z.Q. and Jacobson, R.J. (2000). Using adult female morphological characters for differentiating *Tetranychus urticae* complex (Acari: Tetranychidae) from greenhouse tomato crops in UK. Syst Appl Acarol., 5: 69-76.

Zhao, Y., Bian, J., Cai, L. and Gao, J. (2007). Ecology and control method of *Agrotis ipsilon* in cabbage fields. China vegetables, 9: 52.

Zhao, Y., Zhang, S., Luo, J.Y., Wang, C.Y., Lv, L.M. and Cui, J.J. (2016). Bacterial communities of the cotton aphid *Aphis gossypii* associated with Bt cotton in northern China. Scientific Reports, 6: 22958-22966.

Zheng, C.Q., Lin, H.R., Luo, L.Y., Chen, X.Y., Lai, Z.J., Yin, Z.G. and Lu, X.Y. (1992). Ramie Cultivar Records in China. Agriculture Press; Beijing, China (In Chinese)

Zheng, J., Gao, J., Zhang, S., Chen, H., Liu, Q. and Yi, K. (2011-2012). Pathogen identification of zebra disease of sisal. Journal of Northeast Agricultural University, website:en.cnki.com.cn/Article_en/CJFDTOTAL-DBDN201112016.htm

Zheng, L., Rujing, Lv, Huang, J., Jiang, D., Liu, X. and Hsiang, T. (2010). Integrated Control of Garlic Leaf Blight Caused by *Stemphylium solani* in China. Canadian Journal of Plant Pathology, 32(2): 135-145.

Zheng, T-Z., Tan, Y-F., Zhao, S., Huang, G-Y., Fan, H. And Ma, B. (1988). Mulberry cultivation, FAO Agriculture Service, Bulletin 73/1, Food and Agriculture Organization of the United Nations, Rome.

Zheng, Z., Nonomura, T., Bóka, K., Matsuda, Y., Visser, R. G. F., Toyoda, H., Kiss, L., and Bai, Y. (2013). Detection and quantification of *Leveillula taurica* growth in pepper leaves. Phytopathology, 103: 623-632.

Zhou, T., Fan, Z.F., Li, H.F. and Wong, S.M. (2006). *Hibiscus chlorotic ringspot virus* p27 and Its Isoforms Affect Symptom Expression and Potentiate Virus Movement in Kenaf (*Hibiscus cannabinus* L.). Molecular Plant-Microbe Interactions, 19(9): 948-957.

Zhou, X., Applebaum, S. and Coll, M. (2000). Overwintering and spring migration in the bollworm *Helicoverpa armigera* (Lepidoptera: Noctuidae) in Israel. Environmental Entomology, 29: 1289-1294.

Zhou, Z.C. and Chen, Y.N. (2004). Forecasting tobacco loss damaged by *Agrotis ipsilon* with moth-rainfall coefficient. Chinese Tobacco Science, 25(2): 13-16.

Zhu, G.P., Bu, W.J., Gao, Y.B. and Liu, G.Q. (2012). Potential geographic distribution of Brown Marmorated Stink Bug invasion (*Halyomorpha halys*). PLoS ONE, 7(2): e31246.

Zhu, S., Liu, T., Tang, Q. and Tang, S. (2012). Physio-ecological and cytological features of ramie from continuous cropping system. J. Hunan Agric. Univ. (Nat. Sci.), 38: 360-365.

Zhu, S., Tang, S., Tang, Q. and Liu, T. (2014). Genome-wide transcriptional changes of ramie (*Boehmeria nivea* L. Gaud) in response to root-lesion nematode infection. Gene, 552 (1): 67-74.

Zhu, X.Q. and Xiao, C.L. (2015). Phylogenetic, morphological and pathogenic characterization of *Alternaria* species associated with fruit rot of Blueberry in California. Phytopathology, 105 (12): 1555-1567.

Zhu, Y. and Qiang, S. (2011). *Curvularia eragrostidis*, a promising mycoherbicide agent for grass weeds. Pest Technology, 5(Special Issue 1): 61-66.

Zhu, Y.C. and Luttrell, R. (2014).Altered Gene Regulation and Potential Association with Metabolic Resistance Development to Imidacloprid in the Tarnished Plant Bug, Lyguslineolaris. Pest Management Science, 71(1): 40-57.

Zimmer, D.E. and Hoes, J.A. (1974). Race 370, a new and dangerous North American race of flax rust. Plant Dis. Rep., 58: 311-313.

Zimmerman, E.C. (1958). *Heliothis zea* (Boddie). pp. 213-215. In: Insects of Hawaii A Manual of the Insects of the Hawaiian Islands, including Enumeration of the Species and Notes of the Origin, Distribution, Hosts, Parasites, etc. Volume 7: Macrolepidoptera. The University Press of Hawaii, Honolulu. P.542.

Zornitsa, B. S., Rossitza, M. R., Ilija, K., Biljana, K. ,Vasilissa, I. M. and Ralitsa, G. G. (2013). Morphological and molecular characterization of *Colletotrichum coccodes* isolated from pepper cultivated in Bulgaria and Macedonia. Jour. Nat. Sci, Matica Srpska Novi Sad, 124: 249-261.

Index

A

A. attenuata, 389
A. brasiliensis, 382
Aphis citri, 443
Aphis citrulli, 443
Aphis cucumeris, 443
Aphis cucurbiti, 444
Aphis minuta, 444
A. nemorum, 272
A. tenuissima, 385
A. tubingensis, 382
Abelmoschus esculentus, 126
Acalypha sp. 450
Acanthospermum spp., 450
Acarus cinnabarinus, 501
Accelerometer, 520, 421
Aceria crotalariae, 168
Acontia xanthophila, 508
Acrosporium abelmoschi, 108
Acrosporium lini, 300
Acrosporium obductum, 109
Acrosternum hilare, 516
Acrosternum hilaris, 516
Aculops cannabicola, 284
Aculops lycopersici 284
Aecidium desmium, 427
Aecidium gossypii, 423
Aenasius arizonensis, 471
Aenasius bambawalei, 45, 46, 470
Aenasius bombawale, 133
Aenasius sp. 470
Aeolothrips fasciatus, 371
African cotton bollworm, 449
Agave attenuata var. *marginata,* 389
Agave sisalana, 377, 386, 387
Agave Snout Weevil, 390
Agave Weevil, 390
Ageratum yellow vein virus 83
AgipMNPV, 362
Agrilus acutus, 134
Agrotis cinereomacula, 357
Agrotis gularis, 357
Agrotis insignata, 356
Agrotis ipsilon, 360, 362
Agrotis ipsilon multiple, 362
Agrotis islandica, 357
Agrotis orthogonia, 349, 358
Agrotis orthogonoides, 358
Albigo humuli, 261
Alcides affaber, 136
Alcidodes affaber, 136
Aleiodes aligharensis, 514
Alfalfa mosaic virus, 84, 85, 269
Alphitomorpha humuli, 261
Alphitomorpha macularis , 261
Alternaria alternata, 181, 182, 385, 389, 400, 401,
Alternaria alternata var alternata, 181
Alternaria alternata var rosicola, 181
Alternaria blight, 314, 399
Alternaria gossypina, 400, 401,402
Alternaria Leaf spot, 181, 183, 399, 402, 403,
Alternaria lini, 314, 315, 317
Alternaria linicola, 314, 315, 317
Alternaria longipedicellata, 400
Alternaria macrospora, 399, 400, 402
Alternaria spp., 316, 403
Alternaria tenuis, 181, 400
Alternaria tenuis f. tenuis, 181
Alternaria, 18, 181, 182, 183, 314, 315, 317, 385, 399, 400, 401, 402, 403
Amaranthus gracilis, 203
Amaranthus spp, 450
American cotton bollworm, 446
Ampelomyces quisqualis, 263
AMV, 84, 85
Anaphes regulus 367
Angled gem, 53
Anguillula arenaria, 209
Angular leaf spot, 183, 184, 395, 399, 412
Anomis sabulifera, 53, 54, 538

Antennariella, 18
Anthracnose, 3, 4, 5, 61, 62, 63, 139, 186, 187, 188, 189, 190, 191, 192, 239, 242, 244, 287, 291, 377, 405, 406
Anthracnose boll rot, 405
Antirrhinum, 84
Apanteles tetragammae, 179
Aphanus hyalinipennis, 456
Aphanus tardus var. hyalipennis, 456
Aphidius colemani, 272
Aphidius nigripes, 348
Aphidoletes aphidimyza, 266, 270
Aphids, 21, 85, 222, 263, 264, 265, 266, 267, 270, 271, 346, 347, 348, 348, 442, 443, 444, 445, 446, 486
Aphis bauhiniae, 443
Aphis cannabis, 269
Aphis fabae, 264, 267
Aphis gossypii, 224, 442, 443
Aphthona aeneomicans, 365
Aphthona euphorbiae, 363, 365, 367
Aphthona virescens, 365
Aphusia speiplena, 512
Apion corchori, 58
Apple Weevil, 462
Aprostocetus minutus, 470
Arcte coerula, 231, 232
Arcte coerulea, 231
Areolate mildew, 395
Argina astrea, 165, 166
Argina cribraria, 166
Army cutworm, 348, 349, 350, 351, 368
Arsenophonus, 445
Artaxa scintillans, 129
Artemisia vulgaris, 273
Ascochyta linicola, 293
Ascochyta nicotianae, 106
Ash weevil, 33, 458, 460, 461, 462,
Asian admiral, 219,
Aspergillopsis nigra, 381
Aspergillus alabamensis, 382
Aspergillus niger, 380, 381, 390
Aspergillus niger, 380, 381, 390
Aster yellows, 338, 339, 340, 341, 343, 344
Aster Yellows Phytoplasma, 339, 340, 341
Athelia rolfsii, 1415, 64, 65, 115, 116, 148, 158, 197, 198, 199,
Atomoscelis hyalinus, 175
Atospora gossypii, 494, 496
Aureobasidium, 18, 324, 325

Aureobasidium lini, 324
Aureobasidium pullulans, 324, 325
Aureobasidium pullulans var. *lini*, 324
Autographa gamma, 368
AYP, 339
Azadirachta indica, 503
Azadirachtin 37, 48, 49, 270, 465
Azadiractin, 37, 233

B

B. subtilis, 398
B. tabaci, 530, 532, 533
B. thuringiensis sub sp. *israelensis*, 235
B. thuringiensis toxin, 449
Bacillus campestris, 2
Bacillus subtilis, 75, 119, 250
Bacillus thurigiensis, 362
Bacillus thuringiensis serovar *kurstaki*, 510
Bacillus thuringiensis var. *kurstaki*, 37
Bacterial blight, 244, 395, 397, 398, 407, 512
Bacterial boll rot, 406
Bacterial leaf blight, 1
Bacterial leaf spot, 1, 246
Bacterial Wilt, 88
Bacterium campestris, 2
Baculovirus heliothis, 449
Barathra brassicae, 349, 368
Basal stem blight, 292
BCTV, 344
Bean strain of tobacco mosaic virus, 160
Beauveria bassiana, 46, 56, 127, 130, 457
Beet curly top virus, 344
Begomovirus, 30, 31, 82, 83, 122, 124, 164, 204, 205
Begomoviruses, 30, 31, 82, 83, 125, 126, 164
Bemisia tabaci, 31, 83, 529, 530
Bhang Aphid, 268, 269, 270, 271, 272
Bhindi leaf roller, 463
Bihar hairy caterpillar, 35
Black arm disease, 395
Black band disease 6
Black Bean Aphid, 267, 268
Black cutworm, 360, 362, 363
Black drongo bird, 355
Black foot, 92, 93
Black heart, 417, 418
Black leg 89, 90, 91, 93, 96
Black root rot, 417
Black shank, 92, 93, 95

Black thrips, 375
Blacklight trap 518
Blossom blight 309
Blue dwarf virus, 342, 344
Boehmeria australis, 231
Boehmeria nivea, 130, 181, 191, 230, 231, 232
Boeremia (*Phoma*) leaf bligh,t 105
Boeremia exigua, 106, 111, 293
Boeremia exigua var. *linicola,* 293
Boermia exigua var. *coffeae,* 106
Boermia exigua var. *exigua,* 106
Bole rot, 380, 382, 383, 387, 390
Boll rot, 386, 395, 399, 402, 403, 404, 405, 406, 489, 496, 516
Bombax ceiba, 496
Bombyx pylotis, 166
Bombyx spinula, 361
Botryobasidium rolfsii, 14, 64, 115, 148, 197

Botryobasidium solani, 319
Botryodiplodia phaseoli, 26
Botryodiplodia theobromae, 7
Botrytis, 80, 254, 255, , 256, 311, 336, 421
Botrytis cinerea, 254, 255, 256, 336
Botrytis cinerea f. *cinerea,* 254
Botrytis cinerea f. *coffeae,* 254
Botrytis cinerea var. *cinerea,* 254
Botrytis cinerea var. *dianthi,* 254
Botys multilinealis, 464
Botys otysalis, 464
Bouteloua spp. 425, 426
Brachysporium eragrostidis, 195
Bracon brevicornis, 167, 173
Bracon Kirkpatrick, 492
Brassica napus, 165
Brinckochrysa scelestes, 514
Brinjal Grey weevil, 460
Brown marmorated stink bug, 515, 521, 522, 523
Brown root rot, 192, 194
Brown stem blight, 314
Brown stink bug, 515, 518, 520
Brumoides suturalis, 471
Brumus suturalis, 45, 230
Buchnera, 445
Buprestis acuta, 134
Burkholderia cepacia, 253

C

C. aestuans, 38
C. amaranticolor, 84
C. circinans, 243
C. destructivum, 289
C. fuscum, 289
C. gloeosporioides, 5, 189, 191, 192
C. higginsianum, 187, 188, 289
C. lineola, 243
C. medicaginea, 146, 147
C. pacificus, 483
C. pilolobus, 9
C. retusa, 145, 146, 149, 168
C. ruderalis, 269
C. sativa, 269, 273
C. spectabilis, 157, 218
C. spinaciae, 243
C. tridens, 9
C. americae-borealis, 289
Cabbage thrips, 372
Caeoma hibisciatum, 423
Cajanus cajan, 351, 450
Caliothrips indicus, 369, 375
Campylomma livida, 477, 480, 481
Campylomma morosa, 481
Candidatus Phytoplasma asteris, 341
Cannabis, 239, 240, 245, 246, 247, 248, 251, 252, 253, 254, 255, 258, 260, 264, 265, 267, 268, 269, 270, 271, 273, 276, 279, 281, 282, 284, 286
Cannabis aphid, 268, 270
Cannabis ruderalis, 248, 279
Cannabis sativa, 239, 245, 246, 251, 273, 284
Capnodium, 18
Capsicum annuum, 429
Capsus flavonotatus, 486
Capsus lineolaris, 486
Capsus oblineatus, 486
Capsus strigulatus, 486
Carmine spider mite, 500, 501
Carmovirus, 84
Ceiba pentandra, 496
Ceratobasidium solani, 319
Ceratocystis fimbriata, 152
Ceratocystis fumbriata f. *fimbriata,* 152
Ceratocystis moniliformis f. *coffeae,* 152

Ceratocystis wilt, 151
Ceratostomella fimbriata, 152
Cercospora boehmeriae, 184
Cercospora boehmeriana, 184
Cercospora cassicola, 18
Cercospora corchori, 9, 18, 19
Cercospora crotalariae, 142
Cercospora crotalariae-junceae, 142
Cercospora fukuii, 184
Cercospora gossypina, 413
Cercospora hibisci-esculenti, 70, 101
Cercospora hibisci-sabdariffae, 70, 101
Cercospora Leaf spot, 141, 183, 413
Cercospora malayensis, 69, 70, 71, 100, 101
Cercospora melonis, 19, 104, 408
Cercospora vignicola, 19, 104, 408
Cercospora vignicola, 19, 104, 408
Cercosporella gossypii, 410
Cerosypha gossypii, 444
Cerotelium desmium, 427
Cerotelium gossypii, 427
Chalcaspis arizonensis, 470
Charcoal rot, 23, 24, 25, 26, 29, 63
Charrinia diplodiella, 120
Cheilomenes sexmaculata, 222
Cheiloneurus sp., 470
Chenopodium, 84, 161
Chenopodium quinoa, 84
Chilomenes sexmaculata, 222
Chinavia hilare, 516
Chinavia hilaris, 515, 516
Chloridea armigera, 450
Chloridea obsoleta, 450
ChNRV1, 188
Choanephora americana, 150
Choanephora cucurbitarum, 150
Choanephora heterospora, 150
Choanephora infundibulifera f., 151
Cucurbitarum, 150, 151
Choanephora leaf blight, 149
Chorizagrotis auxiliaris, 349
Chromobacterium subtsugae, 270
Chrysoperla carnea, 45, 270, 272, 445, 553
Cimex cingulata, 496
Citronella, 514
Cladosporium, 18
CLCuBwV, 164
CLCuMB, 82, 125, 126
CLCuMuB, 122, 123
CLCuMuV, 122, 123

CLCuV, 124, 532
Cleome sp., 450
Clerodendrum viscosum, 48
Climbing cutworms, 367, 368
Clisosporium diplodiella, 120
Coccinella novemnotata franciscana, 445
Coccinella undecimpunctata, 445
Cochliobolus eragrostidis, 195
Cocytodes coerulea, 231
CoGMV, 30, 31, 205
Collar rot, 63, 64, 89, 92, 197
Collar rot and wilt, 92
Colletotrichum agaves, 377, 378
Colletotrichum atramentarium, 240
Colletotrichum bakeri, 242
Colletotrichum brassicae, 242
Colletotrichum coccodes, 157, 239, 240, 378, 379
Colletotrichum corchori, 4
Colletotrichum corchorum, 4, 5
Colletotrichum corchorum-capsularis, 4
Colletotrichum crotalariae, 157
Colletotrichum curvatum, 139, 140, 157
Colletotrichum dematium, 239, 242,
Colletotrichum gossypii, 405
Colletotrichum hibisci, 61
Colletotrichum hibisci-cannabini, 61
Colletotrichum higginsianum, 186, 187, 188
Colletotrichum higginsianum non-segmented dsRNA virus1, 188
Colletotrichum lini, 288
Colletotrichum linicola, 288,
Colletotrichum lysimachiae, 242
Colletotrichum pucciniophilum, 242
Colletotrichum sanguisorbae, 242
Colletotrichum sp., 62
Colletotrichum tabacum, 190
Colletotrichum truncatum, 140
Colletotrichum volutella, 142
Colletotrichum boehmeriae, 189
Colletotrichum gloeosporioides, 189, 190, 546
Colletotrichum higginsianum, 187, 188
Coniella diplodiella, 120, 121
Coniella hibisci, 72, 101, 102, 103, 540
Coniella musaiaensis var. *hibisci*, 102
Coniella petrakii, 120
Coniothyrium diplodiella, 120
Contarinia gossypii, 471, 474
Conyza bonariensis, 203

Corchorus capsularis, 463
Corchorus golden mosaic virus, 30, 31, 205
Corchorus olitorius, 1, 3, 8, 19, 57, 173
Corchorus spp., 35, 8
Corchorus tridens, 38
Corn earworm, 446, 449
Corticium rolfsii, 64, 197, 577, 14
Corticium sasakii, 319
Corticum rolfsii, 115, 148
Corynespora cassiicola, 18, 19, 103, 104, 408
Corynesporaleaf blight,
Corynespora mazei, 104
Corynespora melonis, 20, 408
Corynespora vignicola, 20, 408
Cotton anthocyanosis virus, 443
Cotton Aphid, 442, 443, 444, 445
Cotton blue disease, 443
Cotton bollworm, 446, 449
Cotton curliness virus, 443
Cotton Flower Bud Fly, 472
Cotton Flower-bud Maggot, 474
Cotton Gall Midge, 474
Cotton grey weevil, 458, 460
Cotton leaf curl Begomovirus, 124
Cotton leafcurl Burecoala virus,
Cotton leaf curl Multan beta satellite, 82, 125
Cotton leaf curl Multan virus, 122
Cotton leaf curl virus, 83, 124, 530
Cotton leaf roll and purple wilt viruses, 443
Cotton Leaf Roller, 463, 465
Cotton mealybug, 42, 130, 133, 446, 471, 537
Cotton midge, 472
Cotton necrosis disease, 431
Cotton Red Maggot, 476
Cotton red spider, 500
Cotton root-knot nematode, 206
Cotton rust, 422, 424, 425, 426, 427
Cotton seed bug, 455, 457
Cotton spotted bollworm, 506
Cotton stainer, 494, 496, 498, 499, 500, 553
Cotton stainers, 494, 499, 500
Cotton thrips, 527, 432
Cottony soft rot, 78
Cowpea chlorotic spot virus, 78
Cowpea mosaic virus, 160
Cowpea strain of tobacco mosaic virus, 160
Cowpea Yellow mosaic virus, 160
Crazy cotton, 485
Cremastus sp., 179
Creontiades biseratense, 477, 479

Creontiades dilutus, 477, 480, 482
Creontiades signatus, 483
Crimson-speckled flunkey, 171
Crimson-speckled moth, 171
Crinkle, 30, 43, 342, 343, 344, 439, 467, 474
Crotalaria albida, 146
Crotalaria mucronata mosaic virus, 160
Crotalaria pod borer, 545, 165
Crotalaria aegyptiaca,
Crotalaria juncea, 139, 145, 157, 166, 168, 169, 177, 218
Crotalaria spectabilis,
Crown rot, 92, 93, 96
Cryptolaemus montrouzieri, 45, 46, 132
Cucurbita maxima, 15
Curly top, 344
Curvularia eragrostidis,
Curvularia leaf blight, 194, 546
Cuscuta sp., 344
Cuscuta subinclusa, 123
Cydia delineana, 279
Cydia sinana, 279
Cylindrophora albedinis, 117
Cymbopogon martini, 514
Cymus cincticornis, 456

D

Dacrydium roridum, 67, 415
Dahiya disease, 410
Dalpada brevis, 521
Dalpata remota,
Damping off, 73, 74, 76, 91, 113, 139, 141, 148, 149, 157, 158, 174, 243, 254, 287, 292, 314, 335, 336
Damping-off, 74, 292, 314, 335, 336, 158
Dasineura gossypii, 471, 472
Dasineura lini, 352
Dasineura sampaina, 352
Datura spp., 450
Deiopea dulcis, 166
Deiopeia pulchella var *candida*, 171
Dematophora necatrix, 201
Depressaria gossypiella, 487, 489
Desetangsia humuli, 261
Diacrisia obliqua, 36
Diaeretiella rapae, 266
Dicaeoma hibisciatum, 423
Dieback, 4, 53, 61, 78, 335
Digitalis, 84

Diphorodon cannabis, 269
Diplodia boll rot, 405
Diplodia corchori, 7
Dothiorella phaseoli, 26
Downy mildew, 248, 249, 250, 271
Drino inconspicua, 173
Dulichos enation mosaic virus, 160
Dusky Cotton Bug, 455, 457, 458
Dysdercus cingulatus, 495, 496, 497, 550
Dysdercus koenigii, 494
Dysdercus megalopygus, 496
Dysdercus spp., 494
Dysdercus suturellus, 498, 499

E

E. Cichracearum, 299
E. lini, 299
Earias anthophilana, 508
Earias fabia, 512
Earias frondosana, 508
Earias huegeli, 512
Earias insulana, 506, 508, 510
Earias siliquana, 508
Earias simillima, 508
Earias smaragdinana, 508
Earias tristrigosa, 508
Earias vittella, 453, 511, 512
Eggplant ash weevil, 460
Egyptian bollworm, 506
Egyptian stem borer, 506
Elasmus homonae, 179
Elasmus sp. 353
Elleniaim portunitas, 175
Endoconidiophora fimbriata, 152
Ephestia gossypiella, 489
Epiccocum nigrum, 318
Epilachna beetle, 222
Epitrtranycus athaea, 282
Eremothecium gossypii, 495
Eriophid mite, 167
Erysiphe cichoracearum, 144, 300
Erysiphe cichoracearum f. *apocyni,* 300
Erysiphe diffusa, 145
Erysiphe humuli, 261
Erysiphe lini, 303
Erysiphe macularis, 261
Erysiphe orontii, 300
Erysiphe polygoni, 299, 303
Erysiphe tabaci, 300

Erysiphe taurica, 109, 258
Euoidium abelmoschi, 108
Euoidium lini, 300
Euonymus europaeus, 268
Euonymus spp. 267
Euproctis scintillans, 129
Eurasian hemp moth, 278
European corn borer, 273, 274, 275, 276
Eurytoma sp., 253
Euschistus servus, 407, 515, 518, 519
Euschistus servus euschistoides, 519
Euschistus servus servus, 518, 519
Eutetranychus cinnabarinus, 501
Eutetranychus cucurbitacearum, 501
Eutetranychus dianthica, 501
Eutettix tenellus, 345
Euthrips gobsypii
Euxoa auxiliaries, 368
Euxoa ochrogaster, 349, 356
Eye Rot, 66

F

F. lateritium, 405
F. macrophylla, 377
F. moniliforme, 405
F. oxysporum, 118, 119, 252, 253, 329, 331, 333, 405, 435
F. semitectum, 405
F. solani, 405
Ferrisia virgata, 38
Fibroidium abelmoschi, 107
Field thrips, 372, 373, 374
Flax Bollworm, 354, 355
Flax crinkle virus, 342
Flax flea beetle, 363, 364, 365, 366, 367
Flax thrips, 369, 372
Flea Beetle, 126, 127, 128, 169, 276, 277, 278, 363, 364, 365, 366
Floodplain cutworm, 360
Flower Bud Maggot, 472, 473, 474
Foliar blight, 97
Foot and stem rot, 92, 541
Foot rot, 147, 292
Formica subsericea, 271
FOV, 435, 436, 437, 438
Frankliniella schultzei, 527, 528
Frankliniella schultzei nigra, 528
Fulcrifera tricentra, 177
Fumago, 18

Furcraea, 377
Fusarium albedinis, 177
Fusarium apii, 117
Fusarium batatas, 117
Fusarium boll rot, 405
Fusarium bulbigenum, 117
Fusarium camptoceras, 111, 113
Fusarium lini, 328
Fusarium nygamai, 111, 112
Fusarium oxysporfum
Fusarium oxysporum f *apii*, 117
Fusarium oxysporum f. *lini*, 328
Fusarium oxysporum f. sp. *vasinfectum*, 434, 435, 436
Fusarium oxysporum f.sp. *cannabis*, 251
Fusarium oxysporum f.sp. *lini*, 327, 328, 548
Fusarium roseum, 405
Fusarium udum, 154, 155, 156, 157, 543
Fusarium udum var *pusillum*, 155
Fusarium udum var *solani*, 155
Fusarium vasinfectum, 434
Fusarium wilt, 89, 116, 154, 157, 251, 252, 327, 328, 332, 333, 433, 436, 437, 438, 543
Fusarium equiseti, 385
Fusidium udum, 155
Fusisporium udum, 155

G

G. barbadense, 395, 398, 399, 419, 437, 441, 513
Gelechia gossypiella, 489
Gelechia umbripennis, 489
Gelechiella gossypiella, 489
Geocoris sp., 480
Gibberella indica, 155
Gibberella nygamai, 111, 112
Gliocladium roseum, 257
Gliocladium sp. ,253
Gliocladium virens, 419, 149
Globisporangium megalacanthum, 337
Globisporangium ultimum, 74
Globisporangium ultimum var., 74 *sporangiferum*, 74, 75
Globisporangium ultimum var. *ultimum*, 74
Gloeosporium affine, 190
Gloeosporium fructigenum, 190
Gloeosporium lini, 288
Glomerella cingulata, 190

Glomus intraradices, 253
Golovinomyces cichoracearum, 12, 544, 144, 143, 144, 544
Golovinomyces orontii, 299, 300, 304
Gomphrena, 84, 450,
Gomphrena celosioides, 450
Goniozus sp. 179
Gonitis marginata, 54
Gonitis propingua, 54
Gonitis sabulifera, 54
Gossypium hirsutum var. *punctatum*, 398
Gram pod borer, 449
Grapholita compositella, 281
Grapholita delineana, 279
Gray mould, 336
Graymold, 253
Greasy cutworm, 360
Green mirid, 477
Green Peach Aphid, 264, 265, 266, 268
Green stink bug, 406, 515, 516, 517, 518, 520
Green weevil, 461
Grey mildew, 409, 410, 412
Grey mould, 253, 254, 256
Grey weevil, 33, 458, 460, 461

H

H. cannabinus, 63, 71, 72
H. Sabdariffa, 72, 85, 87, 93, 96, 99, 101, 102, 103
H. sabdariffa var. *altissima*, 133
Hairy caterpillar, 35, 37, 38, 128, 171, 172, 173, 542, 544,
Haltica attenuata, 277
Halyomorpha halys, 515, 521, 559
Hamiltonella, 445
HaNPV, 454
Haritalodes derogata, 227, 464
Harmonia axyridis, 272
HCRSV, 84, 85
Head blight, 254
Helianthus annus, 144
Helicotylenchus mucronatus, 206
Helicoverpa armigera, 449, 450, 454, 482
Helicoverpa armigera nuclear polyhedrosis virus, 454
Helicoverpa commoni, 450
Helicoverpa spp. 480
Helicoverpa stombleri, 447

Helicoverpa zea, 446, 447, 448
Heliothis armigera, 450
Heliothis intensiva, 355
Heliothis lugubris, 355
Heliothis NPV, 449
Heliothis ochracea, 447
Heliothis ononidis, 354
Heliothis ononis, 354
Heliothis rama, 450
Heliothis stombleri, 447
Heliothis umbrosus, 447
Heliothis zea, 447
Heliothrips indicus, 375
Helminthosporium cassiaecola, 104
Helminthosporium cassiicola, 408
Helminthosporium papaya, 20, 104, 408
Helminthosporium papayae, 104, 408
Helminthosporium vignae, 20, 104, 408
Helminthosporium vignicola, 20
Helminthosporium warpuriae, 104
Hemp aphid, 268
Hemp borer, 278, 279
Hemp flea beetle, 169, 276, 277
Hemp leaf rollers, 278
Hemp Louse, 268
Hemp mosaic virus, 160, 161, 269
Hemp moth, 278, 281
Hemp Russet Mite, 284, 286
Hemp seed eaters, 278
Heterodera arenaria, 209
Heterodera schachtii, 242, 244
Hibiscus cannabinus, 61, 463, 488
Hibiscus chlorotic ringspot virus, 84, 85
Hibiscus esculentus, 135, 507
hibiscus mealybug, 130, 230
Hibiscus mutabilis, 97
Hibiscus rosa-sinensis, 122
Hibiscus schizopetalus, 89
Hibiscuss, 84
Hippodamia convergens, 270, 445
Hippodamia convergens, 270, 445
Hololepta yucateca, 394
Hop, 65, 67, 70, 73, 75, 77, 79, 86, 88, 90, 94, 97, 99, 102, 104, 106, 262, 263, 269, 270, 271, 271, 272, 273, 276, 277, 279, 288, 293, 296, 300, 305, 310, 315
Hop Aphid, 270, 271, 272
Hop flea beetle, 276
Hoplolaimus indicus, 206
Howardula phyllotretae, 367

Humulus japonicas
Humulus lupulus, 260, 271, 273
Hyalopeplus linefer, 477
Hymenoscyphus sclerotiorum, 79, 310
Hyperaspis maindroni, 471
Hypoxylon necatrix, 201
Hyptis, 363

I

Indian red admiral, 219, 221, 547,
Indian tomato leaf curl virus, 164
Ipomea carnea, 363
Isaria fumosorosea, 498
IToLCV, 164

J

Jute, 1, 2, 3, 4, 5, 6, 8, 9, 10, 13, 14, 16, 17, 18, 19, 21, 22, 23, 25, 28, 29, 30, 31, 32, 33, 35, 36, 37, 38, 40, 42, 43, 46, 49, 50, 52, 53, 54, 55, 56, 57, 58, 59, 64, 66, 86, 87, 126, 133, 161, 173, 224, 235, 462, 463, 496
jute hairy caterpillar, 35, 37
Jute semilooper, 53, 54, 56
Jute stem-weevil, 57

K

Kabatiella lini, 324, 325
Kenaf, 32, 61, 62, 324, 325
Kenaf leaf curl virus, 82
KLCuV, 82
Korogwe leaf spot, 383, 384, 385
Kuehneola desmium, 427

L

L. usitatissimum, 304, 314
Lantana, 363
Large flea beetle, 363, 365, 366
Lasiodiplodia boll rot
Lasiodiplodia theobromae, 7, 405
Laspeyresia crocopa, 177
Laspeyresia delineana (Walker), 279
Laspeyresia pseudonectis, 177
Laspeyresia tricentra, 177
Leaf beetle, 224

Leaf blight, 1, 15, 72, 97, 105, 149, 194, 407, 409, 429
Leaf curl, 82, 83, 121, 122, 124, 125, 163, 164, 165, 204, 443, 501, 530, 532
Leaf fleck, 98, 107
Leaf hopper, 163, 344
Leaf roller, 225, 227, 278, 279, 463, 465
Leaf spot, 8, 18, 19, 69, 70, 71, 72, 84, 98, 100, 101, 103, 105, 119, 120, 141, 142, 181, 182, 183, 184, 242, 244, 246, 249, 282, 288, 377, 383, 384, 385, 387, 388, 395, 399, 402, 403, 413, 415, 417, 429, 508, 512
Leaf spot of hemp, 244, 246
*Lecythea lin*i, 305
Lepidiota mansueta, 236
Lepidiota sp., 236
Lesion nematode, 212, 213, 214, 215
Leucothallia macularis, 261
Leveillula taurica, 108, 109, 258, 259, 299
Leveillula taurica f. *cannabis*, 258
Limacinula, 18
Limothrips alli, 525
Linseed blossom midge, 351
Linseed bud fly, 351
Linseed thrips, 369, 375
Linum grandiflorum, 314
Linum usitatissimum, 287, 292, 293
Litchi weevil, 462
Longitarsus belgaumensis, 170
Longitarsus parvulus, 363, 365, 367
Loxostege sticticalis, 368
Ludwigia leaf distortion beta satellite, 82, 125
LuLDB, 82, 125
Lycopersicon esculentum, 378
Lycopersicum esculentum, 429
Lygaeus cingulatus, 496
Lygus lineolaris, 484, 486
Lygus pratensis var. *rubidus*, 846
Lysiphlebus fabarum, 268
Lysiphlebus testaceipes, 445

M

M. arenarea, 86, 87
M. javanica, 32, 86, 87, 210
M. thamesi, 206, 208, 209, 206, 208
Maconellicoccus hirsutus, 46, 130, 131, 230
Macrocentrus, 281
Macrophoma corchori, 23, 26
Macrophoma phaseoli, 26

Macrophoma phaseolina, 26
Macrophomina phaseoli, 26, 29, 64, 90, 158, 418
Macrophomina phaseoli var. *Indica*, 26
Macrophomina phaseolina, 23, 24, 26, 29, 90, 158, 418
Macrophomina philippinensis, 26
Macrophomina rot, 158
Macrosiphon solanifolii, 346
Macrosiphum amygdaloides, 346
Macrosiphum cyprissiae var. *cucurbitae*, 346
Macrosiphum euphorbiae, 346
Macrosporium gossypium, 400
Macrosporium macrosporum, 400
Macrosteles fascifrons, 339, 340, 343
Macrosteles quadrilineatus, 339, 340
Malachra capitata, 126
Malva vein clearing Potyvirus, 124
Malva vein clearing virus, 124
Malva verticillata, 71
Malvaviscus concinnus, 104
Mamestra configurata, 349, 368
Marijuana, 251, 264, 279
Marshalliella unicolor, 175
Mealy bug, 38, 40, 42, 43, 44, 45, 46, 130, 132, 133, 230, 231, 466, 467, 468, 470, 471
Melampsora lini, 305, 306
Melampsora lini var. *lini*, 305
Melia azadirachta, 49, 363
Melicleptria septentrionalis, 354
Meliola, 18
Meloidogyne arenaria, 208, 209
Meloidogyne arenaria arenaria, 209
Meloidogyne arenaria thamesi, 209
Meloidogyne incognita, 32, 36, 206, 436
Meloidogyne javanica, 32, 33
Meloidogyne species, 36, 87, 210, 211
Meloidogyne spp., 32, 86, 435
Meloidogyne thamesi, 209
Melon aphid, 442
Menochilus quadriplagiatus, 232
Menochilus sexmaculatus, 232
Mesta, 32, 35, 61, 63, 66, 70, 76, 81, 82, 84, 85, 87, 101, 121, 124, 125, 126, 127, 128, 130, 133, 135, 136, 230, 463, 488
Mesta Hairy caterpillar, 128
Mesta stem weevil, 135
Mesta yellow vein mosaic Bahraich virus, 82, 125
Mesta yellow vein mosaic virus, 82, 124, 125
Metarhizium anisopliae, 457, 498

Mexican sisal weevil, 390
MeYVMBV, 82, 125
MeYVMV, 82, 124, 125
Micra partita, 512
Micraspis discolour, 224
Microplitis croceipes, 448
Microsphaera diffusa, 145
Microstroma lini, 324
Microtermes obesi, 235
Microtermes sp., 234
Miller moth, 348, 350
Mirid bug, 477, 478, 479, 480, 482, 484
Mites, 46, 47, 48, 50, 51, 52, 129, 167, 168, 169, 232, 234, 235, 282, 283, 284, 285, 286, 457, 480, 500, 501, 502, 503, 504, 506, 529
Moissonia importunitas, 175
Moissoniaim portunitas, 175
Moringa olifera, 461
Morion georgiae, 394
Mucor cucurbitarum, 150
Mycosphaerella, 142, 296, 410, 413, 414
Mycosphaerella areola, 410
Mycosphaerella crotalariae, 142
Mycosphaerella gossypina, 413
Mycosphaerella Leaf spot, 414
Mycosphaerella linicola, 296
Mycosphaerella linicola var. *linicola,* 296
Myllocerus angustifrons, 462
Myllocerus discolor, 462, 463
Myllocerus fausti, 459
Myllocerus maculosus, 459
Myllocerus marmoratus, 459
Myllocerus pistor, 459
Myllocerus subfasciatus, 460, 461
Myllocerus undatus, 459
Myllocerus undecimpustulatus, 458, 459
Myllocerus undecimpustulatus undatus, 459
Myllocerus viridanus, 461, 462
Myrothecium advena, 67, 415
Myrothecium advena var *terricola,* 67
Myrothecium leaf spot, 415, 417
Myrothecium roridum, 67, 68, 415, 416
Myrothecium roridum var. *apiculatum,* 67
Myrothecium roridum var. *eichhorniae,* 67
Myrothecium roridum var. *roridum,* 67
Myrothecium roridum var. *violae,* 67
Myzus cannabis, 269

Myzus humuli, 271
Myzus persicae, 264

N

Nectarophora ascepiadis, 346
Neem extract, 457
Neem oil, 48, 52, 128, 130, 145, 286, 414, 515, 527, 533
Neemguard, 260
Nematicides, 33, 212, 215, 216, 219, 233, 438
Nematode, 32, 33, 85, 86, 87, 205, 206, 207, 208, 209, 211, 212, 213, 214 215, 216, 217, 218, 219, 244, 338, 367, 429, 435, 436, 437, 438, 440, 492
Nematospora gossypii, 494
Nephus regularis, 471
Nephus sp., 230
Nerium oleander, 363
Neurospora crassa, 311
Nezara hilaris, 516
Nezara viridula, 406, 417
Nigredo decorata, 147
Nigrospora oryzae, 404
Nilaparvata lugens, 234
Nipaecoccus viridis, 38, 40
Nisotra orbiculata, 127
Noctua armigera, 450
Noctua fabia, 512
Noctua ochrogaster, 356
Noctua pulchra, 171
Noctua suffusa, 360
Noctua ypsilon, 361
Nozemia nicotianae, 387
NPV, 56, 201, 362, 368, 449, 454
Nuclear polyhedrosis virus, 56, 368, 454
Nupserha bicolor, 174
Nygmia scintillans, 129

O

Oat blue dwarf marafivirus, 343
Oat blue dwarf virus, 342, 344
OBDV, 342, 343
Oidiopsis taurica, 109, 258
Oidium abelmoschi, 107, 108
Oidium erysiphoides, 10

Oidium lini, 299, 300
Oidium obductum, 109
OkMV, 124
Okra mosaic Tymovirus, 124
Okra mosaic virus, 124
Okra yellow mosaic Mexico virus, 125, 126
OkYMMV, 125, 126
Old World bollworm, 449
Oligonychus coffeae, 47
Onion thrips, 432, 524
Ophiostoma fimbriatum, 152
Orbilia obscura, 26
Ostrinia nubilalis, 274
Ovatisporangium vexans, 193
Oxycarenus cruralis, 456
Oxycarenus hyalinipennis, 456
Oxycarenus laetus, 456
Oxycarenus leucopterus, 456
Ozonium omnivorum, 420

P

P. aphanidermatum, 158
P. brachyurus, 206
P. celtidis, 250
P. cepacia, 322
P. cofleae, 206
P. cubensis, 250
P. ferrugunea, 262
P. filipendule, 262
P. fluorescens, 398, 403
P. nephrolepidis, 185
P. pouzolziae, 185
P. syringae pv. *alisalensis*, 245
Pachnephorus bretinghami, 224
Pachystachys spicata, 225
Padomyia setosa, 173
Paecilomyces lilacinus, 203, 218
Pale Western Cutworm, 358
Pantoea agglomerans, 406, 519
Pachystachys spicata, 225
Paralongidorus sp., 206
Paramyrothecium leaf spot, 417
Paramyrothecium roridum, 67, 68, 415, 416
Paraphorodon cannabis, 269
Paraphorodon omeishanensis, 269
Parthenium, 363, 431, 433
Pasmo, 294, 295, 297, 298, 299
Patania sabinusalis, 229
Patania silicalis, 229

Patania sp., 229
Patania spp., 225, 229
Pectinophora gossypiella, 453, 487, 489
Pellicularia rolfsii, 14, 64, 115, 148, 197
Pentatoma hilaris, 516
Peronoplasmopara cannabina, 249
Peronospora cannabina, 249, 250
Peziza sclerotiorum, 79, 310
Phakopsora desmium, 427
Phakopsora gossypii, 427
Phalaena (Noctua) astrea, 166
Phalaena astrea, 166
Phalaena cribraria, 166
Phalaena derogata, 464
Phalaena idonea, 361
Phalaena ipsilon, 360
Phanurus sp., 173
Phaseolus, 84, 177, 450
Phenacoccus cevalliae, 468
Phenacoccus gossypiphilous, 468
Phenacoccus hirsutus, 38, 131, 230
Phenacoccus solenopsis, 38, 42, 130, 133, 466, 468
Pheromone, 132, 272, 273, 275, 347, 357, 362, 393, 394, 454, 484, 492, 493, 510, 518, 523
Pheromone trap, 362, 393, 494, 484, 492, 510, 514, 518, 523,
Philadelphius spp., 267
Phloeophthora nicotianae, 93
Phloeospora ulmi, 297
Phlyctema linicola, 296
Phoma diplodiella, 120
Phoma exigua, 106, 107, 111, 293
Phoma exigua var *exigua*, 106, 107
Phoma exigua var. linicola, 293
Phoma exigua f.sp. *linicola*, 293
Phoma herbarum, 385
Phoma linicola, 293, 355
Phoma spp., 77
Phomopsis perexigua, 106
Phonoctonus spp., 498
Phorodon asacola, 269
Phorodon cannabis, 269
Phorodon humuli, 264, 270, 271, 272
Phorodon persicae, 264
Phyllody, 162, 339, 341
Phyllosticta, 97, 98, 106, 142, 190
Phyllosticta araliae, 190
Phyllosticta asclepiadearum, 190
Phyllosticta crotalariae, 142

Phyllosticta hibiscina, 97
Phyllosticta hibiscini, 97
Phyllosticta hydrangea, 106
Phymatotrichopsis omnivora, 420
Phymatotrichopsis root rot, 417, 419
Phymatotrichum omnivorum, 420
Physoderma corchori, 22
Phytomonas campestris, 2
Phytophthora boehmeriae, 404
Phytophthora boll rot, 404
Phytophthora melongenae, 93
Phytophthora nicotianae, 93, 94, 386, 387, 404
Phytophthora nicotianae var *nicotianae*, 387
Phytophthora nicotianae var. *parasitica*, 93, 387
Phytophthora parasitica, 93, 94
Phytophthora parasitica var *sabdariffae*, 93, 94
Phytophthora parasitica var. *nicotianae*, 93
Phytophthora parasitica var. *parasitica*, 93
Phytophthora parasitica var. *rhei*, 94
Phytophthora tabaci, 94
Phytophthora terrestris, 93
Phytoplasma, 162, 163, 338, 339, 340, 341
Phytopthora lycopersici, 387
Phytopthora parasitica var. *nicotianae*, 387
Phytopthora tabaci, 387
Phytopythium vexans, 193, 184
Phytoseiulus persimilis, 283, 286
Pilidiella diplodiella, 120, 121
Pilocrocis ramentalis, 225, 226
Pink and green potato aphid, 345
Pink bollworm, 453, 487, 488, 491, 492, 493
Pink hibiscus mealybug, 130, 230
Pink mealybug, 130
Pionnotes uda, 155
Plant lice, 263
Platyedra gossypiella, 489
Plectosphaera atractina, 190
Pleuroptya sp., 229
Plusiaori chalcea, 368
Podosphaera fuliginea, 10, 11, 12
Podosphaera fusca, 12
Podosphaera macularis, 258, 260, 261
Podosphaera sp, 110
Podosphaera xanthii, 10, 11, 12
Polygonum hydropiper, 49
Polyphagotarsonemus latus, 50
Polyspora lini, 234

Poppiocapsidea biseratense, 477, 478
Porricondyla gossypii, 471, 476
Potato aphid, 345, 346, 347, 348
Powdery mildew, 10, 11, 12, 13, 107, 108, 109, 110, 143, 144, 145, 146, 258, 259, 260, 261, 262, 263, 299, 300, 301, 302, 303, 304
Pratylenchus coffeae, 212
Pratylenchus elachistus, 206
Pratylenchus mahogani, 212
Pratylenchus musicola, 212
Pratylenchus penetrans, 338
Prosopis farcta, 203
Prothesia scintillans, 129
Prunus domestica, 271
Prunus spp., 264, 271
Psallus impictus, 175
Pseudocercospora boehmeriae, 184
Pseudocercospora boehmeriigena, 184
Pseudococcus filamentosus var. *corymbatus*, 38, 40
Pseudocochliobolus eragrostidis, 195
Pseudomonas campestris, 2
Pseudomonas cannabina, 245
Pseudomonas fluorescens, 18, 96, 116
Pseudomonas malvacearum, 396
Pseudomonas solanacearum, 88
Pseudomonas sp., 322
Pseudomonas syringae pv. *cannabina*, 245
Pseudoperonospora cannabina, 249, 250
Pseudoperonospora humuli, 249, 271
Psychotria nervosa, 15
Psylliodes attenuata, 277
Psylliodes attenuatus, 277
Psylliodes japonica, 277
Puccinia cacabata, 424, 425
Puccinia muhlenbergiae, 423
Puccinia schedonnardii, 422
Puccinia stakmanii, 425
Pullularia pullulans var. *lini*, 324
Pyralis glabralis, 274
Pyralis nubilalis, 274
Pyrameis asakurae, 220
Pyrameis atalanta, 220
Pyrameis buana, 220
Pyrameis callirhoe, 220
Pyrameis horishanus, 220
Pyrameis occidentalis, 220
Pyrausta nubilalis, 274
Pyrausta nubilalis f. *fanalis*, 274
Pyrethrins, 270

Pythium allantocladon, 193
Pythium butleri, 157
Pythium complectens, 193
Pythium delicense, 73
Pythium indicum, 73
Pythium megalacanthum, 337
Pythium piperinum, 193
Pythium sp, 157, 335
Pythium spp., 335
Pythium ultimum, 74, 75
Pythium ultimum var. *sporangiferum*, 75
Pythium ultimum var. *ultimum*, 74, 75
Pythium vexans, 793, 194
Pythium vexans var. *minutum*, 193
Pythium vexans var. *vexans*, 193

Q

Quarantine, 211, 246, 281424, 426, 428, 448, 452, 470, 491, 529

R

R. bataticola, 157, 418
Rachiplusia nu, 368
Ragmus importunitas, 175
Ragmus morosus, 481
Ralstonia solanacearum, 88, 89
Ramie mosaic, 203, 204, 205
Ramie mosaic begomovirus, 204
Ramie mosaic virus, 204, 205
Ramie moth, 231, 233, 234
Ramie MV Begomovirus, 204
RamMV, 204, 205, 547
Ramularia areola, 410
Ramularia areolata, 410
Ramularia arisaematis, 190
Ramularia gossypii, 410, 411
Ramulariopsis gossypii, 410
Red Cotton Bugs, 494
Red mite/ Red spider mite, 46
Red spider mite, 46, 48, 49, 500
Red-backed cutworm, 356, 358
Reniform nematode, 217, 218, 219
Reynoutrias achaliensis, 145
Rhinyptia meridionalis, 236, 237
Rhinyptia puncticollis, 236
Rhizoctonia bataticola, 26, 418
Rhizoctonia solani, 23, 111, 114, 157, 158, 242, 244, 318, 319

Rhizopus nigricans, 508
Rhopalocystis nigra, 381
Rhopalomyces elegans var *cucurbitarium*, 150
Rhynchophorus asperulus, 391
RnPV1-W8, 201
Rogas aligharensis, 514
Root and stem rot, 92, 147 149
Root Knot Disease, 32, 33
Root knot nematode, 436, 437, 438
Root lesion nematode, 212, 214
Root rot, 23, 63, 73, 74, 76, 89, 91, 116, 119, 157, 192, 194, 200, 202, 203, 212, 318, 319, 322, 417, 419
Root rot and wilt, 76, 89, 91, 116, 119
Root-knot nematode, 86, 87, 206, 208, 211, 437
Roselle, 61, 67, 72, 85, 86, 87, 88, 89, 90, 91, 92, 93, 96, 97, 98, 99, 100, 101, 102, 103, 105, 107, 108, 109, 113, 115, 116117, 118, 119, 121, 123, 124, 125, 126, 127, 130, 132, 230, 488
Rosellinia necatrix partiti virus 1-W8, 201
Rosellinia necatrix, 200, 201, 202
Rostrella coffeae, 152
Rotylenchulus sp., 217
Rullia prostrata, 268
Rust, 146, 147, 304, 306, 307, 308, 309, 422, 423, 424, 425, 426, 427, 428

S

S. apiicola, 297
S. coffeicola, 149, 198, 199
S. populicola, 149, 198, 199
S. rolfsii var. *delphini*, 149
Salvia greggii, 293
Scarce bordered straw worm, 449
Sclerotinia boll rot, 404
Sclerotinia libertiana, 79, 310
Sclerotinia sclerotiorum, 79, 80, 310, 311, 404
Sclerotinia stem rot, 309, 312, 313
Sclerotium rolfsii, 14, 15, 17, 64,65, 115, 148, 149, 157, 158, 197, 198
Sclerotium rot, 158, 197
Sclerotium varium, 79, 310
Scorch, 121, 145, 122, 327, 337, 377
Scorias, 18
Scutellonema brachyurum, 206
Scymnus pallidicollis, 132
Scyphophorus acupunctatus, 390, 391

Scyphophorus interstitialis, 382, 391
Scyphophorus interstitialis, 382, 391
Scyphophorus robustior, 391
Seedling blight, 111, 112, 113, 114, 157, 158, 291, 314, 315, 318, 319, 322, 323, 395
Seedling blight and root rot, 318, 322
Selenophoma linicola, 325
Selenosporella sp, 76
Septocylindrium gossypii, 410
Septogloeum linicola, 296
Septoria linicola, 296
Septoriosis, 294
Sesamum orientale, 351
Sesbania thrips, 375
Shoot borer moth, 176
Siphonophora euphorbiae, 346
Sisal weevil, 390, 394
Siteroptes reniformis, 404
Six-spotted zigzag ladybird beetle, 222
Soft Rot, 14, 78, 89, 117, 309
Solanum tuberosum, 377
Somena scintillans, 129
Sooty Mould, 9, 18, 19, 21, 264, 271, 443, 467, 530, 536
Southern blight, 92, 115, 116, 200
Southern root-knot nematode, 206
Southern sunn-hemp mosaic virus, 160
Southwestern cotton rust, 424, 425, 426
Spasm, 294
Sphacelotheca lini, 299
Sphaerella crotalariae, 142
Sphaerella gossypina, 413
Sphaerella linicola, 296
Sphaeronaema fimbriatum, 152
Sphaerotheca fuliginea, 10, 11, 12, 13
Sphaerotheca fuliginea f. *fuliginea*, 10
Sphaerotheca fuliginea var. *fuliginea*, 10
Sphaerotheca fusca, 12
Sphaerotheca humuli, 261
Sphaerotheca macularis, 261
Sphyracephala hearciana, 179
Spider mites, 46, 282, 283, 500, 503, 504
Spilarctia obliqua, 36
Spilarctia obliqua Montana, 36
Spilarctia obliqua obliqua, 36
Spilosoma obliqua, 35, 36
Spiny bollworm, 506, 507, 508, 509, 510, 511
Spiral borer, 133
Spondylocladium maculans, 195
Sporothrix flocculosa, 263

Spotted bollworm, 506, 511, 513
Sri Lanka weevil, 458
Stem break, 64, 139, 295, 323, 326
Stem break and browning, 323, 326
Stem fasciation, 341
Stem Gall, 21
Stem grinder, 176
Stem rot, 23, 24, 29, 64, 72, 78, 89, 92, 93, 95, 96, , 98, 115, 116, 147, 149, 197, 309, 312, 313
Stemphylium botryosum, 316
Stemphylium Leaf Blight, 429
Stemphylium Leaf Spot, 429
Stemphylium solani, 429, 430
Sterigmatocystis nigra, 381
Stibara bicolor, 174
Stink bug, 406, 515, 516, 517, 518, 520, 521, 522, 523
Streptomyces griseoviridis, 75, 253
Stunt nematode, 215, 216
Sunnhemp, 128, 139, 141, 142, 143, 144, 147, 149, 151, 152, 153, 154, 156, 158, 159, 160, 161, 162, 163, 164, 165, 167, 168, 169, 170, 171, 172, 173, 175, 176, 177, 179
Sunnhemp capsid, 175
Sunnhemp flea beetle, 169
Sunnhemp mirid, 175
Sunnhemp mosaic, 159, 161
Sunn-hemp mosaic virus, 160, 161
Sunn-hemp phyllody phytoplasma, 162
Sunn-hemp rosette virus, 160
Sylepta derogata, 225, 227, 464
Syllepte derogata, 227, 463, 464, 465
Systasis dasyneurae, 353

T

T. achaeae, 514
T. claytoni, 215, 216
T. longibrachiatum, 96
T. patula, 218
T. viride, 29, 114, 403
Tagetes erecta, 218
Tagetes patula, 529
Tarnished plant bug, 484, 485, 486, 487
Tarsonemus latus, 49, 50
Tectona grandis, 496
Termite, 234, 235
Tetranychus bimaculatus, 282
Tetranychus cinnabarinus, 500, 501, 504

Tetranychus cucurbitacearum, 501
Tetranychus piger, 504
Tetranychus reetalius, 504
Tetranychus rosarum, 504
Tetranychus russeolus, 504
Tetranychus telarius, 282, 504
Tetranychus urticae, 282, 501, 503, 504
Tetrastichus sp., 353
Texus root rot, 417
Thanatephorus corchori, 319
Thanatephorus cucumeris, 114, 158, 199, 318, 319, 320
Thanatephorus sasakii, 319
Thespesia populnea, 496
Thrips dianthi, 525
Thrips angusticeps, 372
Thrips armeniacus, 370
Thrips linarius, 369, 370
Thrips lini, 369, 370
Thrips palmi, 432
Thrips ponticus, 370
Thrips tenuisetosus, 370
Thyrospora solani, 429
Tiarosporella phaseolina, 90
Tight lock bolls, 405
Tigria taurica, 258
Tinea pulchella, 171
Tip rot, 77, 149
Tobacco budworm, 449
Tobacco streak virus disease, 431
Tobacco streak virus (TSV), 84, 85, 431, 432
Tobamovirus, 160
ToLCHsV, 204
ToLCJV, 82
ToLCuVB, 164
ToLCV, 164
Tomato grub, 449
Tomato leaf curl Hsinchu virus, 204
Tomato leaf curl New Delhi virus, 83
Tomato leaf curl virus, 83, 164
Tomato leaf curl virus Bangalore, 164
Tomato yellow leaf curl virus, 83
Tomato worm, 449
Torenia., 84
Tortrix insulana, 508
Torymus sp., 353
Tossa jute, 1, 2, 5, 6, 21, 28, 33, 34, 35, 38, 40, 43, 53, 56, 58, 463
Trap crop, 33, 233, 448, 460, 484, 517, 520, 529, 532

Trichocladia diffusa, 145
Trichoderma gamsii, 96, 119
Trichoderma lignorum, 253
Trichoderma sp., 257
Trichoderma spp., 200
Trichoderma virens, 119, 441
Trichoderma viride, 29, 96, 114, 318,
Trichoderma harzianum, 18, 91, 119, 149, 242
Trichogramma, 233, 276, 281, 448, 453, 492, 493, 514
Trichogramma australicum, 514
Trichogramma brasiliense,
Trichogramma chilonis, 492
Trichogramma evanescens, 276
Trichogramma spp. 276, 448, 453
Trichogramma wasps, 233
Trichopoda pennipes, 518
Trichospheria spp., 77
Triumfetta rhomboidea, 59
Trombidium sp., 367
Tropical rust, 427, 428
TSV, 84, 85, 431, 432, 433
Twig blight, 149
Two spotted spider mite, 282, 283, 500, 503, 504, 505
Tylenchorhynchus sp., 215
Tylenchus arenarius, 209
Tylenchus coffeae, 212
Tylenchus mahogani, 212
Tylenchus musicola, 212

U

Uredo chloridis-berroi, 425
Uredo chloridis-polydactylidis, 425
Uredo lini, 305
*Urena lobata,*66, 126
Uromyces decoratus, 147
Urtica dioica, 271
Utetheisa pulchella, 171, 173
Utethesia callima dilutior, 171
Utethesia shyama, 172
Utethesia thytea, 172
UV-light trap, 457

V

Vanessa calliroe, 220
Vanessa indica, 220
Vasates cannabicola, 284
Vascular wilt, 116, 154, 433

Verania discolor, 223
Vermicularia atramentaria, 240
Vermicularia gloeosporioides, 190
Verticillium albo-atrum, 439
Verticillium albo-atrum var. *dahliae,* 439
Verticillium albo-atrum var. *chlamydosporale,* 439
Verticillium albo-atrum var. *medium,* 439
Verticillium blight, 338
Verticillium dahliae, 338
Verticillium lecanii, 46, 263
Verticillium ovatum, 439
Verticillium tracheiphilum, 439
Verticillium wilt, 438, 439, 441
Viburnum spp., 267
Vigna, 20, 84, 104, 406
Viral diseases, 81
Virus, 30, 31, 50, 56, 82, 83, 84, 85, 121, 122, 123, 124, 125, 126, 159, 160, 161164, 188, 201, 202, 204, 205, 264, 269, 338, 342, 343, 344, 345, 362, 363, 368, 383, 384, 385, 431, 432, 433, 449, 454, 524, 5247, 530

W

Watery leaf Spot 119
Watery soft rot 309
Whetzelinia sclerotiorum 79, 310
White ants 234
White cane rot 197, 200
White fungus disease 200
White grub 236, 237
White jute 1, 3, 30, 33, 35, 42, 46, 53, 56, 463, 496
White mould 78, 93, 309, 404
White root rot 200, 202, 203
White Stem Rot 78
Whiteflies 21, 31, 123, 530, 531, 533
Wilt 17, 24, 25, 58, 61, 63, 64, 73, 76, 88, 89, 90, 92, 93, 96, 97, 115, 116, 117, 118, 119, 139, 148, 151, 153, 154, 155, 156, 157, 158, 197, 200, 239, 251, 252, 253, 254, 264, 269, 271, 282, 292, 310, 319, 327, 328, 330, 332, 333, 334, 360, 364, 380, 405, 417, 420, 433, 434, 435, 436, 437, 438, 439, 441, 461, 462, 507, 524, 527, 535, 543, 548
Witches' broom 162
Withania somnifera 282

X

Xanthomonas axonopodis pv. *malvacearum* 396
Xanthomonas campestris 1, 2, 247, 396
Xanthomonas campestris pv. *cannabis* 247
Xanthomonas campestris pv. *malvacearum* 396
Xanthomonas campestris pv.*capsularii* 1
Xanthomonas campestris pv.*olitorii* 1
Xanthomonas cannabis pv. *cannabis* 247
Xanthomonas citri subsp. *malvacearum* 396
Xanthomonas leaf spot 246
Xanthomonas malvacearum 396
Xiphinema sp. 206
Xyloma lini 305

Y

Yellow mirid 49, 50, 51, 52
Yellow mosaic disease 30, 31, 205, 537
Yellow tail tussock moth 128
Yellow vein mosaic disease 81, 82, 84, 123, 124, 125, 540
Yellow-headed ravenous weevil 458

Z

Zea mays 273
Zebra disease 385, 386, 387, 388, 389, 349
Zebra leaf rot 385
Zebra leaf spot 385, 387, 388
Zebronia salomealis 464
Zinckenia perfuscalis 226